Fundamental Equations of Dynamics

KINEMATICS

Particle Rectilinear Motion

Variable a

$$a = \frac{dv}{dt}$$

$$v = \frac{ds}{dt}$$

$$a\,ds = v\,dv$$

Constant $a = a_c$

$$v = v_0 + a_c t$$

$$s = s_0 + v_0 t + \tfrac{1}{2} a_c t^2$$

$$v^2 = v_0^2 + 2a_c(s - s_0)$$

Particle Curvilinear Motion

x, y, z Coordinates

$$v_x = \dot{x} \qquad a_x = \ddot{x}$$
$$v_y = \dot{y} \qquad a_y = \ddot{y}$$
$$v_z = \dot{z} \qquad a_z = \ddot{z}$$

r, θ, z Coordinates

$$v_r = \dot{r} \qquad a_r = \ddot{r} - r\dot{\theta}^2$$
$$v_\theta = r\dot{\theta} \qquad a_\theta = r\ddot{\theta} + 2\dot{r}\dot{\theta}$$
$$v_z = \dot{z} \qquad a_z = \ddot{z}$$

n, t, b Coordinates

$$v = \dot{s} \qquad a_t = \dot{v} = v\frac{dv}{ds}$$

$$a_n = \frac{v^2}{\rho} \qquad \rho = \frac{[1 + (dy/dx)^2]^{3/2}}{|d^2y/dx^2|}$$

Relative Motion

$$\mathbf{v}_B = \mathbf{v}_A + \mathbf{v}_{B/A} \qquad \mathbf{a}_B = \mathbf{a}_A + \mathbf{a}_{B/A}$$

Rigid Body Motion About a Fixed Axis

Variable α

$$\alpha = \frac{d\omega}{dt}$$

$$\omega = \frac{d\theta}{dt}$$

$$\omega\,d\omega = \alpha\,d\theta$$

Constant $\alpha = \alpha_c$

$$\omega = \omega_0 + \alpha_c t$$

$$\theta = \theta_0 + \omega_0 t + \tfrac{1}{2}\alpha_c t^2$$

$$\omega^2 = \omega_0^2 + 2\alpha_c(\theta - \theta_0)$$

For Point P

$$s = \theta r \qquad v = \omega r \qquad a_t = \alpha r \qquad a_n = \omega^2 r$$

Relative General Plane Motion—Translating Axes

$$\mathbf{v}_B = \mathbf{v}_A + \mathbf{v}_{B/A(\text{pin})} \qquad \mathbf{a}_B = \mathbf{a}_A + \mathbf{a}_{B/A(\text{pin})}$$

Relative General Plane Motion—Trans. and Rot. Axis

$$\mathbf{v}_B = \mathbf{v}_A + \Omega \times \mathbf{r}_{B/A} \times (\mathbf{v}_{B/A})_{xyz}$$
$$\mathbf{a}_B = \mathbf{a}_A + \Omega \times \mathbf{r}_{B/A} + \Omega \times (\Omega \times \mathbf{r}_{B/A}) +$$
$$2\Omega \times (\mathbf{v}_{B/A})_{xyz} \times (\mathbf{a}_{B/A})_{xyz}$$

KINETICS

Mass Moment of Inertia $\quad I = \displaystyle\int r^2\,dm$

Parallel-Axis Theorem $\quad I = I_G + md^2$

Radius of Gyration $\quad k = \sqrt{\dfrac{I}{m}}$

Equations of Motion

Particle	$\Sigma \mathbf{F} = m\mathbf{a}$
Rigid Body (Plane Motion)	$\Sigma F_x = m(a_G)_x$ $\Sigma F_y = m(a_G)_y$ $\Sigma M_G = I_G\alpha$ or $\Sigma M_P = \Sigma(\mathcal{M}_k)_P$

Principle of Work and Energy

$$T_1 + U_{1-2} = T_2$$

Kinetic Energy

Particle	$T = \tfrac{1}{2}mv^2$
Rigid Body (Plane Motion)	$T = \tfrac{1}{2}mv_G^2 + \tfrac{1}{2}I_G\omega^2$

Work

Variable force	$U_F = \displaystyle\int F\cos\theta\,ds$
Constant force	$U_F = (F_c\cos\theta)\,\Delta s$
Weight	$U_W = -W\,\Delta y$
Spring	$U_s = -(\tfrac{1}{2}ks_2^2 - \tfrac{1}{2}ks_1^2)$
Couple moment	$U_M = M\,\Delta\theta$

Power and Efficiency

$$P = \frac{dU}{dt} = \mathbf{F}\cdot\mathbf{v} \qquad \epsilon = \frac{P_{\text{out}}}{P_{\text{in}}} = \frac{U_{\text{out}}}{U_{\text{in}}}$$

Conservation of Energy Theorem

$$T_1 + V_1 = T_2 + V_2$$

Potential Energy

$$V = V_g + V_e, \text{ where } V_g = \pm Wy, \; V_e = +\tfrac{1}{2}ks^2$$

Principle of Linear Impulse and Momentum

Particle	$m\mathbf{v}_1 + \Sigma\displaystyle\int \mathbf{F}\,dt = m\mathbf{v}_2$
Rigid Body	$m(\mathbf{v}_G)_1 + \Sigma\displaystyle\int \mathbf{F}\,dt = m(\mathbf{v}_G)_2$

Conservation of Linear Momentum

$$\Sigma \, 1\text{syst. } m v2_1 = \Sigma \, 1\text{syst. } m v2_2$$

Coefficient of Restitution $\quad e = \dfrac{(v_B)_2 - (v_A)_2}{(v_A)_1 - (v_B)_1}$

Principle of Angular Impulse and Momentum

Particle	$(\mathbf{H}_O)_1 + \Sigma\displaystyle\int \mathbf{M}_O\,dt = (\mathbf{H}_O)_2$ where $H_O = (d)(mv)$
Rigid Body (Plane motion)	$(\mathbf{H}_G)_1 + \Sigma\displaystyle\int \mathbf{M}_G\,dt = (\mathbf{H}_G)_2$ where $H_G = I_G\omega$ $(\mathbf{H}_O)_1 + \Sigma\displaystyle\int \mathbf{M}_O\,dt = (\mathbf{H}_O)_2$ where $H_O = I_O\omega$

Conservation of Angular Momentum

$$\Sigma(\text{syst. } \mathbf{H})_1 = \Sigma(\text{syst. } \mathbf{H})_2$$

SI Prefixes

Multiple	Exponential Form	Prefix	SI Symbol
1 000 000 000	10^9	giga	G
1 000 000	10^6	mega	M
1 000	10^3	kilo	k

Submultiple			
0.001	10^{-3}	milli	m
0.000 001	10^{-6}	micro	μ
0.000 000 001	10^{-9}	nano	n

Conversion Factors (FPS) to (SI)

Quantity	Unit of Measurement (FPS)	Equals	Unit of Measurement (SI)
Force	lb		4.4482 N
Mass	slug		14.5938 kg
Length	ft		0.3048 m

Conversion Factors (FPS)

1 ft = 12 in. (inches)

1 mi. (mile) = 5280 ft

1 kip (kilopound) = 1000 lb

1 ton = 2000 lb

PRINCIPLES OF STATICS AND DYNAMICS

ENGINEERING MECHANICS

PRINCIPLES OF STATICS AND DYNAMICS

R. C. Hibbeler

Upper Saddle River, NJ 07458

Library of Congress Cataloging-in-Publication Data on File

Vice President and Editorial Director, ECS: *Marcia Horton*
Associate Editor: *Dee Bernhard*
Executive Managing Editor: *Vince O'Brien*
Managing Editor: *David A. George*
Production Editor: *Daniel Sandin*
Director of Creative Services: *Paul Belfanti*
Manager of Electronic Composition and Digital Content: *Allyson Graesser*
Assistant Manager of Electronic Composition and Digital Content: *William Johnson*
Art Director: *Jonathan Boylan*
Electronic Composition: *Lawrence La Raia*
Art Editor: *Xiaohong Zhu*
Manufacturing Buyer: *Lisa McDowell*
Senior Marketing Manager: *Holly Stark*

About the Cover: Left: The forces within this truss bridge must be determined if they are to be properly designed. Photo by R. C. Hibbeler. Right: In order to properly design the loop of this roller coaster it is necessary to ensure that the cars have enough energy to be able to make the loop without leaving the tracks. Image courtesy of Corbis.

 © 2006 by R.C. Hibbeler
Published by Pearson Prentice Hall
Pearson Education, Inc.
Upper Saddle River, New Jersey 07458

Printed in the United States of America

10 9 8 7 6 5 4 3 2 1

ISBN 0-13-187256-7

Pearson Education Ltd., *London*
Pearson Education Australia Pty. Ltd., *Sydney*
Pearson Education Singapore, Pte. Ltd.
Pearson Education North Asia Ltd., *Hong Kong*
Pearson Education Canada, Inc., *Toronto*
Pearson Educación de Mexico, S.A. de C.V.
Pearson Education—Japan, *Tokyo*
Pearson Education Malaysia, Pte. Ltd.
Pearson Education, Inc., *Upper Saddle River, New Jersey*

TO THE STUDENT

With the hope that this work will stimulate an interest in Engineering Mechanics and provide an acceptable guide to its understanding.

Hibbeler's *Principles of Statics* and *Principles of Dynamics* series is a unique text/web system designed with two goals in mind: 1) to provide students with a lower priced, paperback mechanics textbook and 2) to provide instructors with a better, more secure way to assign homework.

Paperback Text with On-Line Homework Assignments

These paperback texts contain all the topics, explanations, and examples of the best-selling Hibbeler hardcover texts; no mechanics content has been deleted. Students who use these texts will have every equation, example, description, etc., available for study during the course and to keep as long-term reference material. However, almost all problem material has been deleted from the printed text; only a section of end of chapter review problems remain.

Instead of accessing homework problems printed in the text, students simply access their homework assignments on-line through Hibbeler's OneKey Course. Hibbeler's OneKey course contains a bank of approximately 3000 editable *Statics* and *Dynamics* problems that instructors can use to create their assignments. Instructors simply choose the homework problems they want, customize them if they choose, and then post these problems to the secure assignment area for their class. Instructors are provided with Mathcad versions of problem solutions so they can quickly generate custom solutions for their custom problems. Or, instructors can use PHGradeAssist—Prentice Hall's on-line, algorithmic homework generator—to create assignments. PHGradeAssist automatically generates a unique version of each problem for students and grades their answer.

OneKey: Custom Homework, Secure Solutions, and Much More

OneKey is all you need

Hibbeler's OneKey course—availble at www.prenhall.com/onekey—offers over 3000 *Statics* and *Dynamics* problems that you can personalize and post for your student assignments. Editing the values in a problem guarantees a fresh problem for your students. Then, use solutions powered by Mathcad to generate your own personal solution, and if you choose, post the solutions for your students on-line. OneKey also contains PHGradeAssist—an on-line assessment tool with approximately 600 algorithmic test bank problems. PHGA generates unique problems for students, grades the answer, and tracks student results automatically.

You'll find Hibbeler's OneKey course contains much more to help you and your students, including

- Student Hints: Each problem contains a student hint that you may choose to provide to your students. This hint is also fully editable should you wish to change it.

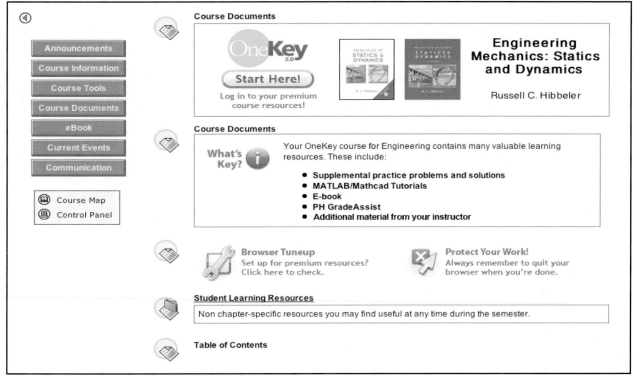

A sample OneKey homepage.

- Active Book: A complete, online HTML version of the textbook students can use and refer to while completing homework and assignments.
- An extra bank of practice problems with solutions.
- Complete bank of .jpg images.
- Complete set of PowerPoint Slides.
- Active Learning slides—Perfect for classroom response systems.
- Mathcad and MATLAB tutorials.
- Animations and Simulations
- Math review tutorials.
- Mechanics visualization software—Ideal for in-class demonstrations

Instructors should visit www.prenhall.com/onekey and/or contact their local sales rep to register and receive more information. You may also send an email requesting information to engineering@prenhall.com.

Hibbeler's OneKey course is available free with any new Hibbeler text: Contact your local Prentice Hall sales rep or email engineering@prenhall.com for ordering isbns and course management system options. It is also available as a standalone item for student purchase at either your bookstore or on-line at www.prenhall.com/onekey.

How to Get Started

For Instructors

To use this text, you need to create your own unique OneKey course, where students will retrieve their homework problems and any other course materials you provide. To do so, you will need a OneKey access code. This will have either come bundled with the text or is available through either your local Prentice Hall Sales Rep, by request at www.prenhall.com/onekey, or by emailing engineering@prenhall.com.

Once you have your access code, go to www.prenhall.com/onekey and follow instructions for instructors to register and establish your course. Complete on-line help is available through OneKey. You'll also find a "How To" FAQ in the Hibbeler OneKey Instructor Resource Center, which quickly explains how to create/save/post problem sets, solutions, and student hints. For general help, you can also contact the support staff at 1-800-677-6337.

Distribute your course information to your students in class. Students use this identifier to find your assignments at the site. Complete help is available on-line for students as well as through our support staff at 1-800-677-6337.

For Students

You should keep the Access card that comes with new copies of the text since it contains complete instructions for registering for your course. Once you register, you simply log in to access these course materials as you need them. If you have difficulty registering, or logging in, either report the problem on-line as described on your access card, or call our support staff at 1-800-677-6337. If you need to purchase an access code, you'll find instructions for ordering one at www.prenhall.com/onekey. Before ordering, you should verify with your instructor which course management system you use.

You can always visit www.prenhall.com/onekey for information and help getting started.

The main purpose of this book is to provide the student with a clear and thorough presentation of the theory and applications of engineering mechanics. To achieve this objective, the author has by no means worked alone; to a large extent, this series, through 10 editions, has been shaped by the comments and suggestions of hundreds of reviewers in the teaching profession as well as many of the author's students.

New Features

Some unique features used throughout this tenth edition include the following:

- **Illustrations.** Throughout the book, new photorealistic illustrations have been added that provide a strong connection to the 3-D nature of engineering. In addition, particular attention has been placed on providing a view of any physical object, its dimensions, and the vectors applied to it in a manner that can be easily understood.

- **Problems.** The problems sets in the Hibbeler OneKey Course have been revised so that instructors can select both design and analysis problems having a wide range of difficulty. Apart from the author, two other professionals have checked all the problems for clarity and accuracy of the solutions. At the end of some chapters, design projects are included.

- **Review Material.** New end-of-chapter review sections have been added to help students recall and study key chapter points.

Of course, the hallmarks of the book remain the same: Where necessary, a strong emphasis is placed on drawing a free-body diagram, and the importance of selecting an appropriate coordinate system, and associated sign convention for vector components is stressed when the equations of mechanics are applied.

Contents

Statics

The subject of Statics is covered in the first 11 chapters, in which the principles are applied first to simple, then to more complicated situations. Most often, each principle is applied first to a particle, then to a rigid body subjected to a coplanar system of forces, and finally to a general case of three-dimensional force systems acting on a rigid body.

Chapter 1 begins with an introduction to mechanics and a discussion of units. The notation of a vector and the properties of a concurrent force system are introduced in Chapter 2. This theory is then applied to the equilibrium of a particle in Chapter 3. Chapter 4 contains a general discussion of both concentrated and distributed force systems and the methods used to simplify them. The principles of rigid-body equilibrium are developed in Chapter 5 and then applied to specific problems involving the equilibrium of trusses, frames, and machines in Chapter 6, and to the analysis of internal forces in beams and cables in Chapter 7. Applications to problems involving frictional forces are discussed in Chapter 8, and topics related to the center of gravity and centroid are treated in Chapter 9. If time permits, sections concerning more advanced topics, indicated by stars (★) may be covered. Most of these topics are included in Chapter 10 (area and mass moments of inertia) and Chapter 11 (virtual work and potential energy). Note that this material also provides a suitable reference for basic principles when it is discussed in more advanced courses.

Alternative Coverage. At the discretion of the instructor, some of the material may be presented in a different sequence with no loss of continuity. For example, it is possible to introduce the concept of a force and all the necessary methods of vector analysis by first covering Chapter 2 and Section 4.2. Then after covering the rest of Chapter 4 (force and moment systems), the equilibrium methods of Chapters 3 and 5 can be discussed.

Dynamics

The subject of Dynamics is presented in the last 11 chapters, the kinematics of a particle is discussed in Chapter 12, followed by a discussion of particle kinetics in Chapter 13 (equation of motion), Chapter 14 (work and energy), and Chapter 15 (impulse and momentum). The concepts of particle dynamics contained in these four chapters are then summarized in a "review" section, and the student is given the chance to identify and solve a variety of problems. A similar sequence of presentation is given for the planar motion of a rigid body: Chapter 16 (planar kinematics), Chapter 17 (equations of motion), Chapter 18 (work and energy), and Chapter 19 (impulse and momentum), followed by a summary and review set of problems for these chapters.

If time permits, some of the material involving three-dimensional regid-body motion may be included in the course. The kinematics and kinetics of this motion are discussed in Chapters 20 and 21, respectively. Chapter 22 (vibrations) may be included if the student has the necessary mathematical background. Sections of the book which are considered to be beyond the scope of the basic dynamics course are indicated by a star (★) and may be omitted. Note that this material also provides a suitable reference for basic principles when it is discussed in more advanced courses.

Alternative Coverage. At the discretion of the instructor, it is possible to cover Chapter 12 through 19 in the following order with no loss in continuity: Chapers 12 and 16 (kinematics), Chapters 13 and 17 (equations of motion), Chapters 14 and 18 (work and energy), and Chapters 15 and 19 (impulse and momentum).

Special Features

Organization and Approach.   The contents of each chapter are organized into well-defined sections that contain an explanation of specific topics and illustrative example problems. The topics within each section are placed into subgroups defined by boldface titles. The purpose of this is to present a structured method for introducing each new definition or concept and to make the book convenient for later reference and review.

Chapter Contents.   Each chapter begins with an illustration demonstrating a broad-range application of the material within the chapter. A bulleted list of the chapter contents is provided to give a general overview of the material that will be covered.

Free-Body Diagrams.   The first step to solving most mechanics problems requires drawing a diagram. By doing so, the student forms the habit of tabulating the necessary data while focusing on the physical aspects of the problem and its associated geometry. If this step is performed correctly, applying the relevant equations of mechanics becomes somewhat methodical since the data can be taken directly from the diagram. This step is particularly important when solving equilibrium problems, and for this reason drawing free-body diagrams is strongly emphasized throughout the book. In particular, special sections and examples are devoted to show how to draw free-body diagrams.

Procedures for Analysis.   Found after many of the sections of the book, this unique feature provides the student with a logical and orderly method to follow when applying the theory. The example problems are solved using this outlined method in order to clarify its numerical application. It is to be understood, however, that once the relevant principles have been mastered and enough confidence and judgment have been obtained, the student can then develop his or her own procedures for solving problems.

Photographs.   Many photographs are used throughout the book to explain how the principles of mechanics apply to real-world situations. In some sections, photographs have been used to show how engineers must first make an idealized model for analysis and then proceed to draw a free-body diagram of this model in order to apply the theory.

Important Points.   This feature provides a review or summary of the most important concepts in a section and highlights the most significant points that should be realized when applying the theory to solve problems.

Conceptual Understanding. Through the use of photographs placed throughout the book, theory is applied in a simplified way in order to illustrate some of its more important conceptual features and instill the physical meaning of many of the terms used in the equations. These simplified applications increase interest in the subject matter and better prepare the student to understand the examples and solve problems.

Example Problems. All the example problems are presented in a concise manner and in a style that is easy to understand.

Homework Problems Instead of accessing homework problems printed in the text, students simply access their homework assignments on-line through Hibbeler's OneKey Course. Hibbeler's OneKey course contains a bank of approximately 3000 editable *Statics* and *Dynamics* problems that instructors can use to create their assignments. Instructors simply choose the homework problems they want, customize them if they choose, and then post these problems to the secure assignment area for their class. Instructors are provided with Mathcad versions of problem solutions so they can quickly generate custom solutions for their custom problems. Or, instructors can use PHGradeAssist—Prentice Hall's on-line, algorithmic homework generator—to create assignments. PHGradeAssist automatically generates a unique version of each problem for students and grades their answer.

Types of problems in OneKey include:

- **Free-Body Diagram Problems.** These assignments will impress upon the student the importance of mastering this skill as a requirement for a complete solution of any equilibrium problem.

- **General Analysis and Design Problems.** The majority of problems in the book depict realistic situations encountered in engineering practice. Some of these problems come from actual products used in industry and are stated as such. It is hoped that this realism will both stimulate the student's interest in engineering mechanics and provide a means for developing the skill to reduce any such problem from its physical description to a model or symbolic representation to which the principles of mechanics may be applied.

- **Computer Problems.** An effort has been made to include some problems that may be solved using a numerical procedure executed on either a desktop computer or a programmable pocket calculator. Suitable numerical techniques along with associated computer programs are given in Appendix B. The intent here is to broaden the student's capacity for using other forms of mathematical analysis without sacrificing the time needed to focus on the application of the principles of mechanics.

- **Design Projects.** At the end of some of the chapters, design projects have been included. It is felt that this type of assignment should be given only after the student has developed a basic understanding of the subject matter. These projects focus on solving a problem by specifying the geometry of a structure or mechanical object needed for a specific purpose. A force analysis is required and, in many cases, safety and cost issues must be addressed

Chapter Reviews. New chapter review sections summarize key points of the chapter, often in bulleted lists.

Appendices. The appendices provide a source of mathematical formula and numerical analysis needed to solve the problems in the book. Appendix C provides a set of problems typically found on the Fundamentals of Engineering Examination. By providing a partial solution to all the problems, the student is given a chance to further practice his or her skills.

Supplements

Hibbeler's robust supplements package supports students and instructors.

OneKey: Custom Homework, Secure Solutions, and Much More

Hibbeler's OneKey course—availble at www.prenhall.com/onekey—offers over 3000 *Statics* and *Dynamics* problems that you can personalize and post for your student assignments. Editing the values in a problem guarantees a fresh problem for your students. Then, use solutions powered by Mathcad to generate your own personal solution, and if you choose, post the solutions for your students on-line. OneKey also contains PHGradeAssist—an on-line assessment tool with approximately 600 algorithmic test bank problems. PHGA generates unique problems for students, grades the answer, and tracks student results automatically.

You'll find Hibbeler's OneKey course contains much more to help you and your students, including

- Student Hints: Each problem contains a student hint that you may choose to provide to your students. This hint is also fully editable should you wish to change it.

- Active Book: A complete, online HTML version of the textbook students can use and refer to while completing homework and assignments.

- An extra bank of practice problems with solutions.

- Complete bank of .jpg images.

- Complete set of PowerPoint Slides.

- Active Learning slides—Perfect for classroom response systems.

- Mathcad and MATLAB tutorials.

- Animations and Simulations

- Math review tutorials.

- Mechanics visualization software—Ideal for in-class demonstrations

Instructors should visit www.prenhall.com/onekey and/or contact their local sales rep to register and receive more information. You may also send an email requesting information to engineering@prenhall.com.

Hibbeler's OneKey course is available free with any new Hibbeler text: Contact your local Prentice Hall sales rep or email engineering@prenhall.com for ordering isbns and course management system options. It is also available as a standalone item for student purchase at either your bookstore or on-line at www.prenhall.com/onekey.

Instructor's Resource CD-ROM and Instructor Access Code This supplement offers visual resources in CD-ROM format. These resources are also found on the Hibbeler OneKey Course.

Study Pack *Statics* (0-13-141209-4) and *Dynamics* (0-13-141680-4) Improved for the tenth edition, this supplement now contains chapter-by-chapter study materials, a Free-Body Diagram Workbook and access to a separate Practice Problems Website.

Student Study Guide *Statics* (0-13-141211-6) *Dynamics* (0-13-141679-0) Students may purchase a Study Guide containing more worked problems. Problems are partially solved and designed to help guide students through difficult topics.

Acknowledgments

This text was derived from the tenth edition of Hibbeler's *Engineering Mechanics* series.

The author endeavored to write this book so that it will appeal to both the student and instructor. Through the years, many people have helped in its development, and I will always be grateful for their valued suggestions and comments. Specifically, I wish to personally thank the following individuals who have contributed their comments to this edition of *Statics* and *Dynamics*:

Paul Heyliger, *Colorado State University*
Kenneth Sawyers, *Lehigh University*
John Oyler, *University of Pittsburgh*
Glenn Beltz, *University of California—Santa Barbara*
Johannes Gessler, *Colorado State University*
Wilfred Nixon, *University of Iowa*
Jonathan Russell, *U.S. Coast Guard Academy*
Robert Hinks, *Arizona State University*
Cap. Mark Orwat, *U.S. Military Academy, West Point*
Cetin Cetinyaka, *Clarkson University*
Jack Xin, *Kansas State University*
Pierre Julien, *Colorado State University*
Stephen Bechtel, *Ohio State University*
W. A. Curtain, *Brown University*
Robert Oakberg, *Montana State University*
Richard Bennett, *University of Tennessee*

A particular note of thanks is also given to Professors Will Liddell, Jr. and Henry Kuhlman for their specific help. A special note of thanks is given to the accuracy checkers, Scott Hendricks of VPI and Karim Nohra of the University of South Florida, who diligently checked all of the text and problems. I should also like to acknowledge the proofreading assistance of my wife, Conny (Cornelie), during the time it has taken to prepare this manuscript for publication.

Lastly, many thanks are extended to all my students and to members of the teaching profession who have freely taken the time to send me their suggestions and comments. Since this list is too long to mention, it is hoped that those who have given help in this manner will accept this anonymous recognition.

I would greatly appreciate hearing from you if at any time you have any comments, suggestions, or problems related to any matters regarding this edition.

Russell Charles Hibbeler
hibbeler@bellsouth.net

C O N T E N T S

Principles of Statics

5

Equilibrium of a Rigid Body   131

7

Internal Forces   221

6

Structural Analysis   177

8

Friction   259

C O N T E N T S

Principles of Dynamics

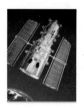

PRINCIPLES OF STATICS

The design of this rocket and gantry structure requires a basic knowledge of both statics and dynamics, which forms the subject matter of engineering mechanics.

General Principles

- To provide an introduction to the basic quantities and idealizations of mechanics.
- To give a statement of Newton's Laws of Motion and Gravitation.
- To review the principles for applying the SI system of units.
- To examine the standard procedures for performing numerical calculations.
- To present a general guide for solving problems.

1.1 Mechanics

Mechanics can be defined as that branch of the physical sciences concerned with the state of rest or motion of bodies that are subjected to the action of forces. In general, this subject is subdivided into three branches: *rigid-body mechanics, deformable-body mechanics*, and *fluid mechanics*. This book treats only rigid-body mechanics since it forms a suitable basis for the design and analysis of many types of structural, mechanical, or electrical devices encountered in engineering. Also, rigid-body mechanics provides part of the necessary background for the study of the mechanics of deformable bodies and the mechanics of fluids.

Rigid-body mechanics is divided into two areas: statics and dynamics. *Statics* deals with the equilibrium of bodies, that is, those that are either at rest or move with a constant velocity; whereas *dynamics* is concerned with the accelerated motion of bodies. Although statics can be considered as a special case of dynamics, in which the acceleration is zero, statics deserves separate treatment in engineering education since many objects are designed with the intention that they remain in equilibrium.

Historical Development.   The subject of statics developed very early in history because the principles involved could be formulated simply from measurements of geometry and force. For example, the writings of Archimedes (287–212 B.C.) deal with the principle of the lever. Studies of the pulley, inclined plane, and wrench are also recorded in ancient writings—at times when the requirements of engineering were limited primarily to building construction.

Since the principles of dynamics depend on an accurate measurement of time, this subject developed much later. Galileo Galilei (1564–1642) was one of the first major contributors to this field. His work consisted of experiments using pendulums and falling bodies. The most significant contributions in dynamics, however, were made by Isaac Newton (1642–1727), who is noted for his formulation of the three fundamental laws of motion and the law of universal gravitational attraction. Shortly after these laws were postulated, important techniques for their application were developed by Euler, D'Alembert, Lagrange, and others.

1.2   Fundamental Concepts

Before we begin our study of engineering mechanics, it is important to understand the meaning of certain fundamental concepts and principles.

Basic Quantities.   The following four quantities are used throughout mechanics.

Length.   *Length* is needed to locate the position of a point in space and thereby describe the size of a physical system. Once a standard unit of length is defined, one can then quantitatively define distances and geometric properties of a body as multiples of the unit length.

Time.   *Time* is conceived as a succession of events. Although the principles of statics are time independent, this quantity does play an important role in the study of dynamics.

Mass.   *Mass* is a property of matter by which we can compare the action of one body with that of another. This property manifests itself as a gravitational attraction between two bodies and provides a quantitative measure of the resistance of matter to a change in velocity.

Force.   In general, *force* is considered as a "push" or "pull" exerted by one body on another. This interaction can occur when there is direct contact between the bodies, such as a person pushing on a wall, or it can occur through a distance when the bodies are physically separated. Examples of the latter type include gravitational, electrical, and magnetic forces. In any case, a force is completely characterized by its magnitude, direction, and point of application.

Idealizations.   Models or idealizations are used in mechanics in order to simplify application of the theory. A few of the more important idealizations will now be defined. Others that are noteworthy will be discussed at points where they are needed.

Particle.   A *particle* has a mass, but a size that can be neglected. For example, the size of the earth is insignificant compared to the size of its orbit, and therefore the earth can be modeled as a particle when studying its orbital motion. When a body is idealized as a particle, the principles of mechanics reduce to a rather simplified form since the geometry of the body will not be involved in the analysis of the problem.

Rigid Body.   A *rigid body* can be considered as a combination of a large number of particles in which all the particles remain at a fixed distance from one another both before and after applying a load. As a result, the material properties of any body that is assumed to be rigid will not have to be considered when analyzing the forces acting on the body. In most cases the actual deformations occurring in structures, machines, mechanisms, and the like are relatively small, and the rigid-body assumption is suitable for analysis.

Concentrated Force.   A *concentrated force* represents the effect of a loading which is assumed to act at a point on a body. We can represent a load by a concentrated force, provided the area over which the load is applied is very small compared to the overall size of the body. An example would be the contact force between a wheel and the ground.

Newton's Three Laws of Motion.   The entire subject of rigid-body mechanics is formulated on the basis of Newton's three laws of motion, the validity of which is based on experimental observation. They apply to the motion of a particle as measured from a nonaccelerating reference frame. With reference to Fig. 1–1, they may be briefly stated as follows.

First Law.   A particle originally at rest, or moving in a straight line with constant velocity, will remain in this state provided the particle is *not* subjected to an unbalanced force.

Second Law.   A particle acted upon by an *unbalanced force* **F** experiences an acceleration **a** that has the same direction as the force and a magnitude that is directly proportional to the force.* If **F** is applied to a particle of mass m, this law may be expressed mathematically as

$$\mathbf{F} = m\mathbf{a} \tag{1–1}$$

Third Law.   The mutual forces of action and reaction between two particles are equal, opposite, and collinear.

*Stated another way, the unbalanced force acting on the particle is proportional to the time rate of change of the particle's linear momentum.

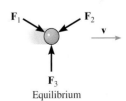

Equilibrium

Accelerated motion

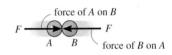

Action — reaction

Fig. 1–1

Newton's Law of Gravitational Attraction. Shortly after formulating his three laws of motion, Newton postulated a law governing the gravitational attraction between any two particles. Stated mathematically,

$$F = G\frac{m_1 m_2}{r^2} \tag{1-2}$$

where F = force of gravitation between the two particles

         G = universal constant of gravitation; according to experimental evidence, $G = 66.73(10^{-12})$ m^3/(kg · s^2)

  m_1, m_2 = mass of each of the two particles

        r = distance between the two particles

Weight. According to Eq. 1–2, any two particles or bodies have a mutual attractive (gravitational) force acting between them. In the case of a particle located at or near the surface of the earth, however, the only gravitational force having any sizable magnitude is that between the earth and the particle. Consequently, this force, termed the *weight*, will be the only gravitational force considered in our study of mechanics.

From Eq. 1–2, we can develop an approximate expression for finding the weight W of a particle having a mass $m_1 = m$. If we assume the earth to be a nonrotating sphere of constant density and having a mass $m_2 = M_e$, then if r is the distance between the earth's center and the particle, we have

$$W = G\frac{mM_e}{r^2}$$

Letting $g = GM_e/r^2$ yields

$$\boxed{W = mg} \tag{1–3}$$

By comparison with $\mathbf{F} = m\mathbf{a}$, we term g the acceleration due to gravity. Since it depends on r, it can be seen that the weight of a body is *not* an absolute quantity. Instead, its magnitude is determined from where the measurement was made. For most engineering calculations, however, g is determined at sea level and at a latitude of 45°, which is considered the "standard location."

1.3   Units of Measurement

The four basic quantities—force, mass, length and time—are not all independent from one another; in fact, they are *related* by Newton's second law of motion, $\mathbf{F} = m\mathbf{a}$. Because of this, the *units* used to measure these quantities cannot *all* be selected arbitrarily. The equality $\mathbf{F} = m\mathbf{a}$ is maintained only if three of the four units, called *base units*, are *arbitrarily defined* and the fourth unit is then *derived* from the equation.

SI Units. The International System of units, abbreviated SI after the French "Système International d'Unités," is a modern version of the metric system which has received worldwide recognition. As shown in Table 1–1, the SI system specifies length in meters (m), time in seconds (s), and mass in kilograms (kg). The unit of force, called a newton (N), is *derived* from $\mathbf{F} = m\mathbf{a}$. Thus, 1 newton is equal to a force required to give 1 kilogram of mass an acceleration of 1 m/s^2 ($\text{N} = \text{kg} \cdot \text{m/s}^2$).

If the weight of a body located at the "standard location" is to be determined in newtons, then Eq. 1–3 must be applied. Here $g = 9.806\ 65 \text{ m/s}^2$; however, for calculations, the value $g = 9.81 \text{ m/s}^2$ will be used. Thus,

$$W = mg \qquad (g = 9.81 \text{ m/s}^2) \qquad (1\text{--}4)$$

Therefore, a body of mass 1 kg has a weight of 9.81 N, a 2-kg body weighs 19.62 N, and so on, Fig. 1–2*a*.

U.S. Customary. In the U.S. Customary system of units (FPS) length is measured in feet (ft), force in pounds (lb), and time in seconds (s), Table 1–1. The unit of mass, called a *slug*, is *derived* from $\mathbf{F} = m\mathbf{a}$. Hence, 1 slug is equal to the amount of matter accelerated at 1 ft/s^2 when acted upon by a force of 1 lb ($\text{slug} = \text{lb} \cdot \text{s}^2/\text{ft}$).

In order to determine the mass of a body having a weight measured in pounds, we must apply Eq. 1–3. If the measurements are made at the "standard location," then $g = 32.2 \text{ ft/s}^2$ will be used for calculations. Therefore,

$$m = \frac{W}{g} \qquad (g = 32.2 \text{ ft/s}^2) \qquad (1\text{--}5)$$

And so a body weighing 32.2 lb has a mass of 1 slug, a 64.4-lb body has a mass of 2 slugs, and so on, Fig. 1–2*b*.

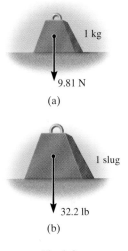

(a)

(b)

Fig. 1–2

TABLE 1–1 • Systems of Units

Name	Length	Time	Mass	Force
International System of Units (SI)	meter (m)	second (s)	kilogram (kg)	newton* (N) $\left(\dfrac{\text{kg} \cdot \text{m}}{\text{s}^2}\right)$
U.S. Customary (FPS)	foot (ft)	second (s)	slug* $\left(\dfrac{\text{lb} \cdot \text{s}^2}{\text{ft}}\right)$	pound (lb)

*Derived unit.

Conversion of Units. Table 1–2 provides a set of direct conversion factors between FPS and SI units for the basic quantities. Also, in the FPS system, recall that 1 ft = 12 in. (inches), 5280 ft = 1 mi (mile), 1000 lb = 1 kip (kilo-pound), and 2000 lb = 1 ton.

TABLE 1–2 • Conversion Factors

Quantity	Unit of Measurement (FPS)	Equals	Unit of Measurement (SI)
Force	lb		4.448 2 N
Mass	slug		14.593 8 kg
Length	ft		0.304 8 m

1.4 The International System of Units

The SI system of units is used extensively in this book since it is intended to become the worldwide standard for measurement. Consequently, the rules for its use and some of its terminology relevant to mechanics will now be presented.

Prefixes. When a numerical quantity is either very large or very small, the units used to define its size may be modified by using a prefix. Some of the prefixes used in the SI system are shown in Table 1–3. Each represents a multiple or submultiple of a unit which, if applied successively, moves the decimal point of a numerical quantity to every third place. * For example, 4 000 000 N = 4 000 kN(kilo-newton) = 4MN (mega-newton), or 0.005m = 5 mm (milli-meter). Notice that the SI system does not include the multiple deca (10) or the submultiple centi (0.01), which form part of the metric system. Except for some volume and area measurements, the use of these prefixes is to be avoided in science and engineering.

TABLE 1–3 • Prefixes

	Exponential Form	Prefix	SI Symbol
Multiple			
1 000 000 000	10^9	giga	G
1 000 000	10^6	mega	M
1 000	10^3	kilo	k
Submultiple			
0.001	10^{-3}	milli	m
0.000 001	10^{-6}	micro	μ
0.000 000 001	10^{-9}	nano	n

*The kilogram is the only base unit that is defined with a prefix.

Rules for Use. The following rules are given for the proper use of the various SI symbols:

1. A symbol is *never* written with a plural "s," since it may be confused with the unit for second (s).

2. Symbols are always written in lowercase letters, with the following exceptions: symbols for the two largest prefixes shown in Table 1–3, giga and mega, are capitalized as G and M, respectively; and symbols named after an individual are also capitalized, e.g., N.

3. Quantities defined by several units which are multiples of one another are separated by a *dot* to avoid confusion with prefix notation, as indicated by $N = kg \cdot m/s^2 = kg \cdot m \cdot s^{-2}$. Also, $m \cdot s$ (meter-second), whereas ms (milli-second).

4. The exponential power represented for a unit having a prefix refers to both the unit *and* its prefix. For example, $\mu N^2 = (\mu N)^2 = \mu N \cdot \mu N$. Likewise, mm^2 represents $(mm)^2 = mm \cdot mm$.

5. Physical constants or numbers having several digits on either side of the decimal point should be reported with a *space* between every three digits rather than with a comma; e.g., 73 569.213 427. In the case of four digits on either side of the decimal, the spacing is optional; e.g., 8537 or 8 537. Furthermore, always try to use decimals and avoid fractions; that is, write 15.25 *not* $15\frac{1}{4}$.

6. When performing calculations, represent the numbers in terms of their *base or derived units* by converting all prefixes to powers of 10. The final result should then be expressed using a *single prefix*. Also, after calculation, it is best to keep numerical values between 0.1 and 1000; otherwise, a suitable prefix should be chosen. For example,

$$(50 \text{ kN})(60 \text{ nm}) = [50(10^3) \text{ N}][60(10^{-9}) \text{ m}]$$
$$= 3000(10^{-6}) \text{ N} \cdot \text{m} = 3(10^{-3}) \text{ N} \cdot \text{m} = 3 \text{ mN} \cdot \text{m}$$

7. Compound prefixes should not be used; e.g., $k\mu s$ (kilo-micro-second) should be expressed as ms (milli-second) since $1 \text{ k}\mu\text{s} = 1(10^3)(10^{-6}) \text{ s} = 1(10^{-3}) \text{ s} = 1 \text{ ms}$.

8. With the exception of the base unit the kilogram, in general avoid the use of a prefix in the denominator of composite units. For example, do not write N/mm, but rather kN/m; also, m/mg should be written as Mm/kg.

9. Although not expressed in multiples of 10, the minute, hour, etc., are retained for practical purposes as multiples of the second. Furthermore, plane angular measurement is made using radians (rad). In this book, however, degrees will often be used, where $180° = \pi$ rad.

1.5 Numerical Calculations

Numerical work in engineering practice is most often performed by using handheld calculators and computers. It is important, however, that the answers to any problem be reported with both justifiable accuracy and appropriate significant figures. In this section we will discuss these topics together with some other important aspects involved in all engineering calculations.

Dimensional Homogeneity. The terms of any equation used to describe a physical process must be *dimensionally homogeneous;* that is, each term must be expressed in the same units. Provided this is the case, all the terms of an equation can then be combined if numerical values are substituted for the variables. Consider, for example, the equation $s = vt + \frac{1}{2}at^2$, where, in SI units, s is the position in meters, m, t is time in seconds, s, v is velocity in m/s, and a is acceleration in m/s^2. Regardless of how this equation is evaluated, it maintains its dimensional homogeneity. In the form stated, each of the three terms is expressed in meters [m, (m/s̸)s̸, (m/s̸^{2})s̸^{2},] or solving for a, $a = 2s/t^2 - 2v/t$, the terms are each expressed in units of m/s^2 [m/s^2, m/s^2, (m/s)/s].

Since problems in mechanics involve the solution of dimensionally homogeneous equations, the fact that all terms of an equation are represented by a consistent set of units can be used as a partial check for algebraic manipulations of an equation.

Significant Figures. The accuracy of a number is specified by the number of significant figures it contains. A *significant figure* is any digit, including a zero, provided it is not used to specify the location of the decimal point for the number. For example, the numbers 5604 and 34.52 each have four significant figures. When numbers begin or end with zeros,

Computers are often used in engineering for advanced design and analysis.

however, it is difficult to tell how many significant figures are in the number. Consider the number 400. Does it have one (4), or perhaps two (40), or three (400) significant figures? In order to clarify this situation, the number should be reported using powers of 10. Using *engineering notation*, the exponent is displayed in multiples of three in order to facilitate conversion of SI units to those having an appropriate prefix. Thus, 400 expressed to one significant figure would be $0.4(10^3)$. Likewise, 2500 and 0.00546 expressed to three significant figures would be $2.50(10^3)$ and $5.46(10^{-3})$.

Rounding Off Numbers. For numerical calculations, the accuracy obtained from the solution of a problem generally can never be better than the accuracy of the problem data. This is what is to be expected, but often handheld calculators or computers involve more figures in the answer than the number of significant figures used for the data. For this reason, a calculated result should always be "rounded off" to an appropriate number of significant figures.

To convey appropriate accuracy, the following rules for rounding off a number to n significant figures apply:

* If the $n + 1$ digit is *less than 5*, the $n + 1$ digit and others following it are dropped. For example, 2.326 and 0.451 rounded off to $n = 2$ significant figures would be 2.3 and 0.45.

* If the $n + 1$ digit is equal to 5 with zeros following it, then round off the nth digit to an *even number*. For example, $1.245(10^3)$ and 0.8655 rounded off to $n = 3$ significant figures become $1.24(10^3)$ and 0.866.

* If the $n + 1$ digit is *greater than 5* or equal to 5 with any nonzero digits following it, then increase the nth digit by 1 and drop the $n + 1$ digit and others following it. For example, 0.723 87 and 565.500 3 rounded off to $n = 3$ significant figures become 0.724 and 566.

Calculations. As a general rule, to ensure accuracy of a final result when performing calculations on a pocket calculator, always retain a greater number of digits than the problem data. If possible, try to work out the computations so that numbers which are approximately equal are not subtracted since accuracy is often lost from this calculation.

In engineering we generally round off final answers to *three* significant figures since the data for geometry, loads, and other measurements are often reported with this accuracy.* Consequently, in this book the intermediate calculations for the examples are often worked out to four significant figures and the answers are generally reported to *three* significant figures.

---

*Of course, some numbers, such as π, e, or numbers used in derived formulas are exact and are therefore accurate to an infinite number of significant figures.

EXAMPLE 1.1

Convert 2 km/h to m/s. How many ft/s is this?

Solution
Since 1 km = 1000 m and 1 h = 3600 s, the factors of conversion are arranged in the following order, so that a cancellation of the units can be applied:

$$2 \text{ km/h} = \frac{2 \cancel{\text{km}}}{\cancel{\text{h}}} \left(\frac{1000 \text{ m}}{\cancel{\text{km}}} \right) \left(\frac{1 \cancel{\text{h}}}{3600 \text{ s}} \right)$$

$$= \frac{2000 \text{ m}}{3600 \text{ s}} = 0.556 \text{ m/s} \qquad \text{Ans.}$$

From Table 1–2, 1 ft = 0.3048 m. Thus

$$0.556 \text{ m/s} = \frac{0.556 \cancel{\text{m}}}{\text{s}} \frac{1 \text{ ft}}{0.3048 \cancel{\text{m}}}$$

$$= 1.82 \text{ ft/s} \qquad \text{Ans.}$$

EXAMPLE 1.2

Convert the quantities 300 lb · s and 52 slug/ft³ to appropriate SI units.

Solution
Using Table 1–2, 1 lb = 4.448 2 N.

$$300 \text{ lb} \cdot \text{s} = 300 \cancel{\text{lb}} \cdot \text{s} \left(\frac{4.448 \text{ 2 N}}{\cancel{\text{lb}}} \right)$$

$$= 1334.5 \text{ N} \cdot \text{s} = 1.33 \text{ kN} \cdot \text{s} \qquad \text{Ans.}$$

Also, 1 slug = 14.593 8 kg and 1 ft = 0.304 8 m.

$$52 \text{ slug/ft}^3 = \frac{52 \cancel{\text{slug}}}{\cancel{\text{ft}^3}} \left(\frac{14.593 \text{ 8 kg}}{1 \cancel{\text{slug}}} \right) \left(\frac{1 \cancel{\text{ft}}}{0.304 \text{ 8 m}} \right)^3$$

$$= 26.8(10^3) \text{kg/m}^3$$

$$= 26.8 \text{ Mg/m}^3 \qquad \text{Ans.}$$

E X A M P L E **1.3**

Evaluate each of the following and express with SI units having an appropriate prefix: (a) $(50 \text{ mN})(6 \text{ GN})$, (b) $(400 \text{ mm})(0.6 \text{ MN})^2$, (c) $45 \text{ MN}^3/900 \text{ Gg}$.

Solution

First convert each number to base units, perform the indicated operations, then choose an appropriate prefix (see Rule 6 on p. 9).

Part (a)

$$(50 \text{ mN})(6 \text{ GN}) = [50(10^{-3}) \text{ N}][6(10^9) \text{ N}]$$

$$= 300(10^6) \text{ N}^2$$

$$= 300(10^6) \text{ N}^2 \left(\frac{1 \text{ kN}}{10^3 \text{ N}}\right)\left(\frac{1 \text{ kN}}{10^3 \text{ N}}\right)$$

$$= 300 \text{ kN}^2 \qquad\qquad Ans.$$

Note carefully the convention $\text{kN}^2 = (\text{kN})^2 = 10^6 \text{ N}^2$ (Rule 4 on p. 9).

Part (b)

$$(400 \text{ mm})(0.6 \text{ MN})^2 = [400(10^{-3}) \text{ m}][0.6(10^6) \text{ N}]^2$$

$$= [400(10^{-3}) \text{ m}][0.36(10^{12}) \text{ N}^2]$$

$$= 144(10^9) \text{ m} \cdot \text{N}^2$$

$$= 144 \text{ Gm} \cdot \text{N}^2 \qquad\qquad Ans.$$

We can also write

$$144(10^9)\text{m} \cdot \text{N}^2 = 144(10^9)\text{m} \cdot \text{N}^2\left(\frac{1 \text{ MN}}{10^6 \text{ N}}\right)\left(\frac{1 \text{ MN}}{10^6 \text{ N}}\right)$$

$$= 0.144 \text{ m} \cdot \text{MN}^2$$

Part (c)

$$45 \text{ MN}^3/900 \text{ Gg} = \frac{45(10^6 \text{ N})^3}{900(10^6) \text{ kg}}$$

$$= 0.05(10^{12}) \text{ N}^3/\text{kg}$$

$$= 0.05(10^{12}) \text{ N}^3\left(\frac{1 \text{ kN}}{10^3 \text{ N}}\right)^3 \frac{1}{\text{kg}}$$

$$= 0.05(10^3) \text{ kN}^3/\text{kg}$$

$$= 50 \text{ kN}^3/\text{kg} \qquad\qquad Ans.$$

Here we have used Rules 4 and 8 on p. 9.

1.6 General Procedure for Analysis

When solving problems, do the work as neatly as possible. Being neat generally stimulates clear and orderly thinking, and vice versa.

The most effective way of learning the principles of engineering mechanics is to *solve problems*. To be successful at this, it is important to always present the work in a *logical* and *orderly manner*, as suggested by the following sequence of steps:

1. Read the problem carefully and try to correlate the actual physical situation with the theory studied.
2. Draw any necessary diagrams and tabulate the problem data.
3. Apply the relevant principles, generally in mathematical form.
4. Solve the necessary equations algebraically as far as practical, then, making sure they are dimensionally homogeneous, use a consistent set of units and complete the solution numerically. Report the answer with no more significant figures than the accuracy of the given data.
5. Study the answer with technical judgment and common sense to determine whether or not it seems reasonable.

IMPORTANT POINTS

- Statics is the study of bodies that are at rest or move with constant velocity.
- A particle has a mass but a size that can be neglected.
- A rigid body does not deform under load.
- Concentrated forces are assumed to act at a point on a body.
- Newton's three laws of motion should be memorized.
- Mass is a property of matter that does not change from one location to another.
- Weight refers to the gravitational attraction of the earth on a body or quantity of mass. Its magnitude depends upon the elevation at which the mass is located.
- In the SI system the unit of force, the newton, is a derived unit. The meter, second, and kilogram are base units.
- Prefixes G, M, k, m, μ, n are used to represent large and small numerical quantities. Their exponential size should be known, along with the rules for using the SI units.
- Perform numerical calculations to several significant figures and then report the final answer to three significant figures.
- Algebraic manipulations of an equation can be checked in part by verifying that the equation remains dimensionally homogeneous.
- Know the rules for rounding off numbers.

This communications tower is stabilized by cables that exert forces at the points of connection. In this chapter, we will show how to determine the magnitude and direction of the resultant force at each point.

Force Vectors

- To show how to add forces and resolve them into components using the Parallelogram Law.
- To express force and position in Cartesian vector form and explain how to determine the vector's magnitude and direction.
- To introduce the dot product in order to determine the angle between two vectors or the projection of one vector onto another.

2.1 Scalars and Vectors

Most of the physical quantities in mechanics can be expressed mathematically by means of scalars and vectors.

Scalar. A quantity characterized by a positive or negative number is called a *scalar*. For example, mass, volume, and length are scalar quantities often used in statics. In this book, scalars are indicated by letters in italic type, such as the scalar A.

Vector. A *vector* is a quantity that has both a magnitude and a direction. In statics the vector quantities frequently encountered are position, force, and moment. For handwritten work, a vector is generally represented by a letter with an arrow written over it, such as $\vec{A}$. The magnitude is designated $|\vec{A}|$ or simply A. In this book vectors will be symbolized in boldface type; for example, **A** is used to designate the vector "A." Its magnitude, which is always a positive quantity, is symbolized in italic type, written as $|A|$, or simply A when it is understood that A is a positive scalar.

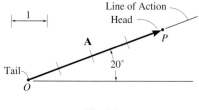

Line of Action
Head

A

P

20°

Tail

O

Fig. 2–1

A vector is represented graphically by an arrow, which is used to define its magnitude, direction, and sense. The *magnitude* of the vector is the length of the arrow, the *direction* is defined by the angle between a reference axis and the arrow's line of action, and the *sense* is indicated by the arrowhead. For example, the vector **A** shown in Fig. 2–1 has a magnitude of 4 units, a direction which is 20° measured counterclockwise from the horizontal axis, and a sense which is upward and to the right. The point *O* is called the *tail* of the vector, the point *P* the *tip* or *head*.

2.2 Vector Operations

A

−**A**

Vector **A** and its negative counterpart

Fig. 2–2

Multiplication and Division of a Vector by a Scalar. The product of vector **A** and scalar a, yielding $a\mathbf{A}$, is defined as a vector having a magnitude $|a\mathbf{A}|$. The *sense* of $a\mathbf{A}$ is the *same* as **A** provided a is *positive*; it is *opposite* to **A** if a is *negative*. In particular, the negative of a vector is formed by multiplying the vector by the scalar (-1), Fig. 2–2. Division of a vector by a scalar can be defined using the laws of multiplication, since $\mathbf{A}/a = (1/a)\mathbf{A}, a \neq 0$. Graphic examples of these operations are shown in Fig. 2–3.

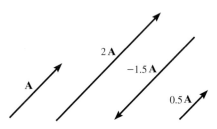

2 **A**

−1.5 **A**

A

0.5 **A**

Scalar Multiplication and Division

Fig. 2–3

Vector Addition. Two vectors **A** and **B** such as force or position, Fig. 2–4*a*, may be added to form a "resultant" vector $\mathbf{R} = \mathbf{A} + \mathbf{B}$ by using the *parallelogram law*. To do this, **A** and **B** are joined at their tails, Fig. 2–4*b*. Parallel lines drawn from the head of each vector intersect at a common point, thereby forming the adjacent sides of a parallelogram. As shown, the resultant **R** is the diagonal of the parallelogram, which extends from the tails of **A** and **B** to the intersection of the lines.

We can also add **B** to **A** using a *triangle construction*, which is a special case of the parallelogram law, whereby vector **B** is added to vector **A** in a "head-to-tail" fashion, i.e., by connecting the head of **A** to the tail of **B**, Fig. 2–4*c*. The resultant **R** extends from the tail of **A** to the head of **B**. In a similar manner, **R** can also be obtained by adding **A** to **B**, Fig. 2–4*d*. By comparison, it is seen that vector addition is commutative; in other words, the vectors can be added in either order, i.e., $\mathbf{R} = \mathbf{A} + \mathbf{B} = \mathbf{B} + \mathbf{A}$.

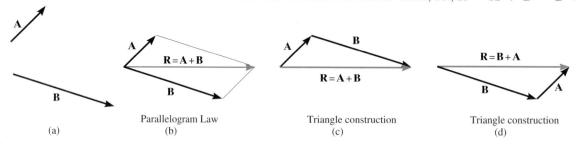

A

B

(a)

A

B

$\mathbf{R} = \mathbf{A} + \mathbf{B}$

Parallelogram Law
(b)

A

B

$\mathbf{R} = \mathbf{A} + \mathbf{B}$

Triangle construction
(c)

$\mathbf{R} = \mathbf{B} + \mathbf{A}$

B

A

Triangle construction
(d)

Vector Addition

Fig. 2–4

As a special case, if the two vectors **A** and **B** are *collinear*, i.e., both have the same line of action, the parallelogram law reduces to an *algebraic* or *scalar addition* $R = A + B$, as shown in Fig. 2–5.

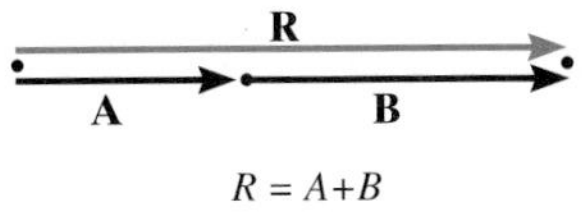

Vector Subtraction.   The resultant *difference* between two vectors **A** and **B** of the same type may be expressed as

$$\mathbf{R}' = \mathbf{A} - \mathbf{B} = \mathbf{A} + (-\mathbf{B})$$

This vector sum is shown graphically in Fig. 2–6. Subtraction is therefore defined as a special case of addition, so the rules of vector addition also apply to vector subtraction.

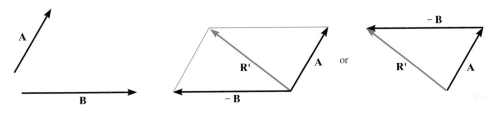

$R = A+B$

Addition of collinear vectors

Fig. 2–5

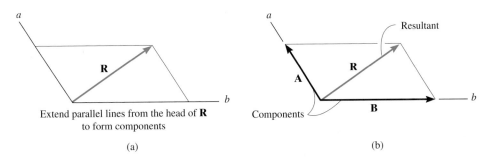

Parallelogram law

Vector Subtraction

Triangle construction

Fig. 2–6

Resolution of Vector.   A vector may be resolved into two "components" having known lines of action by using the parallelogram law. For example, if **R** in Fig. 2–7a is to be resolved into components acting along the lines *a* and *b*, one starts at the *head* of **R** and extends a line *parallel* to *a* until it intersects *b*. Likewise, a line parallel to *b* is drawn from the *head* of **R** to the point of intersection with *a*, Fig. 2–7a. The two components **A** and **B** are then drawn such that they extend from the tail of **R** to the points of intersection, as shown in Fig. 2–7b.

Extend parallel lines from the head of **R** to form components

(a)

Components

(b)

Resolution of a vector

Fig. 2–7

2.3 Vector Addition of Forces

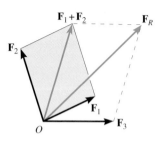

Fig. 2–8

Experimental evidence has shown that a force is a vector quantity since it has a specified magnitude, direction, and sense and it adds according to the parallelogram law. Two common problems in statics involve either finding the resultant force, knowing its components, or resolving a known force into two components. As described in Sec. 2.2, both of these problems require application of the parallelogram law.

If more than two forces are to be added, successive applications of the parallelogram law can be carried out in order to obtain the resultant force. For example, if three forces $\mathbf{F}_1$, $\mathbf{F}_2$, $\mathbf{F}_3$ act at a point O, Fig. 2–8, the resultant of any two of the forces is found—say, $\mathbf{F}_1 + \mathbf{F}_2$—and then this resultant is added to the third force, yielding the resultant of all three forces; i.e., $\mathbf{F}_R = (\mathbf{F}_1 + \mathbf{F}_2) + \mathbf{F}_3$. Using the parallelogram law to add more than two forces, as shown here, often requires extensive geometric and trigonometric calculation to determine the numerical values for the magnitude and direction of the resultant. Instead, problems of this type are easily solved by using the "rectangular-component method," which is explained in Sec. 2.4.

If we know the forces $\mathbf{F}_a$ and $\mathbf{F}_b$ that the two chains a and b exert on the hook, we can find their resultant force $\mathbf{F}_c$ by using the parallelogram law. This requires drawing lines parallel to a and b from the heads of $\mathbf{F}_a$ and $\mathbf{F}_b$ as shown thus forming a parallelogram.

In a similar manner, if the force $\mathbf{F}_c$ along chain c is known, then its two components $\mathbf{F}_a$ and $\mathbf{F}_b$, that act along a and b, can be determined from the parallelogram law. Here we must start at the head of $\mathbf{F}_c$ and construct lines parallel to a and b, thereby forming the parallelogram.

PROCEDURE FOR ANALYSIS

Problems that involve the addition of two forces can be solved as follows:

Parallelogram Law.

- Make a sketch showing the vector addition using the parallelogram law.
- Two "component" forces add according to the parallelogram law, yielding a *resultant* force that forms the diagonal of the parallelogram.
- If a force is to be resolved into *components* along two axes directed from the tail of the force, then start at the head of the force and construct lines parallel to the axes, thereby forming the parallelogram. The sides of the parallelogram represent the components.
- Label all the known and unknown force magnitudes and the angles on the sketch and identify the two unknowns.

Trigonometry.

- Redraw a half portion of the parallelogram to illustrate the triangular head-to-tail addition of the components.
- The magnitude of the resultant force can be determined from the law of cosines, and its direction is determined from the law of sines, Fig. 2–9.
- The magnitude of two force components are determined from the law of sines, Fig. 2–9.

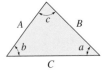

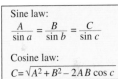

Sine law:
$$\frac{A}{\sin a} = \frac{B}{\sin b} = \frac{C}{\sin c}$$

Cosine law:
$$C = \sqrt{A^2 + B^2 - 2AB \cos c}$$

Fig. 2–9

IMPORTANT POINTS

- A scalar is a positive or negative number.
- A vector is a quantity that has magnitude, direction, and sense.
- Multiplication or division of a vector by a scalar will change the magnitude of the vector. The sense of the vector will change if the scalar is negative.
- As a special case, if the vectors are collinear, the resultant is formed by an algebraic or scalar addition.

E X A M P L E   2.1

The screw eye in Fig. 2–10a is subjected to two forces, $\mathbf{F}_1$ and $\mathbf{F}_2$. Determine the magnitude and direction of the resultant force.

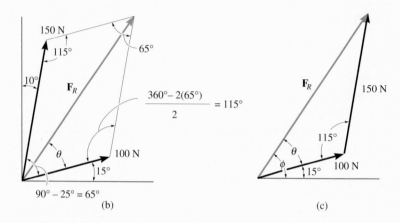

(a)

(b)

(c)

Fig. 2-10

Solution

Parallelogram Law. The parallelogram law of addition is shown in Fig. 2–10b. The two unknowns are the magnitude of $\mathbf{F}_R$ and the angle θ (theta).

Trigonometry. From Fig. 2–10b, the vector triangle, Fig. 2–10c, is constructed. F_R is determined by using the law of cosines:

$$F_R = \sqrt{(100 \text{ N})^2 + (150 \text{ N})^2 - 2(100 \text{ N})(150 \text{ N}) \cos 115°}$$
$$= \sqrt{10\,000 + 22\,500 - 30\,000(-0.4226)} = 212.6 \text{ N}$$
$$= 213 \text{ N} \qquad\qquad\qquad Ans.$$

The angle θ is determined by applying the law of sines, using the computed value of F_R.

$$\frac{150 \text{ N}}{\sin \theta} = \frac{212.6 \text{ N}}{\sin 115°}$$

$$\sin \theta = \frac{150 \text{ N}}{212.6 \text{ N}}(0.9063)$$

$$\theta = 39.8°$$

Thus, the direction ϕ (phi) of $\mathbf{F}_R$, measured from the horizontal, is

$$\phi = 39.8° + 15.0° = 54.8° \; \angle^\phi \qquad\qquad Ans.$$

EXAMPLE 2.2

Resolve the 200-lb force acting on the pipe, Fig. 2–11a, into components in the (a) x and y directions, and (b) x' and y directions.

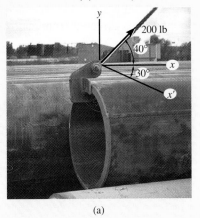

(a)

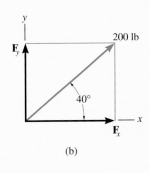

(b)

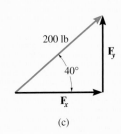

(c)

Fig. 2–11

Solution

In each case the parallelogram law is used to resolve **F** into its two components, and then the vector triangle is constructed to determine the numerical results by trigonometry.

Part (a). The vector addition $\mathbf{F} = \mathbf{F}_x + \mathbf{F}_y$ is shown in Fig. 2–11b. In particular, note that the length of the components is scaled along the x and y axes by first constructing lines from the tip of **F** parallel to the axes in accordance with the parallelogram law. From the vector triangle, Fig. 2–11c,

$$F_x = 200 \text{ lb} \cos 40° = 153 \text{ lb} \qquad Ans.$$
$$F_y = 200 \text{ lb} \sin 40° = 129 \text{ lb} \qquad Ans.$$

Part (b). The vector addition $\mathbf{F} = \mathbf{F}_{x'} + \mathbf{F}_y$ is shown in Fig. 2–11d. Note carefully how the parallelogram is constructed. Applying the law of sines and using the data listed on the vector triangle, Fig. 2–11e, yields

$$\frac{F_{x'}}{\sin 50°} = \frac{200 \text{ lb}}{\sin 60°}$$

$$F_{x'} = 200 \text{ lb}\left(\frac{\sin 50°}{\sin 60°}\right) = 177 \text{ lb} \qquad Ans.$$

$$\frac{F_y}{\sin 70°} = \frac{200 \text{ lb}}{\sin 60°}$$

$$F_y = 200 \text{ lb}\left(\frac{\sin 70°}{\sin 60°}\right) = 217 \text{ lb} \qquad Ans.$$

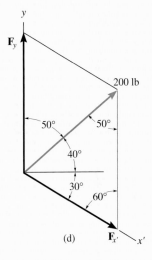

(d)

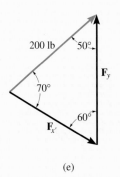

(e)

E X A M P L E 2.3

The force **F** acting on the frame shown in Fig. 2–12a has a magnitude of 500 N and is to be resolved into two components acting along members AB and AC. Determine the angle θ, measured *below* the horizontal, so that the component $\mathbf{F}_{AC}$ is directed from A toward C and has a magnitude of 400 N.

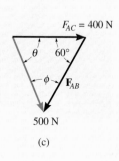

(b)

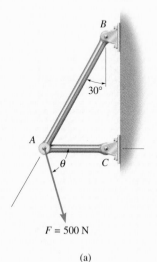

Fig. 2–12

(a)

$F = 500$ N

(c)

Solution

By using the parallelogram law, the vector addition of the two components yielding the resultant is shown in Fig. 2–12b. Note carefully how the resultant force is resolved into the two components $\mathbf{F}_{AB}$ and $\mathbf{F}_{AC}$, which have specified lines of action. The corresponding vector triangle is shown in Fig. 2–12c.

The angle ϕ can be determined by using the law of sines:

$$\frac{400 \text{ N}}{\sin \phi} = \frac{500 \text{ N}}{\sin 60°}$$

$$\sin \phi = \left(\frac{400 \text{ N}}{500 \text{ N}}\right) \sin 60° = 0.6928$$

$$\phi = 43.9°$$

Hence,

$$\theta = 180° - 60° - 43.9° = 76.1° \quad \text{\searrow}_\theta \qquad Ans.$$

Using this value for θ, apply the law of cosines or the law of sines and show that $\mathbf{F}_{AB}$ has a magnitude of 561 N.

Notice that **F** can also be directed at an angle θ *above* the horizontal, as shown in Fig. 2–12d, and still produce the required component $\mathbf{F}_{AC}$. Show that in this case $\theta = 16.1°$ and $F_{AB} = 161$ N.

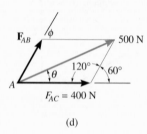

(d)

E X A M P L E   **2.4**

The ring shown in Fig. 2–13a is subjected to two forces, F_1 and F_2. If it is required that the resultant force have a magnitude of 1 kN and be directed vertically downward, determine (a) the magnitudes of F_1 and F_2 provided $\theta = 30°$, and (b) the magnitudes of F_1 and F_2 if F_2 is to be a minimum.

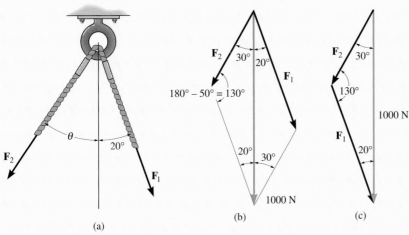

(a)

(b)

(c)

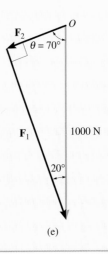

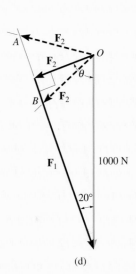

(d)

(e)

Fig. 2–13

Solution

Part (a). A sketch of the vector addition according to the parallelogram law is shown in Fig. 2–13b. From the vector triangle constructed in Fig. 2–13c, the unknown magnitudes F_1 and F_2 are determined by using the law of sines:

$$\frac{F_1}{\sin 30°} = \frac{1000 \text{ N}}{\sin 130°}$$

$$F_1 = 653 \text{ N} \qquad\qquad Ans.$$

$$\frac{F_2}{\sin 20°} = \frac{1000 \text{ N}}{\sin 130°}$$

$$F_2 = 446 \text{ N} \qquad\qquad Ans.$$

Part (b). If θ is not specified, then by the vector triangle, Fig. 2–13d, F_2 may be added to F_1 in various ways to yield the resultant 1000-N force. In particular, the *minimum* length or magnitude of F_2 will occur when its line of action is *perpendicular to F_1*. Any other direction, such as OA or OB, yields a larger value for F_2. Hence, when $\theta = 90° - 20° = 70°$, F_2 is minimum. From the triangle shown in Fig. 2–13e, it is seen that

$$F_1 = 1000 \sin 70° \text{N} = 940 \text{ N} \qquad\qquad Ans.$$

$$F_2 = 1000 \cos 70° \text{N} = 342 \text{ N} \qquad\qquad Ans.$$

2.4 Addition of a System of Coplanar Forces

When the resultant of more than two forces has to be obtained, it is easier to find the components of each force along specified axes, add these components algebraically, and then form the resultant, rather than form the resultant of the forces by successive application of the parallelogram law as discussed in Sec. 2.3.

In this section we will resolve each force into its rectangular components $\mathbf{F}_x$ and $\mathbf{F}_y$, which lie along the x and y axes, respectively, Fig. 2–14a. Although the axes are horizontal and vertical, they may in general be directed at any inclination, as long as they remain perpendicular to one another, Fig. 2–14b. In either case, by the parallelogram law, we require

$$\mathbf{F} = \mathbf{F}_x + \mathbf{F}_y$$

and

$$\mathbf{F}' = \mathbf{F}'_x + \mathbf{F}'_y$$

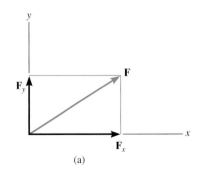

(a)

As shown in Fig. 2–14, the sense of direction of each force component is represented *graphically* by the *arrowhead*. For *analytical* work, however, we must establish a notation for representing the directional sense of the rectangular components. This can be done in one of two ways.

Scalar Notation.   Since the x and y axes have designated positive and negative directions, the magnitude and directional sense of the rectangular components of a force can be expressed in terms of *algebraic scalars*. For example, the components of $\mathbf{F}$ in Fig. 2–14a can be represented by positive scalars F_x and F_y since their sense of direction is along the *positive x* and *y* axes, respectively. In a similar manner, the components of $\mathbf{F}'$ in Fig. 2–14b are F'_x and $-F'_y$. Here the y component is negative, since $\mathbf{F}'_y$ is directed along the negative y axis.

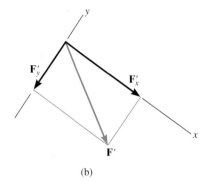

(b)

Fig. 2–14

It is important to keep in mind that this scalar notation is to be used only for computational purposes, not for graphical representations in figures. Throughout the book, the *head of a vector arrow* in any figure indicates the sense of the vector *graphically*; algebraic signs are not used for this purpose. Thus, the vectors in Figs. 2–14a and 2–14b are designated by using boldface (vector) notation.* Whenever italic symbols are written near vector arrows in figures, they indicate the *magnitude* of the vector, which is *always* a *positive* quantity.

---

*Negative signs are used only in figures with boldface notation when showing equal but opposite pairs of vectors as in Fig. 2–2.

Cartesian Vector Notation. It is also possible to represent the components of a force in terms of Cartesian unit vectors. When we do this the methods of vector algebra are easier to apply, and we will see that this becomes particularly advantageous for solving problems in three dimensions.

In two dimensions the *Cartesian unit vectors* **i** and **j** are used to designate the *directions* of the *x* and *y* axes, respectively, Fig. 2–15a.* These vectors have a dimensionless magnitude of unity, and their sense (or arrowhead) will be described analytically by a plus or minus sign, depending on whether they are pointing along the positive or negative *x* or *y* axis.

As shown in Fig. 2–15a, the *magnitude* of each component of **F** is *always a positive quantity*, which is represented by the (positive) scalars F_x and F_y. Therefore, having established notation to represent the magnitude and the direction of each vector component, we can express **F** in Fig. 2–15a as the *Cartesian vector*,

$$\mathbf{F} = F_x\mathbf{i} + F_y\mathbf{j}$$

And in the same way, **F′** in Fig. 2–15b can be expressed as

$$\mathbf{F}' = F'_x\mathbf{i} + F'_y(-\mathbf{j})$$

or simply

$$\mathbf{F}' = F'_x\mathbf{i} - F'_y\mathbf{j}$$

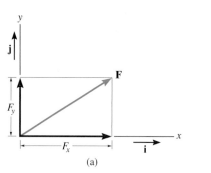

(a)

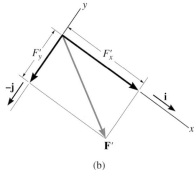

(b)

Fig. 2–15

*For handwritten work, unit vectors are usually indicated using a circumflex, e.g., $\hat{i}$ and $\hat{j}$.

Coplanar Force Resultants. Either of the two methods just described can be used to determine the resultant of several *coplanar forces*. To do this, each force is first resolved into its x and y components, and then the respective components are added using *scalar algebra* since they are collinear. The resultant force is then formed by adding the resultants of the x and y components using the parallelogram law. For example, consider the three concurrent forces in Fig. 2–16a, which have x and y components as shown in Fig. 2–16b. To solve this problem using *Cartesian vector notation*, each force is first represented as a Cartesian vector, i.e.,

$$\mathbf{F}_1 = F_{1x}\mathbf{i} + F_{1y}\mathbf{j}$$
$$\mathbf{F}_2 = -F_{2x}\mathbf{i} + F_{2y}\mathbf{j}$$
$$\mathbf{F}_3 = F_{3x}\mathbf{i} - F_{3y}\mathbf{j}$$

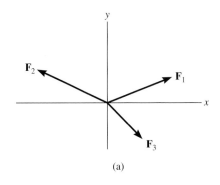

(a)

The vector resultant is therefore

$$\begin{aligned}
\mathbf{F}_R &= \mathbf{F}_1 + \mathbf{F}_2 + \mathbf{F}_3 \\
&= F_{1x}\mathbf{i} + F_{1y}\mathbf{j} - F_{2x}\mathbf{i} + F_{2y}\mathbf{j} + F_{3x}\mathbf{i} - F_{3y}\mathbf{j} \\
&= (F_{1x} - F_{2x} + F_{3x})\mathbf{i} + (F_{1y} + F_{2y} - F_{3y})\mathbf{j} \\
&= (F_{Rx})\mathbf{i} + (F_{Ry})\mathbf{j}
\end{aligned}$$

If *scalar notation* is used, then, from Fig. 2–16b, since x is positive to the right and y is positive upward, we have

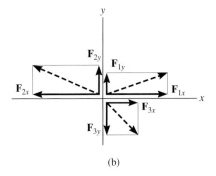

(b)

$$(\xrightarrow{+}) \qquad\qquad F_{Rx} = F_{1x} - F_{2x} + F_{3x}$$
$$(+\uparrow) \qquad\qquad F_{Ry} = F_{1y} + F_{2y} - F_{3y}$$

These results are the *same* as the $\mathbf{i}$ and $\mathbf{j}$ components of $\mathbf{F}_R$ determined above.

In the general case, the x and y components of the resultant of any number of coplanar forces can be represented symbolically by the algebraic sum of the x and y components of all the forces, i.e.,

$$\boxed{\begin{aligned} F_{Rx} &= \Sigma F_x \\ F_{Ry} &= \Sigma F_y \end{aligned}} \qquad\qquad (2\text{–}1)$$

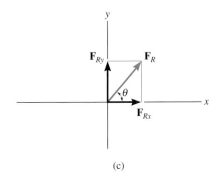

(c)

Fig. 2–16

When applying these equations, it is important to use the *sign convention* established for the components; and that is, components having a directional sense along the positive coordinate axes are considered positive scalars, whereas those having a directional sense along the negative coordinate axes are considered negative scalars. If this convention is followed, then the signs of the resultant components will specify the sense of these components. For example, a positive result indicates that the component has a directional sense which is in the positive coordinate direction.

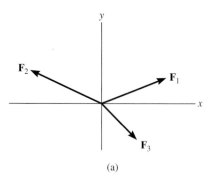

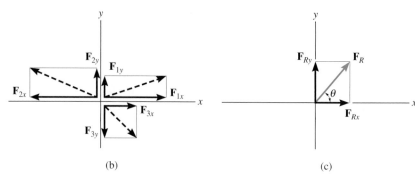

(a)   (b)   (c)

Fig. 2–16

The resultant force of the four cable forces acting on the supporting bracket can be determined by adding algebraically the separate x and y components of each cable force. This resultant $\mathbf{F}_R$ produces the *same pulling effect* on the bracket as all four cables.

Once the resultant components are determined, they may be sketched along the x and y axes in their proper directions, and the resultant force can be determined from vector addition, as shown in Fig. 2–16c. From this sketch, the magnitude of $\mathbf{F}_R$ is then found from the Pythagorean theorem; that is,

$$F_R = \sqrt{F_{Rx}^2 + F_{Ry}^2}$$

Also, the direction angle θ, which specifies the orientation of the force, is determined from trigonometry:

$$\theta = \tan^{-1}\left|\frac{F_{Ry}}{F_{Rx}}\right|$$

The above concepts are illustrated numerically in the examples which follow.

IMPORTANT POINTS

- The resultant of several coplanar forces can easily be determined if an x, y coordinate system is established and the forces are resolved along the axes.
- The direction of each force is specified by the angle its line of action makes with one of the axes, or by a sloped triangle.
- The orientation of the x and y axes is arbitrary, and their positive direction can be specified by the Cartesian unit vectors $\mathbf{i}$ and $\mathbf{j}$.
- The x and y components of the *resultant force* are simply the algebraic addition of the components of all the coplanar forces.
- The magnitude of the resultant force is determined from the Pythagorean theorem, and when the components are sketched on the x and y axes, the direction can be determined from trigonometry.

EXAMPLE 2.5

Determine the x and y components of $\mathbf{F}_1$ and $\mathbf{F}_2$ acting on the boom shown in Fig. 2–17a. Express each force as a Cartesian vector.

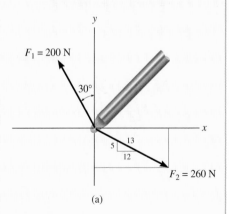

$F_1 = 200$ N

$30°$

x

$F_2 = 260$ N

(a)

Solution

Scalar Notation. By the parallelogram law, $\mathbf{F}_1$ is resolved into x and y components, Fig. 2–17b. The magnitude of each component is determined by trigonometry. Since $\mathbf{F}_{1x}$ acts in the $-x$ direction, and $\mathbf{F}_{1y}$ acts in the $+y$ direction, we have

$$F_{1x} = -200 \sin 30° \text{ N} = -100 \text{ N} = 100 \text{ N} \leftarrow \qquad Ans.$$
$$F_{1y} = 200 \cos 30° \text{ N} = 173 \text{ N} = 173 \text{ N} \uparrow \qquad Ans.$$

The force $\mathbf{F}_2$ is resolved into its x and y components as shown in Fig. 2–17c. Here the *slope* of the line of action for the force is indicated. From this "slope triangle" we could obtain the angle θ, e.g., $\theta = \tan^{-1}(\frac{5}{12})$, and then proceed to determine the magnitudes of the components in the same manner as for $\mathbf{F}_1$. An easier method, however, consists of using proportional parts of similar triangles, i.e.,

$$\frac{F_{2x}}{260 \text{ N}} = \frac{12}{13} \qquad F_{2x} = 260 \text{ N}\left(\frac{12}{13}\right) = 240 \text{ N}$$

Similarly,

$$F_{2y} = 260 \text{ N}\left(\frac{5}{13}\right) = 100 \text{ N}$$

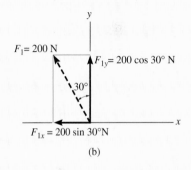

y

$F_1 = 200$ N

$F_{1y} = 200 \cos 30°$ N

$30°$

x

$F_{1x} = 200 \sin 30°$ N

(b)

Notice that the magnitude of the *horizontal component*, F_{2x}, was obtained by multiplying the force magnitude by the ratio of the *horizontal leg* of the slope triangle divided by the hypotenuse; whereas the magnitude of the *vertical component*, F_{2y}, was obtained by multiplying the force magnitude by the ratio of the *vertical leg* divided by the hypotenuse. Hence, using scalar notation,

$$F_{2x} = 240 \text{ N} = 240 \text{ N} \rightarrow \qquad Ans.$$
$$F_{2y} = -100 \text{ N} = 100 \text{ N} \downarrow \qquad Ans.$$

Cartesian Vector Notation. Having determined the magnitudes and directions of the components of each force, we can express each force as a Cartesian vector.

$$\mathbf{F}_1 = \{-100\mathbf{i} + 173\mathbf{j}\} \text{ N} \qquad Ans.$$
$$\mathbf{F}_2 = \{240\mathbf{i} - 100\mathbf{j}\} \text{ N} \qquad Ans.$$

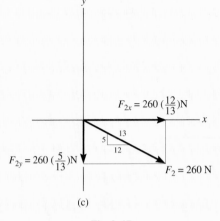

y

$F_{2x} = 260 \left(\frac{12}{13}\right)$N

x

$F_{2y} = 260 \left(\frac{5}{13}\right)$N

$F_2 = 260$ N

(c)

Fig. 2–17

E X A M P L E 2.6

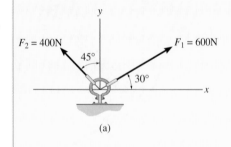

(a)

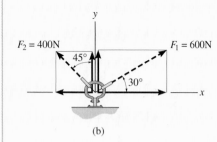

(b)

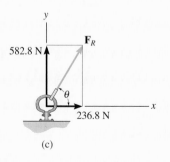

(c)

Fig. 2–18

The link in Fig. 2–18a is subjected to two forces $\mathbf{F}_1$ and $\mathbf{F}_2$. Determine the magnitude and orientation of the resultant force.

Solution I

Scalar Notation. This problem can be solved by using the parallelogram law; however, here we will resolve each force into its x and y components, Fig. 2–18b, and sum these components algebraically. Indicating the "positive" sense of the x and y force components alongside each equation, we have

$$\xrightarrow{+} F_{Rx} = \Sigma F_x; \qquad F_{Rx} = 600 \cos 30° \text{ N} - 400 \sin 45° \text{ N}$$
$$= 236.8 \text{ N} \rightarrow$$
$$+\uparrow F_{Ry} = \Sigma F_y; \qquad F_{Ry} = 600 \sin 30° \text{ N} + 400 \cos 45° \text{ N}$$
$$= 582.8 \text{ N}\uparrow$$

The resultant force, shown in Fig. 2–18c, has a *magnitude* of

$$F_R = \sqrt{(236.8 \text{ N})^2 + (582.8 \text{ N})^2}$$
$$= 629 \text{ N} \qquad \qquad \textit{Ans.}$$

From the vector addition, Fig. 2–18c, the direction angle θ is

$$\theta = \tan^{-1}\left(\frac{582.8\text{N}}{236.8\text{N}}\right) = 67.9° \qquad \textit{Ans.}$$

Solution II

Cartesian Vector Notation. From Fig. 2–18b, each force is expressed as a Cartesian vector

$$\mathbf{F}_1 = \{600 \cos 30°\mathbf{i} + 600 \sin 30°\mathbf{j}\} \text{ N}$$
$$\mathbf{F}_2 = \{-400 \sin 45°\mathbf{i} + 400 \cos 45°\mathbf{j}\} \text{ N}$$

Thus,

$$\mathbf{F}_R = \mathbf{F}_1 + \mathbf{F}_2 = (600 \cos 30° \text{ N} - 400 \sin 45° \text{ N})\mathbf{i}$$
$$+ (600 \sin 30° \text{ N} + 400 \cos 45° \text{ N})\mathbf{j}$$
$$= \{236.8\mathbf{i} + 582.8\mathbf{j}\} \text{ N}$$

The magnitude and direction of $\mathbf{F}_R$ are determined in the same manner as shown above.

Comparing the two methods of solution, note that use of scalar notation is more efficient since the components can be found *directly*, without first having to express each force as a Cartesian vector before adding the components. Later we will show that Cartesian vector analysis is very beneficial for solving three-dimensional problems.

EXAMPLE 2.7

The end of the boom O in Fig. 2–19a is subjected to three concurrent and coplanar forces. Determine the magnitude and orientation of the resultant force.

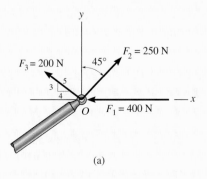

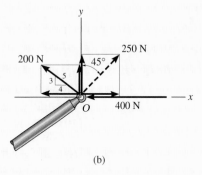

(a)   (b)

Fig. 2–19

Solution

Each force is resolved into its x and y components, Fig. 2–19b. Summing the x components, we have

$$\xrightarrow{+} F_{Rx} = \Sigma F_x; \qquad F_{Rx} = -400\ \text{N} + 250\sin 45°\ \text{N} - 200\left(\tfrac{4}{5}\right)\ \text{N}$$

$$= -383.2\ \text{N} = 383.2\ \text{N} \leftarrow$$

The negative sign indicates that F_{Rx} acts to the left, i.e., in the negative x direction as noted by the small arrow. Summing the y components yields

$$+\uparrow F_{Ry} = \Sigma F_y; \qquad F_{Ry} = 250\cos 45°\ \text{N} + 200\left(\tfrac{3}{5}\right)\ \text{N}$$

$$= 296.8\ \text{N} \uparrow$$

The resultant force, shown in Fig. 2–19c, has a *magnitude* of

$$F_R = \sqrt{(-383.2\text{N})^2 + (296.8\text{N})^2}$$

$$= 485\ \text{N} \qquad\qquad Ans.$$

From the vector addition in Fig. 2–19c, the direction angle θ is

$$\theta = \tan^{-1}\left(\frac{296.8}{383.2}\right) = 37.8° \qquad\qquad Ans.$$

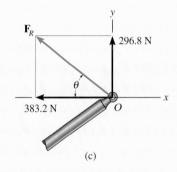

(c)

Note how convenient it is to use this method, compared to two applications of the parallelogram law.

2.5 Cartesian Vectors

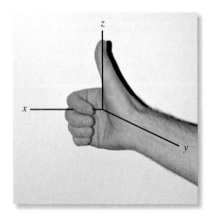

Right-handed coordinate system.

Fig. 2–20

The operations of vector algebra, when applied to solving problems in *three dimensions*, are greatly simplified if the vectors are first represented in Cartesian vector form. In this section we will present a general method for doing this; then in the next section we will apply this method to solving problems involving the addition of forces. Similar applications will be illustrated for the position and moment vectors given in later sections of the book.

Right-Handed Coordinate System. A right-handed coordinate system will be used for developing the theory of vector algebra that follows. A rectangular or Cartesian coordinate system is said to be *right-handed* provided the thumb of the right hand points in the direction of the positive z axis when the right-hand fingers are curled about this axis and directed from the positive x toward the positive y axis, Fig. 2–20. Furthermore, according to this rule, the z axis for a two-dimensional problem as in Fig. 2–19 would be directed outward, perpendicular to the page.

Rectangular Components of a Vector. A vector **A** may have one, two, or three rectangular components along the *x, y, z* coordinate axes, depending on how the vector is oriented relative to the axes. In general, though, when **A** is directed within an octant of the *x, y, z* frame, Fig. 2–21, then by two successive applications of the parallelogram law, we may resolve the vector into components as $\mathbf{A} = \mathbf{A'} + \mathbf{A}_z$ and then $\mathbf{A'} = \mathbf{A}_x + \mathbf{A}_y$. Combining these equations, **A** is represented by the vector sum of its *three* rectangular components,

$$\mathbf{A} = \mathbf{A}_x + \mathbf{A}_y + \mathbf{A}_z \tag{2-2}$$

Unit Vector. The direction of **A** can be specified using a unit vector. This vector is so named since it has a magnitude of 1. If **A** is a vector having a magnitude $A \neq 0$, then the unit vector having the *same direction* as **A** is represented by

$$\mathbf{u}_A = \frac{\mathbf{A}}{A} \tag{2-3}$$

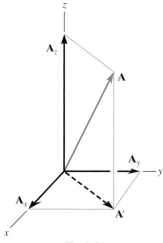

Fig. 2–21

So that

$$\mathbf{A} = A\mathbf{u}_A \tag{2-4}$$

Since **A** is of a certain type, e.g., a force vector, it is customary to use the proper set of units for its description. The magnitude A also has this same set of units; hence, from Eq. 2–3, the *unit vector will be dimensionless* since the units will cancel out. Equation 2–4 therefore indicates that vector **A** may be expressed in terms of both its magnitude and direction *separately*; i.e., A (a positive scalar) defines the *magnitude* of **A**, and $\mathbf{u}_A$ (a dimensionless vector) defines the *direction* and sense of **A**, Fig. 2–22.

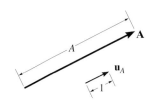

Fig. 2–22

Cartesian Unit Vectors. In three dimensions, the set of Cartesian unit vectors, **i**, **j**, **k**, is used to designate the directions of the *x, y, z* axes respectively. As stated in Sec. 2.4, the *sense* (or arrowhead) of these vectors will be described analytically by a plus or minus sign, depending on whether they are pointing along the positive or negative *x, y,* or *z* axes. The positive Cartesian unit vectors are shown in Fig. 2–23.

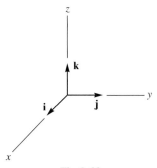

Fig. 2–23

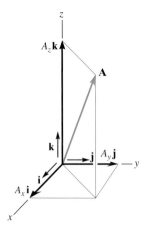

Fig. 2–24

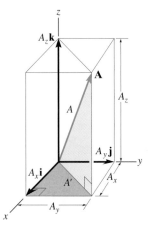

Fig. 2–25

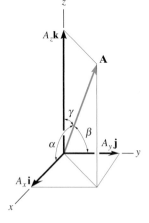

Fig. 2–26

Cartesian Vector Representation. Since the three components of **A** in Eq. 2–2 act in the positive **i**, **j**, and **k** directions, Fig. 2–24, we can write **A** in Cartesian vector form as

$$A = A_x\mathbf{i} + A_y\mathbf{j} + A_z\mathbf{k} \qquad (2\text{–}5)$$

There is a distinct advantage to writing vectors in this manner. Note that the *magnitude* and *direction* of each *component vector* are *separated*, and as a result this will simplify the operations of vector algebra, particularly in three dimensions.

Magnitude of a Cartesian Vector. It is always possible to obtain the magnitude of **A** provided it is expressed in Cartesian vector form. As shown in Fig. 2–25, from the colored right triangle, $A = \sqrt{A'^2 + A_z^2}$, and from the shaded right triangle, $A' = \sqrt{A_x^2 + A_y^2}$. Combining these equations yields

$$A = \sqrt{A_x^2 + A_y^2 + A_z^2} \qquad (2\text{–}6)$$

*Hence, the magnitude of **A** is equal to the positive square root of the sum of the squares of its components.*

Direction of a Cartesian Vector. The *orientation* of **A** is defined by the *coordinate direction angles* α (alpha), β (beta), and γ (gamma), measured between the *tail* of **A** and the *positive x, y, z* axes located at the tail of **A**, Fig. 2–26. Note that regardless of where **A** is directed, each of these angles will be between 0° and 180°.

To determine α, β, and γ, consider the projection of **A** onto the *x, y, z* axes, Fig. 2–27. Referring to the blue colored right triangles shown in each figure, we have

$$\cos\alpha = \frac{A_x}{A} \qquad \cos\beta = \frac{A_y}{A} \qquad \cos\gamma = \frac{A_z}{A} \qquad (2\text{–}7)$$

These numbers are known as the *direction cosines* of **A**. Once they have been obtained, the coordinate direction angles α, β, γ can then be determined from the inverse cosines.

An easy way of obtaining the direction cosines of **A** is to form a unit vector in the direction of **A**, Eq. 2–3. Provided **A** is expressed in Cartesian vector form, $A = A_x\mathbf{i} + A_y\mathbf{j} + A_z\mathbf{k}$ (Eq. 2–5), then

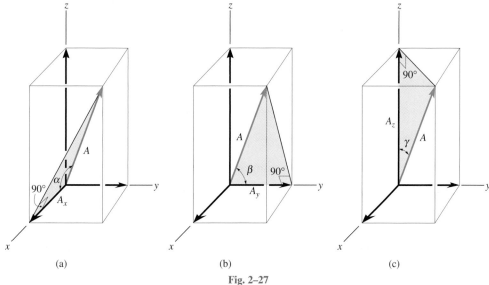

(a)   (b)   (c)

Fig. 2–27

$$\mathbf{u}_A = \frac{\mathbf{A}}{A} = \frac{A_x}{A}\mathbf{i} + \frac{A_y}{A}\mathbf{j} + \frac{A_z}{A}\mathbf{k} \qquad (2\text{--}8)$$

where $A = \sqrt{A_x^2 + A_y^2 + A_z^2}$ (Eq. 2–6). By comparison with Eqs. 2–7, it is seen that *the* **i, j, k** *components of* $\mathbf{u}_A$ *represent the direction cosines of* **A**, i.e.,

$$\mathbf{u}_A = \cos \alpha \mathbf{i} + \cos \beta \mathbf{j} + \cos \gamma \mathbf{k} \qquad (2\text{--}9)$$

Since the magnitude of a vector is equal to the positive square root of the sum of the squares of the magnitudes of its components, and $\mathbf{u}_A$ has a magnitude of 1, then from Eq. 2–9 an important relation between the direction cosines can be formulated as

$$\cos^2 \alpha + \cos^2 \beta + \cos^2 \gamma = 1 \qquad (2\text{--}10)$$

Provided vector **A** lies in a known octant, this equation can be used to determine one of the coordinate direction angles if the other two are known.

Finally, if the magnitude and coordinate direction angles of **A** are given, **A** may be expressed in Cartesian vector form as

$$\begin{aligned} \mathbf{A} &= A\mathbf{u}_A \\ &= A \cos \alpha \mathbf{i} + A \cos \beta \mathbf{j} + A \cos \gamma \mathbf{k} \\ &= A_x \mathbf{i} + A_y \mathbf{j} + A_z \mathbf{k} \end{aligned} \qquad (2\text{--}11)$$

2.6   Addition and Subtraction of Cartesian Vectors

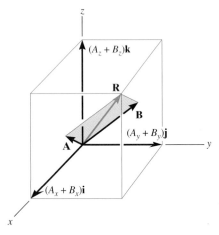

Fig. 2–28

The vector operations of addition and subtraction of two or more vectors are greatly simplified if the vectors are expressed in terms of their Cartesian components. For example, if $\mathbf{A} = A_x\mathbf{i} + A_y\mathbf{j} + A_z\mathbf{k}$ and $\mathbf{B} = B_x\mathbf{i} + B_y\mathbf{j} + B_z\mathbf{k}$, Fig. 2–28, then the resultant vector, $\mathbf{R}$, has components which represent the scalar sums of the $\mathbf{i}, \mathbf{j}, \mathbf{k}$ components of $\mathbf{A}$ and $\mathbf{B}$, i.e.,

$$\mathbf{R} = \mathbf{A} + \mathbf{B} = (A_x + B_x)\mathbf{i} + (A_y + B_y)\mathbf{j} + (A_z + B_z)\mathbf{k}$$

Vector subtraction, being a special case of vector addition, simply requires a scalar subtraction of the respective $\mathbf{i}, \mathbf{j}, \mathbf{k}$ components of either $\mathbf{A}$ or $\mathbf{B}$. For example,

$$\mathbf{R}' = \mathbf{A} - \mathbf{B} = (A_x - B_x)\mathbf{i} + (A_y - B_y)\mathbf{j} + (A_z - B_z)\mathbf{k}$$

Concurrent Force Systems.   If the above concept of vector addition is generalized and applied to a system of several concurrent forces, then the force resultant is the vector sum of all the forces in the system and can be written as

$$\mathbf{F}_R = \Sigma \mathbf{F} = \Sigma F_x\mathbf{i} + \Sigma F_y\mathbf{j} + \Sigma F_z\mathbf{k} \qquad (2\text{–}12)$$

Here ΣF_x, ΣF_y, and ΣF_z represent the algebraic sums of the respective x, y, z or $\mathbf{i}, \mathbf{j}, \mathbf{k}$ components of each force in the system.

The examples which follow illustrate numerically the methods used to apply the above theory to the solution of problems involving force as a vector quantity.

The force $\mathbf{F}$ that the tie-down rope exerts on the ground support at O is directed along the rope. Using the local x, y, z axes, the coordinate direction angles α, β, γ can be measured. The cosines of their values form the components of a unit vector $\mathbf{u}$ which acts in the direction of the rope. If the force has a magnitude F, then the force can be written in Cartesian vector form, as $\mathbf{F} = F\mathbf{u} = F \cos \alpha\mathbf{i} + F \cos \beta\mathbf{j} + F \cos \gamma\mathbf{k}$.

IMPORTANT POINTS

- Cartesian vector analysis is often used to solve problems in three dimensions.
- The positive direction of the x, y, z axes are defined by the Cartesian unit vectors **i**, **j**, **k**, respectively.
- The *magnitude* of a Cartesian vector is $A = \sqrt{A_x^2 + A_y^2 + A_z^2}$.
- The *direction* of a Cartesian vector is specified using coordinate direction angles which the tail of the vector makes with the positive x, y, z axes, respectively. The components of the unit vector $\mathbf{u} = \mathbf{A}/A$ represent the direction cosines of α, β, γ. Only two of the angles α, β, γ have to be specified. The third angle is determined from the relationship $\cos^2 \alpha + \cos^2 \beta + \cos^2 \gamma = 1$.
- To find the *resultant* of a concurrent force system, express each force as a Cartesian vector and add the **i**, **j**, **k** components of all the forces in the system.

E X A M P L E 2.8

Express the force **F** shown in Fig. 2–29 as a Cartesian vector.

Solution

Since only two coordinate direction angles are specified, the third angle α must be determined from Eq. 2–10; i.e.,

$$\cos^2 \alpha + \cos^2 \beta + \cos^2 \gamma = 1$$
$$\cos^2 \alpha + \cos^2 60° + \cos^2 45° = 1$$
$$\cos \alpha = \sqrt{1 - (0.5)^2 - (0.707)^2} = \pm 0.5$$

Hence, two possibilities exist, namely,

$$\alpha = \cos^{-1}(0.5) = 60° \quad \text{or} \quad \alpha = \cos^{-1}(-0.5) = 120°$$

By inspection of Fig. 2–29, it is necessary that $\alpha = 60°$, since $\mathbf{F}_x$ is in the $+x$ direction.

Using Eq. 2–11, with $F = 200$ N, we have

$$\mathbf{F} = F \cos \alpha \mathbf{i} + F \cos \beta \mathbf{j} + F \cos \gamma \mathbf{k}$$
$$= (200 \cos 60° \text{ N})\mathbf{i} + (200 \cos 60° \text{ N})\mathbf{j} + (200 \cos 45° \text{ N})\mathbf{k}$$
$$= \{100.0\mathbf{i} + 100.0\mathbf{j} + 141.4\mathbf{k}\}\text{N} \qquad \qquad Ans.$$

By applying Eq. 2–6, note that indeed the magnitude of $F = 200$ N.

$$F = \sqrt{F_x^2 + F_y^2 + F_z^2}$$
$$= \sqrt{(100.0)^2 + (100.0)^2 + (141.4)^2} = 200 \text{ N}$$

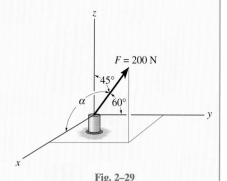

Fig. 2–29

E X A M P L E 2.9

Determine the magnitude and the coordinate direction angles of the resultant force acting on the ring in Fig. 2–30*a*.

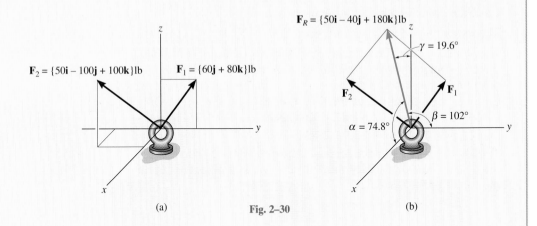

(a)   **Fig. 2–30**   (b)

Solution

Since each force is represented in Cartesian vector form, the resultant force, shown in Fig. 2–30*b*, is

$$\mathbf{F}_R = \Sigma\mathbf{F} = \mathbf{F}_1 + \mathbf{F}_2 = \{60\mathbf{j} + 80\mathbf{k}\} \text{ lb} + \{50\mathbf{i} - 100\mathbf{j} + 100\mathbf{k}\} \text{ lb}$$
$$= \{50\mathbf{i} - 40\mathbf{j} + 180\mathbf{k}\} \text{ lb}$$

The magnitude of $\mathbf{F}_R$ is found from Eq. 2–6, i.e.,

$$F_R = \sqrt{(50)^2 + (-40)^2 + (180)^2} = 191.0$$
$$= 191 \text{ lb} \qquad \qquad \textit{Ans.}$$

The coordinate direction angles α, β, γ are determined from the components of the unit vector acting in the direction of $\mathbf{F}_R$.

$$\mathbf{u}_{F_R} = \frac{\mathbf{F}_R}{F_R} = \frac{50}{191.0}\mathbf{i} - \frac{40}{191.0}\mathbf{j} + \frac{180}{191.0}\mathbf{k}$$
$$= 0.2617\mathbf{i} - 0.2094\mathbf{j} + 0.9422\mathbf{k}$$

so that

$$\cos\alpha = 0.2617 \qquad \alpha = 74.8° \qquad \textit{Ans.}$$
$$\cos\beta = -0.2094 \qquad \beta = 102° \qquad \textit{Ans.}$$
$$\cos\gamma = 0.9422 \qquad \gamma = 19.6° \qquad \textit{Ans.}$$

These angles are shown in Fig. 2–30*b*. In particular, note that $\beta > 90°$ since the $\mathbf{j}$ component of $\mathbf{u}_{F_R}$ is negative.

E X A M P L E 2.10

Express the force $\mathbf{F}_1$, shown in Fig. 2–31a as a Cartesian vector.

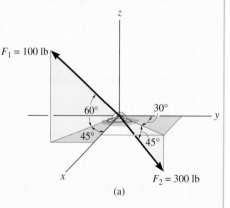

(a)

Solution

The angles of 60° and 45° defining the direction of $\mathbf{F}_1$ are *not* coordinate direction angles. The two successive applications of the parallelogram law needed to resolve $\mathbf{F}_1$ into its x, y, z components are shown in Fig. 2–31b. By trigonometry, the magnitudes of the components are

$$F_{1z} = 100 \sin 60° \text{ lb} = 86.6 \text{ lb}$$
$$F' = 100 \cos 60° \text{ lb} = 50 \text{ lb}$$
$$F_{1x} = 50 \cos 45° \text{ lb} = 35.4 \text{ lb}$$
$$F_{1y} = 50 \sin 45° \text{ lb} = 35.4 \text{ lb}$$

Realizing that $\mathbf{F}_{1y}$ has a direction defined by $-\mathbf{j}$, we have

$$\mathbf{F}_1 = \{35.4\mathbf{i} - 35.4\mathbf{j} + 86.6\mathbf{k}\} \text{ lb} \qquad \text{Ans.}$$

To show that the magnitude of this vector is indeed 100 lb, apply Eq. 2–6,

$$F_1 = \sqrt{F_{1x}^2 + F_{1y}^2 + F_{1z}^2}$$

$$= \sqrt{(35.4)^2 + (-35.4)^2 + (86.6)^2} = 100 \text{ lb}$$

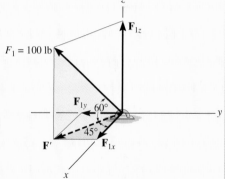

(b)

If needed, the coordinate direction angles of $\mathbf{F}_1$ can be determined from the components of the unit vector acting in the direction of $\mathbf{F}_1$. Hence,

$$\mathbf{u}_1 = \frac{\mathbf{F}_1}{F_1} = \frac{F_{1x}}{F_1}\mathbf{i} + \frac{F_{1y}}{F_1}\mathbf{j} + \frac{F_{1z}}{F_1}\mathbf{k}$$

$$= \frac{35.4}{100}\mathbf{i} - \frac{35.4}{100}\mathbf{j} + \frac{86.6}{100}\mathbf{k}$$

$$= 0.354\mathbf{i} - 0.354\mathbf{j} + 0.866\mathbf{k}$$

so that

$$\alpha_1 = \cos^{-1}(0.354) = 69.3°$$
$$\beta_1 = \cos^{-1}(-0.354) = 111°$$
$$\gamma_1 = \cos^{-1}(0.866) = 30.0°$$

These results are shown in Fig. 2–31c.

Using this same method, show that $\mathbf{F}_2$ in Fig. 2–31a can be written in Cartesian vector form as

$$\mathbf{F}_2 = \{106\mathbf{i} + 184\mathbf{j} - 212\mathbf{k}\} \text{ N} \qquad \text{Ans.}$$

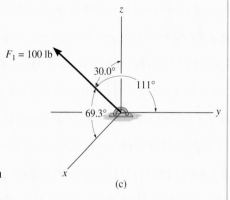

(c)

Fig. 2–31

EXAMPLE **2.11**

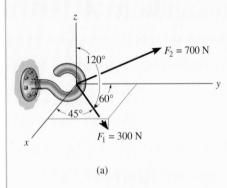

(a)

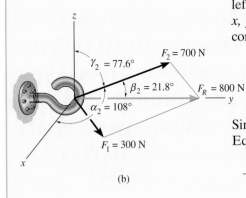

(b)

Fig. 2–32

Two forces act on the hook shown in Fig. 2–32a. Specify the coordinate direction angles of $\mathbf{F}_2$ so that the resultant force $\mathbf{F}_R$ acts along the positive y axis and has a magnitude of 800 N.

Solution

To solve this problem, the resultant force $\mathbf{F}_R$ and its two components, $\mathbf{F}_1$ and $\mathbf{F}_2$, will each be expressed in Cartesian vector form. Then, as shown in Fig. 2–32b, it is necessary that $\mathbf{F}_R = \mathbf{F}_1 + \mathbf{F}_2$.

Applying Eq. 2–11,

$$\mathbf{F}_1 = F_1 \cos \alpha_1 \mathbf{i} + F_1 \cos \beta_1 \mathbf{j} + F_1 \cos \gamma_1 \mathbf{k}$$
$$= 300 \cos 45° \, \mathbf{N i} + 300 \cos 60° \, \mathbf{N j} + 300 \cos 120° \, \mathbf{k}$$
$$= \{212.1\mathbf{i} + 150\mathbf{j} - 150\mathbf{k}\} \, \text{N}$$
$$\mathbf{F}_2 = F_{2x}\mathbf{i} + F_{2y}\mathbf{j} + F_{2z}\mathbf{k}$$

Since the resultant force $\mathbf{F}_R$ has a magnitude of 800 N and acts in the $+\mathbf{j}$ direction.

$$\mathbf{F}_R = (800 \text{ N})(+\mathbf{j}) = \{800\mathbf{j}\} \text{ N}$$

We require

$$\mathbf{F}_R = \mathbf{F}_1 + \mathbf{F}_2$$
$$800\mathbf{j} = 212.1\mathbf{i} + 150\mathbf{j} - 150\mathbf{k} + F_{2x}\mathbf{i} + F_{2y}\mathbf{j} + F_{2z}\mathbf{k}$$
$$800\mathbf{j} = (212.1 + F_{2x})\mathbf{i} + (150 + F_{2y})\mathbf{j} + (-150 + F_{2z})\mathbf{k}$$

To satisfy this equation, the corresponding $\mathbf{i}$, $\mathbf{j}$, $\mathbf{k}$ components on the left and right sides must be equal. This is equivalent to stating that the x, y, z components of $\mathbf{F}_R$ must be equal to the corresponding x, y, z components of $(\mathbf{F}_1 + \mathbf{F}_2)$. Hence,

$$0 = 212.1 + F_{2x} \qquad F_{2x} = -212.1 \text{ N}$$
$$800 = 150 + F_{2y} \qquad F_{2y} = 650 \text{ N}$$
$$0 = -150 + F_{2z} \qquad F_{2z} = 150 \text{ N}$$

Since the magnitudes of $\mathbf{F}_2$ and its components are known, we can use Eq. 2–11 to determine α_2, β_2, γ_2.

$$-212.1 = 700 \cos \alpha_2; \qquad \alpha_2 = \cos^{-1}\left(\frac{-212.1}{700}\right) = 108° \qquad \textit{Ans.}$$

$$650 = 700 \cos \beta_2; \qquad \beta_2 = \cos^{-1}\left(\frac{650}{700}\right) = 21.8° \qquad \textit{Ans.}$$

$$150 = 700 \cos \gamma_2; \qquad \gamma_2 = \cos^{-1}\left(\frac{150}{700}\right) = 77.6° \qquad \textit{Ans.}$$

These results are shown in Fig. 2–32b.

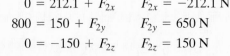

2.7 Position Vectors

In this section we will introduce the concept of a position vector. It will be shown that this vector is of importance in formulating a Cartesian force vector directed between any two points in space. Later, in Chapter 4, we will use it for finding the moment of a force.

x, y, z **Coordinates.**   Throughout the book we will use a *right-handed* coordinate system to reference the location of points in space. Furthermore, we will use the convention followed in many technical books, and that is to require the positive *z* axis to be directed *upward* (the zenith direction) so that it measures the height of an object or the altitude of a point. The *x, y* axes then lie in the horizontal plane, Fig. 2–33. Points in space are located relative to the origin of coordinates, *O*, by successive measurements along the *x, y, z* axes. For example, in Fig. 2–33 the coordinates of point *A* are obtained by starting at *O* and measuring $x_A = +4$ m along the *x* axis, $y_A = +2$ m along the *y* axis, and $z_A = -6$ m along the *z* axis. Thus, $A(4, 2, -6)$. In a similar manner, measurements along the *x, y, z* axes from *O* to *B* yield the coordinates of *B*, i.e., $B(0, 2, 0)$. Also notice that $C(6, -1, 4)$.

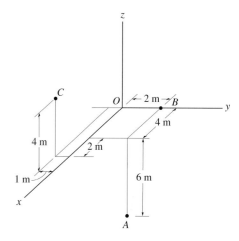

Fig. 2–33

Position Vector.   The *position vector* **r** is defined as a fixed vector which locates a point in space relative to another point. For example, if **r** extends from the origin of coordinates, *O*, to point $P(x, y, z)$, Fig. 2–34*a*, then **r** can be expressed in Cartesian vector form as

$$\mathbf{r} = x\mathbf{i} + y\mathbf{j} + z\mathbf{k}$$

Note how the head-to-tail vector addition of the three components yields vector *r*, Fig. 2–34*b*. Starting at the origin *O*, one travels *x* in the $+\mathbf{i}$ direction, then *y* in the $+\mathbf{j}$ direction, and finally *z* in the $+\mathbf{k}$ direction to arrive at point $P(x, y, z)$.

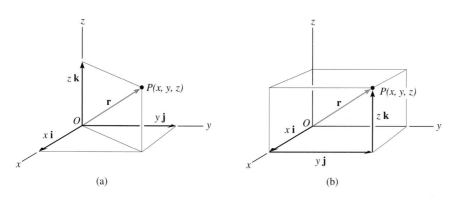

(a)

(b)

Fig. 2–34

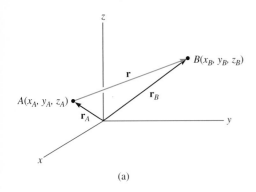

(a)

In the more general case, the position vector may be directed from point A to point B in space, Fig. 2–35a. As noted, this vector is also designated by the symbol $\mathbf{r}$. As a matter of convention, however, we will *sometimes* refer to this vector with *two subscripts* to indicate from and to the point where it is directed. Thus, $\mathbf{r}$ can also be designated as $\mathbf{r}_{AB}$. Also, note that $\mathbf{r}_A$ and $\mathbf{r}_B$ in Fig. 2–35a are referenced with only one subscript since they extend from the origin of coordinates.

From Fig. 2–35a, by the head-to-tail vector addition, we require

$$\mathbf{r}_A + \mathbf{r} = \mathbf{r}_B$$

Solving for $\mathbf{r}$ and expressing $\mathbf{r}_A$ and $\mathbf{r}_B$ in Cartesian vector form yields

$$\mathbf{r} = \mathbf{r}_B - \mathbf{r}_A = (x_B\mathbf{i} + y_B\mathbf{j} + z_B\mathbf{k}) - (x_A\mathbf{i} + y_A\mathbf{j} + z_A\mathbf{k})$$

or

$$\mathbf{r} = (x_B - x_A)\mathbf{i} + (y_B - y_A)\mathbf{j} + (z_B - z_A)\mathbf{k} \qquad (2\text{–}13)$$

Thus, the $\mathbf{i}$, $\mathbf{j}$, $\mathbf{k}$ *components of the position vector* $\mathbf{r}$ *may be formed by taking the coordinates of the tail of the vector,* $A(x_A, y_A, z_A)$, *and subtracting them from the corresponding coordinates of the head,* $B(x_B, y_B, z_B)$. Again note how the head-to-tail addition of these three components yields $\mathbf{r}$, i.e., going from A to B, Fig. 2–35b, one first travels $(x_B - x_A)$ in the $+\mathbf{i}$ direction, then $(y_B - y_A)$ in the $+\mathbf{j}$ direction, and finally $(z_B - z_A)$ in the $+\mathbf{k}$ direction.

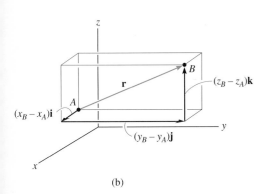

(b)

Fig. 2–35

The length and direction of cable AB used to support the stack can be determined by measuring the coordinates of points A and B using the x, y, z axes. The position vector $\mathbf{r}$ along the cable can then be established. The magnitude r represents the length of the cable, and the direction of the cable is defined by α, β, γ, which are determined from the components of the unit vector found from the position vector, $\mathbf{u} = \mathbf{r}/r$.

EXAMPLE **2.12**

An elastic rubber band is attached to points A and B as shown in Fig. 2–36a. Determine its length and its direction measured from A toward B.

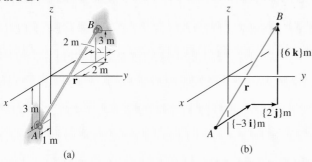

(a)

(b)

Fig. 2–36

Solution

We first establish a position vector from A to B, Fig. 2–36b. In accordance with Eq. 2–13, the coordinates of the tail $A(1 \text{ m}, 0, -3 \text{ m})$ are subtracted from the coordinates of the head $B(-2 \text{ m}, 2 \text{ m}, 3 \text{ m})$, which yields

$$\mathbf{r} = [-2 \text{ m} - 1 \text{ m}]\mathbf{i} + [2 \text{ m} - 0]\mathbf{j} + [3 \text{ m} - (-3 \text{ m})]\mathbf{k}$$
$$= \{-3\mathbf{i} + 2\mathbf{j} + 6\mathbf{k}\} \text{ m}$$

These components of $\mathbf{r}$ can also be determined *directly* by realizing from Fig. 2–36a that they represent the direction and distance one must go along each axis in order to move from A to B, i.e., along the x axis $\{-3\mathbf{i}\}$ m, along the y axis $\{2\mathbf{j}\}$ m, and finally along the z axis $\{6\mathbf{k}\}$ m.

The magnitude of $\mathbf{r}$ represents the length of the rubber band.

$$r = \sqrt{(-3)^2 + (2)^2 + (6)^2} = 7 \text{ m} \qquad \textit{Ans.}$$

Formulating a unit vector in the direction of $\mathbf{r}$, we have

$$\mathbf{u} = \frac{\mathbf{r}}{r} = \frac{-3}{7}\mathbf{i} + \frac{2}{7}\mathbf{j} + \frac{6}{7}\mathbf{k}$$

The components of this unit vector yield the coordinate direction angles

$$\alpha = \cos^{-1}\left(\frac{-3}{7}\right) = 115° \qquad \textit{Ans.}$$

$$\beta = \cos^{-1}\left(\frac{2}{7}\right) = 73.4° \qquad \textit{Ans.}$$

$$\gamma = \cos^{-1}\left(\frac{6}{7}\right) = 31.0° \qquad \textit{Ans.}$$

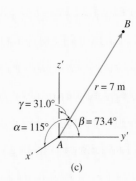

(c)

These angles are measured from the *positive axes* of a localized coordinate system placed at the tail of $\mathbf{r}$, point A, as shown in Fig. 2–36c.

2.8   Force Vector Directed along a Line

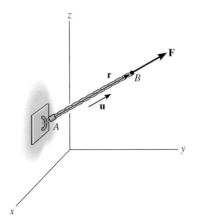

Fig. 2–37

Quite often in three-dimensional statics problems, the direction of a force is specified by two points through which its line of action passes. Such a situation is shown in Fig. 2–37, where the force **F** is directed along the cord *AB*. We can formulate **F** as a Cartesian vector by realizing that it has the *same direction* and *sense* as the position vector **r** directed from point *A* to point *B* on the cord. This common direction is specified by the *unit vector* **u** = **r**/*r*. Hence,

$$\mathbf{F} = F\mathbf{u} = F\left(\frac{\mathbf{r}}{r}\right)$$

Although we have represented **F** symbolically in Fig. 2–37, note that it has *units of force*, unlike **r**, which has units of length.

The force **F** acting along the chain can be represented as a Cartesian vector by first establishing *x, y, z* axes and forming a position vector **r** along the length of the chain, then finding the corresponding unit vector **u** = **r**/*r* that defines the direction of both the chain and the force. Finally, the magnitude of the force is combined with its direction, **F** = *F***u**.

IMPORTANT POINTS

- A position vector locates one point in space relative to another point.
- The easiest way to formulate the components of a position vector is to determine the distance and direction that must be traveled along the *x, y, z* directions—going from the tail to the head of the vector.
- A force **F** acting in the direction of a position vector **r** can be represented in Cartesian form if the unit vector **u** of the position vector is determined and this is multiplied by the magnitude of the force, i.e., **F** = *F***u** = *F*(**r**/*r*).

EXAMPLE 2.13

The man shown in Fig. 2–38a pulls on the cord with a force of 70 lb. Represent this force, acting on the support A, as a Cartesian vector and determine its direction.

Solution

Force **F** is shown in Fig. 2–38b. The *direction* of this vector, **u**, is determined from the position vector **r**, which extends from A to B, Fig. 2–38b. The coordinates of the end points of the cord are A(0, 0, 30 ft) and B(12 ft, −8 ft, 6 ft). Forming the position vector by subtracting the corresponding x, y, and z coordinates of A from those of B, we have

$$\mathbf{r} = (12 \text{ ft} - 0)\mathbf{i} + (-8 \text{ ft} - 0)\mathbf{j} + (6 \text{ ft} - 30 \text{ ft})\mathbf{k}$$
$$= \{12\mathbf{i} - 8\mathbf{j} - 24\mathbf{k}\} \text{ ft}$$

This result can also be determined *directly* by noting in Fig. 2–38a, that one must go from A $\{-24\mathbf{k}\}$ ft, then $\{-8\mathbf{j}\}$ ft, and finally $\{12\mathbf{i}\}$ ft to get to B.

The magnitude of **r**, which represents the *length* of cord AB, is

$$r = \sqrt{(12 \text{ ft})^2 + (-8 \text{ ft})^2 + (-24 \text{ ft})^2} = 28 \text{ ft}$$

Forming the unit vector that defines the direction and sense of both **r** and **F** yields

$$\mathbf{u} = \frac{\mathbf{r}}{r} = \frac{12}{28}\mathbf{i} - \frac{8}{28}\mathbf{j} - \frac{24}{28}\mathbf{k}$$

Since **F** has a *magnitude* of 70 lb and a *direction* specified by **u**, then

$$\mathbf{F} = F\mathbf{u} = 70 \text{ lb} \left(\frac{12}{28}\mathbf{i} - \frac{8}{28}\mathbf{j} - \frac{24}{28}\mathbf{k} \right)$$
$$= \{30\mathbf{i} - 20\mathbf{j} - 60\mathbf{k}\} \text{ lb} \qquad Ans.$$

The coordinate direction angles are measured between **r** (or **F**) and the *positive axes* of a localized coordinate system with origin placed at A, Fig. 2–38b. From the components of the unit vector:

$$\alpha = \cos^{-1}\left(\frac{12}{28}\right) = 64.6° \qquad Ans.$$

$$\beta = \cos^{-1}\left(\frac{-8}{28}\right) = 107° \qquad Ans.$$

$$\gamma = \cos^{-1}\left(\frac{-24}{28}\right) = 149° \qquad Ans.$$

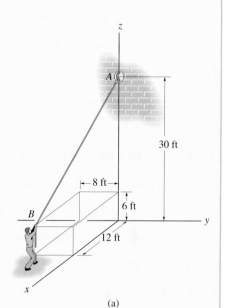

30 ft

8 ft

6 ft

12 ft

(a)

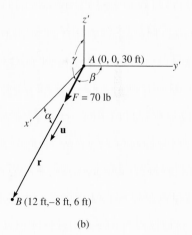

A (0, 0, 30 ft)

F = 70 lb

B (12 ft,−8 ft, 6 ft)

(b)

Fig. 2–38

E X A M P L E   2.14

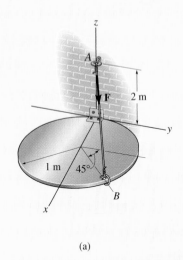

(a)

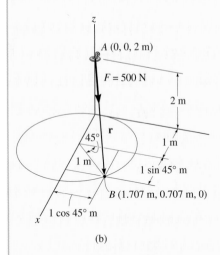

(b)

Fig. 2–39

The circular plate in Fig. 2–39a is partially supported by the cable AB. If the force of the cable on the hook at A is $F = 500$ N, express **F** as a Cartesian vector.

Solution

As shown in Fig. 2–39b, **F** has the same direction and sense as the position vector **r**, which extends from A to B. The coordinates of the end points of the cable are $A(0, 0, 2$ m$)$ and $B(1.707$ m, 0.707 m, $0)$, as indicated in the figure. Thus,

$$\mathbf{r} = (1.707 \text{ m} - 0)\mathbf{i} + (0.707 \text{ m} - 0)\mathbf{j} + (0 - 2 \text{ m})\mathbf{k}$$

$$= \{1.707\mathbf{i} + 0.707\mathbf{j} - 2\mathbf{k}\} \text{ m}$$

Note how one can calculate these components *directly* by going from A, $\{-2\mathbf{k}\}$ m along the z axis, then $\{1.707\mathbf{i}\}$ m along the x axis, and finally $\{0.707\mathbf{j}\}$ m along the y axis to get to B.

The magnitude of **r** is

$$r = \sqrt{(1.707)^2 + (0.707)^2 + (-2)^2} = 2.723 \text{ m}$$

Thus,

$$\mathbf{u} = \frac{\mathbf{r}}{r} = \frac{1.707}{2.723}\mathbf{i} + \frac{0.707}{2.723}\mathbf{j} - \frac{2}{2.723}\mathbf{k}$$

$$= 0.6269\mathbf{i} + 0.2597\mathbf{j} - 0.7345\mathbf{k}$$

Since $F = 500$ N and **F** has the direction **u**, we have

$$\mathbf{F} = F\mathbf{u} = 500 \text{ N}(0.6269\mathbf{i} + 0.2597\mathbf{j} - 0.7345\mathbf{k})$$

$$= \{313\mathbf{i} + 130\mathbf{j} - 367\mathbf{k}\} \text{ N} \qquad \qquad Ans.$$

Using these components, notice that indeed the magnitude of **F** is 500 N; i.e.,

$$F = \sqrt{(313)^2 + (130)^2 + (-367)^2} = 500 \text{ N}$$

Show that the coordinate direction angle $\gamma = 137°$, and indicate this angle on the figure.

EXAMPLE 2.15

The roof is supported by cables as shown in the photo. If the cables exert forces $F_{AB} = 100 \text{ N}$ and $F_{AC} = 120 \text{ N}$ on the wall hook at A as shown in Fig. 2–40*a*, determine the magnitude of the resultant force acting at A.

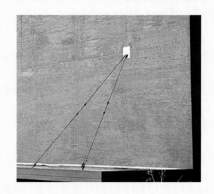

Solution

The resultant force $\mathbf{F}_R$ is shown graphically in Fig. 2–40*b*. We can express this force as a Cartesian vector by first formulating $\mathbf{F}_{AB}$ and $\mathbf{F}_{AC}$ as Cartesian vectors and then adding their components. The directions of $\mathbf{F}_{AB}$ and $\mathbf{F}_{AC}$ are specified by forming unit vectors $\mathbf{u}_{AB}$ and $\mathbf{u}_{AC}$ along the cables. These unit vectors are obtained from the associated position vectors $\mathbf{r}_{AB}$ and $\mathbf{r}_{AC}$. With reference to Fig. 2–40*b*, for $\mathbf{F}_{AB}$ we have

$$\mathbf{r}_{AB} = (4 \text{ m} - 0)\mathbf{i} + (0 - 0)\mathbf{j} + (0 - 4 \text{ m})\mathbf{k}$$

$$= \{4\mathbf{i} - 4\mathbf{k}\} \text{ m}$$

$$r_{AB} = \sqrt{(4)^2 + (-4)^2} = 5.66 \text{ m}$$

$$\mathbf{F}_{AB} = 100 \text{ N}\left(\frac{\mathbf{r}_{AB}}{r_{AB}}\right) = 100 \text{ N}\left(\frac{4}{5.66}\mathbf{i} - \frac{4}{5.66}\mathbf{k}\right)$$

$$\mathbf{F}_{AB} = \{70.7\mathbf{i} - 70.7\mathbf{k}\} \text{ N}$$

For $\mathbf{F}_{AC}$ we have

$$\mathbf{r}_{AC} = (4 \text{ m} - 0)\mathbf{i} + (2 \text{ m} - 0)\mathbf{j} + (0 - 4 \text{ m})\mathbf{k}$$

$$= \{4\mathbf{i} + 2\mathbf{j} - 4\mathbf{k}\} \text{ m}$$

$$r_{AC} = \sqrt{(4)^2 + (2)^2 + (-4)^2} = 6 \text{ m}$$

$$\mathbf{F}_{AC} = 120 \text{ N}\left(\frac{\mathbf{r}_{AC}}{r_{AC}}\right) = 120 \text{ N}\left(\frac{4}{6}\mathbf{i} + \frac{2}{6}\mathbf{j} - \frac{4}{6}\mathbf{k}\right)$$

$$= \{80\mathbf{i} + 40\mathbf{j} - 80\mathbf{k}\} \text{ N}$$

The resultant force is therefore

$$\mathbf{F}_R = \mathbf{F}_{AB} + \mathbf{F}_{AC} = \{70.7\mathbf{i} - 70.7\mathbf{k}\} \text{ N} + \{80\mathbf{i} + 40\mathbf{j} - 80\mathbf{k}\} \text{ N}$$

$$= \{150.7\mathbf{i} + 40\mathbf{j} - 150.7\mathbf{k}\} \text{ N}$$

The magnitude of $\mathbf{F}_R$ is thus

$$F_R = \sqrt{(150.7)^2 + (40)^2 + (-150.7)^2}$$

$$= 217 \text{ N} \qquad \qquad \qquad \qquad \qquad \qquad \qquad Ans.$$

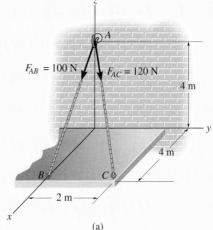

(a)

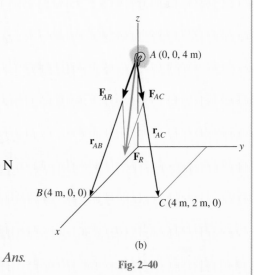

(b)

Fig. 2–40

2.9 Dot Product

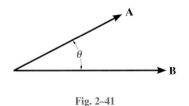

Fig. 2–41

Occasionally in statics one has to find the angle between two lines or the components of a force parallel and perpendicular to a line. In two dimensions, these problems can readily be solved by trigonometry since the geometry is easy to visualize. In three dimensions, however, this is often difficult, and consequently vector methods should be employed for the solution. The dot product defines a particular method for "multiplying" two vectors and is used to solve the above-mentioned problems.

The *dot product* of vectors **A** and **B**, written **A** · **B**, and read "**A** dot **B**," is defined as the product of the magnitudes of **A** and **B** and the cosine of the angle θ between their tails, Fig. 2–41. Expressed in equation form,

$$\mathbf{A} \cdot \mathbf{B} = AB \cos \theta \qquad (2\text{–}14)$$

where $0° \leq \theta \leq 180°$. The dot product is often referred to as the *scalar product* of vectors since the result is a *scalar* and not a vector.

Laws of Operation

1. Commutative law:

$$\mathbf{A} \cdot \mathbf{B} = \mathbf{B} \cdot \mathbf{A}$$

2. Multiplication by a scalar:

$$a(\mathbf{A} \cdot \mathbf{B}) = (a\mathbf{A}) \cdot \mathbf{B} = \mathbf{A} \cdot (a\mathbf{B}) = (\mathbf{A} \cdot \mathbf{B})a$$

3. Distributive law:

$$\mathbf{A} \cdot (\mathbf{B} + \mathbf{D}) = (\mathbf{A} \cdot \mathbf{B}) + (\mathbf{A} \cdot \mathbf{D})$$

It is easy to prove the first and second laws by using Eq. 2–14. The proof of the distributive law is left as an exercise (see Prob. 2–109).

Cartesian Vector Formulation. Equation 2–14 may be used to find the dot product for each of the Cartesian unit vectors. For example, $\mathbf{i} \cdot \mathbf{i} = (1)(1) \cos 0° = 1$ and $\mathbf{i} \cdot \mathbf{j} = (1)(1) \cos 90° = 0$. In a similar manner,

$$\mathbf{i} \cdot \mathbf{i} = 1 \qquad \mathbf{j} \cdot \mathbf{j} = 1 \qquad \mathbf{k} \cdot \mathbf{k} = 1$$
$$\mathbf{i} \cdot \mathbf{j} = 0 \qquad \mathbf{i} \cdot \mathbf{k} = 0 \qquad \mathbf{k} \cdot \mathbf{j} = 0$$

These results should not be memorized; rather, it should be clearly understood how each is obtained.

Consider now the dot product of two general vectors **A** and **B** which are expressed in Cartesian vector form. We have

$$\mathbf{A} \cdot \mathbf{B} = (A_x\mathbf{i} + A_y\mathbf{j} + A_z\mathbf{k}) \cdot (B_x\mathbf{i} + B_y\mathbf{j} + B_z\mathbf{k})$$
$$= A_xB_x(\mathbf{i} \cdot \mathbf{i}) + A_xB_y(\mathbf{i} \cdot \mathbf{j}) + A_xB_z(\mathbf{i} \cdot \mathbf{k})$$
$$+ A_yB_x(\mathbf{j} \cdot \mathbf{i}) + A_yB_y(\mathbf{j} \cdot \mathbf{j}) + A_yB_z(\mathbf{j} \cdot \mathbf{k})$$
$$+ A_zB_x(\mathbf{k} \cdot \mathbf{i}) + A_zB_y(\mathbf{k} \cdot \mathbf{j}) + A_zB_z(\mathbf{k} \cdot \mathbf{k})$$

Carrying out the dot-product operations, the final result becomes

$$\mathbf{A} \cdot \mathbf{B} = A_x B_x + A_y B_y + A_z B_z \qquad (2\text{--}15)$$

Thus, to determine the dot product of two Cartesian vectors, multiply their corresponding x, y, z components and sum their products algebraically. Since the result is a scalar, be careful *not* to include any unit vectors in the final result.

Applications. The dot product has two important applications in mechanics.

1. *The angle formed between two vectors or intersecting lines.* The angle θ between the tails of vectors **A** and **B** in Fig. 2–41 can be determined from Eq. 2–14 and written as

$$\theta = \cos^{-1}\left(\frac{\mathbf{A} \cdot \mathbf{B}}{AB}\right) \qquad 0° \le \theta \le 180°$$

   Here $\mathbf{A} \cdot \mathbf{B}$ is found from Eq. 2–15. In particular, notice that if $\mathbf{A} \cdot \mathbf{B} = 0, \theta = \cos^{-1} 0 = 90°$, so that **A** will be *perpendicular* to **B**.

2. *The components of a vector parallel and perpendicular to a line.* The component of vector **A** parallel to or collinear with the line aa' in Fig. 2–42 is defined by $\mathbf{A}_{\parallel}$, where $A_{\parallel} = A \cos \theta$. This component is sometimes referred to as the *projection* of **A** onto the line, since a right angle is formed in the construction. If the *direction* of the line is specified by the unit vector **u**, then, since $u = 1$, we can determine $A_{\parallel}$ directly from the dot product (Eq. 2–14); i.e.,

$$A_{\parallel} = A \cos \theta = \mathbf{A} \cdot \mathbf{u}$$

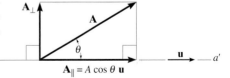

Fig. 2–42

*Hence, the scalar projection of **A** along a line is determined from the dot product of **A** and the unit vector **u** which defines the direction of the line.* Notice that if this result is positive, then $\mathbf{A}_{\parallel}$ has a directional sense which is the same as **u**, whereas if $A_{\parallel}$ is a negative scalar, then $\mathbf{A}_{\parallel}$ has the opposite sense of direction to **u**. The component $\mathbf{A}_{\parallel}$ represented as a *vector* is therefore

$$\mathbf{A}_{\parallel} = A \cos \theta \; \mathbf{u} = (\mathbf{A} \cdot \mathbf{u})\mathbf{u}$$

The component of **A** which is *perpendicular* to line aa' can also be obtained, Fig. 2–42. Since $\mathbf{A} = \mathbf{A}_{\parallel} + \mathbf{A}_{\perp}$, then $\mathbf{A}_{\perp} = \mathbf{A} - \mathbf{A}_{\parallel}$. There are two possible ways of obtaining $A_{\perp}$. One way would be to determine θ from the dot product, $\theta = \cos^{-1}(\mathbf{A} \cdot \mathbf{u}/A)$, then $A_{\perp} = A \sin \theta$. Alternatively, if $A_{\parallel}$ is known, then by the Pythagorean theorem we can also write $A_{\perp} = \sqrt{A^2 - A_{\parallel}^2}$.

The angle θ which is made between the rope and the connecting beam A can be determined by using the dot product. Simply formulate position vectors or unit vectors along the beam, $\mathbf{u}_A = \mathbf{r}_A/r_A$, and along the rope, $\mathbf{u}_r = \mathbf{r}_r/r_r$. Since θ is defined between the tails of these vectors we can solve for θ using $\theta = \cos^{-1}(\mathbf{r}_A \cdot \mathbf{r}_r/r_A r_r) = \cos^{-1} \mathbf{u}_A \cdot \mathbf{u}_r$.

If the rope exerts a force $\mathbf{F}$ on the joint, the projection of this force along beam A can be determined by first defining the *direction of the beam* using the unit vector $\mathbf{u}_A = \mathbf{r}_A/r_A$ and then formulating the force as a Cartesian vector $\mathbf{F} = F(\mathbf{r}_r/r_r) = F\mathbf{u}_r$. Applying the dot product, the projection is $F_{\parallel} = \mathbf{F} \cdot \mathbf{u}_A$.

IMPORTANT POINTS

- The dot product is used to determine the angle between two vectors or the projection of a vector in a specified direction.

- If the vectors $\mathbf{A}$ and $\mathbf{B}$ are expressed in Cartesian form, the dot product is determined by multiplying the respective x, y, z scalar components together and algebraically adding the results, i.e., $\mathbf{A} \cdot \mathbf{B} = A_x B_x + A_y B_y + A_z B_z$.

- From the definition of the dot product, the angle formed between the tails of vectors $\mathbf{A}$ and $\mathbf{B}$ is $\theta = \cos^{-1}(\mathbf{A} \cdot \mathbf{B}/AB)$.

- The magnitude of the projection of vector $\mathbf{A}$ along a line whose direction is specified by $\mathbf{u}$ is determined from the dot product $A_{\parallel} = \mathbf{A} \cdot \mathbf{u}$.

E X A M P L E **2.16**

The frame shown in Fig. 2–43a is subjected to a horizontal force $\mathbf{F} = \{300\mathbf{j}\}$ N. Determine the magnitude of the components of this force parallel and perpendicular to member AB.

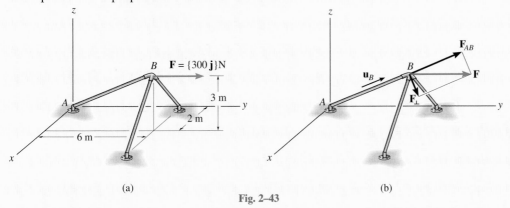

Fig. 2–43

Solution

The magnitude of the component of $\mathbf{F}$ along AB is equal to the dot product of $\mathbf{F}$ and the unit vector $\mathbf{u}_B$, which defines the direction of AB, Fig. 2–43b. Since

$$\mathbf{u}_B = \frac{\mathbf{r}_B}{r_B} = \frac{2\mathbf{i} + 6\mathbf{j} + 3\mathbf{k}}{\sqrt{(2)^2 + (6)^2 + (3)^2}} = 0.286\mathbf{i} + 0.857\mathbf{j} + 0.429\mathbf{k}$$

then

$$\begin{aligned} F_{AB} = F \cos\theta &= \mathbf{F} \cdot \mathbf{u}_B = (300\mathbf{j}) \cdot (0.286\mathbf{i} + 0.857\mathbf{j} + 0.429\mathbf{k}) \\ &= (0)(0.286) + (300)(0.857) + (0)(0.429) \\ &= 257.1 \text{ N} \qquad\qquad\qquad\qquad\qquad\qquad\qquad \textit{Ans.} \end{aligned}$$

Since the result is a positive scalar, $\mathbf{F}_{AB}$ has the same sense of direction as $\mathbf{u}_B$, Fig. 2–43b.

Expressing $\mathbf{F}_{AB}$ in Cartesian vector form, we have

$$\begin{aligned} \mathbf{F}_{AB} = F_{AB}\mathbf{u}_B &= (257.1 \text{ N})(0.286\mathbf{i} + 0.857\mathbf{j} + 0.429\mathbf{k}) \\ &= \{73.5\mathbf{i} + 220\mathbf{j} + 110\mathbf{k}\} \text{ N} \qquad\qquad \textit{Ans.} \end{aligned}$$

The perpendicular component, Fig. 2–43b, is therefore

$$\begin{aligned} \mathbf{F}_\perp = \mathbf{F} - \mathbf{F}_{AB} &= 300\mathbf{j} - (73.5\mathbf{i} + 220\mathbf{j} + 110\mathbf{k}) \\ &= \{-73.5\mathbf{i} + 80\mathbf{j} - 110\mathbf{k}\} \text{ N} \end{aligned}$$

Its magnitude can be determined either from this vector or from the Pythagorean theorem, Fig. 2–43b:

$$\begin{aligned} F_\perp &= \sqrt{F^2 - F^2_{AB}} \\ &= \sqrt{(300 \text{ N})^2 - (257.1 \text{ N})^2} \\ &= 155 \text{ N} \qquad\qquad\qquad\qquad\qquad \textit{Ans.} \end{aligned}$$

EXAMPLE 2.17

The pipe in Fig. 2–44a is subjected to the force of $F = 80$ lb. Determine the angle θ between **F** and the pipe segment BA, and the magnitudes of the components of **F**, which are parallel and perpendicular to BA.

Solution

Angle θ. First we will establish position vectors from B to A and B to C. Then we will determine the angle θ between the tails of these two vectors.

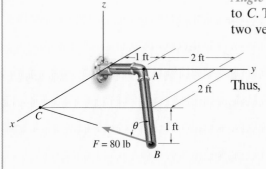

$$\mathbf{r}_{BA} = \{-2\mathbf{i} - 2\mathbf{j} + 1\mathbf{k}\} \text{ ft}$$
$$\mathbf{r}_{BC} = \{-3\mathbf{j} + 1\mathbf{k}\} \text{ ft}$$

Thus,

$$\cos \theta = \frac{\mathbf{r}_{BA} \cdot \mathbf{r}_{BC}}{\mathbf{r}_{BA}\mathbf{r}_{BC}} = \frac{(-2)(0) + (-2)(-3) + (1)(1)}{3\sqrt{10}}$$
$$= 0.7379$$
$$\theta = 42.5° \qquad\qquad Ans.$$

Components of F. The force **F** is resolved into components as shown in Fig. 2–44b. Since $F_{BA} = \mathbf{F} \cdot \mathbf{u}_{BA}$, we must first formulate the unit vector along BA and force **F** as Cartesian vectors.

$$\mathbf{u}_{BA} = \frac{\mathbf{r}_{BA}}{r_{BA}} = \frac{(-2\mathbf{i} - 2\mathbf{j} + 1\mathbf{k})}{3} = -\frac{2}{3}\mathbf{i} - \frac{2}{3}\mathbf{j} + \frac{1}{3}\mathbf{k}$$

$$\mathbf{F} = 80 \text{ lb}\left(\frac{\mathbf{r}_{BC}}{r_{BC}}\right) = 80\left(\frac{-3\mathbf{j} + 1\mathbf{k}}{\sqrt{10}}\right) = -75.89\mathbf{j} + 25.30\mathbf{k}$$

Thus,

$$F_{BA} = \mathbf{F} \cdot \mathbf{u}_{BA} = (-75.89\mathbf{j} + 25.30\mathbf{k}) \cdot \left(-\frac{2}{3}\mathbf{i} - \frac{2}{3}\mathbf{j} + \frac{1}{3}\mathbf{k}\right)$$
$$= 0 + 50.60 + 8.43$$
$$= 59.0 \text{ lb} \qquad\qquad Ans.$$

Since θ was calculated in Fig. 2–44b, this same result can also be obtained directly from trigonometry.

$$F_{BA} = 80 \cos 42.5° \text{ lb} = 59.0 \text{ lb} \qquad\qquad Ans.$$

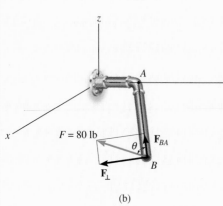

(b)

Fig. 2–44

The perpendicular component can be obtained by trigonometry,

$$F_\perp = F \sin \theta$$
$$= 80 \sin 42.5° \text{ lb}$$
$$= 54.0 \text{ lb} \qquad\qquad Ans.$$

Or, by the Pythagorean theorem,

$$F_\perp = \sqrt{F^2 - F_{BA}^2} = \sqrt{(80)^2 - (59.0)^2}$$
$$= 54.0 \text{ lb} \qquad\qquad Ans.$$

CHAPTER REVIEW

- **Parallelogram Law.** Two vectors add according to the parallelogram law. The *components* form the sides of the parallelogram and the *resultant* is the diagonal. To obtain the components or the resultant, show how the vectors add by the tip-to-tail addition using the triangle rule, and then use the law of sines and the law of cosines to calculate their values.

- **Cartesian Vectors.** A vector can be resolved into its Cartesian components along the x, y, z axes so that $\mathbf{F} = F_x\mathbf{i} + F_y\mathbf{j} + F_z\mathbf{k}$.

  The magnitude of $\mathbf{F}$ is determined from $F = \sqrt{F_x^2 + F_y^2 + F_z^2}$ and the coordinate direction angles α, β, γ are determined by formulating a unit vector in the direction of $\mathbf{F}$, that is $\mathbf{u} = (F_x/F)\mathbf{i} + (F_y/F)\mathbf{j} + (F_z/F)\mathbf{k}$. The components of $\mathbf{u}$ represent $\cos\alpha, \cos\beta, \cos\gamma$. These three angles are related by $\cos^2\alpha + \cos^2\beta + \cos^2\gamma = 1$, so that only two of the three angles are independent of one another.

- **Force and Position Vectors.** A position vector is directed between two points. It can be formulated by finding the distance and the direction one has to travel along the x, y, z axes from one point (the tail) to the other point (the tip). If the line of action of a force passes through these two points, then it acts in the same direction $\mathbf{u}$ as the position vector. The force can be expressed as a Cartesian vector using $\mathbf{F} = F\mathbf{u} = F(\mathbf{r}/r)$.

- **Dot Product.** The dot product between two vectors $\mathbf{A}$ and $\mathbf{B}$ is defined by $\mathbf{A} \cdot \mathbf{B} = AB\cos\theta$. If $\mathbf{A}$ and $\mathbf{B}$ are expressed as Cartesian vectors, then $\mathbf{A} \cdot \mathbf{B} = A_xB_x + A_yB_y + A_zB_z$. In statics the dot product is used to determine the angle between the tails of the vectors, $\theta = \cos^{-1}(\mathbf{A} \cdot \mathbf{B}/AB)$. It is also used to determine the projected component of a vector $\mathbf{A}$ onto an axis defined by its unit vector $\mathbf{u}$, so that $A = A\cos\theta = \mathbf{A} \cdot \mathbf{u}$.

REVIEW PROBLEMS

2-1. Determine the magnitude and coordinate direction angles of $\mathbf{F}_3$ so that the resultant of the three forces acts along the positive y axis and has a magnitude of 600 lb.

2-2. Determine the magnitude and coordinate direction angles of $\mathbf{F}_3$ so that the resultant of the three forces is zero.

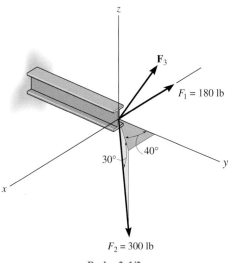

$\mathbf{F}_3$

$F_1 = 180$ lb

$40°$

$30°$

y

x

$F_2 = 300$ lb

Probs. 2–1/2

*2-4. The force $\mathbf{F}$ has a magnitude of 80 lb and acts at the midpoint C of the thin rod. Express the force as a Cartesian vector.

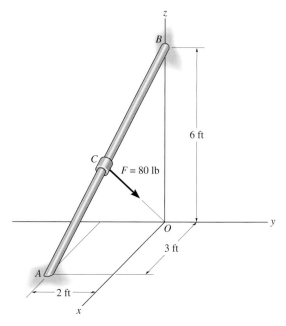

z

B

6 ft

C

$F = 80$ lb

O

3 ft

y

A

2 ft

x

Prob. 2–4

2-3. Determine the design angle θ $(\theta < 90°)$ between the two struts so that the 500-lb horizontal force has a component of 600-lb directed from A toward C. What is the component of force acting along member BA?

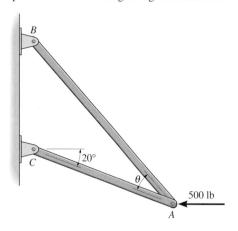

B

$20°$

C

θ

500 lb

A

Prob. 2–3

2-5. Two forces $\mathbf{F}_1$ and $\mathbf{F}_2$ act on the hook. If their lines of action are at an angle θ apart and the magnitude of each force is $F_1 = F_2 = F$, determine the magnitude of the resultant force $\mathbf{F}_R$ and the angle between $\mathbf{F}_R$ and $\mathbf{F}_1$.

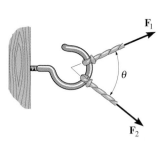

$\mathbf{F}_1$

θ

$\mathbf{F}_2$

Prob. 2–5

2-6. Determine the angles θ and ϕ between the wire segments.

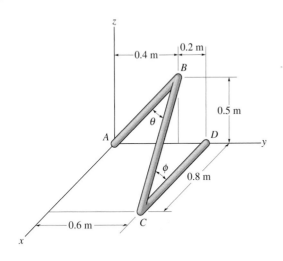

Prob. 2–6

2-7. Determine the magnitudes of the projected components of the force $\mathbf{F} = \{60\mathbf{i} + 12\mathbf{j} - 40\mathbf{k}\}$ N in the direction of the cables AB and AC.

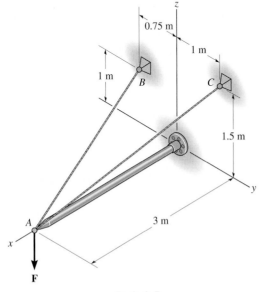

Prob. 2–7

*2-8. Determine the magnitude of the projected component of the 100-lb force acting along the axis BC of the pipe.

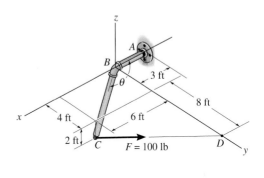

Prob. 2–8

2-9. The boat is to be pulled onto the shore using two ropes. If the resultant force is to be 80 lb, directed along the keel aa, as shown, determine the magnitudes of forces $\mathbf{T}$ and $\mathbf{P}$ acting in each rope and the angle θ so that $\mathbf{P}$ is a *minimum*. $\mathbf{T}$ acts at 30° from the keel as shown.

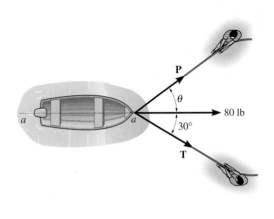

Prob. 2–9

Whenever cables are used for hoisting loads, they must be selected so that they do not fail when they are placed at their points of attachment. In this chapter, we will show how to calculate cable loadings for such cases.

Equilibrium of a Particle

- To introduce the concept of the free-body diagram for a particle.
- To show how to solve particle equilibrium problems using the equations of equilibrium.

3.1   Condition for the Equilibrium of a Particle

A particle is in *equilibrium* provided it is at rest if originally at rest or has a constant velocity if originally in motion. Most often, however, the term "equilibrium" or, more specifically, "static equilibrium" is used to describe an object at rest. To maintain equilibrium, it is *necessary* to satisfy Newton's first law of motion, which requires the *resultant force* acting on a particle to be equal to *zero*. This condition may be stated mathematically as

$$\Sigma \mathbf{F} = \mathbf{0} \qquad (3\text{–}1)$$

where $\Sigma \mathbf{F}$ is the vector *sum of all the forces* acting on the particle.

Not only is Eq. 3–1 a necessary condition for equilibrium, it is also a *sufficient* condition. This follows from Newton's second law of motion, which can be written as $\Sigma \mathbf{F} = m\mathbf{a}$. Since the force system satisfies Eq. 3–1, then $m\mathbf{a} = \mathbf{0}$, and therefore the particle's acceleration $\mathbf{a} = \mathbf{0}$. Consequently the particle indeed moves with constant velocity or remains at rest.

3.2 The Free-Body Diagram

To apply the equation of equilibrium, we must account for *all* the known and unknown forces ($\Sigma \mathbf{F}$) which act *on* the particle. The best way to do this is to draw the particle's *free-body diagram*. This diagram is simply a sketch which shows the particle "free" from its surroundings with *all* the forces that act *on* it.

Before presenting a formal procedure as to how to draw a free-body diagram, we will first consider two types of connections often encountered in particle equilibrium problems.

Springs. If a *linear elastic spring* is used for support, the length of the spring will change in direct proportion to the force acting on it. A characteristic that defines the "elasticity" of a spring is the *spring constant* or *stiffness k*. The magnitude of force exerted on a linear elastic spring which has a stiffness k and is deformed (elongated or compressed) a distance s, measured from its *unloaded* position, is

$$F = ks \qquad (3\text{--}2)$$

Here s is determined from the difference in the spring's deformed length l and its undeformed length l_0, i.e., $s = l - l_0$. If s is positive, $\mathbf{F}$ "pulls" on the spring; whereas if s is negative, $\mathbf{F}$ must "push" on it. For example, the spring shown in Fig. 3–1 has an undeformed length $l_0 = 0.4$ m and stiffness $k = 500$ N/m. To stretch it so that $l = 0.6$ m, a force $F = ks = (500 \text{ N/m})(0.6 \text{ m} - 0.4 \text{ m}) = 100$ N is needed. Likewise, to compress it to a length $l = 0.2$ m, a force $F = ks = (500 \text{ N/m})(0.2 \text{ m} - 0.4 \text{ m}) = -100$ N is required, Fig. 3–1.

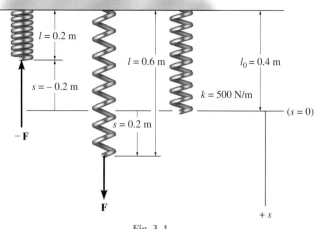

Fig. 3–1

Cables and Pulleys. Throughout this book, except in Sec. 7.4, all cables (or cords) are assumed to have negligible weight and they cannot stretch. Also, a cable can support *only* a tension or "pulling" force, and this force always acts in the direction of the cable. In Chapter 5 it will be shown that the tension force developed in a *continuous cable* which passes over a frictionless pulley must have a *constant* magnitude to keep the cable in equilibrium. Hence, for any angle θ, shown in Fig. 3–2, the cable is subjected to a constant tension T throughout its length.

Cable is in tension

Fig. 3–2

PROCEDURE FOR DRAWING A FREE-BODY DIAGRAM

Since we must account for *all the forces acting on the particle* when applying the equations of equilibrium, the importance of first drawing a free-body diagram cannot be overemphasized. To construct a free-body diagram, the following three steps are necessary.

Draw Outlined Shape. Imagine the particle to be *isolated* or cut "free" from its surroundings by drawing its outlined shape.

Show All Forces. Indicate on this sketch *all* the forces that act *on the particle*. These forces can be *active forces*, which tend to set the particle in motion, or they can be *reactive forces* which are the result of the constraints or supports that tend to prevent motion. To account for all these forces, it may help to trace around the particle's boundary, carefully noting each force acting on it.

Identify Each Force. The forces that are *known* should be labeled with their proper magnitudes and directions. Letters are used to represent the magnitudes and directions of forces that are unknown.

The bucket is held in equilibrium by the cable, and instinctively we know that the force in the cable must equal the weight of the bucket. By drawing a free-body diagram of the bucket we can understand why this is so. This diagram shows that there are only two forces *acting on the bucket*, namely, its weight **W** and the force **T** of the cable. For equilibrium, the resultant of these forces must be equal to zero, and so $T = W$. The important point is that by *isolating the bucket* the unknown cable force **T** becomes "exposed" and must be considered as a requirement for equilibrium.

Consider the spool having a weight W which is suspended from the crane boom. If we wish to obtain the forces in cables AB and AC then we can consider the free-body diagram of the ring at A since these forces act on the ring. Here the cables AD exert a resultant force of **W** on the ring and the condition of equilibrium is used to obtain $\mathbf{T}_B$ and $\mathbf{T}_C$.

EXAMPLE 3.1

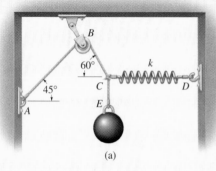

(a)

The sphere in Fig. 3–3a has a mass of 6 kg and is supported as shown. Draw a free-body diagram of the sphere, the cord CE, and the knot at C.

Solution

Sphere. By inspection, there are only two forces acting on the sphere, namely, its weight and the force of cord CE. The sphere has a weight of 6 kg $(9.81 \text{ m/s}^2) = 58.9$ N. The free-body diagram is shown in Fig. 3–3b.

Cord CE. When the cord CE is isolated from its surroundings, its free-body diagram shows only two forces acting on it, namely, the force of the sphere and the force of the knot, Fig. 3–3c. Notice that $\mathbf{F}_{CE}$ shown here is equal but opposite to that shown in Fig. 3–3b, a consequence of Newton's third law. Also, $\mathbf{F}_{CE}$ and $\mathbf{F}_{EC}$ pull on the cord and keep it in tension so that it doesn't collapse. For equilibrium, $F_{CE} = F_{EC}$.

Knot. The knot at C is subjected to three forces, Fig. 3–3d. They are caused by the cords CBA and CE and the spring CD. As required the free-body diagram shows all these forces labeled with their magnitudes and directions. It is important to recognize that the weight of the sphere does not directly act on the knot. Instead, the cord CE subjects the knot to this force.

$\mathbf{F}_{CE}$ (Force of cord CE acting on sphere)

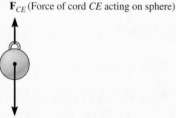

58.9N (Weight or gravity acting on sphere)

(b)

$\mathbf{F}_{EC}$ (Force of knot acting on cord CE)

$\mathbf{F}_{CE}$ (Force of sphere acting on cord CE)

(c)

$\mathbf{F}_{CBA}$ (Force of cord CBA acting on knot)

60° C

$\mathbf{F}_{CD}$ (Force of spring acting on knot)

$\mathbf{F}_{CE}$ (Force of cord CE acting on knot)

(d)

Fig. 3–3

3.3 Coplanar Force Systems

If a particle is subjected to a system of coplanar forces that lie in the x–y plane, Fig. 3–4, then each force can be resolved into its $\mathbf{i}$ and $\mathbf{j}$ components. For equilibrium, Eq. 3–1 can be written as

$$\Sigma \mathbf{F} = \mathbf{0}$$

$$\Sigma F_x \mathbf{i} + \Sigma F_y \mathbf{j} = \mathbf{0}$$

For this vector equation to be satisfied, both the x and y components must be equal to zero. Hence,

$$\boxed{\begin{aligned} \Sigma F_x &= 0 \\ \Sigma F_y &= 0 \end{aligned}}$$

(3–3)

These *scalar equations of equilibrium* require that the *algebraic sum* of the x and y components of all the forces acting on the particle be equal to zero. As a result, Eqs. 3–3 can be solved for at most two unknowns, generally represented as angles and magnitudes of forces shown on the particle's free-body diagram.

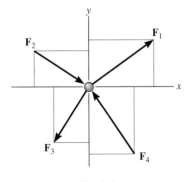

Fig. 3–4

Scalar Notation. Since each of the two equilibrium equations requires the resolution of vector components along a specified x or y axis, we will use scalar notation to represent the components when applying these equations. When doing this, the sense of direction for each component is accounted for by an *algebraic sign* which corresponds to the arrowhead direction of the component along each axis. If a force has an *unknown magnitude*, then the arrowhead sense of the force on the free-body diagram can be *assumed*. Since the magnitude of a force is *always positive*, then if the *solution* yields a *negative scalar*, this indicates that the sense of the force acts in the opposite direction.

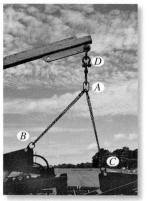

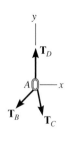

The chains exert three forces on the ring at A. The ring will not move, or will move with constant velocity, provided the summation of these forces along the x and along the y axis on the free-body diagram is zero. If one of the three forces is known, the magnitudes of the other two forces can be obtained from the two equations of equilibrium.

Fig. 3–5

For example, consider the free-body diagram of the particle subjected to the two forces shown in Fig. 3–5. Here it is *assumed* that the *unknown force* **F** acts to the right to maintain equilibrium. Applying the equation of equilibrium along the x axis, we have

$$\xrightarrow{+} \Sigma F_x = 0; \qquad\qquad +F + 10\ \text{N} = 0$$

Both terms are "positive" since both forces act in the positive x direction. When this equation is solved, $F = -10\ \text{N}$. Here the *negative sign* indicates that **F** must act to the left to hold the particle in equilibrium, Fig. 3–5. Notice that if the $+x$ axis in Fig. 3–5 was directed to the left, both terms in the above equation would be negative, but again, after solving, $F = -10\ \text{N}$, indicating again **F** would be directed to the left.

PROCEDURE FOR ANALYSIS

Coplanar force equilibrium problems for a particle can be solved using the following procedure.

Free-Body Diagram.

- Establish the x, y axes in any suitable orientation.
- Label all the known and unknown force magnitudes and directions on the diagram.
- The sense of a force having an unknown magnitude can be assumed.

Equations of Equilibrium.

- Apply the equations of equilibrium $\Sigma F_x = 0$ and $\Sigma F_y = 0$.
- Components are positive if they are directed along a positive axis, and negative if they are directed along a negative axis.
- If more than two unknowns exist and the problem involves a spring, apply $F = ks$ to relate the spring force to the deformation s of the spring.
- If the solution yields a negative result, this indicates the sense of the force is the reverse of that shown on the free-body diagram.

EXAMPLE 3.2

Determine the tension in cables AB and AD for equilibrium of the 250-kg engine shown in Fig. 3–6a.

Solution

Free-Body Diagram. To solve this problem, we will investigate the equilibrium of the ring at A because this "particle" is subjected to the forces of both cables AB and AD. First, however, note that the engine has a weight $(250 \text{ kg})(9.81 \text{ m/s}^2) = 2.452 \text{ kN}$ which is supported by cable CA. Therefore, as shown in Fig. 3–6b, there are three concurrent forces *acting on the ring*. The forces $\mathbf{T}_B$ and $\mathbf{T}_D$ have unknown magnitudes but known directions, and cable AC exerts a downward force on A equal to 2.452 kN.

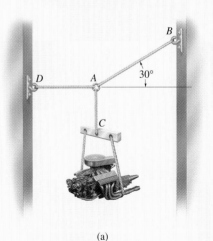

(a)

Fig. 3–6

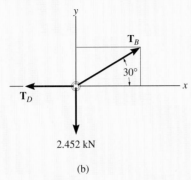

(b)

Equations of Equilibrium. The two unknown magnitudes T_B and T_D can be obtained from the two scalar equations of equilibrium, $\Sigma F_x = 0$ and $\Sigma F_y = 0$. To apply these equations, the x, y axes are established on the free-body diagram and $\mathbf{T}_B$ must be resolved into its x and y components. Thus,

$$\xrightarrow{+} \Sigma F_x = 0; \qquad T_B \cos 30° - T_D = 0 \qquad (1)$$
$$+\uparrow \Sigma F_y = 0; \qquad T_B \sin 30° - 2.452 \text{ kN} = 0 \qquad (2)$$

Solving Eq. 2 for T_B and substituting into Eq. 1 to obtain T_D yields

$$T_B = 4.90 \text{ kN} \qquad \qquad \textit{Ans.}$$
$$T_D = 4.25 \text{ kN} \qquad \qquad \textit{Ans.}$$

The accuracy of these results, of course, depends on the accuracy of the data, i.e., measurements of geometry and loads. For most engineering work involving a problem such as this, the data as measured to three significant figures would be sufficient. Also, note that here we have neglected the weights of the cables, a reasonable assumption since they would be small in comparison with the weight of the engine.

E X A M P L E 3.3

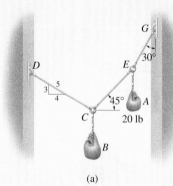

(a)

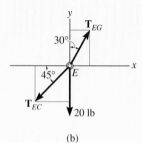

(b)

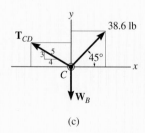

(c)

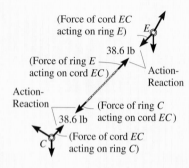

(d)

Fig. 3–7

If the sack at A in Fig. 3–7a has a weight of 20 lb, determine the weight of the sack at B and the force in each cord needed to hold the system in the equilibrium position shown.

Solution

Since the weight of A is known, the unknown tension in the two cords EG and EC can be determined by investigating the equilibrium of the ring at E. Why?

Free-Body Diagram. There are three forces acting on E, as shown in Fig. 3–7b.

Equations of Equilibrium. Establishing the x, y axes and resolving each force onto its x and y components using trigonometry, we have

$$\xrightarrow{+} \Sigma F_x = 0; \qquad T_{EG} \sin 30° - T_{EC} \cos 45° = 0 \qquad (1)$$
$$+\uparrow \Sigma F_y = 0; \qquad T_{EG} \cos 30° - T_{EC} \sin 45° - 20 \text{ lb} = 0 \qquad (2)$$

Solving Eq. 1 for T_{EG} in terms of T_{EC} and substituting the result into Eq. 2 allows a solution for T_{EC}. One then obtains T_{EG} from Eq. 1. The results are

$$T_{EC} = 38.6 \text{ lb} \qquad\qquad Ans.$$
$$T_{EG} = 54.6 \text{ lb} \qquad\qquad Ans.$$

Using the calculated result for T_{EC}, the equilibrium of the ring at C can now be investigated to determine the tension in CD and the weight of B.

Free-Body Diagram. As shown in Fig. 3–7c, $T_{EC} = 38.6$ lb "pulls" on C. The reason for this becomes clear when one draws the free-body diagram of cord CE and applies both equilibrium and the principle of action, equal but opposite force reaction (Newton's third law), Fig. 3–7d.

Equations of Equilibrium. Establishing the x, y axes and noting the components of $\mathbf{T}_{CD}$ are proportional to the slope of the cord as defined by the 3–4–5 triangle, we have

$$\xrightarrow{+} \Sigma F_x = 0; \qquad 38.6 \cos 45° \text{ lb} - \left(\tfrac{4}{5}\right)T_{CD} = 0 \qquad (3)$$
$$\xrightarrow{+} \Sigma F_y = 0; \qquad \left(\tfrac{3}{5}\right)T_{CD} + 38.6 \sin 45° \text{ lb} - W_B = 0 \qquad (4)$$

Solving Eq. 3 and substituting the result into Eq. 4 yields

$$T_{CD} = 34.2 \text{ lb} \qquad\qquad Ans.$$
$$W_B = 47.8 \text{ lb} \qquad\qquad Ans.$$

EXAMPLE **3.4**

Determine the required length of cord AC in Fig. 3–8a so that the 8-kg lamp is suspended in the position shown. The *undeformed* length of spring AB is $l'_{AB} = 0.4$ m, and the spring has a stiffness of $k_{AB} = 300$ N/m.

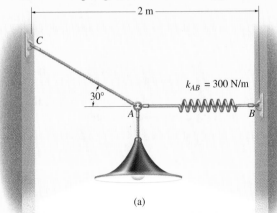

(a)

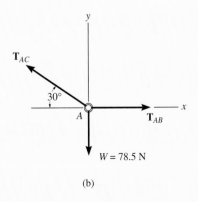

(b)

Fig. 3–8

Solution

If the force in spring AB is known, the stretch of the spring can be found using $F = ks$. From the problem geometry, it is then possible to calculate the required length of AC.

Free-Body Diagram. The lamp has a weight $W = 8(9.81) = 78.5$ N. The free-body diagram of the ring at A is shown in Fig. 3–8b.

Equations of Equilibrium. Using the x, y axes,

$$\xrightarrow{+} \Sigma F_x = 0; \qquad T_{AB} - T_{AC} \cos 30° = 0$$
$$+\uparrow \Sigma F_y = 0; \qquad T_{AC} \sin 30° - 78.5 \text{ N} = 0$$

Solving, we obtain

$$T_{AC} = 157.0 \text{ N}$$
$$T_{AB} = 136.0 \text{ N}$$

The stretch of spring AB is therefore

$$T_{AB} = k_{AB}s_{AB}; \qquad 136.0 \text{ N} = 300 \text{ N/m}(s_{AB})$$
$$s_{AB} = 0.453 \text{ m}$$

so the stretched length is

$$l_{AB} = l'_{AB} + s_{AB}$$
$$l_{AB} = 0.4 \text{ m} + 0.453 \text{ m} = 0.853 \text{ m}$$

The horizontal distance from C to B, Fig. 3–8a, requires

$$2 \text{ m} = l_{AC} \cos 30° + 0.853 \text{ m}$$
$$l_{AC} = 1.32 \text{ m} \qquad\qquad\qquad \text{Ans.}$$

3.4 Three-Dimensional Force Systems

For particle equilibrium we require

$$\Sigma \mathbf{F} = \mathbf{0} \qquad (3\text{--}4)$$

If the forces are resolved into their respective $\mathbf{i}, \mathbf{j}, \mathbf{k}$ components, Fig. 3–9, then we have

$$\Sigma F_x \mathbf{i} + \Sigma F_y \mathbf{j} + \Sigma F_z \mathbf{k} = \mathbf{0}$$

To ensure equilibrium, we must therefore require that the following three scalar component equations be satisfied:

$$\begin{aligned} \Sigma F_x &= 0 \\ \Sigma F_y &= 0 \\ \Sigma F_z &= 0 \end{aligned} \qquad (3\text{--}5)$$

These equations represent the *algebraic sums* of the x, y, z force components acting on the particle. Using them we can solve for at most three unknowns, generally represented as angles or magnitudes of forces shown on the particle's free-body diagram.

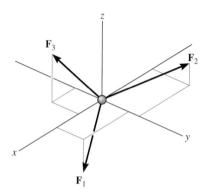

Fig. 3–9

The ring at A is subjected to the force from the hook as well as forces from each of the three chains. If the electromagnet and its load has a weight W, then the hook force will be W, and the three scalar equations of equilibrium can be applied to the free-body diagram of the ring in order to determine the chain forces, $\mathbf{F}_B$, $\mathbf{F}_C$ and $\mathbf{F}_D$.

PROCEDURE FOR ANALYSIS

Three-dimensional force equilibrium problems for a particle can be solved using the following procedure.

Free-Body Diagram.

- Establish the x, y, z axes in any suitable orientation.
- Label all the known and unknown force magnitudes and directions on the diagram.
- The sense of a force having an unknown magnitude can be assumed.

Equations of Equilibrium.

- Use the scalar equations of equilibrium, $\Sigma F_x = 0$, $\Sigma F_y = 0$, $\Sigma F_z = 0$, in cases where it is easy to resolve each force into its x, y, z components.
- If the three-dimensional geometry appears difficult, then first express each force as a Cartesian vector, substitute these vectors into $\Sigma \mathbf{F} = \mathbf{0}$, and then set the $\mathbf{i}$, $\mathbf{j}$, $\mathbf{k}$ components equal to zero.
- If the solution yields a negative result, this indicates the sense of the force is the reverse of that shown on the free-body diagram.

E X A M P L E 3.5

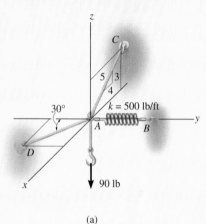

(a)

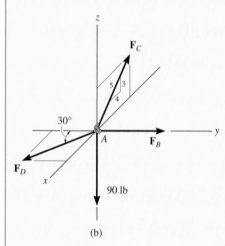

(b)

Fig. 3–10

A 90-lb load is suspended from the hook shown in Fig. 3–10a. The load is supported by two cables and a spring having a stiffness $k = 500$ lb/ft. Determine the force in the cables and the stretch of the spring for equilibrium. Cable AD lies in the x–y plane and cable AC lies in the x–z plane.

Solution

The stretch of the spring can be determined once the force in the spring is determined.

Free-Body Diagram. The connection at A is chosen for the equilibrium analysis since the cable forces are concurrent at this point. The free-body diagram is shown in Fig. 3–10b.

Equations of Equilibrium. By inspection, each force can easily be resolved into its x, y, z components, and therefore the three scalar equations of equilibrium can be directly applied. Considering components directed along the positive axes as "positive," we have

$$\Sigma F_x = 0; \qquad\qquad F_D \sin 30° - \tfrac{4}{5}F_C = 0 \qquad\qquad (1)$$
$$\Sigma F_y = 0; \qquad\qquad -F_D \cos 30° + F_B = 0 \qquad\qquad (2)$$
$$\Sigma F_z = 0; \qquad\qquad \tfrac{3}{5}F_C - 90 \text{ lb} = 0 \qquad\qquad (3)$$

Solving Eq. 3 for F_C, then Eq. 1 for F_D, and finally Eq. 2 for F_B, yields

$$F_C = 150 \text{ lb} \qquad\qquad Ans.$$
$$F_D = 240 \text{ lb} \qquad\qquad Ans.$$
$$F_B = 208 \text{ lb} \qquad\qquad Ans.$$

The stretch of the spring is therefore

$$F_B = ks_{AB}$$
$$208 \text{ lb} = 500 \text{ lb/ft}(s_{AB})$$
$$s_{AB} = 0.416 \text{ ft} \qquad\qquad Ans.$$

E X A M P L E 3.6

Determine the magnitude and coordinate direction angles of force **F** in Fig. 3–11a that are required for equilibrium of particle O.

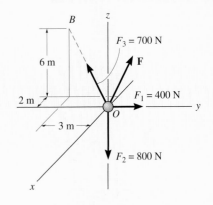

(a)

Solution

Free-Body Diagram. Four forces act on particle O, Fig. 3–11b.

Equations of Equilibrium. Each of the forces can be expressed in Cartesian vector form, and the equations of equilibrium can be applied to determine the x, y, z components of **F**. Noting that the coordinates of B are $B(-2\text{ m}, -3\text{ m}, 6\text{ m})$, we have

$$\mathbf{F}_1 = \{400\mathbf{j}\}\text{ N}$$
$$\mathbf{F}_2 = \{-800\mathbf{k}\}\text{ N}$$
$$\mathbf{F}_3 = F_3\left(\frac{\mathbf{r}_B}{r_B}\right) = 700\text{ N}\left[\frac{-2\mathbf{i} - 3\mathbf{j} + 6\mathbf{k}}{\sqrt{(-2)^2 + (-3)^2 + (6)^2}}\right]$$
$$= \{-200\mathbf{i} - 300\mathbf{j} + 600\mathbf{k}\}\text{ N}$$
$$\mathbf{F} = F_x\mathbf{i} + F_y\mathbf{j} + F_z\mathbf{k}$$

For equilibrium

$$\Sigma\mathbf{F} = \mathbf{0}; \qquad\qquad \mathbf{F}_1 + \mathbf{F}_2 + \mathbf{F}_3 + \mathbf{F} = \mathbf{0}$$
$$400\mathbf{j} - 800\mathbf{k} - 200\mathbf{i} - 300\mathbf{j} + 600\mathbf{k} + F_x\mathbf{i} + F_y\mathbf{j} + F_z\mathbf{k} = \mathbf{0}$$

Equating the respective **i**, **j**, **k** components to zero, we have

$$\Sigma F_x = 0; \qquad\quad -200 + F_x = 0 \qquad F_x = 200\text{ N}$$
$$\Sigma F_y = 0; \qquad\quad 400 - 300 + F_y = 0 \qquad F_y = -100\text{ N}$$
$$\Sigma F_z = 0; \qquad\quad -800 + 600 + F_z = 0 \qquad F_z = 200\text{ N}$$

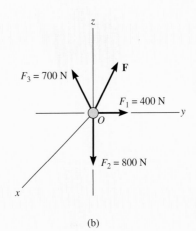

(b)

Thus,

$$\mathbf{F} = \{200\mathbf{i} - 100\mathbf{j} + 200\mathbf{k}\}\text{ N}$$
$$F = \sqrt{(200)^2 + (-100)^2 + (200)^2} = 300\text{ N} \qquad\qquad Ans.$$
$$\mathbf{u}_F = \frac{\mathbf{F}}{F} = \frac{200}{300}\mathbf{i} - \frac{100}{300}\mathbf{j} + \frac{200}{300}\mathbf{k}$$
$$\alpha = \cos^{-1}\left(\frac{200}{300}\right) = 48.2° \qquad\qquad Ans.$$
$$\beta = \cos^{-1}\left(\frac{-100}{300}\right) = 109° \qquad\qquad Ans.$$
$$\gamma = \cos^{-1}\left(\frac{200}{300}\right) = 48.2° \qquad\qquad Ans.$$

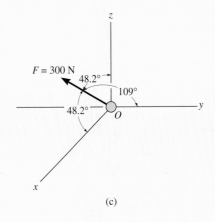

(c)

Fig. 3–11

The magnitude and correct direction of **F** are shown in Fig. 3–11c.

EXAMPLE **3.7**

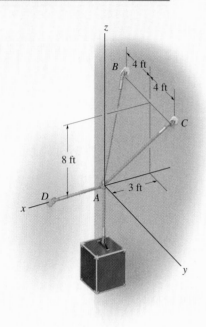

(a)

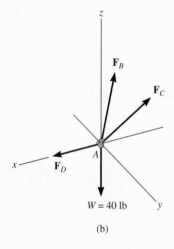

$W = 40$ lb

(b)

Fig. 3–12

Determine the force developed in each cable used to support the 40-lb crate shown in Fig. 3–12a.

Solution

Free-Body Diagram. As shown in Fig. 3–12b, the free-body diagram of point A is considered in order to "expose" the three unknown forces in the cables.

Equations of Equilibrium. First we will express each force in Cartesian vector form. Since the coordinates of points B and C are B(−3 ft, −4 ft, 8 ft) and C(−3 ft, 4 ft, 8 ft), we have

$$\mathbf{F}_B = F_B \left[\frac{-3\mathbf{i} - 4\mathbf{j} + 8\mathbf{k}}{\sqrt{(-3)^2 + (-4)^2 + (8)^2}} \right]$$

$$= -0.318 F_B \mathbf{i} - 0.424 F_B \mathbf{j} + 0.848 F_B \mathbf{k}$$

$$\mathbf{F}_C = F_C \left[\frac{-3\mathbf{i} + 4\mathbf{j} + 8\mathbf{k}}{\sqrt{(-3)^2 + (4)^2 + (8)^2}} \right]$$

$$= -0.318 F_C \mathbf{i} + 0.424 F_C \mathbf{j} + 0.484 F_C \mathbf{k}$$

$$\mathbf{F}_D = F_D \mathbf{i}$$

$$\mathbf{W} = \{-40\mathbf{k}\} \text{ lb}$$

Equilibrium requires

$$\Sigma \mathbf{F} = \mathbf{0}; \qquad \mathbf{F}_B + \mathbf{F}_C + \mathbf{F}_D + \mathbf{W} = \mathbf{0}$$

$$-0.318 F_B \mathbf{i} - 0.424 F_B \mathbf{j} + 0.848 F_B \mathbf{k} - 0.318 F_C \mathbf{i} + 0.424 F_C \mathbf{j}$$

$$+ 0.848 F_C \mathbf{k} + F_D \mathbf{i} - 40 \mathbf{k} = \mathbf{0}$$

Equating the respective **i**, **j**, **k** components to zero yields

$$\Sigma F_x = 0; \qquad -0.318 F_B - 0.318 F_C + F_D = 0 \qquad (1)$$

$$\Sigma F_y = 0; \qquad -0.424 F_B + 0.424 F_C = 0 \qquad (2)$$

$$\Sigma F_z = 0; \qquad 0.848 F_B + 0.848 F_C - 40 = 0 \qquad (3)$$

Equation 2 states that $F_B = F_C$. Thus, solving Eq. 3 for F_B and F_C and substituting the result into Eq. 1 to obtain F_D, we have

$$F_B = F_C = 23.6 \text{ lb} \qquad \qquad Ans.$$

$$F_D = 15.0 \text{ lb} \qquad \qquad Ans.$$

EXAMPLE 3.8

The 100-kg crate shown in Fig. 3–13a is supported by three cords, one of which is connected to a spring. Determine the tension in cords AC and AD and the stretch of the spring.

Solution

Free-Body Diagram. The force in each of the cords can be determined by investigating the equilibrium of point A. The free-body diagram is shown in Fig. 3–13b. The weight of the crate is $W = 100(9.81) = 981$ N.

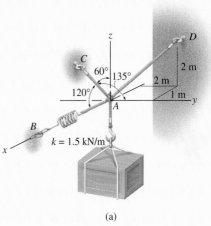

(a)

Equations of Equilibrium. Each vector on the free-body diagram is first expressed in Cartesian vector form. Using Eq. 2–11 for $\mathbf{F}_C$ and noting point $D(-1 \text{ m}, 2 \text{ m}, 2 \text{ m})$ for $\mathbf{F}_D$, we have

$$\mathbf{F}_B = F_B\mathbf{i}$$

$$\mathbf{F}_C = F_C \cos 120°\mathbf{i} + F_C \cos 135°\mathbf{j} + F_C \cos 60°\mathbf{k}$$

$$= -0.5F_C\mathbf{i} - 0.707F_C\mathbf{j} + 0.5F_C\mathbf{k}$$

$$\mathbf{F}_D = F_D\left[\frac{-1\mathbf{i} + 2\mathbf{j} + 2\mathbf{k}}{\sqrt{(-1)^2 + (2)^2 + (2)^2}}\right]$$

$$= -0.333F_D\mathbf{i} + 0.667F_D\mathbf{j} + 0.667F_D\mathbf{k}$$

$$\mathbf{W} = \{-981\mathbf{k}\} \text{ N}$$

Equilibrium requires

$$\Sigma\mathbf{F} = \mathbf{0}; \qquad \mathbf{F}_B + \mathbf{F}_C + \mathbf{F}_D + \mathbf{W} = \mathbf{0}$$

$$F_B\mathbf{i} - 0.5F_C\mathbf{i} - 0.707F_C\mathbf{j} + 0.5F_C\mathbf{k} - 0.333F_D\mathbf{i} + 0.667F_D\mathbf{j}$$
$$+ 0.667F_D\mathbf{k} - 981\mathbf{k} = \mathbf{0}$$

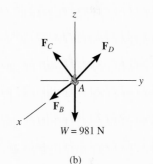

(b)

Fig. 3–13

Equating the respective $\mathbf{i}$, $\mathbf{j}$, $\mathbf{k}$ components to zero,

$$\Sigma F_x = 0; \qquad F_B - 0.5F_C - 0.333F_D = 0 \qquad (1)$$

$$\Sigma F_y = 0; \qquad -0.707F_C + 0.667F_D = 0 \qquad (2)$$

$$\Sigma F_z = 0; \qquad 0.5F_C + 0.667F_D - 981 = 0 \qquad (3)$$

Solving Eq. 2 for F_D in terms of F_C and substituting into Eq. 3 yields F_C. F_D is determined from Eq. 2. Finally, substituting the results into Eq. 1 gives F_B. Hence,

$$F_C = 813 \text{ N} \qquad\qquad Ans.$$

$$F_D = 862 \text{ N} \qquad\qquad Ans.$$

$$F_B = 693.7 \text{ N}$$

The stretch of the spring is therefore

$$F = ks; \qquad\qquad 693.7 = 1500s$$

$$s = 0.462 \text{ m} \qquad\qquad Ans.$$

CHAPTER REVIEW

- ***Equilibrium.*** When a particle is at rest or moves with constant velocity, it is in equilibrium. This requires that all the forces acting on the particle form a zero force resultant. In order to account for all the forces, it is necessary to draw a free-body diagram. This diagram is an outlined shape of the particle that shows all the forces, listed with their known or unknown magnitudes and directions.

- ***Two Dimensions.*** The two scalar equations of force equilibrium $\Sigma F_x = 0$ and $\Sigma F_y = 0$ can be applied when referenced from an established x, y coordinate system. If the solution for a force magnitude yields a negative scalar, then the force acts in the opposite direction to that shown on the free-body diagram. If the problem involves a linear elastic spring then the stretch or compression s of the spring can be related to the force applied to it using $F = ks$.

- ***Three Dimensions.*** Since three-dimensional geometry can be difficult to visualize, the equilibrium equation $\Sigma \mathbf{F} = \mathbf{0}$ should be applied using a Cartesian vector analysis. This requires first expressing each force on the free-body diagram as a Cartesian vector. When the forces are summed and set equal to zero, then the $\mathbf{i}$, $\mathbf{j}$, and $\mathbf{k}$ components are also zero, so that $\Sigma F_x = 0$, $\Sigma F_y = 0$ and $\Sigma F_z = 0$.

REVIEW PROBLEMS

3-1. The pipe is held in place by the vice. If the bolt exerts a force of 50 lb on the pipe in the direction shown, determine the forces F_A and F_B that the smooth contacts at A and B exert on the pipe.

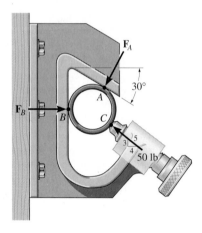

Prob. 3–1

3-2. When y is zero, the springs sustain a force of 60 lb. Determine the magnitude of the applied vertical forces $\mathbf{F}$ and $-\mathbf{F}$ required to pull point A away from point B a distance of $y = 2$ ft. The ends of cords CAD and CBD are attached to rings at C and D.

*3-3. When y is zero, the springs are each stretched 1.5 ft. Determine the distance y if a force of $F = 60$ lb is applied to points A and B as shown. The ends of cords CAD and CBD are attached to rings at C and D.

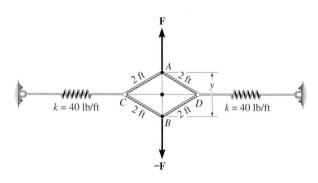

Probs. 3–2/3

3-4. Romeo tries to reach Juliet by climbing with constant velocity up a rope which is knotted at point A. Any of the three segments of the rope can sustain a maximum force of 2 kN before it breaks. Determine if Romeo, who has a mass of 65 kg, can climb the rope, and if so, can he along with his Juliet, who has a mass of 60 kg, climb down with constant velocity?

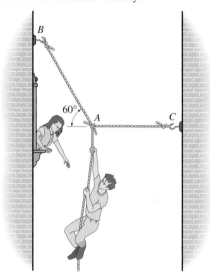

Prob. 3–4

■ **3-5.** Determine the magnitudes of forces $\mathbf{F}_1$, $\mathbf{F}_2$, and $\mathbf{F}_3$ necessary to hold the force $\mathbf{F} = \{-9\mathbf{i} - 8\mathbf{j} - 5\mathbf{k}\}$ kN in equilibrium.

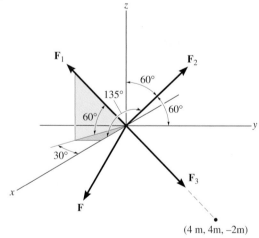

Prob. 3–5

3-6. The man attempts to pull the log at *C* by using the three ropes. Determine the direction θ in which he should pull on his rope with a force of 80 lb, so that he exerts a maximum force on the log. What is the force on the log for this case? Also, determine the direction in which he should pull in order to maximize the force in the rope attached to *B*. What is this maximum force?

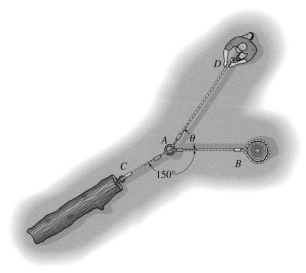

Prob. 3–6

*■3-7. The ring of negligible size is subjected to a vertical force of 200 lb. Determine the required length *l* of cord *AC* such that the tension acting in *AC* is 160 lb. Also, what is the force acting in cord *AB*? *Hint:* Use the equilibrium condition to determine the required angle θ for attachment, then determine *l* using trigonometry applied to $\triangle ABC$.

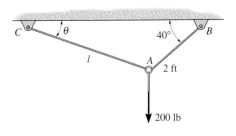

Prob. 3–7

3-8. Determine the maximum weight of the engine that can be supported without exceeding a tension of 450 lb in chain *AB* and 480 lb in chain *AC*.

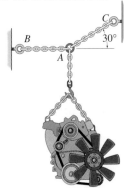

Prob. 3–8

3-9. Determine the force in each cable needed to support the 500-lb load.

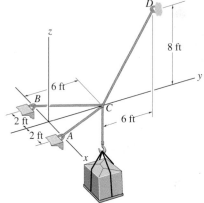

Prob. 3–9

3-10. The joint of a space frame is subjected to four member forces. Member *OA* lies in the $x - y$ plane and member *OB* lies in the $y - z$ plane. Determine the forces acting in each of the members required for equilibrium of the joint.

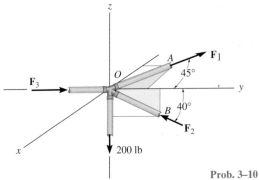

Prob. 3–10

This utility pole is subjected to many forces, caused by the cables and the weight of the transformer. In some cases, it is important to be able to simplify this system to a single resultant force and specify where this resultant acts on the pole.

Force System Resultants

- To discuss the concept of the moment of a force and show how to calculate it in two and three dimensions.
- To provide a method for finding the moment of a force about a specified axis.
- To define the moment of a couple.
- To present methods for determining the resultants of nonconcurrent force systems.
- To indicate how to reduce a simple distributed loading to a resultant force having a specified location.

4.1 Moment of a Force—Scalar Formulation

The *moment* of a force about a point or axis provides a measure of the tendency of the force to cause a body to rotate about the point or axis. For example, consider the horizontal force $\mathbf{F}_x$, which acts perpendicular to the handle of the wrench and is located a distance d_y from point O, Fig. 4–1a. It is seen that this force tends to cause the pipe to turn about the z axis. The larger the force or the distance d_y, the greater the turning effect. This tendency for rotation caused by $\mathbf{F}_x$ is sometimes called a *torque*, but most often it is called the *moment of a force* or simply the *moment* $(\mathbf{M}_O)_z$. Note that the *moment axis* (z) is perpendicular to the shaded plane (x–y) which contains both $\mathbf{F}_x$ and d_y and that this axis intersects the plane at point O.

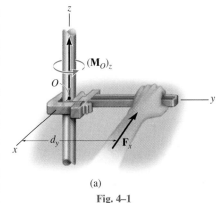

(a)

Fig. 4–1

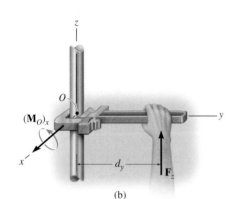

(b)

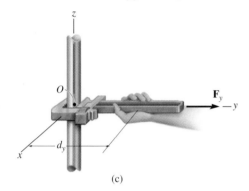

(c)

Fig. 4–1

Now consider applying the force $\mathbf{F}_z$ to the wrench, Fig. 4–1b. This force will *not* rotate the pipe about the z axis. Instead, it tends to rotate it about the x axis. Keep in mind that although it may not be possible to actually "rotate" or turn the pipe in this manner, $\mathbf{F}_z$ still creates the *tendency* for rotation and so the moment $(\mathbf{M}_O)_x$ is produced. As before, the force and distance d_y lie in the shaded plane $(y$–$z)$ which is perpendicular to the moment axis (x). Lastly, if a force $\mathbf{F}_y$ is applied to the wrench, Fig. 4–1c, no moment is produced about point O. This results in a lack of turning since the line of action of the force passes through O and therefore no tendency for rotation is possible.

We will now generalize the above discussion and consider the force $\mathbf{F}$ and point O which lie in a shaded plane as shown in Fig. 4–2a. The moment $\mathbf{M}_O$ about point O, or about an axis passing through O and perpendicular to the plane, is a *vector quantity* since it has a specified magnitude and direction.

Magnitude. The magnitude of M_O is

$$M_O = Fd \tag{4–1}$$

where d is referred to as the *moment arm* or perpendicular distance from the axis at point O to the line of action of the force. Units of moment magnitude consist of force times distance, e.g., N·m or lb·ft.

Direction. The direction of $\mathbf{M}_O$ will be specified by using the "right-hand rule." To do this, the fingers of the right hand are curled such that they follow the sense of rotation, which would occur if the force could rotate about point O, Fig. 4–2a. The *thumb* then *points* along the *moment axis* so that it gives the direction and sense of the moment vector, which is *upward* and *perpendicular* to the shaded plane containing $\mathbf{F}$ and d.

In three dimensions, $\mathbf{M}_O$ is illustrated by a vector arrow with a curl on it to *distinguish* it from a force vector, Fig. 4–2a. Many problems in mechanics, however, involve coplanar force systems that may be conveniently viewed in two dimensions. For example, a two-dimensional view of Fig. 4–2a is given in Fig. 4–2b. Here $\mathbf{M}_O$ is simply represented by the (counterclockwise) curl, which indicates the action of $\mathbf{F}$. The arrowhead on this curl is used to show the *sense of rotation* caused by $\mathbf{F}$. Using the right-hand rule, however, realize that the direction and sense of the moment vector in Fig. 4–2b are specified by the thumb, which points *out* of the page since the fingers follow the curl. In particular, notice that *this curl or sense of rotation can always be determined by observing in which direction the force would "orbit" about point O* (counterclockwise in Fig. 4–2b). In two dimensions we will often refer to finding the moment of a force "about a point" (O). Keep in mind, however, that the moment *always acts about an axis* which is perpendicular to the plane containing $\mathbf{F}$ and d, and this axis intersects the plane at the point (O), Fig. 4–2a.

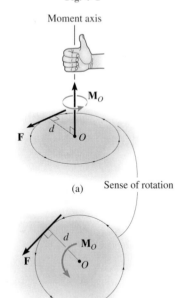

(a)

(b)

Fig. 4–2

Resultant Moment of a System of Coplanar Forces. If a system of forces lies in an x–y plane, then the moment produced by each force about point O will be directed along the z axis, Fig. 4–3. Consequently, the resultant moment $\mathbf{M}_{R_O}$ of the system can be determined by simply adding the moments of all forces *algebraically* since all the moment vectors are collinear. We can write this vector sum symbolically as

$$\downarrow + M_{R_O} = \Sigma F d \qquad (4\text{--}2)$$

Here the counterclockwise curl written alongside the equation indicates that, by the scalar sign convention, the moment of any force will be positive if it is directed along the $+z$ axis, whereas a negative moment is directed along the $-z$ axis.

The following examples illustrate numerical application of Eqs. 4–1 and 4–2.

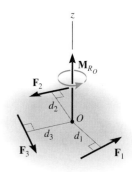

Fig. 4–3

By pushing down on the pry bar the load on the ground at A can be lifted. The turning effect, caused by the applied force, is due to the moment about A. To produce this moment with minimum effort we instinctively know that the force should be applied to the *end* of the bar; however, the *direction* in which this force is applied is also important. This is because moment is the product of the force and the moment arm. Notice that when the force is at an angle $\theta < 90°$, then the moment arm distance is *shorter* than when the force is applied perpendicular to the bar $\theta = 90°$, i.e., $d' < d$. Hence the greatest moment is produced when the force is farthest from point A and applied perpendicular to the axis of the bar so as to maximize the moment arm.

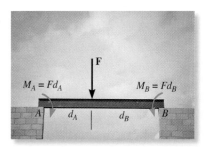

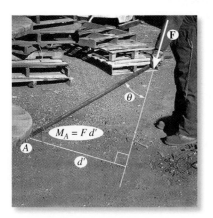

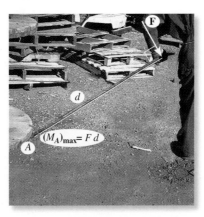

The moment of a force does not always cause a rotation. For example, the force $\mathbf{F}$ tends to rotate the beam clockwise about its support at A with a moment $M_A = Fd_A$. The actual rotation would occur if the support at B were removed. In the same manner, $\mathbf{F}$ creates a tendency to rotate the beam counterclockwise about B with a moment $M_B = Fd_B$. Here the support at A prevents the rotation.

E X A M P L E   4.1

For each case illustrated in Fig. 4–4, determine the moment of the force about point O.

Solution *(Scalar Analysis)*

The line of action of each force is extended as a dashed line in order to establish the moment arm d. Also illustrated is the tendency of rotation of the member as caused by the force. Furthermore, the orbit of the force is shown as a colored curl. Thus,

Fig. 4–4a	$M_O = (100\,\text{N})(2\,\text{m}) = 200\,\text{N}\cdot\text{m}\ \downarrow$		*Ans.*
Fig. 4–4b	$M_O = (50\,\text{N})(0.75\,\text{m}) = 37.5\,\text{N}\cdot\text{m}\ \downarrow$		*Ans.*
Fig. 4–4c	$M_O = (40\,\text{lb})(4\,\text{ft} + 2\cos 30°\,\text{ft}) = 229\,\text{lb}\cdot\text{ft}\ \downarrow$		*Ans.*
Fig. 4–4d	$M_O = (60\,\text{lb})(1\sin 45°\,\text{ft}) = 42.4\,\text{lb}\cdot\text{ft}\ \uparrow$		*Ans.*
Fig. 4–4e	$M_O = (7\,\text{kN})(4\,\text{m} - 1\,\text{m}) = 21.0\,\text{kN}\cdot\text{m}\ \uparrow$		*Ans.*

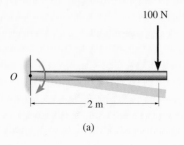

(a)

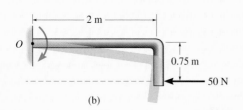

(b)

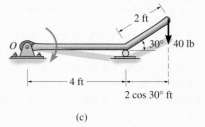

(c)

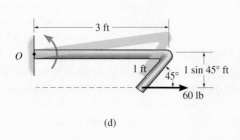

(d)

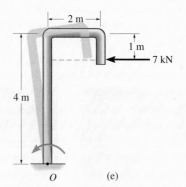

(e)

Fig. 4–4

E X A M P L E   **4.2**

Determine the moments of the 800-N force acting on the frame in Fig. 4–5 about points A, B, C, and D.

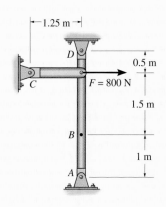

Solution (Scalar Analysis)
In general, $M = Fd$, where d is the moment arm or *perpendicular distance* from the *point* on the moment axis to the *line of action* of the force. Hence,

$M_A = 800 \text{ N}(2.5 \text{ m}) = 2000 \text{ N} \cdot \text{m} \downdownarrows$ *Ans.*

$M_B = 800 \text{ N}(1.5 \text{ m}) = 1200 \text{ N} \cdot \text{m} \downdownarrows$ *Ans.*

$M_C = 800 \text{ N}(0) = 0$ (line of action of **F** passes through C)   *Ans.*

$M_D = 800 \text{ N}(0.5 \text{ m}) = 400 \text{ N} \cdot \text{m} \upharpoonleft$ *Ans.*

The curls indicate the sense of rotation of the moment, which is defined by the direction the force orbits about each point.

Fig. 4–5

E X A M P L E   **4.3**

Determine the resultant moment of the four forces acting on the rod shown in Fig. 4–6 about point O.

Solution
Assuming that positive moments act in the $+\mathbf{k}$ direction, i.e., counterclockwise, we have

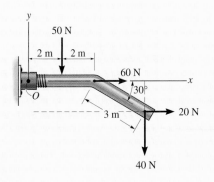

$\downdownarrows +M_{R_O} = \Sigma Fd;$

$\qquad M_{R_O} = -50 \text{ N}(2\text{m}) + 60 \text{ N}(0) + 20 \text{ N}(3 \sin 30° \text{ m})$

$\qquad\qquad -40 \text{ N}(4 \text{ m} + 3 \cos 30° \text{ m})$

$\qquad M_{R_O} = -334 \text{ N} \cdot \text{m} = 334 \text{ N} \cdot \text{m} \downdownarrows$ *Ans.*

For this calculation, note how the moment-arm distances for the 20-N and 40-N forces are established from the extended (dashed) lines of action of each of these forces.

Fig. 4–6

4.2 Cross Product

The moment of a force will be formulated using Cartesian vectors in the next section. Before doing this, however, it is first necessary to expand our knowledge of vector algebra and introduce the cross-product method of vector multiplication.

The *cross product* of two vectors **A** and **B** yields the vector **C**, which is written

$$\mathbf{C} = \mathbf{A} \times \mathbf{B}$$

and is read "**C** equals **A** cross **B**."

Magnitude. The *magnitude* of **C** is defined as the product of the magnitudes of **A** and **B** and the sine of the angle θ between their tails $(0° \leq \theta \leq 180°)$. Thus, $C = AB \sin \theta$.

Direction. Vector **C** has a *direction* that is perpendicular to the plane containing **A** and **B** such that **C** is specified by the right-hand rule; i.e., curling the fingers of the right hand from vector **A** (cross) to vector **B**, the thumb then points in the direction of **C**, as shown in Fig. 4–7.

Knowing both the magnitude and direction of **C**, we can write

$$\mathbf{C} = \mathbf{A} \times \mathbf{B} = (AB \sin \theta)\mathbf{u}_C \qquad (4\text{–}3)$$

where the scalar $AB \sin \theta$ defines the *magnitude* of **C** and the unit vector $\mathbf{u}_C$ defines the *direction* of **C**. The terms of Eq. 4–3 are illustrated graphically in Fig. 4–8.

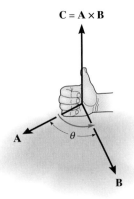

$C = \mathbf{A} \times \mathbf{B}$

Fig. 4–7

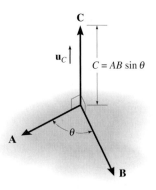

Fig. 4–8

Laws of Operation.

1. The commutative law is *not* valid; i.e.,

$$\mathbf{A} \times \mathbf{B} \neq \mathbf{B} \times \mathbf{A}$$

Rather,

$$\mathbf{A} \times \mathbf{B} = -\mathbf{B} \times \mathbf{A}$$

This is shown in Fig. 4–9 by using the right-hand rule. The cross product $\mathbf{B} \times \mathbf{A}$ yields a vector that acts in the opposite direction to $\mathbf{C}$; i.e., $\mathbf{B} \times \mathbf{A} = -\mathbf{C}$.

2. Multiplication by a scalar:

$$a(\mathbf{A} \times \mathbf{B}) = (a\mathbf{A}) \times \mathbf{B} = \mathbf{A} \times (a\mathbf{B}) = (\mathbf{A} \times \mathbf{B})a$$

This property is easily shown since the magnitude of the resultant vector ($|a| AB \sin \theta$) and its direction are the same in each case.

3. The distributive law:

$$\mathbf{A} \times (\mathbf{B} + \mathbf{D}) = (\mathbf{A} \times \mathbf{B}) + (\mathbf{A} \times \mathbf{D})$$

The proof of this identity is left as an exercise (see Prob. 4–1). It is important to note that *proper order* of the cross products must be maintained, since they are not commutative.

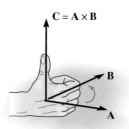

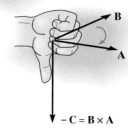

$$\mathbf{C} = \mathbf{A} \times \mathbf{B}$$

$$-\mathbf{C} = \mathbf{B} \times \mathbf{A}$$

Fig. 4–9

Cartesian Vector Formulation. Equation 4–3 may be used to find the cross product of a pair of Cartesian unit vectors. For example, to find $\mathbf{i} \times \mathbf{j}$, the *magnitude* of the resultant vector is $(i)(j)(\sin 90°) = (1)(1)(1) = 1$, and its *direction* is determined using the right-hand rule. As shown in Fig. 4–10, the resultant vector points in the $+\mathbf{k}$ direction. Thus, $\mathbf{i} \times \mathbf{j} = (1)\mathbf{k}$. In a similar manner,

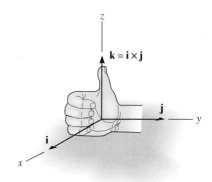

$$\mathbf{k} = \mathbf{i} \times \mathbf{j}$$

Fig. 4–10

$$
\begin{array}{lll}
\mathbf{i} \times \mathbf{j} = \mathbf{k} & \mathbf{i} \times \mathbf{k} = -\mathbf{j} & \mathbf{i} \times \mathbf{i} = 0 \\
\mathbf{j} \times \mathbf{k} = \mathbf{i} & \mathbf{j} \times \mathbf{i} = -\mathbf{k} & \mathbf{j} \times \mathbf{j} = 0 \\
\mathbf{k} \times \mathbf{i} = \mathbf{j} & \mathbf{k} \times \mathbf{j} = -\mathbf{i} & \mathbf{k} \times \mathbf{k} = 0
\end{array}
$$

These results should *not* be memorized; rather, it should be clearly understood how each is obtained by using the right-hand rule and the definition of the cross product. A simple scheme shown in Fig. 4–11 is helpful for obtaining the same results when the need arises. If the circle is constructed as shown, then "crossing" two unit vectors in a *counterclockwise* fashion around the circle yields the *positive* third unit vector; e.g., $\mathbf{k} \times \mathbf{i} = \mathbf{j}$. Moving *clockwise*, a *negative* unit vector is obtained; e.g., $\mathbf{i} \times \mathbf{k} = -\mathbf{j}$.

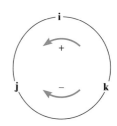

Fig. 4–11

Consider now the cross product of two general vectors **A** and **B** which are expressed in Cartesian vector form. We have

$$\mathbf{A} \times \mathbf{B} = (A_x\mathbf{i} + A_y\mathbf{j} + A_z\mathbf{k}) \times (B_x\mathbf{i} + B_y\mathbf{j} + B_z\mathbf{k})$$
$$= A_xB_x(\mathbf{i} \times \mathbf{i}) + A_xB_y(\mathbf{i} \times \mathbf{j}) + A_xB_z(\mathbf{i} \times \mathbf{k})$$
$$+ A_yB_x(\mathbf{j} \times \mathbf{i}) + A_yB_y(\mathbf{j} \times \mathbf{j}) + A_yB_z(\mathbf{j} \times \mathbf{k})$$
$$+ A_zB_x(\mathbf{k} \times \mathbf{i}) + A_zB_y(\mathbf{k} \times \mathbf{j}) + A_zB_z(\mathbf{k} \times \mathbf{k})$$

Carrying out the cross-product operations and combining terms yields

$$\mathbf{A} \times \mathbf{B} = (A_yB_z - A_zB_y)\mathbf{i} - (A_xB_z - A_zB_x)\mathbf{j} + (A_xB_y - A_yB_x)\mathbf{k} \qquad (4\text{–}4)$$

This equation may also be written in a more compact determinant form as

$$\mathbf{A} \times \mathbf{B} = \begin{vmatrix} \mathbf{i} & \mathbf{j} & \mathbf{k} \\ A_x & A_y & A_z \\ B_x & B_y & B_z \end{vmatrix} \qquad (4\text{–}5)$$

Thus, to find the cross product of any two Cartesian vectors **A** and **B**, it is necessary to expand a determinant whose first row of elements consists of the unit vectors **i**, **j**, and **k** and whose second and third rows represent the *x, y, z* components of the two vectors **A** and **B**, respectively.*

---

*A determinant having three rows and three columns can be expanded using three minors, each of which is multiplied by one of the three terms in the first row. There are four elements in each minor, e.g.,

$$\begin{vmatrix} A_{11} & A_{12} \\ A_{21} & A_{22} \end{vmatrix}$$

By *definition*, this notation represents the terms $(A_{11}A_{22} - A_{12}A_{21})$, which is simply the product of the two elements of the arrow slanting downward to the right $(A_{11}A_{22})$ *minus* the product of the two elements intersected by the arrow slanting downward to the left $(A_{12}A_{21})$. For a 3 × 3 determinant, such as Eq. 4–5, the three minors can be generated in accordance with the following scheme:

For element **i**: $\begin{vmatrix} \mathbf{i} & \mathbf{j} & \mathbf{k} \\ A_x & A_y & A_z \\ B_x & B_y & B_z \end{vmatrix} = \mathbf{i}(A_yB_z - A_zB_y)$

For element **j**: $\begin{vmatrix} \mathbf{i} & \mathbf{j} & \mathbf{k} \\ A_x & A_y & A_z \\ B_x & B_y & B_z \end{vmatrix} = -\mathbf{j}(A_xB_z - A_zB_x)$

For element **k**: $\begin{vmatrix} \mathbf{i} & \mathbf{j} & \mathbf{k} \\ A_x & A_y & A_z \\ B_x & B_y & B_z \end{vmatrix} = \mathbf{k}(A_xB_y - A_yB_x)$

Adding the results and noting that the **j** element *must include the minus sign* yields the expanded form of **A** × **B** given by Eq. 4–4.

4.3 Moment of a Force—Vector Formulation

The moment of a force **F** about point O, or actually about the moment axis passing through O and perpendicular to the plane containing O and **F**, Fig. 4–12a, can be expressed using the vector cross product, namely,

$$\mathbf{M}_O = \mathbf{r} \times \mathbf{F} \qquad\qquad (4\text{–}6)$$

Here **r** represents a position vector drawn *from O* to *any point* lying on the line of action of **F**. We will now show that indeed the moment $\mathbf{M}_O$, when determined by this cross product, has the proper magnitude and direction.

Magnitude. The magnitude of the cross product is defined from Eq. 4–3 as $M_O = rF \sin \theta$, where the angle θ is measured between the *tails* of **r** and **F**. To establish this angle, **r** must be treated as a sliding vector so that θ can be constructed properly, Fig. 4–12b. Since the moment arm $d = r \sin \theta$, then

$$M_O = rF \sin \theta = F(r \sin \theta) = Fd$$

which agrees with Eq. 4–1.

Direction. The direction and sense of $\mathbf{M}_O$ in Eq. 4–6 are determined by the right-hand rule as it applies to the cross product. Thus, extending **r** to the dashed position and curling the right-hand fingers from **r** toward **F**, "**r** cross **F**," the thumb is directed upward or perpendicular to the plane containing **r** and **F** and this is in the *same direction* as $\mathbf{M}_O$, the moment of the force about point O, Fig. 4–12b. Note that the "curl" of the fingers, like the curl around the moment vector, indicates the sense of rotation caused by the force. Since the cross product is not commutative, it is important that the *proper order* of **r** and **F** be maintained in Eq. 4–6.

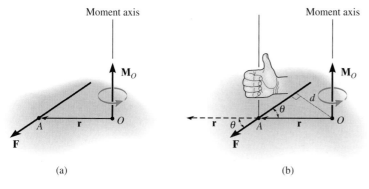

(a)

(b)

Fig. 4–12

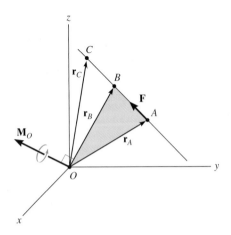

Fig. 4–13

Principle of Transmissibility. Consider the force **F** applied at point A in Fig. 4–13. The moment created by **F** about O is $\mathbf{M}_O = \mathbf{r}_A \times \mathbf{F}$; however, it was shown that "**r**" can extend from O to *any point* on the line of action of **F**. Consequently, **F** may be applied at point B or C, and the same moment $\mathbf{M}_O = \mathbf{r}_B \times \mathbf{F} = \mathbf{r}_C \times \mathbf{F}$ will be computed. As a result, **F** has the properties of a *sliding vector* and can therefore act at *any point along its line of action* and still create the same moment about point O. We refer to this as the *principle of transmissibility*, and we will discuss this property further in Sec. 4.7.

Cartesian Vector Formulation. If we establish x, y, z coordinate axes, then the position vector **r** and force **F** can be expressed as Cartesian vectors, Fig. 4–14. Applying Eq. 4–5 we have

$$\mathbf{M}_O = \mathbf{r} \times \mathbf{F} = \begin{vmatrix} \mathbf{i} & \mathbf{j} & \mathbf{k} \\ r_x & r_y & r_z \\ F_x & F_y & F_z \end{vmatrix} \qquad (4\text{–}7)$$

where

r_x, r_y, r_z represent the x, y, z components of the position vector drawn from point O to *any point* on the line of action of the force

F_x, F_y, F_z represent the x, y, z components of the force vector

If the determinant is expanded, then like Eq. 4–4 we have

$$\mathbf{M}_O = (r_y F_z - r_z F_y)\mathbf{i} - (r_x F_z - r_z F_x)\mathbf{j} + (r_x F_y - r_y F_x)\mathbf{k} \quad (4\text{–}8)$$

The physical meaning of these three moment components becomes evident by studying Fig. 4–14a. For example, the **i** component of $\mathbf{M}_O$ is

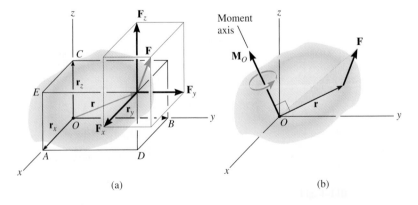

(a)   (b)

Fig. 4–14

determined from the moments of $\mathbf{F}_x$, $\mathbf{F}_y$, and $\mathbf{F}_z$ about the x axis. In particular, note that $\mathbf{F}_x$ does *not* create a moment or tendency to cause turning about the x axis since this force is *parallel* to the x axis. The line of action of $\mathbf{F}_y$ passes through point E, and so the magnitude of the moment of $\mathbf{F}_y$ about point A on the x axis is $r_z F_y$. By the right-hand rule this component acts in the negative $\mathbf{i}$ direction. Likewise, $\mathbf{F}_z$ contributes a moment component of $r_y F_z \mathbf{i}$. Thus, $(M_O)_x = (r_y F_z - r_z F_y)$ as shown in Eq. 4–8. As an exercise, establish the $\mathbf{j}$ and $\mathbf{k}$ components of $\mathbf{M}_O$ in this manner and show that indeed the expanded form of the determinant, Eq. 4–8, represents the moment of $\mathbf{F}$ about point O. Once $\mathbf{M}_O$ is determined, realize that it will always be *perpendicular* to the shaded plane containing vectors $\mathbf{r}$ and $\mathbf{F}$, Fig. 4–14b.

It will be shown in Example 4.4 that the computation of the moment using the cross product has a distinct advantage over the scalar formulation when solving problems in *three dimensions*. This is because it is generally easier to establish the position vector $\mathbf{r}$ to the force, rather than determining the moment-arm distance d that must be directed *perpendicular* to the line of action of the force.

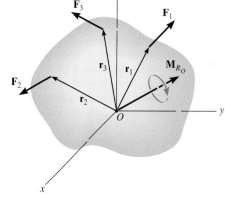

Fig. 4–15

Resultant Moment of a System of Forces. If a body is acted upon by a system of forces, Fig. 4–15, the resultant moment of the forces about point O can be determined by vector addition resulting from successive applications of Eq. 4–6. This resultant can be written symbolically as

$$\mathbf{M}_{R_o} = \Sigma(\mathbf{r} \times \mathbf{F}) \qquad (4\text{–}9)$$

and is shown in Fig. 4–15.

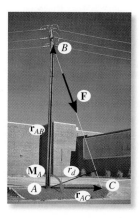

If we pull on cable BC with a force $\mathbf{F}$ at *any point along the cable*, the moment of this force about the base of the utility pole at A will always be the *same*. This is a consequence of the principle of transmissibility. Note that the moment arm, or perpendicular distance from A to the cable, is r_d, and so $M_A = r_d F$. In three dimensions this distance is often difficult to determine, and so we can use the vector cross product to obtain the moment in a more direct manner. For example, $\mathbf{M}_A = \mathbf{r}_{AB} \times \mathbf{F} = \mathbf{r}_{AC} \times \mathbf{F}$. As required, both of these vectors are directed from point A to a point on the line of action of the force.

E X A M P L E 4.4

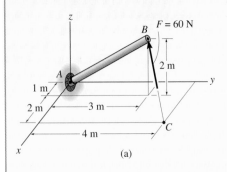

(a)

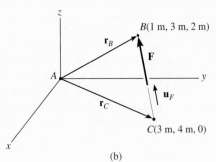

(b)

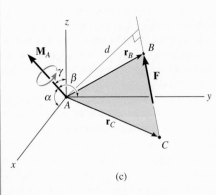

(c)

Fig. 4–16

The pole in Fig. 4–16a is subjected to a 60-N force that is directed from C to B. Determine the magnitude of the moment created by this force about the support at A.

Solution *(Vector Analysis)*

As shown in Fig. 4–16b, either one of two position vectors can be used for the solution, since $\mathbf{M}_A = \mathbf{r}_B \times \mathbf{F}$ or $\mathbf{M}_A = \mathbf{r}_C \times \mathbf{F}$. The position vectors are represented as

$$\mathbf{r}_B = \{1\mathbf{i} + 3\mathbf{j} + 2\mathbf{k}\} \text{ m} \quad \text{and} \quad \mathbf{r}_C = \{3\mathbf{i} + 4\mathbf{j}\} \text{ m}$$

The force has a magnitude of 60 N and a direction specified by the unit vector $\mathbf{u}_F$, directed from C to B. Thus,

$$\mathbf{F} = (60 \text{ N})\mathbf{u}_F = (60 \text{ N}) \left[\frac{(1-3)\mathbf{i} + (3-4)\mathbf{j} + (2-0)\mathbf{k}}{\sqrt{(-2)^2 + (-1)^2 + (2)^2}} \right]$$

$$= \{-40\mathbf{i} - 20\mathbf{j} + 40\mathbf{k}\} \text{ N}$$

Substituting into the determinant formulation, Eq. 4–7, and following the scheme for determinant expansion as stated in the footnote on page 120, we have

$$\mathbf{M}_A = \mathbf{r}_B \times \mathbf{F} = \begin{vmatrix} \mathbf{i} & \mathbf{j} & \mathbf{k} \\ 1 & 3 & 2 \\ -40 & -20 & 40 \end{vmatrix}$$

$$= [3(40)-2(-20)]\mathbf{i}-[1(40)-2(-40)]\mathbf{j} + [1(-20)-3(-40)]\mathbf{k}$$

or

$$\mathbf{M}_A = \mathbf{r}_C \times \mathbf{F} = \begin{vmatrix} \mathbf{i} & \mathbf{j} & \mathbf{k} \\ 3 & 4 & 0 \\ -40 & -20 & 40 \end{vmatrix}$$

$$= [4(40)-0(-20)]\mathbf{i}-[3(40)-0(-40)]\mathbf{j} + [3(-20)-4(-40)]\mathbf{k}$$

In both cases,

$$\mathbf{M}_A = \{160\mathbf{i} - 120\mathbf{j} + 100\mathbf{k}] \text{ N} \cdot \text{m}$$

The *magnitude* of $\mathbf{M}_A$ is therefore

$$M_A = \sqrt{(160)^2 + (-120)^2 + (100)^2} = 224 \text{ N} \cdot \text{m} \qquad \textit{Ans.}$$

As expected, $\mathbf{M}_A$ acts perpendicular to the shaded plane containing vectors $\mathbf{F}$, $\mathbf{r}_B$, and $\mathbf{r}_C$, Fig. 4–16c. (How would you find its coordinate direction angles $\alpha = 44.3°$, $\beta = 122°$, $\gamma = 63.4°$?) Had this problem been worked using a scalar approach, where $M_A = Fd$, notice the difficulty that can arise in obtaining the moment arm d.

EXAMPLE 4.5

Three forces act on the rod shown in Fig. 4–17a. Determine the resultant moment they create about the flange at O and determine the coordinate direction angles of the moment axis.

Solution

Position vectors are directed from point O to each force as shown in Fig. 4–17b. These vectors are

$$\mathbf{r}_A = \{5\mathbf{j}\} \text{ ft}$$
$$\mathbf{r}_B = \{4\mathbf{i} + 5\mathbf{j} - 2\mathbf{k}\} \text{ ft}$$

The resultant moment about O is therefore

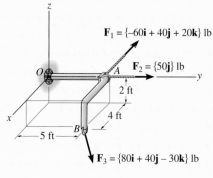

$\mathbf{F}_1 = \{-60\mathbf{i} + 40\mathbf{j} + 20\mathbf{k}\}$ lb

$\mathbf{F}_2 = \{50\mathbf{j}\}$ lb

2 ft

4 ft

5 ft

$\mathbf{F}_3 = \{80\mathbf{i} + 40\mathbf{j} - 30\mathbf{k}\}$ lb

(a)

$$
\begin{aligned}
\mathbf{M}_{R_O} &= \Sigma(\mathbf{r} \times \mathbf{F}) \\
&= \mathbf{r}_A \times \mathbf{F}_1 + \mathbf{r}_A \times \mathbf{F}_2 + \mathbf{r}_B \times \mathbf{F}_3 \\
&= \begin{vmatrix} \mathbf{i} & \mathbf{j} & \mathbf{k} \\ 0 & 5 & 0 \\ -60 & 40 & 20 \end{vmatrix} + \begin{vmatrix} \mathbf{i} & \mathbf{j} & \mathbf{k} \\ 0 & 5 & 0 \\ 0 & 50 & 0 \end{vmatrix} + \begin{vmatrix} \mathbf{i} & \mathbf{j} & \mathbf{k} \\ 4 & 5 & -2 \\ 80 & 40 & -30 \end{vmatrix} \\
&= [5(20) - 40(0)]\mathbf{i} - [0\mathbf{j}] + [0(40) - (-60)(5)]\mathbf{k} + [0\mathbf{i} - 0\mathbf{j} + 0\mathbf{k}] \\
&\quad + [5(-30) - (40)(-2)]\mathbf{i} - [4(-30) - 80(-2)]\mathbf{j} + [4(40) - 80(5)]\mathbf{k} \\
&= \{30\mathbf{i} - 40\mathbf{j} + 60\mathbf{k}\} \text{ lb} \cdot \text{ft} \qquad\qquad\qquad Ans.
\end{aligned}
$$

The moment axis is directed along the line of action of $\mathbf{M}_{R_O}$. Since the magnitude of this moment is

$$M_{R_O} = \sqrt{(30)^2 + (-40)^2 + (60)^2} = 78.10 \text{ lb} \cdot \text{ft}$$

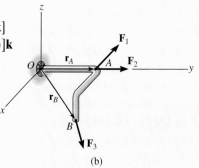

(b)

the unit vector which defines the direction of the moment axis is

$$\mathbf{u} = \frac{\mathbf{M}_{R_O}}{M_{R_O}} = \frac{30\mathbf{i} - 40\mathbf{j} + 60\mathbf{k}}{78.10} = 0.3841\mathbf{i} - 0.5121\mathbf{j} + 0.7682\mathbf{k}$$

Therefore, the coordinate direction angles of the moment axis are

$$\cos \alpha = 0.3841; \qquad \alpha = 67.4° \qquad\qquad Ans.$$
$$\cos \beta = -0.5121; \qquad \beta = 121° \qquad\qquad Ans.$$
$$\cos \gamma = 0.7682; \qquad \gamma = 39.8° \qquad\qquad Ans.$$

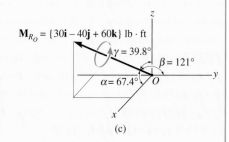

$\mathbf{M}_{R_O} = \{30\mathbf{i} - 40\mathbf{j} + 60\mathbf{k}\}$ lb · ft

$\gamma = 39.8°$

$\beta = 121°$

$\alpha = 67.4°$

(c)

Fig. 4–17

These results are shown in Fig. 4–17c. Realize that the three forces tend to cause the rod to rotate about this axis in the manner shown by the curl indicated on the moment vector.

4.4 Principle of Moments

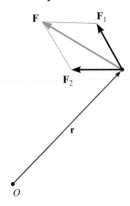

Fig. 4–18

A concept often used in mechanics is the *principle of moments*, which is sometimes referred to as *Varignon's theorem* since it was originally developed by the French mathematician Varignon (1654 –1722). It states that *the moment of a force about a point is equal to the sum of the moments of the force's components about the point.* The proof follows directly from the distributive law of the vector cross product. To show this, consider the force $\mathbf{F}$ and two of its rectangular components, where $\mathbf{F} = \mathbf{F}_1 + \mathbf{F}_2$, Fig. 4–18. We have

$$\mathbf{M}_O = \mathbf{r} \times \mathbf{F}_1 + \mathbf{r} \times \mathbf{F}_2 = \mathbf{r} \times (\mathbf{F}_1 + \mathbf{F}_2) = \mathbf{r} \times \mathbf{F}$$

This concept has important applications to the solution of problems and proofs of theorems that follow, since it is often easier to determine the moments of a force's components rather than the moment of the force itself.

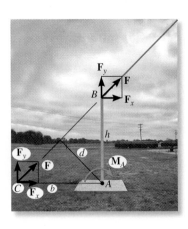

The guy cable exerts a force $\mathbf{F}$ on the pole and this creates a moment about the base at A of $M_A = Fd$. If the force is replaced by its two components $\mathbf{F}_x$ and $\mathbf{F}_y$ at point B where the cable acts on the pole, then the sum of the moments of these two components about A will yield the *same* resultant moment. For the calculation $\mathbf{F}_y$ will create zero moment about A and so $M_A = F_x h$. This is an application of the *principle of moments*. In addition we can apply the *principle of transmissibility* and slide the force to where its line of action intersects the ground at C. In this case $\mathbf{F}_x$ will create zero moment about A, and so $M_A = F_y b$.

IMPORTANT POINTS

- The moment of a force indicates the tendency of a body to turn about an axis passing through a specific point O.

- Using the right-hand rule, the sense of rotation is indicated by the fingers, and the thumb is directed along the moment axis, or line of action of the moment.

- The magnitude of the moment is determined from $M_O = Fd$, where d is the perpendicular or shortest distance from point O to the line of action of the force $\mathbf{F}$.

- In three dimensions use the vector cross product to determine the moment, i.e., $\mathbf{M}_O = \mathbf{r} \times \mathbf{F}$. Remember that $\mathbf{r}$ is directed *from* point O to *any point* on the line of action of $\mathbf{F}$.

- The principle of moments states that the moment of a force about a point is equal to the sum of the moments of the force's components about the point. This is a very convenient method to use in two dimensions.

EXAMPLE 4.6

A 200-N force acts on the bracket shown in Fig. 4–19a. Determine the moment of the force about point A.

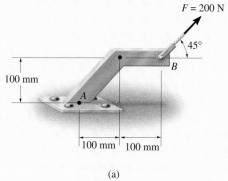

(a)

Fig. 4–19

Solution I

The moment arm d can be found by trigonometry, using the construction shown in Fig. 4–19b. From the right triangle BCD,

$$CB = d = 100 \cos 45° = 70.71 \text{ mm} = 0.070\,71 \text{ m}$$

Thus,

$$M_A = Fd = 200 \text{ N}(0.070\,71 \text{ m}) = 14.1 \text{ N} \cdot \text{m}\,\raise{2pt}{\curvearrowleft}$$

According to the right-hand rule, $\mathbf{M}_A$ is directed in the $+\mathbf{k}$ direction since the force tends to rotate or orbit *counterclockwise* about point A. Hence, reporting the moment as a Cartesian vector, we have

$$\mathbf{M}_A = \{14.1\mathbf{k}\} \text{ N} \cdot \text{m} \qquad\qquad Ans.$$

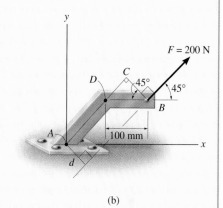

(b)

Solution II

The 200-N force may be resolved into x and y components, as shown in Fig. 4–19c. In accordance with the principle of moments, the moment of $\mathbf{F}$ computed about point A is equivalent to the sum of the moments produced by the two force components. Assuming counterclockwise rotation as positive, i.e., in the $+\mathbf{k}$ direction, we can apply Eq. 4–2 ($M_A = \Sigma Fd$), in which case

$$\zeta+M_A = (200 \sin 45° \text{ N})(0.20 \text{ m}) - (200 \cos 45° \text{ N})(0.10 \text{ m})$$
$$= 14.1 \text{ N} \cdot \text{m}\,\raise{2pt}{\curvearrowleft}$$

Thus

$$\mathbf{M}_A = \{14.1\mathbf{k}\} \text{ N} \cdot \text{m} \qquad\qquad Ans.$$

By comparison, it is seen that Solution II provides a more *convenient method* for analysis than Solution I since the moment arm for each component force is easier to establish.

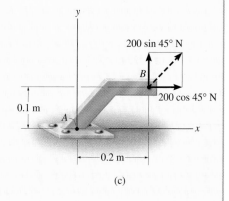

(c)

EXAMPLE 4.7

The force **F** acts at the end of the angle bracket shown in Fig. 4–20a. Determine the moment of the force about point O.

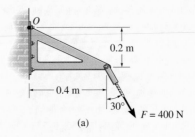

(a)

Solution I *(Scalar Analysis)*

The force is resolved into its x and y components as shown in Fig. 4–20b, and the moments of the components are computed about point O. Taking positive moments as counterclockwise, i.e., in the $+\mathbf{k}$ direction, we have

$$\zeta + M_O = 400 \sin 30° \, \text{N}(0.2 \, \text{m}) - 400 \cos 30° \, \text{N}(0.4 \, \text{m})$$
$$= -98.6 \, \text{N} \cdot \text{m} = 98.6 \, \text{N} \cdot \text{m} \, \downarrow$$

or

$$\mathbf{M}_O = \{-98.6\mathbf{k}\} \, \text{N} \cdot \text{m} \qquad\qquad \textit{Ans.}$$

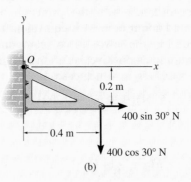

(b)

Fig. 4–20

Solution II *(Vector Analysis)*

Using a Cartesian vector approach, the force and position vectors shown in Fig. 4–20c can be represented as

$$\mathbf{r} = \{0.4\mathbf{i} - 0.2\mathbf{j}\} \text{ m}$$
$$\mathbf{F} = \{400 \sin 30°\mathbf{i} - 400 \cos 30°\mathbf{j}\} \text{ N}$$
$$= \{200.0\mathbf{i} - 346.4\mathbf{j}\} \text{ N}$$

The moment is therefore

$$\mathbf{M}_O = \mathbf{r} \times \mathbf{F} = \begin{vmatrix} \mathbf{i} & \mathbf{j} & \mathbf{k} \\ 0.4 & -0.2 & 0 \\ 200.0 & -346.4 & 0 \end{vmatrix}$$
$$= 0\mathbf{i} - 0\mathbf{j} + [0.4(-346.4) - (-0.2)(200.0)]\mathbf{k}$$
$$= \{-98.6\mathbf{k}\} \text{ N} \cdot \text{m} \qquad\qquad\qquad Ans.$$

By comparison, it is seen that the scalar analysis (Solution I) provides a more *convenient method* for analysis than Solution II since the direction of the moment and the moment arm for each component force are easy to establish. Hence, this method is generally recommended for solving problems displayed in two dimensions. On the other hand, Cartesian vector analysis is generally recommended only for solving three-dimensional problems, where the moment arms and force components are often more difficult to determine.

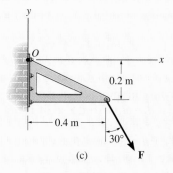

(c)

Fig. 4–20

4.5 Moment of a Force about a Specified Axis

Recall that when the moment of a force is computed about a point, the moment and its axis are *always* perpendicular to the plane containing the force and the moment arm. In some problems it is important to find the *component* of this moment along a *specified axis* that passes through the point. To solve this problem either a scalar or vector analysis can be used.

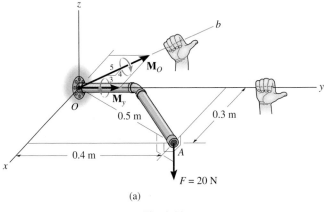

(a)

Fig. 4–21

Scalar Analysis. As a numerical example of this problem, consider the pipe assembly shown in Fig. 4–21a, which lies in the horizontal plane and is subjected to the vertical force of $F = 20$ N applied at point A. The moment of this force about point O has a *magnitude* of $M_O = (20 \text{ N})(0.5 \text{ m}) = 10 \text{ N} \cdot \text{m}$, and a *direction* defined by the right-hand rule, as shown in Fig. 4–21a. This moment tends to turn the pipe about the Ob axis. For practical reasons, however, it may be necessary to determine the *component* of M_O about the y axis, M_y, since this component tends to unscrew the pipe from the flange at O. From Fig. 4–21a, M_y has a magnitude of $M_y = \frac{3}{5}(10 \text{ N} \cdot \text{m}) = 6 \text{ N} \cdot \text{m}$ and a sense of direction shown by the vector resolution. Rather than performing this *two-step* process of first finding the moment of the force about point O and then resolving the moment along the y axis, it is also possible to solve this problem *directly*. To do so, it is necessary to determine the perpendicular or moment-arm distance from the line of action of $\mathbf{F}$ to the y axis. From Fig. 4–21a this distance is 0.3 m. Thus the *magnitude* of the moment of the force about the y axis is again $M_y = 0.3(20 \text{ N}) = 6 \text{ N} \cdot \text{m}$, and the *direction* is determined by the right-hand rule as shown.

In general, then, *if the line of action of a force* $\mathbf{F}$ *is perpendicular to any specified axis aa*, the magnitude of the moment of $\mathbf{F}$ about the axis can be determined from the equation

$$M_a = F d_a \qquad (4\text{-}10)$$

Here d_a is the *perpendicular or shortest distance* from the force line of action to the axis. The direction is determined from the thumb of the right hand when the fingers are curled in accordance with the direction of rotation as produced by the force. In particular, realize that a *force will not contribute a moment about a specified axis if the force line of action is parallel to the axis or its line of action passes through the axis.*

If a horizontal force **F** is applied to the handle of the flex-headed wrench, it tends to turn the socket at A about the z axis. This effect is caused by the moment of **F** about the z axis. The *maximum moment* is determined when the wrench is in the horizontal plane so that full leverage from the handle can be achieved, i.e., $(M_z)_{max} = Fd$. If the handle is not in the horizontal position, then the moment about the z axis is determined from $M_z = Fd'$, where d' is the perpendicular distance from the force line of action to the axis. We can also determine this moment by first finding the moment of **F** about A, $M_A = Fd$, then finding the projection or component of this moment along z, i.e., $M_z = M_A \cos \theta$.

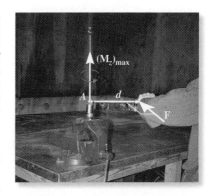

Vector Analysis.   The previous two-step solution of first finding the moment of the force about a point on the axis and then finding the projected component of the moment about the axis can also be performed using a vector analysis, Fig. 4–21b. Here the moment about point O is first determined from $\mathbf{M}_O = \mathbf{r}_A \times \mathbf{F} = (0.3\mathbf{i} + 0.4\mathbf{j}) \times (-20\mathbf{k}) = \{-8\mathbf{i} + 6\mathbf{j}\}\text{N} \cdot \text{m}$. The component or projection of this moment along the y axis is then determined from the dot product (Sec. 2.9). Since the unit vector for this axis (or line) is $\mathbf{u}_a = \mathbf{j}$, then $M_y = \mathbf{M}_O \cdot \mathbf{u}_a = (-8\mathbf{i} + 6\mathbf{j}) \cdot \mathbf{j} = 6\,\text{N} \cdot \text{m}$. This result, of course, is to be expected, since it represents the **j** component of $\mathbf{M}_O$.

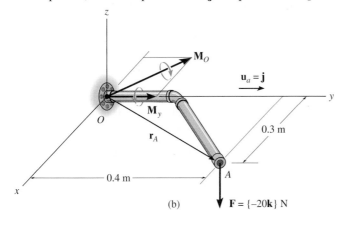

(b)

$\mathbf{F} = \{-20\mathbf{k}\}$ N

Fig. 4–21

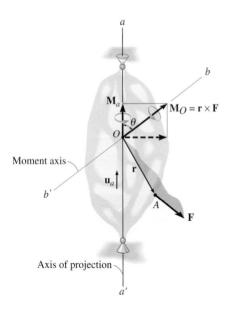

a

b

$\mathbf{M}_a$

$\mathbf{M}_O = \mathbf{r} \times \mathbf{F}$

θ

O

Moment axis

$\mathbf{u}_a$

$\mathbf{r}$

b'

A

$\mathbf{F}$

Axis of projection

a'

Fig. 4–22

A vector analysis such as this is particularly advantageous for finding the moment of a force about an axis when the force components or the appropriate moment arms are difficult to determine. For this reason, the above two-step process will now be generalized and applied to a body of arbitrary shape. To do so, consider the body in Fig. 4–22, which is subjected to the force **F** acting at point *A*. Here we wish to determine the effect of **F** in tending to rotate the body about the *aa'* axis. This tendency for rotation is measured by the moment component **M**_a. To determine **M**_a we first compute the moment of **F** about any *arbitrary point O* that lies on the *aa'* axis. In this case, **M**_O is expressed by the cross product $\mathbf{M}_O = \mathbf{r} \times \mathbf{F}$, where **r** is directed from *O* to *A*. Here **M**_O acts along the moment axis *bb'*, and so the component or projection of **M**_O onto the *aa'* axis is then **M**_a. The *magnitude* of **M**_a is determined by the dot product, $M_a = M_O \cos\theta = \mathbf{M}_O \cdot \mathbf{u}_a$ where **u**_a is a unit vector that defines the direction of the *aa'* axis. Combining these two steps as a general expression, we have $M_a = (\mathbf{r} \times \mathbf{F}) \cdot \mathbf{u}_a$. Since the dot product is commutative, we can also write

$$M_a = \mathbf{u}_a \cdot (\mathbf{r} \times \mathbf{F})$$

In vector algebra, this combination of dot and cross product yielding the scalar M_a is called the *triple scalar product*. Provided *x, y, z* axes are established and the Cartesian components of each of the vectors can be determined, then the triple scalar product may be written in determinant form as

$$M_a = (u_{a_x}\mathbf{i} + u_{a_y}\mathbf{j} + u_{a_z}\mathbf{k}) \cdot \begin{vmatrix} \mathbf{i} & \mathbf{j} & \mathbf{k} \\ r_x & r_y & r_z \\ F_x & F_y & F_z \end{vmatrix}$$

or simply

$$M_a = \mathbf{u}_a \cdot (\mathbf{r} \times \mathbf{F}) = \begin{vmatrix} u_{a_x} & u_{a_y} & u_{a_z} \\ r_x & r_y & r_z \\ F_x & F_y & F_z \end{vmatrix} \tag{4–11}$$

where

$u_{a_x}, u_{a_y}, u_{a_z}$ represent the *x, y, z* components of the unit vector defining the direction of the *aa'* axis

r_x, r_y, r_z represent the *x, y, z* components of the position vector drawn from *any point O* on the *aa'* axis to *any point A* on the line of action of the force

F_x, F_y, F_z represent the *x, y, z* components of the force vector.

When M_a is evaluated from Eq. 4–11, it will yield a positive or negative scalar. The sign of this scalar indicates the sense of direction of $\mathbf{M}_a$ along the aa' axis. If it is positive, then $\mathbf{M}_a$ will have the same sense as $\mathbf{u}_a$, whereas if it is negative, then $\mathbf{M}_a$ will act opposite to $\mathbf{u}_a$.

Once M_a is determined, we can then express $\mathbf{M}_a$ as a Cartesian vector, namely,

$$\mathbf{M}_a = M_a\mathbf{u}_a = [\mathbf{u}_a \cdot (\mathbf{r} \times \mathbf{F})]\mathbf{u}_a \qquad (4\text{–}12)$$

Finally, if the resultant moment of a series of forces is to be computed about the aa' axis, then the moment components of each force are added together *algebraically*, since each component lies along the same axis. Thus the magnitude of $\mathbf{M}_a$ is

$$M_a = \Sigma[\mathbf{u}_a \cdot (\mathbf{r} \times \mathbf{F})] = \mathbf{u}_a \cdot \Sigma(\mathbf{r} \times \mathbf{F})$$

The examples which follow illustrate a numerical application of the above concepts.

Wind blowing on the face of this traffic sign creates a resultant force $\mathbf{F}$ that tends to tip the sign over due to the moment $\mathbf{M}_A$ created about the $a - a$ axis. The moment of $\mathbf{F}$ about a point A that lies on the axis is $\mathbf{M}_A = \mathbf{r} \times \mathbf{F}$. The projection of this moment along the axis, whose direction is defined by the unit vector $\mathbf{u}_a$, is $M_a = \mathbf{u}_a \cdot (\mathbf{r} \times \mathbf{F})$. Had this moment been calculated using scalar methods, then the perpendicular distance from the force line of action to the $a - a$ axis would have to be determined, which in this case would be a more difficult task.

IMPORTANT POINTS

- The moment of a force about a specified axis can be determined provided the perpendicular distance d_a from *both* the force line of action and the axis can be determined. $M_a = Fd_a$.
- If vector analysis is used, $M_a = \mathbf{u}_a \cdot (\mathbf{r} \times \mathbf{F})$, where $\mathbf{u}_a$ defines the direction of the axis and $\mathbf{r}$ is directed from *any point* on the axis to *any point* on the line of action of the force.
- If M_a is calculated as a negative scalar, then the sense of direction of $\mathbf{M}_a$ is opposite to $\mathbf{u}_a$.
- The moment $\mathbf{M}_a$ expressed as a Cartesian vector is determined from $\mathbf{M}_a = M_a\mathbf{u}_a$.

EXAMPLE 4.8

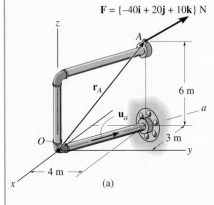

(a)

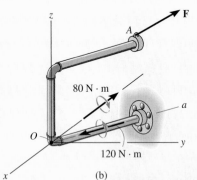

(b)

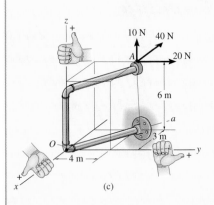

(c)

Fig. 4–23

The force $\mathbf{F} = \{-40\mathbf{i} + 20\mathbf{j} + 10\mathbf{k}\}$ N acts at point A shown in Fig. 4–23a. Determine the moments of this force about the x and a axes.

Solution I (Vector Analysis)
We can solve this problem by using the position vector $\mathbf{r}_A$. Why? Since $\mathbf{r}_A = \{-3\mathbf{i} + 4\mathbf{j} + 6\mathbf{k}\}$ m and $\mathbf{u}_x = \mathbf{i}$, then applying Eq. 4–11,

$$M_x = \mathbf{i} \cdot (\mathbf{r}_A \times \mathbf{F}) = \begin{vmatrix} 1 & 0 & 0 \\ -3 & 4 & 6 \\ -40 & 20 & 10 \end{vmatrix}$$

$$= 1[4(10)-6(20)]-0[(-3)(10)-6(-40)]+0[(-3)(20)-4(-40)]$$

$$= -80 \text{ N} \cdot \text{m} \qquad \qquad Ans.$$

The negative sign indicates that the sense of $\mathbf{M}_x$ is opposite to $\mathbf{i}$.
  We can compute M_a also using $\mathbf{r}_A$ because $\mathbf{r}_A$ extends from a point on the a axis to the force. Also, $\mathbf{u}_a = -\frac{3}{5}\mathbf{i} + \frac{4}{5}\mathbf{j}$. Thus,

$$M_a = \mathbf{u}_a \cdot (\mathbf{r}_A \times \mathbf{F}) = \begin{vmatrix} -\frac{3}{5} & \frac{4}{5} & 0 \\ -3 & 4 & 6 \\ -40 & 20 & 10 \end{vmatrix}$$

$$= -\tfrac{3}{5}[4(10)-6(20)]-\tfrac{4}{5}[(-3)(10)-6(-40)]+0[(-3)(20)-4(-40)]$$

$$= -120 \text{ N} \cdot \text{m} \qquad \qquad Ans.$$

What does the negative sign indicate?
  The moment components are shown in Fig. 4–23b.

Solution II (Scalar Analysis)
Since the force components and moment arms are easy to determine for computing M_x, a scalar analysis can be used to solve this problem. Referring to Fig. 4–23c, only the 10-N and 20-N forces contribute moments about the x axis. (The line of action of the 40-N force is parallel to this axis and hence its moment about the x axis is zero.) Using the right-hand rule, the algebraic sum of the moment components about the x axis is therefore

$$M_x = (10 \text{ N})(4 \text{ m}) - (20 \text{ N})(6 \text{ m}) = -80 \text{ N} \cdot \text{m} \qquad Ans.$$

Although not required here, note also that

$$M_y = (10 \text{ N})(3 \text{ m}) - (40 \text{ N})(6 \text{ m}) = -210 \text{ N} \cdot \text{m}$$
$$M_z = (40 \text{ N})(4 \text{ m}) - (20 \text{ N})(3 \text{ m}) = 100 \text{ N} \cdot \text{m}$$

  If we were to determine M_a by this scalar method, it would require much more effort since the force components of 40 N and 20 N are *not perpendicular* to the direction of the a axis. The vector analysis yields a more direct solution.

EXAMPLE 4.9

The rod shown in Fig. 4–24a is supported by two brackets at A and B. Determine the moment $\mathbf{M}_{AB}$ produced by $\mathbf{F} = \{-600\mathbf{i}+200\mathbf{j}-300\mathbf{k}\}$ N, which tends to rotate the rod about the AB axis.

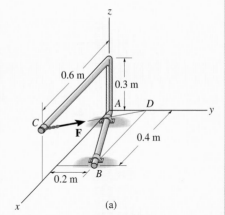

0.6 m
0.3 m
A D
C
F
0.4 m
0.2 m B

x (a)

Solution

A vector analysis using $M_{AB} = \mathbf{u}_B \cdot (\mathbf{r} \times \mathbf{F})$ will be considered for the solution since the moment arm or perpendicular distance from the line of action of $\mathbf{F}$ to the AB axis is difficult to determine. Each of the terms in the equation will now be identified.

Unit vector $\mathbf{u}_B$ defines the direction of the AB axis of the rod, Fig. 4–24b, where

$$\mathbf{u}_B = \frac{\mathbf{r}_B}{r_B} = \frac{0.4\mathbf{i} + 0.2\mathbf{j}}{\sqrt{(0.4)^2 + (0.2)^2}} = 0.894\mathbf{i} + 0.447\mathbf{j}$$

Vector $\mathbf{r}$ is directed from *any point* on the AB axis to *any point* on the line of action of the force. For example, position vectors $\mathbf{r}_C$ and $\mathbf{r}_D$ are suitable, Fig. 4–24b. (Although not shown, $\mathbf{r}_{BC}$ or $\mathbf{r}_{BD}$ can also be used.) For simplicity, we choose $\mathbf{r}_D$, where

$$\mathbf{r}_D = \{0.2\mathbf{j}\} \text{ m}$$

The force is

$$\mathbf{F} = \{-600\mathbf{i} + 200\mathbf{j} - 300\mathbf{k}\} \text{ N}$$

Substituting these vectors into the determinant form and expanding, we have

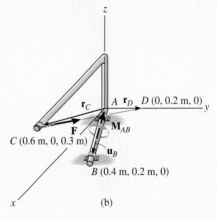

$\mathbf{r}_C$ A $\mathbf{r}_D$ D (0, 0.2 m, 0)
y
F $\mathbf{M}_{AB}$
C (0.6 m, 0, 0.3 m)
$\mathbf{u}_B$
B (0.4 m, 0.2 m, 0)

x (b)

Fig. 4–24

$$M_{AB} = \mathbf{u}_B \cdot (\mathbf{r}_D \times \mathbf{F}) = \begin{vmatrix} 0.894 & 0.447 & 0 \\ 0 & 0.2 & 0 \\ -600 & 200 & -300 \end{vmatrix}$$

$$= 0.894[0.2(-300) - 0(200)] - 0.447[0(-300) - 0(-600)] +$$
$$0[0(200) - 0.2(-600)]$$

$$= -53.67 \text{ N} \cdot \text{m}$$

The negative sign indicates that the sense of $\mathbf{M}_{AB}$ is opposite to that of $\mathbf{u}_B$.

Expressing $\mathbf{M}_{AB}$ as a Cartesian vector yields

$$\mathbf{M}_{AB} = M_{AB}\mathbf{u}_B = (-53.67 \text{ N} \cdot \text{m})(0.894\mathbf{i} + 0.447\mathbf{j})$$
$$= \{-48.0\mathbf{i} - 24.0\mathbf{j}\} \text{ N} \cdot \text{m} \qquad\qquad Ans.$$

The result is shown in Fig. 4–24b.

Note that if axis AB is defined using a unit vector directed from B toward A, then in the above formulation $-\mathbf{u}_B$ would have to be used. This would lead to $M_{AB} = +53.67 \text{ N} \cdot \text{m}$. Consequently, $\mathbf{M}_{AB} = M_{AB}(-\mathbf{u}_B)$, and the above result would again be determined.

4.6 Moment of a Couple

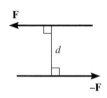

Fig. 4–25

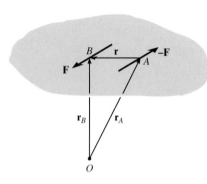

Fig. 4–26

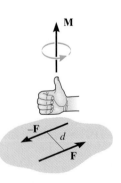

Fig. 4–27

A *couple* is defined as two parallel forces that have the same magnitude, have opposite directions, and are separated by a perpendicular distance d, Fig. 4–25. Since the resultant force is zero, the only effect of a couple is to produce a rotation or tendency of rotation in a specified direction.

The moment produced by a couple is called a *couple moment*. We can determine its value by finding the sum of the moments of both couple forces about *any* arbitrary point. For example, in Fig. 4–26, position vectors $\mathbf{r}_A$ and $\mathbf{r}_B$ are directed from point O to points A and B lying on the line of action of $-\mathbf{F}$ and $\mathbf{F}$. The couple moment computed about O is therefore

$$\mathbf{M} = \mathbf{r}_A \times (-\mathbf{F}) + \mathbf{r}_B \times \mathbf{F}$$

Rather than sum the moments of both forces to determine the couple moment, it is simpler to take moments about a point lying on the line of action of one of the forces. If point A is chosen, then the moment of $-\mathbf{F}$ about A is zero, and we have

$$\mathbf{M} = \mathbf{r} \times \mathbf{F} \tag{4–13}$$

The fact that we obtain the *same result* in both cases can be demonstrated by noting that in the first case we can write $\mathbf{M} = (\mathbf{r}_B - \mathbf{r}_A) \times \mathbf{F}$; and by the triangle rule of vector addition, $\mathbf{r}_A + \mathbf{r} = \mathbf{r}_B$ or $\mathbf{r} = \mathbf{r}_B - \mathbf{r}_A$, so that upon substitution we obtain Eq. 4–13. This result indicates that a couple moment is a *free vector*, i.e., it can act at *any point* since $\mathbf{M}$ depends *only* upon the position vector $\mathbf{r}$ directed *between* the forces and *not* the position vectors $\mathbf{r}_A$ and $\mathbf{r}_B$, directed from the arbitrary point O to the forces. This concept is therefore unlike the moment of a force, which requires a definite point (or axis) about which moments are determined.

Scalar Formulation. The moment of a couple, $\mathbf{M}$, Fig. 4–27, is defined as having a *magnitude* of

$$M = Fd \tag{4–14}$$

where F is the magnitude of one of the forces and d is the perpendicular distance or moment arm between the forces. The *direction* and sense of the couple moment are determined by the right-hand rule, where the thumb indicates the direction when the fingers are curled with the sense of rotation caused by the two forces. In all cases, $\mathbf{M}$ acts perpendicular to the plane containing these forces.

Vector Formulation. The moment of a couple can also be expressed by the vector cross product using Eq. 4–13, i.e.,

$$\mathbf{M} = \mathbf{r} \times \mathbf{F} \tag{4–15}$$

Application of this equation is easily remembered if one thinks of taking the moments of both forces about a point lying on the line of action of one of the forces. For example, if moments are taken about point A in Fig. 4–26, the moment of $-\mathbf{F}$ is *zero* about this point, and the moment or **F** is defined from Eq. 4–15. Therefore, in the formulation **r** is crossed with the force **F** to which it is directed.

Equivalent Couples. Two couples are said to be equivalent if they produce the same moment. Since the moment produced by a couple is always perpendicular to the plane containing the couple forces, it is therefore necessary that the forces of equal couples lie either in the same plane or in planes that are *parallel* to one another. In this way, the direction of each couple moment will be the same, that is, perpendicular to the parallel planes.

Resultant Couple Moment. Since couple moments are free vectors, they may be applied at any point P on a body and added vectorially. For example, the two couples acting on different planes of the body in Fig. 4–28a may be replaced by their corresponding couple moments **M**$_1$ and **M**$_2$, Fig. 4–28b, and then these free vectors may be moved to the *arbitrary point P* and added to obtain the resultant couple moment **M**$_R$ = **M**$_1$ + **M**$_2$, shown in Fig. 4–28c.

If more than two couple moments act on the body, we may generalize this concept and write the vector resultant as

$$\mathbf{M}_R = \Sigma(\mathbf{r} \times \mathbf{F}) \tag{4–16}$$

These concepts are illustrated numerically in the examples which follow. In general, problems projected in two dimensions should be solved using a scalar analysis since the moment arms and force components are easy to compute.

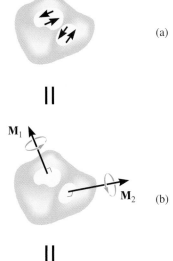

(a)

||

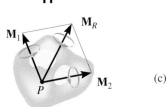

(b)

||

(c)

Fig. 4–28

The frictional forces of the floor on the blades of the concrete finishing machine create a couple moment **M**$_c$ on the machine that tends to turn it. An equal but opposite couple moment must be applied by the hands of the operator to prevent the turning. Here the couple moment, $M_c = Fd$, is applied on the handle, although it could be applied at any other point on the machine.

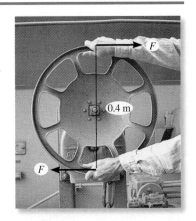

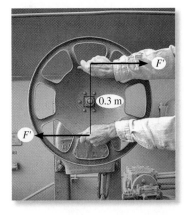

A moment of 12 N · m is needed to turn the shaft connected to the center of the wheel. To do this it is efficient to apply a couple since this effect produces a pure rotation. The couple forces can be made as small as possible by placing the hands on the *rim* of the wheel, where the spacing is 0.4 m. In this case 12 N · m = $F(0.4$ m), F = 30 N. An equivalent couple moment of 12 N · m can be produced if one grips the wheel within the inner hub, although here much larger forces are needed. If the distance between the hands becomes 0.3 m, then 12 N · m = $F'(0.3)$, F' = 40 N. Also, realize that if the wheel was connected to the shaft at a point other than at its center, the wheel would still turn when the forces are applied since the 12-N · m couple moment is a *free vector*.

IMPORTANT POINTS

- A couple moment is produced by two noncollinear forces that are equal but opposite. Its effect is to produce pure rotation, or tendency for rotation in a specified direction.

- A couple moment is a free vector, and as a result it causes the same effect of rotation on a body regardless of where the couple moment is applied to the body.

- The moment of the two couple forces can be computed about *any point*. For convenience, this point is often chosen on the line of action of one of the forces in order to eliminate the moment of this force about the point.

- In three dimensions the couple moment is often determined using the vector formulation, $\mathbf{M} = \mathbf{r} \times \mathbf{F}$, where $\mathbf{r}$ is directed from *any point* on the line of action of one of the forces to *any point* on the line of action of the other force $\mathbf{F}$.

- A resultant couple moment is simply the vector sum of all the couple moments of the system.

EXAMPLE 4.10

A couple acts on the gear teeth as shown in Fig. 4–29a. Replace it by an equivalent couple having a pair of forces that act through points *A* and *B*.

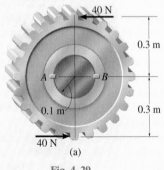
(a)

Fig. 4–29

(b)

(c)

Solution (Scalar Analysis)

The couple has a magnitude of $M = Fd = 40(0.6) = 24 \text{ N} \cdot \text{m}$ and a direction that is out of the page since the forces tend to rotate counterclockwise. $\mathbf{M}$ is a free vector, and so it can be placed at any point on the gear, Fig. 4–29b. To preserve the counterclockwise rotation of $\mathbf{M}$, *vertical* forces acting through points *A* and *B* must be directed as shown in Fig. 4–29c. The magnitude of each force is

$$M = Fd \qquad 24 \text{ N} \cdot \text{m} = F(0.2 \text{ m})$$

$$F = 120 \text{ N} \qquad\qquad Ans.$$

EXAMPLE 4.11

Determine the moment of the couple acting on the member shown in Fig. 4–30a.

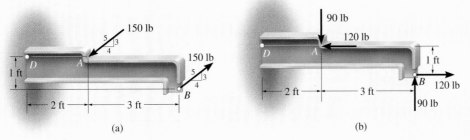

(a)   (b)

Solution (Scalar Analysis)

Here it is somewhat difficult to determine the perpendicular distance between the forces and compute the couple moment as $M = Fd$. Instead, we can resolve each force into its horizontal and vertical components, $F_x = \frac{4}{5}(150 \text{ lb}) = 120 \text{ lb}$ and $F_y = \frac{3}{5}(150 \text{ lb}) = 90 \text{ lb}$, Fig. 4–30b, and then use the principle of moments. The couple moment can be determined about *any point*. For example, if point D is chosen, we have for all four forces,

$$\downarrow + M = 120 \text{ lb}(0 \text{ ft}) - 90 \text{ lb}(2 \text{ ft}) + 90 \text{ lb}(5 \text{ ft}) + 120 \text{ lb}(1 \text{ ft})$$
$$= 390 \text{ lb} \cdot \text{ft} \uparrow \qquad\qquad\qquad Ans.$$

It is easier, however, to determine the moments about point A or B in order to *eliminate* the moment of the forces acting at the moment point. For point A, Fig. 4-30b, we have

$$\downarrow + M = 90 \text{ lb}(3 \text{ ft}) + 120 \text{ lb}(1 \text{ ft})$$
$$= 390 \text{ lb} \cdot \text{ft} \uparrow \qquad\qquad\qquad Ans.$$

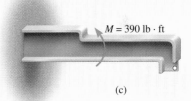

(c)

Fig. 4–30

Show that one obtains this same result if moments are summed about point B. Notice also that the couple in Fig. 4–30a can be replaced by *two* couples in Fig. 4–30b. Using $M = Fd$, one couple has a moment of $M_1 = 90 \text{ lb}(3 \text{ ft}) = 270 \text{ lb} \cdot \text{ft}$ and the other has a moment of $M_2 = 120 \text{ lb}(1 \text{ ft}) = 120 \text{ lb} \cdot \text{ft}$. By the right-hand rule, both couple moments are counterclockwise and are therefore directed out of the page. Since these couples are free vectors, they can be moved to any point and added, which yields $M = 270 \text{ lb} \cdot \text{ft} + 120 \text{ lb} \cdot \text{ft} = 390 \text{ lb} \cdot \text{ft} \uparrow$, the same result determined above. **M** is a free vector and can therefore act at any point on the member, Fig. 4–30c. Also, realize that the external effect, such as the support reactions on the member, will be the *same* if the member supports the couple, Fig. 4–30a, or the couple moment, Fig. 4–30c.

EXAMPLE 4.12

Determine the couple moment acting on the pipe shown in Fig. 4–31a. Segment AB is directed 30° below the x-y plane.

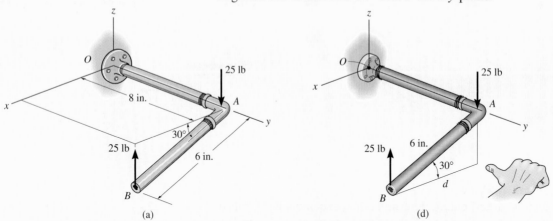

(a)

(d)

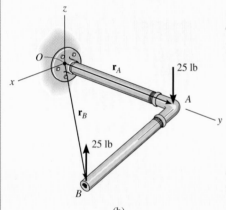

(b)

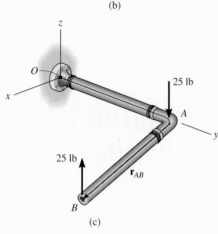

(c)

Fig. 4–31

Solution I *(Vector Analysis)*

The moment of the two couple forces can be found about *any point*. If point O is considered, Fig. 4–31b, we have

$$\mathbf{M} = \mathbf{r}_A \times (-25\mathbf{k}) + \mathbf{r}_B \times (25\mathbf{k})$$
$$= (8\mathbf{j}) \times (-25\mathbf{k}) + (6\cos 30°\mathbf{i} + 8\mathbf{j} - 6\sin 30°\mathbf{k}) \times (25\mathbf{k})$$
$$= -200\mathbf{i} - 129.9\mathbf{j} + 200\mathbf{i}$$
$$= \{-130\mathbf{j}\}\ \text{lb}\cdot\text{in.} \qquad\qquad Ans.$$

It is *easier* to take moments of the couple forces about a point lying on the line of action of one of the forces, e.g., point A, Fig. 4–31c. In this case the moment of the force A is zero, so that

$$\mathbf{M} = \mathbf{r}_{AB} \times (25\mathbf{k})$$
$$= (6\cos 30°\mathbf{i} - 6\sin 30°\mathbf{k}) \times (25\mathbf{k})$$
$$= \{-130\mathbf{j}\}\ \text{lb}\cdot\text{in.} \qquad\qquad Ans.$$

Solution II *(Scalar Analysis)*

Although this problem is shown in three dimensions, the geometry is simple enough to use the scalar equation $M = Fd$. The perpendicular distance between the lines of action of the forces is $d = 6\cos 30° = 5.20$ in., Fig. 4–31d. Hence, taking moments of the forces about either point A or B yields

$$M = Fd = 25\ \text{lb}(5.20\ \text{in.}) = 129.9\ \text{lb}\cdot\text{in.}$$

Applying the right-hand rule, $\mathbf{M}$ acts in the $-\mathbf{j}$ direction. Thus,

$$\mathbf{M} = \{-130\mathbf{j}\}\ \text{lb}\cdot\text{in.} \qquad\qquad Ans.$$

EXAMPLE 4.13

Replace the two couples acting on the pipe column in Fig. 4–32a by a resultant couple moment.

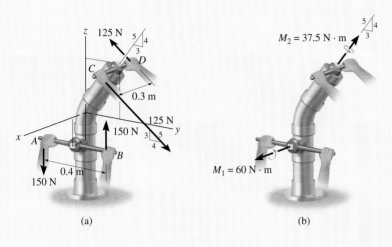

Fig. 4–32

Solution *(Vector Analysis)*

The couple moment $\mathbf{M}_1$, developed by the forces at A and B, can easily be determined from a scalar formulation.

$$M_1 = Fd = 150\,\text{N}(0.4\,\text{m}) = 60\,\text{N}\cdot\text{m}$$

By the right-hand rule, $\mathbf{M}_1$ acts in the $+\mathbf{i}$ direction, Fig. 4–32b. Hence,

$$\mathbf{M}_1 = \{60\mathbf{i}\}\,\text{N}\cdot\text{m}$$

Vector analysis will be used to determine $\mathbf{M}_2$, caused by forces at C and D. If moments are computed about point D, Fig. 4–32a, $\mathbf{M}_2 = \mathbf{r}_{DC} \times \mathbf{F}_C$, then

$$\begin{aligned}
\mathbf{M}_2 = \mathbf{r}_{DC} \times \mathbf{F}_C &= (0.3\mathbf{i}) \times [125(\tfrac{4}{5})\mathbf{j} - 125(\tfrac{3}{5})\mathbf{k}]\\
&= (0.3\mathbf{i}) \times [100\mathbf{j} - 75\mathbf{k}] = 30(\mathbf{i} \times \mathbf{j}) - 22.5(\mathbf{i} \times \mathbf{k})\\
&= \{22.5\mathbf{j} + 30\mathbf{k}\}\,\text{N}\cdot\text{m}
\end{aligned}$$

Try to establish $\mathbf{M}_2$ by using a scalar formulation, Fig. 4–32b.

Since $\mathbf{M}_1$ and $\mathbf{M}_2$ are free vectors, they may be moved to some arbitrary point P and added vectorially, Fig. 4–32c. The resultant couple moment becomes

$$\mathbf{M}_R = \mathbf{M}_1 + \mathbf{M}_2 = \{60\mathbf{i} + 22.5\mathbf{j} + 30\mathbf{k}\}\,\text{N}\cdot\text{m} \qquad Ans.$$

4.7 Equivalent System

A force has the effect of both translating and rotating a body, and the amount by which it does so depends upon where and how the force is applied. In the next section we will discuss the method used to *simplify* a system of forces and couple moments acting on a body to a single resultant force and couple moment acting at a specified point *O*. To do this, however, it is necessary that the force and couple moment system produce the *same* "external" effects of translation and rotation of the body as their resultants. When this occurs these two sets of loadings are said to be *equivalent*.

In this section we wish to show how to maintain this equivalency when a single force is applied to a specific point on a body and when it is located at another point *O*. Two cases for the location of point *O* will now be considered.

Point *O* Is On the Line of Action of the Force. Consider the body shown in Fig. 4–33*a*, which is subjected to the force **F** applied to point *A*. In order to apply the force to point *O* without altering the external effects on the body, we will first apply equal but opposite forces **F** and −**F** at *O*, as shown in Fig. 4–33*b*. The two forces indicated by the slash across them can be canceled, leaving the force at point *O* as required, Fig. 4–33*c*. By using this construction procedure, an *equivalent system* has been maintained between each of the diagrams, as shown by the equal signs. Note, however, that the force has simply been "transmitted" along its line of action, from point *A*, Fig. 4–33*a*, to point *O*, Fig. 4–33*c*. In other words, the force can be considered as a *sliding vector* since it can act at any point *O* along its line of action. In Sec. 4.3 we referred to this concept as the *principle of transmissibility*. It is important to realize that only the *external effects*, such as the body's motion or the forces needed to support the body if it is stationary, remain *unchanged* after **F** is moved. Certainly the *internal effects* depend on where **F** is located. For example, when **F** acts at *A*, the internal forces in the body have a high intensity around *A*; whereas movement of **F** away from this point will cause these internal forces to decrease.

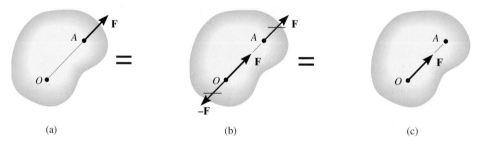

(a) (b) (c)

Fig. 4–33

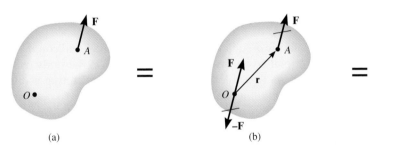
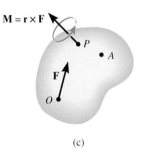

$$M = r \times F$$

(a)   (b)   (c)

Fig. 4–34

Point *O* Is Not On the Line of Action of the Force. This case is shown in Fig. 4–34*a*, where **F** is to be moved to point *O* without altering the external effects on the body. Following the same procedure as before, we first apply equal but opposite forces **F** and −**F** at point *O*, Fig. 4–34*b*. Here the two forces indicated by a slash across them form a couple which has a moment that is perpendicular to **F** and is defined by the cross product **M** = **r** × **F**. Since the couple moment is a *free vector*, it may be applied at *any point P* on the body as shown in Fig. 4–34*c*. In addition to this couple moment, **F** now acts at point *O* as required.

To summarize these concepts, when the point on the body is *on the line of action of the force*, simply transmit or slide the force along its line of action to the point. When the point is not on the line of action of the force, then move the force to the point and add a couple moment anywhere to the body. This couple moment is found by taking the moment of the force about the point. When these rules are carried out, equivalent external effects will be produced.

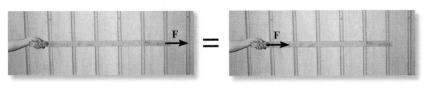

Consider the effects on the hand when a stick of negligible weight supports a force **F** at its end. When the force is applied horizontally, the same force is felt at the grip, regardless of where it is applied along its line of action. This is a consequence of the principle of transmissibility.

When the force is applied vertically it causes both a downward force **F** to be felt at the grip and a clockwise couple moment or twist of $M = Fd$. These same effects are felt if **F** is applied at the grip and **M** is applied anywhere on the stick. In both cases the systems are equivalent.

4.8 Resultants of a Force and Couple System

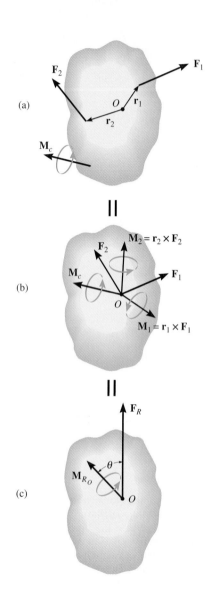

(a)

||

(b)

||

(c)

Fig. 4–35

When a rigid body is subjected to a *system* of forces and couple moments, it is often simpler to study the external effects on the body by *replacing* the system by an equivalent single resultant force acting at a specified point O and a resultant couple moment. To show how to determine these resultants we will consider the rigid body in Fig. 4–35*a* and use the concepts discussed in the previous section. Since point O is not on the line of action of the forces, an equivalent effect is produced if the forces are moved to point O *and* the corresponding couple moments $\mathbf{M}_1 = \mathbf{r}_1 \times \mathbf{F}_1$ and $\mathbf{M}_2 = \mathbf{r}_2 \times \mathbf{F}_2$ are applied to the body. Furthermore, the couple moment $\mathbf{M}_c$ is simply moved to point O since it is a free vector. These results are shown in Fig. 4–35*b*. By vector addition, the resultant force is $\mathbf{F}_R = \mathbf{F}_1 + \mathbf{F}_2$, and the resultant couple moment is $\mathbf{M}_{R_O} = \mathbf{M}_c + \mathbf{M}_1 + \mathbf{M}_2$, Fig. 4–35*c*. Since equivalency is maintained between the diagrams in Fig. 4–35, each force and couple system will cause the *same external effects*, i.e., the same translation and rotation of the body. Note that both the magnitude and direction of $\mathbf{F}_R$ are independent of the location of point O; however, $\mathbf{M}_{R_O}$ depends upon this location since the moments $\mathbf{M}_1$ and $\mathbf{M}_2$ are determined using the position vectors $\mathbf{r}_1$ and $\mathbf{r}_2$. Also note that $\mathbf{M}_{R_O}$ is a free vector and can act at *any point* on the body, although point O is generally chosen as its point of application.

The above method of simplifying any force and couple moment system to a resultant force acting at point O and a resultant couple moment can be generalized and represented by application of the following two equations.

$$\mathbf{F}_R = \Sigma\mathbf{F}$$
$$\mathbf{M}_{R_O} = \Sigma\mathbf{M}_c + \Sigma\mathbf{M}_O \qquad (4\text{–}17)$$

The first equation states that the resultant force of the system is equivalent to the sum of all the forces; and the second equation states that the resultant couple moment of the system is equivalent to the sum of all the couple moments $\Sigma\mathbf{M}_c$, plus the moments about point O of all the forces $\Sigma\mathbf{M}_O$. If the force system lies in the $x\text{–}y$ plane and any couple moments are perpendicular to this plane, that is along the z axis, then the above equations reduce to the following three scalar equations.

$$F_{R_x} = \Sigma F_x$$
$$F_{R_y} = \Sigma F_y$$
$$M_{R_O} = \Sigma M_c + \Sigma M_O \qquad (4\text{–}18)$$

Note that the resultant force $\mathbf{F}_R$ is equivalent to the vector sum of its two components $\mathbf{F}_{R_x}$ and $\mathbf{F}_{R_y}$.

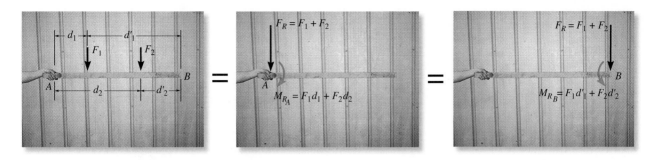

If the two forces acting on the stick are replaced by an equivalent resultant force and couple moment at point A, or by the equivalent resultant force and couple moment at point B, then in each case the hand must provide the same resistance to translation and rotation in order to keep the stick in the horizontal position. In other words, the external effects on the stick are the *same* in each case.

PROCEDURE FOR ANALYSIS

The following points should be kept in mind when applying Eqs. 4–17 or 4–18.

- Establish the coordinate axes with the origin located at the point O and the axes having a selected orientation.

Force Summation.

- If the force system is *coplanar*, resolve each force into its x and y components. If a component is directed along the positive x or y axis, it represents a positive scalar; whereas if it is directed along the negative x or y axis, it is a negative scalar.
- In three dimensions, represent each force as a Cartesian vector before summing the forces.

Moment Summation.

- When determining the moments of a *coplanar* force system about point O, it is generally advantageous to use the principle of moments, i.e., determine the moments of the components of each force rather than the moment of the force itself.
- In three dimensions use the vector cross product to determine the moment of each force about the point. Here the position vectors extend from point O to any point on the line of action of each force.

EXAMPLE 4.14

Replace the forces acting on the brace shown in Fig. 4–36a by an equivalent resultant force and couple moment acting at point A.

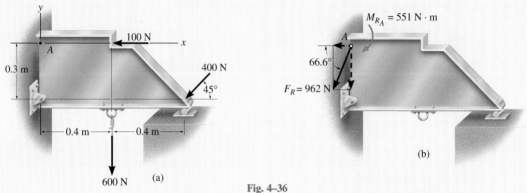

Fig. 4–36

Solution (*Scalar Analysis*)

The principle of moments will be applied to the 400-N force, whereby the moments of its two rectangular components will be considered.

Force Summation. The resultant force has x and y components of

$$\xrightarrow{+} F_{R_x} = \Sigma F_x; \quad F_{R_x} = -100 \text{ N} - 400 \cos 45°\text{N} = -382.8 \text{ N} = 382.8 \text{ N} \leftarrow$$

$$+\uparrow F_{R_y} = \Sigma F_y; \quad F_{R_y} = -600 \text{ N} - 400 \sin 45°\text{N} = -882.8 \text{ N} = 882.8 \text{ N}\downarrow$$

As shown in Fig. 4–36b, $\mathbf{F}_R$ has a magnitude of

$$F_R = \sqrt{(F_{R_x})^2 + (F_{R_y})^2} = \sqrt{(382.8)^2 + (882.8)^2} = 962 \text{ N} \quad \textit{Ans.}$$

and a direction of

$$\theta = \tan^{-1}\left(\frac{F_{R_y}}{F_{R_x}}\right) = \tan^{-1}\left(\frac{882.8}{382.8}\right) = 66.6° \quad _{\theta}\!\!\nearrow \quad \textit{Ans.}$$

Moment Summation. The resultant couple moment $\mathbf{M}_{R_A}$ is determined by summing the moments of the forces about point A. Assuming that positive moments act counterclockwise, i.e., in the $+\mathbf{k}$ direction, we have

$$\zeta + M_{R_A} = \Sigma M_A;$$

$$M_{R_A} = 100 \text{ N}(0) - 600 \text{ N}(0.4 \text{ m}) - (400 \sin 45°\text{N})(0.8 \text{ m})$$

$$- (400 \cos 45°\text{N})(0.3 \text{ m})$$

$$= -551 \text{ N} \cdot \text{m} = 551 \text{ N} \cdot \text{m} \,\rfloor \quad \textit{Ans.}$$

In conclusion, when $\mathbf{M}_{R_A}$ and $\mathbf{F}_R$ act on the brace at point A, Fig. 4–36b, they will produce the *same* external effect or reactions at the supports as that produced by the force system in Fig. 4–36a.

EXAMPLE 4.15

A structural member is subjected to a couple moment $\mathbf{M}$ and forces $\mathbf{F}_1$ and $\mathbf{F}_2$ as shown in Fig. 4–37a. Replace this system by an equivalent resultant force and couple moment acting at its base, point O.

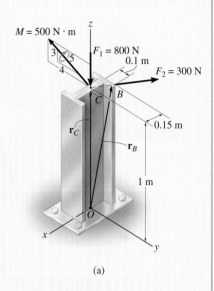

(a)

Solution *(Vector Analysis)*

The three-dimensional aspects of the problem can be simplified by using a Cartesian vector analysis. Expressing the forces and couple moment as Cartesian vectors, we have

$$\mathbf{F}_1 = \{-800\mathbf{k}\}\ \text{N}$$

$$\mathbf{F}_2 = (300\ \text{N})\mathbf{u}_{CB} = (300\ \text{N})\left(\frac{\mathbf{r}_{CB}}{r_{CB}}\right)$$

$$= 300\left[\frac{-0.15\mathbf{i} + 0.1\mathbf{j}}{\sqrt{(-0.15)^2 + (0.1)^2}}\right] = \{-249.6\mathbf{i} + 166.4\mathbf{j}\}\ \text{N}$$

$$\mathbf{M} = -500(\tfrac{4}{5})\mathbf{j} + 500(\tfrac{3}{5})\mathbf{k} = \{-400\mathbf{j} + 300\mathbf{k}\}\ \text{N}\cdot\text{m}$$

Force Summation.

$$\mathbf{F}_R = \Sigma\mathbf{F}; \qquad \mathbf{F}_R = \mathbf{F}_1 + \mathbf{F}_2 = -800\mathbf{k} - 249.6\mathbf{i} + 166.4\mathbf{j}$$

$$= \{-249.6\mathbf{i} + 166.4\mathbf{j} - 800\mathbf{k}\}\ \text{N} \qquad\qquad Ans.$$

Moment Summation.

$$\mathbf{M}_{R_O} = \Sigma\mathbf{M}_C + \Sigma\mathbf{M}_O$$

$$\mathbf{M}_{R_O} = \mathbf{M} + \mathbf{r}_C \times \mathbf{F}_1 + \mathbf{r}_B \times \mathbf{F}_2$$

$$\mathbf{M}_{R_O} = (-400\mathbf{j} + 300\mathbf{k}) + (1\mathbf{k}) \times (-800\mathbf{k}) + \begin{vmatrix} \mathbf{i} & \mathbf{j} & \mathbf{k} \\ -0.15 & 0.1 & 1 \\ -249.6 & 166.4 & 0 \end{vmatrix}$$

$$= (-400\mathbf{j} + 300\mathbf{k}) + (\mathbf{0}) + (-166.4\mathbf{i} - 249.6\mathbf{j})$$

$$= \{-166\mathbf{i} - 650\mathbf{j} + 300\mathbf{k}\}\ \text{N}\cdot\text{m} \qquad\qquad Ans.$$

The results are shown in Fig. 4–37b.

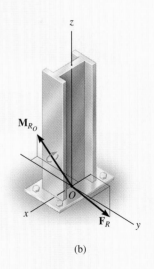

(b)

Fig. 4–37

4.9 Further Reduction of a Force and Couple System

Simplification to a Single Resultant Force. Consider now a special case for which the system of forces and couple moments acting on a rigid body, Fig. 4–38a, reduces at point O to a resultant force $\mathbf{F}_R = \Sigma\mathbf{F}$ and resultant couple moment $\mathbf{M}_{R_O} = \Sigma\mathbf{M}_O$, which are *perpendicular* to one another, Fig. 4–38b. Whenever this occurs, we can further simplify the force and couple moment system by moving $\mathbf{F}_R$ to another point P, located either on or off the body so that no resultant couple moment has to be applied to the body, Fig. 4–38c. In other words, if the force and couple moment system in Fig. 4–38a is reduced to a resultant system at point P, only the force resultant will have to be applied to the body, Fig. 4–38c.

The location of point P, measured from point O, can always be determined provided $\mathbf{F}_R$ and $\mathbf{M}_{R_O}$ are known, Fig. 4–38b. As shown in Fig. 4–38c, P must lie on the *bb* axis, which is perpendicular to both the line of action of $\mathbf{F}_R$ and the *aa* axis. This point is chosen such that the distance d satisfies the scalar equation $M_{R_O} = F_R d$ or $d = M_{R_O}/F_R$. With $\mathbf{F}_R$ so located, it will produce the same external effects on the body as the force and couple moment system in Fig. 4–38a, or the force and couple moment resultants in Fig. 4–38b.

If a system of forces is either concurrent, coplanar, or parallel, it can always be reduced, as in the above case, to a single resultant force $\mathbf{F}_R$ acting through. This is because in each of these cases $\mathbf{F}_R$ and $\mathbf{M}_{R_O}$ will always be perpendicular to each other when the force system is simplified at *any* point O.

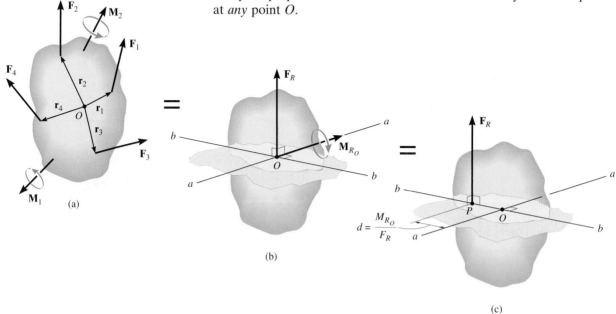

Fig. 4–38

Concurrent Force Systems. A concurrent force system has been treated in detail in Chapter 2. Obviously, all the forces act at a point for which there is no resultant couple moment, so the point P is automatically specified, Fig. 4–39.

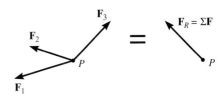

Fig. 4–39

Coplanar Force Systems. Coplanar force systems, which may include couple moments directed perpendicular to the plane of the forces as shown in Fig. 4–40a, can be reduced to a single resultant force, because when each force in the system is moved to any point O in the x–y plane, it produces a couple moment that is *perpendicular* to the plane, i.e., in the $\pm\mathbf{k}$ direction. The resultant moment $\mathbf{M}_{R_O} = \Sigma\mathbf{M} + \Sigma(\mathbf{r} \times \mathbf{F})$ is thus perpendicular to the resultant force $\mathbf{F}_R$. Fig. 4–40b; and so $\mathbf{F}_R$ can be positioned a distance d from O so as to create this same moment $\mathbf{M}_{R_O}$ about O, Fig. 4–40c.

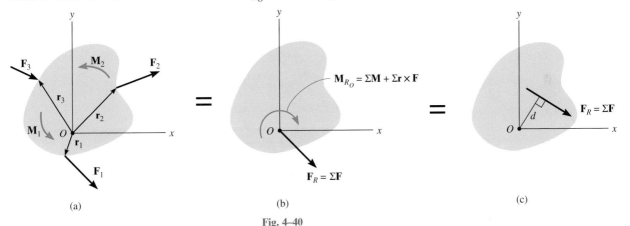

(a)   (b)   (c)

Fig. 4–40

Parallel Force Systems. Parallel force systems, which can include couple moments that are perpendicular to the forces, as shown in Fig. 4–41a, can be reduced to a single resultant force because when each force is moved to any point O in the x–y plane, it produces a couple moment that has components only about the x and y axes. The resultant moment $\mathbf{M}_{R_O} = \Sigma\mathbf{M}_O + \Sigma(\mathbf{r} \times \mathbf{F})$ is thus perpendicular to the resultant force $\mathbf{F}_R$, Fig. 4–41b; and so $\mathbf{F}_R$ can be moved to a point a distance d away so that it produces the same moment about O.

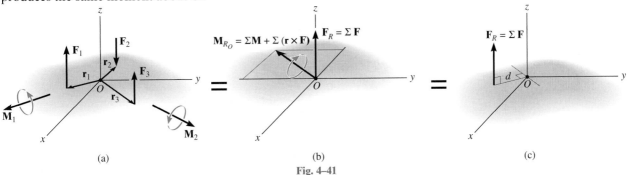

(a)   (b)   (c)

Fig. 4–41

The three parallel forces acting on the stick can be replaced by a single resultant force F_R acting at a distance d from the grip. To be equivalent we require the resultant force to equal the sum of the forces, $F_R = F_1 + F_2 + F_3$, and to find the distance d the moment of the resultant force about the grip must be equal to the moment of all the forces about the grip, $F_R d = F_1 d_1 + F_2 d_2 + F_3 d_3$.

PROCEDURE FOR ANALYSIS

The technique used to reduce a coplanar or parallel force system to a single resultant force follows a similar procedure outlined in the previous section.

- Establish the x, y, z axes and locate the resultant force $\mathbf{F}_R$ an arbitrary distance away from the origin of the coordinates.

Force Summation.

- The resultant force is equal to the sum of all the forces in the system.

- For a coplanar force system, resolve each force into its x and y components. Positive components are directed along the positive x and y axes, and negative components are directed along the negative x and y axes.

Moment Summation.

- The moment of the resultant force about point O is equal to the sum of all the couple moments in the system plus the moments about point O of all the forces in the system.

- This moment condition is used to find the location of the resultant force from point O.

Reduction to a Wrench. In the general case, the force and couple moment system acting on a body, Fig. 4–35a, will reduce to a single resultant force $\mathbf{F}_R$ and couple moment $\mathbf{M}_{R_O}$ at O which are *not* perpendicular. Instead, $\mathbf{F}_R$ will act at an angle θ from $\mathbf{M}_{R_O}$, Fig. 4–35c. As shown in Fig. 4–42a, however, $\mathbf{M}_{R_O}$ may be resolved into two components: one perpendicular, $\mathbf{M}_\perp$, and the other parallel $\mathbf{M}_\parallel$, to the line of action of $\mathbf{F}_R$. As in the previous discussion, the perpendicular component $\mathbf{M}_\perp$ may be *eliminated* by moving $\mathbf{F}_R$ to point P, as shown in Fig. 4–42b. This point lies on axis bb, which is perpendicular to both $\mathbf{M}_{R_O}$ and $\mathbf{F}_R$. In order to maintain an equivalency of loading, the distance from O to P is $d = M_\perp/F_R$. Furthermore, when $\mathbf{F}_R$ is applied at P, the moment of $\mathbf{F}_R$ tending to cause rotation of the body *about O* is in the *same direction* as $\mathbf{M}_\perp$, Fig. 4–42a. Finally, since $\mathbf{M}_\parallel$ is a free vector, it may be moved to P so that it is collinear with $\mathbf{F}_R$, Fig. 4–42c. This combination of a collinear force and couple moment is called a *wrench* or *screw*. The *axis of the wrench* has the same line of action as the force. Hence, the wrench tends to cause both a translation along and a rotation about this axis. Comparing Fig. 4–42a to Fig. 4–42c, it is seen that a general force and couple moment system acting on a body can be reduced to a wrench. The axis of the wrench and the point through which this axis passes can always be determined.

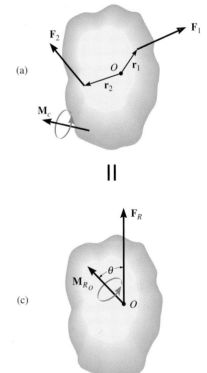

(a)

(c)

Fig. 4–35, (Repeated)

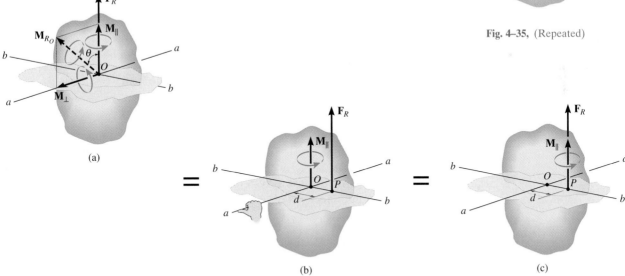

Fig. 4–42

E X A M P L E 4.16

The beam AE in Fig. 4–43a is subjected to a system of coplanar forces. Determine the magnitude, direction, and location on the beam of a resultant force which is equivalent to the given system of forces measured from E.

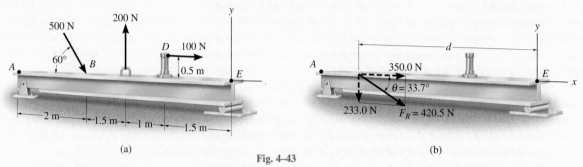

Fig. 4–43

Solution

The origin of coordinates is located at point E as shown in Fig. 4–43a.

Force Summation. Resolving the 500-N force into x and y components and summing the force components yields

$$\xrightarrow{+} F_{R_x} = \Sigma F_x; \quad F_{R_x} = 500 \cos 60° \text{ N} + 100 \text{ N} = 350.0 \text{ N} \rightarrow$$
$$+\uparrow F_{R_y} = \Sigma F_y; \quad F_{R_y} = -500 \sin 60° \text{ N} + 200 \text{ N} = -233.0 \text{ N}$$
$$= 233.0 \text{ N} \downarrow$$

The magnitude and direction of the resultant force are established from the vector addition shown in Fig. 4–43b. We have

$$F_R = \sqrt{(350.0)^2 + (233.0)^2} = 420.5 \text{ N} \qquad \qquad Ans.$$

$$\theta = \tan^{-1}\left(\frac{233.0}{350.0}\right) = 33.7° \searrow_\theta \qquad \qquad Ans.$$

Moment Summation. Moments will be summed about point E. Hence, from Figs. 4–43a and 4–43b, we require the moments of the components of $\mathbf{F}_R$ (or the moment of $\mathbf{F}_R$) about point E to equal the moments of the force system about E. Assuming positive moments are counterclockwise, we have

$$\zeta+M_{R_E} = \Sigma M_E$$
$$233.0 \text{ N}(d)+350.0 \text{ N}(0) = (500 \sin 60° \text{ N})(4 \text{ m})+(500 \cos 60° \text{ N})(0)$$
$$- (100 \text{ N})(0.5 \text{ m}) - (200 \text{ N})(2.5 \text{ m})$$

$$d = \frac{1182.1}{233.0} = 5.07 \text{ m} \qquad \qquad Ans.$$

Note that using a clockwise sign convention would yield this same result. Since d is *positive*, $\mathbf{F}_R$ acts to the left of E as shown. Try to solve this problem by summing moments about point A and show $d' = 0.927$ m, measured to the right of A.

E X A M P L E 4.17

The jib crane shown in Fig. 4–44a is subjected to three coplanar forces. Replace this loading by an equivalent resultant force and specify where the resultant's line of action intersects the column AB and boom BC.

Solution

Force Summation. Resolving the 250-lb force into x and y components and summing the force components yields

$$\xrightarrow{+} F_{R_x} = \Sigma F_x;\quad F_{R_x} = -250\ \text{lb}(\tfrac{3}{5}) - 175\ \text{lb} = -325\ \text{lb} = 325\ \text{lb} \leftarrow$$
$$+\uparrow F_{R_y} = \Sigma F_y;\quad F_{R_y} = -250\ \text{lb}(\tfrac{4}{5}) - 60\ \text{lb} = -260\ \text{lb} = 260\ \text{lb} \downarrow$$

As shown by the vector addition in Fig. 4–44b,

$$F_R = \sqrt{(325)^2 + (260)^2} = 416\ \text{lb} \qquad \textit{Ans.}$$

$$\theta = \tan^{-1}\left(\frac{260}{325}\right) = 38.7°\ {}^\theta\!\nearrow \quad \textit{Ans.}$$

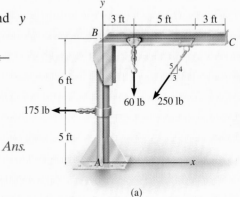

(a)

Moment Summation. Moments will be summed about the arbitrary point A. Assuming the line of action of $\mathbf{F}_R$ intersects AB, Fig. 4–44b, we require the moment of the components of $\mathbf{F}_R$ in Fig. 4–44b about A to equal the moments of the force system in Fig. 4–44a about A; i.e.,

$$\zeta + M_{R_A} = \Sigma M_A;\qquad 325\ \text{lb}(y) + 260\ \text{lb}(0)$$
$$= 175\ \text{lb}(5\ \text{ft}) - 60\ \text{lb}(3\ \text{ft}) + 250\ \text{lb}(\tfrac{3}{5})(11\ \text{ft}) - 250\ \text{lb}(\tfrac{4}{5})(8\ \text{ft})$$
$$y = 2.29\ \text{ft} \qquad \textit{Ans.}$$

By the principle of transmissibility, $\mathbf{F}_R$ can also be treated as intersecting BC, Fig. 4–44b, in which case we have

$$\zeta + M_{R_A} = \Sigma M_A;\qquad 325\ \text{lb}(11\ \text{ft}) - 260\ \text{lb}(x)$$
$$= 175\ \text{lb}(5\ \text{ft}) - 60\ \text{lb}(3\ \text{ft}) + 250\ \text{lb}(\tfrac{3}{5})(11\ \text{ft}) - 250\ \text{lb}(\tfrac{4}{5})(8\ \text{ft})$$
$$x = 10.9\ \text{ft} \qquad \textit{Ans.}$$

We can also solve for these positions by assuming $\mathbf{F}_R$ acts at the arbitrary point (x, y) on its line of action, Fig. 4–44b. Summing moments about point A yields

$$\zeta + M_{R_A} = \Sigma M_A;\qquad 325\ \text{lb}(y) - 260\ \text{lb}(x)$$
$$= 175\ \text{lb}(5\ \text{ft}) - 60\ \text{lb}(3\ \text{ft}) + 250\ \text{lb}(\tfrac{3}{5})(11\ \text{ft}) - 250\ \text{lb}(\tfrac{4}{5})(8\ \text{ft})$$
$$325y - 260x = 745$$

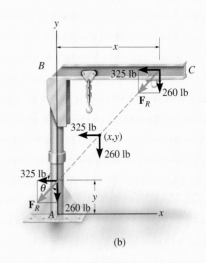

(b)

Fig. 4–44

which is the equation of the colored dashed line in Fig. 4–44b. To find the points of intersection with the crane along AB, set $x = 0$, then $y = 2.29$ ft, and along BC set $y = 11$ ft, then $x = 10.9$ ft.

EXAMPLE 4.18

The slab in Fig. 4–45a is subjected to four parallel forces. Determine the magnitude and direction of a resultant force equivalent to the given force system and locate its point of application on the slab.

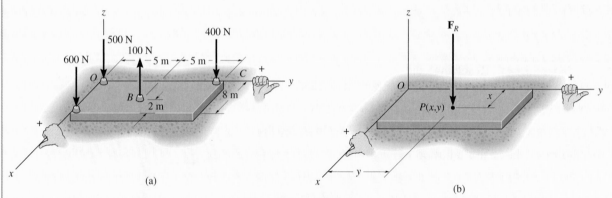

Fig. 4–45

Solution (Scalar Analysis)

Force Summation. From Fig. 4–45a, the resultant force is

$$+\uparrow F_R = \Sigma F; \quad F_R = -600\text{ N} + 100\text{ N} - 400\text{ N} - 500\text{ N}$$
$$= -1400\text{ N} = 1400\text{ N} \downarrow \qquad \qquad Ans.$$

Moment Summation. We require the moment about the x axis of the resultant force, Fig. 4–45b, to be equal to the sum of the moments about the x axis of all the forces in the system, Fig. 4–45a. The moment arms are determined from the y coordinates since these coordinates represent the *perpendicular distances* from the x axis to the lines of action of the forces. Using the right-hand rule, where positive moments act in the $+\mathbf{i}$ direction, we have

$$M_{R_x} = \Sigma M_x;$$
$$-(1400\text{ N})y = 600\text{ N}(0) + 100\text{ N}(5\text{ m}) - 400\text{ N}(10\text{ m}) + 500\text{ N}(0)$$
$$-1400y = -3500 \qquad y = 2.50\text{ m} \qquad \qquad Ans.$$

In a similar manner, assuming that positive moments act in the $+\mathbf{j}$ direction, a moment equation can be written about the y axis using moment arms defined by the x coordinates of each force.

$$M_{R_y} = \Sigma M_y;$$
$$(1400\text{ N})x = 600\text{ N}(8\text{ m}) - 100\text{ N}(6\text{ m}) + 400\text{ N}(0) + 500\text{ N}(0)$$
$$1400x = 4200 \qquad x = 3.00\text{ m} \qquad \qquad Ans.$$

Hence, a force of $F_R = 1400$ N placed at point $P(3.00\text{ m}, 2.50\text{ m})$ on the slab, Fig. 4–45b, is equivalent to the parallel force system acting on the slab in Fig. 4–45a.

E X A M P L E **4.19**

Three parallel bolting forces act on the rim of the circular cover plate in Fig. 4–46a. Determine the magnitude and direction of a resultant force equivalent to the given force system and locate its point of application, P, on the cover plate.

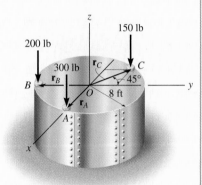

(a)

Solution (*Vector Analysis*)

Force Summation. From Fig. 4–46a, the force resultant $\mathbf{F}_R$ is

$$\mathbf{F}_R = \Sigma\mathbf{F}; \qquad \mathbf{F}_R = -300\mathbf{k} - 200\mathbf{k} - 150\mathbf{k}$$
$$= \{-650\mathbf{k}\}\ \text{lb} \qquad\qquad\qquad Ans.$$

Moment Summation. Choosing point O as a reference for computing moments and assuming that $\mathbf{F}_R$ acts at a point $P(x, y)$, Fig. 4–46b, we require

$$\mathbf{M}_{R_O} = \Sigma\mathbf{M}_O;$$
$$\mathbf{r} \times \mathbf{F}_R = \mathbf{r}_A \times (-300\mathbf{k}) + \mathbf{r}_B \times (-200\mathbf{k}) + \mathbf{r}_C \times (-150\mathbf{k})$$
$$(x\mathbf{i} + y\mathbf{j}) \times (-650\mathbf{k}) = (8\mathbf{i}) \times (-300\mathbf{k}) + (-8\mathbf{j}) \times (-200\mathbf{k})$$
$$+ (-8\sin 45°\mathbf{i} + 8\cos 45°\mathbf{j}) \times (-150\mathbf{k})$$
$$650x\mathbf{j} - 650y\mathbf{i} = 2400\mathbf{j} + 1600\mathbf{i} - 848.5\mathbf{j} - 848.5\mathbf{i}$$

Equating the corresponding $\mathbf{j}$ and $\mathbf{i}$ components yields

$$650x = 2400 - 848.5 \qquad\qquad (1)$$
$$-650y = 1600 - 848.5 \qquad\qquad (2)$$

Solving these equations, we obtain the coordinates of point P,

$$x = 2.39\ \text{ft} \qquad y = -1.16\ \text{ft} \qquad\qquad Ans.$$

The negative sign indicates that it was wrong to have assumed a $+y$ position for $\mathbf{F}_R$ as shown in Fig. 4–46b.

It is also possible to establish Eqs. 1 and 2 directly by summing moments about the y and x axes. Using the right-hand rule we have

$$M_{R_y} = \Sigma M_y; \qquad 650x = 300\ \text{lb}\ (8\ \text{ft}) - 150\ \text{lb}\ (8\sin 45°\ \text{ft})$$
$$M_{R_x} = \Sigma M_x; \qquad -650y = 200\ \text{lb}\ (8\ \text{ft}) - 150\ \text{lb}\ (8\cos 45°\ \text{ft})$$

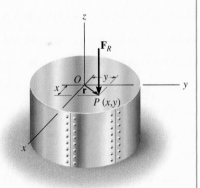

(b)

Fig. 4–46

4.10 Reduction of a Simple Distributed Loading

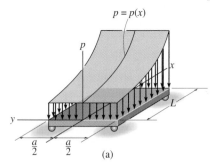

(a)

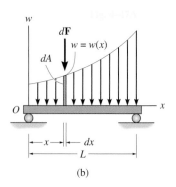

(b)

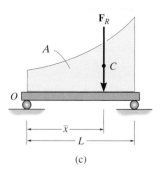

(c)

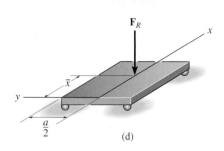

(d)

Fig. 4–47

In many situations a very large surface area of a body may be subjected to *distributed loadings* such as those caused by wind, fluids, or simply the weight of material supported over the body's surface. The *intensity* of these loadings at each point on the surface is defined as the *pressure p* (force per unit area), which can be measured in units of lb/ft^2 or pascals (Pa), where $1\ Pa = 1\ N/m^2$.

In this section we will consider the most common case of a distributed pressure loading, which is *uniform* along one axis of a flat rectangular body upon which the loading is applied.* An example of such a loading is shown in Fig. 4–47a. The direction of the intensity of the pressure load is indicated by arrows shown on the *load-intensity diagram*. The entire loading on the plate is therefore a system of parallel forces, infinite in number and each acting on a separate differential area of the plate. Here the *loading function*, $p = p(x)$ Pa, is only a function of x since the pressure is uniform along the y axis. If we multiply $p = p(x)$ by the *width a* m of the plate, we obtain $w = [p(x)\ N/m^2]a\ m = w(x)\ N/m$. This loading function, shown in Fig. 4–47b, is a measure of load distribution along the line $y = 0$ which is in the plane of symmetry of the loading, Fig. 4–47a. As noted, it is measured as a force per unit length, rather than a force per unit area. Consequently, the load-intensity diagram for $w = w(x)$ can be represented by a system of *coplanar* parallel forces, shown in two dimensions in Fig. 4–47b. Using the methods of Sec. 4.9, this system of forces can be simplified to a single resultant force $\mathbf{F}_R$ and its location $\bar{x}$ can be specified, Fig. 4–47c.

Magnitude of Resultant Force. From Eq. 4–17 ($F_R = \Sigma F$), the magnitude of $\mathbf{F}_R$ is equivalent to the sum of all the forces in the system. In this case integration must be used since there is an infinite number of parallel forces $d\mathbf{F}$ acting along the plate, Fig. 4–47b. Since $d\mathbf{F}$ is acting on an element of length dx and $w(x)$ is a force per unit length, then at the location x, $dF = w(x)\ dx = dA$. In other words, the magnitude of $d\mathbf{F}$ is determined from the colored differential *area dA* under the loading curve. For the entire plate length,

$$+\downarrow F_R = \Sigma F; \qquad \boxed{F_R = \int_L w(x)\ dx = \int_A dA = A} \qquad (4\text{–}19)$$

Hence, the magnitude of the resultant force is equal to the total area A under the loading diagram $w = w(x)$, Fig. 4–47c.

*The more general case of a nonuniform surface loading acting on a body is considered in Sec. 9.5.

Location of Resultant Force.   Applying Eq. 4–17 ($M_{R_O} = \Sigma M_O$), the location $\bar{x}$ of the line of action of $\mathbf{F}_R$ can be determined by equating the moments of the force resultant and the force distribution about point O (the y axis). Since $d\mathbf{F}$ produces a moment of $x\, dF = x\, w(x)\, dx$ about O, Fig. 4–47b, then for the entire plate, Fig. 4-47c,

$$\curvearrowright + M_{R_O} = \Sigma M_O; \qquad \bar{x}F_R = \int_L x\, w(x)\, dx$$

Solving for $\bar{x}$, using Eq. 4–19, we can write

$$\bar{x} = \frac{\displaystyle\int_L x\, w(x)\, dx}{\displaystyle\int_L w(x)\, dx} = \frac{\displaystyle\int_A x\, dA}{\displaystyle\int_A dA} \qquad (4\text{--}20)$$

This equation represents the x coordinate for the geometric center or *centroid* of the *area* under the distributed-loading diagram $w(x)$. *Therefore, the resultant force has a line of action which passes through the centroid C (geometric center) of the area defined by the distributed-loading diagram $w(x)$*, Fig. 4–47c.

Once $\bar{x}$ is determined, $\mathbf{F}_R$ by symmetry passes through point $(\bar{x}, 0)$ on the surface of the plate, Fig. 4–47d. If we now consider the three-dimensional pressure loading $p(x)$, Fig. 4–47a, we can therefore conclude that *the resultant force has a magnitude equal to the volume under the distributed-loading curve $p = p(x)$ and a line of action which passes through the centroid (geometric center) of this volume.* Detailed treatment of the integration techniques for computing the centroids of volumes or areas is given in Chapter 9. In many cases, however, the distributed-loading diagram is in the shape of a rectangle, triangle, or some other simple geometric form. The centroids for such common shapes do not have to be determined from Eq. 4–20; rather, they can be obtained directly from the tabulation given on the inside back cover.

The beam supporting this stack of lumber is subjected to a *uniform* distributed loading, and so the load-intensity diagram has a rectangular shape. If the load intensity is w_0, then the resultant force is determined from the area of the rectangle, $F_R = w_0 b$. The line of action of this force passes through the centroid or center of this area, $\bar{x} = a + b/2$. This resultant is equivalent to the distributed load, and so both loadings produce the same "external" effects or support reactions on the beam.

IMPORTANT POINTS

- Distributed loadings are defined by using a loading function $w = w(x)$ that indicates the intensity of the loading along the length of the member. This intensity is measured in N/m or lb/ft.

- The external effects caused by a coplanar distributed load acting on a body can be represented by a single resultant force.

- The resultant force is equivalent to the *area* under the distributed loading diagram, and has a line of action that passes through the *centroid* or geometric center of this area.

EXAMPLE 4.20

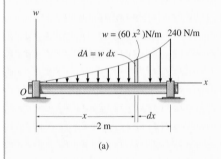

$w = (60\,x^2\,)$N/m   240 N/m

$dA = w\,dx$

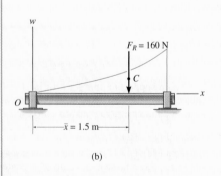

(a)

$F_R = 160$ N

$\bar{x} = 1.5$ m

(b)

Fig. 4–48

Determine the magnitude and location of the equivalent resultant force acting on the shaft in Fig. 4–48a.

Solution

Since $w = w(x)$ is given, this problem will be solved by integration. The colored differential area element $dA = w\,dx = 60x^2\,dx$. Applying Eq. 4–19, by summing these elements from $x = 0$ to $x = 2$ m, we obtain the resultant force $\mathbf{F}_R$.

$$F_R = \Sigma F;$$

$$F_R = \int_A dA = \int_0^2 60x^2\,dx = 60\left[\frac{x^3}{3}\right]_0^2 = 60\left[\frac{2^3}{3} - \frac{0^3}{3}\right]$$

$$= 160 \text{ N} \hspace{2cm} Ans.$$

Since the element of area dA is located an arbitrary distance x from O, the location $\bar{x}$ of $\mathbf{F}_R$ *measured from* O, Fig. 4–48b, is determined from Eq. 4–20.

$$\bar{x} = \frac{\int_A x\,dA}{\int_A dA} = \frac{\int_0^2 x(60x^2)dx}{160} = \frac{60\left[\frac{x^4}{4}\right]_0^2}{160} = \frac{60\left[\frac{2^4}{4} - \frac{0^4}{4}\right]}{160}$$

$$= 1.5 \text{ m} \hspace{2cm} Ans.$$

These results may be checked by using the table on the inside back cover, where it is shown that for an exparabolic area of length a, height b, and shape shown in Fig. 4–48a,

$$A = \frac{ab}{3} = \frac{2\text{ m}(240 \text{ N/m})}{3} = 160 \text{ N and } \bar{x} = \frac{3}{4}a = \frac{3}{4}(2 \text{ m}) = 1.5 \text{ m}$$

E X A M P L E **4.21**

A distributed loading of $p = 800x$ Pa acts over the top surface of the beam shown in Fig. 4–49a. Determine the magnitude and location of the equivalent resultant force.

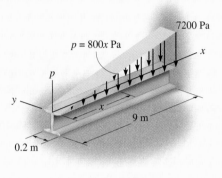

7200 Pa

$p = 800x$ Pa

x

y

x

9 m

0.2 m

(a)

Fig. 4–49

Solution

The loading function $p = 800x$ Pa indicates that the load intensity varies uniformly from $p = 0$ at $x = 0$ to $p = 7200$ Pa at $x = 9$ m. Since the intensity is uniform along the width of the beam (the y axis), the loading may be viewed in two dimensions as shown in Fig. 4–49b. Here

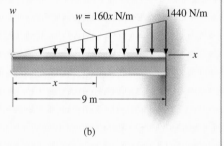

w

$w = 160x$ N/m

1440 N/m

x

x

9 m

(b)

$$w = (800x \text{ N/m}^2)(0.2 \text{ m})$$
$$= (160x) \text{ N/m}$$

At $x = 9$ m, note that $w = 1440$ N/m. Although we may again apply Eqs. 4–19 and 4–20 as in Example 4.20, it is simpler to use the table on the inside back cover.

The magnitude of the resultant force is equivalent to the area under the triangle.

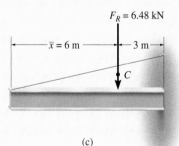

$F_R = 6.48$ kN

$\bar{x} = 6$ m

3 m

C

(c)

$$F_R = \tfrac{1}{2}(9 \text{ m})(1440 \text{ N/m}) = 6480 \text{ N} = 6.48 \text{ kN} \qquad Ans.$$

The line of action of $\mathbf{F}_R$ passes through the *centroid C* of the triangle. Hence,

$$\bar{x} = 9 \text{ m} - \tfrac{1}{3}(9 \text{ m}) = 6 \text{ m} \qquad Ans.$$

The results are shown in Fig. 4–49c.

We may also view the resultant $\mathbf{F}_R$ as *acting* through the *centroid* of the *volume* of the loading diagram $p = p(x)$ in Fig. 4–49a. Hence $\mathbf{F}_R$ intersects the x–y plane at the point (6 m, 0). Furthermore, the magnitude of $\mathbf{F}_R$ is equal to the volume under the loading diagram; i.e.,

$$F_R = V = \tfrac{1}{2}(7200 \text{ N/m}^2)(9 \text{ m})(0.2 \text{ m}) = 6.48 \text{ kN} \qquad Ans.$$

E X A M P L E **4.22**

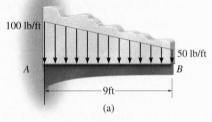

(a)

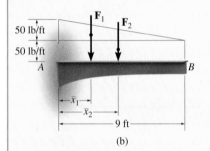

(b)

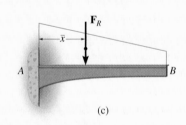

(c)

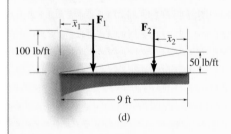

(d)

Fig. 4–50

The granular material exerts the distributed loading on the beam as shown in Fig. 4–50a. Determine the magnitude and location of the equivalent resultant of this load.

Solution

The area of the loading diagram is a *trapezoid*, and therefore the solution can be obtained directly from the area and centroid formulas for a trapezoid listed on the inside back cover. Since these formulas are not easily remembered, instead we will solve this problem by using "composite" areas. In this regard, we can divide the trapezoidal loading into a rectangular and triangular loading as shown in Fig. 4–50b. The magnitude of the force represented by each of these loadings is equal to its associated *area*,

$$F_1 = \tfrac{1}{2}(9 \text{ ft})(50 \text{ lb/ft}) = 225 \text{ lb}$$
$$F_2 = (9 \text{ ft})(50 \text{ lb/ft}) = 450 \text{ lb}$$

The lines of action of these parallel forces act through the *centroid* of their associated areas and therefore intersect the beam at

$$\bar{x}_1 = \tfrac{1}{3}(9 \text{ ft}) = 3 \text{ ft}$$
$$\bar{x}_2 = \tfrac{1}{2}(9 \text{ ft}) = 4.5 \text{ ft}$$

The two parallel forces $\mathbf{F}_1$ and $\mathbf{F}_2$ can be reduced to a single resultant $\mathbf{F}_R$. The magnitude of $\mathbf{F}_R$ is

$$+\downarrow F_R = \Sigma F; \qquad F_R = 225 + 450 = 675 \text{ lb} \qquad \textit{Ans.}$$

With reference to point A, Fig. 4–50b and 4–50c, we can find the location of $\mathbf{F}_R$. We require

$$\curvearrowleft + M_{R_A} = \Sigma M_A; \qquad \bar{x}(675) = 3(225) + 4.5(450)$$
$$\bar{x} = 4 \text{ ft} \qquad \textit{Ans.}$$

Note: The trapezoidal area in Fig. 4–50a can also be divided into two triangular areas as shown in Fig. 4–50d. In this case

$$F_1 = \tfrac{1}{2}(9 \text{ ft})(100 \text{ lb/ft}) = 450 \text{ lb}$$
$$F_2 = \tfrac{1}{2}(9 \text{ ft})(50 \text{ lb/ft}) = 225 \text{ lb}$$

and

$$\bar{x}_1 = \tfrac{1}{3}(9 \text{ ft}) = 3 \text{ ft}$$
$$\bar{x}_2 = \tfrac{1}{3}(9 \text{ ft}) = 3 \text{ ft}$$

Using these results, show that again $F_R = 675$ lb and $\bar{x} = 4$ ft.

CHAPTER REVIEW

- *Moment of a Force.* A force produces a turning effect about a point O that does not lie on its line of action. In scalar form, the moment *magnitude* is $M_O = Fd$, where d is the moment arm or perpendicular distance from point O to the line of action of the force. The *direction* of the moment is defined using the right-hand rule. Rather than finding d, it is normally easier to resolve the force into its x and y components, determine the moment of each component about the point, and then sum the results. Since three-dimensional geometry is generally more difficult to visualize, the vector cross product can be used to determine the moment, $\mathbf{M}_O = \mathbf{r} \times \mathbf{F}$, where $\mathbf{r}$ is a position vector that extends from point O to any point on the line of action of $\mathbf{F}$.

- *Moment about a Specified Axis.* If the moment of a force is to be determined about an arbitrary axis, then the projection of the moment onto the axis must be obtained. Provided the distance d_a that is perpendicular to *both* the line of action of the force and the axis can be determined, then the moment of the force about the axis is simply $M_a = F\, d_a$. If this distance d_a cannot be found, then the vector triple product should be used, where $M_a = \mathbf{u}_a \cdot \mathbf{r} \times \mathbf{F}$. Here $\mathbf{u}_a$ is the unit vector that specifies the direction of the axis and $\mathbf{r}$ is a position vector that is directed from any point on the axis to any point on the line of action of the force.

- *Couple Moment.* A couple consists of two equal but opposite forces that act a perpendicular distance d apart. Couples tend to produce a rotation without translation. The moment of the couple is determined from $M = Fd$, and its direction is established using the right-hand rule. If the vector cross product is used to determine the moment of the couple then $\mathbf{M} = \mathbf{r} \times \mathbf{F}$. Here $\mathbf{r}$ extends from any point on the line of action of one of the forces to any point on the line of action of the force $\mathbf{F}$ used in the cross product.

- *Reduction of a Force and Couple System.* Any system of forces and couples can be reduced to a single resultant force and resultant couple moment acting at a point. The resultant force is the sum of all the forces in the system, and the resultant couple moment is equal to the sum of all the forces and couple moments about the point. Further simplification to a single resultant force is possible provided the force system is *concurrent, coplanar*, or *parallel*. For this case, to find the location of the resultant force from a point, it is necessary to equate the moment of the resultant force about the point to the moment of the forces and couples in the system about the same point. Doing this for any *other type* of force system would yield a *wrench*, which consists of the resultant force and a resultant collinear couple moment.

- *Distributed Loading.* A simple distributed loading can be replaced by a *resultant force*, which is equivalent to the *area* under the loading curve. This resultant has a line of action that passes through the *centroid* or geometric center of the area or volume under the loading diagram.

REVIEW PROBLEMS

4-1. Determine the coordinate direction angles α, β, γ of **F**, which is applied to the end A of the pipe assembly, so that the moment of **F** about O is zero.

4-2. Determine the moment of the force **F** about point O. The force has coordinate direction angles of $\alpha = 60°$, $\beta = 120°$, $\gamma = 45°$. Express the result as a Cartesian vector.

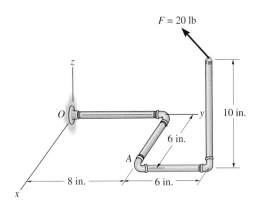

Probs. 4–1/2

4-3. If it takes a force of $F = 125$ lb to pull the nail out, determine the smallest vertical force **P** that must be applied to the handle of the crowbar. *Hint:* This requires the moment of **F** about point A to be equal to the moment of **P** about A. Why?

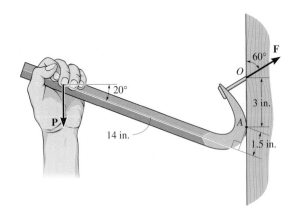

Prob. 4–3

*__**4-4.**__ Determine the moment of the force $\mathbf{F}_c$ about the door hinge at A. Express the result as a Cartesian vector.

4-5. Determine the magnitude of the moment of the force $\mathbf{F}_c$ about the hinged axis aa of the door.

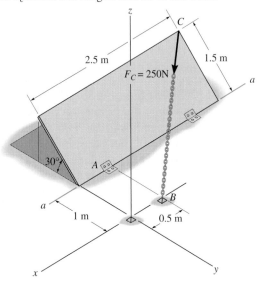

Probs. 4–4/5

4-6. Determine the resultant couple moment of the two couples that act on the assembly. Member OB lies in the x-z plane.

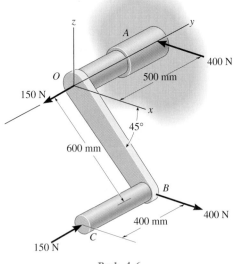

Prob. 4–6

4-7. Replace the force **F** having a magnitude of $F = 50$ lb and acting at point A by an equivalent force and couple moment at point C.

4-9. The horizontal 30-N force acts on the handle of the wrench. Determine the moment of this force about point O. Specify the coordinate direction angles α, β, γ of the moment axis.

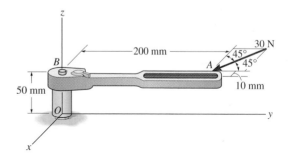

Prob. 4–9

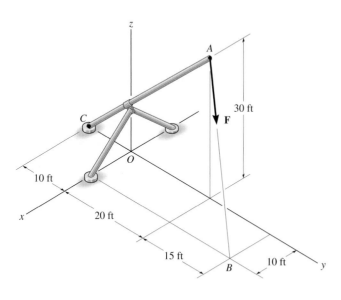

Prob. 4–7

*4-8.** The horizontal 30-N force acts on the handle of the wrench. What is the magnitude of the moment of this force about the z axis?

4-10. The forces and couple moments that are exerted on the toe and heel plates of a snow ski are $\mathbf{F}_t = \{-50\mathbf{i}+80\mathbf{j}-158\mathbf{k}\}$ N, $\mathbf{M}_t = \{-6\mathbf{i}+4\mathbf{j}+2\mathbf{k}\}$ N · m, and $\mathbf{F}_h = \{-20\mathbf{i}+60\mathbf{j}-250\mathbf{k}\}$ N, $\mathbf{M}_h = \{-20\mathbf{i}+8\mathbf{j}+3\mathbf{k}\}$ N · m, respectively. Replace this system by an equivalent force and couple moment acting at point P. Express the results in Cartesian vector form.

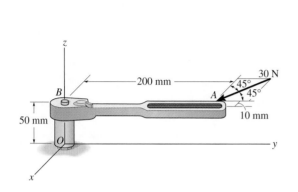

Prob. 4–8

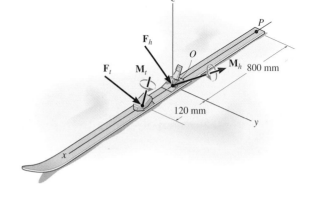

Prob. 4–10

The tower crane is subjected to its weight and the load it supports.
In order to calculate the support reactions for the crane, it is necessary to
apply the principles of equilibrium.

CHAPTER 5

Equilibrium of a Rigid Body

CHAPTER OBJECTIVES

- To develop the equations of equilibrium for a rigid body.
- To introduce the concept of the free-body diagram for a rigid body.
- To show how to solve rigid body equilibrium problems using the equations of equilibrium.

5.1 Conditions for Rigid-Body Equilibrium

In this section we will develop both the necessary and sufficient conditions required for equilibrium of a rigid body. To do this, consider the rigid body in Fig. 5–1a, which is fixed in the x, y, z reference and is either at rest or moves with the reference at constant velocity. A free-body diagram of the arbitrary ith particle of the body is shown in Fig. 5–1b. There are two types of forces which act on it. The resultant *internal force*, $\mathbf{f}_i$, is caused by interactions with adjacent particles. The resultant *external force* $\mathbf{F}_i$ represents, for example, the effects of gravitational, electrical, magnetic, or contact forces between the ith particle and adjacent bodies or particles *not* included within the body. If the particle is in equilibrium, then applying Newton's first law we have

$$\mathbf{F}_i + \mathbf{f}_i = \mathbf{0}$$

When the equation of equilibrium is applied to each of the other particles of the body, similar equations will result. If all these equations are added together *vectorially*, we obtain

$$\Sigma\mathbf{F}_i + \Sigma\mathbf{f}_i = \mathbf{0}$$

The summation of the internal forces will equal zero since the internal forces between particles within the body will occur in equal

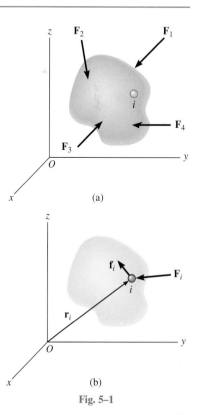

(a)

(b)

Fig. 5–1

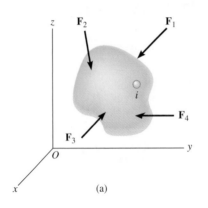

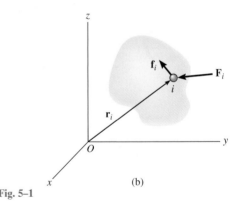

(a)   (b)

Fig. 5–1

but opposite collinear pairs, Newton's third law. Consequently, only the sum of the *external forces* will remain; and therefore, letting $\Sigma \mathbf{F}_i = \Sigma \mathbf{F}$, the above equation can be written as

$$\Sigma \mathbf{F} = \mathbf{0}$$

Let us now consider the moments of the forces acting on the ith particle about the arbitrary point O, Fig. 5–1b. Using the above particle equilibrium equation and the distributive law of the vector cross product we have

$$\mathbf{r}_i \times (\mathbf{F}_i + \mathbf{f}_i) = \mathbf{r}_i \times \mathbf{F}_i + \mathbf{r}_i \times \mathbf{f}_i = \mathbf{0}$$

Similar equations can be written for the other particles of the body, and adding them together vectorially, we obtain

$$\Sigma \mathbf{r}_i \times \mathbf{F}_i + \Sigma \mathbf{r}_i \times \mathbf{f}_i = \mathbf{0}$$

The second term is zero since, as stated above, the internal forces occur in equal but opposite collinear pairs, and therefore the resultant moment of each pair of forces about point O is zero. Hence, using the notation $\Sigma \mathbf{M}_O = \Sigma \mathbf{r}_i \times \mathbf{F}_i$, we have

$$\Sigma \mathbf{M}_O = \mathbf{0}$$

Hence the two *equations of equilibrium* for a rigid body can be summarized as follows:

$$\boxed{\begin{array}{l} \Sigma \mathbf{F} = \mathbf{0} \\ \Sigma \mathbf{M}_O = \mathbf{0} \end{array}}$$

(5–1)

These equations require that a rigid body will remain in equilibrium provided the sum of all the *external forces* acting on the body is equal to zero and the sum of the moments of the external forces about a point is equal to zero. The fact that these conditions are *necessary* for equilibrium has now been proven. They are also *sufficient* for maintaining equilibrium. To show this, let us assume that the body is in equilibrium and the force system acting on the body satisfies Eqs. 5–1. Suppose that an *additional force* $\mathbf{F}'$ is applied to the body. As a result, the equilibrium equations become

$$\Sigma \mathbf{F} + \mathbf{F}' = \mathbf{0}$$
$$\Sigma \mathbf{M}_O + \mathbf{M}'_O = \mathbf{0}$$

where $\mathbf{M}'_O$ is the moment of $\mathbf{F}'$ about O. Since $\Sigma \mathbf{F} = \mathbf{0}$ and $\Sigma \mathbf{M}_O = \mathbf{0}$, then we require $\mathbf{F}' = \mathbf{0}$ (also $\mathbf{M}'_O = \mathbf{0}$). Consequently, the additional force $\mathbf{F}'$ is not required, and indeed Eqs. 5–1 are also sufficient conditions for maintaining equilibrium.

Many types of engineering problems involve symmetric loadings and can be solved by projecting all the forces acting on a body onto a single plane. Hence, in the next section, the equilibrium of a body subjected to a *coplanar* or *two-dimensional force system* will be considered. Ordinarily the geometry of such problems is not very complex, so a scalar solution is suitable for analysis. The more general discussion of rigid bodies subjected to *three-dimensional force systems* is given in the latter part of this chapter. It will be seen that many of these types of problems can best be solved by using vector analysis.

Equilibrium in Two Dimensions

5.2 Free-Body Diagrams

Successful application of the equations of equilibrium requires a complete specification of *all* the known and unknown external forces that act *on* the body. The best way to account for these forces is to draw the body's free-body diagram. This diagram is a sketch of the outlined shape of the body, which represents it as being *isolated* or "free" from its surroundings, i.e., a "free body." On this sketch it is necessary to show *all* the forces and couple moments that the surroundings exert *on the body* so that these effects can be accounted for when the equations of equilibrium are applied. For this reason, *a thorough understanding of how to draw a free-body diagram is of primary importance for solving problems in mechanics.*

TABLE 5–1 • Supports for Rigid Bodies Subjected to Two-Dimensional Force Systems

Types of Connection	Reaction	Number of Unknowns
(1) cable		One unknown. The reaction is a tension force which acts away from the member in the direction of the cable.
(2) weightless link	or	One unknown. The reaction is a force which acts along the axis of the link.
(3) roller		One unknown. The reaction is a force which acts perpendicular to the surface at the point of contact.
(4) roller or pin in confined smooth slot	or	One unknown. The reaction is a force which acts perpendicular to the slot.
(5) rocker		One unknown. The reaction is a force which acts perpendicular to the surface at the point of contact.
(6) smooth contacting surface		One unknown. The reaction is a force which acts perpendicular to the surface at the point of contact.
(7) member pin connected to collar on smooth rod	or	One unknown. The reaction is a force which acts perpendicular to the rod.

continued

TABLE 5–1 • *Continued*

Types of Connection	Reaction	Number of Unknowns
(8) 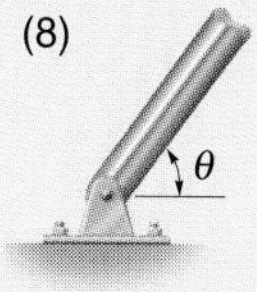 smooth pin or hinge	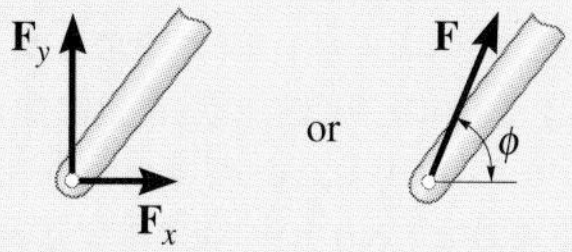 F_y ↑ ↑ F_x or F ϕ	Two unknowns. The reactions are two components of force, or the magnitude and direction ϕ of the resultant force. Note that ϕ and θ are not necessarily equal [usually not, unless the rod shown is a link as in (2)].
(9) 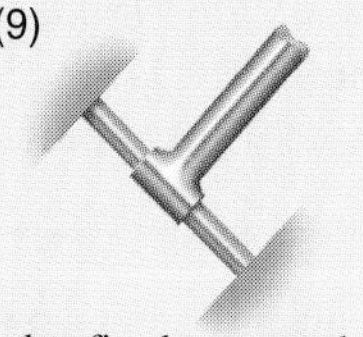 member fixed connected to collar on smooth rod	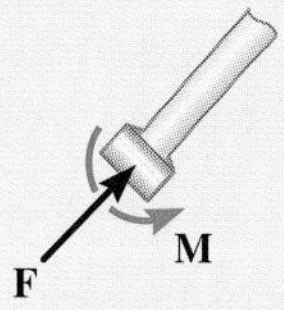 F M	Two unknowns. The reactions are the couple moment and the force which acts perpendicular to the rod.
(10) 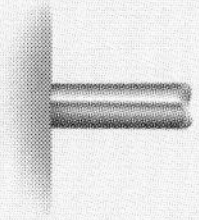 fixed support	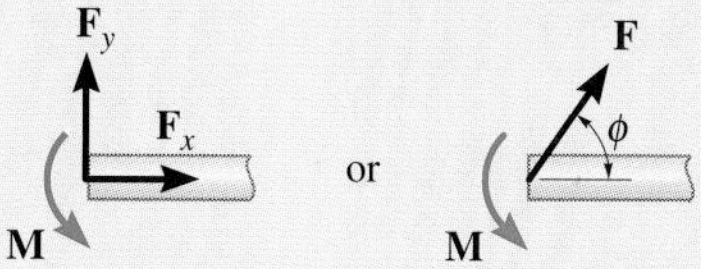 F_y ↑ F_x M or F ϕ M	Three unknowns. The reactions are the couple moment and the two force components, or the couple moment and the magnitude and direction ϕ of the resultant force.

Support Reactions.   Before presenting a formal procedure as to how to draw a free-body diagram, we will first consider the various types of reactions that occur at supports and points of support between bodies subjected to coplanar force systems. *As a general rule, if a support prevents the translation of a body in a given direction, then a force is developed on the body in that direction. Likewise, if rotation is prevented, a couple moment is exerted on the body.*

For example, let us consider three ways in which a horizontal member, such as a beam, is supported at its end. One method consists of a *roller* or cylinder, Fig. 5–2a. Since this support only prevents the beam from *translating* in the vertical direction, the roller can only exert a *force* on the beam in this direction, Fig. 5–2b.

The beam can be supported in a more restrictive manner by using a *pin* as shown in Fig. 5–3a. The pin passes through a hole in the beam and two leaves which are fixed to the ground. Here the pin can prevent *translation* of the beam in *any direction* ϕ, Fig. 5–3b, and so the pin must exert a *force* **F** on the beam in this direction. For purposes of analysis, it is generally easier to represent this resultant force **F** by its two components $\mathbf{F}_x$ and $\mathbf{F}_y$, Fig. 5–3c. If F_x and F_y are known, then F and ϕ can be calculated.

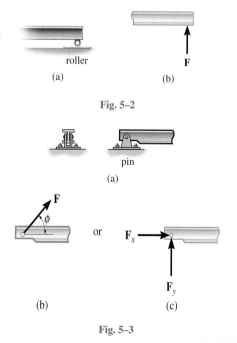

roller

(a)

F

(b)

Fig. 5–2

pin

(a)

F ϕ

or

$\mathbf{F}_x$

$\mathbf{F}_y$

(b)

(c)

Fig. 5–3

fixed support

(a)

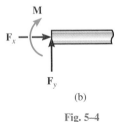

(b)

Fig. 5–4

The most restrictive way to support the beam would be to use a *fixed support* as shown in Fig. 5–4a. This support will prevent both *translation and rotation* of the beam, and so to do this a *force and couple moment* must be developed on the beam at its point of connection, Fig. 5–4b. As in the case of the pin, the force is usually represented by its components $\mathbf{F}_x$ and $\mathbf{F}_y$.

Table 5–1 lists other common types of supports for bodies subjected to coplanar force systems. (In all cases the angle θ is assumed to be known.) Carefully study each of the symbols used to represent these supports and the types of reactions they exert on their contacting members. Although concentrated forces and couple moments are shown in this table, they actually represent the *resultants* of small *distributed surface loads* that exist between each support and its contacting member. It is these *resultants* which will be determined from the equations of equilibrium.

Typical examples of actual supports that are referenced to Table 5–1 are shown in the following sequence of photos.

The cable exerts a force on the bracket in the direction of the cable. (1)

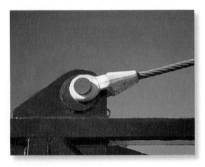

The rocker support for this bridge girder allows horizontal movement so the bridge is free to expand and contract due to temperature. (5)

This concrete girder rests on the ledge that is assumed to act as a smooth contacting surface. (6)

This utility building is pin supported at the top of the column. (8)

The floor beams of this building are welded together and thus form fixed connections. (10)

External and Internal Forces.   Since a rigid body is a composition of particles, both *external* and *internal* loadings may act on it. It is important to realize, however, that if the free-body diagram for the body is drawn, the forces that are *internal* to the body are *not represented* on the free-body diagram. As discussed in Sec. 5.1, these forces always occur in equal but opposite collinear pairs, and therefore their *net effect* on the body is zero.

In some problems, a free-body diagram for a "system" of connected bodies may be used for an analysis. An example would be the free-body diagram of an entire automobile (system) composed of its many parts. Obviously, the connecting forces between its parts would represent *internal forces* which would *not* be included on the free-body diagram of the automobile. To summarize, internal forces act between particles which are contained within the boundary of the free-body diagram. Particles or bodies outside this boundary exert external forces on the system, and these alone must be shown on the free-body diagram.

Weight and the Center of Gravity.   When a body is subjected to a gravitational field, then each of its particles has a specified weight. For the entire body it is appropriate to consider these gravitational forces to be represented as a *system of parallel forces* acting on all the particles contained within the boundary of the body. It was shown in Sec. 4.9 that such a system can be reduced to a single resultant force acting through a specified point. We refer to this force resultant as the *weight* **W** of the body and to the location of its point of application as the *center of gravity*. The methods used for its calculation will be developed in Chapter 9.

In the examples and problems that follow, if the weight of the body is important for the analysis, this force will then be reported in the problem statement. Also, when the body is *uniform* or made of homogeneous material, the center of gravity will be located at the body's *geometric center* or *centroid*; however, if the body is nonhomogeneous or has an unusual shape, then the location of its center of gravity will be given.

Idealized Models.   In order to perform a correct force analysis of any object, it is important to consider a corresponding analytical or idealized model that gives results that approximate as closely as possible the actual situation. To do this, careful choices have to be made so that selection of the type of supports, the material behavior, and the object's dimensions can be justified. This way the engineer can feel confident that any design or analysis will yield results which can be trusted. In complex cases this process may require developing several different models of the object that must be analyzed, but in any case, this selection process requires both skill and experience.

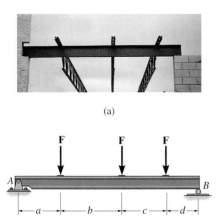

(a)

(b)

Fig. 5–5

To illustrate what is required to develop a proper model, we will now consider a few cases. As shown in Fig. 5–5a, the steel beam is to be used to support the roof joists of a building. For a force analysis it is reasonable to assume the material is rigid since only very small deflections will occur when the beam is loaded. A bolted connection at *A* will allow for any slight rotation that occurs when the load is applied, and so a *pin* can be considered for this support. At *B* a *roller* can be considered since the support offers no resistance to horizontal movement here. Building code requirements are used to specify the roof loading which results in a calculation of the joist loads **F**. These forces will be larger than any actual loading on the beam since they account for extreme loading cases and for dynamic or vibrational effects. The weight of the beam is generally neglected when it is small compared to the load the beam supports. The idealized model of the beam is shown with average dimensions *a*, *b*, *c*, and *d* in Fig. 5–5b.

As a second case, consider the lift boom in Fig. 5–6a. By inspection, it is supported by a pin at *A* and by the hydraulic cylinder *BC*, which can be approximated as a weightless link. The material can be assumed rigid, and with its density known, the weight of the boom and the location of its center of gravity *G* are determined. When a design loading **P** is specified, the idealized model shown in Fig. 5–6b can be used for a force analysis. Average dimensions (not shown) are used to specify the location of the loads and the supports.

Idealized models of specific objects will be given in some of the examples throughout the text. It should be realized, however, that each case represents the reduction of a practical situation using simplifying assumptions like the ones illustrated here.

(a)

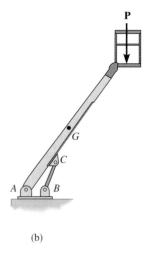

(b)

Fig. 5–6

PROCEDURE FOR DRAWING A FREE-BODY DIAGRAM

To construct a free-body diagram for a rigid body or group of bodies considered as a single system, the following steps should be performed:

Draw Outlined Shape. Imagine the body to be *isolated* or cut "free" from its constraints and connections and draw (sketch) its outlined shape.

Show All Forces and Couple Moments. Identify all the external forces and couple moments that act on the body. Those generally encountered are due to (1) applied loadings, (2) reactions occurring at the supports or at points of contact with other bodies (see Table 5–1), and (3) the weight of the body. To account for all these effects, it may help to trace over the boundary, carefully noting each force or couple moment acting on it.

Identify Each Loading and Give Dimensions. The forces and couple moments that are known should be labeled with their proper magnitudes and directions. Letters are used to represent the magnitudes and direction angles of forces and couple moments that are *unknown*. Establish an *x, y* coordinate system so that these unknowns, A_x, B_y, etc., can be identified. Indicate the dimensions of the body necessary for calculating the moments of forces.

IMPORTANT POINTS

- No equilibrium problem should be solved without *first* drawing the free-body diagram, so as to account for all the forces and couple moments that act on the body.

- If a support *prevents translation* of a body in a particular direction, then the support exerts a *force* on the body in that direction.

- If *rotation is prevented*, then the support exerts a *couple moment* on the body.

- Study Table 5–1.

- Internal forces are never shown on the free-body diagram since they occur in equal but opposite collinear pairs and therefore cancel out.

- The weight of a body is an external force, and its effect is shown as a single resultant force acting through the body's center of gravity *G*.

- *Couple moments* can be placed anywhere on the free-body diagram since they are *free vectors*. *Forces* can act at any point along their lines of action since they are *sliding vectors*.

EXAMPLE 5.1

Draw the free-body diagram of the uniform beam shown in Fig. 5–7a. The beam has a mass of 100 kg.

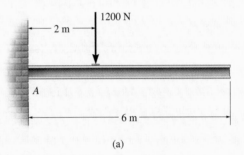

(a)

Solution

The free-body diagram of the beam is shown in Fig. 5–7b. Since the support at A is a fixed wall, there are three reactions acting *on the beam* at A, denoted as $\mathbf{A}_x$, $\mathbf{A}_y$, and $\mathbf{M}_A$ drawn in an arbitrary direction. The magnitudes of these vectors are *unknown*, and their sense has been *assumed*. The weight of the beam, $W = 100(9.81) = 981$ N, acts through the beam's center of gravity G, which is 3 m from A since the beam is uniform.

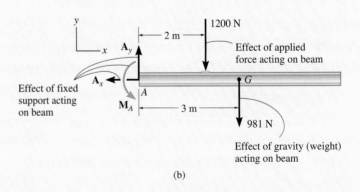

(b)

Fig. 5–7

EXAMPLE 5.2

Draw the free-body diagram of the foot lever shown in Fig. 5–8a. The operator applies a vertical force to the pedal so that the spring is stretched 1.5 in and the force in the short link at B is 20 lb.

(a)

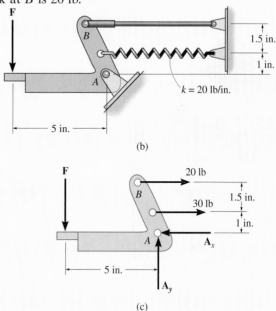

(b)

(c)

Fig. 5–8

Solution

By inspection, the lever is loosely bolted to the frame at A. The rod at B is pinned at its ends and acts as a "short link." After making the proper measurements, the idealized model of the lever is shown in Fig. 5–8b. From this the free-body diagram must be drawn. As shown in Fig. 5–8c, the pin support at A exerts force components A_x and A_y on the lever, each force has a known line of action but unknown magnitude. The link at B exerts a force of 20 lb, acting in the direction of the link. In addition the spring also exerts a horizontal force on the lever. If the stiffness is measured and found to be $k = 20$ lb/in., then since the stretch $s = 1.5$ in., using Eq. 3–2, $F_s = ks = 20$ lb/in. (1.5 in.) $= 30$ lb. Finally, the operator's shoe applies a vertical force of $\mathbf{F}$ on the pedal. The dimensions of the lever are also shown on the free-body diagram, since this information will be useful when computing the moments of the forces. As usual, the senses of the unknown forces at A have been assumed. The correct senses will become apparent after solving the equilibrium equations.

E X A M P L E 5.3

Two smooth pipes, each having a mass of 300 kg, are supported by the forks of the tractor in Fig. 5–9a. Draw the free-body diagrams for each pipe and both pipes together.

(a)

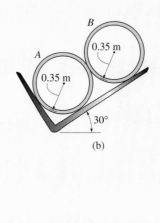

(b)

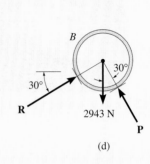

(d)

Effect of *B* acting on *A*

Effect of sloped blade acting on *A*

T

2943 N

Effect of gravity (weight) acting on *A*

F Effect of sloped fork acting on *A*

(c)

Fig. 5–9

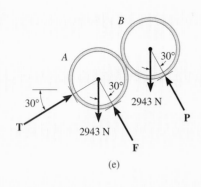

(e)

Solution

The idealized model from which we must draw the free-body diagrams is shown in Fig. 5–9b. Here the pipes are identified, the dimensions have been added, and the physical situation reduced to its simplest form.

The free-body diagram for pipe *A* is shown in Fig. 5–9c. Its weight is $W = 300(9.81) = 2943$ N. Assuming all contacting surfaces are *smooth*, the reactive forces **T**, **F**, **R** act in a direction *normal* to the tangent at their surfaces of contact.

The free-body diagram of pipe *B* is shown in Fig. 5–9d. Can you identify each of the three forces acting *on this pipe*? In particular, note that **R**, representing the force of *A* on *B*, Fig. 5–9d, is equal and opposite to **R** representing the force of *B* on *A*, Fig. 5–9c. This is a consequence of Newton's third law of motion.

The free-body diagram of both pipes combined ("system") is shown in Fig. 5–9e. Here the contact force **R**, which acts between *A* and *B*, is considered as an *internal* force and hence is not shown on the free-body diagram. That is, it represents a pair of equal but opposite collinear forces which cancel each other.

EXAMPLE 5.4

Draw the free-body diagram of the unloaded platform that is suspended off the edge of the oil rig shown in Fig. 5–10a. The platform has a mass of 200 kg.

(a)

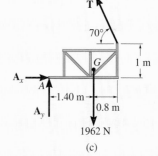

(b)

Fig. 5–10

(c)

Solution

The idealized model of the platform will be considered in two dimensions because by observation the loading and the dimensions are all symmetrical about a vertical plane passing through its center, Fig. 5–10b. Here the connection at A is assumed to be a pin, and the cable supports the platform at B. The direction of the cable and average dimensions of the platform are listed, and the center of gravity G has been determined. It is from this model that we must proceed to draw the free-body diagram, which is shown in Fig. 5–10c. The platform's weight is $200(9.81) = 1962$ N. The force components $\mathbf{A}_x$ and $\mathbf{A}_y$ along with the cable force $\mathbf{T}$ represent the reactions that both pins and both cables exert on the platform, Fig. 5–10a. Consequently, after the solution for these reactions, half their magnitude is developed at A and half is developed at B.

EXAMPLE **5.5**

The free-body diagram of each object in Fig. 5–11 is drawn. Carefully study each solution and identify what each loading represents, as was done in Fig. 5–7b.

Solution

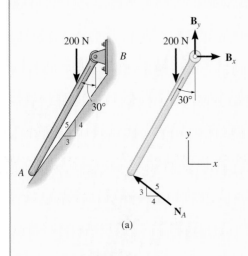

(a)

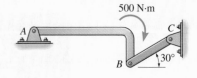

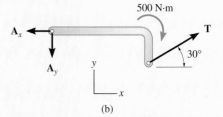

(b)

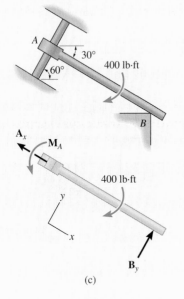

(c)

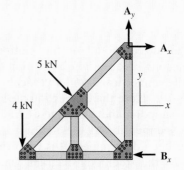

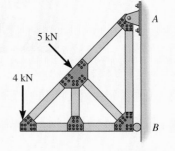

Note: Internal forces of one member on another are equal but opposite collinear forces which are not to be included here since they cancel out.

(d)

Fig. 5–11

5.3 Equations of Equilibrium

In Sec. 5.1 we developed the two equations which are both necessary and sufficient for the equilibrium of a rigid body, namely, $\Sigma \mathbf{F} = \mathbf{0}$ and $\Sigma \mathbf{M}_O = \mathbf{0}$. When the body is subjected to a system of forces, which all lie in the x–y plane, then the forces can be resolved into their x and y components. Consequently, the conditions for equilibrium in two dimensions are

$$\boxed{\begin{aligned} \Sigma F_x &= 0 \\ \Sigma F_y &= 0 \\ \Sigma M_O &= 0 \end{aligned}} \qquad (5\text{–}2)$$

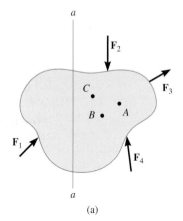

(a)

Here ΣF_x and ΣF_y represent, respectively, the algebraic sums of the x and y components of all the forces acting on the body, and ΣM_O represents the algebraic sum of the couple moments and the moments of all the force components about an axis perpendicular to the x–y plane and passing through the arbitrary point O, which may lie either on or off the body.

Alternative Sets of Equilibrium Equations. Although Eqs. 5–2 are *most often* used for solving coplanar equilibrium problems, two *alternative* sets of three independent equilibrium equations may also be used. One such set is

$$\begin{aligned} \Sigma F_a &= 0 \\ \Sigma M_A &= 0 \\ \Sigma M_B &= 0 \end{aligned} \qquad (5\text{–}3)$$

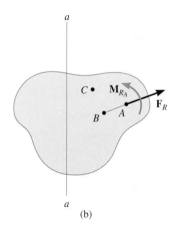

(b)

When using these equations it is required that a line passing through points A and B is *not perpendicular* to the a axis. To prove that Eqs. 5–3 provide the *conditions* for equilibrium, consider the free-body diagram of an arbitrarily shaped body shown in Fig. 5–12a. Using the methods of Sec. 4.8, all the forces on the free-body diagram may be replaced by an equivalent resultant force $\mathbf{F}_R = \Sigma \mathbf{F}$, acting at point A, and a resultant couple moment $\mathbf{M}_{R_A} = \Sigma \mathbf{M}_A$, Fig. 5–12$b$. If $\Sigma M_A = 0$ is satisfied, it is necessary that $\mathbf{M}_{R_A} = \mathbf{0}$. Furthermore, in order that $\mathbf{F}_R$ satisfy $\Sigma F_a = 0$, it must have *no component* along the a axis, and therefore its line of action must be perpendicular to the a axis, Fig. 5–12c. Finally, if it is required that $\Sigma M_B = 0$, where B does not lie on the line of action of $\mathbf{F}_R$, then $\mathbf{F}_R = \mathbf{0}$. Since $\Sigma \mathbf{F} = \mathbf{0}$ and $\Sigma \mathbf{M}_A = \mathbf{0}$, indeed the body in Fig. 5–12a must be in equilibrium.

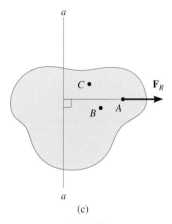

(c)

Fig. 5–12

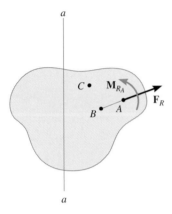

Fig. 5–13

A second alternative set of equilibrium equations is

$$\Sigma M_A = 0$$
$$\Sigma M_B = 0 \qquad\qquad (5\text{–}4)$$
$$\Sigma M_C = 0$$

Here it is necessary that points A, B, and C do not lie on the same line. To prove that these equations, when satisfied, ensure equilibrium, consider the free-body diagram in Fig. 5–13. If $\Sigma M_A = 0$ is to be satisfied, then $\mathbf{M}_{R_A} = \mathbf{0}$. $\Sigma M_B = 0$ is satisfied if the line of action of $\mathbf{F}_R$ passes through point B as shown. Finally, if we require $\Sigma M_C = 0$, where C does not lie on line AB, it is necessary that $\mathbf{F}_R = \mathbf{0}$, and the body in Fig. 5–12a must then be in equilibrium.

PROCEDURE FOR ANALYSIS

Coplanar force equilibrium problems for a rigid body can be solved using the following procedure.

Free-Body Diagram.

- Establish the x, y coordinate axes in any suitable orientation.
- Draw an outlined shape of the body.
- Show all the forces and couple moments acting on the body.
- Label all the loadings and specify their directions relative to the x, y axes. The sense of a force or couple moment having an *unknown* magnitude but known line of action can be *assumed*.
- Indicate the dimensions of the body necessary for computing the moments of forces.

Equations of Equilibrium.

- Apply the moment equation of equilibrium, $\Sigma M_O = 0$, about a point (O) that lies at the intersection of the lines of action of two unknown forces. In this way, the moments of these unknowns are zero about O, and a *direct solution* for the third unknown can be determined.
- When applying the force equilibrium equations, $\Sigma F_x = 0$ and $\Sigma F_y = 0$, orient the x and y axes along lines that will provide the simplest resolution of the forces into their x and y components.
- If the solution of the equilibrium equations yields a negative scalar for a force or couple moment magnitude, this indicates that the sense is opposite to that which was assumed on the free-body diagram.

E X A M P L E   5.6

Determine the horizontal and vertical components of reaction for the beam loaded as shown in Fig. 5–14a. Neglect the weight of the beam in the calculations.

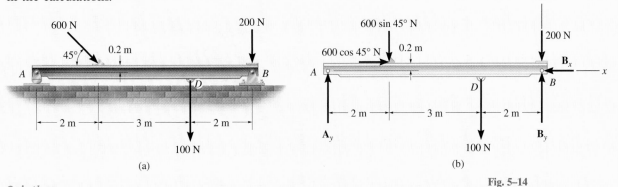

(a)

(b)

Fig. 5–14

Solution

Free-Body Diagram.   Can you identify each of the forces shown on the free-body diagram of the beam, Fig. 5–14b? For simplicity, the 600-N force is represented by its x and y components as shown. Also, note that a 200-N force acts on the beam at B and is independent of the force components $\mathbf{B}_x$ and $\mathbf{B}_y$, which represent the effect of the pin on the beam.

Equations of Equilibrium.   Summing forces in the x direction yields

$$\xrightarrow{+} \Sigma F_x = 0; \qquad 600 \cos 45° \,\text{N} - B_x = 0$$
$$B_x = 424 \,\text{N} \qquad\qquad\qquad Ans.$$

A direct solution for $\mathbf{A}_y$ can be obtained by applying the moment equation $\Sigma M_B = 0$ about point B. For the calculation, it should be apparent that forces 200 N, $\mathbf{B}_x$, and $\mathbf{B}_y$ all create zero moment about B. Assuming counterclockwise rotation about B to be positive (in the $+\mathbf{k}$ direction), Fig. 5–14b, we have

$$\zeta + \Sigma M_B = 0; \qquad 100 \,\text{N}(2 \,\text{m}) + (600 \sin 45° \,\text{N})(5 \,\text{m})$$
$$- (600 \cos 45° \,\text{N})(0.2 \,\text{m}) - A_y(7 \,\text{m}) = 0$$
$$A_y = 319 \,\text{N} \qquad\qquad\qquad Ans.$$

Summing forces in the y direction, using this result, gives

$$+\uparrow \Sigma F_y = 0; \qquad 319 \,\text{N} - 600 \sin 45° \,\text{N} - 100 \,\text{N} - 200 \,\text{N} + B_y = 0$$
$$B_y = 405 \,\text{N} \qquad\qquad\qquad Ans.$$

We can check this result by summing moments about point A.

$$\zeta + \Sigma M_A = 0; \qquad -(600 \sin 45° \,\text{N})(2 \,\text{m}) - (600 \cos 45° \,\text{N})(0.2 \,\text{m})$$
$$-(100 \,\text{N})(5 \,\text{m}) - (200 \,\text{N})(7 \,\text{m}) + B_y(7 \,\text{m}) = 0$$
$$B_y = 405 \,\text{N} \qquad\qquad\qquad Ans.$$

E X A M P L E **5.7**

The cord shown in Fig. 5–15a supports a force of 100 lb and wraps over the frictionless pulley. Determine the tension in the cord at C and the horizontal and vertical components of reaction at pin A.

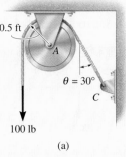

0.5 ft

A

$\theta = 30°$

C

100 lb

(a)

Fig. 5–15

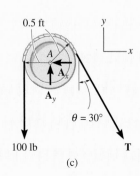

p

30°

100 lb

T

p

A

A_x

A_y

(b)

Solution

Free-Body Diagrams. The free-body diagrams of the cord and pulley are shown in Fig. 5–15b. Note that the principle of action, equal but opposite reaction must be carefully observed when drawing each of these diagrams: the cord exerts an unknown load distribution p along part of the pulley's surface, whereas the pulley exerts an equal but opposite effect on the cord. For the solution, however, it is simpler to *combine* the free-body diagrams of the pulley and the contacting portion of the cord, so that the distributed load becomes *internal* to the system and is therefore eliminated from the analysis, Fig. 5–15c.

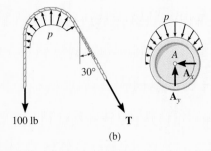

0.5 ft

y

x

A

A_x

A_y

$\theta = 30°$

100 lb

T

(c)

Equations of Equilibrium. Summing moments about point A to eliminate $\mathbf{A}_x$ and $\mathbf{A}_y$, Fig. 5–15c, we have

$$\zeta + \Sigma M_A = 0; \qquad 100\ \text{lb}(0.5\ \text{ft}) - T(0.5\ \text{ft}) = 0$$
$$T = 100\ \text{lb} \qquad\qquad Ans.$$

It is seen that the tension remains *constant* as the cord passes over the pulley. (This of course is true for *any angle* θ at which the cord is directed and for *any radius* r of the pulley.) Using the result for T, a force summation is applied to determine the components of reaction at pin A.

$$\xrightarrow{+} \Sigma F_x = 0; \qquad -A_x + 100 \sin 30°\ \text{lb} = 0$$
$$A_x = 50.0\ \text{lb} \qquad\qquad Ans.$$
$$+\uparrow \Sigma F_y = 0; \qquad A_y - 100\ \text{lb} - 100 \cos 30°\ \text{lb} = 0$$
$$A_y = 187\ \text{lb} \qquad\qquad Ans.$$

E X A M P L E **5.8**

The link shown in Fig. 5–16a is pin-connected at A and rests against a smooth support at B. Compute the horizontal and vertical components of reaction at the pin A.

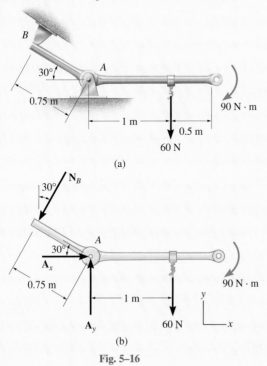

(a)

(b)

Fig. 5–16

Solution

Free-Body Diagram. As shown in Fig. 5–16b, the reaction $\mathbf{N}_B$ is perpendicular to the link at B. Also, horizontal and vertical components of reaction are represented at A.

Equations of Equilibrium. Summing moments about A, we obtain a direct solution for N_B,

$$\zeta + \Sigma M_A = 0; \quad -90 \text{ N} \cdot \text{m} - 60 \text{ N}(1 \text{ m}) + N_B(0.75 \text{ m}) = 0$$
$$N_B = 200 \text{ N}$$

Using this result,

$$\xrightarrow{+} \Sigma F_x = 0; \quad A_x - 200 \sin 30° \text{ N} = 0$$
$$A_x = 100 \text{ N} \qquad \textit{Ans.}$$
$$+\uparrow \Sigma F_y = 0; \quad A_y - 200 \cos 30° \text{ N} - 60 \text{ N} = 0$$
$$A_y = 233 \text{ N} \qquad \textit{Ans.}$$

EXAMPLE 5.9

The box wrench in Fig. 5–17a is used to tighten the bolt at A. If the wrench does not turn when the load is applied to the handle, determine the torque or moment applied to the bolt and the force of the wrench on the bolt.

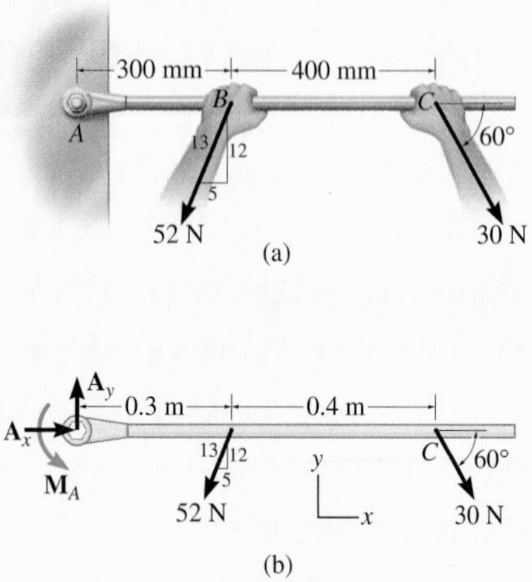

Solution

Free-Body Diagram. The free-body diagram for the wrench is shown in Fig. 5–17b. Since the bolt acts as a "fixed support," it exerts force components $\mathbf{A}_x$ and $\mathbf{A}_y$ and a torque $\mathbf{M}_A$ on the wrench at A.

Equations of Equilibrium.

$$\xrightarrow{+} \Sigma F_x = 0; \qquad A_x - 52(\tfrac{5}{13})\text{N} + 30 \cos 60° \text{ N} = 0$$
$$A_x = 5.00 \text{ N} \qquad\qquad Ans.$$

$$+\uparrow \Sigma F_y = 0; \qquad A_y - 52(\tfrac{12}{13})\text{N} - 30 \sin 60° \text{ N} = 0$$
$$A_y = 74.0 \text{ N} \qquad\qquad Ans.$$

$$\downarrow+\Sigma M_A = 0; \qquad M_A - 52(\tfrac{12}{13})\text{N} (0.3 \text{ m}) - (30 \sin 60° \text{ N})(0.7 \text{ m}) = 0$$
$$M_A = 32.6 \text{ N} \cdot \text{m} \qquad\qquad Ans.$$

Point A was chosen for summing moments because the lines of action of the *unknown* forces $\mathbf{A}_x$ and $\mathbf{A}_y$ pass through this point, and therefore these forces were not included in the moment summation. Realize, however, that $\mathbf{M}_A$ must be *included* in this moment summation. This couple moment is a free vector and represents the twisting resistance of the bolt on the wrench. By Newton's third law, the wrench exerts an equal but opposite moment or torque on the bolt. Furthermore, the resultant force on the wrench is

$$F_A = \sqrt{(5.00)^2 + (74.0)^2} = 74.1 \text{ N} \qquad Ans.$$

Because the force components A_x and A_y were calculated as positive quantities, their directional sense is shown correctly on the free-body diagram in Fig. 5–17b. Hence

$$\theta = \tan^{-1}\frac{74.0 \text{ N}}{5.00 \text{ N}} = 86.1° \; \measuredangle$$

Realize that $\mathbf{F}_A$ acts in the opposite direction on the bolt. Why?

Although only *three* independent equilibrium equations can be written for a rigid body, it is a good practice to *check* the calculations using a fourth equilibrium equation. For example, the above computations may be verified in part by summing moments about point C:

$$\downarrow+\Sigma M_C = 0; \; 52(\tfrac{12}{13})\text{N} (0.4 \text{ m}) + 32.6 \text{ N} \cdot \text{m} - 74.0 \text{ N}(0.7 \text{ m}) = 0$$
$$19.2 \text{ N} \cdot \text{m} + 32.6 \text{ N} \cdot \text{m} - 51.8 \text{ N} \cdot \text{m} = 0$$

(a)

(b)

Fig. 5–17

EXAMPLE 5.10

Placement of concrete from the truck is accomplished using the chute shown in the photos, Fig. 5–18a. Determine the force that the hydraulic cylinder and the truck frame exert on the chute to hold it in the position shown. The chute and wet concrete contained along its length have a uniform weight of 35 lb/ft.

Solution

The idealized model of the chute is shown in Fig. 5–18b. Here the dimensions are given, and it is assumed the chute is pin connected to the frame at A and the hydraulic cylinder BC acts as a short link.

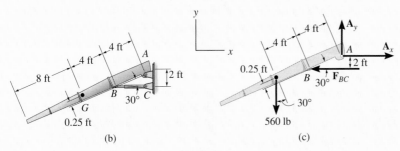

(b)

(c)

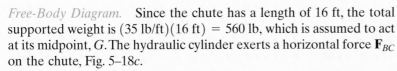

Fig. 5–18

(a)

Free-Body Diagram. Since the chute has a length of 16 ft, the total supported weight is $(35 \text{ lb/ft})(16 \text{ ft}) = 560 \text{ lb}$, which is assumed to act at its midpoint, G. The hydraulic cylinder exerts a horizontal force $\mathbf{F}_{BC}$ on the chute, Fig. 5–18c.

Equations of Equilibrium. A direct solution for $\mathbf{F}_{BC}$ is possible by summing moments about the pin at A. To do this we will use the principle of moments and resolve the weight into components parallel and perpendicular to the chute. We have,

$\zeta + \Sigma M_A = 0;$
$$-F_{BC}(2 \text{ ft}) + 560 \cos 30° \text{ lb}(8 \text{ ft}) + 560 \sin 30° \text{ lb}(0.25 \text{ ft}) = 0$$
$$F_{BC} = 1975 \text{ lb} \qquad \qquad Ans.$$

Summing forces to obtain A_x and A_y, we obtain

$\xrightarrow{+} \Sigma F_x = 0; \qquad -A_x + 1975 \text{ lb} = 0$
$$A_x = 1975 \text{ lb} \qquad \qquad Ans.$$

$+\uparrow \Sigma F_y = 0; \qquad A_y - 560 \text{ lb} = 0$
$$A_y = 560 \text{ lb} \qquad \qquad Ans.$$

To verify this solution we can sum moments about point B.

$\zeta + \Sigma M_B = 0; \qquad -1975 \text{ lb}(2 \text{ ft}) + 560 \text{ lb}(4 \cos 30° \text{ ft}) +$
$$560 \cos 30° \text{lb}(4 \text{ ft}) + 560 \sin 30° \text{ lb}(0.25 \text{ ft}) = 0$$

E X A M P L E 5.11

The uniform smooth rod shown in Fig. 5–19a is subjected to a force and couple moment. If the rod is supported at A by a smooth wall and at B and C either at the top or bottom by rollers, determine the reactions at these supports. Neglect the weight of the rod.

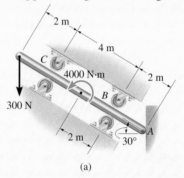

(a)

Solution

Free-Body Diagram. As shown in Fig. 5–19b, all the support reactions act normal to the surface of contact since the contacting surfaces are smooth. The reactions at B and C are shown acting in the positive y′ direction. This assumes that only the rollers located on the bottom of the rod are used for support.

Equations of Equilibrium. Using the x, y coordinate system in Fig. 5–19b, we have

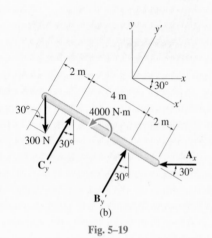

$$\xrightarrow{+} \Sigma F_x = 0; \qquad C_{y'} \sin 30° + B_{y'} \sin 30° - A_x = 0 \qquad (1)$$

$$+\uparrow \Sigma F_y = 0; \qquad -300 \text{ N} + C_{y'} \cos 30° + B_{y'} \cos 30° = 0 \qquad (2)$$

$$\zeta + \Sigma M_A = 0; \qquad -B_{y'}(2 \text{ m}) + 4000 \text{ N} \cdot \text{m} - C_{y'}(6 \text{ m})$$
$$+ (300 \cos 30° \text{ N})(8 \text{ m}) = 0 \qquad (3)$$

When writing the moment equation, it should be noticed that the line of action of the force component 300 sin 30° N passes through point A, and therefore this force is not included in the moment equation.
Solving Eqs. 2 and 3 simultaneously, we obtain

$$B_{y'} = -1000.0 \text{ N} = -1 \text{ kN} \qquad \textit{Ans.}$$
$$C_{y'} = 1346.4 \text{ N} = 1.35 \text{ kN} \qquad \textit{Ans.}$$

Since $B_{y'}$ is a negative scalar, the sense of $\mathbf{B}_{y'}$ is opposite to that shown on the free-body diagram in Fig. 5–19b. Therefore, the top roller at B serves as the support rather than the bottom one. Retaining the negative sign for $B_{y'}$ (Why?) and substituting the results into Eq. 1, we obtain

$$1346.4 \sin 30° \text{ N} - 1000.0 \sin 30° \text{ N} - A_x = 0$$

$$A_x = 173 \text{ N} \qquad \textit{Ans.}$$

Fig. 5–19

EXAMPLE 5.12

The uniform truck ramp shown in Fig. 5–20a has a weight of 400 lb and is pinned to the body of the truck at each end and held in the position shown by the two side cables. Determine the tension in the cables.

(a)

Solution

The idealized model of the ramp, which indicates all necessary dimensions and supports, is shown in Fig. 5–20b. Here the center of gravity is located at the midpoint since the ramp is approximately uniform.

Free-Body Diagram.   Working from the idealized model, the ramp's free-body diagram is shown in Fig. 5–20c.

Equations of Equilibrium.   Summing moments about point A will yield a direct solution for the cable tension. Using the principle of moments, there are several ways of determining the moment of $\mathbf{T}$ about A. If we use x and y components, with $\mathbf{T}$ applied at B, we have

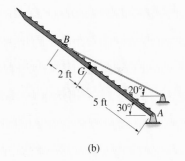

(b)

$$\zeta + \Sigma M_A = 0; \quad -T\cos 20°(7\sin 30° \text{ ft}) + T\sin 20°(7\cos 30° \text{ ft})$$
$$+ 400 \text{ lb}(5\cos 30° \text{ ft}) = 0$$
$$T = 1425 \text{ lb}$$

By the principle of transmissibility, we can locate $\mathbf{T}$ at C, even though this point is not on the ramp, Fig. 5–20c. In this case the horizontal component of $\mathbf{T}$ does not create a moment about A. First we must determine d using the sine law.

$$\frac{d}{\sin 10°} = \frac{7 \text{ ft}}{\sin 20°}; \quad d = 3.554 \text{ ft}$$

$$\zeta + \Sigma M_A = 0; \quad -T\sin 20°(3.554 \text{ ft}) + 400 \text{ lb}(5\cos 30° \text{ ft}) = 0$$

$$T = 1425 \text{ lb}$$

The simplest way to compute the moment of $\mathbf{T}$ about A is to resolve it into components parallel and perpendicular to the ramp at B. Then the moment of the parallel component is zero about A, so that

$$\zeta + \Sigma M_A = 0; \quad -T\sin 10°(7 \text{ ft}) + 400 \text{ lb}(5\cos 30° \text{ ft}) = 0$$

$$T = 1425 \text{ lb}$$

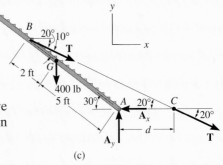

(c)

Since there are two cables supporting the ramp,

$$T' = \frac{T}{2} = 712 \text{ lb} \qquad\qquad Ans.$$

Fig. 5–20

As an exercise, show that $A_x = 1339$ lb and $A_y = 887.4$ lb.

5.4 Two- and Three-Force Members

The solution to some equilibrium problems can be simplified if one is able to recognize members that are subjected to only two or three forces.

Two-Force Members. When a member is subject to *no couple moments* and forces are applied at only two points on a member, the member is called a *two-force member*. An example is shown in Fig. 5–21a. The forces at A and B are summed to obtain their respective *resultants* $\mathbf{F}_A$ and $\mathbf{F}_B$, Fig. 5–21b. These two forces will maintain *translational or force equilibrium* ($\Sigma\mathbf{F} = \mathbf{0}$) provided $\mathbf{F}_A$ is of equal magnitude and opposite direction to $\mathbf{F}_B$. Furthermore, *rotational or moment equilibrium* ($\Sigma\mathbf{M}_O = \mathbf{0}$) is satisfied if $\mathbf{F}_A$ is *collinear* with $\mathbf{F}_B$. As a result, the line of action of both forces is known since it always passes through A and B. Hence, only the force magnitude must be determined or stated. Other examples of two-force members held in equilibrium are shown in Fig. 5–22.

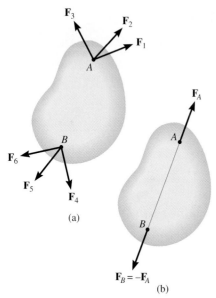

(a)

(b)

Two-force member

Fig. 5–21

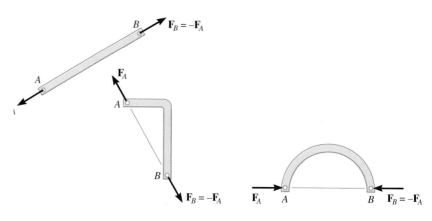

Two-force members

Fig. 5–22

Three-Force Members. If a member is subjected to only three forces, then it is necessary that the forces be either *concurrent* or *parallel* for the member to be in equilibrium. To show the concurrency requirement, consider the body in Fig. 5–23a and suppose that any two of the three forces acting on the body have lines of action that intersect at point O. To satisfy moment equilibrium about O, i.e., $\Sigma M_O = 0$, the third force must also pass through O, which then makes the force system *concurrent*. If two of the three forces are parallel, Fig. 5–23b, the point of concurrency, O, is considered to be at "infinity," and the third force must be parallel to the other two forces to intersect at this "point."

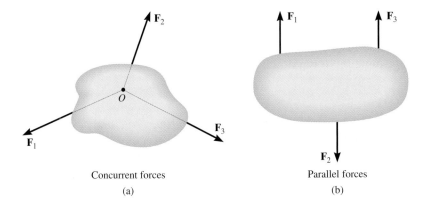

Concurrent forces
(a)

Parallel forces
(b)

Three-force members

Fig. 5–23

Many mechanical elements act as two- or three-force members, and the ability to recognize them in a problem will considerably simplify an equilibrium analysis.

- The bucket link AB on the back-hoe is a typical example of a two-force member since it is pin connected at its ends and, provided its weight is neglected, no other force acts on this member.

- The hydraulic cylinder BC is pin connected at its ends. It is a two-force member. The boom ABD is subjected to the weight of the suspended motor at D, the force of the hydraulic cylinder at B, and the force of the pin at A. If the boom's weight is neglected, it is a three-force member.

- The dump bed of the truck operates by extending the telescopic hydraulic cylinder AB. If the weight of AB is neglected, we can classify it as a two-force member since it is pin connected at its end points.

EXAMPLE 5.13

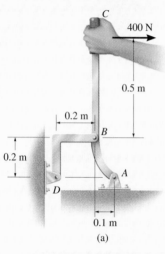

(a)

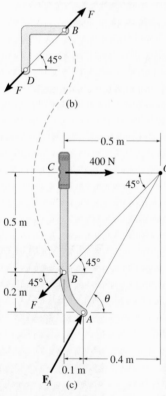

(b)

(c)

Fig. 5–24

The lever *ABC* is pin-supported at *A* and connected to a short link *BD* as shown in Fig. 5–24a. If the weight of the members is negligible, determine the force of the pin on the lever at *A*.

Solution

Free-Body Diagrams. As shown by the free-body diagram, Fig. 5–24b, the short link *BD* is a *two-force member*, so the *resultant forces* at pins *D* and *B* must be equal, opposite, and collinear. Although the magnitude of the force is unknown, the line of action is known since it passes through *B* and *D*.

Lever *ABC* is a *three-force member*, and therefore, in order to satisfy moment equilibrium, the three nonparallel forces acting on it must be concurrent at *O*, Fig. 5–24c. In particular, note that the force *F* on the lever at *B* is equal but opposite to the force *F* acting at *B* on the link. Why? The distance *CO* must be 0.5 m since the lines of action of **F** and the 400-N force are known.

Equations of Equilibrium. By requiring the force system to be concurrent at *O*, since $\Sigma M_O = 0$, the angle θ which defines the line of action of $\mathbf{F}_A$ can be determined from trigonometry,

$$\theta = \tan^{-1}\left(\frac{0.7}{0.4}\right) = 60.3° \quad \measuredangle\theta \qquad \text{Ans.}$$

Using the *x, y* axes and applying the force equilibrium equations, we can obtain F_A and *F*.

$$\overset{+}{\rightarrow} \Sigma F_x = 0; \qquad F_A \cos 60.3° - F \cos 45° + 400\text{ N} = 0$$
$$+\uparrow \Sigma F_y = 0; \qquad F_A \sin 60.3° - F \sin 45° = 0$$

Solving, we get

$$F_A = 1.07 \text{ kN} \qquad \text{Ans.}$$
$$F = 1.32 \text{ kN}$$

Note: We can also solve this problem by representing the force at *A* by its two components $\mathbf{A}_x$ and $\mathbf{A}_y$ and applying $\Sigma M_A = 0, \Sigma F_x = 0, \Sigma F_y = 0$ to the lever. Once A_x and A_y are determined, how would you find F_A and θ?

Equilibrium in Three Dimensions

5.5 Free-Body Diagrams

The first step in solving three-dimensional equilibrium problems, as in the case of two dimensions, is to draw a free-body diagram of the body (or group of bodies considered as a system). Before we show this, however, it is necessary to discuss the types of reactions that can occur at the supports.

Support Reactions. The reactive forces and couple moments acting at various types of supports and connections, when the members are viewed in three dimensions, are listed in Table 5–2. It is important to recognize the symbols used to represent each of these supports and to understand clearly how the forces and couple moments are developed by each support. As in the two-dimensional case, *a force is developed by a support that restricts the translation of the attached member, whereas a couple moment is developed when rotation of the attached member is prevented.* For example, in Table 5–2, the ball-and-socket joint (4) prevents any translation of the connecting member; therefore, a force must act on the member at the point of connection. This force has three components having unknown magnitudes, F_x, F_y, F_z. Provided these components are known, one can obtain the magnitude of force. $F = \sqrt{F_x^2 + F_y^2 + F_z^2}$, and the force's orientation defined by the coordinate direction angles α, β, γ, Eqs. 2–7.* Since the connecting member is allowed to rotate freely about *any* axis, no couple moment is resisted by a ball-and-socket joint.

It should be noted that the *single* bearing supports (5) and (7), the *single* pin (8), and the *single* hinge (9) are shown to support both force and couple-moment components. If, however, these supports are used in conjunction with *other* bearings, pins, or hinges to hold a rigid body in equilibrium and the supports are *properly aligned* when connected to the body, then the *force reactions* at these supports *alone* may be adequate for supporting the body. In other words, the couple moments become redundant and are not shown on the free-body diagram. The reason for this should become clear after studying the examples which follow.

---

*The three unknowns may also be represented as an unknown force magnitude F and two unknown coordinate direction angles. The third direction angle is obtained using the identity $\cos^2 \alpha + \cos^2 \beta + \cos^2 \gamma = 1$, Eq. 2–10.

TABLE 5–2 • Supports for Rigid Bodies Subjected to Three-Dimensional Force Systems

Types of Connection	Reaction	Number of Unknowns
(1) cable	**F**	One unknown. The reaction is a force which acts away from the member in the known direction of the cable.
(2) smooth surface support	**F**	One unknown. The reaction is a force which acts perpendicular to the surface at the point of contact.
(3) roller	**F**	One unknown. The reaction is a force which acts perpendicular to the surface at the point of contact.
(4) ball and socket	F_z F_x F_y	Three unknowns. The reactions are three rectangular force components.
(5) single journal bearing	M_z F_z M_x F_x	Four unknowns. The reactions are two force and two couple-moment components which act perpendicular to the shaft.

continued

TABLE 5-2 • Continued

Types of Connection	Reaction	Number of Unknown
(6) single journal bearing with square shaft		Five unknowns. The reactions are two force and three couple-moment components.
(7) single thrust bearing		Five unknowns. The reactions are three force and two couple-moment components.
(8) single smooth pin		Five unknowns. The reactions are three force and two couple-moment components.
(9) single hinge		Five unknowns. The reactions are three force and two couple-moment components.
(10) fixed support		Six unknowns. The reactions are three force and three couple-moment components.

Typical examples of actual supports that are referenced to Table 5–2 are shown in the following sequence of photos.

This ball-and-socket joint provides a connection for the housing of an earth grader to its frame. (4)

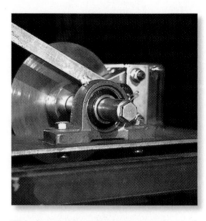

This journal bearing supports the end of the shaft. (5)

This thrust bearing is used to support the drive shaft on a machine. (7)

This pin is used to support the end of the strut used on a tractor. (8)

Free-Body Diagrams. The general procedure for establishing the free-body diagram of a rigid body has been outlined in Sec. 5.2. Essentially it requires first "isolating" the body by drawing its outlined shape. This is followed by a careful *labeling* of *all* the forces and couple moments in reference to an established *x, y, z* coordinate system. As a general rule, *components of reaction* having an *unknown magnitude* are shown acting on the free-body diagram in the *positive sense*. In this way, if any negative values are obtained, they will indicate that the components act in the negative coordinate directions.

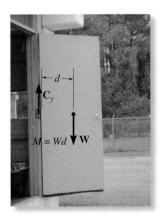 

It is a mistake to support a door using a single hinge since the hinge must develop a force $\mathbf{C}_y$ to support the weight $\mathbf{W}$ of the door and a couple moment $\mathbf{M}$ to support the moment of $\mathbf{W}$, i.e., $M = Wd$. If instead two properly aligned hinges are used, then the weight is carried by both hinges, $A_y + B_y = W$, and the moment of the door is resisted by the two hinge forces $\mathbf{F}_x$ and $-\mathbf{F}_x$. These forces form a couple, such that $F_x d' = Wd$. In other words, no couple moments are produced by the hinges on the door provided they are in *proper alignment*. Instead, the forces $\mathbf{F}_x$ and $-\mathbf{F}_x$ resist the rotation caused by $\mathbf{W}$.

E X A M P L E 5.14

Several examples of objects along with their associated free-body diagrams are shown in Fig. 5–25. In all cases, the *x, y, z* axes are established and the unknown reaction components are indicated in the positive sense. The weight of the objects is neglected.

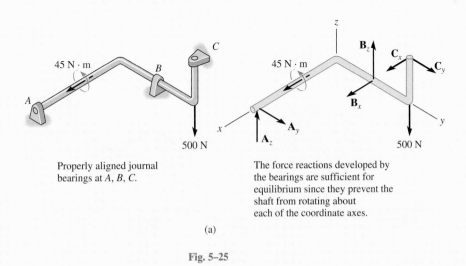

Properly aligned journal bearings at A, B, C.

The force reactions developed by the bearings are sufficient for equilibrium since they prevent the shaft from rotating about each of the coordinate axes.

(a)

Fig. 5–25

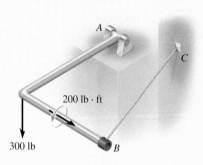

200 lb · ft

300 lb

B

A

C

Pin at *A* and cable *BC*.

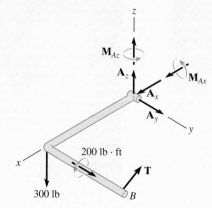

$\mathbf{M}_{Az}$

$\mathbf{A}_z$

$\mathbf{M}_{Ax}$

$\mathbf{A}_x$

$\mathbf{A}_y$

z

y

x

200 lb · ft

300 lb

T

B

(b)

Moment components are developed
by the pin on the rod to prevent
rotation about the *x* and *z* axes.

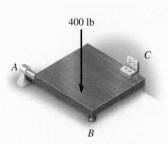

400 lb

A

C

B

Properly aligned journal bearing
at *A* and hinge at *C*. Roller at *B*.

(c)

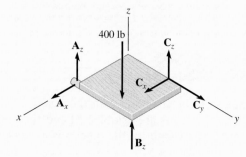

z

400 lb

$\mathbf{A}_z$

$\mathbf{C}_z$

$\mathbf{C}_x$

$\mathbf{A}_x$

$\mathbf{C}_y$

$\mathbf{B}_z$

x

y

Only force reactions are developed by
the bearing and hinge on the plate to
prevent rotation about each coordinate axis.
No moments at the hinge are developed.

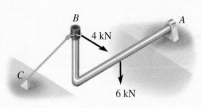

B

4 kN

A

C

6 kN

Thrust bearing at *A* and
cable *BC*

(d)

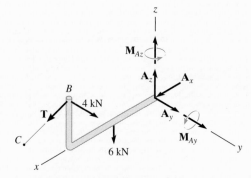

z

$\mathbf{M}_{Az}$

$\mathbf{A}_z$

$\mathbf{A}_x$

B

4 kN

T

$\mathbf{A}_y$

$\mathbf{M}_{Ay}$

C

6 kN

x

y

Moment components are developed
by the bearing on the rod in order to
prevent rotation about the *y* and *z* axes.

Fig. 5–25

5.6 Equations of Equilibrium

As stated in Sec. 5.1, the conditions for equilibrium of a rigid body subjected to a three-dimensional force system require that both the *resultant* force and *resultant* couple moment acting on the body be equal to *zero*.

Vector Equations of Equilibrium. The two conditions for equilibrium of a rigid body may be expressed mathematically in vector form as

$$\Sigma \mathbf{F} = \mathbf{0}$$
$$\Sigma \mathbf{M}_O = \mathbf{0} \tag{5-5}$$

where $\Sigma \mathbf{F}$ is the vector sum of all the external forces acting on the body and $\Sigma \mathbf{M}_O$ is the sum of the couple moments and the moments of all the forces about any point O located either on or off the body.

Scalar Equations of Equilibrium. If all the applied external forces and couple moments are expressed in Cartesian vector form and substituted into Eqs. 5–5, we have

$$\Sigma \mathbf{F} = \Sigma F_x \mathbf{i} + \Sigma F_y \mathbf{j} + \Sigma F_z \mathbf{k} = \mathbf{0}$$
$$\Sigma \mathbf{M}_O = \Sigma M_x \mathbf{i} + \Sigma M_y \mathbf{j} + \Sigma M_z \mathbf{k} = \mathbf{0}$$

Since the $\mathbf{i}, \mathbf{j}$, and $\mathbf{k}$ components are independent from one another, the above equations are satisfied provided

$$\Sigma F_x = 0$$
$$\Sigma F_y = 0 \tag{5-6a}$$
$$\Sigma F_z = 0$$

and

$$\Sigma M_x = 0$$
$$\Sigma M_y = 0 \tag{5-6b}$$
$$\Sigma M_z = 0$$

These *six scalar equilibrium equations* may be used to solve for at most six unknowns shown on the free-body diagram. Equations 5–6a express the fact that the sum of the external force components acting in the x, y, and z directions must be zero, and Eqs. 5–6b require the sum of the moment components about the x, y, and z axes to be zero.

5.7 Constraints for a Rigid Body

To ensure the equilibrium of a rigid body, it is not only necessary to satisfy the equations of equilibrium, but the body must also be properly held or constrained by its supports. Some bodies may have more supports than are necessary for equilibrium, whereas others may not have enough or the supports may be arranged in a particular manner that could cause the body to collapse. Each of these cases will now be discussed.

Redundant Constraints. When a body has redundant supports, that is, more supports than are necessary to hold it in equilibrium, it becomes statically indeterminate. *Statically indeterminate* means that there will be more unknown loadings on the body than equations of equilibrium available for their solution. For example, the two-dimensional problem, Fig. 5–26a, and the three-dimensional problem, Fig. 5–26b, shown together with their free-body diagrams, are both statically indeterminate because of additional support reactions. In the two-dimensional case, there are five unknowns, that is, M_A, A_x, A_y, B_y, and C_y, for which only three equilibrium equations can be written ($\Sigma F_x = 0$, $\Sigma F_y = 0$, and $\Sigma M_O = 0$, Eqs. 5–2). The three-dimensional problem has eight unknowns, for which only six equilibrium equations can be written, Eqs. 5–6. The additional equations needed to solve indeterminate problems of the type shown in Fig. 5–26 are generally obtained from the deformation conditions at the points of support. These equations involve the physical properties of the body which are studied in subjects dealing with the mechanics of deformation, such as "mechanics of materials."*

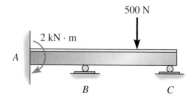

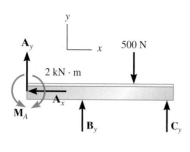

(a)

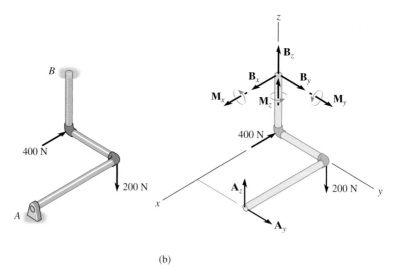

(b)

Fig. 5–26

*See R. C. Hibbeler, *Mechanics of Materials*, 5th edition (Pearson Education/Prentice Hall, Inc., 2003).

Improper Constraints.   In some cases, there may be as many unknown forces on the body as there are equations of equilibrium; however, *instability* of the body can develop because of *improper constraining* by the supports. In the case of three-dimensional problems, the body is improperly constrained if the support reactions *all intersect a common axis*. For two-dimensional problems, this axis is *perpendicular* to the plane of the forces and therefore appears as a point. Hence, when all the reactive forces are *concurrent* at this point, the body is improperly constrained. Examples of both cases are given in Fig. 5–27. From the free-body diagrams it is seen that the summation of moments about the *x* axis, Fig. 5–27a, or point *O*, Fig. 5–27b, will *not* be equal to zero; thus rotation about the *x* axis or point *O* will take place.* Furthermore, in both cases, it becomes *impossible* to solve *completely* for all the unknowns since one can write a moment equation that *does not* involve any of the unknown support reactions, and as a result, this reduces the number of available equilibrium equations by one.

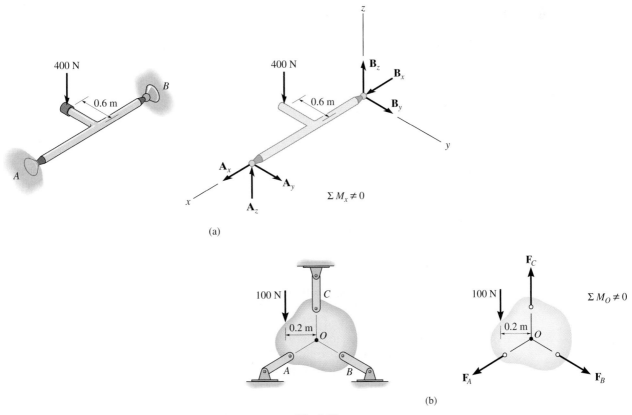

(a)

(b)

Fig. 5–27

*For the three-dimensional problem, $\Sigma M_x = (400 \text{ N})(0.6 \text{ m}) \neq 0$, and for the two-dimensional problem, $\Sigma M_O = (100 \text{ N})(0.2 \text{ m}) \neq 0$.

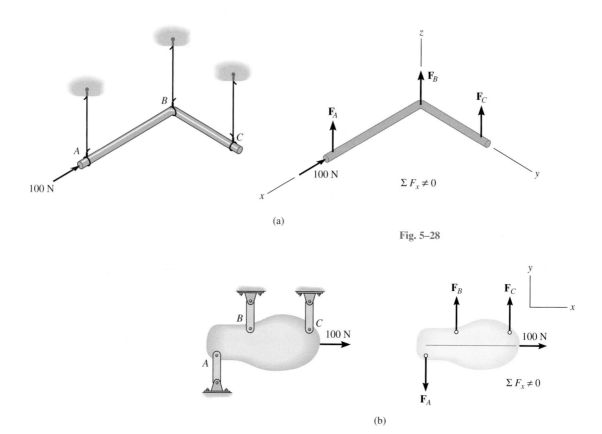

(a)

Fig. 5–28

$\Sigma F_x \neq 0$

(b)

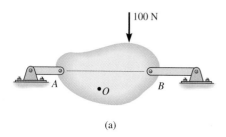

(a)

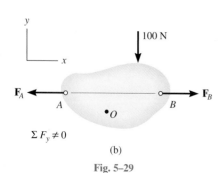

$\Sigma F_y \neq 0$

(b)

Fig. 5–29

Another way in which improper constraining leads to instability occurs when the *reactive forces* are all *parallel*. Three- and two-dimensional examples of this are shown in Fig. 5–28. In both cases, the summation of forces along the x axis will not equal zero.

In some cases, a body may have *fewer* reactive forces than equations of equilibrium that must be satisfied. The body then becomes only *partially constrained*. For example, consider the body shown in Fig. 5–29*a* with its corresponding free-body diagram in Fig. 5–29*b*. If O is a point not located on the line AB, the equation $\Sigma F_x = 0$ gives $F_A = F_B$ and $\Sigma M_O = 0$ and $\Sigma F_y = 0$, however, will not be satisfied for the loading conditions and therefore equilibrium will not be maintained.

Proper constraining therefore requires that (1) the lines of action of the reactive forces do not intersect points on a common axis, and (2) the reactive forces must not all be parallel to one another. When the minimum number of reactive forces is needed to properly constrain the body in question, the problem will be statically determinate, and therefore the equations of equilibrium can be used to determine *all* the reactive forces.

IMPORTANT POINTS

- Always draw the free-body diagram first.
- If a support *prevents translation* of a body in a specific direction, then the support exerts a *force* on the body in that direction.
- If *rotation about an axis is prevented*, then the support exerts a *couple moment* on the body about the axis.
- If a body is subjected to more unknown reactions than available equations of equilibrium, then the problem is *statically indeterminate*.
- To avoid instability of a body require that the lines of action of the reactive forces do not intersect a common axis and are not parallel to one another.

PROCEDURE FOR ANALYSIS

Three-dimensional equilibrium problems for a rigid body can be solved using the following procedure.

Free-Body Diagram

- Draw an outlined shape of the body.
- Show all the forces and couple moments acting on the body.
- Establish the origin of the x, y, z axes at a convenient point and orient the axes so that they are parallel to as many of the external forces and moments as possible.
- Label all the loadings and specify their directions relative to the x, y, z axes. In general, show all the unknown components having a positive sense along the x, y, z axes if the sense cannot be determined.
- Indicate the dimensions of the body necessary for computing the moments of forces.

Equations of Equilibrium

- If the x, y, z force and moment components seem easy to determine, then apply the six scalar equations of equilibrium; otherwise use the vector equations.
- It is not necessary that the set of axes chosen for force summation coincide with the set of axes chosen for moment summation. Also, any set of nonorthogonal axes may be chosen for this purpose.
- Choose the direction of an axis for moment summation such that it intersects the lines of action of as many unknown forces as possible. In this way, the moments of forces passing through points on this axis and forces which are parallel to the axis will then be zero.
- If the solution of the equilibrium equations yields a negative scalar for a force or couple moment magnitude, it indicates that the sense is opposite to that which was assumed on the free-body diagram.

E X A M P L E   5.15

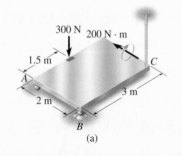

(a)

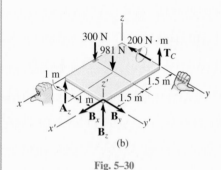

(b)

Fig. 5–30

The homogeneous plate shown in Fig. 5–30a has a mass of 100 kg and is subjected to a force and couple moment along its edges. If it is supported in the horizontal plane by means of a roller at A, a ball-and-socket joint at B, and a cord at C, determine the components of reaction at the supports.

Solution (Scalar Analysis)

Free-Body Diagram.   There are five unknown reactions acting on the plate, as shown in Fig. 5–30b. Each of these reactions is assumed to act in a positive coordinate direction.

Equations of Equilibrium.   Since the three-dimensional geometry is rather simple, a *scalar analysis* provides a *direct solution* to this problem. A force summation along each axis yields

$$\Sigma F_x = 0; \qquad B_x = 0 \qquad\qquad\qquad\qquad Ans.$$

$$\Sigma F_y = 0; \qquad B_y = 0 \qquad\qquad\qquad\qquad Ans.$$

$$\Sigma F_z = 0; \qquad A_z + B_z + T_C - 300\text{ N} - 981\text{ N} = 0 \qquad (1)$$

Recall that the moment of a force about an axis is equal to the product of the force magnitude and the perpendicular distance (moment arm) from the line of action of the force to the axis. The sense of the moment is determined by the right-hand rule. Also, forces that are parallel to an axis or pass through it create no moment about the axis. Hence, summing moments of the forces on the free-body diagram, with positive moments acting along the positive x or y axis, we have

$$\Sigma M_x = 0; \qquad T_C(2\text{ m}) - 981\text{ N}(1\text{ m}) + B_z(2\text{ m}) = 0 \qquad (2)$$

$$\Sigma M_y = 0;$$

$$300\text{ N}(1.5\text{ m}) + 981\text{ N}(1.5\text{ m}) - B_z(3\text{ m}) - A_z(3\text{ m}) - 200\text{ N} \cdot \text{m} = 0 \quad (3)$$

The components of force at B can be eliminated if the x', y', z' axes are used. We obtain

$$\Sigma M_{x'} = 0; \qquad 981\text{ N}(1\text{ m}) + 300\text{ N}(2\text{ m}) - A_z(2\text{ m}) = 0 \qquad (4)$$

$$\Sigma M_{y'} = 0;$$

$$-300\text{ N}(1.5\text{ m}) - 981\text{ N}(1.5\text{ m}) - 200\text{ N} \cdot \text{m} + T_C(3\text{ m}) = 0 \qquad (5)$$

Solving Eqs. 1 through 3 or the more convenient Eqs. 1, 4, and 5 yields

$$A_z = 790\text{ N} \qquad B_z = -217\text{ N} \qquad T_C = 707\text{ N} \qquad Ans.$$

The negative sign indicates that $\mathbf{B}_z$ acts downward.

Note that the solution of this problem does not require the use of a summation of moments about the z axis. The plate is partially constrained since the supports cannot prevent it from turning about the z axis if a force is applied to it in the x–y plane.

E X A M P L E 5.16

The windlass shown in Fig. 5–31a is supported by a thrust bearing at A and a smooth journal bearing at B, which are properly aligned on the shaft. Determine the magnitude of the vertical force **P** that must be applied to the handle to maintain equilibrium of the 100-kg bucket. Also calculate the reactions at the bearings.

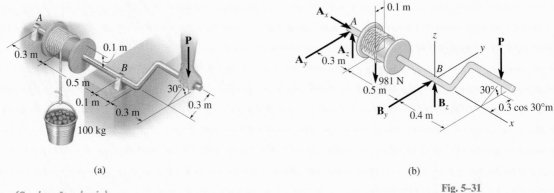

(a)   (b)

Fig. 5–31

Solution *(Scalar Analysis)*

Free-Body Diagram. Since the bearings at A and B are aligned correctly, *only* force reactions occur at these supports, Fig. 5–31b. Why are there no moment reactions?

Equations of Equilibrium. Summing moments about the x axis yields a direct solution for **P**. Why? For a scalar moment summation, it is necessary to determine the moment of each force as the product of the force magnitude and the *perpendicular distance* from the x axis to the line of action of the force. Using the right-hand rule and assuming positive moments act in the +**i** direction, we have

$$\Sigma M_x = 0; \quad 981 \text{ N}(0.1 \text{ m}) - P(0.3 \cos 30°\text{m}) = 0$$
$$P = 377.6 \text{ N} \qquad \qquad Ans.$$

Using this result and summing moments about the y and z axes yields
$$\Sigma M_y = 0;$$
$$-981 \text{ N}(0.5 \text{ m}) + A_z(0.8 \text{ m}) + (377.6 \text{ N})(0.4 \text{ m}) = 0$$
$$A_z = 424.3 \text{ N} \qquad \qquad Ans.$$
$$\Sigma M_z = 0; \quad -A_y(0.8 \text{ m}) = 0 \quad A_y = 0$$

The reactions at B are determined by a force summation using these results.

$$\Sigma F_x = 0; \quad A_x = 0$$
$$\Sigma F_y = 0; \quad 0 + B_y = 0 \quad B_y = 0$$
$$\Sigma F_z = 0; \quad 424.3 - 981 + B_z - 377.6 = 0 \quad B_z = 934 \text{ N} \quad Ans.$$

EXAMPLE 5.17

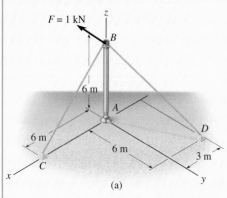

F = 1 kN

6 m

6 m

6 m

3 m

(a)

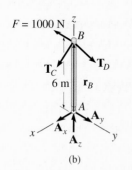

F = 1000 N

6 m $\mathbf{r}_B$

(b)

$\mathbf{F}_B = -\mathbf{F}_A$

(c)

Fig. 5–32

Determine the tension in cables BC and BD and the reactions at the ball-and-socket joint A for the mast shown in Fig. 5–32a.

Solution (Vector Analysis)

Free-Body Diagram. There are five unknown force magnitudes shown on the free-body diagram, Fig. 5–32b.

Equations of Equilibrium. Expressing each force in Cartesian vector form, we have

$$\mathbf{F} = \{-1000\mathbf{j}\}\text{ N}$$
$$\mathbf{F}_A = A_x\mathbf{i} + A_y\mathbf{j} + A_z\mathbf{k}$$
$$\mathbf{T}_C = 0.707T_C\mathbf{i} - 0.707T_C\mathbf{k}$$
$$\mathbf{T}_D = T_D\left(\frac{\mathbf{r}_{BD}}{r_{BD}}\right) = -\frac{3}{9}T_D\mathbf{i} + \frac{6}{9}T_D\mathbf{j} - \frac{6}{9}T_D\mathbf{k}$$

Applying the force equation of equilibrium gives

$$\Sigma\mathbf{F} = \mathbf{0}; \quad \mathbf{F} + \mathbf{F}_A + \mathbf{T}_C + \mathbf{T}_D = \mathbf{0}$$
$$\left(A_x + 0.707T_C - \frac{3}{9}T_D\right)\mathbf{i} + \left(-1000 + A_y + \frac{6}{9}T_D\right)\mathbf{j}$$
$$+ \left(A_z - 0.707T_C - \frac{6}{9}T_D\right)\mathbf{k} = \mathbf{0}$$

$$\Sigma F_x = 0; \quad A_x + 0.707T_C - \frac{3}{9}T_D = 0 \quad (1)$$

$$\Sigma F_y = 0; \quad A_y + \frac{6}{9}T_D - 1000 = 0 \quad (2)$$

$$\Sigma F_z = 0; \quad A_z - 0.707T_C - \frac{6}{9}T_D = 0 \quad (3)$$

Summing moments about point A, we have

$$\Sigma\mathbf{M}_A = \mathbf{0}; \quad \mathbf{r}_B \times (\mathbf{F} + \mathbf{T}_C + \mathbf{T}_D) = \mathbf{0}$$
$$6\mathbf{k} \times (-1000\mathbf{j} + 0.707T_C\mathbf{i} - 0.707T_C\mathbf{k}$$
$$-\frac{3}{9}T_D\mathbf{i} + \frac{6}{9}T_D\mathbf{j} - \frac{6}{9}T_D\mathbf{k}) = \mathbf{0}$$

Evaluating the cross product and combining terms yields

$$(-4T_D + 6000)\mathbf{i} + (4.24T_C - 2T_D)\mathbf{j} = \mathbf{0}$$

$$\Sigma M_x = 0; \quad -4T_D + 6000 = 0 \quad (4)$$

$$\Sigma M_y = 0; \quad 4.24T_C - 2T_D = 0 \quad (5)$$

The moment equation about the z axis, $\Sigma M_z = 0$, is automatically satisfied. Why? Solving Eqs. 1 through 5 we have

$$T_C = 707\text{ N} \qquad T_D = 1500\text{ N} \qquad \qquad Ans.$$
$$A_x = 0\text{ N} \qquad A_y = 0\text{ N} \qquad A_z = 1500\text{ N} \qquad Ans.$$

Since the mast is a two-force member, Fig. 5–32c, note that the value $A_x = A_y = 0$ could have been determined *by inspection*.

EXAMPLE 5.18

Rod AB shown in Fig. 5–33a is subjected to the 200-N force. Determine the reactions at the ball-and-socket joint A and the tension in cables BD and BE.

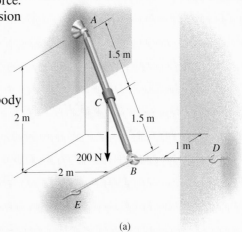

(a)

Solution (Vector Analysis)

Free-Body Diagram. Fig. 5–33b.

Equations of Equilibrium. Representing each force on the free-body diagram in Cartesian vector form, we have

$$\mathbf{F}_A = A_x\mathbf{i} + A_y\mathbf{j} + A_z\mathbf{k}$$
$$\mathbf{T}_E = T_E\mathbf{i}$$
$$\mathbf{T}_D = T_D\mathbf{j}$$
$$\mathbf{F} = \{-200\mathbf{k}\}\ \text{N}$$

Applying the force equation of equilibrium.

$$\Sigma\mathbf{F} = \mathbf{0}; \qquad \mathbf{F}_A + \mathbf{T}_E + \mathbf{T}_D + \mathbf{F} = \mathbf{0}$$
$$(A_x + T_E)\mathbf{i} + (A_y + T_D)\mathbf{j} + (A_z - 200)\mathbf{k} = \mathbf{0}$$

$$\Sigma F_x = 0; \qquad A_x + T_E = 0 \qquad\qquad (1)$$
$$\Sigma F_y = 0; \qquad A_y + T_D = 0 \qquad\qquad (2)$$
$$\Sigma F_z = 0; \qquad A_z - 200 = 0 \qquad\qquad (3)$$

Summing moments about point A yields

$$\Sigma\mathbf{M}_A = \mathbf{0}; \qquad \mathbf{r}_C \times \mathbf{F} + \mathbf{r}_B \times (\mathbf{T}_E + \mathbf{T}_D) = \mathbf{0}$$

Since $\mathbf{r}_C = \frac{1}{2}\mathbf{r}_B$, then

$$(0.5\mathbf{i} + 1\mathbf{j} - 1\mathbf{k}) \times (-200\mathbf{k}) + (1\mathbf{i} + 2\mathbf{j} - 2\mathbf{k}) \times (T_E\mathbf{i} + T_D\mathbf{j}) = \mathbf{0}$$

Expanding and rearranging terms gives

$$(2T_D - 200)\mathbf{i} + (-2T_E + 100)\mathbf{j} + (T_D - 2T_E)\mathbf{k} = \mathbf{0}$$

$$\Sigma M_x = 0; \qquad 2T_D - 200 = 0 \qquad\qquad (4)$$
$$\Sigma M_y = 0; \qquad -2T_E + 100 = 0 \qquad\qquad (5)$$
$$\Sigma M_z = 0; \qquad T_D - 2T_E = 0 \qquad\qquad (6)$$

Solving Eqs. 1 through 6, we get

$$
\begin{aligned}
T_D &= 100\ \text{N} & \textit{Ans.} \\
T_E &= 50\ \text{N} & \textit{Ans.} \\
A_x &= -50\ \text{N} & \textit{Ans.} \\
A_y &= -100\ \text{N} & \textit{Ans.} \\
A_z &= 200\ \text{N} & \textit{Ans.}
\end{aligned}
$$

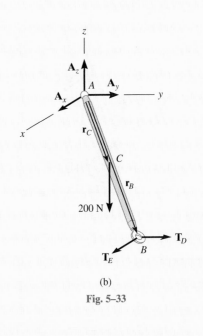

(b)

Fig. 5–33

The negative sign indicates that $\mathbf{A}_x$ and $\mathbf{A}_y$ have a sense which is opposite to that shown on the free-body diagram, Fig. 5–33b.

E X A M P L E 5.19

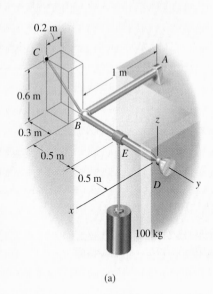

0.2 m

C

1 m

A

0.6 m

0.3 m

B

z

0.5 m

E

0.5 m

D

y

x

100 kg

(a)

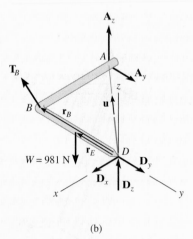

A_z

A

A_y

T_B

z

B

$\mathbf{r}_B$

u

$\mathbf{r}_E$

D

$W = 981$ N

D_y

D_x D_z

x

y

(b)

Fig. 5–34

The bent rod in Fig. 5–34a is supported at *A* by a journal bearing, at *D* by a ball-and-socket joint, and at *B* by means of cable *BC*. Using only *one equilibrium equation*, obtain a direct solution for the tension in cable *BC*. The bearing at *A* is capable of exerting force components only in the *z* and *y* directions since it is properly aligned on the shaft.

Solution (*Vector Analysis*)

Free-Body Diagram. As shown in Fig. 5–34b, there are six unknowns: three force components caused by the ball-and-socket joint, two caused by the bearing, and one caused by the cable.

Equations of Equilibrium. The cable tension $\mathbf{T}_B$ may be obtained *directly* by summing moments about an axis passing through points *D* and *A*. Why? The direction of the axis is defined by the unit vector **u**, where

$$\mathbf{u} = \frac{\mathbf{r}_{DA}}{r_{DA}} = -\frac{1}{\sqrt{2}}\mathbf{i} - \frac{1}{\sqrt{2}}\mathbf{j}$$
$$= -0.707\mathbf{i} - 0.707\mathbf{j}$$

Hence, the sum of the moments about this axis is zero provided

$$\Sigma M_{DA} = \mathbf{u} \cdot \Sigma (\mathbf{r} \times \mathbf{F}) = 0$$

Here **r** represents a position vector drawn from *any point* on the axis *DA* to any point on the line of action of force **F** (see Eq. 4–11). With reference to Fig. 5–34b, we can therefore write

$$\mathbf{u} \cdot (\mathbf{r}_B \times \mathbf{T}_B + \mathbf{r}_E \times \mathbf{W}) = 0$$
$$(-0.707\mathbf{i} - 0.707\mathbf{j}) \cdot [(-1\mathbf{j}) \times (\tfrac{0.2}{0.7}T_B\mathbf{i} - \tfrac{0.3}{0.7}T_B\mathbf{j} + \tfrac{0.6}{0.7}T_B\mathbf{k})$$
$$+ (-0.5\mathbf{j}) \times (-981\mathbf{k})] = \mathbf{0}$$
$$(-0.707\mathbf{i} - 0.707\mathbf{j}) \cdot [(-0.857T_B + 490.5)\mathbf{i} + 0.286T_B\mathbf{k}] = \mathbf{0}$$
$$- 0.707(-0.857T_B + 490.5) + 0 + 0 = 0$$

$$T_B = \frac{490.5}{0.857} = 572 \text{ N} \qquad \qquad Ans.$$

The advantage of using Cartesian vectors for this solution should be noted. It would be especially tedious to determine the perpendicular distance from the *DA* axis to the line of action of $\mathbf{T}_B$ using scalar methods.

Note: In a similar manner, we can obtain $D_z(=490.5 \text{ N})$ by summing moments about an axis passing through *AB*. Also, $A_z(=0)$ is obtained by summing moments about the *y* axis.

CHAPTER REVIEW

- **Free-Body Diagram.** Before analyzing any equilibrium problem it is first necessary to draw a free-body diagram. This is an outlined shape of the body, which shows all the forces and couple moments that act on the body. Remember that a support will exert a *force* on the body in a particular direction if it prevents *translation* of the body in that direction, and it will exert a *couple moment* on the body if it prevents *rotation*. Angles used to resolve forces, and dimensions used to take moments of the forces, should also be shown on the free-body diagram.

- **Two Dimensions.** Normally the three scalar equations of equilibrium, $\Sigma F_x = 0$, $\Sigma F_y = 0$, $\Sigma M_o = 0$, can be applied when solving problems in two dimensions, since the geometry is easy to visualize. For the most direct solution, try to sum forces along an axis that will eliminate as many unknown forces as possible. Sum moments about a point O that passes through the line of action of as many unknown forces as possible.

- **Three Dimensions.** In three dimensions, it is often advantageous to use a Cartesian vector analysis when applying the equations of equilibrium. To do this, first express each known and unknown force and couple moment shown on the free-body diagram as a Cartesian vector. Then set the force summation equal to zero, $\Sigma \mathbf{F} = \mathbf{0}$. Take moments about a point O that lies on the line of action of as many unknown force components as possible. From point O direct position vectors to each force, and then use the cross product to determine the moment of each force. Require $\Sigma \mathbf{M}_o = \Sigma \mathbf{r} \times \mathbf{F} = \mathbf{0}$. The six scalar equations of equilibrium are established by setting the respective $\mathbf{i}$, $\mathbf{j}$, and $\mathbf{k}$ components of these force and moment sums equal to zero.

REVIEW PROBLEMS

5-1. The shaft assembly is supported by two smooth journal bearings A and B and a short link DC. If a couple moment is applied to the shaft as shown, determine the components of force reaction at the bearings and the force in the link. The link lies in a plane parallel to the y–z plane and the bearings are properly aligned on the shaft.

*5-2. Determine the horizontal and vertical components of reaction at the pin A and the reaction at the roller B required to support the truss. Set $F = 600$ N.

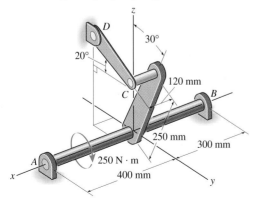

Prob. 5–1

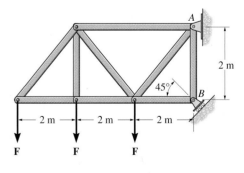

Prob. 5–2

5-3. If the roller at B can sustain a maximum load of 3 kN, determine the largest magnitude of each of the three forces F that can be supported by the truss.

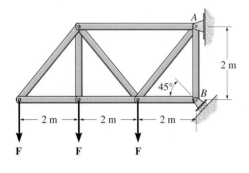

Prob. 5–3

5-5. The symmetrical shelf is subjected to a uniform load of 4 kPa. Support is provided by a bolt (or pin) located at each end A and A' and by the symmetrical brace arms, which bear against the smooth wall on both sides at B and B'. Determine the force resisted by each bolt at the wall and the normal force at B for equilibrium.

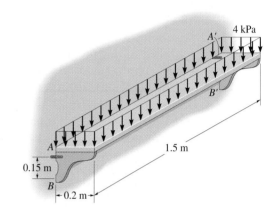

Prob. 5–5

5-4. Determine the normal reaction at the roller A and horizontal and vertical components at pin B for equilibrium of the member.

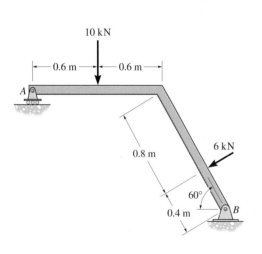

Prob. 5–4

*5-6. Determine the x and z components of reaction at the journal bearing A and the tension in cords BC and BD necessary for equilibrium of the rod.

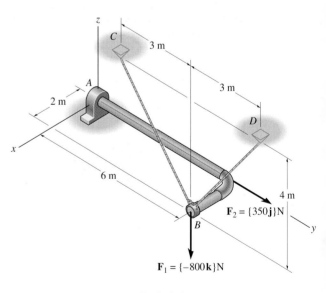

Prob. 5–6

5-7. Determine the reactions at the supports A and B for equilibrium of the beam.

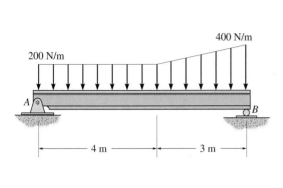

Prob. 5–7

5-9. Determine the x, y, z components of reaction at the fixed wall A. The 150-N force is parallel to the z axis and the 200-N force is parallel to the y axis.

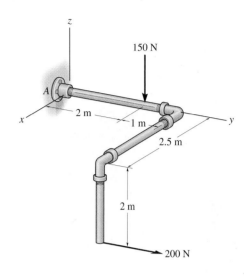

Prob. 5–9

5-8. Determine the x, y, z components of reaction at the ball supports B and C and the ball-and-socket A (not shown) for the uniformly loaded plate.

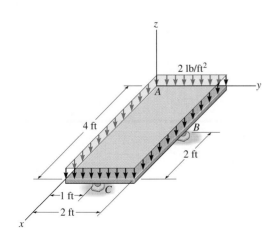

Prob. 5–8

*5-10. The horizontal beam is supported by springs at its ends. If the stiffness of the spring at A is $k_A = 5$ kN/m, determine the required stiffness of the spring at B so that if the beam is loaded with the 800-N force, it remains in the horizontal position both before and after loading.

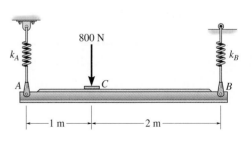

Prob. 5–10

The forces within the members of this truss bridge must be determined if
they are to be properly designed.

Structural Analysis

CHAPTER OBJECTIVES

- To show how to determine the forces in the members of a truss using the method of joints and the method of sections.
- To analyze the forces acting on the members of frames and machines composed of pin-connected members.

6.1 Simple Trusses

A *truss* is a structure composed of slender members joined together at their end points. The members commonly used in construction consist of wooden struts or metal bars. The joint connections are usually formed by bolting or welding the ends of the members to a common plate, called a *gusset plate,* as shown in Fig. 6–1a, or by simply passing a large bolt or pin through each of the members, Fig. 6–1b.

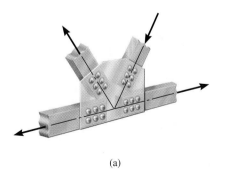

(a)

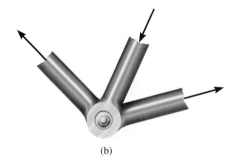

(b)

Fig. 6–1

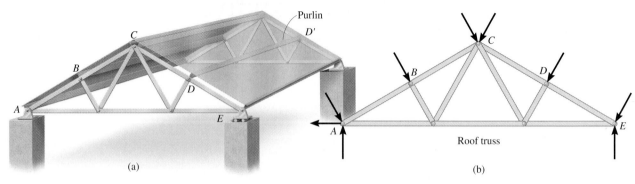

Fig. 6–2

Planar Trusses. *Planar* trusses lie in a single plane and are often used to support roofs and bridges. The truss $ABCDE$, shown in Fig. 6–2a, is an example of a typical roof-supporting truss. In this figure, the roof load is transmitted to the truss *at the joints* by means of a series of *purlins,* such as DD'. Since the imposed loading acts in the same plane as the truss, Fig. 6–2b, the analysis of the forces developed in the truss members is two-dimensional.

In the case of a bridge, such as shown in Fig. 6–3a, the load on the deck is first transmitted to *stringers,* then to *floor beams,* and finally to the *joints B, C,* and *D* of the two supporting side trusses. Like the roof truss, the bridge truss loading is also coplanar, Fig. 6–3b.

When bridge or roof trusses extend over large distances, a rocker or roller is commonly used for supporting one end, e.g., joint *E* in Figs. 6–2a and 6–3a. This type of support allows freedom for expansion or contraction of the members due to temperature or application of loads.

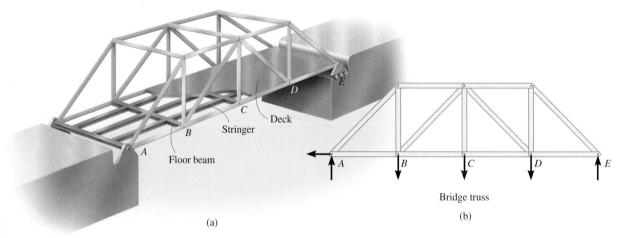

Fig. 6–3

Assumptions for Design. To design both the members and the connections of a truss, it is first necessary to determine the *force* developed in each member when the truss is subjected to a given loading. In this regard, two important assumptions will be made:

1. *All loadings are applied at the joints.* In most situations, such as for bridge and roof trusses, this assumption is true. Frequently in the force analysis the weight of the members is neglected since the forces supported by the members are usually large in comparison with their weight. If the member's weight is to be included in the analysis, it is generally satisfactory to apply it as a vertical force, half of its magnitude applied at each end of the member.

2. *The members are joined together by smooth pins.* In cases where bolted or welded joint connections are used, this assumption is satisfactory provided the center lines of the joining members are *concurrent,* as in Fig. 6–1a.

Because of these two assumptions, *each truss member acts as a two-force member,* and therefore the forces at the ends of the member must be directed along the axis of the member. If the force tends to *elongate* the member, it is a *tensile force* (T), Fig. 6–4a; whereas if it tends to *shorten* the member, it is a *compressive force* (C), Fig. 6–4b. In the actual design of a truss it is important to state whether the nature of the force is tensile or compressive. Often, compression members must be made *thicker* than tension members because of the buckling or column effect that occurs when a member is in compression.

Simple Truss. To prevent collapse, the form of a truss must be rigid. Obviously, the four-bar shape *ABCD* in Fig. 6–5 will collapse unless a diagonal member, such as *AC*, is added for support. The simplest form that is rigid or stable is a *triangle*. Consequently, a *simple truss* is constructed by *starting* with a basic triangular element, such as *ABC* in Fig. 6–6, and connecting two members (*AD* and *BD*) to form an additional element. As each additional element consisting of two members and a joint is placed on the truss, it is possible to construct a simple truss.

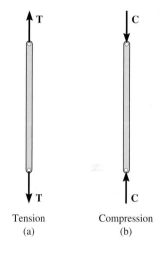

Tension
(a)

Compression
(b)

Fig. 6–4

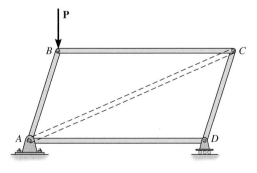

Fig. 6–5

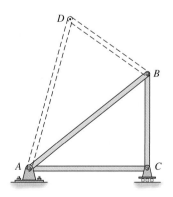

Fig. 6–6

6.2 The Method of Joints

These Howe trusses are used to support the roof of the metal building. Note how the members come together at a common point on the gusset plate and how the roof purlins transmit the load to the joints.

In order to analyze or design a truss, we must obtain the force in each of its members. If we were to consider a free-body diagram of the entire truss, then the forces in the members would be *internal forces,* and they could not be obtained from an equilibrium analysis. Instead, if we consider the equilibrium of a joint of the truss then a member force becomes an *external force* on the joint's free-body diagram, and the equations of equilibrium can be applied to obtain its magnitude. This forms the basis for the *method of joints.*

Because the truss members are all straight two-force members lying in the same plane, the force system acting at each joint is *coplanar and concurrent.* Consequently, rotational or moment equilibrium is automatically satisfied at the joint (or pin), and it is only necessary to satisfy $\Sigma F_x = 0$ and $\Sigma F_y = 0$ to ensure equilibrium.

When using the method of joints, it is *first* necessary to draw the joint's free-body diagram before applying the equilibrium equations. To do this, recall that the *line of action* of each member force acting on the joint is *specified* from the geometry of the truss since the force in a member passes along the axis of the member. As an example, consider the pin at joint *B* of the truss in Fig. 6–7*a.* Three forces act on the pin, namely, the 500-N force and the forces exerted by members *BA* and *BC.* The free-body diagram is shown in Fig. 6–7*b.* As shown, $\mathbf{F}_{BA}$ is "pulling" on the pin, which means that member *BA* is in *tension;* whereas $\mathbf{F}_{BC}$ is "pushing" on the pin, and consequently member *BC* is in *compression.* These effects are clearly demonstrated by isolating the joint with small segments of the member connected to the pin, Fig. 6–7*c.* The pushing or pulling on these small segments indicates the effect of the member being either in compression or tension.

In all cases, the analysis should start at a joint having at least one known force and at most two unknown forces, as in Fig. 6–7*b.* In this way, application of $\Sigma F_x = 0$ and $\Sigma F_y = 0$ yields two algebraic equations which can be solved for the two unknowns. When applying these equations, the correct sense of an unknown member force can be determined using one of two possible methods:

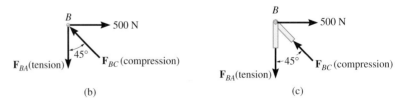

(a)

(b)

(c)

Fig. 6–7

- *Always assume* the *unknown member forces* acting on the joint's free-body diagram to be in *tension,* i.e., "pulling" on the pin. If this is done, then numerical solution of the equilibrium equations will yield *positive scalars for members in tension and negative scalars for members in compression.* Once an unknown member force is found, use its *correct* magnitude and sense (T or C) on subsequent joint free-body diagrams.

- The *correct* sense of direction of an unknown member force can, in many cases, be determined "by inspection." For example, $\mathbf{F}_{BC}$ in Fig. 6–7b must push on the pin (compression) since its horizontal component, $F_{BC} \sin 45°$, must balance the 500-N force ($\Sigma F_x = 0$). Likewise, $\mathbf{F}_{BA}$ is a tensile force since it balances the vertical component, $F_{BC} \cos 45°$ ($\Sigma F_y = 0$). In more complicated cases, the sense of an unknown member force can be *assumed;* then, after applying the equilibrium equations, the assumed sense can be verified from the numerical results. A *positive* answer indicates that the sense is *correct,* whereas a *negative* answer indicates that the sense shown on the free-body diagram must be *reversed.* This is the method we will use in the example problems which follow.

PROCEDURE FOR ANALYSIS

The following procedure provides a typical means for analyzing a truss using the method of joints.

- Draw the free-body diagram of a joint having at least one known force and at most two unknown forces. (If this joint is at one of the supports, then it may be necessary to know the external reactions at the truss support.)

- Use one of the two methods described above for establishing the sense of an unknown force.

- Orient the x and y axes such that the forces on the free-body diagram can be easily resolved into their x and y components and then apply the two force equilibrium equations $\Sigma F_x = 0$ and $\Sigma F_y = 0$. Solve for the two unknown member forces and verify their correct sense.

- Continue to analyze each of the other joints, where again it is necessary to choose a joint having at most two unknowns and at least one known force.

- Once the force in a member is found from the analysis of a joint at one of its ends, the result can be used to analyze the forces acting on the joint at its other end. Remember that a member in *compression* "pushes" on the joint and a member in *tension* "pulls" on the joint.

E X A M P L E 6.1

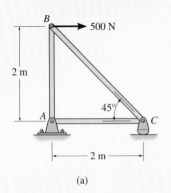

(a)

(b)

(c)

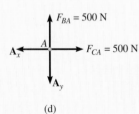

(d)

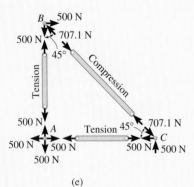

(e)

Fig. 6–8

Determine the force in each member of the truss shown in Fig. 6–8a and indicate whether the members are in tension or compression.

Solution

By inspection of Fig. 6–8a, there are two unknown member forces at joint B, two unknown member forces and an unknown reaction force at joint C, and two unknown member forces and two unknown reaction forces at joint A. Since we should have no more than two unknowns at the joint and at least one known force acting there, we will begin the analysis at joint B.

Joint B. The free-body diagram of the pin at B is shown in Fig. 6–8b. Applying the equations of joint equilibrium, we have

$$\xrightarrow{+} \Sigma F_x = 0; \quad 500\ \text{N} - F_{BC} \sin 45° = 0 \qquad F_{BC} = 707.1\ \text{N (C)} \quad Ans.$$
$$+\uparrow \Sigma F_y = 0; \quad F_{BC} \cos 45° - F_{BA} = 0 \qquad F_{BA} = 500\ \text{N} \quad \text{(T)} \quad Ans.$$

Since the force in member BC has been calculated, we can proceed to analyze joint C in order to determine the force in member CA and the support reaction at the rocker.

Joint C. From the free-body diagram of joint C, Fig. 6–8c, we have

$$\xrightarrow{+} \Sigma F_x = 0; \quad -F_{CA} + 707.1 \cos 45°\text{N} = 0 \quad F_{CA} = 500\ \text{N} \quad \text{(T)} \quad Ans.$$
$$+\uparrow \Sigma F_y = 0; \quad C_y - 707.1 \sin 45°\text{N} = 0 \quad C_y = 500\ \text{N} \qquad Ans.$$

Joint A. Although it is not necessary, we can determine the support reactions at joint A using the results of $F_{CA} = 500\ \text{N}$ and $F_{BA} = 500\ \text{N}$. From the free-body diagram, Fig. 6–8d, we have

$$\xrightarrow{+} \Sigma F_x = 0; \quad 500\ \text{N} - A_x = 0 \quad A_x = 500\ \text{N}$$
$$+\uparrow \Sigma F_y = 0; \quad 500\ \text{N} - A_y = 0 \quad A_y = 500\ \text{N}$$

The results of the analysis are summarized in Fig. 6–8e. Note that the free-body diagram of each pin shows the effects of all the connected members and external forces applied to the pin, whereas the free-body diagram of each member shows only the effects of the end pins on the member.

EXAMPLE 6.2

Determine the forces acting in all the members of the truss shown in Fig. 6–9a.

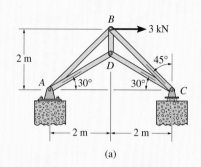

(a)

Solution

By inspection, there are more than two unknowns at each joint. Consequently, the support reactions on the truss must first be determined. Show that they have been correctly calculated on the free-body diagram in Fig. 6–9b. We can now begin the analysis at joint C. Why?

Joint C. From the free-body diagram, Fig. 6–9c,

$$\xrightarrow{+} \Sigma F_x = 0; \qquad -F_{CD}\cos 30° + F_{CB}\sin 45° = 0$$
$$+\uparrow \Sigma F_y = 0; \quad 1.5 \text{ kN} + F_{CD}\sin 30° - F_{CB}\cos 45° = 0$$

These two equations must be solved *simultaneously* for each of the two unknowns. Note, however, that a *direct solution* for one of the unknown forces may be obtained by applying a force summation along an axis that is *perpendicular* to the direction of the other unknown force. For example, summing forces along the y' axis, which is perpendicular to the direction of $\mathbf{F}_{CD}$, Fig. 6–9d, yields a direct solution for F_{CB}.

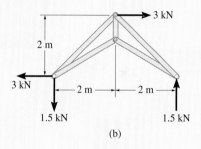

(b)

$$+\nearrow\Sigma F_{y'} = 0;$$

$$1.5\cos 30°\text{kN} - F_{CB}\sin 15° = 0 \quad F_{CB} = 5.02 \text{ kN} \quad \text{(C)} \qquad Ans.$$

In a similar fashion, summing forces along the y'' axis, Fig. 6–9e, yields a direct solution for F_{CD}.

$$+\nearrow\Sigma F_{y''} = 0;$$

$$1.5\cos 45°\text{kN} - F_{CD}\sin 15° = 0 \qquad F_{CD} = 4.10 \text{ kN} \quad \text{(T)} \quad Ans.$$

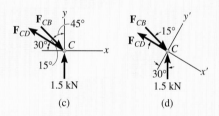

(c)   (d)

Joint D. We can now proceed to analyze joint D. The free-body diagram is shown in Fig. 6–9f.

$$\xrightarrow{+} \Sigma F_x = 0; \quad -F_{DA}\cos 30° + 4.10\cos 30°\text{kN} = 0$$
$$F_{DA} = 4.10 \text{ kN} \quad \text{(T)} \qquad\qquad Ans.$$
$$+\uparrow \Sigma F_y = 0; \qquad F_{DB} - 2(4.10\sin 30°\text{kN}) = 0$$
$$F_{DB} = 4.10 \text{ kN} \quad \text{(T)} \qquad\qquad Ans.$$

The force in the last member, *BA*, can be obtained from joint B or joint A. As an exercise, draw the free-body diagram of joint B, sum the forces in the horizontal direction, and show that $F_{BA} = 0.776$ kN (C).

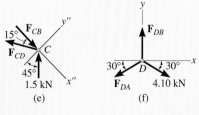

(e)   (f)

Fig. 6–9

E X A M P L E 6.3

Determine the force in each member of the truss shown in Fig. 6–10a. Indicate whether the members are in tension or compression.

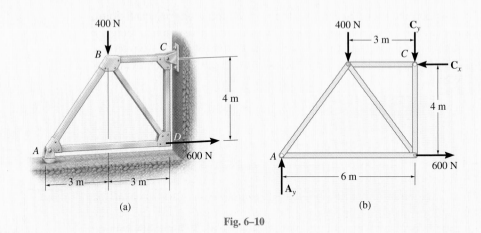

(a)

Fig. 6–10

(b)

Solution

Support Reactions. No joint can be analyzed until the support reactions are determined. Why? A free-body diagram of the entire truss is given in Fig. 6–10b. Applying the equations of equilibrium, we have

$$\xrightarrow{+} \Sigma F_x = 0; \qquad 600 \text{ N} - C_x = 0 \qquad C_x = 600 \text{ N}$$

$$\zeta + \Sigma M_C = 0; \quad -A_y(6 \text{ m}) + 400 \text{ N}(3 \text{ m}) + 600 \text{ N}(4 \text{ m}) = 0$$

$$A_y = 600 \text{ N}$$

$$+\uparrow \Sigma F_y = 0; \qquad 600 \text{ N} - 400 \text{ N} - C_y = 0 \qquad C_y = 200 \text{ N}$$

The analysis can now start at either joint A or C. The choice is arbitrary since there are one known and two unknown member forces acting on the pin at each of these joints.

Joint A (Fig. 6–10c). As shown on the free-body diagram, there are three forces that act on the pin at joint A. The inclination of $\mathbf{F}_{AB}$ is determined from the geometry of the truss. By inspection, can you see why this force is assumed to be compressive and $\mathbf{F}_{AD}$ tensile? Applying the equations of equilibrium, we have

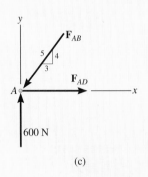

(c)

$$+\uparrow \Sigma F_y = 0; \qquad 600 \text{ N} - \tfrac{4}{5}F_{AB} = 0 \qquad F_{AB} = 750 \text{ N} \quad (\text{C}) \qquad Ans.$$

$$\xrightarrow{+} \Sigma F_x = 0; \quad F_{AD} - \tfrac{3}{5}(750 \text{ N}) = 0 \qquad F_{AD} = 450 \text{ N} \quad (\text{T}) \qquad Ans.$$

Joint D (Fig. 6–10d). The pin at this joint is chosen next since, by inspection of Fig. 6–10a, the force in AD is known and the unknown forces in DB and DC can be determined. Summing forces in the horizontal direction, Fig. 6–10d, we have

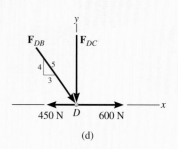

(d)

$$\xrightarrow{+} \Sigma F_x = 0; \quad -450 \text{ N} + \tfrac{3}{5} F_{DB} + 600 \text{ N} = 0 \quad F_{DB} = -250 \text{ N}$$

The negative sign indicates that $\mathbf{F}_{DB}$ acts in the *opposite sense* to that shown in Fig. 6–10d. Hence,

$$F_{DB} = 250 \text{ N} \quad (\text{T}) \qquad\qquad Ans.$$

To determine $\mathbf{F}_{DC}$, we can either correct the sense of $\mathbf{F}_{DB}$ and then apply $\Sigma F_y = 0$, or apply this equation and retain the negative sign for F_{DB}, i.e.,

$$+\uparrow \Sigma F_y = 0; \quad -F_{DC} - \tfrac{4}{5}(-250 \text{ N}) = 0 \qquad F_{DC} = 200 \text{ N} \quad (\text{C}) \quad Ans.$$

Joint C (Fig. 6–10e).
$$\xrightarrow{+} \Sigma F_x = 0; \quad F_{CB} - 600 \text{ N} = 0 \qquad F_{CB} = 600 \text{ N} \quad (\text{C}) \qquad\qquad Ans.$$

$$+\uparrow \Sigma F_y = 0; \quad 200 \text{ N} - 200 \text{ N} \equiv 0 \quad (\text{check})$$

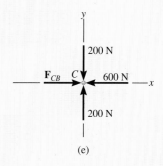

(e)

The analysis is summarized in Fig. 6–10f, which shows the correct free-body diagram for each pin and member.

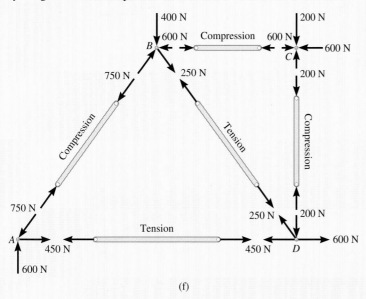

(f)

*The proper sense could have been determined by inspection, prior to applying $\Sigma F_x = 0$.

6.3 Zero-Force Members

Truss analysis using the method of joints is greatly simplified if one is first able to determine those members which support *no loading*. These *zero-force members* are used to increase the stability of the truss during construction and to provide support if the applied loading is changed.

The zero-force members of a truss can generally be determined *by inspection* of each of its joints. For example, consider the truss shown in Fig. 6–11a. If a free-body diagram of the pin at joint A is drawn, Fig. 6–11b, it is seen that members AB and AF are zero-force members. On the other hand, notice that we could not have come to this conclusion if we had considered the free-body diagrams of joints F or B simply because there are five unknowns at each of these joints. In a similar manner, consider the free-body diagram of joint D, Fig. 6–11c. Here again it is seen that DC and DE are zero-force members. As a general rule, *if only two members form a truss joint and no external load or support reaction is applied to the joint, the members must be zero-force members.* The load on the truss in Fig. 6–11a is therefore supported by only five members as shown in Fig. 6–11d.

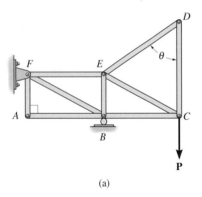

(a)

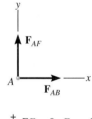

$$\xrightarrow{+} \Sigma F_x = 0; \quad F_{AB} = 0$$
$$+\uparrow \Sigma F_y = 0; \quad F_{AF} = 0$$

(b)

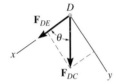

$$+\searrow \Sigma F_y = 0; F_{DC} \sin \theta = 0; \quad F_{DC} = 0 \text{ since } \sin \theta \neq 0$$
$$+\swarrow \Sigma F_x = 0; F_{DE} + 0 = 0; \quad F_{DE} = 0$$

(c)

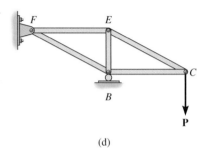

(d)

Fig. 6–11

Now consider the truss shown in Fig. 6–12a. The free-body diagram of the pin at joint D is shown in Fig. 6–12b. By orienting the y axis along members DC and DE and the x axis along member DA, it is seen that DA is a zero-force member. Note that this is also the case for member CA, Fig. 6–12c. In general, *if three members form a truss joint for which two of the members are collinear, the third member is a zero-force member provided no external force or support reaction is applied to the joint.* The truss shown in Fig. 6–12d is therefore suitable for supporting the load **P**.

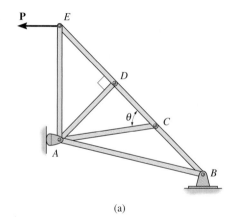

(a)

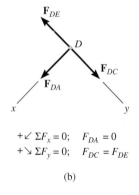

$$+\swarrow \Sigma F_x = 0; \quad F_{DA} = 0$$
$$+\searrow \Sigma F_y = 0; \quad F_{DC} = F_{DE}$$

(b)

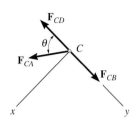

$$+\swarrow \Sigma F_x = 0; \quad F_{CA}\sin\theta = 0; \quad F_{CA} = 0 \text{ since } \sin\theta \neq 0;$$
$$+\searrow \Sigma F_y = 0; \quad F_{CB} = F_{CD}$$

(c)

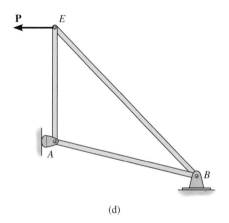

(d)

Fig. 6–12

E X A M P L E **6.4**

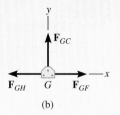

(b)

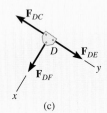

(c)

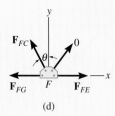

(d)

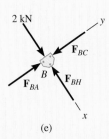

(e)

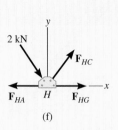

(f)

Using the method of joints, determine all the zero-force members of the *Fink roof truss* shown in Fig. 6–13a. Assume all joints are pin connected.

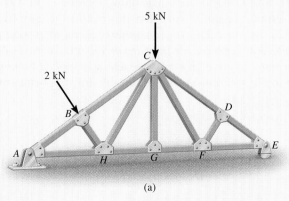

(a)

Fig. 6–13

Solution

Look for joint geometries that have three members for which two are collinear. We have

Joint G (Fig. 6–13b).

$$+\uparrow \Sigma F_y = 0; \qquad F_{GC} = 0 \qquad\qquad\qquad Ans.$$

Realize that we could not conclude that *GC* is a zero-force member by considering joint *C*, where there are five unknowns. The fact that *GC* is a zero-force member means that the 5-kN load at *C* must be supported by members *CB*, *CH*, *CF*, and *CD*.

Joint D (Fig. 6–13c).

$$+\swarrow \Sigma F_x = 0; \qquad F_{DF} = 0 \qquad\qquad\qquad Ans.$$

Joint F (Fig. 6–13d).

$$+\uparrow \Sigma F_y = 0; \quad F_{FC} \cos\theta = 0 \qquad \text{Since } \theta \neq 90°, \qquad F_{FC} = 0 \quad Ans.$$

Note that if joint *B* is analyzed, Fig. 6–13e,

$$+\searrow \Sigma F_x = 0; \quad 2 \text{ kN} - F_{BH} = 0 \quad F_{BH} = 2 \text{ kN} \qquad \text{(C)}$$

Note that F_{HC} must satisfy $\Sigma F_y = 0$, Fig. 6–13f, and therefore *HC* is *not* a zero-force member.

6.4 The Method of Sections

The *method of sections* is used to determine the loadings acting within a body. It is based on the principle that if a body is in equilibrium then any part of the body is also in equilibrium. For example, consider the two truss members shown on the left in Fig. 6–14. If the forces within the members are to be determined, then an imaginary section indicated by the blue line, can be used to cut each member into two parts and thereby "expose" each internal force as "external" to the free-body diagrams shown on the right. Clearly, it can be seen that equilibrium requires that the member in tension (T) be subjected to a "pull," whereas the member in compression (C) is subjected to a "push."

The method of sections can also be used to "cut" or section the members of an entire truss. If the section passes through the truss and the free-body diagram of either of its two parts is drawn, we can then apply the equations of equilibrium to that part to determine the member forces at the "cut section." Since only *three* independent equilibrium equations ($\Sigma F_x = 0$, $\Sigma F_y = 0$, $\Sigma M_O = 0$) can be applied to the isolated part of the truss, try to select a section that, in general, passes through not more than *three* members in which the forces are unknown. For example, consider the truss in Fig. 6–15a. If the force in member *GC* is to be determined, section *aa* would be appropriate. The free-body diagrams of the two parts are shown in Figs. 6–15b and 6–15c. In particular, note that the line of action of each member force is specified from the *geometry* of the truss, since the force in a member passes along its axis. Also, the member forces acting on one part of the truss are equal but opposite to those acting on the other part—Newton's third law. As noted above, members assumed to be in *tension* (*BC* and *GC*) are subjected to a "pull," whereas the member in *compression* (*GF*) is subjected to a "push."

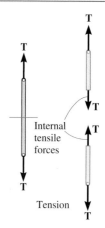

Tension

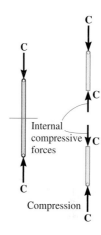

Compression

Fig. 6–14

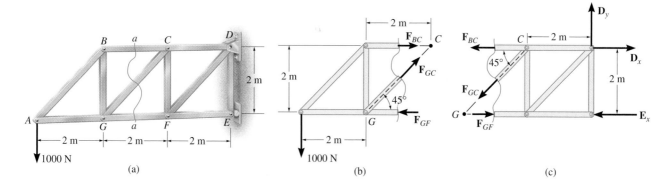

(a) (b) (c)

Fig. 6–15

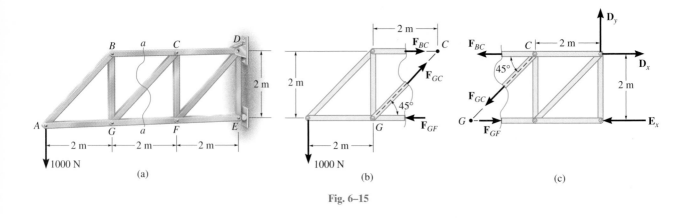

(a)

(b)

(c)

Fig. 6–15

Two Pratt trusses are used to construct this pedestrian bridge.

The three unknown member forces $\mathbf{F}_{BC}$, $\mathbf{F}_{GC}$, and $\mathbf{F}_{GF}$ can be obtained by applying the three equilibrium equations to the free-body diagram in Fig. 6–15b. If, however, the free-body diagram in Fig. 6–15c is considered, the three support reactions $\mathbf{D}_x$, $\mathbf{D}_y$ and $\mathbf{E}_x$ will have to be determined *first*. Why? (This, of course, is done in the usual manner by considering a free-body diagram of the *entire truss.*)

When applying the equilibrium equations, one should consider ways of writing the equations so as to yield a *direct solution* for each of the unknowns, rather than having to solve simultaneous equations. For example, summing moments about C in Fig. 6–15b would yield a direct solution for $\mathbf{F}_{GF}$ since $\mathbf{F}_{BC}$ and $\mathbf{F}_{GC}$ create zero moment about C. Likewise, $\mathbf{F}_{BC}$ can be directly obtained by summing moments about G. Finally, $\mathbf{F}_{GC}$ can be found directly from a force summation in the vertical direction since $\mathbf{F}_{GF}$ and $\mathbf{F}_{BC}$ have no vertical components. This ability to *determine directly* the force in a particular truss member is one of the main advantages of using the method of sections.*

---

*By comparison, if the method of joints were used to determine, say, the force in member GC, it would be necessary to analyze joints A, B, and G in sequence.

As in the method of joints, there are two ways in which one can determine the correct sense of an unknown member force:

- *Always assume* that the unknown member forces at the cut section are in *tension,* i.e., "pulling" on the member. By doing this, the numerical solution of the equilibrium equations will yield *positive scalars for members in tension and negative scalars for members in compression.*

- The correct sense of an unknown member force can in many cases be determined "by inspection." For example, $\mathbf{F}_{BC}$ is a tensile force as represented in Fig. 6–15*b* since moment equilibrium about G requires that $\mathbf{F}_{BC}$ create a moment opposite to that of the 1000-N force. Also, $\mathbf{F}_{GC}$ is tensile since its vertical component must balance the 1000-N force which acts downward. In more complicated cases, the sense of an unknown member force may be *assumed.* If the solution yields a *negative* scalar, it indicates that the force's sense is *opposite* to that shown on the free-body diagram. This is the method we will use in the example problems which follow.

PROCEDURE FOR ANALYSIS

The forces in the members of a truss may be determined by the method of sections using the following procedure.

Free-Body Diagram

- Make a decision as to how to "cut" or section the truss through the members where forces are to be determined.

- Before isolating the appropriate section, it may first be necessary to determine the truss's *external* reactions. Then three equilibrium equations are available to solve for member forces at the cut section.

- Draw the free-body diagram of that part of the sectioned truss which has the least number of forces acting on it.

- Use one of the two methods described above for establishing the sense of an unknown member force.

Equations of Equilibrium

- Moments should be summed about a point that lies at the intersection of the lines of action of two unknown forces, so that the third unknown force is determined directly from the moment equation.

- If two of the unknown forces are *parallel,* forces may be summed *perpendicular* to the direction of these unknowns to determine *directly* the third unknown force.

E X A M P L E 6.5

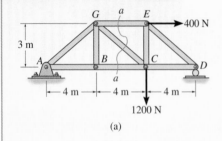

(a)

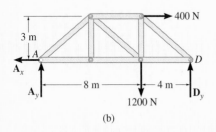

(b)

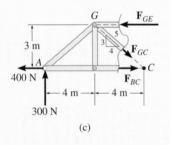

(c)

Fig. 6–16

Determine the force in members *GE*, *GC*, and *BC* of the truss shown in Fig. 6–16a. Indicate whether the members are in tension or compression.

Solution

Section *aa* in Fig. 6–16a has been chosen since it cuts through the *three* members whose forces are to be determined. In order to use the method of sections, however, it is *first* necessary to determine the external reactions at *A* or *D*. Why? A free-body diagram of the entire truss is shown in Fig. 6–16b. Applying the equations of equilibrium, we have

$$\xrightarrow{+} \Sigma F_x = 0; \quad 400 \text{ N} - A_x = 0 \quad A_x = 400 \text{ N}$$

$$\zeta + \Sigma M_A = 0; \quad -1200 \text{ N}(8 \text{ m}) - 400 \text{ N}(3 \text{ m}) + D_y(12 \text{ m}) = 0$$

$$D_y = 900 \text{ N}$$

$$+\uparrow \Sigma F_y = 0; \quad A_y - 1200 \text{ N} + 900 \text{ N} = 0 \quad A_y = 300 \text{ N}$$

Free-Body Diagram. The free-body diagram of the left portion of the sectioned truss is shown in Fig. 6–16c. For the analysis this diagram will be used since it involves the least number of forces.

Equations of Equilibrium. Summing moments about point *G* eliminates $\mathbf{F}_{GE}$ and $\mathbf{F}_{GC}$ and yields a direct solution for F_{BC}.

$$\zeta + \Sigma M_G = 0; \quad -300 \text{ N}(4 \text{ m}) - 400 \text{ N}(3 \text{ m}) + F_{BC}(3 \text{ m}) = 0$$

$$F_{BC} = 800 \text{ N} \quad (\text{T}) \qquad\qquad Ans.$$

In the same manner, by summing moments about point *C* we obtain a direct solution for F_{GE}.

$$\zeta + \Sigma M_C = 0; \quad -300 \text{ N}(8 \text{ m}) + F_{GE}(3 \text{ m}) = 0$$

$$F_{GE} = 800 \text{ N} \quad (\text{C}) \qquad\qquad Ans.$$

Since $\mathbf{F}_{BC}$ and $\mathbf{F}_{GE}$ have no vertical components, summing forces in the *y* direction directly yields F_{GC}, i.e.,

$$+\uparrow \Sigma F_y = 0 \quad 300\text{N} - \tfrac{3}{5}F_{GC} = 0$$

$$F_{GC} = 500 \text{ N} \quad (\text{T}) \qquad\qquad Ans.$$

As an exercise, obtain these results by applying the equations of equilibrium to the free-body diagram of the right portion of the sectioned truss.

EXAMPLE 6.6

Determine the force in member *CF* of the bridge truss shown in Fig. 6–17a. Indicate whether the member is in tension or compression. Assume each member is pin-connected.

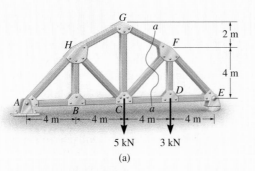

(a)

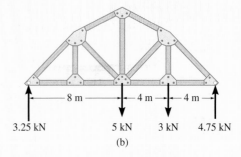

(b)

Fig. 6–17

Solution

Free-Body Diagram. Section *aa* in Fig. 6–17a will be used since this section will "expose" the internal force in member *CF* as "external" on the free-body diagram of either the right or left portion of the truss. It is first necessary, however, to determine the external reactions on either the left or right side. Verify the results shown on the free-body diagram in Fig. 6–17b.

The free-body diagram of the right portion of the truss, which is the easiest to analyze, is shown in Fig. 6–17c. There are three unknowns, F_{FG}, F_{CF}, and F_{CD}.

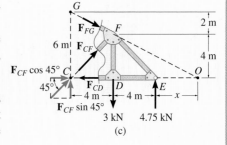

Equations of Equilibrium. The most direct method for solving this problem requires application of the moment equation about a point that eliminates two of the unknown forces. Hence, to obtain $\mathbf{F}_{CF}$, we will eliminate $\mathbf{F}_{FG}$ and $\mathbf{F}_{CD}$ by summing moments about point *O*, Fig. 6–17c. Note that the location of point *O* measured from *E* is determined from proportional triangles, i.e., $4/(4 + x) = 6/(8 + x)$, $x = 4$ m. Or, stated in another manner, the slope of member *GF* has a drop of 2 m to a horizontal distance of 4 m. Since *FD* is 4 m, Fig. 6–17c, then from *D* to *O* the distance must be 8 m.

An easy way to determine the moment of $\mathbf{F}_{CF}$ about point *O* is to use the principle of transmissibility and move $\mathbf{F}_{CF}$ to point *C*, and then resolve $\mathbf{F}_{CF}$ into its two rectangular components. We have

$$\zeta + \Sigma M_O = 0;$$
$$-F_{CF} \sin 45°(12 \text{ m}) + (3 \text{ kN})(8 \text{ m}) - (4.75 \text{ kN})(4 \text{ m}) = 0$$
$$F_{CF} = 0.589 \text{ kN} \quad \text{(C)} \qquad\qquad Ans.$$

EXAMPLE 6.7

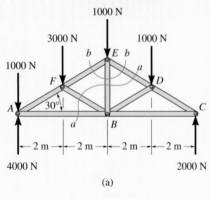

(a)

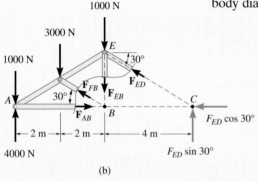

(b)

Fig. 6–18

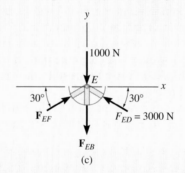

(c)

Determine the force in member EB of the roof truss shown in Fig. 6–18a. Indicate whether the member is in tension or compression.

Solution

Free-Body Diagrams. By the method of sections, any imaginary vertical section that cuts through EB, Fig. 6–18a, will also have to cut through three other members for which the forces are unknown. For example, section aa cuts through ED, EB, FB, and AB. If the components of reaction at A are calculated first ($A_x = 0$, $A_y = 4000$ N) and a free-body diagram of the left side of this section is considered, Fig. 6–18b, it is possible to obtain $\mathbf{F}_{ED}$ by summing moments about B to eliminate the other three unknowns; however, $\mathbf{F}_{EB}$ cannot be determined from the remaining two equilibrium equations. One possible way of obtaining $\mathbf{F}_{EB}$ is first to determine $\mathbf{F}_{ED}$ from section aa, then use this result on section bb, Fig. 6–18a, which is shown in Fig. 6–18c. Here the force system is concurrent and our sectioned free-body diagram is the same as the free-body diagram for the pin at E (method of joints).

Equations of Equilibrium. In order to determine the moment of $\mathbf{F}_{ED}$ about point B, Fig. 6–18b, we will resolve the force into its rectangular components and, by the principle of transmissibility, extend it to point C as shown. The moments of 1000 N, F_{AB}, F_{FB}, F_{EB}, and $F_{ED} \cos 30°$ are all zero about B. Therefore,

$$\zeta + \Sigma M_B = 0; \quad 1000 \text{ N}(4 \text{ m}) + 3000 \text{ N}(2 \text{ m}) - 4000 \text{ N}(4 \text{ m})$$
$$+ F_{ED} \sin 30°(4) = 0$$
$$F_{ED} = 3000 \text{ N} \quad (C)$$

Considering now the free-body diagram of section bb, Fig. 6–18c, we have

$$\overset{+}{\rightarrow} \Sigma F_x = 0; \quad F_{EF} \cos 30° - 3000 \cos 30° \text{ N} = 0$$
$$F_{EF} = 3000 \text{ N} \quad (C)$$
$$+\uparrow \Sigma F_y = 0; \, 2(3000 \sin 30° \text{ N}) - 1000 \text{ N} - F_{EB} = 0$$
$$F_{EB} = 2000 \text{ N} \quad (T) \qquad \qquad Ans.$$

*6.5   Space Trusses

A *space truss* consists of members joined together at their ends to form a stable three-dimensional structure. The simplest element of a space truss is a *tetrahedron*, formed by connecting six members together, as shown in Fig. 6–19. Any additional members added to this basic element would be redundant in supporting the force **P**. A *simple space truss* can be built from this basic tetrahedral element by adding three additional members and a joint, forming a system of multiconnected tetrahedrons.

Assumptions for Design.   The members of a space truss may be treated as two-force members provided the external loading is applied at the joints and the joints consist of ball-and-socket connections. These assumptions are justified if the welded or bolted connections of the joined members intersect at a common point and the weight of the members can be neglected. In cases where the weight of a member is to be included in the analysis, it is generally satisfactory to apply it as a vertical force, half of its magnitude applied at each end of the member.

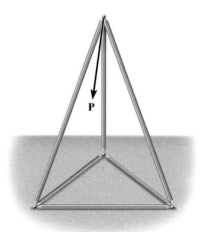

Fig. 6-19

PROCEDURE FOR ANALYSIS

Either the method of joints or the method of sections can be used to determine the forces developed in the members of a simple space truss.

Method of Joints.

Generally, if the forces in *all* the members of the truss must be determined, the method of joints is most suitable for the analysis. When using the method of joints, it is necessary to solve the three scalar equilibrium equations $\Sigma F_x = 0, \Sigma F_y = 0, \Sigma F_z = 0$ at each joint. The solution of many simultaneous equations can be avoided if the force analysis begins at a joint having at least one known force and at most three unknown forces. If the three-dimensional geometry of the force system at the joint is hard to visualize, it is recommended that a Cartesian vector analysis be used for the solution.

Method of Sections.

If only a *few* member forces are to be determined, the method of sections may be used. When an imaginary section is passed through a truss and the truss is separated into two parts, the force system acting on one of the parts must satisfy the *six* scalar equilibrium equations: $\Sigma F_x = 0$, $\Sigma F_y = 0$, $\Sigma F_z = 0$, $\Sigma M_x = 0$, $\Sigma M_y = 0$, $\Sigma M_z = 0$ (Eqs. 5–6). By proper choice of the section and axes for summing forces and moments, many of the unknown member forces in a space truss can be computed *directly*, using a single equilibrium equation.

Typical roof-supporting space truss. Notice the use of ball-and-socket joints for the connections.

EXAMPLE 6.8

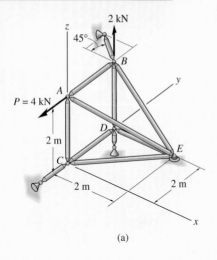

(a)

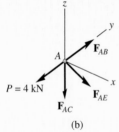

(b)

Determine the forces acting in the members of the space truss shown in Fig. 6–20a. Indicate whether the members are in tension or compression.

Solution

Since there are one known force and three unknown forces acting at joint A, the force analysis of the truss will begin at this joint.

Joint A (Fig. 6–20b). Expressing each force that acts on the free-body diagram of joint A in vector notation, we have

$$\mathbf{P} = \{-4\mathbf{j}\} \text{ kN}, \quad \mathbf{F}_{AB} = F_{AB}\mathbf{j}, \quad \mathbf{F}_{AC} = -F_{AC}\mathbf{k},$$

$$\mathbf{F}_{AE} = F_{AE}\left(\frac{\mathbf{r}_{AE}}{r_{AE}}\right) = F_{AE}(0.577\mathbf{i} + 0.577\mathbf{j} - 0.577\mathbf{k})$$

For equilibrium,

$$\Sigma\mathbf{F} = \mathbf{0}; \qquad \mathbf{P} + \mathbf{F}_{AB} + \mathbf{F}_{AC} + \mathbf{F}_{AE} = \mathbf{0}$$
$$-4\mathbf{j} + F_{AB}\mathbf{j} - F_{AC}\mathbf{k} + 0.577F_{AE}\mathbf{i} + 0.577F_{AE}\mathbf{j} - 0.577F_{AE}\mathbf{k} = \mathbf{0}$$
$$\Sigma F_x = 0; \qquad\qquad 0.577F_{AE} = 0$$
$$\Sigma F_y = 0; \qquad -4 + F_{AB} + 0.577F_{AE} = 0$$
$$\Sigma F_z = 0; \qquad\qquad -F_{AC} - 0.577F_{AE} = 0$$
$$\qquad\qquad F_{AC} = F_{AE} = 0 \qquad\qquad Ans.$$
$$\qquad\qquad F_{AB} = 4 \text{ kN} \quad (\text{T}) \qquad Ans.$$

Since F_{AB} is known, joint B may be analyzed next.

Joint B (Fig. 6–20c).

$$\Sigma F_x = 0; \qquad -R_B \cos 45° + 0.707F_{BE} = 0$$
$$\Sigma F_y = 0; \qquad\qquad -4 + R_B \sin 45° = 0$$
$$\Sigma F_z = 0; \qquad\qquad 2 + F_{BD} - 0.707F_{BE} = 0$$
$$R_B = F_{BE} = 5.66 \text{ kN} \quad (\text{T}), \qquad F_{BD} = 2 \text{ kN} \quad (\text{C}) \qquad Ans.$$

The *scalar* equations of equilibrium may also be applied directly to the force systems on the free-body diagrams of joints D and C since the force components are easily determined. Show that

$$F_{DE} = F_{DC} = F_{CE} = 0 \qquad\qquad Ans.$$

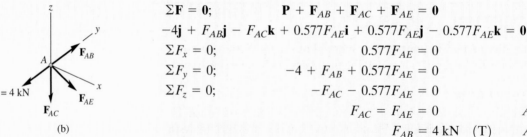

(c)

Fig. 6–20

6.6 Frames and Machines

Frames and machines are two common types of structures which are often composed of pin-connected *multiforce members*, i.e., members that are subjected to more than two forces. *Frames* are generally stationary and are used to support loads, whereas *machines* contain moving parts and are designed to transmit and alter the effect of forces. Provided a frame or machine is properly constrained and contains no more supports or members than are necessary to prevent collapse, the forces acting at the joints and supports can be determined by applying the equations of equilibrium to each member. Once the forces at the joints are obtained, it is then possible to *design* the size of the members, connections, and supports using the theory of mechanics of materials and an appropriate engineering design code.

Free-Body Diagrams.   In order to determine the forces acting at the joints and supports of a frame or machine, the structure must be disassembled and the free-body diagrams of its parts must be drawn. The following important points *must* be observed:

• Isolate each part by drawing its *outlined shape*. Then show all the forces and/or couple moments that act on the part. Make sure to *label* or *identify* each known and unknown force and couple moment with reference to an established *x, y* coordinate system. Also, indicate any dimensions used for taking moments. Most often the equations of equilibrium are easier to apply if the forces are represented by their rectangular components. As usual, the sense of an unknown force or couple moment can be assumed.

• Identify all the two-force members in the structure and represent their free-body diagrams as having two equal but opposite collinear forces acting at their points of application. (See Sec. 5.4.) By recognizing the two-force members, we can avoid solving an unnecessary number of equilibrium equations.

• Forces common to any two *contacting* members act with equal magnitudes but opposite sense on the respective members. If the two members are treated as a *"system" of connected members*, then these forces are *"internal"* and are *not shown* on the *free-body diagram of the system*; however, if the free-body diagram of *each member* is drawn, the forces are *"external"* and *must* be shown on each of the free-body diagrams.

The following examples graphically illustrate application of these points in drawing the free-body diagrams of a dismembered frame or machine. In all cases, the weight of the members is neglected.

E X A M P L E 6.9

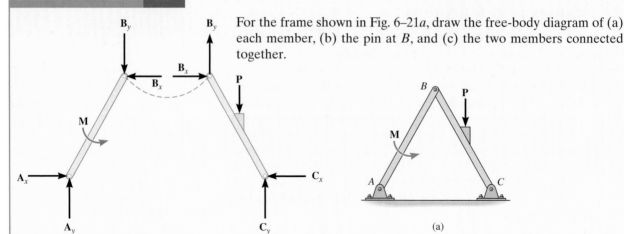

For the frame shown in Fig. 6–21a, draw the free-body diagram of (a) each member, (b) the pin at B, and (c) the two members connected together.

(a)

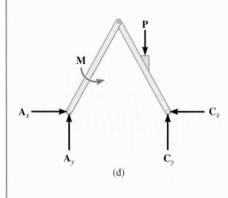

(b)

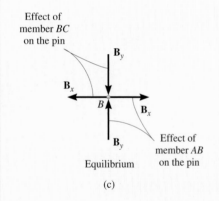

Effect of member BC on the pin

Equilibrium

(c)

(d)

Fig. 6–21

Solution

Part (a). By inspection, members BA and BC are *not* two-force members. Instead, as shown on the free-body diagrams, Fig. 6–21b, BC is subjected to *not* five but *three forces*, namely, the resultant force from pins B and C and the external force **P**. Likewise, AB is subjected to the *resultant* forces from the pins at A and B and the external couple moment **M**.

Part (b). It can be seen in Fig. 6–21a that the pin at B is subjected to only *two forces*, i.e., the force of member BC on the pin and the force of member AB on the pin. For *equilibrium* these forces and therefore their respective components must be equal but opposite, Fig. 6–21c. Notice carefully how Newton's third law is applied between the pin and its contacting members, i.e., the effect of the pin on the two members, Fig. 6–21b, and the equal but opposite effect of the two members on the pin, Fig. 6–21c. Also note that $\mathbf{B}_x$ and $\mathbf{B}_y$, shown equal but opposite in Fig. 6–21b on members AB and BC, is *not* the effect of Newton's third law; instead, this results from the *equilibrium* analysis of the pin, Fig. 6–21c.

Part (c). The free-body diagram of both members connected together, yet removed from the supporting pins at A and C, is shown in Fig. 6–21d. The force components $\mathbf{B}_x$ and $\mathbf{B}_y$ are *not shown* on this diagram since they form equal but opposite collinear pairs of *internal* forces (Fig. 6–21b) and therefore cancel out. Also, to be consistent when later applying the equilibrium equations, the unknown force components at A and C must act in the *same sense* as those shown in Fig. 6–21b. Here the couple moment **M** can be applied at any point on the frame in order to determine the reactions at A and C. Note, however, that it must act on member AB in Fig. 6–21b and *not* on member BC.

EXAMPLE 6.10

A constant tension in the conveyor belt is maintained by using the device shown in Fig. 6–22a. Draw the free-body diagrams of the frame and the cylinder which supports the belt. The suspended block has a weight of W.

(a)

Fig. 6–22

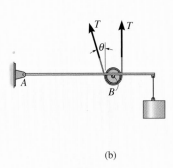

(b)

Solution

The idealized model of the device is shown in Fig. 6–22b. Here the angle θ is assumed to be known. Notice that the tension in the belt is the same on each side of the cylinder, since the cylinder is free to turn. From this model, the free-body diagrams of the frame and cylinder are shown in Figs. 6–22c and 6–22d, respectively. Note that the force that the pin at B exerts on the cylinder can be represented by either its horizontal and vertical components $\mathbf{B}_x$ and $\mathbf{B}_y$, which can be determined by using the force equations of equilibrium applied to the cylinder, or by the two components T, which provide equal but opposite couple moments on the cylinder and thus keep it from turning. Also, realize that once the pin reactions at A have been determined, half of their values act on each side of the frame since pin connections occur on each side, Fig. 6–22a.

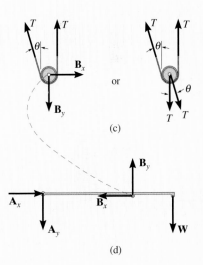

(c)

(d)

E X A M P L E **6.11**

Draw the free-body diagram of each part of the smooth piston and link mechanism used to crush recycled cans, which is shown in Fig. 6–23a.

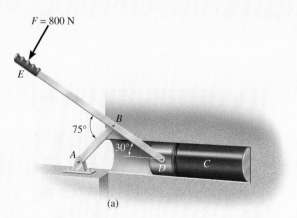

(a)

Solution

By inspection, member *AB* is a two-force member. The free-body diagrams of the parts are shown in Fig. 6–23b. Since the pins at *B* and *D connect only two parts together*, the forces there are shown as equal but opposite on the separate free-body diagrams of their connected members. In particular, four components of force act on the piston: $\mathbf{D}_x$ and $\mathbf{D}_y$ represent the effect of the pin (or lever *EBD*), $\mathbf{N}_w$ is the *resultant force* of the floor, and $\mathbf{P}$ is the resultant compressive force caused by the can *C*.

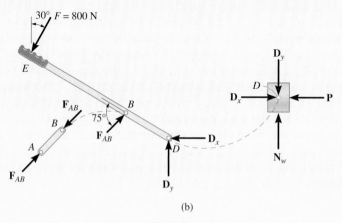

(b)

Fig. 6–23

E X A M P L E **6.12**

For the frame shown in Fig. 6–24*a*, draw the free-body diagrams of (a) the entire frame including the pulleys and cords, (b) the frame without the pulleys and cords, and (c) each of the pulleys.

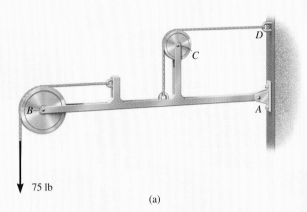

(a)

Solution

Part (a).   When the entire frame including the pulleys and cords is considered, the interactions at the points where the pulleys and cords are connected to the frame become pairs of *internal forces* which cancel each other and therefore are not shown on the free-body diagram, Fig. 6–24*b*.

Part (b).   When the cords and pulleys are removed, their effect *on the frame* must be shown, Fig. 6–24*c*.

Part (c).   The force components $\mathbf{B}_x$, $\mathbf{B}_y$, $\mathbf{C}_x$, $\mathbf{C}_y$ of the pins on the pulleys, Fig. 6–24*d*, are equal but opposite to the force components exerted by the pins on the frame, Fig. 6–24*c*. Why?

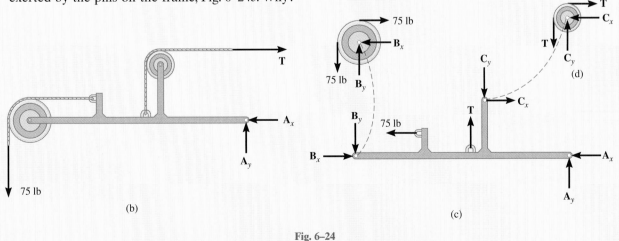

(b)

(c)

(d)

Fig. 6–24

EXAMPLE **6.13**

Draw the free-body diagrams of the bucket and the vertical boom of the back hoe shown in the photo, Fig. 6–25a. The bucket and its contents have a weight W. Neglect the weight of the members.

(a)

Fig. 6–25

Solution

The idealized model of the assembly is shown in Fig. 6–25b. Not shown are the required dimensions and angles that must be obtained, along with the location of the center of gravity G of the load. By inspection, members AB, BC, BE, and HI are all two-force members since they are pin connected at their end points and no other forces act on them. The free-body diagrams of the bucket and the boom are shown in Fig. 6–25c. Note that pin C is subjected to only two forces, the force of link BC and the force of the boom. For equilibrium, these forces must be equal in magnitude but opposite in direction, Fig. 6–25d. The pin at B is subjected to three forces, Fig. 6–25e. The force $\mathbf{F}_{BE}$ is caused by the hydraulic cylinder, and the forces $\mathbf{F}_{BA}$ and $\mathbf{F}_{BC}$ are caused by the links. These three forces are related by the two equations of force equilibrium applied to the pin.

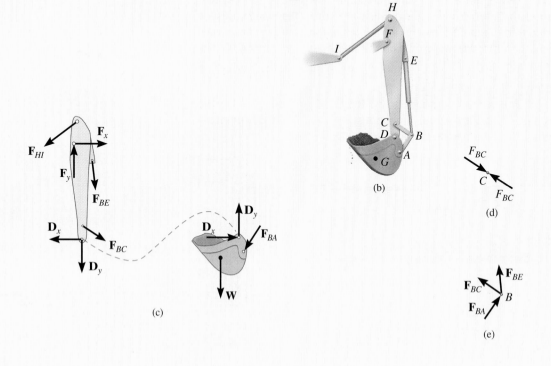

(b)

(c)

(d)

(e)

Before proceeding, it is recommended to cover the solutions to the previous examples and attempt to draw the requested free-body diagrams. When doing so, make sure the work is neat and that all the forces and couple moments are properly labeled.

Equations of Equilibrium. Provided the structure (frame or machine) is properly supported and contains no more supports or members than are necessary to prevent its collapse, then the unknown forces at the supports and connections can be determined from the equations of equilibrium. If the structure lies in the *x*–*y* plane, then for *each* free-body diagram drawn the loading must satisfy $\Sigma F_x = 0$, $\Sigma F_y = 0$, and $\Sigma M_O = 0$. The selection of the free-body diagrams used for the analysis is *completely arbitrary*. They may represent each of the members of the structure, a portion of the structure, or its entirety. For example, consider finding the six components of the pin reactions at *A*, *B*, and *C* for the frame shown in Fig. 6–26*a*. If the frame is dismembered, as it is in Fig. 6–26*b*, these unknowns can be determined by applying the three equations of equilibrium to each of the two members (total of six equations). The free-body diagram of the *entire frame* can also be used for part of the analysis, Fig. 6–26*c*. Hence, if so desired, all six unknowns can be determined by applying the three equilibrium equations to the entire frame, Fig. 6–26*c*, and also to either one of its members. Furthermore, the answers can be checked in part by applying the three equations of equilibrium to the remaining "second" member. In general, then, this problem can be solved by writing *at most* six equilibrium equations using free-body diagrams of the members and/or the combination of connected members. Any more than six equations written would *not* be unique from the original six and would only serve to check the results.

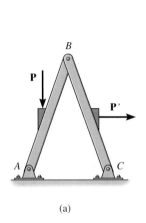

(a)

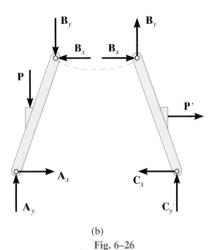

(b)

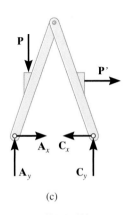

(c)

Fig. 6–26

PROCEDURE FOR ANALYSIS

The joint reactions on frames or machines (structures) composed of multiforce members can be determined using the following procedure.

Free-Body Diagram.

- Draw the free-body diagram of the entire structure, a portion of the structure, or each of its members. The choice should be made so that it leads to the most direct solution of the problem.

- When the free-body diagram of a group of members of a structure is drawn, the forces at the connected parts of this group are internal forces and are not shown on the free-body diagram of the group.

- Forces common to two members which are in contact act with equal magnitude but opposite sense on the respective free-body diagrams of the members.

- Two-force members, regardless of their shape, have equal but opposite collinear forces acting at the ends of the member.

- In many cases it is possible to tell by inspection the proper sense of the unknown forces acting on a member; however, if this seems difficult, the sense can be assumed.

- A couple moment is a free vector and can act at any point on the free-body diagram. Also, a force is a sliding vector and can act at any point along its line of action.

Equations of Equilibrium.

- Count the number of unknowns and compare it to the total number of equilibrium equations that are available. In two dimensions, there are three equilibrium equations that can be written for each member.

- Sum moments about a point that lies at the intersection of the lines of action of as many unknown forces as possible.

- If the solution of a force or couple moment magnitude is found to be negative, it means the sense of the force is the reverse of that shown on the free-body diagrams.

EXAMPLE 6.14

Determine the horizontal and vertical components of force which the pin at C exerts on member CB of the frame in Fig. 6–27a.

Solution I

Free-Body Diagrams. By inspection it can be seen that AB is a two-force member. The free-body diagrams are shown in Fig. 6–27b.

Equations of Equilibrium. The *three unknowns*, C_x, C_y, and F_{AB}, can be determined by applying the three equations of equilibrium to member CB.

$$\zeta+\Sigma M_C = 0;\ 2000\ \text{N}(2\ \text{m}) - (F_{AB}\sin 60°)(4\ \text{m}) = 0\ F_{AB} = 1154.7\ \text{N}$$

$$\overset{+}{\to}\Sigma F_x = 0;\qquad 1154.7\cos 60°\text{N} - C_x = 0\qquad C_x = 577\ \text{N}\qquad Ans.$$

$$+\uparrow\Sigma F_y = 0;\ 1154.7\sin 60°\text{N} - 2000\ \text{N} + C_y = 0\quad C_y = 1000\ \text{N}\quad Ans.$$

Solution II

Free-Body Diagrams. If one does not recognize that AB is a two-force member, then more work is involved in solving this problem. The free-body diagrams are shown in Fig. 6–27c.

Equations of Equilibrium. The *six unknowns*, A_x, A_y, B_x, B_y, C_x, C_y, are determined by applying the three equations of equilibrium to each member.

Member AB

$$\zeta+\Sigma M_A = 0;\qquad B_x(3\sin 60°\ \text{m}) - B_y(3\cos 60°\ \text{m}) = 0 \qquad (1)$$

$$\overset{+}{\to}\Sigma F_x = 0;\qquad A_x - B_x = 0 \qquad (2)$$

$$+\uparrow\Sigma F_y = 0;\qquad A_y - B_y = 0 \qquad (3)$$

Member BC

$$\zeta+\Sigma M_C = 0;\qquad 2000\ \text{N}(2\ \text{m}) - B_y(4\ \text{m}) = 0 \qquad (4)$$

$$\overset{+}{\to}\Sigma F_x = 0;\qquad B_x - C_x = 0 \qquad (5)$$

$$+\uparrow\Sigma F_y = 0;\qquad B_y - 2000\ \text{N} + C_y = 0 \qquad (6)$$

The results for C_x and C_y can be determined by solving these equations in the following sequence: 4, 1, 5, then 6. The results are

$$B_y = 1000\ \text{N}$$
$$B_x = 577\ \text{N}$$
$$C_x = 577\ \text{N} \qquad\qquad Ans.$$
$$C_y = 1000\ \text{N} \qquad\qquad Ans.$$

By comparison, Solution I is simpler since the requirement that F_{AB} in Fig. 6–27b be equal, opposite, and collinear at the ends of member AB automatically satisfies Eqs. 1, 2, and 3 above and therefore eliminates the need to write these equations. *As a result, always identify the two-force members before starting the analysis!*

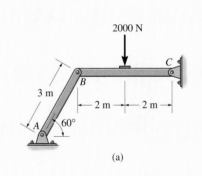

(a)

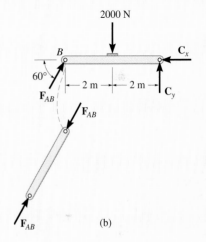

(b)

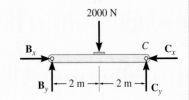

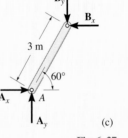

(c)

Fig. 6–27

E X A M P L E 6.15

The compound beam shown in Fig. 6–28a is pin connected at B. Determine the reactions at its supports. Neglect its weight and thickness.

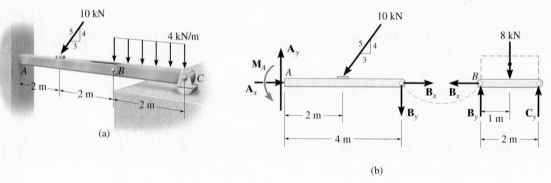

(a)

(b)

Fig. 6–28

Solution

Free-Body Diagrams. By inspection, if we consider a free-body diagram of the entire beam ABC, there will be three unknown reactions at A and one at C. These four unknowns cannot all be obtained from the three equations of equilibrium, and so it will become necessary to dismember the beam into its two segments as shown in Fig. 6–28b.

Equations of Equilibrium. The six unknowns are determined as follows:

Segment BC

$$\xrightarrow{+} \Sigma F_x = 0; \qquad\qquad\qquad\qquad B_x = 0$$

$$\zeta + \Sigma M_B = 0; \qquad -8 \text{ kN}(1 \text{ m}) + C_y(2 \text{ m}) = 0$$

$$+\uparrow \Sigma F_y = 0; \qquad\qquad B_y - 8 \text{ kN} + C_y = 0$$

Segment AB

$$\xrightarrow{+} \Sigma F_x = 0; \qquad\qquad A_x - (10 \text{ kN})(\tfrac{3}{5}) + B_x = 0$$

$$\zeta + \Sigma M_A = 0; \qquad M_A - (10 \text{ kN})(\tfrac{4}{5})(2 \text{ m}) - B_y(4 \text{ m}) = 0$$

$$+\uparrow \Sigma F_y = 0; \qquad\qquad A_y - (10 \text{ kN})(\tfrac{4}{5}) - B_y = 0$$

Solving each of these equations successively, using previously calculated results, we obtain

$$A_x = 6 \text{ kN} \qquad A_y = 12 \text{ kN} \qquad M_A = 32 \text{ kN} \cdot \text{m} \qquad \textit{Ans.}$$

$$B_x = 0 \qquad\qquad B_y = 4 \text{ kN}$$

$$C_y = 4 \text{ kN} \qquad\qquad\qquad\qquad\qquad\qquad\qquad\qquad \textit{Ans.}$$

EXAMPLE 6.16

Determine the horizontal and vertical components of force which the pin at C exerts on member $ABCD$ of the frame shown in Fig. 6–29a.

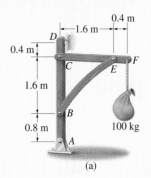

(a)

Solution

Free-Body Diagrams. By inspection, the three components of reaction that the supports exert on $ABCD$ can be determined from a free-body diagram of the entire frame, Fig. 6–29b. Also, the free-body diagram of each frame member is shown in Fig. 6–29c. Notice that member BE is a two-force member. As shown by the colored dashed lines, the forces at B, C, and E have equal magnitudes but opposite directions on the separate free-body diagrams.

Equations of Equilibrium. The six unknowns A_x, A_y, F_B, C_x, C_y, and D_x will be determined from the equations of equilibrium applied to the entire frame and then to member CEF. We have

Entire Frame

$$\zeta+\Sigma M_A = 0; \quad -981 \text{ N}(2 \text{ m}) + D_x(2.8 \text{ m}) = 0 \qquad D_x = 700.7 \text{ N}$$

$$\xrightarrow{+} \Sigma F_x = 0; \qquad A_x - 700.7 \text{ N} = 0 \qquad A_x = 700.7 \text{ N}$$

$$+\uparrow\Sigma F_y = 0; \qquad A_y - 981 \text{ N} = 0 \qquad A_y = 981 \text{ N}$$

Member CEF

$$\zeta+\Sigma M_C = 0; \quad -981 \text{ N}(2 \text{ m}) - (F_B \sin 45°)(1.6 \text{ m}) = 0$$
$$F_B = -1734.2 \text{ N}$$

$$\xrightarrow{+} \Sigma F_x = 0; \qquad -Cx - (-1734.2 \cos 45° \text{ N}) = 0$$
$$C_x = 1226 \text{ N} \qquad\qquad Ans.$$

$$+\uparrow\Sigma F_y = 0; \quad C_y - (-1734.2 \sin 45° \text{ N}) - 981 \text{ N} = 0$$
$$C_y = -245 \text{ N} \qquad\qquad Ans.$$

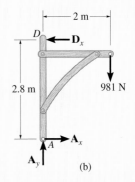

(b)

Since the magnitudes of $\mathbf{F}_B$ and $\mathbf{C}_y$ were calculated as negative quantities, they were assumed to be acting in the wrong sense on the free-body diagrams, Fig. 6–29c. The correct sense of these forces might have been determined "by inspection" *before* applying the equations of equilibrium to member CEF. As shown in Fig. 6–29c, moment equilibrium about point E on member CEF indicates that $\mathbf{C}_y$ must actually act *downward* to counteract the moment created by the 981-N force about E. Similarly, summing moments about C, it is seen that the vertical component of $\mathbf{F}_B$ must actually act *upward*, and so $\mathbf{F}_B$ must act upward and to the right.

The above calculations can be checked by applying the three equilibrium equations to member $ABCD$, Fig. 6–29c.

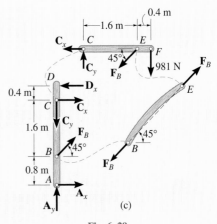

(c)

Fig. 6–29

EXAMPLE 6.17

The smooth disk shown in Fig. 6–30a is pinned at D and has a weight of 20 lb. Neglecting the weights of the other members, determine the horizontal and vertical components of reaction at pins B and D.

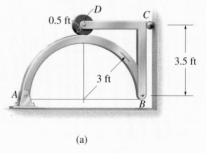

(a)

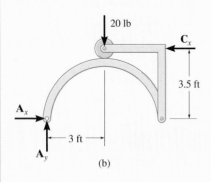

(b)

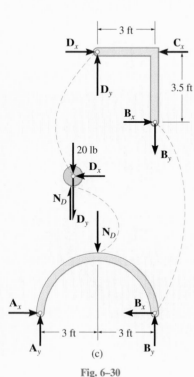

(c)

Fig. 6–30

Solution

Free-Body Diagrams. By inspection, the three components of reaction at the supports can be determined from a free-body diagram of the entire frame, Fig. 6–30b. Also, free-body diagrams of the members are shown in Fig. 6–30c.

Equations of Equilibrium. The eight unknowns can of course be obtained by applying the eight equilibrium equations to each member—three to member AB, three to member BCD, and two to the disk. (Moment equilibrium is automatically satisfied for the disk.) If this is done, however, all the results can be obtained only from a simultaneous solution of some of the equations. (Try it and find out.) To avoid this situation, it is best to first determine the three support reactions on the *entire* frame; then, using these results, the remaining five equilibrium equations can be applied to two other parts in order to solve successively for the other unknowns.

Entire Frame

$$\zeta + \Sigma M_A = 0; \quad -20\ \text{lb}(3\ \text{ft}) + C_x(3.5\ \text{ft}) = 0 \quad C_x = 17.1\ \text{lb}$$
$$\xrightarrow{+} \Sigma F_x = 0; \quad A_x - 17.1\ \text{lb} = 0 \quad A_x = 17.1\ \text{lb}$$
$$+\uparrow \Sigma F_y = 0; \quad A_y - 20\ \text{lb} = 0 \quad A_y = 20\ \text{lb}$$

Member AB

$$\xrightarrow{+} \Sigma F_x = 0; \quad 17.1\ \text{lb} - B_x = 0 \quad B_x = 17.1\ \text{lb} \qquad Ans.$$
$$\zeta + \Sigma M_B = 0; \quad -20\ \text{lb}(6\ \text{ft}) + N_D(3\ \text{ft}) = 0 \quad N_D = 40\ \text{lb}$$
$$+\uparrow \Sigma F_y = 0; \quad 20\ \text{lb} - 40\ \text{lb} + B_y = 0 \quad B_y = 20\ \text{lb} \qquad Ans.$$

Disk

$$\xrightarrow{+} \Sigma F_x = 0; \quad D_x = 0 \qquad \qquad Ans.$$
$$+\uparrow \Sigma F_y = 0; \quad 40\ \text{lb} - 20\ \text{lb} - D_y = 0 \quad D_y = 20\ \text{lb} \qquad Ans.$$

E X A M P L E 6.18

Determine the tension in the cables and also the force **P** required to support the 600-N force using the frictionless pulley system shown in Fig. 6–31*a*.

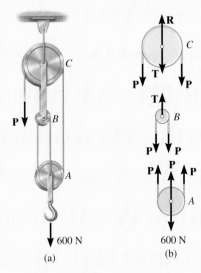

Fig. 6–31

Solution

Free-Body Diagram.   A free-body diagram of each pulley *including* its pin and a portion of the contacting cable is shown in Fig. 6–31*b*. Since the cable is *continuous* and the pulleys are frictionless, the cable has a *constant tension P* acting throughout its length (see Example 5.7). The link connection between pulleys *B* and *C* is a two-force member, and therefore it has an unknown tension *T* acting on it. Notice that the *principle of action, equal but opposite reaction* must be carefully observed for forces **P** and **T** when the *separate* free-body diagrams are drawn.

Equations of Equilibrium.   The three unknowns are obtained as follows:

Pulley A

$+\uparrow \Sigma F_y = 0;$ $3P - 600 \text{ N} = 0$ $P = 200 \text{ N}$ *Ans.*

Pulley B

$+\uparrow \Sigma F_y = 0;$ $T - 2P = 0$ $T = 400 \text{ N}$ *Ans.*

Pulley C

$+\uparrow \Sigma F_y = 0;$ $R - 2P - T = 0$ $R = 800 \text{ N}$ *Ans.*

EXAMPLE 6.19

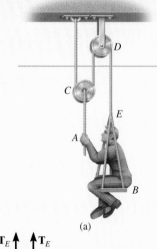

(a)

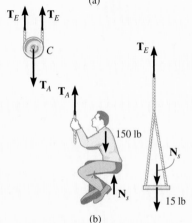

(b)

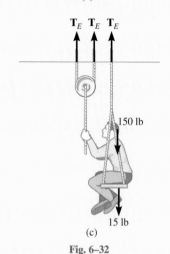

(c)

Fig. 6–32

A man having a weight of 150 lb supports himself by means of the cable and pulley system shown in Fig. 6–32*a*. If the seat has a weight of 15 lb, determine the force that he must exert on the cable at *A* and the force he exerts on the seat. Neglect the weight of the cables and pulleys.

Solution I

Free-Body Diagrams. The free-body diagrams of the man, seat, and pulley *C* are shown in Fig. 6–32*b*. The *two* cables are subjected to tensions T_A and T_E, respectively. The man is subjected to three forces: his weight, the tension T_A of cable *AC*, and the reaction N_s of the seat.

Equations of Equilibrium. The three unknowns are obtained as follows:

Man

$$+\uparrow \Sigma F_y = 0; \qquad T_A + N_s - 150 \text{ lb} = 0 \qquad (1)$$

Seat

$$+\uparrow \Sigma F_y = 0; \qquad T_E - N_s - 15 \text{ lb} = 0 \qquad (2)$$

Pulley C

$$+\uparrow \Sigma F_y = 0; \qquad 2T_E - T_A = 0 \qquad (3)$$

Here T_E can be determined by adding Eqs. 1 and 2 to eliminate N_s and then using Eq. 3. The other unknowns are then obtained by resubstitution of T_E.

$$T_A = 110 \text{ lb} \qquad \qquad Ans.$$
$$T_E = 55 \text{ lb}$$
$$N_s = 40 \text{ lb} \qquad \qquad Ans.$$

Solution II

Free-Body Diagrams. By using the blue section shown in Fig. 6–32*a*, the man, pulley, and seat can be considered as a *single system*, Fig. 6–32*c*. Here N_s and T_A are *internal* forces and hence are not included on this "combined" free-body diagram.

Equations of Equilibrium. Applying $\Sigma F_y = 0$ yields a *direct* solution for T_E.

$$+\uparrow \Sigma F_y = 0; \qquad 3T_E - 15 \text{ lb} - 150 \text{ lb} = 0 \qquad T_E = 55 \text{ lb}$$

The other unknowns can be obtained from Eqs. 2 and 3.

EXAMPLE 6.20

The hand exerts a force of 8 lb on the grip of the spring compressor shown in Fig. 6–33a. Determine the force in the spring needed to maintain equilibrium of the mechanism.

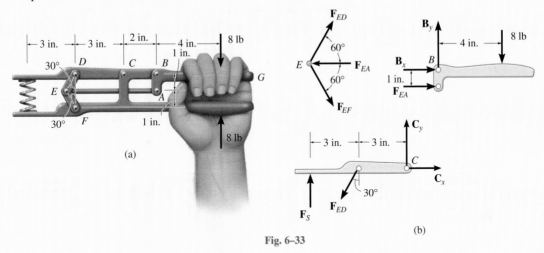

Fig. 6–33

Solution

Free-Body Diagrams. By inspection, members *EA, ED,* and *EF* are all two-force members. The free-body diagrams for parts *DC* and *ABG* are shown in Fig. 6–33b. The pin at *E* has also been included here since *three* force interactions occur on this pin. They represent the effects of members *ED, EA,* and *EF.* Note carefully how equal and opposite force reactions occur between each of the parts.

Equations of Equilibrium. By studying the free-body diagrams, the most direct way to obtain the spring force is to apply the equations of equilibrium in the following sequence:

Lever ABG

$$\zeta + \Sigma M_B = 0; \quad F_{EA}(1 \text{ in.}) - 8 \text{ lb}(4 \text{ in.}) = 0 \quad F_{EA} = 32 \text{ lb}$$

Pin E

$$+\uparrow \Sigma F_y = 0; \quad F_{ED} \sin 60° - F_{EF} \sin 60° = 0 \quad F_{ED} = F_{EF} = F$$
$$\xrightarrow{+} \Sigma F_x = 0; \quad 2F \cos 60° - 32 \text{ lb} = 0 \quad F = 32 \text{ lb}$$

Arm DC

$$\zeta + \Sigma M_C = 0; \quad -F_s(6 \text{ in.}) + 32 \cos 30° \text{ lb}(3 \text{ in.}) = 0$$
$$F_s = 13.9 \text{ lb} \qquad \qquad Ans.$$

EXAMPLE 6.21

The 100-kg block is held in equilibrium by means of the pulley and continuous cable system shown in Fig. 6–34a. If the cable is attached to the pin at *B*, compute the forces which this pin exerts on each of its connecting members.

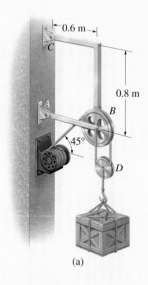

(a)

Fig. 6–34

Solution

Free-Body Diagrams. A free-body diagram of each member of the frame is shown in Fig. 6–34b. By inspection, members *AB* and *CB* are two-force members. Furthermore, the cable must be subjected to a force of 490.5 N in order to hold pulley *D* and the block in equilibrium. A free-body diagram of the pin at *B* is needed since *four interactions* occur at this pin. These are caused by the attached cable (490.5 N), member *AB* ($\mathbf{F}_{AB}$), member *CB* ($\mathbf{F}_{CB}$), and pulley *B* ($\mathbf{B}_x$ and $\mathbf{B}_y$).

Equations of Equilibrium. Applying the equations of force equilibrium to pulley *B*, we have

$$\xrightarrow{+} \Sigma F_x = 0; \quad B_x - 490.5 \cos 45° \text{ N} = 0 \quad B_x = 346.8 \text{ N} \qquad Ans.$$

$$+\uparrow \Sigma F_y = 0; \quad B_y - 490.5 \sin 45° \text{ N} - 490.5 \text{ N} = 0$$

$$B_y = 837.3 \text{ N} \qquad Ans.$$

Using these results, equilibrium of the pin requires that

$$+\uparrow \Sigma F_y = 0; \quad \tfrac{4}{5}F_{CB} - 837.3 \text{ N} - 490.5 \text{ N} \qquad F_{CB} = 1660 \text{ N} \qquad Ans.$$

$$\xrightarrow{+} \Sigma F_x = 0; \quad F_{AB} - \tfrac{3}{5}(1660 \text{ N}) - 346.8 \text{ N} = 0 \quad F_{AB} = 1343 \text{ N} \qquad Ans.$$

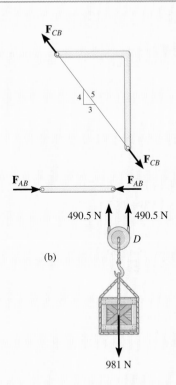

(b)

It may be noted that the two-force member *CB* is subjected to bending as caused by the force $\mathbf{F}_{CB}$. From the standpoint of design, it would be better to make this member *straight* (from *C* to *B*) so that the force $\mathbf{F}_{CB}$ would create only tension in the member.

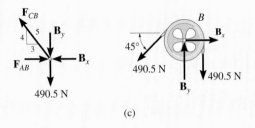

(c)

Fig. 6–34

CHAPTER REVIEW

- *Truss Analysis.* A simple truss consists of triangular elements connected together by pin joints. The forces within it members can be determined by assuming the members are all two-force members, connected concurrently at each joint.

- *Method of Joints.* If a truss is in equilibrium, then each of its joints is also in equilibrium. For a coplanar truss, the concurrent force system at each joint must satisfy force equilibrium, $\Sigma F_x = 0$, $\Sigma F_y = 0$. To obtain a numerical solution for the forces in the members, select a joint that has a free-body diagram with at most two unknown forces and one known force. (This may require first finding the reactions at the supports.) Once a member force is determined, use its value and apply it to an adjacent joint. Remember that forces that are found to *pull* on the joint are in *tension*, and those that *push* on the joint are in *compression*. To avoid a simultaneous solution of two equations, try to sum forces in a direction that is perpendicular to one of the unknowns. This will allow a direct solution for the other unknown. To further simplify the analysis, first identify all the zero-force members.

- *Method of Sections.* If a truss is in equilibrium, then each section of the truss is also in equilibrium. Pass a section through the member whose force is to be determined. Then draw the free-body diagram of the sectioned part having the least number of forces on it. Sectioned members subjected to *pulling* are in *tension*, and those that are subjected to *pushing* are in *compression*. If the force system is coplanar, then three equations of equilibrium are available to determine the unknowns. If possible, sum forces in a direction that is perpendicular to two of the three unknown forces. This will yield a direct solution for the third force. Likewise, sum moments about a point that passes through the line of action of two of the three unknown forces, so that the third unknown force can be determined directly.

- *Frames and Machines.* The forces acting at the joints of a frame or machine can be determined by drawing the free-body diagrams of each of its members or parts. The principle of action-reaction should be carefully observed when drawing these forces on each adjacent member or pin. For a coplanar force system, there are three equilibrium equations available for each member.

REVIEW PROBLEMS

*6-1. Determine the resultant forces at pins B and C on member ABC of the four-member frame.

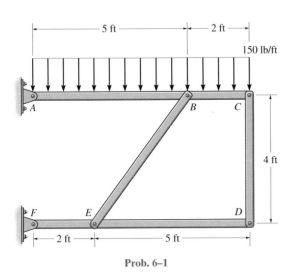

Prob. 6–1

6-3. Determine the horizontal and vertical components of force at pins A and C of the two-member frame.

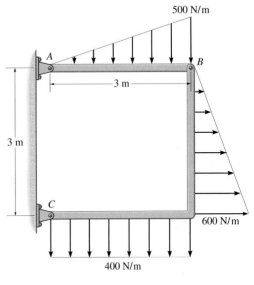

Prob. 6–3

6-2. The mechanism consists of identical meshed gears A and B and arms which are fixed to the gears. The spring attached to the ends of the arms has an unstretched length of 100 mm and a stiffness of $k = 250$ N/m. If a torque of $M = 6$ N·m is applied to gear A, determine the angle θ through which each arm rotates. The gears are each pinned to fixed supports at their centers.

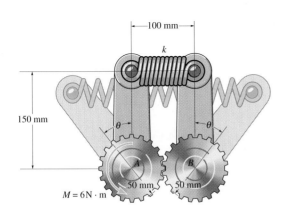

Prob. 6–2

6-4. The spring has an unstretched length of 0.3 m. Determine the angle θ for equilibrium if the uniform links each have a mass of 5 kg.

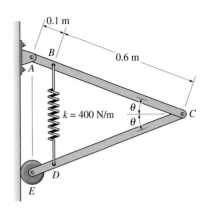

Prob. 6–4

*6-5. The spring has an unstretched length of 0.3 m. Determine the mass m of each uniform link if the angle $\theta = 20°$ for equilibrium.

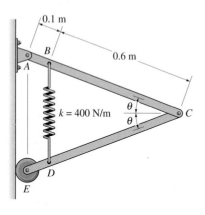

Prob. 6–5

6-7. Determine the horizontal and vertical components of force that pins A and B exert on the two-member frame. Set $F = 500$ N.

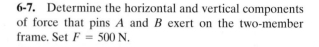
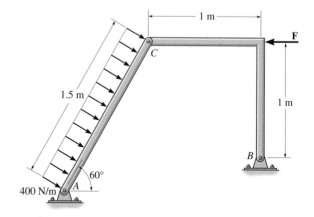

Prob. 6–7

6-6. Determine the horizontal and vertical components of force that the pins A and B exert on the two-member frame. Set $F = 0$.

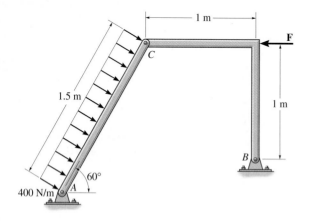

Prob. 6–6

6-8. The two-bar mechanism consists of a lever arm AB and smooth link CD, which has a fixed collar at its end C and a roller at the other end D. Determine the force **P** needed to hold the lever in the position θ. The spring has a stiffness k and unstretched length $2L$. The roller contacts either the top or bottom portion of the horizontal guide.

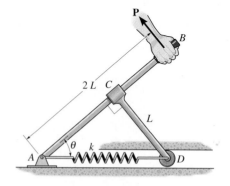

Prob. 6–8

*6-9. Determine the force in each member of the truss and state if the members are in tension or compression.

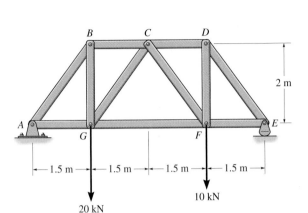

2 m

A
G
F
E

1.5 m — 1.5 m — 1.5 m — 1.5 m

10 kN

20 kN

Prob. 6–9

6-10. Determine the force in members AB, AD, and AC of the space truss and state if the members are in tension or compression.

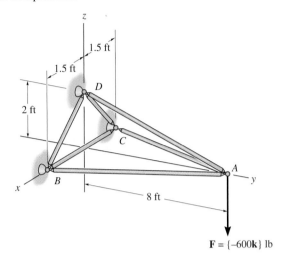

$\mathbf{F} = \{-600\mathbf{k}\}$ lb

Prob. 6–10

DESIGN PROJECTS

6–1D   DESIGN OF A BRIDGE TRUSS

A bridge having a horizontal top cord is to span between the two piers A and B having an arbitrary height. It is required that a pin-connected truss be used, consisting of steel members bolted together to steel gusset plates, such as the one shown in the figure. The end supports are assumed to be a pin at A and a roller at B. A vertical loading of 5 kN is to be supported within the middle 3m of the span. This load can be applied in part to several joints on the top cord within this region, or to a single joint at the middle of the top cord. The force of the wind and the weight of the members are to be neglected.

Assume the maximum tensile force in each member cannot exceed 4.25 kN; and regardless of the length of the member, the maximum compressive force cannot exceed 3.5 kN. Design the most economical truss that will support the loading. The members cost $3.50/m, and the gusset plates cost $8.00 each. Submit your cost analysis for the materials, along with a scaled drawing of the truss, identifying on this drawing the tensile and compressive force in each member. Also, include your calculations of the complete force analysis.

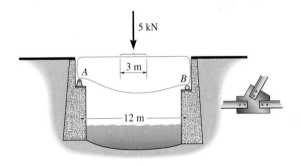

5 kN

3 m

A   B

12 m

Prob. 6–1D

6–2D DESIGN OF A CART LIFT

A hand cart is used to move a load from one loading dock to another. Any dock will have a different elevation relative to the bed of a truck that backs up to it. It is necessary that the loading platform on the hand cart will bring the load resting on it up to the elevation of each truck bed as shown. The maximum elevation difference between the frame of the hand cart and a truck bed is 1 ft. Design a hand-operated mechanical system that will allow the load to be lifted this distance from the frame of the hand cart. Assume the operator can exert a (comfortable) force of 20 lb to make the lift, and that the maximum load, centered on the loading platform, is 400 lb. Submit a scaled drawing of your design, and explain how it works based on a force analysis.

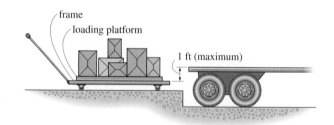

Prob. 6–2D

6–3D DESIGN OF A PULLEY SYSTEM

The steel beam AB, having a length of 5 m and a mass of 700 kg is to be hoisted in its horizontal position to a height of 4 m. Design a pulley-and-rope system, which can be suspended from the overhead beam CD, that will allow a single worker to hoist the beam. Assume that the maximum (comfortable) force that he can apply to the rope is 180 N. Submit a drawing of your design, specify its approximate material cost, and discuss the safety aspects of its operation. Rope costs $1.25/m and each pulley costs $3.00.

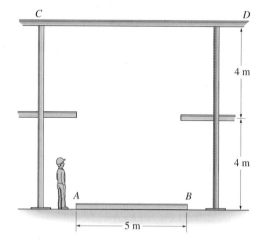

Prob. 6–3D

6–4D   DESIGN OF A TOOL USED TO POSITION A SUSPENDED LOAD

Heavy loads are suspended from an overhead pulley and each load must be positioned over a depository. Design a tool that can be used to shorten or lengthen the pulley cord AB a small amount in order to make the location adjustment. Assume the worker can apply a maximum (comfortable) force of 25 lb to the tool, and the maximum force allowed in cord AB is 500 lb. Submit a scaled drawing of the tool, and a brief paragraph to explain how it works using a force analysis. Include a discussion on the safety aspects of its use.

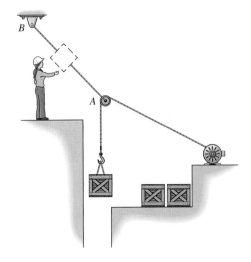

Prob. 6–4D

6–5D   DESIGN OF A FENCE-POST REMOVER

A farmer wishes to remove several fence posts. Each post is buried 18 in. in the ground and will require a maximum vertical pulling force of 175 lb to remove it. He can use his truck to develop the force, but he needs to devise a method for their removal without breaking the posts. Design a method that can be used, considering that the only materials available are a strong rope and several pieces of wood having various sizes and lengths. Submit a sketch of your design and discuss the safety and reliability of its use. Also, provide a force analysis to show how it works and why it will cause minimal damage to a post when it is removed.

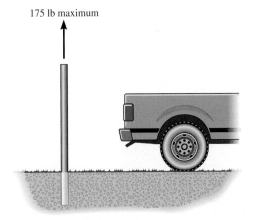

175 lb maximum

Prob. 6–5D

The design and analysis of any structural member requires knowledge of the internal loadings acting within it, not only when it is in place and subjected to service loads, but also when it is being hoisted as shown here. In this chapter, we will discuss how engineers determine these loadings.

Internal Forces

CHAPTER OBJECTIVES

- To show how to use the method of sections for determining the internal loadings in a member.
- To generalize this procedure by formulating equations that can be plotted so that they describe the internal shear and moment throughout a member.
- To analyze the forces and study the geometry of cables supporting a load.

7.1 Internal Forces Developed in Structural Members

The design of any structural or mechanical member requires an investigation of the loading acting within the member in order to be sure the material can resist this loading. These internal loadings can be determined by using the *method of sections*. To illustrate the procedure, consider the "simply supported" beam shown in Fig. 7–1a, which is subjected to the forces $\mathbf{F}_1$ and $\mathbf{F}_2$ and the *support reactions* $\mathbf{A}_x$, $\mathbf{A}_y$, and $\mathbf{B}_y$, Fig. 7–1b. If the *internal loadings* acting on the cross section at C are to be determined, then an imaginary section is passed through the beam, cutting it into two segments. By doing this the internal loadings at the section become *external* on the free-body diagram of each segment,

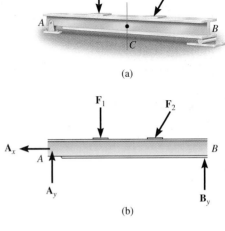

(a)

(b)

Fig. 7–1

221

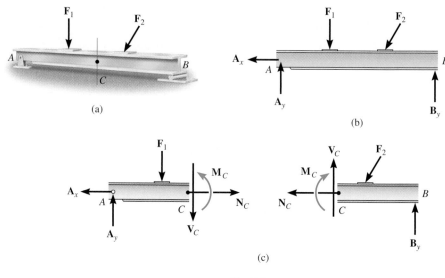

(a)

(b)

(c)

Fig. 7–1

Fig. 7–1c. Since both segments (AC and CB) were in equilibrium *before* the beam was sectioned, equilibrium of each segment is maintained provided rectangular force components $\mathbf{N}_C$ and $\mathbf{V}_C$ and a resultant couple moment $\mathbf{M}_C$ are developed at the section. Note that these loadings must be equal in magnitude and opposite in direction on each of the segments (Newton's third law). The magnitude of each of these loadings can now be determined by applying the three equations of equilibrium to either segment AC or CB. A *direct solution* for $\mathbf{N}_C$ is obtained by applying $\Sigma F_x = 0$; $\mathbf{V}_C$ is obtained directly from $\Sigma F_y = 0$; and $\mathbf{M}_C$ is determined by summing moments about point C, $\Sigma M_C = 0$, in order to eliminate the moments of the unknowns $\mathbf{N}_C$ and $\mathbf{V}_C$.

To save on material the beams used to support the roof of this shelter were tapered since the roof loading will produce a larger internal moment at the beams' centers than at their ends.

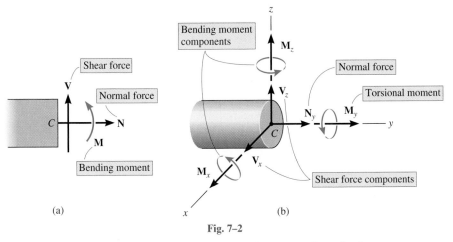

(a)   (b)

Fig. 7–2

In mechanics, the force components **N**, acting normal to the beam at the cut section, and **V**, acting tangent to the section, are termed the *normal or axial force* and the *shear force*, respectively. The couple moment **M** is referred to as the *bending moment*, Fig. 7–2a. In three dimensions, a general internal force and couple moment resultant will act at the section. The *x, y, z* components of these loadings are shown in Fig. 7–2b. Here N_y is the *normal force*, and V_x and V_z are *shear force components*. M_y is a *torsional or twisting moment*, and M_x and M_z are *bending moment components*. For most applications, these *resultant loadings* will act at the geometric center or centroid (*C*) of the section's cross-sectional area. Although the magnitude for each loading generally will be different at various points along the axis of the member, the method of sections can always be used to determine their values.

Free-Body Diagrams. Since frames and machines are composed of *multiforce members*, each of these members will generally be subjected to internal normal, shear, and bending loadings. For example, consider the frame shown in Fig. 7–3a. If the blue section is passed through the frame to determine the internal loadings at points *H*, *G*, and *F*, the resulting free-body diagram of the top portion of this section is shown in Fig. 7–3b. At each point where a member is sectioned there is an unknown normal force, shear force, and bending moment. As a result, we cannot apply the *three* equations of equilibrium to this section in order to obtain these *nine unknowns*.*Instead, to solve this problem we must *first dismember* the frame and determine the reactions at the connections of the members using the techniques of Sec. 6.6. Once this is done, *each member* may then be sectioned at its appropriate point, and the three equations of equilibrium can be applied to determine **N**, **V**, and **M**. For example, the free-body diagram of segment *DG*, Fig. 7–3c, can be used to determine the internal loadings at *G* provided the reactions of the pin, D_x and D_y, are known.

*Recall that this method of analysis worked well for trusses since truss members are *straight two-force members* which support only an axial or normal load.

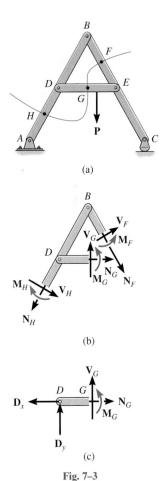

(a)

(b)

(c)

Fig. 7–3

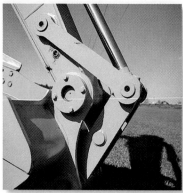

In each case, the link on the backhoe is a two-force member. In the top photo it is subjected to both bending and axial load at its center. By making the member straight, as in the bottom photo, then only an axial force acts within the member.

PROCEDURE FOR ANALYSIS

The method of sections can be used to determine the internal loadings at a specific location in a member using the following procedure.

Support Reactions.

- Before the member is "cut" or sectioned, it may first be necessary to determine the member's support reactions, so that the equilibrium equations are used only to solve for the internal loadings when the member is sectioned.

- If the member is part of a frame or machine, the reactions at its connections are determined using the methods of Sec. 6.6.

Free-Body Diagram.

- Keep all distributed loadings, couple moments, and forces acting on the member in their *exact locations*, then pass an imaginary section through the member, perpendicular to its axis at the point where the internal loading is to be determined.

- After the section is made, draw a free-body diagram of the segment that has the least number of loads on it, and indicate the *x, y, z* components of the force and couple moment resultants at the section.

- If the member is subjected to a *coplanar* system of forces, only **N**, **V**, and **M** act at the section.

- In many cases it may be possible to tell by inspection the proper sense of the unknown loadings; however, if this seems difficult, the sense can be assumed.

Equations of Equilibrium.

- Moments should be summed at the section about axes passing through the *centroid* or geometric center of the member's cross-sectional area in order to eliminate the unknown normal and shear forces and thereby obtain direct solutions for the moment components.

- If the solution of the equilibrium equations yields a negative scalar, the assumed sense of the quantity is opposite to that shown on the free-body diagram.

EXAMPLE 7.1

The bar is fixed at its end and is loaded as shown in Fig. 7–4a. Determine the internal normal force at points B and C.

Solution

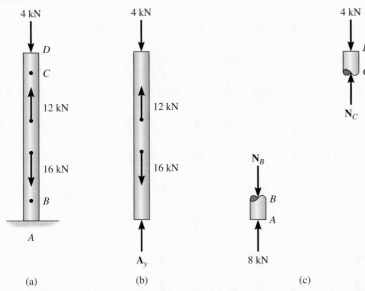

Fig. 7–4

Support Reactions. A free-body diagram of the entire bar is shown in Fig. 7–4b. By inspection, only a normal force A_y acts at the fixed support since the loads are applied symmetrically along the bar's axis. ($A_x = 0$, $M_A = 0$.)

$+\uparrow \Sigma F_y = 0;$ $A_y - 16 \text{ kN} + 12 \text{ kN} - 4 \text{ kN} = 0$ $A_y = 8 \text{ kN}$

Free-Body Diagrams. The internal forces at B and C will be found using the free-body diagrams of the sectioned bar shown in Fig. 7–4c. No shear or moment act on the sections since they are not required for equilibrium. In particular, segments AB and DC will be chosen here, since they contain the *least* number of forces.

Equations of Equilibrium.

Segment AB

$+\uparrow \Sigma F_y = 0;$ $8 \text{ kN} - N_B = 0$ $N_B = 8 \text{ kN}$ *Ans.*

Segment DC

$+\uparrow \Sigma F_y = 0;$ $N_C - 4 \text{ kN} = 0$ $N_C = 4 \text{ kN}$ *Ans.*

Try working this problem in the following manner: Determine N_B from segment BD. (Note that this approach *does not require* solution for the support reaction at A.) Using the result for N_B, isolate segment BC to determine N_C.

EXAMPLE 7.2

The circular shaft is subjected to three concentrated torques as shown in Fig. 7–5a. Determine the internal torques at points B and C.

Solution

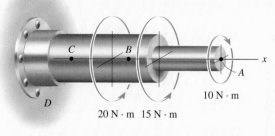

(a)

Support Reactions. Since the shaft is subjected only to collinear torques, a torque reaction occurs at the support, Fig. 7–5b. Using the right-hand rule to define the positive directions of the torques, we require

$$\Sigma M_x = 0; \quad -10 \text{ N} \cdot \text{m} + 15 \text{ N} \cdot \text{m} + 20 \text{ N} \cdot \text{m} - T_D = 0$$

$$T_D = 25 \text{ N} \cdot \text{m}$$

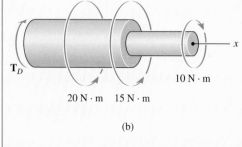

(b)

Fig. 7–5

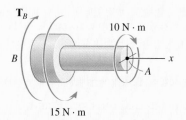

(c)

Free-Body Diagrams. The internal torques at B and C will be found using the free-body diagrams of the shaft segments AB and CD shown in Fig. 7–5c.

Equations of Equilibrium. Applying the equation of moment equilibrium along the shaft's axis, we have

    Segment AB

$$\Sigma M_x = 0; \quad -10 \text{ N} \cdot \text{m} + 15 \text{ N} \cdot \text{m} - T_B = 0 \quad T_B = 5 \text{ N} \cdot \text{m} \quad \textit{Ans.}$$

    Segment CD

$$\Sigma M_x = 0; \quad T_C - 25 \text{ N} \cdot \text{m} = 0 \quad T_C = 25 \text{ N} \cdot \text{m} \quad \textit{Ans.}$$

Try to solve for T_C by using segment CA. Note that this approach *does not require* a solution for the support reaction at D.

E X A M P L E   **7.3**

The beam supports the loading shown in Fig. 7–6a. Determine the internal normal force, shear force, and bending moment acting just to the left, point B, and just to the right, point C, of the 6-kN force.

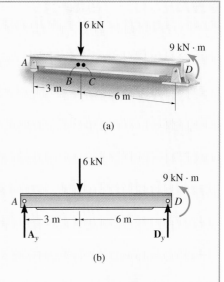

(a)

Solution

Support Reactions.   The free-body diagram of the beam is shown in Fig. 7–6b. When determining the *external reactions*, realize that the 9-kN · m couple moment is a free vector and therefore it can be placed *anywhere* on the free-body diagram of the entire beam. Here we will only determine $\mathbf{A}_y$, since segments AB and AC will be used for the analysis.

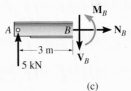

(b)

$$\zeta + \Sigma M_D = 0; \quad 9 \text{ kN} \cdot \text{m} + (6 \text{ kN})(6 \text{ m}) - A_y(9 \text{ m}) = 0$$

$$A_y = 5 \text{ kN}$$

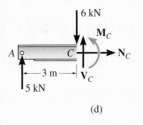

(c)

Free-Body Diagrams.   The free-body diagrams of the left segments AB and AC of the beam are shown in Figs. 7–6c and 7–6d. In this case the 9-kN · m couple moment is *not included* on these diagrams since it must be kept in its *original position* until *after* the section is made and the appropriate body is isolated. In other words, the free-body diagrams of the left segments of the beam do not show the couple moment since this moment does not actually act on these segments.

(d)

Equations of Equilibrium.

Fig. 7–6

Segment AB

$$\xrightarrow{+} \Sigma F_x = 0; \qquad\qquad N_B = 0 \qquad\qquad\qquad Ans.$$
$$+\uparrow \Sigma F_y = 0; \quad 5 \text{ kN} - V_B = 0 \qquad V_B = 5 \text{ kN} \quad Ans.$$
$$\zeta + \Sigma M_B = 0; \quad -(5 \text{ kN})(3 \text{ m}) + M_B = 0 \quad M_B = 15 \text{ kN} \cdot \text{m} \; Ans.$$

Segment AC

$$\xrightarrow{+} \Sigma F_x = 0; \qquad\qquad N_C = 0 \qquad\qquad\qquad Ans.$$
$$+\uparrow \Sigma F_y = 0; \quad 5 \text{ kN} - 6 \text{ kN} + V_C = 0 \qquad V_C = 1 \text{ kN} \quad Ans.$$
$$\zeta + \Sigma M_C = 0; \quad -(5 \text{ kN})(3 \text{ m}) + M_C = 0 \qquad M_C = 15 \text{ kN} \cdot \text{m} \quad Ans.$$

Here the moment arm for the 5-kN force in both cases is approximately 3 m since B and C are "almost" coincident.

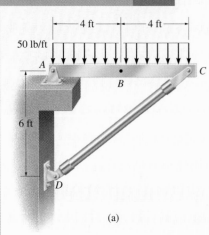

(a)

Determine the internal normal force, shear force, and bending moment acting at point B of the two-member frame shown in Fig. 7–7a.

Solution

Support Reactions.   A free-body diagram of each member is shown in Fig. 7–7b. Since CD is a two-force member, the equations of equilibrium need to be applied only to member AC.

$\zeta + \Sigma M_A = 0$; $-400\text{ lb}(4\text{ ft}) + (\frac{3}{5})F_{DC}(8\text{ ft}) = 0$ $F_{DC} = 333.3\text{ lb}$

$\xrightarrow{+} \Sigma F_x = 0$; $-A_x + (\frac{4}{5})(333.3\text{ lb}) = 0$ $A_x = 266.7\text{ lb}$

$+\uparrow \Sigma F_y = 0$; $A_y - 400\text{ lb} + \frac{3}{5}(333.3\text{ lb}) = 0$ $A_y = 200\text{ lb}$

Free-Body Diagrams.   Passing an imaginary section perpendicular to the axis of member AC through point B yields the free-body diagrams of segments AB and BC shown in Fig. 7–7c. When constructing these diagrams it is important to keep the distributed loading exactly as it is until *after* the section is made. Only then can it be replaced by a single resultant force. Why? Also, notice that $\mathbf{N}_B$, $\mathbf{V}_B$, and $\mathbf{M}_B$ act with equal magnitude but opposite direction on each segment—Newton's third law.

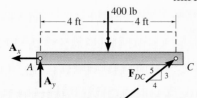

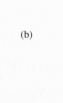

(b)

Equations of Equilibrium.   Applying the equations of equilibrium to segment AB, we have

$\xrightarrow{+} \Sigma F_x = 0$; $N_B - 266.7\text{ lb} = 0$ $N_B = 267\text{ lb}$ *Ans.*

$+\uparrow \Sigma F_y = 0$; $200\text{ lb} - 200\text{ lb} - V_B = 0$ $V_B = 0$ *Ans.*

$\zeta + \Sigma M_B = 0$; $M_B - 200\text{ lb}(4\text{ ft}) + 200\text{ lb}(2\text{ ft}) = 0$

$$M_B = 400\text{ lb} \cdot \text{ft}$$ *Ans.*

As an exercise, try to obtain these same results using segment BC.

Fig. 7–7

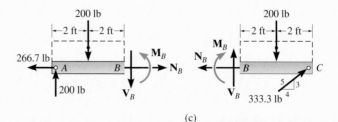

(c)

E X A M P L E **7.5**

Determine the normal force, shear force, and bending moment acting at point E of the frame loaded as shown in Fig. 7–8a.

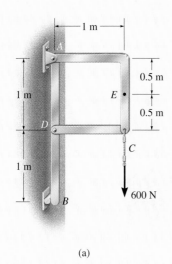

(a)

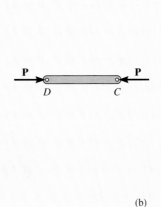

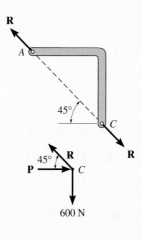

(b)

Solution

Support Reactions.   By inspection, members AC and CD are two-force members, Fig. 7–8b. In order to determine the internal loadings at E, we must first determine the force **R** at the end of member AC. To do this we must analyze the equilibrium of the pin at C. Why? Summing forces in the vertical direction on the pin, Fig. 7–8b, we have

$+\uparrow\Sigma F_y = 0;$ $R \sin 45° - 600 \text{ N} = 0$ $R = 848.5 \text{ N}$

Free-Body Diagram.   The free-body diagram of segment CE is shown in Fig. 7–8c.

Equations of Equilibrium.

$\xrightarrow{+}\Sigma F_x = 0;$ $848.5 \cos 45° \text{ N} - V_E = 0$ $V_E = 600 \text{ N } Ans.$

$+\uparrow\Sigma F_y = 0;$ $-848.5 \sin 45° \text{ N} + N_E = 0$ $N_E = 600 \text{ N } Ans.$

$\zeta+\Sigma M_E = 0;$ $848.5 \cos 45° \text{ N}(0.5 \text{ m}) - M_E = 0$ $M_E = 300 \text{ N} \cdot \text{m } Ans.$

These results indicate a poor design. Member AC should be *straight* (from A to C) so that bending within the member is *eliminated*. If AC is straight then the internal force would only create tension in the member. See Example 6.21.

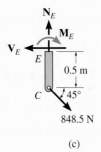

(c)

Fig. 7–8

The uniform sign shown in Fig. 7–9a has a mass of 650 kg and is supported on the fixed column. Design codes indicate that the expected maximum uniform wind loading that will occur in the area where it is located is 900 Pa. Determine the internal loadings at A.

(a)

Solution

The idealized model for the sign is shown in Fig. 7–9b. Here the necessary dimensions are indicated. We can consider the free-body diagram of a section above point A since it does not involve the support reactions.

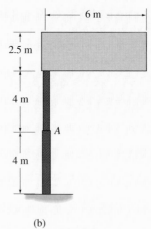

(b)

Fig. 7–9

Free-Body Diagram. The sign has a weight of $W = 650(9.81) = 6.376$ kN, and the wind creates a resultant force of $F_w = 900$ N/m^2(6m)(2.5m) $= 13.5$ kN perpendicular to the face of the sign. These loadings are shown on the free-body diagram, Fig. 7–9c.

Equations of Equilibrium. Since the problem is three dimensional, a vector analysis will be used.

$$\Sigma F = 0; \qquad F_A - 13.5i - 6.376k = 0$$
$$F_A = \{13.5i + 6.38k\} \text{ kN} \qquad\qquad Ans.$$

$$\Sigma M_A = 0; \qquad M_A + r \times (F_w + W) = 0$$

$$M_A + \begin{vmatrix} i & j & k \\ 0 & 3 & 5.25 \\ -13.5 & 0 & 6.376 \end{vmatrix} = 0$$

$$M_A = \{-19.1i + 70.9j + 40.5k\} \text{ kN} \cdot \text{m} \qquad Ans.$$

Here $F_{A_z} = \{6.38k\}$ kN represents the normal force N, whereas $F_{A_x} = \{13.5i\}$ kN is the shear force. Also, the torsional moment is $M_{A_z} = \{40.5k\}$ kN $\cdot$ m, and the bending moment is determined from its components $M_{A_x} = \{-19.1i\}$ kN $\cdot$ m and $M_{A_y} = \{-70.9j\}$ kN $\cdot$ m; i.e., $M_b = \sqrt{M_x^2 + M_y^2}$.

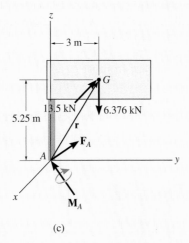

(c)

Fig. 7–9

*7.2   Shear and Moment Equations and Diagrams

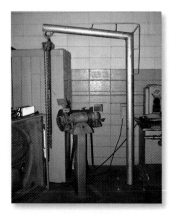

The designer of this shop crane realized the need for additional reinforcement around the joint in order to prevent severe internal bending of the joint when a large load is suspended from the chain hoist.

Beams are structural members which are designed to support loadings applied perpendicular to their axes. In general, beams are long, straight bars having a constant cross-sectional area. Often they are classified as to how they are supported. For example, a *simply supported beam* is pinned at one end and roller-supported at the other, Fig. 7–10, whereas a *cantilevered beam* is fixed at one end and free at the other. The actual design of a beam requires a detailed knowledge of the *variation* of the internal shear force V and bending moment M acting at *each point* along the axis of the beam. After this force and bending-moment analysis is complete, one can then use the theory of mechanics of materials and an appropriate engineering design code to determine the beam's required cross-sectional area.

The *variations* of V and M as functions of the position x along the beam's axis can be obtained by using the method of sections discussed in Sec. 7.1. Here, however, it is necessary to section the beam at an arbitrary distance x from one end rather than at a specified point. If the results are plotted, the graphical variations of V and M as functions of x are termed the *shear diagram* and *bending-moment diagram*, respectively.

In general, the internal shear and bending-moment functions generally will be discontinuous, or their slopes will be discontinuous at points where a distributed load changes or where concentrated forces or couple moments are applied. Because of this, these functions must be determined for *each segment* of the beam located between any two discontinuities of loading. For example, sections located at x_1, x_2, and x_3 will have to be used to describe the variation of V and M throughout the length of the beam in Fig. 7–10. These functions will be valid *only* within regions from O to a for x_1, from a to b for x_2, and from b to L for x_3.

The internal normal force will not be considered in the following discussion for two reasons. In most cases, the loads applied to a beam act perpendicular to the beam's axis and hence produce only an internal shear force and bending moment. For design purposes, the beam's resistance to shear, and particularly to bending, is more important than its ability to resist a normal force.

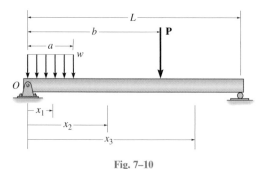

Fig. 7–10

Sign Convention. Before presenting a method for determining the shear and bending moment as functions of x and later plotting these functions (shear and bending-moment diagrams), it is first necessary to establish a *sign convention* so as to define a "positive" and "negative" shear force and bending moment acting in the beam. [This is analogous to assigning coordinate directions x positive to the right and y positive upward when plotting a function $y = f(x)$.] Although the choice of a sign convention is arbitrary, here we will choose the one used for the majority of engineering applications. It is illustrated in Fig. 7–11. Here the positive directions are denoted by an internal *shear force* that causes *clockwise rotation* of the member on which it acts, and by an internal *moment* that causes *compression or pushing on the upper part* of the member. Also, positive moment would tend to bend the member if it were elastic, concave upward. Loadings that are opposite to these are considered negative.

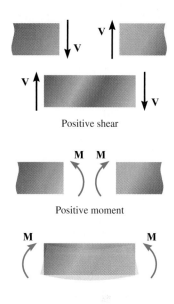

Positive shear

Positive moment

Beam sign convention

Fig. 7–11

PROCEDURE FOR ANALYSIS

The shear and bending-moment diagrams for a beam can be constructed using the following procedure.

Support Reactions.

- Determine all the reactive forces and couple moments acting on the beam and resolve all the forces into components acting perpendicular and parallel to the beam's axis.

Shear and Moment Functions.

- Specify separate coordinates x having an origin at the beam's *left end* and extending to regions of the beam *between* concentrated forces and/or couple moments, or where there is no discontinuity of distributed loading.

- Section the beam perpendicular to its axis at each distance x and draw the free-body diagram of one of the segments. Be sure **V** and **M** are shown acting in their *positive sense*, in accordance with the sign convention given in Fig. 7–11.

- The shear V is obtained by summing forces perpendicular to the beam's axis.

- The moment M is obtained by summing moments about the sectioned end of the segment.

Shear and Moment Diagrams.

- Plot the shear diagram (V versus x) and the moment diagram (M versus x). If computed values of the functions describing V and M are *positive*, the values are plotted above the x axis, whereas *negative* values are plotted below the x axis.

- Generally, it is convenient to plot the shear and bending-moment diagrams directly below the free-body diagram of the beam.

E X A M P L E 7.7

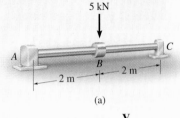

5 kN

A

B

C

2 m

2 m

(a)

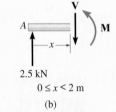

V

A

M

x

2.5 kN

$0 \leq x < 2$ m

(b)

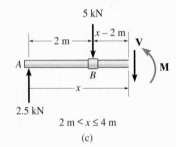

5 kN

2 m

x − 2 m

V

A

B

M

x

2.5 kN

2 m $< x \leq 4$ m

(c)

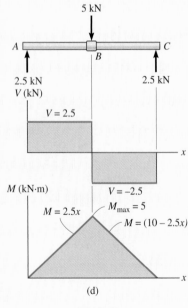

5 kN

A

B

C

2.5 kN

2.5 kN

V (kN)

V = 2.5

x

M (kN·m)

V = −2.5

M = 2.5x

M_{max} = 5

M = (10 − 2.5x)

x

(d)

Fig. 7–12

Draw the shear and bending-moment diagrams for the shaft shown in Fig. 7–12a. The support at A is a thrust bearing and the support at C is a journal bearing.

Solution

Support Reactions. The support reactions have been computed, as shown on the shaft's free-body diagram, Fig. 7–12d.

Shear and Moment Functions. The shaft is sectioned at an arbitrary distance x from point A, extending within the region AB, and the free-body diagram of the left segment is shown in Fig. 7–12b. The unknowns **V** and **M** are assumed to act in the *positive sense* on the right-hand face of the segment according to the established sign convention. Why? Applying the equilibrium equations yields

$$+\uparrow \Sigma F_y = 0; \qquad\qquad V = 2.5 \text{ kN} \qquad\qquad (1)$$

$$\downarrow + \Sigma M = 0; \qquad\qquad M = 2.5x \text{ kN} \cdot \text{m} \qquad\qquad (2)$$

A free-body diagram for a left segment of the shaft extending a distance x within the region BC is shown in Fig. 7–12c. As always, **V** and **M** are shown acting in the positive sense. Hence,

$$+\uparrow \Sigma F_y = 0; \qquad 2.5 \text{ kN} - 5 \text{ kN} - V = 0$$
$$V = -2.5 \text{ kN} \qquad\qquad (3)$$

$$\downarrow + \Sigma M = 0; \qquad M + 5 \text{ kN}(x - 2\text{m}) - 2.5 \text{ kN}(x) = 0$$
$$M = (10 - 2.5x) \text{ kN} \cdot \text{m} \qquad\qquad (4)$$

Shear and Moment Diagrams. When Eqs. 1 through 4 are plotted within the regions in which they are valid, the shear and bending-moment diagrams shown in Fig. 7–12d are obtained. The shear diagram indicates that the internal shear force is always 2.5 kN (positive) within shaft segment AB. Just to the right of point B, the shear force changes sign and remains at a constant value of −2.5 kN for segment BC. The moment diagram starts at zero, increases linearly to point B at x = 2 m, where M_{max} = 2.5 kN(2 m) = 5 kN · m, and thereafter decreases back to zero.

It is seen in Fig. 7–12d that the graph of the shear and moment diagrams is discontinuous at points of concentrated force, i.e., points A, B, and C. For this reason, as stated earlier, it is necessary to express both the shear and bending-moment functions separately for regions between concentrated loads. It should be realized, however, that all loading discontinuities are mathematical, arising from the *idealization of a concentrated force and couple moment*. Physically, loads are always applied over a finite area, and if the load variation could actually be accounted for, the shear and bending-moment diagrams would then be continuous over the shaft's entire length.

EXAMPLE **7.8**

Draw the shear and bending-moment diagrams for the beam shown in Fig. 7–13a.

Solution

Support Reactions. The support reactions have been computed as shown on the beam's free-body diagram, Fig. 7–13c.

Shear and Moment Functions. A free-body diagram for a left segment of the beam having a length x is shown in Fig. 7–13b. The distributed loading acting on this segment has an intensity of $\frac{2}{3}x$ at its end and is replaced by a resultant force *after* the segment is isolated as a free-body diagram. The *magnitude* of the resultant force is equal to $\frac{1}{2}(x)(\frac{2}{3}x) = \frac{1}{3}x^2$. This force *acts through the centroid* of the distributed loading area, a distance $\frac{1}{3}x$ from the right end. Applying the two equations of equilibrium yields

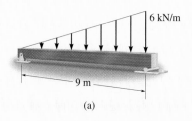
(a)

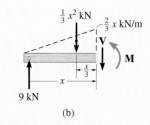
(b)

$$+\uparrow \Sigma F_y = 0; \qquad 9 - \frac{1}{3}x^2 - V = 0$$

$$V = \left(9 - \frac{x^2}{3}\right) \text{kN} \qquad (1)$$

$$\zeta + \Sigma M = 0; \qquad M + \frac{1}{3}x^2\left(\frac{x}{3}\right) - 9x = 0$$

$$M = \left(9x - \frac{x^3}{9}\right) \text{kN} \cdot \text{m} \qquad (2)$$

Shear and Moment Diagrams. The shear and bending-moment diagrams shown in Fig. 7–13c are obtained by plotting Eqs. 1 and 2. The point of *zero shear* can be found using Eq. 1:

$$V = 9 - \frac{x^2}{3} = 0$$

$$x = 5.20 \text{ m}$$

It will be shown in Sec. 7.3 that this value of x happens to represent the point on the beam where the *maximum moment* occurs. Using Eq. (2), we have

$$M_{\text{max}} = \left(9(5.20) - \frac{(5.20)^3}{9}\right) \text{kN} \cdot \text{m}$$

$$= 31.2 \text{ kN} \cdot \text{m}$$

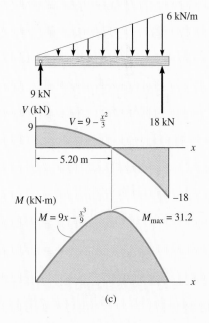
(c)

Fig. 7–13

*7.3 Relations between Distributed Load, Shear, and Moment

In cases where a beam is subjected to several concentrated forces, couple moments, and distributed loads, the method of constructing the shear and bending-moment diagrams discussed in Sec. 7.2 may become quite tedious. In this section a simpler method for constructing these diagrams is discussed—a method based on differential relations that exist between the load, shear, and bending moment.

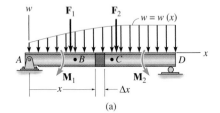

(a)

Distributed Load. Consider the beam AD shown in Fig. 7–14a, which is subjected to an arbitrary load $w = w(x)$ and a series of concentrated forces and couple moments. In the following discussion, the *distributed load* will be considered *positive* when the *loading acts downward* as shown. A free-body diagram for a small segment of the beam having a length Δx is chosen at a point x along the beam which is *not* subjected to a concentrated force or couple moment, Fig. 7–14b. Hence any results obtained will not apply at points of concentrated loading. The internal shear force and bending moment shown on the free-body diagram are assumed to act in the *positive sense* according to the established sign convention. Note that both the shear force and moment acting on the right-hand face must be increased by a small, finite amount in order to keep the segment in equilibrium. The distributed loading has been replaced by a resultant force $\Delta F = w(x)\,\Delta x$ that acts at a fractional distance $k\,(\Delta x)$ from the right end, where $0 < k < 1$ [for example, if $w(x)$ is *uniform*, $k = \frac{1}{2}$]. Applying the equations of equilibrium, we have

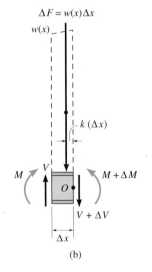

(b)

Fig. 7–14

$$+\uparrow \Sigma F_y = 0; \qquad V - w(x)\,\Delta x - (V + \Delta V) = 0$$
$$\Delta V = -w(x)\,\Delta x$$
$$\zeta + \Sigma M_O = 0; \quad -V\Delta x - M + w(x)\,\Delta x[k(\Delta x)] + (M + \Delta M) = 0$$
$$\Delta M = V\Delta x - w(x)k(\Delta x)^2$$

Dividing by Δx and taking the limit as $\Delta x \to 0$, these two equations become

$$\frac{dV}{dx} = -w(x)$$

$$\begin{array}{c}\text{Slope of} \\ \text{shear diagram}\end{array} = \begin{array}{c}\text{Negative of distributed} \\ \text{load intensity}\end{array}$$

(7–1)

$$\frac{dM}{dx} = V$$

$$\begin{array}{c}\text{Slope of} \\ \text{moment diagram}\end{array} = \text{Shear}$$

(7–2)

These two equations provide a convenient means for plotting the shear and moment diagrams for a beam. At a specific point in a beam, Eq. 7–1 states that the *slope of the shear diagram is equal to the negative of the intensity of the distributed load*, while Eq. 7–2 states that the *slope of the moment diagram is equal to the shear*. In particular, if the shear is equal to zero, $dM/dx = 0$, and therefore *a point of zero shear corresponds to a point of maximum (or possibly minimum) moment*.

Equations 7–1 and 7–2 may also be rewritten in the form $dV = -w(x)\,dx$ and $dM = V\,dx$. Noting that $w(x)\,dx$ and $V\,dx$ represent differential areas under the distributed-loading and shear diagrams, respectively, we can integrate these areas between two points B and C along the beam, Fig. 7–14a, and write

This concrete beam is used to support the roof. Its size and the placement of steel reinforcement within it can be determined once the shear and moment diagrams have been established.

$$\Delta V_{BC} = -\int w(x)\,dx$$

$$\begin{array}{c} \text{Change} \\ \text{in shear} \end{array} = \begin{array}{c} \text{Negative of area under} \\ \text{loading curve} \end{array}$$

(7–3)

and

$$\Delta M_{BC} = \int V\,dx$$

$$\begin{array}{c} \text{Change} \\ \text{in moment} \end{array} = \begin{array}{c} \text{Area under} \\ \text{shear diagram} \end{array}$$

(7–4)

Equation 7–3 states that the *change in shear between points B and C is equal to the negative of the area under the distributed-loading curve between these points*. Similarly, from Eq. 7–4, the *change in moment between B and C is equal to the area under the shear diagram within region BC*. Because two integrations are involved, first to determine the change in shear, Eq. 7–3, then to determine the change in moment, Eq. 7–4, we can state that if the loading curve $w = w(x)$ is a polynomial of degree n, then $V = V(x)$ will be a curve of degree $n + 1$, and $M = M(x)$ will be a curve of degree $n + 2$.

As stated previously, the above equations do not apply at points where a *concentrated* force or couple moment acts. These two special cases create *discontinuities* in the shear and moment diagrams, and as a result, each deserves separate treatment.

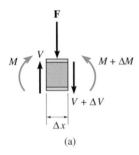

(a)

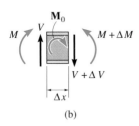

(b)

Fig. 7–15

Force. A free-body diagram of a small segment of the beam in Fig. 7–14a, taken from under one of the forces, is shown in Fig. 7–15a. Here it can be seen that force equilibrium requires

$$+\uparrow \Sigma F_y = 0; \qquad\qquad \Delta V = -F \qquad\qquad (7\text{–}5)$$

Thus, the *change in shear is negative*, so that on the shear diagram the shear will "jump" *downward when* **F** *acts downward* on the beam. Likewise, the jump in shear (ΔV) is upward when **F** acts upward.

Couple Moment. If we remove a segment of the beam in Fig. 7–14a that is located at the couple moment, the free-body diagram shown in Fig. 7–15b results. In this case letting $\Delta x \rightarrow 0$, moment equilibrium requires

$$\zeta + \Sigma M = 0; \qquad\qquad \Delta M = M_0 \qquad\qquad (7\text{–}6)$$

Thus, the *change in moment is positive*, or the moment diagram will "jump" *upward if* **M**$_0$ *is clockwise*. Likewise, the jump ΔM is downward when **M**$_0$ is counterclockwise.

The examples which follow illustrate application of the above equations for the construction of the shear and moment diagrams. After working through these examples, it is recommended that Examples 7–7 and 7–8 be solved using this method.

Each outrigger such as AB supporting this crane acts as a beam which is fixed to the frame of the crane at one end and subjected to a force **F** on the footing at its other end. A proper design requires that the outrigger is able to resist its maximum internal shear and moment. The shear and moment diagrams indicate that the shear will be constant throughout its length and the maximum moment occurs at the support A.

IMPORTANT POINTS

- The slope of the shear diagram is equal to the negative of the intensity of the distributed loading, where positive distributed loading is downward, i.e., $dV/dx = -w(x)$.

- If a concentrated force acts downward on the beam, the shear will jump downward by the amount of the force.

- The change in the shear ΔV between two points is equal to *the negative of the area* under the distributed-loading curve between the points.

- The slope of the moment diagram is equal to the shear, i.e., $dM/dx = V$.

- The change in the moment ΔM between two points is equal to the *area* under the shear diagram between the two points.

- If a *clockwise* couple moment acts on the beam, the shear will not be affected, however, the moment diagram will jump *upward* by the amount of the moment.

- Points of *zero shear* represent points of *maximum or minimum moment* since $dM/dx = 0$.

EXAMPLE 7.9

Draw the shear and moment diagrams for the beam shown in Fig. 7–16a.

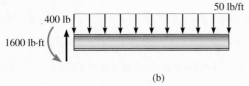

(b)

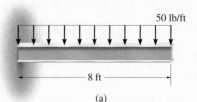

(a)

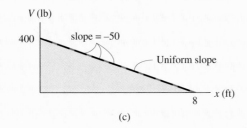

(c)

Fig. 7–16

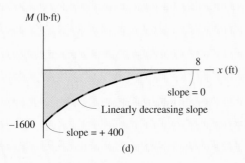

(d)

Solution

Support Reactions. The reactions at the fixed support have been calculated and are shown on the free-body diagram of the beam, Fig. 7–16b.

Shear Diagram. The shear at the end points is plotted first, Fig. 7–16c. From the sign convention, Fig. 7–11, $V = +400$ at $x = 0$ and $V = 0$ at $x = 8$. Since $dV/dx = -w = -50$, a straight, *negative* sloping line connects the end points.

Moment Diagram. From our sign convention, Fig. 7–11, the moments at the beam's end points, $M = -1600$ at $x = 0$ and $M = 0$ at $x = 8$, are plotted first, Fig. 7–16d. Successive values of shear taken from the shear diagram, Fig. 7–16c, indicate that the *slope $dM/dx = V$* of the moment diagram, Fig. 7–16d, is always positive yet *linearly decreasing* from $dM/dx = 400$ at $x = 0$ to $dM/dx = 0$ at $x = 8$. Thus, due to the integrations, w a constant yields V a sloping line (first-degree curve) and M a parabola (second-degree curve).

E X A M P L E   7.10

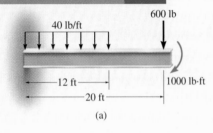

(a)

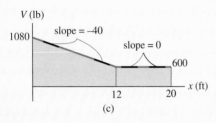

(b)

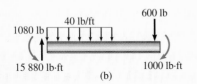

(c)

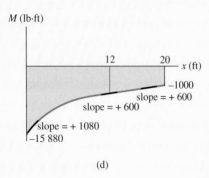

(d)

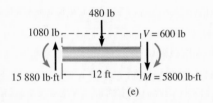

(e)

Fig. 7–17

Draw the shear and moment diagrams for the cantilevered beam shown in Fig. 7–17a.

Solution

Support Reactions. The reactions at the fixed support have been calculated and are shown on the free-body diagram of the beam, Fig. 7–17b.

Shear Diagram. Using the established sign convention, Fig. 7–11, the shear at the ends of the beam is plotted first; i.e., $x = 0$, $V = +1080$; $x = 20$, $V = +600$, Fig. 7–17c.

Since the uniform distributed load is downward and *constant*, the slope of the shear diagram is $dV/dx = -w = -40$ for $0 \le x < 12$ as indicated.

The magnitude of shear at $x = 12$ is $V = +600$. This can be determined by first finding the area under the load diagram between $x = 0$ and $x = 12$. This represents the change in shear. That is, $\Delta V = -\int w(x)\, dx = -40(12) = -480$. Thus $V|_{x=12} = V|_{x=0} + (-480) = 1080 - 480 = 600$. Also, we can obtain this value by using the method of sections, Fig. 7–17e, where for equilibrium $V = +600$.

Since the load between $12 < x \le 20$ is $w = 0$, the slope $dV/dx = 0$ as indicated. This brings the shear to the required value of $V = +600$ at $x = 20$.

Moment Diagram. Again, using the established sign convention, the moments at the ends of the beam are plotted first; i.e., $x = 0$, $M = -15\,880$; $x = 20$, $M = -1000$, Fig. 7–17d.

Each value of shear gives the slope of the moment diagram since $dM/dx = V$. As indicated, at $x = 0$, $dM/dx = +1080$; and at $x = 12$, $dM/dx = +600$. For $0 \le x < 12$, specific values of the shear diagram are positive but linearly decreasing. Hence, the moment diagram is parabolic with a linearly decreasing positive slope.

The magnitude of moment at $x = 12$ is -5800. This can be found by first determining the trapezoidal area under the shear diagram, which represents the change in moment, $\Delta M = \int V dx = 600(12) + \frac{1}{2}(1080 - 600)(12) = +10080$. Thus, $M|_{x=12} = M|_{x=0} + 10080 = -15\,880 + 10\,080 = -5800$. The more "basic" method of sections can also be used, where equilibrium at $x = 12$ requires $M = -5800$, Fig. 7–17e.

The moment diagram has a constant slope for $12 < x \le 20$ since, from the shear diagram, $dM/dx = V = +600$. This brings the value of $M = -1000$ at $x = 20$, as required.

E X A M P L E   7.11

Draw the shear and moment diagrams for the shaft in Fig. 7–18a. The support at A is a thrust bearing and the support at B is a journal bearing.

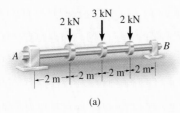

(a)

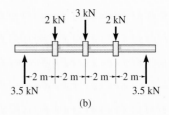

(b)

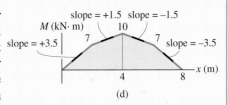

(c)

Solution

Support Reactions. The reactions at the supports are shown on the free-body diagram in Fig. 7–18b.

Shear Diagram. The end points $x = 0$, $V = +3.5$ and $x = 8$, $V = -3.5$ are plotted first, as shown in Fig. 7–18c.

Since there is no distributed load on the shaft, the slope of the shear diagram throughout the shaft's length is zero; i.e., $dV/dx = -w = 0$. There is a discontinuity or "jump" of the shear diagram, however, at each concentrated force. From Eq. 7–5, $\Delta V = -F$, the change in shear is negative when the force acts downward and positive when the force acts upward. Stated another way, the "jump" follows the force, i.e., a downward force causes a downward jump, and vice versa. Thus, the 2-kN force at $x = 2$ m changes the shear from 3.5 kN to 1.5 kN; the 3-kN force at $x = 4$ m changes the shear from 1.5 kN to -1.5 kN, etc. We can *also* obtain numerical values for the shear at a specified point in the shaft by using the method of sections, as for example, $x = 2^+$ m, $V = 1.5$ kN in Fig. 7–18e.

Moment Diagram. The end points $x = 0$, $M = 0$ and $x = 8$, $M = 0$ are plotted first, as shown in Fig. 7–18d.

Since the shear is constant in each region of the shaft, the moment diagram has a corresponding constant positive or negative slope as indicated on the diagram. Numerical values for the change in moment at any point can be computed from the *area* under the shear diagram. For example, at $x = 2$ m, $\Delta M = \int V \, dx = 3.5(2) = 7$. Thus, $M|_{x=2} = M|_{x=0} + 7 = 0 + 7 = 7$. Also, by the method of sections, we can determine the moment at a specified point, as for example, $x = 2^+$ m, $M = 7$ kN·m, Fig. 7–18e.

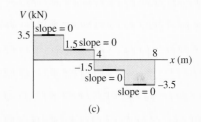

(d)

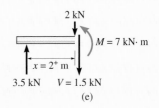

(e)

Fig. 7–18

EXAMPLE 7.12

Sketch the shear and moment diagrams for the beam shown in Fig. 7–19a.

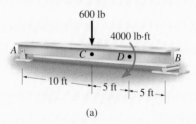

(a)

Solution

Support Reactions. The reactions are calculated and indicated on the free-body diagram, Fig. 7–19b.

Shear Diagram. As in Example 7.11, the shear diagram can be constructed by "following the load" on the free-body diagram. In this regard, beginning at A, the reaction is up so $V_A = +100$ lb, Fig. 7–19c. No load acts between A and C, so the shear remains constant; i.e., $dV/dx = -w(x) = 0$. At C the 600-lb force acts downward, so the shear jumps down 600 lb, from 100 lb to -500 lb. Again the shear is constant (no load) and ends at -500 lb, point B. Notice that no jump or discontinuity in shear occurs at D, the point where the 4000-lb·ft couple moment is applied, Fig. 7–19b. This is because, for force equilibrium, $\Delta V = 0$ in Fig. 7–15b.

Moment Diagram. The moment at each end of the beam is zero. These two points are plotted first, Fig. 7–19d. The slope of the moment diagram from A to C is constant since $dM/dx = V = +100$. The value

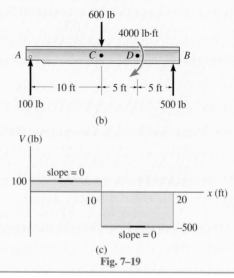

(b)

(c)

Fig. 7–19

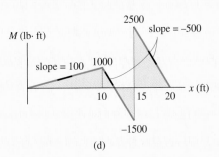

(d)

of the moment at C can be determined by the method of sections, Fig. 7–19e where $M_C = +1000 \text{ lb} \cdot \text{ft}$; or by first computing the rectangular area under the shear diagram between A and C to obtain the change in moment $\Delta M_{AC} = (100 \text{ lb})(10 \text{ ft}) = 1000 \text{ lb} \cdot \text{ft}$. Since $M_A = 0$, then $M_C = 0 + 1000 \text{ lb} \cdot \text{ft} = 1000 \text{ lb} \cdot \text{ft}$. From C to D the slope of the moment diagram is $dM/dx = V = -500$, Fig. 7–19c. The area under the shear diagram between points C and D is $\Delta M_{CD} = (-500 \text{ lb})(5 \text{ ft}) = -2500 \text{ lb} \cdot \text{ft}$, so that $M_D = M_C + \Delta M_{CD}$ $= 1000 - 2500 = -1500 \text{ lb} \cdot \text{ft}$. A jump in the moment diagram occurs at point D, which is caused by the concentrated couple moment of $4000 \text{ lb} \cdot \text{ft}$. From Eq. 7–6, the jump is *positive* since the couple moment is *clockwise*. Thus, at $x = 15^+$ ft, the moment is $M_D = -1500 + 4000 = 2500 \text{ lb} \cdot \text{ft}$. This value can *also* be determined by the method of sections, Fig. 7–19f. From point D the slope of $dM/dx = -500$ is maintained until the diagram closes to zero at B, Fig. 7–19d.

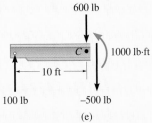

(e)

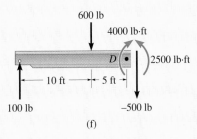

(f)

Fig. 7–19

*7.4 Cables

Flexible cables and chains are often used in engineering structures for support and to transmit loads from one member to another. When used to support suspension bridges and trolley wheels, cables form the main load-carrying element of the structure. In the force analysis of such systems, the weight of the cable itself may be neglected because it is often small compared to the load it carries. On the other hand, when cables are used as transmission lines and guys for radio antennas and derricks, the cable weight may become important and must be included in the structural analysis. Three cases will be considered in the analysis that follows: (1) a cable subjected to concentrated loads; (2) a cable subjected to a distributed load; and (3) a cable subjected to its own weight. Regardless of which loading conditions are present, provided the loading is coplanar with the cable, the requirements for equilibrium are formulated in an identical manner.

When deriving the necessary relations between the force in the cable and its slope, we will make the assumption that the cable is *perfectly flexible* and *inextensible*. Due to its flexibility, the cable offers no resistance to bending, and therefore, the tensile force acting in the cable is always tangent to the cable at points along its length. Being inextensible, the cable has a constant length both before and after the load is applied. As a result, once the load is applied, the geometry of the cable remains fixed, and the cable or a segment of it can be treated as a rigid body.

Cable Subjected to Concentrated Loads.

When a cable of negligible weight supports several concentrated loads, the cable takes the form of several straight-line segments, each of which is subjected to a constant tensile force. Consider, for example, the cable shown in Fig. 7–20, where the distances h, L_1, L_2, and L_3 and the loads $\mathbf{P}_1$ and $\mathbf{P}_2$ are known. The problem here is to determine the *nine unknowns* consisting of the tension in each of the *three* segments, the *four* components of reaction at A and B, and the sags y_C and y_D at the *two* points C and D. For the solution we can write *two* equations of force equilibrium at each of points A, B, C, and D. This results in a total of *eight equations*.* To complete the solution, it will be necessary to know something about the geometry of the cable in order to obtain the necessary ninth equation. For example, if the cable's total *length L* is specified, then the Pythagorean theorem can be used to relate each of the three segmental lengths, written in terms of h, y_C, y_D, L_1, L_2, and L_3, to the total length L. Unfortunately, this type of problem cannot be solved easily by hand. Another possibility, however, is to specify one of the sags, either y_C or y_D, instead of the cable length. By doing this, the equilibrium equations are then sufficient for obtaining the unknown forces and the remaining sag. Once the sag at each point of loading is obtained, the length of the cable can be determined by trigonometry. The following example illustrates a procedure for performing the equilibrium analysis for a problem of this type.

Each of the cable segments remains approximately straight as they support the weight of these traffic lights.

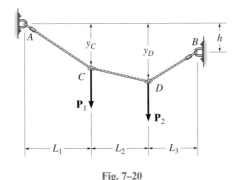

Fig. 7–20

*As will be shown in the following example, the eight equilibrium equations can *also* be written for the entire cable, or any part thereof. But *no more* than *eight* equations are available.

EXAMPLE 7.13

Determine the tension in each segment of the cable shown in Fig. 7–21a.

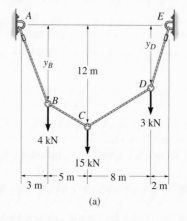

(a)

Solution

By inspection, there are four unknown external reactions (A_x, A_y, E_x, and E_y) and four unknown cable tensions, one in each cable segment. These eight unknowns along with the two unknown sags y_B and y_D can be determined from *ten* available equilibrium equations. One method is to apply these equations as force equilibrium ($\Sigma F_x = 0$, $\Sigma F_y = 0$) to each of the five points A through E. Here, however, we will take a more direct approach.

Consider the free-body diagram for the entire cable, Fig. 7–21b. Thus,

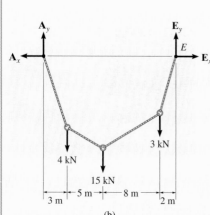

(b)

$$\xrightarrow{+} \Sigma F_x = 0; \qquad -A_x + E_x = 0$$

$$\zeta + \Sigma M_E = 0; \quad -A_y(18\,\text{m}) + 4\,\text{kN}(15\,\text{m}) + 15\,\text{kN}(10\,\text{m}) + 3\,\text{kN}(2\,\text{m}) = 0$$

$$A_y = 12\,\text{kN}$$

$$+\uparrow \Sigma F_y = 0; \quad 12\,\text{kN} - 4\,\text{kN} - 15\,\text{kN} - 3\,\text{kN} + E_y = 0$$

$$E_y = 10\,\text{kN}$$

Since the sag $y_C = 12$ m is known, we will now consider the leftmost section, which cuts cable BC, Fig. 7–21c.

$$\zeta + \Sigma M_C = 0; \quad A_x(12\,\text{m}) - 12\,\text{kN}(8\,\text{m}) + 4\,\text{kN}(5\,\text{m}) = 0$$

$$A_x = E_x = 6.33\,\text{kN}$$

$$\xrightarrow{+} \Sigma F_x = 0; \qquad T_{BC}\cos\theta_{BC} - 6.33\,\text{kN} = 0$$

$$+\uparrow \Sigma F_y = 0; \qquad 12\,\text{kN} - 4\,\text{kN} - T_{BC}\sin\theta_{BC} = 0$$

Thus,

$$\theta_{BC} = 51.6°$$

$$T_{BC} = 10.2\,\text{kN} \qquad \qquad Ans.$$

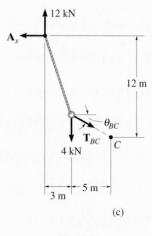

(c)

Fig. 7–21

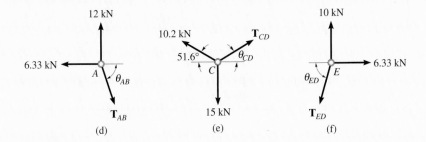

(d)   (e)   (f)

Proceeding now to analyze the equilibrium of points A, C, and E in sequence, we have

Point A.   (Fig. 7–21d)

$\xrightarrow{+} \Sigma F_x = 0;$ $T_{AB} \cos \theta_{AB} - 6.33 \text{ kN} = 0$

$+\uparrow \Sigma F_y = 0;$ $-T_{AB} \sin \theta_{AB} + 12 \text{ kN} = 0$

$$\theta_{AB} = 62.2°$$
$$T_{AB} = 13.6 \text{ kN} \qquad\qquad Ans.$$

Point C.   (Fig. 7–21e)

$\xrightarrow{+} \Sigma F_x = 0;$ $T_{CD} \cos \theta_{CD} - 10.2 \cos 51.6° \text{ kN} = 0$

$+\uparrow \Sigma F_y = 0;$ $T_{CD} \sin \theta_{CD} + 10.2 \sin 51.6° \text{ kN} - 15 \text{ kN} = 0$

$$\theta_{CD} = 47.9°$$
$$T_{CD} = 9.44 \text{ kN} \qquad\qquad Ans.$$

Point E.   (Fig. 7–21f)

$\xrightarrow{+} \Sigma F_x = 0;$ $6.33 \text{ kN} - T_{ED} \cos \theta_{ED} = 0$

$+\uparrow \Sigma F_y = 0;$ $10 \text{ kN} - T_{ED} \sin \theta_{ED} = 0$

$$\theta_{ED} = 57.7°$$
$$T_{ED} = 11.8 \text{ kN} \qquad\qquad Ans.$$

By comparison, the maximum cable tension is in segment AB since this segment has the greatest slope (θ) and it is required that for any left-hand cable segment the horizontal component $T \cos \theta = A_x$ (a constant). Also, since the slope angles that the cable segments make with the horizontal have now been determined, it is possible to determine the sags y_B and y_D, Fig. 7–21a, using trigonometry.

Cable Subjected to a Distributed Load. Consider the weightless cable shown in Fig. 7–22a, which is subjected to a loading function $w = w(x)$ *as measured in the x direction.* The free-body diagram of a small segment of the cable having a length Δs is shown in Fig. 7–22b. Since the tensile force in the cable changes continuously in both magnitude and direction along the cable's length, this change is denoted on the free-body diagram by ΔT. The distributed load is represented by its resultant

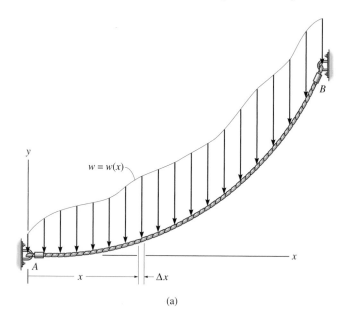

(a)

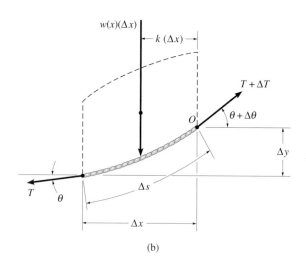

(b)

Fig. 7–22

force $w(x)(\Delta x)$, which acts at a fractional distance $k(\Delta x)$ from point O, where $0 < k < 1$. Applying the equations of equilibrium yields

$$\xrightarrow{+} \Sigma F_x = 0; \qquad -T\cos\theta + (T + \Delta T)\cos(\theta + \Delta\theta) = 0$$
$$+\uparrow \Sigma F_y = 0; \qquad -T\sin\theta - w(x)(\Delta x) + (T + \Delta T)\sin(\theta + \Delta\theta) = 0$$
$$\zeta + \Sigma M_O = 0; \qquad w(x)(\Delta x)k(\Delta x) - T\cos\theta\,\Delta y + T\sin\theta\,\Delta x = 0$$

Dividing each of these equations by Δx and taking the limit as $\Delta x \to 0$, and hence $\Delta y \to 0$, $\Delta\theta \to 0$, and $\Delta T \to 0$, we obtain

$$\frac{d(T\cos\theta)}{dx} = 0 \qquad (7\text{–}7)$$

The cable and suspenders are used to support the uniform load of a gas pipe which crosses the river.

$$\frac{d(T\sin\theta)}{dx} - w(x) = 0 \qquad (7\text{–}8)$$

$$\frac{dy}{dx} = \tan\theta \qquad (7\text{–}9)$$

Integrating Eq. 7–7, we have

$$T\cos\theta = \text{constant} = F_H \qquad (7\text{–}10)$$

Here F_H represents the horizontal component of tensile force at *any point* along the cable.

Integrating Eq. 7–8 gives

$$T\sin\theta = \int w(x)\,dx \qquad (7\text{–}11)$$

Dividing Eq. 7–11 by Eq. 7–10 eliminates T. Then, using Eq. 7–9, we can obtain the slope

$$\tan\theta = \frac{dy}{dx} = \frac{1}{F_H}\int w(x)\,dx$$

Performing a second integration yields

$$y = \frac{1}{F_H}\int \left(\int w(x)\,dx\right)dx \qquad (7\text{–}12)$$

This equation is used to determine the curve for the cable, $y = f(x)$. The horizontal force component F_H and the two constants, say C_1 and C_2, resulting from the integration are determined by applying the boundary conditions for the cable.

EXAMPLE 7.14

The cable of a suspension bridge supports half of the uniform road surface between the two columns at A and B, as shown in Fig. 7–23a. If this distributed loading is w_0, determine the maximum force developed in the cable and the cable's required length. The span length L and, sag h are known.

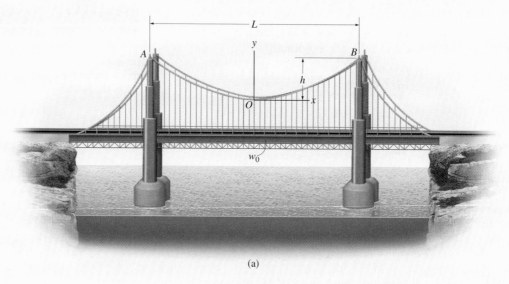

(a)

Fig. 7–23

Solution

We can determine the unknowns in the problem by first finding the curve that defines the shape of the cable by using Eq. 7–12. For reasons of symmetry, the origin of coordinates has been placed at the cable's center. Noting that $w(x) = w_0$, we have

$$y = \frac{1}{F_H} \int \left(\int w_0 \, dx \right) dx$$

Performing the two integrations gives

$$y = \frac{1}{F_H} \left(\frac{w_0 x^2}{2} + C_1 x + C_2 \right) \tag{1}$$

The constants of integration may be determined by using the boundary conditions $y = 0$ at $x = 0$ and $dy/dx = 0$ at $x = 0$. Substituting into Eq. 1 yields $C_1 = C_2 = 0$. The curve then becomes

$$y = \frac{w_0}{2F_H} x^2 \tag{2}$$

This is the equation of a *parabola*. The constant F_H may be obtained by using the boundary condition $y = h$ at $x = L/2$. Thus,

$$F_H = \frac{w_0 L^2}{8h} \qquad (3)$$

Therefore, Eq. 2 becomes

$$y = \frac{4h}{L^2} x^2 \qquad (4)$$

Since F_H is known, the tension in the cable may be determined using Eq. 7–10, written as $T = F_H/\cos\theta$. For $0 \le \theta < \pi/2$, the maximum tension will occur when θ is *maximum*, i.e., at point B, Fig. 7–23a. From Eq. 2, the slope at this point is

$$\left.\frac{dy}{dx}\right|_{x=L/2} = \tan\theta_{max} = \left.\frac{w_0}{F_H} x\right|_{x=L/2}$$

or

$$\theta_{max} = \tan^{-1}\left(\frac{w_0 L}{2F_H}\right) \qquad (5)$$

Therefore,

$$T_{max} = \frac{F_H}{\cos(\theta_{max})} \qquad (6)$$

Using the triangular relationship shown in Fig. 7–23b, which is based on Eq. 5, Eq. 6 may be written as

$$T_{max} = \frac{\sqrt{4F_H^2 + w_0^2 L^2}}{2}$$

Substituting Eq. 3 into the above equation yields

$$T_{max} = \frac{w_0 L}{2}\sqrt{1 + \left(\frac{L}{4h}\right)^2} \qquad \textit{Ans.}$$

For a differential segment of cable length ds, we can write

$$ds = \sqrt{(dx)^2 + (dy)^2} = \sqrt{1 + \left(\frac{dy}{dx}\right)^2}\, dx$$

Hence, the total length of the cable, $\mathscr{L}$, can be determined by integration. Using Eq. 4, we have

$$\mathscr{L} = \int ds = 2\int_0^{L/2}\sqrt{1 + \left(\frac{8h}{L^2}x\right)^2}\, dx \qquad (7)$$

Integrating yields

$$\mathscr{L} = \frac{L}{2}\left[\sqrt{1 + \left(\frac{4h}{L}\right)^2} + \frac{L}{4h}\sinh^{-1}\left(\frac{4h}{L}\right)\right] \qquad \textit{Ans.}$$

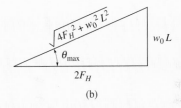

(b)

Fig. 7–23

Over time the forces the cables exert on this telephone pole have caused it to tilt. Proper bracing should be required.

Cable Subjected to Its Own Weight. When the weight of the cable becomes important in the force analysis, the loading function along the cable becomes a function of the arc length s rather than the projected length x. A generalized loading function $w = w(s)$ acting along the cable is shown in Fig. 7–24a. The free-body diagram for a segment of the cable is shown in Fig. 7–24b. Applying the equilibrium equations to the force system on this diagram, one obtains relationships identical to those given by Eqs. 7–7 through 7–9, but with ds replacing dx. Therefore, it may be shown that

$$T \cos \theta = F_H$$

$$T \sin \theta = \int w(s) \, ds \tag{7–13}$$

$$\frac{dy}{dx} = \frac{1}{F_H} \int w(s) \, ds \tag{7–14}$$

To perform a direct integration of Eq. 7–14, it is necessary to replace dy/dx by ds/dx. Since

$$ds = \sqrt{dx^2 + dy^2}$$

then

$$\frac{dy}{dx} = \sqrt{\left(\frac{ds}{dx}\right)^2 - 1}$$

Therefore,

$$\frac{ds}{dx} = \left\{ 1 + \frac{1}{F_H^2} \left(\int w(s) \, ds \right)^2 \right\}^{1/2}$$

Separating the variables and integrating yields

$$x = \int \frac{ds}{\left\{ 1 + \dfrac{1}{F_H^2} \left(\int w(s) \, ds \right)^2 \right\}^{1/2}} \tag{7–15}$$

The two constants of integration, say C_1 and C_2, are found using the boundary conditions for the cable.

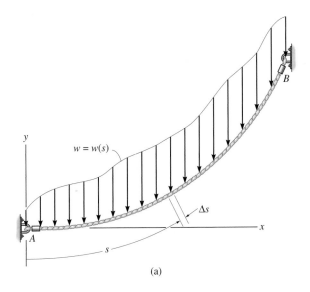

(a)

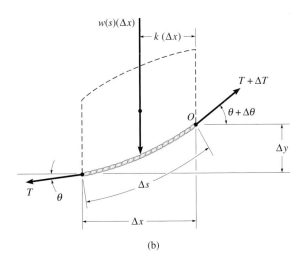

(b)

Fig. 7–24

EXAMPLE 7.15

Determine the deflection curve, the length, and the maximum tension in the uniform cable shown in Fig. 7–25. The cable weighs $w_0 = 5$ N/m.

Solution

For reasons of symmetry, the origin of coordinates is located at the center of the cable. The deflection curve is expressed as $y = f(x)$. We can determine it by first applying Eq. 7–15, where $w(s) = w_0$.

$$x = \int \frac{ds}{[1 + (1/F_H^2)(\int w_0 \, ds)^2]^{1/2}}$$

Integrating the term under the integral sign in the denominator, we have

$$x = \int \frac{ds}{[1 + (1/F_H^2)(w_0 s + C_1)^2]^{1/2}}$$

Substituting $u = (1/F_H)(w_0 s + C_1)$ so that $du = (w_0/F_H) \, ds$, a second integration yields

$$x = \frac{F_H}{w_0}(\sinh^{-1} u + C_2)$$

or

$$x = \frac{F_H}{w_0}\left\{ \sinh^{-1}\left[\frac{1}{F_H}(w_0 s + C_1) \right] + C_2 \right\} \tag{1}$$

To evaluate the constants note that, from Eq. 7–14,

$$\frac{dy}{dx} = \frac{1}{F_H}\int w_0 \, ds \quad \text{or} \quad \frac{dy}{dx} = \frac{1}{F_H}(w_0 s + C_1)$$

Since $dy/dx = 0$ at $s = 0$, then $C_1 = 0$. Thus,

$$\frac{dy}{dx} = \frac{w_0 s}{F_H} \tag{2}$$

The constant C_2 may be evaluated by using the condition $s = 0$ at $x = 0$ in Eq. 1, in which case $C_2 = 0$. To obtain the deflection curve, solve for s in Eq. 1, which yields

$$s = \frac{F_H}{w_0}\sinh\left(\frac{w_0}{F_H}x \right) \tag{3}$$

Now substitute into Eq. 2, in which case

$$\frac{dy}{dx} = \sinh\left(\frac{w_0}{F_H}x \right)$$

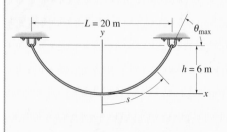

Fig. 7–25

L = 20 m

y

θ_{max}

h = 6 m

s

x

Hence

$$y = \frac{F_H}{w_0}\cosh\left(\frac{w_0}{F_H}x\right) + C_3 \tag{4}$$

If the boundary condition $y = 0$ at $x = 0$ is applied, the constant $C_3 = -F_H/w_0$, and therefore the deflection curve becomes

$$y = \frac{F_H}{w_0}\left[\cosh\left(\frac{w_0}{F_H}x\right) - 1\right]$$

This equation defines the shape of a *catenary curve*. The constant F_H is obtained by using the boundary condition that $y = h$ at $x = L/2$, in which case

$$h = \frac{F_H}{w_0}\left[\cosh\left(\frac{w_0 L}{2F_H}\right) - 1\right] \tag{5}$$

Since $w_0 = 5$ N/m, $h = 6$ m, and $L = 20$ m, Eqs. 4 and 5 become

$$y = \frac{F_H}{5\text{ N/m}}\left[\cosh\left(\frac{5\text{ N/m}}{F_H}x\right) - 1\right] \tag{6}$$

$$6\text{ m} = \frac{F_H}{5\text{ N/m}}\left[\cosh\left(\frac{50\text{ N}}{F_H}\right) - 1\right] \tag{7}$$

Equation 7 can be solved for F_H by using a trial-and-error procedure. The result is

$$F_H = 45.9\text{ N}$$

and therefore the deflection curve, Eq. 6, becomes

$$y = 9.19[\cosh(0.109x) - 1]\text{ m} \qquad \textit{Ans.}$$

Using Eq. 3, with $x = 10$ m, the half-length of the cable is

$$\frac{\mathcal{L}}{2} = \frac{45.9\text{ N}}{5\text{ N/m}}\sinh\left[\frac{5\text{ N/m}}{45.9\text{ N}}(10\text{ m})\right] = 12.1\text{ m}$$

Hence,

$$\mathcal{L} = 24.2\text{ m} \qquad \textit{Ans.}$$

Since $T = F_H/\cos\theta$, Eq. 7–13, the maximum tension occurs when θ is maximum, i.e., at $s = \mathcal{L}/2 = 12.1$ m. Using Eq. 2 yields

$$\left.\frac{dy}{dx}\right|_{s=12.1\text{ m}} = \tan\theta_{max} = \frac{5\text{ N/m}(12.1\text{ m})}{45.9\text{ N}} = 1.32$$

$$\theta_{max} = 52.8°$$

Thus,

$$T_{max} = \frac{F_H}{\cos\theta_{max}} = \frac{45.9\text{ N}}{\cos 52.8°} = 75.9\text{ N} \qquad \textit{Ans.}$$

CHAPTER REVIEW

- **Internal Loadings.** If a coplanar force system acts on a member, then in general a resultant internal *normal force N, shear force V*, and *bending moment M* will act at any cross section along the member. These resultants are determined using the method of sections. To find them, the member is sectioned at the point where the internal loadings are to be determined. A free-body diagram of one of the sectioned parts is then drawn. The normal force is determined by summing forces normal to the cross section. The shear force is found by summing forces tangent to the cross section, and the bending moment is found by summing moments about the centroid of the cross-sectional area. If the member is subjected to a three-dimensional loading, then, in general, a *torsional loading* will also act on the cross section. It can be determined by summing moments about an axis that is perpendicular to the cross section and passes through its centroid.

- **Shear and Moment Diagrams as Functions of *x*.** To construct the shear and moment diagrams for a member, it is necessary to section the member at an arbitrary point, located a distance *x* from one end. The unknown shear and moment are indicated on the cross section in the positive direction according to the established sign convention. Application of the equilibrium equations will give these loadings as a function of *x*, which can then be plotted. If the external loading consists of changes in the distributed load, or a series of concentrated forces and couple moments act on the member, then different expressions for V and M must be determined within regions between these different loadings.

- **Graphical Methods for Establishing Shear and Moment Diagrams.** It is possible to plot the shear and moment diagrams quickly by using differential relationships that exist between the distributed loading w and V and M. The slope of the shear diagram is equal to the distributed loading at any point, $dV/dx = -w$; and the slope of the moment diagram is equal to the shear at any point, $V = dM/dx$. Also, the change in shear between any two points is equal to the area under the distributed loading between the points, $\Delta V = \int w \, dx$, and the change in the moment is equal to the area under the shear diagram between the points, $\Delta M = \int V \, dx$.

- **Cables.** When a flexible and inextensible cable is subjected to a series of concentrated forces, then the analysis of the cable can be performed by using the equations of equilibrium applied to free-body diagrams of either segments or points of application of the loading. If external distributed loads or the weight of the cable are to be considered, then the forces and shape of the cable must be determined by first analyzing the forces on a differential segment of the cable and then integrating this result.

REVIEW PROBLEMS

*∎7-1. A 100-lb cable is attached between two points at a distance 50 ft apart having equal elevations. If the maximum tension developed in the cable is 75 lb, determine the length of the cable and the sag.

7-2. Determine the distance *a* between the supports in terms of the beam's length *L* so that the moment in the *symmetric* beam is zero at the beam's center.

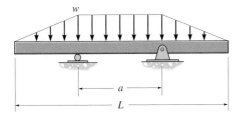

Prob. 7–2

7-3. Draw the shear and moment diagrams for the beam.

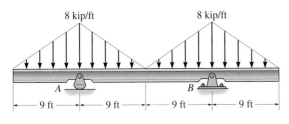

Prob. 7–3

7-4. Draw the shear and moment diagrams for the beam *ABC*.

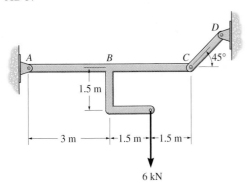

Prob. 7–4

*7-5. Draw the shear and moment diagrams for the beam.

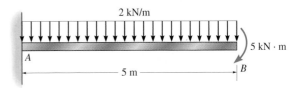

Prob. 7–5

7-6. Determine the normal force, shear force, and moment at points *B* and *C* of the beam.

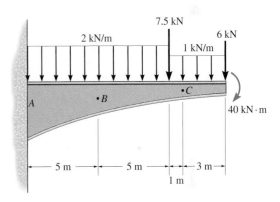

Prob. 7–6

7-7. A chain is suspended between points at the same elevation and spaced a distance of 60 ft apart. If it has a weight of 0.5 lb/ft and the sag is 3 ft, determine the maximum tension in the chain.

7-8. Draw the shear and moment diagrams for the beam.

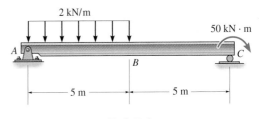

Prob. 7–8

The effective design of a brake system, such as the one for this bicycle, requires efficient capacity for the mechanism to resist frictional forces. In this chapter we will study the nature of friction and show how these forces are considered in engineering analysis.

Friction

- To introduce the concept of dry friction and show how to analyze the equilibrium of rigid bodies subjected to this force.
- To present specific applications of frictional force analysis on wedges, screws, belts, and bearings.
- To investigate the concept of rolling resistance.

8.1   Characteristics of Dry Friction

The heat generated by the abrasive action of friction can be noticed when using this grinder to sharpen a metal blade.

Friction may be defined as a force of resistance acting on a body which prevents or retards slipping of the body relative to a second body or surface with which it is in contact. This force always acts *tangent* to the surface at points of contact with other bodies and is directed so as to oppose the possible or existing motion of the body relative to these points.

In general, two types of friction can occur between surfaces. *Fluid friction* exists when the contacting surfaces are separated by a film of fluid (gas or liquid). The nature of fluid friction is studied in fluid mechanics since it depends upon knowledge of the velocity of the fluid and the fluid's ability to resist shear force. In this book only the effects of *dry friction* will be presented. This type of friction is often called *Coulomb friction* since its characteristics were studied extensively by C. A. Coulomb in 1781. Specifically, dry friction occurs between the contacting surfaces of bodies in the absence of a lubricating fluid.

259

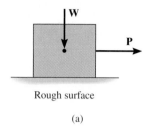

Rough surface

(a)

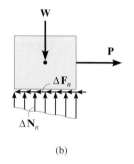

(b)

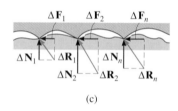

(c)

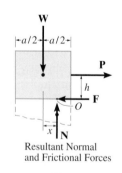

Resultant Normal
and Frictional Forces

(d)

Fig. 8–1

Theory of Dry Friction. The theory of dry friction can best be explained by considering what effects are caused by pulling horizontally on a block of uniform weight **W** which is resting on a rough horizontal surface, Fig. 8–1a. To properly develop a full understanding of the nature of friction, it is necessary to consider the surfaces of contact to be *nonrigid or deformable.* The other portion of the block, however, will be considered rigid. As shown on the free-body diagram of the block, Fig. 8–1b, the floor exerts a *distribution* of both *normal force* $\Delta \mathbf{N}_n$ and *frictional force* $\Delta \mathbf{F}_n$ along the contacting surface. For equilibrium, the normal forces must act *upward* to balance the block's weight **W**, and the frictional forces act to the left to prevent the applied force **P** from moving the block to the right. Close examination of the contacting surfaces between the floor and block reveals how these frictional and normal forces develop, Fig. 8–1c. It can be seen that many microscopic irregularities exist between the two surfaces and, as a result, reactive forces $\Delta \mathbf{R}_n$ are developed at each of the protuberances.* These forces act at all points of contact, and, as shown, each reactive force contributes both a frictional component $\Delta \mathbf{F}_n$ and a normal component $\Delta \mathbf{N}_n$.

Equilibrium. For simplicity in the following analysis, the effect of the *distributed* normal and frictional loadings will be indicated by their *resultants* **N** and **F**, which are represented on the free-body diagram as shown in Fig. 8-1d. Clearly, the distribution of $\Delta \mathbf{F}_n$ in Fig. 8–1b indicates that **F** always acts *tangent to the contacting surface, opposite* to the direction of **P**. On the other hand, the normal force **N** is determined from the distribution of $\Delta \mathbf{N}_n$ in Fig. 8–1b and is directed upward to balance the block's weight **W**. Notice that **N** acts a distance x to the right of the line of action of **W**, Fig. 8–1d. This location, which coincides with the centroid or geometric center of the loading diagram in Fig. 8–1b, is necessary in order to balance the "tipping effect" caused by **P**. For example, if **P** is applied at a height h from the surface, Fig. 8–1d, then moment equilibrium about point O is satisfied if $Wx = Ph$ or $x = Ph/W$. In particular, the block will be on the verge of *tipping* if **N** acts at the right corner of the block, $x = a/2$.

---

*Besides mechanical interactions as explained here, which is referred to as a classical approach, a detailed treatment of the nature of frictional forces must also include the effects of temperature, density, cleanliness, and atomic or molecular attraction between the contacting surfaces. See J. Krim, *Scientific American*, October, 1996.

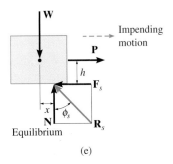

Equilibrium

(e)

Impending Motion. In cases where h is small or the surfaces of contact are rather "slippery," the frictional force $\mathbf{F}$ may *not* be great enough to balance $\mathbf{P}$, and consequently the block will tend to slip *before* it can tip. In other words, as P is slowly increased, F correspondingly increases until it attains a certain *maximum value* F_s, called the *limiting static frictional force*, Fig. 8–1e. When this value is reached, the block is in *unstable equilibrium* since any further increase in P will cause deformations and fractures at the points of surface contact, and consequently the block will begin to move. Experimentally, it has been determined that the limiting static frictional force F_s is *directly proportional* to the resultant normal force N. This may be expressed mathematically as

$$F_s = \mu_s N \qquad (8\text{–}1)$$

where the constant of proportionality, μ_s (mu "sub" s), is called the *coefficient of static friction.*

Thus, when the block is on the *verge of sliding*, the normal force $\mathbf{N}$ and frictional force $\mathbf{F}_s$ combine to create a resultant $\mathbf{R}_s$, Fig. 8–1e. The angle ϕ_s that $\mathbf{R}_s$ makes with $\mathbf{N}$ is called the *angle of static friction*. From the figure,

$$\phi_s = \tan^{-1}\left(\frac{F_s}{N}\right) = \tan^{-1}\left(\frac{\mu_s N}{N}\right) = \tan^{-1}\mu_s$$

Tabular Values of μ_s. Typical values for μ_s, found in many engineering handbooks, are given in Table 8–1. Although this coefficient is generally less than 1, be aware that in some cases it is possible, as in the case of aluminum on aluminum, for μ_s to be greater than 1. Physically this means, of course, that in this case the frictional force is greater than the corresponding normal force. Furthermore, it should be noted that μ_s is dimensionless and depends only on the characteristics of the two surfaces in contact. A wide range of values is given for each value of μ_s since experimental testing was done under variable conditions of roughness and cleanliness of the contacting surfaces. For applications, therefore, it is important that both caution and judgment be exercised when selecting a coefficient of friction for a given set of conditions. When a more accurate calculation of F_s is required, the coefficient of friction should be determined directly by an experiment that involves the two materials to be used.

TABLE 8–1
Typical Values for μ_s

Contact Materials	Coefficient of Static Friction (μ_s)
Metal on ice	0.03–0.05
Wood on wood	0.30–0.70
Leather on wood	0.20–0.50
Leather on metal	0.30–0.60
Aluminum on aluminum	1.10–1.70

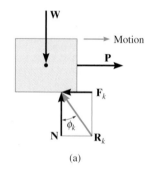

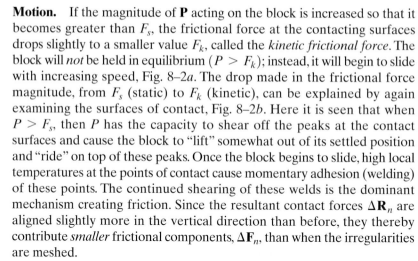

(a)

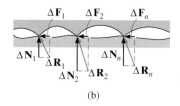

(b)

Fig. 8–2

Motion. If the magnitude of **P** acting on the block is increased so that it becomes greater than F_s, the frictional force at the contacting surfaces drops slightly to a smaller value F_k, called the *kinetic frictional force*. The block will *not* be held in equilibrium ($P > F_k$); instead, it will begin to slide with increasing speed, Fig. 8–2a. The drop made in the frictional force magnitude, from F_s (static) to F_k (kinetic), can be explained by again examining the surfaces of contact, Fig. 8–2b. Here it is seen that when $P > F_s$, then P has the capacity to shear off the peaks at the contact surfaces and cause the block to "lift" somewhat out of its settled position and "ride" on top of these peaks. Once the block begins to slide, high local temperatures at the points of contact cause momentary adhesion (welding) of these points. The continued shearing of these welds is the dominant mechanism creating friction. Since the resultant contact forces $\Delta\mathbf{R}_n$ are aligned slightly more in the vertical direction than before, they thereby contribute *smaller* frictional components, $\Delta\mathbf{F}_n$, than when the irregularities are meshed.

Experiments with sliding blocks indicate that the magnitude of the resultant frictional force $\mathbf{F}_k$ is directly proportional to the magnitude of the resultant normal force $\mathbf{N}$. This may be expressed mathematically as

$$F_k = \mu_k N \qquad (8\text{–}2)$$

Here the constant of proportionality, μ_k, is called the *coefficient of kinetic friction*. Typical values for μ_k are approximately 25 percent *smaller* than those listed in Table 8–1 for μ_s.

As shown in Fig. 8–2a, in this case, the resultant $\mathbf{R}_k$ has a line of action defined by ϕ_k. This angle is referred to as the *angle of kinetic friction*, where

$$\phi_k = \tan^{-1}\left(\frac{F_k}{N}\right) = \tan^{-1}\left(\frac{\mu_k N}{N}\right) = \tan^{-1}\mu_k$$

By comparison, $\phi_s \geq \phi_k$.

The above effects regarding friction can be summarized by reference to the graph in Fig. 8–3, which shows the variation of the frictional force F versus the applied load P. Here the frictional force is categorized in three different ways: namely, F is a *static-frictional force* if equilibrium is maintained; F is a *limiting static-frictional force* F_s when it reaches a maximum value needed to maintain equilibrium; and finally, F is termed a *kinetic-frictional force* F_k when sliding occurs at the contacting surface. Notice also from the graph that for very large values of P or for high speeds, because of aerodynamic effects, F_k and likewise μ_k begin to decrease.

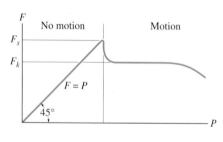

Fig. 8–3

Characteristics of Dry Friction.   As a result of *experiments* that pertain to the foregoing discussion, the following rules which apply to bodies subjected to dry friction may be stated.

- The frictional force acts *tangent* to the contacting surfaces in a direction *opposed* to the *relative motion* or tendency for motion of one surface against another.

- The maximum static frictional force F_s that can be developed is independent of the area of contact, provided the normal pressure is not very low nor great enough to severely deform or crush the contacting surfaces of the bodies.

- The maximum static frictional force is generally greater than the kinetic frictional force for any two surfaces of contact. However, if one of the bodies is moving with a *very low velocity* over the surface of another, F_k becomes approximately equal to F_s, i.e., $\mu_s \approx \mu_k$.

- When *slipping* at the surface of contact is *about to occur*, the maximum static frictional force is proportional to the normal force, such that $F_s = \mu_s N$.

- When *slipping* at the surface of contact is *occurring*, the kinetic frictional force is proportional to the normal force, such that $F_k = \mu_k N$.

8.2   Problems Involving Dry Friction

If a rigid body is in equilibrium when it is subjected to a system of forces that includes the effect of friction, the force system must satisfy not only the equations of equilibrium but *also* the laws that govern the frictional forces.

Types of Friction Problems.   In general, there are three types of mechanics problems involving dry friction. They can easily be classified once the free-body diagrams are drawn and the total number of unknowns are identified and compared with the total number of available equilibrium equations. Each type of problem will now be explained and illustrated graphically by examples. In all these cases the geometry and dimensions for the problem are assumed to be known.

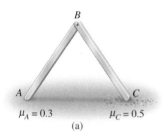

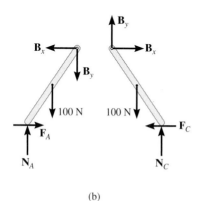

Fig. 8–4

Equilibrium. Problems in this category are strictly equilibrium problems which require *the total number of unknowns to be equal to the total number of available equilibrium equations*. Once the frictional forces are determined from the solution, however, their numerical values must be checked to be sure they satisfy the inequality $F \leq \mu_s N$; otherwise, slipping will occur and the body will not remain in equilibrium. A problem of this type is shown in Fig. 8–4a. Here we must determine the frictional forces at A and C to check if the equilibrium position of the two-member frame can be maintained. If the bars are uniform and have known weights of 100 N each, then the free-body diagrams are as shown in Fig. 8–4b. There are six unknown force components which can be determined *strictly* from the six equilibrium equations (three for each member). Once F_A, N_A, F_C, and N_C are determined, then the bars will remain in equilibrium provided $F_A \leq 0.3 N_A$ and $F_C \leq 0.5 N_C$ are satisfied.

Impending Motion at All Points. In this case *the total number of unknowns will equal the total number of available equilibrium equations plus the total number of available frictional equations, $F = \mu N$*. In particular, if *motion is impending* at the points of contact, then $F_s = \mu_s N$; whereas if the body is *slipping*, then $F_k = \mu_k N$. For example, consider the problem of finding the smallest angle θ at which the 100-N bar in Fig. 8–5a can be placed against the wall without slipping. The free-body diagram is shown in Fig. 8–5b. Here there are *five* unknowns: F_A, N_A, F_B, N_B, θ. For the solution there are *three* equilibrium equations and *two* static frictional equations which apply at *both* points of contact, so that $F_A = 0.3 N_A$ and $F_B = 0.4 N_B$.

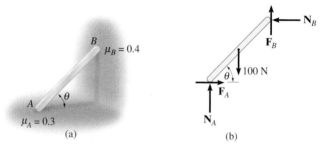

Fig. 8–5

Impending Motion at Some Points. Here *the total number of unknowns will be less than the number of available equilibrium equations plus the total number of frictional equations or conditional equations for tipping.* As a result, several possibilities for motion or impending motion will exist and the problem will involve a determination of the kind of motion which actually occurs. For example, consider the two-member frame shown in Fig. 8–6a. In this problem we wish to determine the horizontal force P needed to cause movement. If each member has a weight of 100 N, then the free-body diagrams are as shown in Fig. 8–6b. There are *seven* unknowns: N_A, F_A, N_C, F_C, B_x, B_y, P. For a unique solution we must satisfy the *six* equilibrium equations (three for each member) and only *one* of two possible static frictional equations. This means that as P increases it will either cause slipping at A and no slipping at C, so that $F_A = 0.3N_A$ and $F_C \leq 0.5N_C$; or slipping occurs at C and no slipping at A, in which case $F_C = 0.5N_C$ and $F_A \leq 0.3N_A$. The actual situation can be determined by calculating P for each case and then choosing the case for which P is *smaller*. If in both cases the *same value* for P is calculated, which in practice would be highly improbable, then slipping at both points occurs simultaneously; i.e., the *seven unknowns* will satisfy *eight equations*.

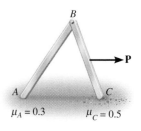

$\mu_A = 0.3$ $\mu_C = 0.5$

(a)

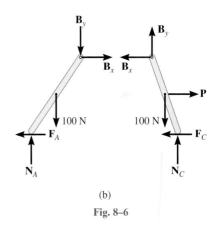

(b)

Fig. 8–6

Consider pushing on the uniform crate that has a weight W and sits on the rough surface. As shown on the first free-body diagram, if the magnitude of **P** is small, the crate will remain in equilibrium. As P increases the crate will either be on the verge of slipping on the surface ($F = \mu_s W$), or if the surface is very rough (large μ_s) then the resultant normal force will shift to the corner, $x = b/2$, as shown on the second free-body diagram, and the crate will tip over. The crate has a greater chance of tipping if **P** is applied at a greater height h above the surface, or if the crate's width b is smaller.

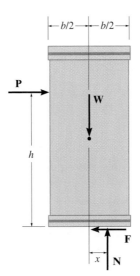

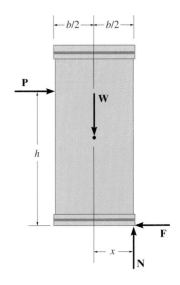

Equilibrium Versus Frictional Equations. It was stated earlier that the frictional force *always* acts so as to either oppose the relative motion or impede the motion of a body over its contacting surface. Realize, however, that we can *assume* the sense of the frictional force in problems which require F to be an "equilibrium force" and satisfy the inequality $F < \mu_s N$. The correct sense is made known *after* solving the equations of equilibrium for F. For example, if F is a negative scalar the sense of $\mathbf{F}$ is the reverse of that which was assumed. This convenience of *assuming* the sense of $\mathbf{F}$ is possible because the equilibrium equations equate to zero the *components of vectors* acting in the *same direction*. In cases where the frictional equation $F = \mu N$ is used in the solution of a problem, however, the convenience of *assuming* the sense of $\mathbf{F}$ is *lost*, since the frictional equation relates only the *magnitudes* of two perpendicular vectors. Consequently, $\mathbf{F}$ *must always* be shown acting with its *correct sense* on the free-body diagram whenever the frictional equation is used for the solution of a problem.

PROCEDURE FOR ANALYSIS

Equilibrium problems involving dry friction can be solved using the following procedure.

Free-Body Diagrams.

- Draw the necessary free-body diagrams, and unless it is stated in the problem that impending motion or slipping occurs, *always* show the frictional forces as unknowns; i.e., *do not assume* $F = \mu N$.

- Determine the number of unknowns and compare this with the number of available equilibrium equations.

- If there are more unknowns than equations of equilibrium, it will be necessary to apply the frictional equation at some, if not all, points of contact to obtain the extra equations needed for a complete solution.

- If the equation $F = \mu N$ is to be used, it will be necessary to show $\mathbf{F}$ acting in the proper direction on the free-body diagram.

Equations of Equilibrium and Friction.

- Apply the equations of equilibrium and the necessary frictional equations (or conditional equations if tipping is possible) and solve for the unknowns.

- If the problem involves a three-dimensional force system such that it becomes difficult to obtain the force components or the necessary moment arms, apply the equations of equilibrium using Cartesian vectors.

EXAMPLE **8.1**

The uniform crate shown in Fig. 8–7a has a mass of 20 kg. If a force $P = 80$ N is applied to the crate, determine if it remains in equilibrium. The coefficient of static friction is $\mu = 0.3$.

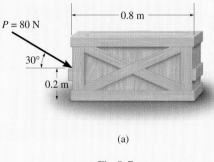

(a)

Fig. 8–7

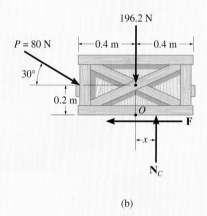

(b)

Solution

Free-Body Diagram. As shown in Fig. 8–7b, the *resultant* normal force $\mathbf{N}_C$ must act a distance x from the crate's center line in order to counteract the tipping effect caused by $\mathbf{P}$. There are *three unknowns,* F, N_C, and x, which can be determined strictly from the *three* equations of equilibrium.

Equations of Equilibrium.

$\xrightarrow{+} \Sigma F_x = 0;$ $\qquad$ $80 \cos 30° \text{ N} - F = 0$

$+\uparrow \Sigma F_y = 0;$ $\qquad$ $-80 \sin 30° \text{ N} + N_C - 196.2 \text{ N} = 0$

$\zeta+\Sigma M_O = 0;$ $\quad$ $80 \sin 30° \text{ N}(0.4 \text{ m}) - 80 \cos 30° \text{ N}(0.2 \text{ m}) + N_C(x) = 0$

Solving,

$$F = 69.3 \text{ N}$$
$$N_C = 236 \text{ N}$$
$$x = -0.00908 \text{ m} = -9.08 \text{ mm}$$

Since x is negative it indicates the *resultant* normal force acts (slightly) to the *left* of the crate's center line. No tipping will occur since $x \le 0.4$ m. Also, the *maximum* frictional force which can be developed at the surface of contact is $F_{max} = \mu_s N_C = 0.3(236 \text{ N}) = 70.8$ N. Since $F = 69.3$ N < 70.8 N, the crate will *not slip*, although it is very close to doing so.

E X A M P L E   **8.2**

It is observed that when the bed of the dump truck is raised to an angle of $\theta = 25°$ the vending machines begin to slide off the bed, Fig. 8–8a. Determine the static coefficient of friction between them and the surface of the truck.

(a)

Solution

An idealized model of a vending machine resting on the bed of the truck is shown in Fig. 8–8b. The dimensions have been measured and the center of gravity has been located. We will assume that the machine weighs W.

Free-Body Diagram. As shown in Fig. 8–8c, the dimension x is used to locate the position of the resultant normal force N. There are four unknowns, N, F, μ_s, and x.

Equations of Equilibrium.

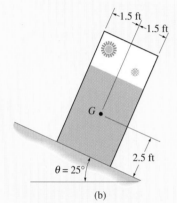

(b)

$$+\searrow\Sigma F_x = 0; \qquad W \sin 25° - F = 0 \qquad (1)$$
$$+\nearrow\Sigma F_y = 0; \qquad N - W \cos 25° = 0 \qquad (2)$$
$$\downarrow+\Sigma M_O = 0; \quad -W \sin \theta \,(2.5 \text{ ft}) + W \cos \theta \,(x) = 0 \qquad (3)$$

Since slipping impends at $\theta = 25°$, using the first two equations, we have

$$F_s = \mu_s N; \qquad W \sin 25° = \mu_s \,(W \cos 25°)$$
$$\mu_s = \tan 25° = 0.466 \qquad \qquad Ans.$$

The angle of $\theta = 25°$ is referred to as the *angle of repose*, and by comparison, it is equal to the angle of static friction $\theta = \phi_s$. Notice from the calculation that θ is independent of the weight of the vending machine, and so knowing θ provides a convenient method for determining the coefficient of static friction.

From Eq. 3, with $\theta = 25°$, we find $x = 1.17$ ft. Since 1.17 ft < 1.5 ft, indeed the vending machine will slip before it can tip as observed in Fig. 8–8a.

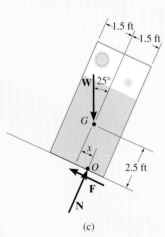

(c)

Fig. 8–8

EXAMPLE 8.3

The uniform rod having a weight W and length l is supported at its ends against the surface at A and B in Fig. 8–9a. If the rod is on the verge of slipping when $\theta = 30°$, determine the coefficient of static friction μ_s at A and B. Neglect the thickness of the rod for the calculation.

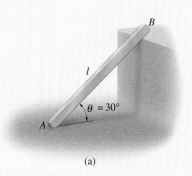

(a)

Solution

Free-Body Diagram. As shown in Fig. 8–9b, there are *five* unknowns: F_A, N_A, F_B, N_B, and μ_s. These can be determined from the *three* equilibrium equations and *two* frictional equations applied at points A and B. The frictional forces must be drawn with their correct sense so that they oppose the tendency for motion of the rod. Why? (Refer to p. 386.)

Equations of Friction and Equilibrium. Writing the frictional equations,

$$F = \mu_s N; \qquad F_A = \mu_s N_A$$
$$F_B = \mu_s N_B$$

Using these results and applying the equations of equilibrium yields

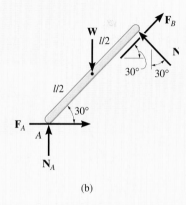

(b)

Fig. 8–9

$$\xrightarrow{+} \Sigma F_x = 0; \quad \mu_s N_A + \mu_s N_B \cos 30° - N_B \sin 30° = 0 \qquad (1)$$
$$+\uparrow \Sigma F_y = 0; \quad N_A - W + N_B \cos 30° + \mu_s N_B \sin 30° = 0 \qquad (2)$$
$$\zeta+\Sigma M_A = 0; \quad N_B l - W\left(\frac{l}{2}\right)\cos 30° = 0 \qquad (3)$$
$$N_B = 0.4330\ W$$

From Eqs. 1 and 2,

$$\mu_s N_A = 0.2165\ W - (0.3750\ W)\mu_s$$
$$N_A = 0.6250\ W - (0.2165\ W)\mu_s$$

By division,

$$0.6250\mu_s - 0.2165\mu_s^2 = 0.2165 - 0.375\mu_s$$

or,

$$\mu_s^2 - 4.619\ \mu_s + 1 = 0$$

Solving for the smallest root,

$$\mu_s = 0.228 \qquad\qquad Ans.$$

EXAMPLE 8.4

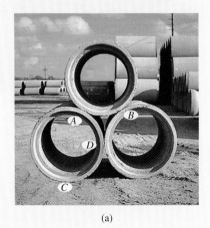

(a)

The concrete pipes are stacked in the yard as shown in Fig. 8–10a. Determine the minimum coefficient of static friction at each point of contact so that the pile does not collapse.

Solution

Free-body Diagrams Recognize that the coefficient of static friction between two pipes, at A and B, and between a pipe and the ground, at C, will be different since the contacting surfaces are different. We will assume each pipe has an outer radius r and weight W. The free-body diagrams for two of the pipes are shown in Fig. 8–10b. There are six unknowns, N_A, F_A, N_B, F_B, N_C, F_C. (Note that when collapse is about to occur the normal force at D is zero.) Since only the six equations of equilibrium are necessary to obtain the unknowns, the sense of direction of the frictional forces can be verified from the solution.

Equations of Equilibrium. For the top pipe we have

$$\curvearrowright + \Sigma M_O = 0; \quad -F_A(r) + F_B(r) = 0; F_A = F_B = F$$

$$\xrightarrow{+} \Sigma F_x = 0; \quad N_A \sin 30° - F \cos 30° - N_B \sin 30° + F \cos 30° = 0$$

$$N_A = N_B = N$$

$$+\uparrow \Sigma F_y = 0; \quad 2N \cos 30° + 2F \sin 30° - W = 0 \qquad (1)$$

For the bottom pipe, using $F_A = F$ and $N_A = N$, we have,

$$\curvearrowright + \Sigma M_{O'} = 0; \quad F_C(r) - F(r) = 0; \quad F_C = F$$

$$\xrightarrow{+} \Sigma F_x = 0; \quad -N \sin 30° + F \cos 30° + F = 0 \qquad (2)$$

$$+\uparrow \Sigma F_y = 0; \quad N_C - W - N \cos 30° - F \sin 30° = 0 \qquad (3)$$

From Eq. 2, $F = 0.2679\, N$, so that between the pipes

$$(\mu_s)_{\min} = \frac{F}{N} = 0.268 \qquad \qquad Ans.$$

Using this result in Eq. 1,

$$N = 0.5\, W$$

From Eq. 3,

$$N_C - W - (0.5\, W) \cos 30° - 0.2679\,(0.5\, W) \sin 30° = 0$$

$$N_C = 1.5\, W$$

At the ground, the smallest required coefficient of static friction would be

$$(\mu'_s)_{\min} = \frac{F}{N_C} = \frac{0.2679(0.5\, W)}{1.5\, W} = 0.0893 \qquad Ans.$$

Hence a greater coefficient of static friction is required between the pipes than that required at the ground; and so it is likely that if slipping would occur between the pipes the bottom two pipes would roll away from one another without slipping as the top pipe falls downward.

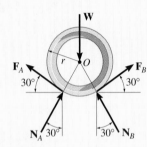

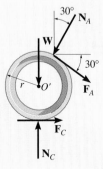

(b)

Fig. 8–10

EXAMPLE 8.5

Beam AB is subjected to a uniform load of 200 N/m and is supported at B by post BC, Fig. 8–11a. If the coefficients of static friction at B and C are $\mu_B = 0.2$ and $\mu_C = 0.5$, determine the force $\mathbf{P}$ needed to pull the post out from under the beam. Neglect the weight of the members and the thickness of the post.

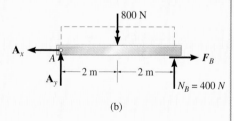

(a)

Solution

Free-Body Diagrams. The free-body diagram of beam AB is shown in Fig. 8–11b. Applying $\Sigma M_A = 0$, we obtain $N_B = 400$ N. This result is shown on the free-body diagram of the post, Fig. 8–11c. Referring to this member, the *four* unknowns F_B, P, F_C, and N_C are determined from the *three* equations of equilibrium and *one* frictional equation applied either at B or C.

Equations of Equilibrium and Friction.

$$\xrightarrow{+} \Sigma F_x = 0; \qquad P - F_B - F_C = 0 \qquad (1)$$
$$+\uparrow \Sigma F_y = 0; \qquad N_C - 400 \text{ N} = 0 \qquad (2)$$
$$\zeta + \Sigma M_C = 0; \qquad -P(0.25 \text{ m}) + F_B(1 \text{ m}) = 0 \qquad (3)$$

(Post Slips Only at B) This requires $F_C \leq \mu_C N_C$ and

$$F_B = \mu_B N_B; \qquad F_B = 0.2(400 \text{ N}) = 80 \text{ N}$$

Using this result and solving Eqs. 1 through 3, we obtain

$$P = 320 \text{ N}$$
$$F_C = 240 \text{ N}$$
$$N_C = 400 \text{ N}$$

Since $F_C = 240$ N $> \mu_C N_C = 0.5(400$ N$) = 200$ N, the other case of movement must be investigated.

(Post Slips Only at C.) Here $F_B \leq \mu_B N_B$ and

$$F_C = \mu_C N_C; \qquad F_C = 0.5 N_C \qquad (4)$$

Solving Eqs. 1 through 4 yields

$$P = 267 \text{ N} \qquad \qquad Ans.$$
$$N_C = 400 \text{ N}$$
$$F_C = 200 \text{ N}$$
$$F_B = 66.7 \text{ N}$$

Obviously, this case occurs first since it requires a *smaller* value for P.

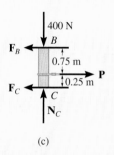

(b)

(c)

Fig. 8–11

EXAMPLE 8.6

Determine the normal force P that must be exerted on the rack to begin pushing the 100-kg pipe shown in Fig. 8–12a up the 20° incline. The coefficients of static friction at the points of contact are $(\mu_s)_A = 0.15$, and $(\mu_s)_B = 0.4$.

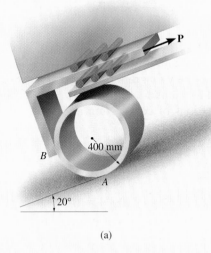

(a)

Solution

Free-Body Diagram.   As shown in Fig. 8–12b, the rack must exert a force P on the pipe due to force equilibrium in the x direction. There are four unknowns P, F_A, N_A, and F_B acting on the pipe Fig. 8–12c. These can be determined from the *three* equations of equilibrium and *one* frictional equation, which apply either at A or B. If slipping begins to occur only at B, the pipe will begin to roll up the incline; whereas if slipping occurs only at A, the pipe will begin to *slide* up the incline. Here we must find N_B.

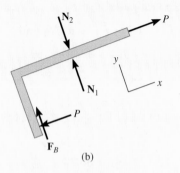

(b)

Fig. 8–12

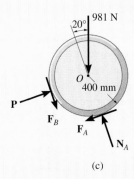

(c)

Fig. 8–12

Equations of Equilibrium and Friction (for Fig. 8–12c)

$$+\nearrow \Sigma F_x = 0; \qquad -F_A + P - 981 \sin 20° \text{ N} = 0 \qquad (1)$$

$$+\nwarrow \Sigma F_y = 0; \qquad N_A - F_B - 981 \cos 20° \text{ N} = 0 \qquad (2)$$

$$\downarrow + \Sigma M_O = 0; \qquad F_B(400 \text{ mm}) - F_A(400 \text{ mm}) = 0 \qquad (3)$$

(Pipe Rolls up Incline.) In this case $F_A \leq 0.15 N_A$ and

$$(F_s)_B = (\mu_s)_B N_B; \qquad F_B = 0.4P \qquad (4)$$

The direction of the frictional force at B must be specified correctly. Why? Since the spool is being forced up the incline, $\mathbf{F}_B$ acts downward to prevent any clockwise rolling motion of the pipe, Fig. 8–12c. Solving Eqs. 1 through 4, we have

$$N_A = 1146 \text{ N} \quad F_A = 224 \text{ N} \quad F_B = 224 \text{ N} \quad P = 559 \text{ N}$$

The assumption regarding no slipping at A should be checked.

$$F_A \leq (\mu_s)_A N_A; \quad 224 \text{ N} \overset{?}{\leq} 0.15(1146 \text{ N}) = 172 \text{ N}$$

The inequality does *not apply*, and therefore slipping occurs at A and not at B. Hence, the other case of motion will occur.

(Pipe Slides up Incline.) In this case, $P \leq 0.4 N_B$ and

$$(F_s)_A = (\mu_s)_A N_A; \qquad F_A = 0.15 N_A \qquad (5)$$

Solving Eqs. 1 through 3 and 5 yields

$$N_A = 1085 \text{ N} \quad F_A = 163 \text{ N} \quad F_B = 163 \text{ N} \quad P = 498 \text{ N} \qquad \textit{Ans.}$$

The validity of the solution ($P = 498$ N) can be checked by testing the assumption that indeed no slipping occurs at B.

$$F_B \leq (\mu_s)_B P; \quad 163 \text{ N} < 0.4(498 \text{ N}) = 199 \text{ N} \qquad \text{(check)}$$

8.3 Wedges

Wedges are often used to adjust the elevation of structural or mechanical parts. Also, they provide stability for objects such as this tank.

A *wedge* is a simple machine which is often used to transform an applied force into much larger forces, directed at approximately right angles to the applied force. Also, wedges can be used to give small displacements or adjustments to heavy loads.

Consider, for example, the wedge shown in Fig. 8–13a, which is used to *lift* a block of weight **W** by applying a force **P** to the wedge. Free-body diagrams of the block and wedge are shown in Fig. 8–13b. Here we have excluded the weight of the wedge since it is usually *small* compared to the weight of the block. Also, note that the frictional forces **F**$_1$ and **F**$_2$ must oppose the motion of the wedge. Likewise, the frictional force **F**$_3$ of the wall on the block must act downward so as to oppose the block's upward motion. The locations of the resultant normal forces are not important in the force analysis since neither the block nor wedge will "tip." Hence the moment equilibrium equations will not be considered. There are seven unknowns consisting of the applied force **P**, needed to cause motion of the wedge, and six normal and frictional forces. The seven available equations consist of two force equilibrium equations ($\Sigma F_x = 0$, $\Sigma F_y = 0$) applied to the wedge and block (four equations total) and the frictional equation $F = \mu N$ applied at each surface of contact (three equations total).

If the block is to be *lowered*, the frictional forces will all act in a sense opposite to that shown in Fig. 8–13b. The applied force **P** will act to the right as shown if the coefficient of friction is very *small* or the wedge angle θ is *large*. Otherwise, **P** may have the reverse sense of direction in order to *pull* on the wedge to remove it. If **P** is *not applied*, or **P** = **0**, and friction forces hold the block in place, then the wedge is referred to as *self-locking*.

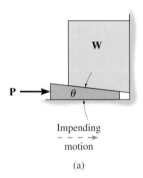

(a)

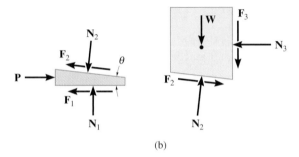

(b)

Fig. 8–13

E X A M P L E   **8.7**

The uniform stone in Fig. 8–14a has a mass of 500 kg and is held in the horizontal position using a wedge at B. If the coefficient of static friction is $\mu_s = 0.3$ at, the surfaces of contact, determine the minimum force **P** needed to remove the wedge. Is the wedge self-locking? Assume that the stone does not slip at A.

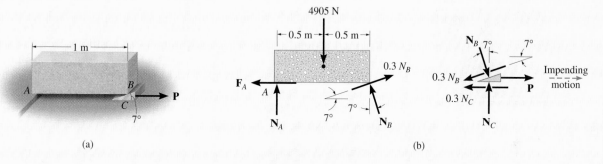

Fig. 8–14

Solution

The minimum force P requires $F = \mu_s N$ at the surfaces of contact with the wedge. The free-body diagrams of the stone and wedge are shown in Fig. 8–14b. On the wedge the friction force opposes the motion, and on the stone at A, $F_A \leq \mu_s N_A$, since slipping does not occur there. There are five unknowns F_A, N_A, N_B, N_C, and P. Three equilibrium equations for the stone and two for the wedge are available for solution. From the free-body diagram of the stone,

$\downarrow + \Sigma M_A = 0;$ $-4905\ \text{N}(0.5\ \text{m}) + (N_B \cos 7°\ \text{N})(1\ \text{m})$
$$+(0.3 N_B \sin 7°\ \text{N})(1\ \text{m}) = 0$$
$$N_B = 2383.1\ \text{N}$$

Using this result for the wedge, we have

$\xrightarrow{+} \Sigma F_x = 0;$ $2383.1 \sin 7°\ \text{N} - 0.3(2383.1 \cos 7°\ \text{N}) + P - 0.3 N_C = 0$
$+\uparrow \Sigma F_y = 0;$
$$N_C - 2383.1 \cos 7°\ \text{N} - 0.3(2383.1 \sin 7°\ \text{N}) = 0$$
$$N_C = 2452.5\ \text{N}$$
$$P = 1154.9\ \text{N} = 1.15\ \text{kN} \qquad\qquad Ans.$$

Since P is positive, indeed the wedge must be pulled out. If P was zero, the wedge would remain in place (self-locking) and the frictional forces developed at B and C would satisfy $F_B < \mu_s N_B$ and $F_C < \mu_s N_C$.

8.4 Frictional Forces on Screws

Square-threaded screws find applications on valves, jacks, and vises, where particularly large forces must be developed along the axis of the screw.

In most cases screws are used as fasteners; however, in many types of machines they are incorporated to transmit power or motion from one part of the machine to another. A *square-threaded screw* is commonly used for the latter purpose, especially when large forces are applied along its axis. In this section we will analyze the forces acting on square-threaded screws. The analysis of other types of screws, such as the V-thread, is based on these same principles.

A *screw* may be thought of simply as an inclined plane or wedge wrapped around a cylinder. A nut initially at position A on the screw shown in Fig. 8–15a will move up to B when rotated 360° around the screw. This rotation is equivalent to translating the nut up an inclined plane of height l and length $2\pi r$, where r is the mean radius of the thread, Fig. 8–15b. The rise l for a single revolution is referred to as the *lead* of the screw, where the *lead angle* is given by $\theta = \tan^{-1}(l/2\pi r)$.

Frictional Analysis. When a screw is subjected to large axial loads, the frictional forces developed on the thread become important if we are to determine the moment $\mathbf{M}$* needed to turn the screw. Consider, for example, the square-threaded jack screw shown in Fig. 8–16, which supports the vertical load $\mathbf{W}$. The reactive forces of the jack to this load are actually distributed over the circumference of the screw thread in contact with the screw hole in the jack, that is, within region h shown in Fig. 8–16. For simplicity, this portion of thread can be imagined as being unwound from the screw and represented as a simple block resting on an inclined plane having the screw's lead angle θ, Fig. 8–17a. Here the inclined plane represents the inside *supporting thread* of the jack base. Three forces act on the block or screw. The force $\mathbf{W}$ is the total axial load applied to the screw. The horizontal force $\mathbf{S}$ is caused by the applied moment $\mathbf{M}$, such that by summing moments about the axis of the screw, $M = Sr$, where r is the screw's mean radius. As a result of $\mathbf{W}$ and $\mathbf{S}$, the inclined plane exerts a resultant force $\mathbf{R}$ on the block, which is shown to have components acting normal, $\mathbf{N}$, and tangent, $\mathbf{F}$, to the contacting surfaces.

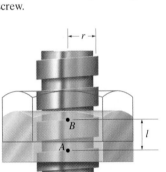

(a)

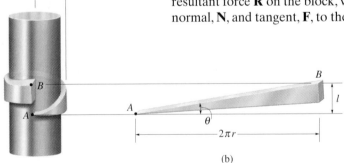

(b)

Fig. 8–15

*For applications, $\mathbf{M}$ is developed by applying a horizontal force $\mathbf{P}$ at a right angle to the end of a lever that would be fixed to the screw.

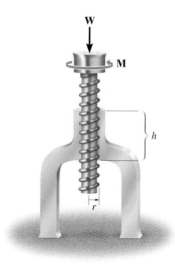

Fig. 8–16

Upward Screw Motion. Provided M is great enough, the screw (and hence the block) can either be brought to the verge of upward impending motion or motion can be occurring. Under these conditions, **R** acts at an angle $(\theta + \phi)$ from the vertical as shown in Fig. 8–17a, where $\phi = \tan^{-1}(F/N) = \tan^{-1}(\mu N/N) = \tan^{-1}\mu$. Applying the two force equations of equilibrium to the block, we obtain

$$\xrightarrow{+} \Sigma F_x = 0; \qquad S - R\sin(\theta + \phi) = 0$$
$$+\uparrow \Sigma F_y = 0; \qquad R\cos(\theta + \phi) - W = 0$$

Eliminating R and solving for S, then substituting this value into the equation $M = Sr$, yields

$$\boxed{M = Wr\tan(\theta + \phi)} \qquad (8\text{–}3)$$

As indicated, M is the moment necessary to cause upward impending motion of the screw, provided $\phi = \phi_s = \tan^{-1}\mu_s$ (the angle of static friction). If ϕ is replaced by $\phi_k = \tan^{-1}\mu_k$ (the angle of kinetic friction), Eq. 8–3 will give a smaller value M necessary to maintain uniform upward motion of the screw.

Downward Screw Motion $(\theta > \phi)$. If the surface of the screw is very *slippery*, it may be possible for the screw to rotate downward if the magnitude of the moment is reduced to, say, $M' < M$. As shown in Fig. 8–17b, this causes the effect of **M′** to become **S′**, and it requires the angle ϕ (ϕ_s or ϕ_k) to lie on the opposite side of the normal n to the plane supporting the block, such that $\theta > \phi$. For this case, Eq. 8–3 becomes

$$\boxed{M' = Wr\tan(\theta - \phi)} \qquad (8\text{–}4)$$

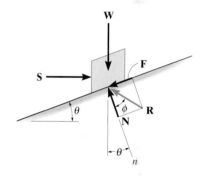

Upward screw motion
(a)

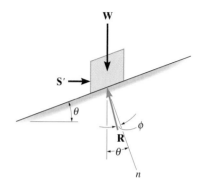

Downward screw motion $(\theta > \phi)$
(b)

Fig. 8–17

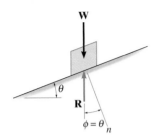

Self-locking screw ($\theta = \phi$)
(on the verge of rotating downward)

(c)

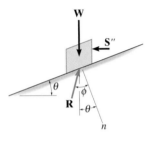

Downward screw motion ($\theta < \phi$)

(d)

Self-Locking Screw. If the moment **M** (or its effect **S**) is *removed*, the screw will remain *self-locking;* i.e., it will support the load **W** by *friction forces alone* provided $\phi \geq \theta$. To show this, consider the necessary limiting case when $\phi = \theta$, Fig. 8–17c. Here vertical equilibrium is maintained since **R** is vertical and thus balances **W**.

Downward Screw Motion ($\theta < \phi$). When the surface of the screw is *very rough*, the screw will not rotate downward as stated above. Instead, the direction of the applied moment must be *reversed* in order to cause the motion. The free-body diagram shown in Fig. 8–17d is representative of this case. Here **S″** is caused by the applied (reverse) moment **M″**. Hence Eq. 8–3 becomes

$$M'' = Wr \tan(\phi - \theta) \qquad (8\text{–}5)$$

Each of the above cases should be thoroughly understood before proceeding to solve problems.

Fig. 8–17

EXAMPLE 8.8

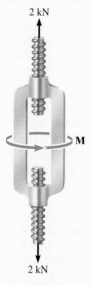

2 kN

M

2 kN

Fig. 8–18

The turnbuckle shown in Fig. 8–18 has a square thread with a mean radius of 5 mm and a lead of 2 mm. If the coefficient of static friction between the screw and the turnbuckle is $\mu_s = 0.25$, determine the moment **M** that must be applied to draw the end screws closer together. Is the turnbuckle self-locking?

Solution

The moment may be obtained by using Eq. 8–3. Why? Since friction at *two screws* must be overcome, this requires

$$M = 2[Wr \tan(\theta + \phi)] \tag{1}$$

Here $W = 2000$ N, $r = 5$ mm, $\phi_s = \tan^{-1} \mu_s = \tan^{-1}(0.25) = 14.04°$, and $\theta = \tan^{-1}(l/2\pi r) = \tan^{-1}(2 \text{ mm}/[2\pi(5 \text{ mm})]) = 3.64°$. Substituting these values into Eq. 1 and solving gives

$$M = 2[(2000 \text{ N})(5 \text{ mm}) \tan(14.04° + 3.64°)]$$
$$= 6374.7 \text{ N} \cdot \text{mm} = 6.37 \text{ N} \cdot \text{m} \qquad \qquad Ans.$$

When the moment is *removed*, the turnbuckle will be self-locking; i.e., it will not unscrew since $\phi_s > \theta$.

8.5 Frictional Forces on Flat Belts

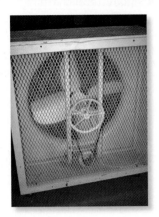

Flat or V-belts are often used to transmit the torque developed by a motor to a fan or blower.

Whenever belt drives or band brakes are designed, it is necessary to determine the frictional forces developed between the belt and its contacting surface. In this section we will analyze the frictional forces acting on a flat belt, although the analysis of other types of belts, such as the V-belt, is based on similar principles.

Here we will consider the flat belt shown in Fig. 8–19a, which passes over a fixed curved surface, such that the total angle of belt to surface contact in radians is β and the coefficient of friction between the two surfaces is μ. We will determine the tension T_2 in the belt which is needed to pull the belt counterclockwise over the surface and thereby overcome both the frictional forces at the surface of contact and the known tension T_1. Obviously, $T_2 > T_1$.

Frictional Analysis. A free-body diagram of the belt segment in contact with the surface is shown in Fig. 8–19b. Here the normal force **N** and the frictional force **F**, acting at different points along the belt, will vary both in magnitude and direction. Due to this *unknown* force distribution, the analysis of the problem will proceed on the basis of initially studying the forces acting on a differential element of the belt.

A free-body diagram of an element having a length ds is shown in Fig. 8–19c. Assuming either impending motion or motion of the belt, the magnitude of the frictional force $dF = \mu\, dN$. This force opposes the sliding motion of the belt and thereby increases the magnitude of the tensile force acting in the belt by dT. Applying the two force equations of equilibrium, we have

$$\searrow + \Sigma F_x = 0; \qquad T \cos\left(\frac{d\theta}{2}\right) + \mu\, dN - (T + dT)\cos\left(\frac{d\theta}{2}\right) = 0$$

$$+ \nearrow \Sigma F_y = 0; \qquad dN - (T + dT)\sin\left(\frac{d\theta}{2}\right) - T \sin\left(\frac{d\theta}{2}\right) = 0$$

Motion or impending motion of belt relative to surface

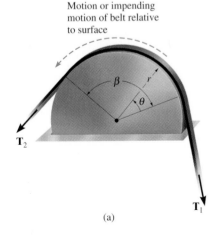

(a)

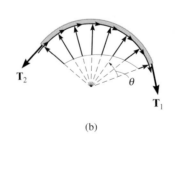

(b)

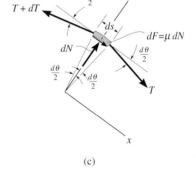

(c)

Fig. 8–19

Since $d\theta$ is of *infinitesimal size*, $\sin(d\theta/2)$ and $\cos(d\theta/2)$ can be replaced by $d\theta/2$ and 1, respectively. Also, the *product* of the two infinitesimals dT and $d\theta/2$ may be neglected when compared to infinitesimals of the first order. The above two equations therefore reduce to

$$\mu \, dN = dT$$

and

$$dN = T \, d\theta$$

Eliminating dN yields

$$\frac{dT}{T} = \mu \, d\theta$$

Integrating this equation between all the points of contact that the belt makes with the drum, and noting that $T = T_1$ at $\theta = 0$ and $T = T_2$ at $\theta = \beta$, yields

$$\int_{T_1}^{T_2} \frac{dT}{T} = \mu \int_0^\beta d\theta$$

$$\ln\frac{T_2}{T_1} = \mu\beta$$

Solving for T_2, we obtain

$$\boxed{T_2 = T_1 e^{\mu\beta}} \qquad\qquad (8\text{–}6)$$

where T_2, T_1 = belt tensions; T_1 opposes the direction of motion (or impending motion) of the belt measured relative to the surface, while T_2 acts in the direction of the relative belt motion (or impending motion); because of friction, $T_2 > T_1$.

μ = coefficient of static or kinetic friction between the belt and the surface of contact

β = angle of belt to surface contact, measured in radians

e = 2.718..., base of the natural logarithm

Note that T_2 is *independent* of the *radius* of the drum and instead it is a function of the angle of belt to surface contact, β. Furthermore, as indicated by the integration, this equation is valid for flat belts placed on *any shape* of contacting surface. For application, Eq. 8–6 is valid only when *impending motion* or *motion* occurs.

EXAMPLE 8.9

The maximum tension that can be developed in the cord shown in Fig. 8–20a is 500 N. If the pulley at A is free to rotate and the coefficient of static friction at the fixed drums B and C is $\mu_s = 0.25$, determine the largest mass of the cylinder that can be lifted by the cord. Assume that the force **F** applied at the end of the cord is directed vertically downward, as shown.

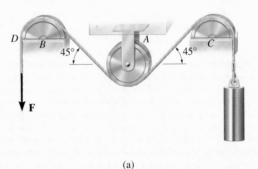

(a)

Solution

Lifting the cylinder, which has a weight $W = mg$, causes the cord to move counterclockwise over the drums at B and C; hence, the maximum tension T_2 in the cord occurs at D. Thus, $T_2 = 500$ N. A section of the cord passing over the drum at B is shown in Fig. 8–20b. Since $180° = \pi$ rad, the angle of contact between the drum and the cord is $\beta = (135°/180°)\pi = 3\pi/4$ rad. Using Eq. 8–6, we have

$$T_2 = T_1 e^{\mu_s \beta}; \qquad 500 \text{ N} = T_1 e^{0.25[(3/4)\pi]}$$

Hence,

$$T_1 = \frac{500 \text{ N}}{e^{0.25[(3/4)\pi]}} = \frac{500 \text{ N}}{1.80} = 277.4 \text{ N}$$

Since the pulley at A is free to rotate, equilibrium requires that the tension in the cord remains the *same* on both sides of the pulley.

The section of the cord passing over the drum at C is shown in Fig. 8–20c. The weight $W < 277.4$ N. Why? Applying Eq. 8–6, we obtain

$$T_2 = T_1 e^{\mu_s \beta}; \qquad 277.4 \text{ N} = W e^{0.25[(3/4)\pi]}$$
$$W = 153.9 \text{ N}$$

so that

$$m = \frac{W}{g} = \frac{153.9 \text{ N}}{9.81 \text{ m/s}^2}$$
$$= 15.7 \text{ kg} \qquad\qquad Ans.$$

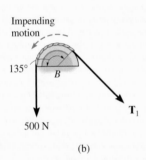

(b)

(c)

Fig. 8–20

*8.6 Frictional Forces on Collar Bearings, Pivot Bearings, and Disks

Pivot and *collar bearings* are commonly used in machines to support an *axial load* on a rotating shaft. These two types of support are shown in Fig. 8–21. Provided the bearings are not lubricated, or are only partially lubricated, the laws of dry friction may be applied to determine the moment **M** needed to turn the shaft when it supports an axial force **P**.

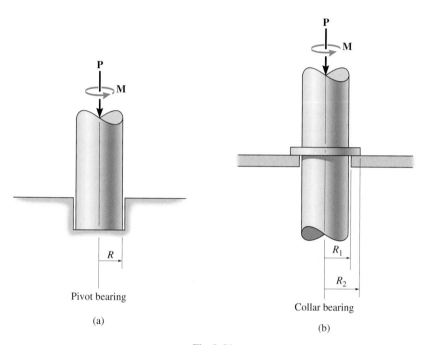

Pivot bearing

(a)

Collar bearing

(b)

Fig. 8–21

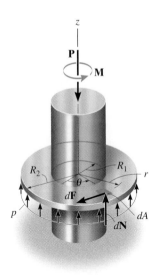

Fig. 8–22

Frictional Analysis. The collar bearing on the shaft shown in Fig. 8–22 is subjected to an axial force **P** and has a total bearing or contact area $\pi(R_2^2 - R_1^2)$. In the following analysis, the normal pressure p is considered to be *uniformly distributed* over this area—a reasonable assumption provided the bearing is new and evenly supported. Since $\Sigma F_z = 0$, then p, measured as a force per unit area, is $p = P/\pi(R_2^2 - R_1^2)$.

The moment needed to cause impending rotation of the shaft can be determined from moment equilibrium about the z axis. A small area element $dA = (r\,d\theta)(dr)$, shown in Fig. 8–22, is subjected to both a normal force $dN = p\,dA$ and an associated frictional force,

$$dF = \mu_s\,dN = \mu_s p\,dA = \frac{\mu_s P}{\pi(R_2^2 - R_1^2)}\,dA$$

The normal force does not create a moment about the z axis of the shaft; however, the frictional force does; namely, $dM = r\,dF$. Integration is needed to compute the total moment created by all the frictional forces acting on differential areas dA. Therefore, for impending rotational motion,

$$\Sigma M_z = 0; \qquad\qquad M - \int_A r\,dF = 0$$

Substituting for dF and dA and integrating over the entire bearing area yields

$$M = \int_{R_1}^{R_2}\int_0^{2\pi} r\left[\frac{\mu_s P}{\pi(R_2^2 - R_1^2)}\right](r\,d\theta\,dr) = \frac{\mu_s P}{\pi(R_2^2 - R_1^2)}\int_{R_1}^{R_2} r^2\,dr \int_0^{2\pi} d\theta$$

or

$$M = \tfrac{2}{3}\mu_s P\left(\frac{R_2^3 - R_1^3}{R_2^2 - R_1^2}\right) \qquad\qquad (8\text{–}7)$$

This equation gives the magnitude of moment required for impending rotation of the shaft. The frictional moment developed at the end of the shaft, when it is *rotating* at constant speed, can be found by substituting μ_k for μ_s in Eq. 8–7.

When $R_2 = R$ and $R_1 = 0$, as in the case of a pivot bearing, Fig. 8–21a, Eq. 8–7 reduces to

$$M = \tfrac{2}{3}\mu_s PR \qquad\qquad (8\text{–}8)$$

Recall from the initial assumption that both Eqs. 8–7 and 8–8 apply only for bearing surfaces subjected to *constant pressure*. If the pressure is not uniform, a variation of the pressure as a function of the bearing area must be determined before integrating to obtain the moment. The following example illustrates this concept.

Frictional forces acting on the disk of this sanding machine must be overcome by the torque developed by the motor which turns it.

E X A M P L E 8.10

The uniform bar shown in Fig. 8–23a has a total mass m. If it is assumed that the normal pressure acting at the contacting surface varies linearly along the length of the bar as shown, determine the couple moment **M** required to rotate the bar. Assume that the bar's width a is negligible in comparison to its length l. The coefficient of static friction is equal to μ_s.

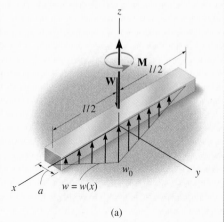

(a)

Solution

A free-body diagram of the bar is shown in Fig. 8–23b. Since the bar has a total weight of $W = mg$, the intensity w_0 of the distributed lead at the center ($x = 0$) is determined from vertical force equilibrium, Fig. 8–23a.

$$+\uparrow \Sigma F_z = 0; \quad -mg + 2\left[\frac{1}{2}\left(\frac{l}{2}\right)w_0\right] = 0 \quad w_0 = \frac{2\,mg}{l}$$

Since $w = 0$ at $x = l/2$, the distributed load expressed as a function of x is

$$w = w_0\left(1 - \frac{2\,x}{l}\right) = \frac{2\,mg}{l}\left(1 - \frac{2\,x}{l}\right)$$

The magnitude of the normal force acting on a segment of area having a length dx is therefore

$$dN = w\,dx = \frac{2\,mg}{l}\left(1 - \frac{2\,x}{l}\right)dx$$

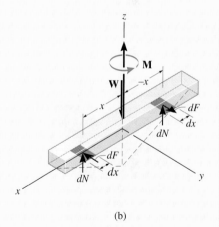

(b)

Fig. 8–23

The magnitude of the frictional force acting on the same element of area is

$$dF = \mu_s\,dN = \frac{2\,\mu_s mg}{l}\left(1 - \frac{2\,x}{l}\right)dx$$

Hence, the moment created by this force about the z axis is

$$dM = x\,dF = \frac{2\,\mu_s mg}{l}x\left(1 - \frac{2\,x}{l}\right)dx$$

The summation of moments about the z axis of the bar is determined by integration, which yields

$$\Sigma M_z = 0; \quad M - 2\int_0^{l/2}\frac{2\,\mu_s mg}{l}x\left(1 - \frac{2\,x}{l}\right)dx = 0$$

$$M = \frac{4\,\mu_s mg}{l}\left(\frac{x^2}{2} - \frac{2\,x^3}{3\,l}\right)\Big|_0^{l/2}$$

$$M = \frac{\mu_s mgl}{6} \qquad \qquad Ans.$$

8.7 Frictional Forces on Journal Bearings

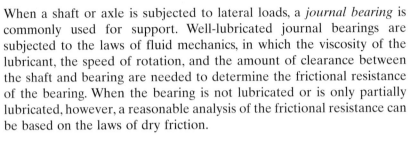

When a shaft or axle is subjected to lateral loads, a *journal bearing* is commonly used for support. Well-lubricated journal bearings are subjected to the laws of fluid mechanics, in which the viscosity of the lubricant, the speed of rotation, and the amount of clearance between the shaft and bearing are needed to determine the frictional resistance of the bearing. When the bearing is not lubricated or is only partially lubricated, however, a reasonable analysis of the frictional resistance can be based on the laws of dry friction.

Frictional Analysis. A typical journal-bearing support is shown in Fig. 8–24a. As the shaft rotates in the direction shown in the figure, it rolls up against the wall of the bearing to some point A where slipping occurs. If the lateral load acting at the end of the shaft is **P**, it is necessary that the bearing reactive force **R** acting at A be equal and opposite to **P**, Fig. 8–24b. The moment needed to maintain constant rotation of the shaft can be found by summing moments about the z axis of the shaft; i.e.,

$$\Sigma M_z = 0; \qquad\qquad M - (R \sin \phi_k)r = 0$$

or

$$M = Rr \sin \phi_k \tag{8–9}$$

where ϕ_k is the angle of kinetic friction defined by $\tan \phi_k = F/N = \mu_k N/N = \mu_k$. In Fig. 8–24c, it is seen that $r \sin \phi_k = r_f$. The dashed circle with radius r_f is called the *friction circle*, and as the shaft rotates, the reaction **R** will always be tangent to it. If the bearing is partially lubricated, μ_k is small, and therefore $\mu_k = \tan \phi_k \approx \sin \phi_k \approx \phi_k$. Under these conditions, a reasonable *approximation* to the moment needed to overcome the frictional resistance becomes

$$M \approx Rr\mu_k \tag{8–10}$$

The following example illustrates a common application of this analysis.

Unwinding the cable from this spool requires overcoming friction from the supporting shaft.

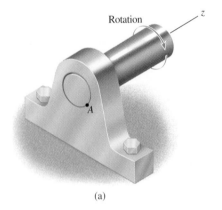

(a)

Fig. 8–24

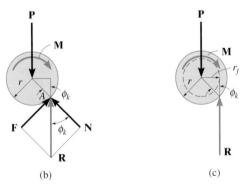

(b) (c)

EXAMPLE 8.11

The 100-mm-diameter pulley shown in Fig. 8–25a fits loosely on a 10-mm-diameter shaft for which the coefficient of static friction is $\mu_s = 0.4$. Determine the minimum tension T in the belt needed to (a) raise the 100-kg block and (b) lower the block. Assume that no slipping occurs between the belt and pulley and neglect the weight of the pulley.

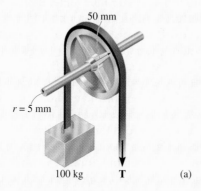

50 mm

$r = 5$ mm

100 kg **T** (a)

Solution

Part (a) A free-body diagram of the pulley is shown in Fig. 8–25b. When the pulley is subjected to belt tensions of 981 N each, it makes contact with the shaft at point P_1. As the tension T is *increased*, the pulley will roll around the shaft to point P_2 before motion impends. From the figure, the friction circle has a radius $r_f = r \sin \phi_s$. Using the simplification that $\sin \phi_s \approx$ (tan $\phi_s \approx \phi_s$), then $r_f \approx r\mu_s = (5 \text{ mm})(0.4) = 2$ mm, so that summing moments about P_2 gives

$$\zeta + \Sigma M_{P_2} = 0; \qquad 981 \text{ N}(52 \text{ mm}) - T(48 \text{ mm}) = 0$$
$$T = 1063 \text{ N} = 1.06 \text{ kN} \qquad \qquad Ans.$$

If a more exact analysis is used, then $\phi_s = \tan^{-1} 0.4 = 21.8°$. Thus, the radius of the friction circle would be $r_f = r \sin \phi_s = 5 \sin 21.8° = 1.86$ mm. Therefore,

$$\zeta + \Sigma M_{P_2} = 0;$$
$$981 \text{ N}(50 \text{ mm} + 1.86 \text{ mm}) - T(50 \text{ mm} - 1.86 \text{ mm}) = 0$$
$$T = 1057 \text{ N} = 1.06 \text{ kN} \qquad \qquad Ans.$$

Part (b) When the block is lowered, the resultant force **R** acting on the shaft passes through point P_3, as shown in Fig. 8–25c. Summing moments about this point yields

$$\zeta + \Sigma M_{P_3} = 0; \qquad 981 \text{ N}(48 \text{ mm}) - T(52 \text{ mm}) = 0$$
$$T = 906 \text{ N} \qquad \qquad Ans.$$

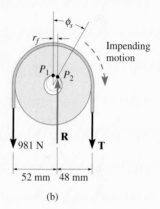

ϕ_s

r_f

Impending motion

P_1 P_2

981 N **R** **T**

52 mm 48 mm

(b)

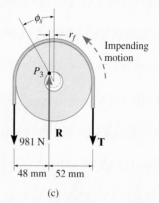

ϕ_s

r_f

Impending motion

P_3

981 N **R** **T**

48 mm 52 mm

(c)

Fig. 8–25

*8.8 Rolling Resistance

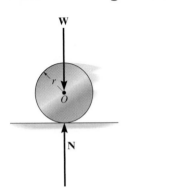

Rigid surface of contact

(a)

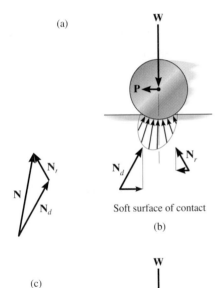

Soft surface of contact

(b)

(c)

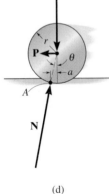

(d)

Fig. 8–26

If a *rigid* cylinder of weight **W** rolls at constant velocity along a *rigid* surface, the normal force exerted by the surface on the cylinder acts at the tangent point of contact, as shown in Fig. 8–26a. Under these conditions, provided the cylinder does not encounter frictional resistance from the air, motion will continue indefinitely. Actually, however, no materials are perfectly rigid, and therefore the reaction of the surface on the cylinder consists of a distribution of normal pressure. For example, consider the cylinder to be made of a very hard material, and the surface on which it rolls to be relatively soft. Due to its weight, the cylinder compresses the surface underneath it, Fig. 8–26b. As the cylinder rolls, the surface material in front of the cylinder *retards* the motion since it is being *deformed*, whereas the material in the rear is *restored* from the deformed state and therefore tends to *push* the cylinder forward. The normal pressures acting on the cylinder in this manner are represented in Fig. 8–26b by their resultant forces $\mathbf{N}_d$ and $\mathbf{N}_r$. Unfortunately, the magnitude of the force of *deformation*, $\mathbf{N}_d$, and its horizontal component is *always greater* than that of *restoration*, $\mathbf{N}_r$, and consequently a horizontal driving force **P** must be applied to the cylinder to maintain the motion. Fig. 8–26b.*

Rolling resistance is caused primarily by this effect, although it is also, to a smaller degree, the result of surface adhesion and relative microsliding between the surfaces of contact. Because the actual force **P** needed to overcome these effects is difficult to determine, a simplified method will be developed here to explain one way engineers have analyzed this phenomenon. To do this, we will consider the resultant of the *entire* normal pressure, $\mathbf{N} = \mathbf{N}_d + \mathbf{N}_r$, acting on the cylinder, Fig. 8–26c. As shown in Fig. 8–26d, this force acts at an angle θ with the vertical. To keep the cylinder in equilibrium, i.e., rolling at a constant rate, it is necessary that **N** be *concurrent* with the driving force **P** and the weight **W**. Summing moments about point A gives $Wa = P(r \cos \theta)$. Since the deformations are generally very small in relation to the cylinder's radius, $\cos \theta \approx 1$; hence,

$$Wa \approx Pr$$

or

$$\boxed{P \approx \frac{Wa}{r}} \qquad (8\text{–}11)$$

---

*Actually, the deformation force $\mathbf{N}_d$ causes *energy* to be stored in the material as its magnitude is increased, whereas the restoration force $\mathbf{N}_r$, as its magnitude is decreased, allows some of this energy to be released. The remaining energy is *lost* since it is used to heat up the surface, and if the cylinder's weight is very large, it accounts for permanent deformation of the surface. Work must be done by the horizontal force **P** to make up for this loss.

The distance a is termed the *coefficient of rolling resistance,* which has the dimension of length. For instance, $a \approx 0.5$ mm for a wheel rolling on a rail, both of which are made of mild steel. For hardened steel ball bearings on steel, $a \approx 0.1$ mm. Experimentally, though, this factor is difficult to measure, since it depends on such parameters as the rate of rotation of the cylinder, the elastic properties of the contacting surfaces, and the surface finish. For this reason, little reliance is placed on the data for determining a. The analysis presented here does, however, indicate why a heavy load (W) offers greater resistance to motion (P) than a light load under the same conditions. Furthermore, since the ratio Wa/r is generally very small compared to $\mu_k W$, the force needed to *roll* the cylinder over the surface will be much less than that needed to *slide* the cylinder across the surface. Hence, the analysis indicates why roller or ball bearings are often used to minimize the frictional resistance between moving parts.

Rolling resistance of railroad wheels on the rails is small since steel is very stiff. By comparison, the rolling resistance of the wheels of a tractor in a wet field is very large.

A 10-kg steel wheel shown in Fig. 8–27a has a radius of 100 mm and rests on an inclined plans made of wood. If θ is increased so that the wheel begins to roll-down the incline with constant velocity when $\theta = 1.2°$, determine the coefficient of rolling resistance.

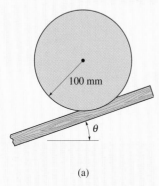

(a)

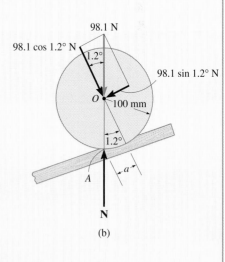

(b)

Fig. 8–27

Solution

As shown on the free-body diagram, Fig. 8–27b, when the wheel has impending motion, the normal reaction $\mathbf{N}$ acts at point A defined by the dimension a. Resolving the weight into components parallel and perpendicular to the incline, and summing moments about point A, yields (approximately)

$$\zeta + \Sigma M_A = 0; \quad 98.1 \cos 1.2° \, N(a) - 98.1 \sin 1.2° \, N(100 \text{ mm}) = 0$$

Solving, we obtain

$$a = 2.09 \text{ mm} \qquad\qquad Ans.$$

CHAPTER REVIEW

- *Dry Friction.* Frictional forces exist at rough surfaces of contact. They act on a body so as to oppose the motion or tendency of motion of the body. A static friction force approaches a maximum value of $F_s = \mu_s N$, where μ_s is the *coefficient of static friction*. In this case motion between the contacting surfaces is about to impend. If slipping occurs, then the friction force remains essentially constant and equal to a value of $F_k = \mu_k N$. Here μ_k is the *coefficient of kinetic friction*. The solution of a problem involving friction requires first drawing the free-body diagram of the body. If the unknowns cannot be determined strictly from the equations of equilibrium, and the possibility of slipping can occur, then the friction equation should be applied at the appropriate points of contact in order to complete the solution. It may also be possible for slender objects to tip over, and this situation should also be investigated.

- *Wedges, Screws, Belts, and Bearings.* A frictional analysis of these objects can be performed by applying the friction equation at the points of contact and then using the equations of equilibrium to relate the frictional force to the other external forces acting on the object. By combining the resulting equations, the force of friction can then be eliminated from the analysis, so that the force needed to overcome the effects of friction can be determined.

- *Rolling Resistance.* The resistance of a wheel to roll over a surface is caused by *deformation* between the two materials of contact. This effect causes the resultant normal force acting on the rolling body to be inclined so that it provides a component that acts in the opposite direction of the force causing the motion. The effect is characterized using the *coefficient of rolling resistance*, which is determined from experiment.

REVIEW PROBLEMS

8-1. A single force **P** is applied to the handle of the drawer. If friction is neglected at the bottom side and the coefficient of static friction along the sides is $\mu_s = 0.4$, determine the largest spacing s between the symmetrically placed handles so that the drawer does not bind at the corners A and B when the force **P** is applied to one of the handles.

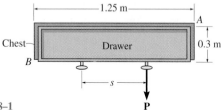

Prob. 8–1

8-2. The truck has a mass of 1.25 Mg and a center of mass at G. Determine the greatest load it can pull if (a) the truck has rear-wheel drive while the front wheels are free to roll, and (b) the truck has four-wheel drive. The coefficient of static friction between the wheels and the ground is $\mu_s = 0.5$, and between the crate and the ground, it is $\mu_s' = 0.4$.

*8-3. Solve Prob. 8-2 if the truck and crate are traveling up a 10° incline.

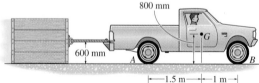

Probs. 8–2/3

8-4. The cam or short link is pinned at A and is used to hold mops or brooms against a wall. If the coefficient of static friction between the broomstick and the cam is $\mu_s = 0.2$, determine if it is possible to support the broom having a weight W. The surface at B is smooth. Neglect the weight of the cam.

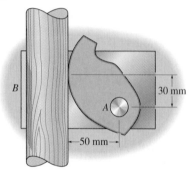

Prob. 8–4

8-5. The carton clamp on the forklift has a coefficient of static friction of $\mu_s = 0.5$ with any cardboard carton, whereas a cardboard carton has a coefficient of static friction of $\mu_s' = 0.4$ with any other cardboard carton. Compute the smallest horizontal force P the clamp must exert on the sides of a carton so that two cartons A and B each weighing 30 lb can be lifted. What smallest clamping force P' is required to lift three 30-lb cartons? The third carton C is placed between A and B.

Prob. 8–5

8-6. The tractor pulls on the fixed tree stump. Determine the torque that must be applied by the engine to the rear wheels to cause them to slip. The front wheels are free to roll. The tractor weighs 3500 lb and has a center of gravity at G. The coefficient of static friction between the rear wheels and the ground is $\mu_s = 0.5$.

*8-7. The tractor pulls on the fixed tree stump. If the coefficient of static friction between the rear wheels and the ground is $\mu_s = 0.6$, determine if the rear wheels slip or the front wheels lift off the ground as the engine provides torque to the rear wheels. What is the torque needed to cause the motion? The front wheels are free to roll. The tractor weighs 2500 lb and has a center of gravity at G.

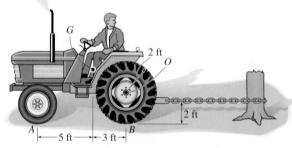

Probs. 8–6/7

DESIGN PROJECTS

8–1D DESIGN OF A ROPE-AND-PULLEY SYSTEM FOR PULLING A CRATE UP AN INCLINE.

A large 300-kg packing crate is to be hoisted up the 25° incline. The coefficient of static friction between the incline and the crate is $\mu_s = 0.5$, and the coefficient of kinetic friction is $\mu_k = 0.4$. Using a system of ropes and pulleys, design a method that will allow a single worker to pull the crate up the ramp. Pulleys can be attached to any point on the wall AB. Assume the worker can exert a maximum (comfortable) pull of 200 N on a rope. Submit a drawing of your design and a force analysis to show how it operates. Estimate the material cost required for its construction. Assume rope costs $0.75/m and a pulley costs $1.80.

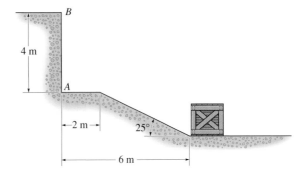

Fig. 8–1D

8–2D DESIGN OF A DEVICE FOR LIFTING STAINLESS-STEEL PIPES.

Stainless-steel pipes are stacked vertically in a manufacturing plant and are to be moved by an overhead crane from one point to another. The pipes have inner diameters ranging from 100 mm $\leq d \leq$ 250 mm and the maximum mass of any pipe is 500 kg. Design a device that can be connected to the hook and used to lift each pipe. The device should be made of structural steel and should be able to grip the pipe only from its inside surface, since the outside surface is required not to be scratched or damaged. Assume the smallest coefficient of static friction between the two steels is $\mu_s = 0.25$. Submit a scaled drawing of your device, along with a brief explanation of how it works based on a force analysis.

Fig. 8–2D

8–3D DESIGN OF A TOOL USED TO TURN PLASTIC PIPE.

PVC plastic is often used for sewer pipe. If the outer diameter of any pipe ranges from 4 in. $\leq d \leq$ 8 in., design a tool that can be used by a worker in order to turn the pipe when it is subjected to a maximum anticipated ground resistance of 80 lb · ft. The device is to be made of steel and should be designed so that it does not cut into the pipe and leave any significant marks on its surface. Assume a worker can apply a maximum (comfortable) force of 40 lb, and take the minimum coefficient of static friction between the PVC and the steel to be $\mu_s = 0.35$. Submit a scaled drawing of the device, and a brief paragraph to explain how it works based on a force analysis.

Fig. 8–3D

When a pressure vessel is designed, it is important to be able to determine the center of gravity of its component parts, calculate its volume and surface area, and reduce three-dimensional distributed loadings to their resultants. These topics are discussed in this chapter.

Center of Gravity and Centroid

- To discuss the concept of the center of gravity, center of mass, and the centroid.
- To show how to determine the location of the center of gravity and centroid for a system of discrete particles and a body of arbitrary shape.
- To use the theorems of Pappus and Guldinus for finding the area and volume for a surface of revolution.
- To present a method for finding the resultant of a general distributed loading and show how it applies to finding the resultant of a fluid.

9.1 Center of Gravity and Center of Mass for a System of Particles

Center of Gravity.   The *center of gravity G* is a point which locates the resultant weight of a system of particles. To show how to determine this point consider the system of *n* particles fixed within a region of space as shown in Fig. 9–1a. The weights of the particles comprise a system of parallel forces* which can be replaced by a single (equivalent) resultant weight having the defined point *G* of application. To find the $\bar{x}, \bar{y}, \bar{z}$ coordinates of *G*, we must use the principles outlined in Sec. 4.9.

---

*This is not true in the exact sense, since the weights are not parallel to each other; rather they are all *concurrent* at the earth's center. Furthermore, the acceleration of gravity *g* is actually different for each particle since it depends on the distance from the earth's center to the particle. For all practical purposes, however, both of these effects can generally be neglected.

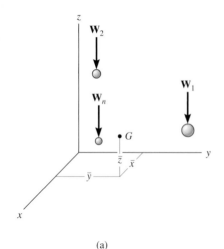

(a)

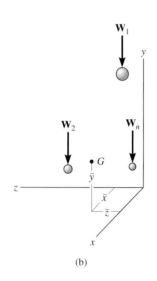

(b)

Fig. 9–1

This requires that the resultant weight be equal to the total weight of all n particles; that is,

$$W_R = \Sigma W$$

The sum of the moments of the weights of all the particles about the x, y, and z axes is then equal to the moment of the resultant weight about these axes. Thus, to determine the $\bar{x}$ coordinate of G, we can sum moments about the y axis. This yields

$$\bar{x}W_R = \tilde{x}_1W_1 + \tilde{x}_2W_2 + \cdots + \tilde{x}_nW_n$$

Likewise, summing moments about the x axis, we can obtain the $\bar{y}$ coordinate; i.e.,

$$\bar{y}W_R = \tilde{y}_1W_1 + \tilde{y}_2W_2 + \cdots + \tilde{y}_nW_n$$

Although the weights do not produce a moment about the z axis, we can obtain the $\bar{z}$ coordinate of G by imagining the coordinate system, with the particles fixed in it, as being rotated 90° about the x (or y) axis, Fig. 9–1b. Summing moments about the x axis, we have

$$\bar{z}W_R = \tilde{z}_1W_1 + \tilde{z}_2W_2 + \cdots + \tilde{z}_nW_n$$

We can generalize these formulas, and write them symbolically in the form

$$\bar{x} = \frac{\Sigma \tilde{x}W}{\Sigma W} \qquad \bar{y} = \frac{\Sigma \tilde{y}W}{\Sigma W} \qquad \bar{z} = \frac{\Sigma \tilde{z}W}{\Sigma W} \qquad (9\text{–}1)$$

Here

$\bar{x}, \bar{y}, \bar{z}$ represent the coordinates of the center of gravity G of the system of particles.

$\tilde{x}, \tilde{y}, \tilde{z}$ represent the coordinates of each particle in the system.

ΣW is the resultant sum of the weights of all the particles in the system.

These equations are easily remembered if it is kept in mind that they simply represent a balance between the sum of the moments of the weights of each particle of the system and the moment of the *resultant* weight for the system.

Center of Mass. To study problems concerning the motion of *matter* under the influence of force, i.e., dynamics, it is necessary to locate a point called the *center of mass*. Provided the acceleration due to gravity g for every particle is constant, then $W = mg$. Substituting into Eqs. 9–1 and canceling g from both the numerator and denominator yields

$$\bar{x} = \frac{\Sigma \tilde{x}m}{\Sigma m} \qquad \bar{y} = \frac{\Sigma \tilde{y}m}{\Sigma m} \qquad \bar{z} = \frac{\Sigma \tilde{z}m}{\Sigma m} \qquad (9\text{–}2)$$

By comparison, then, the location of the center of gravity *coincides* with that of the center of mass.* Recall, however, that particles have "weight" only when under the influence of a gravitational attraction, whereas the center of mass is independent of gravity. For example, it would be meaningless to define the center of gravity of a system of particles representing the planets of our solar system, while the center of mass of this system is important.

9.2 Center of Gravity, Center of Mass, and Centroid for a Body

Center of Gravity.   A rigid body is composed of an infinite number of particles, and so if the principles used to determine Eqs. 9–1 are applied to the system of particles composing a rigid body, it becomes necessary to use integration rather than a discrete summation of the terms. Considering the arbitrary particle located at $(\tilde{x}, \tilde{y}, \tilde{z})$ and having a weight dW, Fig. 9–2, the resulting equations are

$$\bar{x} = \frac{\displaystyle\int \tilde{x}\, dW}{\displaystyle\int dW} \qquad \bar{y} = \frac{\displaystyle\int \tilde{y}\, dW}{\displaystyle\int dW} \qquad \bar{z} = \frac{\displaystyle\int \tilde{z}\, dW}{\displaystyle\int dW} \qquad (9\text{–}3)$$

In order to apply these equations properly, the differential weight dW must be expressed in terms of its associated volume dV. If γ represents the *specific weight* of the body, measured as a weight per unit volume, then $dW = \gamma\, dV$ and therefore

$$\bar{x} = \frac{\displaystyle\int_V \tilde{x}\gamma\, dV}{\displaystyle\int_V \gamma\, dV} \qquad \bar{y} = \frac{\displaystyle\int_V \tilde{y}\gamma\, dV}{\displaystyle\int_V \gamma\, dV} \qquad \bar{z} = \frac{\displaystyle\int_V \tilde{z}\gamma\, dV}{\displaystyle\int_V \gamma\, dV} \qquad (9\text{–}4)$$

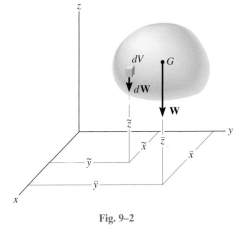

Fig. 9–2

Here integration must be performed throughout the entire volume of the body.

Center of Mass.   The *density* ρ, or mass per unit volume, is related to γ by the equation $\gamma = \rho g$, where g is the acceleration due to gravity. Substituting this relationship into Eqs. 9–4 and canceling g from both the numerators and denominators yields similar equations (with ρ replacing γ) that can be used to determine the body's *center of mass*.

---

*This is true as long as the gravity field is assumed to have the same magnitude and direction everywhere. That assumption is appropriate for most engineering applications, since gravity does not vary appreciably between, for instance, the bottom and the top of a building.

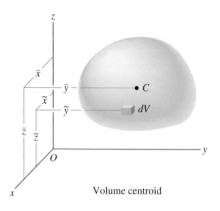

Volume centroid

Fig. 9–3

Centroid. The *centroid C* is a point which defines the *geometric center* of an object. Its location can be determined from formulas similar to those used to determine the body's center of gravity or center of mass. In particular, if the material composing a body is uniform or *homogeneous*, the *density or specific weight* will be *constant* throughout the body, and therefore this term will factor out of the integrals and *cancel* from both the numerators and denominators of Eqs. 9–4. The resulting formulas define the centroid of the body since they are independent of the body's weight and instead depend only on the body's geometry. Three specific cases will be considered.

Volume. If an object is subdivided into volume elements dV, Fig. 9–3, the location of the centroid $C(\overline{x}, \overline{y}, \overline{z})$ for the volume of the object can be determined by computing the "moments" of the elements about each of the coordinate axes. The resulting formulas are

$$\overline{x} = \frac{\int_V \widetilde{x}\, dV}{\int_V dV} \qquad \overline{y} = \frac{\int_V \widetilde{y}\, dV}{\int_V dV} \qquad \overline{z} = \frac{\int_V \widetilde{z}\, dV}{\int_V dV} \qquad (9\text{–}5)$$

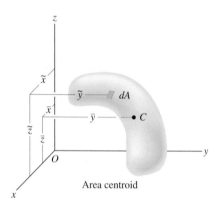

Area centroid

Fig. 9–4

Area. In a similar manner, the centroid for the surface area of an object, such as a plate or shell, Fig. 9–4, can be found by subdividing the area into differential elements dA and computing the "moments" of these area elements about each of the coordinate axes, namely,

$$\overline{x} = \frac{\int_A \widetilde{x}\, dA}{\int_A dA} \qquad \overline{y} = \frac{\int_A \widetilde{y}\, dA}{\int_A dA} \qquad \overline{z} = \frac{\int_A \widetilde{z}\, dA}{\int_A dA} \qquad (9\text{–}6)$$

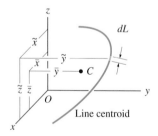

Line centroid

Fig. 9–5

Line. If the geometry of the object, such as a thin rod or wire, takes the form of a line, Fig. 9–5, the balance of moments of the differential elements dL about each of the coordinate axes yields

$$\overline{x} = \frac{\int_L \widetilde{x}\, dL}{\int_L dL} \qquad \overline{y} = \frac{\int_L \widetilde{y}\, dL}{\int_L dL} \qquad \overline{z} = \frac{\int_L \widetilde{z}\, dL}{\int_L dL} \qquad (9\text{–}7)$$

Remember that when applying Eqs. 9–4 through 9–7 it is best to choose a coordinate system that simplifies as much as possible the equation used to describe the object's boundary. For example, polar coordinates are generally appropriate for areas having circular boundaries. Also, the terms $\widetilde{x}, \widetilde{y}, \widetilde{z}$ in the equations refer to the "moment arms" or coordinates of the *center of gravity or centroid for the differential element* used. If possible, this differential element should be chosen such that it has a differential size or thickness in only *one direction*. When this is done, only a single integration is required to cover the entire region.

Symmetry. The *centroids* of some shapes may be partially or completely specified by using conditions of *symmetry*. In cases where the shape has an axis of symmetry, the centroid of the shape will lie along that axis. For example, the centroid C for the line shown in Fig. 9–6 must lie along the y axis since for every elemental length dL at a distance $+\widetilde{x}$ to the right of the y axis there is an identical element at a distance $-\widetilde{x}$ to the left. The total moment for all the elements about the axis of symmetry will therefore cancel; i.e., $\int \widetilde{x}\, dL = 0$ (Eq. 9–7), so that $\bar{x} = 0$. In cases where a shape has two or three axes of symmetry, it follows that the centroid lies at the intersection of these axes, Fig. 9–7 and Fig. 9–8.

Integration must be used to determine the location of the center of gravity of this goal post.

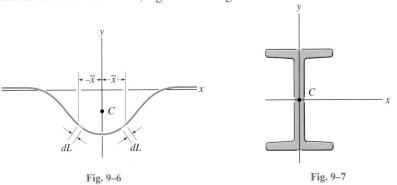

Fig. 9–6

Fig. 9–7

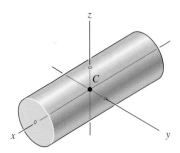

Fig. 9–8

IMPORTANT POINTS

- The centroid represents the geometric center of a body. This point coincides with the center of mass or the center of gravity only if the material composing the body is uniform or homogeneous.

- Formulas used to locate the center of gravity or the centroid simply represent a balance between the sum of moments of all the parts of the system and the moment of the "resultant" for the system.

- In some cases the centroid is located at a point that is not on the object, as in the case of a ring, where the centroid is at its center. Also, this point will lie on any axis of symmetry for the body.

PROCEDURE FOR ANALYSIS

The center of gravity or centroid of an object or shape can be determined by single integrations using the following procedure.

Differential Element.

- Select an appropriate coordinate system, specify the coordinate axes, and then choose a differential element for integration.
- For lines the element dL is represented as a differential line segment.
- For areas the element dA is generally a rectangle having a finite length and differential width.
- For volumes the element dV is either a circular disk having a finite radius and differential thickness, or a shell having a finite length and radius and a differential thickness.
- Locate the element at an arbitrary point (x, y, z) on the curve that defines the shape.

Size and Moment Arms.

- Express the length dL, area dA, or volume dV of the element in terms of the coordinates of the curve used to define the geometric shape.
- Determine the coordinates or moment arms $\tilde{x}, \tilde{y}, \tilde{z}$ for the centroid or center of gravity of the element.

Integrations.

- Substitute the formulations for $\tilde{x}, \tilde{y}, \tilde{z}$ and dL, dA, or dV into the appropriate equations (Eqs. 9–4 through 9–7) and perform the integrations.[*]
- Express the function in the integrand in terms of the *same variable as the differential thickness of the element* in order to perform the integration.
- The limits of the integral are defined from the two extreme locations of the element's differential thickness, so that when the elements are "summed" or the integration performed, the entire region is covered.

[*]Formulas for integration are given in Appendix A.

EXAMPLE 9.1

Locate the centroid of the rod bent into the shape of a parabolic arc, shown in Fig. 9–9.

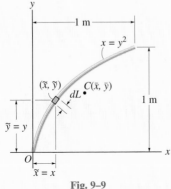

Fig. 9–9

Solution

Differential Element.   The differential element is shown in Fig. 9–9. It is located on the curve at the *arbitrary point* (x, y).

Area and Moment Arms.   The differential length of the element dL can be expressed in terms of the differentials dx and dy by using the Pythagorean theorem.

$$dL = \sqrt{(dx)^2 + (dy)^2} = \sqrt{\left(\frac{dx}{dy}\right)^2 + 1}\, dy$$

Since $x = y^2$, then $dx/dy = 2y$. Therefore, expressing dL in terms of y and dy, we have

$$dL = \sqrt{(2y)^2 + 1}\, dy$$

The centroid is located at $\tilde{x} = x, \tilde{y} = y$.

Integrations.   Applying Eqs. 9–7 and integrating with respect to y using the formulas in Appendix A, we have

$$\bar{x} = \frac{\int_L \tilde{x}\, dL}{\int_L dL} = \frac{\int_0^1 x\sqrt{4y^2 + 1}\, dy}{\int_0^1 \sqrt{4y^2 + 1}\, dy} = \frac{\int_0^1 y^2\sqrt{4y^2 + 1}\, dy}{\int_0^1 \sqrt{4y^2 + 1}\, dy}$$

$$= \frac{0.6063}{1.479} = 0.410 \text{ m} \qquad\qquad Ans.$$

$$\bar{y} = \frac{\int_L \tilde{y}\, dL}{\int_L dL} = \frac{\int_0^1 y\sqrt{4y^2 + 1}\, dy}{\int_0^1 \sqrt{4y^2 + 1}\, dy} = \frac{0.8484}{1.479} = 0.574 \text{ m} \qquad Ans.$$

E X A M P L E   9.2

Locate the centroid of the circular wire segment shown in Fig. 9–10.

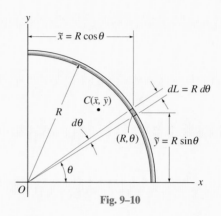

Fig. 9–10

Solution

Polar coordinates will be used to solve this problem since the arc is circular.

Differential Element. A differential circular arc is selected as shown in the figure. This element intersects the curve at (R, θ).

Length and Moment Arm. The differential length of the element is $dL = R\, d\theta$, and its centroid is located at $\tilde{x} = R \cos\theta$ and $\tilde{y} = R \sin\theta$.

Integrations. Applying Eqs. 9–7 and integrating with respect to θ, we obtain

$$\bar{x} = \frac{\displaystyle\int_L \tilde{x}\, dL}{\displaystyle\int_L dL} = \frac{\displaystyle\int_0^{\pi/2} (R\cos\theta)R\, d\theta}{\displaystyle\int_0^{\pi/2} R\, d\theta} = \frac{R^2 \displaystyle\int_0^{\pi/2} \cos\theta\, d\theta}{R \displaystyle\int_0^{\pi/2} d\theta} = \frac{2R}{\pi} \qquad Ans.$$

$$\bar{y} = \frac{\displaystyle\int_L \tilde{y}\, dL}{\displaystyle\int_L dL} = \frac{\displaystyle\int_0^{\pi/2} (R\sin\theta)R\, d\theta}{\displaystyle\int_0^{\pi/2} R\, d\theta} = \frac{R^2 \displaystyle\int_0^{\pi/2} \sin\theta\, d\theta}{R \displaystyle\int_0^{\pi/2} d\theta} = \frac{2R}{\pi} \qquad Ans.$$

EXAMPLE 9.3

Determine the distance $\bar{y}$ from the x axis to the centroid of the area of the triangle shown in Fig. 9–11.

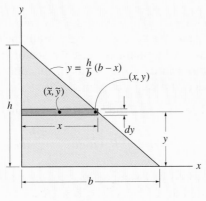

Fig. 9–11

Solution

Differential Element. Consider a rectangular element having thickness dy which intersects the boundary at (x, y), Fig. 9–11.

Area and Moment Arms. The area of the element is $dA = x\,dy = \dfrac{b}{h}(h - y)\,dy$, and its centroid is located a distance $\tilde{y} = y$ from the x axis.

Integrations. Applying the second of Eqs. 9–6 and integrating with respect to y yields

$$\bar{y} = \frac{\displaystyle\int_A \tilde{y}\,dA}{\displaystyle\int_A dA} = \frac{\displaystyle\int_0^h y\frac{b}{h}(h - y)\,dy}{\displaystyle\int_0^h \frac{b}{h}(h - y)\,dy} = \frac{\frac{1}{6}bh^2}{\frac{1}{2}bh}$$

$$= \frac{h}{3} \qquad\qquad\qquad Ans.$$

E X A M P L E 9.4

Locate the centroid for the area of a quarter circle shown in Fig. 9–12a.

Solution I

Differential Element. Polar coordinates will be used since the boundary is circular. We choose the element in the shape of a *triangle*, Fig. 9–12a. (Actually the shape is a circular sector; however, neglecting higher-order differentials, the element becomes triangular.) The element intersects the curve at point (R, θ).

Area and Moment Arms. The area of the element is

$$dA = \tfrac{1}{2}(R)(R \, d\theta) = \frac{R^2}{2} d\theta$$

and using the results of Example 9.3, the centroid of the (triangular) element is located at $\tilde{x} = \tfrac{2}{3}R \cos \theta$, $\tilde{y} = \tfrac{2}{3}R \sin \theta$.

Integrations. Applying Eqs. 9–6 and integrating with respect to θ, we obtain

$$\overline{x} = \frac{\displaystyle\int_A \tilde{x} \, dA}{\displaystyle\int_A dA} = \frac{\displaystyle\int_0^{\pi/2} \left(\frac{2}{3}R \cos \theta\right) \frac{R^2}{2} d\theta}{\displaystyle\int_0^{\pi/2} \frac{R^2}{2} d\theta}$$

$$= \frac{\left(\dfrac{2}{3}R\right) \displaystyle\int_0^{\pi/2} \cos \theta \, d\theta}{\displaystyle\int_0^{\pi/2} d\theta} = \frac{4R}{3\pi} \qquad \textit{Ans.}$$

$$\overline{y} = \frac{\displaystyle\int_A \tilde{y} \, dA}{\displaystyle\int_A dA} = \frac{\displaystyle\int_0^{\pi/2} \left(\frac{2}{3}R \sin \theta\right) \frac{R^2}{2} d\theta}{\displaystyle\int_0^{\pi/2} \frac{R^2}{2} d\theta}$$

$$= \frac{\left(\dfrac{2}{3}R\right) \displaystyle\int_0^{\pi/2} \sin \theta \, d\theta}{\displaystyle\int_0^{\pi/2} d\theta} = \frac{4R}{3\pi} \qquad \textit{Ans.}$$

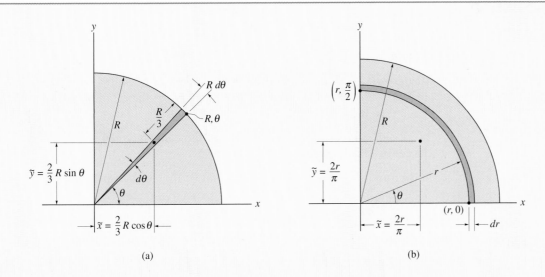

Fig. 9–12

Solution II

Differential Element.   The differential element may be chosen in the form of a *circular arc* having a thickness dr as shown in Fig. 9–12*b*. The element intersects the axes at points $(r, 0)$ and $(r, \pi/2)$.

Area and Moment Arms.   The area of the element is $dA = (2\pi r/4)\,dr$. Since the centroid of a 90° circular arc was determined in Example 9.2, then for the element $\widetilde{x} = 2r/\pi$, $\widetilde{y} = 2r/\pi$.

Integrations.   Using Eqs. 9–6 and integrating with respect to r, we obtain

$$\overline{x} = \frac{\displaystyle\int_A \widetilde{x}\,dA}{\displaystyle\int_A dA} = \frac{\displaystyle\int_0^R \frac{2r}{\pi}\left(\frac{2\pi r}{4}\right)dr}{\displaystyle\int_0^R \frac{2\pi r}{4}\,dr} = \frac{\displaystyle\int_0^R r^2\,dr}{\dfrac{\pi}{2}\displaystyle\int_0^R r\,dr} = \frac{4\,R}{3\,\pi} \qquad Ans.$$

$$\overline{y} = \frac{\displaystyle\int_A \widetilde{y}\,dA}{\displaystyle\int_A dA} = \frac{\displaystyle\int_0^R \frac{2r}{\pi}\left(\frac{2\pi r}{4}\right)dr}{\displaystyle\int_0^R \frac{2\pi r}{4}\,dr} = \frac{\displaystyle\int_0^R r^2\,dr}{\dfrac{\pi}{2}\displaystyle\int_0^R r\,dr} = \frac{4\,R}{3\,\pi} \qquad Ans.$$

E X A M P L E   9.5

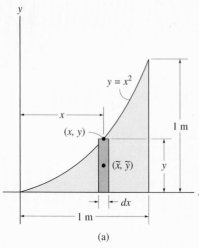

(a)

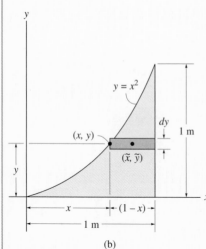

(b)

Fig. 9–13

Locate the centroid of the area shown in Fig. 9–13a.

Solution I

Differential Element.   A differential element of thickness dx is shown in Fig. 9–13a. The element intersects the curve at the *arbitrary point* (x, y), and so it has a height y.

Area and Moment Arms.   The area of the element is $dA = y\,dx$, and its centroid is located at $\widetilde{x} = x$, $\widetilde{y} = y/2$.

Integrations.   Applying Eqs. 9–6 and integrating with respect to x yields

$$\bar{x} = \frac{\displaystyle\int_A \widetilde{x}\,dA}{\displaystyle\int_A dA} = \frac{\displaystyle\int_0^1 xy\,dx}{\displaystyle\int_0^1 y\,dx} = \frac{\displaystyle\int_0^1 x^3\,dx}{\displaystyle\int_0^1 x^2\,dx} = \frac{0.250}{0.333} = 0.75 \text{ m} \qquad Ans.$$

$$\bar{y} = \frac{\displaystyle\int_A \widetilde{y}\,dA}{\displaystyle\int_A dA} = \frac{\displaystyle\int_0^1 (y/2)y\,dx}{\displaystyle\int_0^1 y\,dx} = \frac{\displaystyle\int_0^1 (x^2/2)x^2\,dx}{\displaystyle\int_0^1 x^2\,dx} = \frac{0.100}{0.333} = 0.3 \text{ m } Ans.$$

Solution II

Differential Element.   The differential element of thickness dy is shown in Fig. 9–13b. The element intersects the curve at the *arbitrary point* (x, y), and so it has a length $(1 - x)$.

Area and Moment Arms.   The area of the element is $dA = (1 - x)\,dy$, and its centroid is located at

$$\widetilde{x} = x + \left(\frac{1 - x}{2}\right) = \frac{1 + x}{2}, \quad \widetilde{y} = y$$

Integrations.   Applying Eqs. 9–6 and integrating with respect to y, we obtain

$$\widetilde{x} = \frac{\displaystyle\int_A \widetilde{x}\,dA}{\displaystyle\int_A dA} = \frac{\displaystyle\int_0^1 [(1 + x)/2](1 - x)\,dy}{\displaystyle\int_0^1 (1 - x)\,dy} = \frac{\dfrac{1}{2}\displaystyle\int_0^1 (1 - y)\,dy}{\displaystyle\int_0^1 (1 - \sqrt{y})\,dy} = \frac{0.250}{0.333} = 0.75 \text{ m} \qquad Ans.$$

$$\bar{y} = \frac{\displaystyle\int_A \widetilde{y}\,dA}{\displaystyle\int_A dA} = \frac{\displaystyle\int_0^1 y(1 - x)\,dy}{\displaystyle\int_0^1 (1 - x)\,dy} = \frac{\displaystyle\int_0^1 (y - y^{3/2})\,dy}{\displaystyle\int_0^1 (1 - \sqrt{y})\,dy} = \frac{0.100}{0.333} = 0.3 \text{ m} \qquad Ans.$$

EXAMPLE 9.6

Locate the $\bar{x}$ centroid of the shaded area bounded by the two curves $y = x$ and $y = x^2$, Fig. 9–14.

Solution I

Differential Element. A differential element of thickness dx is shown in Fig. 9–14a. The element intersects the curves at *arbitrary points* (x, y_1) and (x, y_2), and so it has a height $(y_2 - y_1)$.

Area and Moment Arm. The area of the element is $dA = (y_2 - y_1)\, dx$, and its centroid is located at $\tilde{x} = x$.

Integration. Applying Eq. 9–6, we have

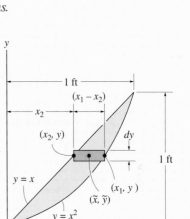

(a)

$$\bar{x} = \frac{\int_A \tilde{x}\, dA}{\int_A dA} = \frac{\int_0^1 x(y_2 - y_1)\, dx}{\int_0^1 (y_2 - y_1)\, dx} = \frac{\int_0^1 x(x - x^2)\, dx}{\int_0^1 (x - x^2)\, dx} = \frac{\frac{1}{12}}{\frac{1}{6}} = 0.5 \text{ ft } Ans.$$

Solution II

Differential Element. A differential element having a thickness dy is shown in Fig. 9–14b. The element intersects the curves at the *arbitrary points* (x_2, y) and (x_1, y), and so it has a length $(x_1 - x_2)$.

Area and Moment Arm. The area of the element is $dA = (x_1 - x_2)\, dy$, and its centroid is located at

$$\tilde{x} = x_2 + \frac{x_1 - x_2}{2} = \frac{x_1 + x_2}{2}$$

Integration. Applying Eq. 9–6, we have

(b)

Fig. 9–14

$$\bar{x} = \frac{\int_A \tilde{x}\, dA}{\int_A dA} = \frac{\int_0^1 [(x_1 + x_2)/2](x_1 - x_2)\, dy}{\int_0^1 (x_1 - x_2)\, dy} = \frac{\int_0^1 [(\sqrt{y} + y)/2](\sqrt{y} - y)\, dy}{\int_0^1 (\sqrt{y} - y)\, dy}$$

$$= \frac{\frac{1}{2}\int_0^1 (y - y^2)\, dy}{\int_0^1 (\sqrt{y} - y)\, dy} = \frac{\frac{1}{12}}{\frac{1}{6}} = 0.5 \text{ ft} \qquad\qquad Ans.$$

E X A M P L E   **9.7**

Locate the $\bar{y}$ centroid for the paraboloid of revolution, which is generated by revolving the shaded area shown in Fig. 9–15*a* about the *y* axis.

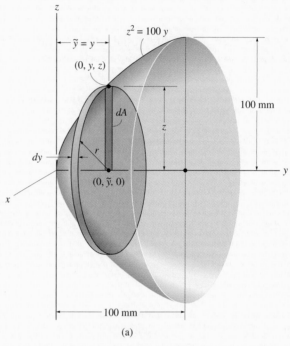

(a)

Fig. 9–15

Solution I

Differential Element. An element having the shape of a *thin disk* is chosen, Fig. 9–15*a*. This element has a thickness *dy*. In this "disk" method of analysis, the element of planar area, *dA*, is always taken *perpendicular* to the axis of revolution. Here the element intersects the generating curve at the *arbitrary point* (0, *y*, *z*), and so its radius is $r = z$.

Area and Moment Arm. The volume of the element is $dV = (\pi z^2)\,dy$, and its centroid is located at $\widetilde{y} = y$.

Integration Applying the second of Eqs. 9–5 and integrating with respect to *y* yields

$$\bar{y} = \frac{\displaystyle\int_V \widetilde{y}\,dV}{\displaystyle\int_V dV} = \frac{\displaystyle\int_0^{100} y(\pi z^2)\,dy}{\displaystyle\int_0^{100} (\pi z^2)\,dy} = \frac{100\pi \displaystyle\int_0^{100} y^2\,dy}{100\pi \displaystyle\int_0^{100} y\,dy} = 66.7 \text{ mm} \quad Ans.$$

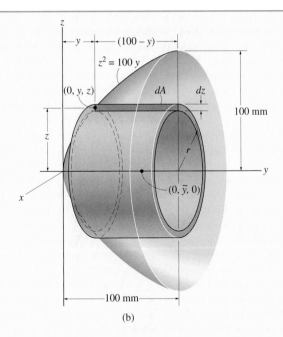

(b)

Solution II

Differential Element.   As shown in Fig. 9–15b, the volume element can be chosen in the form of a *thin cylindrical shell*, where the shell's thickness is dz. In this "shell" method of analysis, the element of planar area, dA, is always taken *parallel* to the axis of revolution. Here the element intersects the generating curve at point $(0, y, z)$, and so the radius of the shell is $r = z$.

Area and Moment Arm.   The volume of the element is $dV = 2\pi r\, dA = 2\pi z(100 - y)\, dz$, and its centroid is located at $\widetilde{y} = y + (100 - y)/2 = (100 + y)/2$.

Integrations.   Applying the second of Eqs. 9–5 and integrating with respect to z yields

$$\overline{y} = \frac{\displaystyle\int_V \widetilde{y}\, dV}{\displaystyle\int_V dV} = \frac{\displaystyle\int_0^{100} [(100 + y)/2]2\pi z(100 - y)\, dz}{\displaystyle\int_0^{100} 2\pi z(100 - y)\, dz}$$

$$= \frac{\pi \displaystyle\int_0^{100} z(10^4 - 10^{-4}z^4)\, dz}{2\pi \displaystyle\int_0^{100} z(100 - 10^{-2}z^2)\, dz} = 66.7 \text{ mm} \qquad \textit{Ans.}$$

EXAMPLE 9.8

Determine the location of the center of mass of the cylinder shown in Fig. 9–16a if its density varies directly with its distance from the base, i.e., $\rho = 200z$ kg/m^3.

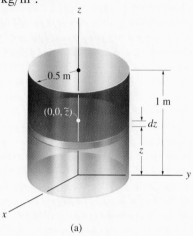

(a)

Solution

For reasons of material symmetry,

$$\bar{x} = \bar{y} = 0 \qquad \qquad Ans.$$

Differential Element. A disk element of radius 0.5 m and thickness dz is chosen for integration, Fig. 9–16a, since the *density of the entire element is constant* for a given value of z. The element is located along the z axis at the *arbitrary point* $(0, 0, z)$.

Volume and Moment Arm. The volume of the element is $dV = \pi(0.5)^2 \, dz$, and its centroid is located at $\tilde{z} = z$.

Integrations. Using an equation similar to the third of Eqs. 9–4 and integrating with respect to z, noting that $\rho = 200z$, we have

$$\bar{z} = \frac{\int_v \tilde{z}\rho \, dV}{\int_v \rho \, dV} = \frac{\int_0^1 z(200z)\pi(0.5)^2 \, dz}{\int_0^1 (200z)\pi(0.5)^2 \, dz}$$

$$= \frac{\int_0^1 z^2 \, dz}{\int_0^1 z \, dz} = 0.667 \text{ m} \qquad \qquad Ans.$$

Note: It is not possible to use a shell element for integration such as shown in Fig. 9–16b since the density of the material composing the shell would *vary* along the shell's height and hence the location of $\tilde{z}$ for the element cannot be specified.

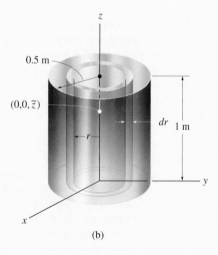

(b)

Fig. 9–16

9.3 Composite Bodies

A *composite body* consists of a series of connected "simpler" shaped bodies, which may be rectangular, triangular, semicircular, etc. Such a body can often be sectioned or divided into its composite parts and, provided the *weight* and location of the center of gravity of each of these parts are known, we can eliminate the need for integration to determine the center of gravity for the entire body. The method for doing this requires treating each composite part like a particle and following the

In order to determine the force required to tip over this concrete barrier it is first necessary to determine the location of its center of gravity.

procedure outlined in Sec. 9.1. Formulas analogous to Eqs. 9–1 result since we must account for a finite number of weights. Rewriting these formulas, we have

$$\overline{x} = \frac{\Sigma \widetilde{x} W}{\Sigma W} \qquad \overline{y} = \frac{\Sigma \widetilde{y} W}{\Sigma W} \qquad \overline{z} = \frac{\Sigma \widetilde{z} W}{\Sigma W} \qquad (9\text{–}8)$$

Here

$\overline{x}, \overline{y}, \overline{z}$ represent the coordinates of the center of gravity G of the composite body.

$\widetilde{x}, \widetilde{y}, \widetilde{z}$ represent the coordinates of the center of gravity of each composite part of the body.

ΣW is the sum of the weights of all the composite parts of the body, or simply the total weight of the body.

When the body has a *constant density or specific weight*, the center of gravity *coincides* with the centroid of the body. The centroid for composite lines, areas, and volumes can be found using relations analogous to Eqs. 9–8; however, the W's are replaced by L's, A's, and V's, respectively. Centroids for common shapes of lines, areas, shells, and volumes that often make up a composite body are given in the table on the inside back cover.

PROCEDURE FOR ANALYSIS

The location of the center of gravity of a body or the centroid of a composite geometrical object represented by a line, area, or volume can be determined using the following procedure.

Composite Parts.

- Using a sketch, divide the body or object into a finite number of composite parts that have simpler shapes.

- If a composite part has a *hole*, or a geometric region having no material, then consider the composite part without the hole and consider the hole as an *additional* composite part having *negative* weight or size.

Moment Arms.

- Establish the coordinate axes on the sketch and determine the coordinates $\widetilde{x}, \widetilde{y}, \widetilde{z}$ of the center of gravity or centroid of each part.

Summations.

- Determine $\overline{x}, \overline{y}, \overline{z}$ by applying the center of gravity equations, Eqs. 9–8, or the analogous centroid equations.

- If an object is *symmetrical* about an axis, the centroid of the object lies on this axis.

If desired, the calculations can be arranged in tabular form, as indicated in the following three examples.

E X A M P L E 9.9

Locate the centroid of the wire shown in Fig. 9–17a.

Solution

Composite Parts. The wire is divided into three segments as shown in Fig. 9–17b.

Moment Arms. The location of the centroid for each piece is determined and indicated in the figure. In particular, the centroid of segment ① is determined either by integration or by using the table on the inside back cover.

Summations. The calculations are tabulated as follows:

Segment	L (mm)	$\widetilde{x}$ (mm)	$\widetilde{y}$ (mm)	$\widetilde{z}$ (mm)	$\widetilde{x}L$ (mm²)	$\widetilde{y}L$ (mm²)	$\widetilde{z}L$ (mm²)
1	$\pi(60) = 188.5$	60	−38.2	0	11 310	−7200	0
2	40	0	20	0	0	0	0
3	20	0	40	−10	0	800	−200
	$\Sigma L = 248.5$				$\Sigma\widetilde{x}L = 11\,310$	$\Sigma\widetilde{y}L = -5600$	$\Sigma\widetilde{z}L = -200$

Thus,

$$\overline{x} = \frac{\Sigma\widetilde{x}L}{\Sigma L} = \frac{11310}{248.5} = 45.5 \text{ mm} \qquad Ans.$$

$$\overline{y} = \frac{\Sigma\widetilde{y}L}{\Sigma L} = \frac{-5600}{248.5} = -22.5 \text{ mm} \qquad Ans.$$

$$\overline{z} = \frac{\Sigma\widetilde{z}L}{\Sigma L} = \frac{-200}{248.5} = -0.805 \text{ mm} \qquad Ans.$$

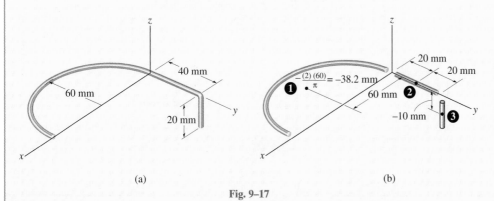

(a)   (b)

Fig. 9–17

E X A M P L E **9.10**

Locate the centroid of the plate area shown in Fig. 9–18a.

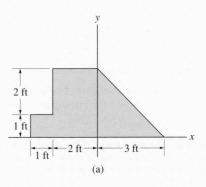

(a)

Fig. 9–18

Solution

Composite Parts. The plate is divided into three segments as shown in Fig. 9–18b. Here the area of the small rectangle ③ is considered "negative" since it must be subtracted from the larger one ②.

Moment Arms. The centroid of each segment is located as indicated in the figure. Note that the $\tilde{x}$ coordinates of ② and ③ are *negative*.

Summations. Taking the data from Fig. 9–18b, the calculations are tabulated as follows:

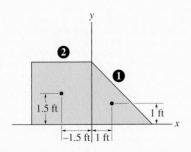

(b)

Segment	A (ft²)	$\tilde{x}$ (ft)	$\tilde{y}$ (ft)	$\tilde{x}A$ (ft³)	$\tilde{y}A$ (ft³)
1	$\frac{1}{2}(3)(3) = 4.5$	1	1	4.5	4.5
2	$(3)(3) = 9$	−1.5	1.5	−13.5	13.5
3	$-(2)(1) = -2$	−2.5	2	5	−4
	$\Sigma A = 11.5$			$\Sigma \tilde{x}A = -4$	$\Sigma \tilde{y}A = 14$

Thus,

$$\bar{x} = \frac{\Sigma \tilde{x}A}{\Sigma A} = \frac{-4}{11.5} = -0.348 \text{ ft} \qquad Ans.$$

$$\bar{y} = \frac{\Sigma \tilde{y}A}{\Sigma A} = \frac{14}{11.5} = 1.22 \text{ ft} \qquad Ans.$$

E X A M P L E   **9.11**

Locate the center of mass of the composite assembly shown in Fig. 9–19a. The conical frustum has a density of $\rho_c = 8\text{ Mg/m}^3$, and the hemisphere has a density of $\rho_h = 4\text{ Mg/m}^3$. There is a 25-mm radius cylindrical hole in the center.

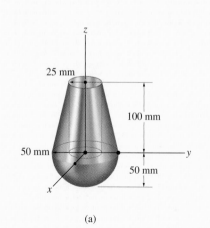

Solution

Composite Parts. The assembly can be thought of as consisting of four segments as shown in Fig. 9–19b. For the calculations, ③ and ④ must be considered as "negative" volumes in order that the four segments, when added together, yield the total composite shape shown in Fig. 9–19a.

Moment Arm. Using the table on the inside back cover, the computations for the centroid $\tilde{z}$ of each piece are shown in the figure.

Summations. Because of *symmetry*, note that

$$\bar{x} = \bar{y} = 0 \qquad Ans.$$

Since $W = mg$ and g is constant, the third of Eqs. 9–8 becomes $\bar{z} = \Sigma\tilde{z}m/\Sigma m$. The mass of each piece can be computed from $m = \rho V$ and used for the calculations. Also, $1\text{ Mg/m}^3 = 10^{-6}\text{ kg/mm}^3$, so that

(a)

Fig. 9–19

Segment	m (kg)	$\tilde{z}$ (mm)	$\tilde{z}m$ (kg·mm)
1	$8(10^{-6})(\frac{1}{3})\pi(50)^2(200) = 4.189$	50	209.440
2	$4(10^{-6})(\frac{2}{3})\pi(50)^3 = 1.047$	-18.75	-19.635
3	$-8(10^{-6})(\frac{1}{3})\pi(25)^2(100) = -0.524$	$100 + 25 = 125$	-65.450
4	$-8(10^{-6})\pi(25)^2(100) = -1.571$	50	-78.540
	$\Sigma m = 3.141$		$\Sigma\tilde{z}m = 45.815$

Thus,

$$\tilde{z} = \frac{\Sigma\tilde{z}m}{\Sigma m} = \frac{45.815}{3.141} = 14.6\text{ mm} \qquad Ans.$$

200 mm

$\dfrac{200\text{ mm}}{4} = 50\text{ mm}$

50 mm

50 mm

❶

50 mm

❷

$-\frac{3}{8}(50) = -18.75\text{ mm}$

❸

100 mm

$\dfrac{100\text{ mm}}{4} = 25\text{ mm}$

25 mm

100 mm

50 mm

25 mm

100 mm

❹

50 mm

(b)

*9.4 Theorems of Pappus and Guldinus

The two *theorems of Pappus and Guldinus*, which were first developed by Pappus of Alexandria during the third century A.D. and then restated at a later time by the Swiss mathematician Paul Guldin or Guldinus (1577–1643), are used to find the surface area and volume of any object of revolution.

A *surface area of revolution* is generated by revolving a *plane curve* about a nonintersecting fixed axis in the plane of the curve; whereas a *volume of revolution* is generated by revolving a *plane area* about a nonintersecting fixed axis in the plane of the area. For example, if the *line AB* shown in Fig. 9–20 is rotated about a fixed axis, it generates the *surface area* of a cone (less the area of the base); if the triangular *area ABC* shown in Fig. 9–21 is rotated about the axis, it generates the *volume* of a cone.

The statements and proofs of the theorems of Pappus and Guldinus follow. The proofs require that the generating curves and areas do *not* cross the axis about which they are rotated; otherwise, two sections on either side of the axis would generate areas or volumes having opposite signs and hence cancel each other.

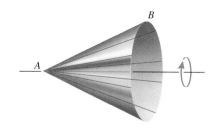

Fig. 9–20

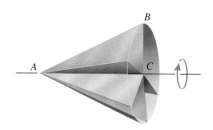

Fig. 9–21

Surface Area. *The area of a surface of revolution equals the product of the length of the generating curve and the distance traveled by the centroid of the curve in generating the surface area.*

Proof. When a differential length dL of the curve shown in Fig. 9–22 is revolved about an axis through a distance $2\pi r$, it generates a ring having a surface area $dA = 2\pi r\, dL$. The entire surface area, generated by revolving the entire curve about the axis, is therefore $A = 2\pi \int_L r\, dL$. This equation may be simplified, however, by noting that the location r of the centroid for the line of total length L can be determined from an equation having the form of Eqs. 9–7, namely, $\int_L r\, dL = \bar{r}L$. Thus, the total surface area becomes $A = 2\pi\bar{r}L$. In general, though, if the line does not undergo a complete revolution, then,

$$A = \theta \bar{r} L \tag{9–9}$$

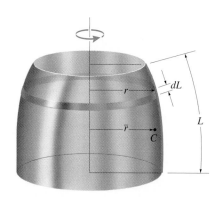

Fig. 9–22

where A = surface area of revolution

θ = angle of revolution measured in radians, $\theta \le 2\pi$

$\bar{r}$ = perpendicular distance from the axis of revolution to the centroid of the generating curve

L = length of the generating curve

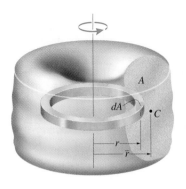

Fig. 9–23

Volume. *The volume of a body of revolution equals the product of the generating area and the distance traveled by the centroid of the area in generating the volume.*

Proof. When the differential area dA shown in Fig. 9–23 is revolved about an axis through a distance $2\pi r$, it generates a ring having a volume $dV = 2\pi r\, dA$. The entire volume, generated by revolving A about the axis, is therefore $V = 2\pi \int_V r\, dA$. Here the integral can be eliminated by using an equation analogous to Eqs. 9–6, $\int_V r\, dA = \bar{r}A$, where $\bar{r}$ locates the centroid C of the generating area A. The volume becomes $V = 2\pi \bar{r}A$. In general, though,

$$V = \theta \bar{r} A \qquad (9\text{–}10)$$

The surface area and the amount of water that can be stored in this water tank can be determined by using the theorems of Pappus and Guldinus.

where V = volume of revolution

θ = angle of revolution measured in radians, $\theta \leq 2\pi$

$\bar{r}$ = perpendicular distance from the axis of revolution to the centroid of the generating area

A = generating area

Composite Shapes. We may also apply the above two theorems to lines or areas that may be composed of a series of composite parts. In this case the total surface area or volume generated is the addition of the surface areas or volumes generated by each of the composite parts. Since each part undergoes the *same* angle of revolution, θ, and the distance from the axis of revolution to the centroid of each composite part is $\tilde{r}$, then

$$A = \theta \Sigma(\tilde{r}L) \qquad (9\text{–}11)$$

and

$$V = \theta \Sigma(\tilde{r}A) \qquad (9\text{–}12)$$

The amount of roofing material used on this storage building can be estimated by using the theorem of Pappus and Guldinus to determine its surface area.

Application of the above theorems is illustrated numerically in the following example.

E X A M P L E   9.12

Show that the surface area of a sphere is $A = 4\pi R^2$ and its volume is $V = \frac{4}{3}\pi R^3$.

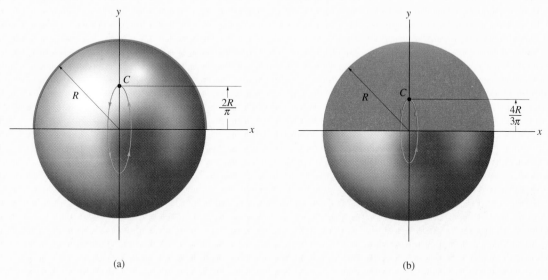

(a)   (b)

Fig. 9–24

Solution

Surface Area. The surface area of the sphere in Fig. 9–24a is generated by rotating a semicircular *arc* about the x axis. Using the table on the inside back cover, it is seen that the centroid of this arc is located at a distance $\bar{r} = 2R/\pi$ from the x axis of rotation. Since the centroid moves through an angle of $\theta = 2\pi$ rad in generating the sphere, then applying Eq. 9–9 we have

$$A = \theta \bar{r} L; \qquad A = 2\pi \left(\frac{2R}{\pi} \right) \pi R = 4\pi R^2 \qquad \text{Ans.}$$

Volume. The volume of the sphere is generated by rotating the semicircular *area* in Fig. 9–24b about the x axis. Using the table on the inside back cover to locate the centroid of the area, i.e., $\bar{r} = 4R/3\pi$, and applying Eq. 9–10, we have

$$V = \theta \bar{r} A; \qquad V = 2\pi \left(\frac{4R}{3\pi} \right) \left(\frac{1}{2}\pi R^2 \right) = \frac{4}{3}\pi R^3 \qquad \text{Ans.}$$

*9.5 Resultant of a General Distributed Loading

In Sec. 4.10, we discussed the method used to simplify a distributed loading that is uniform along an axis of a rectangular surface. In this section we will generalize this method to include surfaces that have an arbitrary shape and are subjected to a variable load distribution. As a specific application, in Sec. 9.6 we will find the resultant loading acting on the surface of a body that is submerged in a fluid.

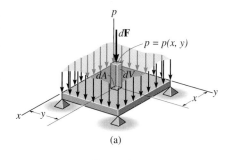

(a)

Pressure Distribution over a Surface. Consider the flat plate shown in Fig. 9–25a, which is subjected to the loading function $p = p(x, y)$ Pa, where Pa (pascal) $= 1$ N/m². Knowing this function, we can determine the force dF acting on the differential area dA m² of the plate, located at the arbitrary point (x, y). This force magnitude is simply $dF = [p(x, y) \text{ N/m}^2](dA \text{ m}^2) = [p(x, y) \, dA]$ N. The entire loading on the plate is therefore represented as a system of *parallel forces* infinite in number and each acting on a separate differential area dA. This system will now be simplified to a single resultant force $\mathbf{F}_R$ acting through a unique point $(\overline{x}, \overline{y})$ on the plate, Fig. 9–25b.

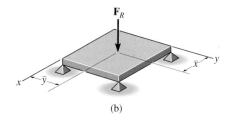

(b)

Fig. 9–25

Magnitude of Resultant Force. To determine the *magnitude* of $\mathbf{F}_R$, it is necessary to sum each of the differential forces dF acting over the plate's *entire surface area A*. This sum may be expressed mathematically as an integral:

$$F_R = \Sigma F; \qquad \boxed{F_R = \int_A p(x, y) \, dA = \int_V dV} \qquad (9\text{–}13)$$

Here $p(x, y) \, dA = dV$, the colored differential *volume element* shown in Fig. 9–25a. Therefore, the result indicates that the *magnitude of the resultant force is equal to the total volume under the distributed-loading diagram.*

Location of Resultant Force. The location $(\overline{x}, \overline{y})$ of $\mathbf{F}_R$ is determined by setting the moments of $\mathbf{F}_R$ equal to the moments of all the forces dF about the respective y and x axes: From Figs. 9–25a and 9–25b, using Eq. 9–13, this results in

$$\boxed{\overline{x} = \frac{\displaystyle\int_A x p(x, y) \, dA}{\displaystyle\int_A p(x, y) \, dA} = \frac{\displaystyle\int_V x \, dV}{\displaystyle\int_V dV} \qquad \overline{y} = \frac{\displaystyle\int_A y p(x, y) \, dA}{\displaystyle\int_A p(x, y) \, dA} = \frac{\displaystyle\int_V y \, dV}{\displaystyle\int_V dV}} \qquad (9\text{–}14)$$

Hence, it can be seen that the *line of action of the resultant force passes through the geometric center or centroid of the volume under the distributed loading diagram.*

*9.6 Fluid Pressure

According to Pascal's law, a fluid at rest creates a pressure p at a point that is the *same* in *all* directions. The magnitude of p, measured as a force per unit area, depends on the specific weight γ or mass density ρ of the fluid and the depth z of the point from the fluid surface.* The relationship can be expressed mathematically as

$$p = \gamma z = \rho g z \qquad (9\text{–}15)$$

where g is the acceleration due to gravity. Equation 9–15 is valid only for fluids that are assumed *incompressible*, as in the case of most liquids. Gases are compressible fluids, and since their density changes significantly with both pressure and temperature, Eq. 9–15 cannot be used.

To illustrate how Eq. 9–15 is applied, consider the submerged plate shown in Fig. 9–26. Three points on the plate have been specified. Since point B is at depth z_1 from the liquid surface, the *pressure* at this point has a magnitude $p_1 = \gamma z_1$. Likewise, points C and D are both at depth z_2; hence, $p_2 = \gamma z_2$. In all cases, the pressure acts *normal* to the surface area dA located at the specified point. Using Eq. 9-15 and the results of Sec. 9.5, it is possible to determine the resultant force caused by a liquid pressure distribution and specify its location on the surface of a submerged plate. Three different shapes of plates will now be considered.

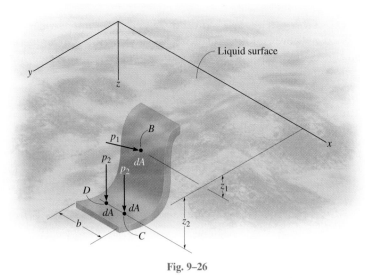

Fig. 9–26

*In particular, for water $\gamma = 62.4$ lb/ft^3, or $\gamma = \rho g = 9810$ N/m^3 since $\rho = 1000$ kg/m^3 and $g = 9.81$ m/s^2.

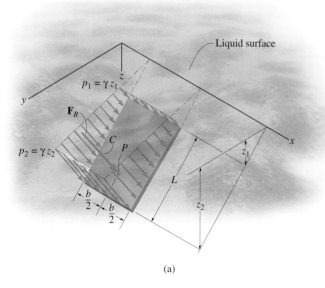

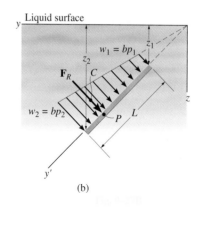

(b)

(a)

Fig. 9–27

Flat Plate of Constant Width.   A flat rectangular plate of constant width, which is submerged in a liquid having a specific weight γ, is shown in Fig. 9–27a. The plane of the plate makes an angle with the horizontal, such that its top edge is located at a depth z_1 from the liquid surface and its bottom edge is located at a depth z_2. Since pressure varies linearly with depth, Eq. 9–15, the distribution of pressure over the plate's surface is represented by a trapezoidal volume having an intensity of $p_1 = \gamma z_1$ at depth z_1 and $p_2 = \gamma z_2$ at depth z_2. As noted in Sec. 9.5, the magnitude of the *resultant force* $\mathbf{F}_R$ is equal to the *volume* of this loading diagram and $\mathbf{F}_R$ has a *line of action* that passes through the volume's centroid C. Hence, $\mathbf{F}_R$ does *not* act at the centroid of the plate; rather, it acts at point P, called the *center of pressure*.

Since the plate has a *constant width*, the loading distribution may also be viewed in two dimensions, Fig. 9–27b. Here the loading intensity is measured as force/length and varies linearly from $w_1 = bp_1 = b\gamma z_1$ to $w_2 = bp_2 = b\gamma z_2$. The magnitude of $\mathbf{F}_R$ in this case equals the trapezoidal *area*, and $\mathbf{F}_R$ has a *line of action* that passes through the area's *centroid* C. For numerical applications, the area and location of the centroid for a trapezoid are tabulated on the inside back cover.

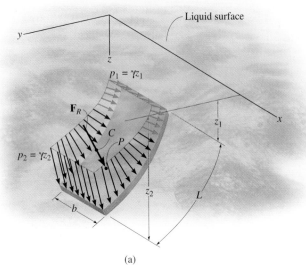

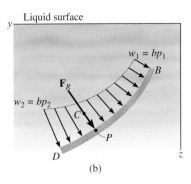

(b)

Fig. 9–28

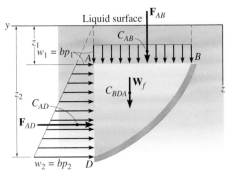

Fig. 9–29

Curved Plate of Constant Width. When the submerged plate is curved, the pressure acting normal to the plate continually changes direction, and therefore calculation of the magnitude of $\mathbf{F}_R$ and its location P is more difficult than for a flat plate. Three- and two-dimensional views of the loading distribution are shown in Figs. 9–28a and 9–28b, respectively. Here integration can be used to determine both F_R and the location of the centroid C or center of pressure P.

A simpler method exists, however, for calculating the magnitude of $\mathbf{F}_R$ and its location along a curved (or flat) plate having a *constant width*. This method requires separate calculations for the horizontal and vertical *components* of $\mathbf{F}_R$. For example, the distributed loading acting on the curved plate DB in Fig. 9–28b can be represented by the *equivalent loading* shown in Fig. 9–29. Here the plate supports the weight of liquid W_f contained within the block BDA. This force has a magnitude $W_f = (\gamma b)(\text{area}_{BDA})$ and acts through the centroid of BDA. In addition, there are the pressure distributions caused by the liquid acting along the vertical and horizontal sides of the block. Along the vertical side AD, the force $\mathbf{F}_{AD}$ has a magnitude that equals the area under the trapezoid and acts through the centroid C_{AD} of this area. The distributed loading along the horizontal side AB is constant since all points lying in this plane are at the same depth from the surface of the liquid. The magnitude of $\mathbf{F}_{AB}$ is simply the area of the rectangle. This force acts through the area's centroid C_{AB} or the midpoint of AB. Summing the three forces in Fig. 9–29 yields $\mathbf{F}_R = \Sigma \mathbf{F} = \mathbf{F}_{AD} + \mathbf{F}_{AB} + \mathbf{W}_f$, which is shown in Fig. 9–28. Finally, the location of the center of pressure P on the plate is determined by applying the equation $M_{R_O} = \Sigma M_O$, which states that the moment of the resultant force about a convenient reference point O, such as D or B, in Fig. 9–28, is equal to the sum of the moments of the three forces in Fig. 9–29 about the same point.

Flat Plate of Variable Width.   The pressure distribution acting on the surface of a submerged plate having a variable width is shown in Fig. 9–30. The resultant force of this loading equals the volume described by the plate area as its base and linearly varying pressure distribution as its altitude. The shaded element shown in Fig. 9–30 may be used if integration is chosen to determine this volume. The element consists of a rectangular strip of area $dA = x \, dy'$ located at a depth z below the liquid surface. Since a uniform pressure $p = \gamma z$ (force/area) acts on dA, the magnitude of the differential force $d\mathbf{F}$ is equal to $dF = dV = p \, dA = \gamma z(x \, dy')$. Integrating over the entire volume yields Eq. 9–13; i.e.,

$$F_R = \int_A p \, dA = \int_V dV = V$$

The resultant force of the water and its location on the elliptical back plate of this tank truck must be determined by integration.

From Eq. 9–14, the centroid of V defines the point through which $\mathbf{F}_R$ acts. The center of pressure, which lies on the surface of the plate just below C, has coordinates $P(\overline{x}, \overline{y}')$ defined by the equations

$$\overline{x} = \frac{\displaystyle\int_V \tilde{x} \, dV}{\displaystyle\int_V dV} \qquad \overline{y}' = \frac{\displaystyle\int_V \tilde{y}' \, dV}{\displaystyle\int_V dV}$$

This point should *not* be mistaken for the centroid of the plate's *area*.

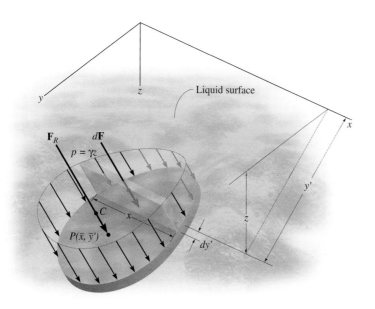

Fig. 9–30

EXAMPLE 9.13

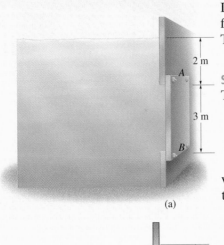

(a)

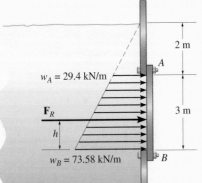

(b)

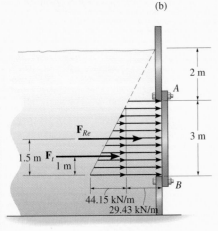

(c)

Fig. 9–31

Determine the magnitude and location of the resultant hydrostatic force acting on the submerged rectangular plate AB shown in Fig. 9–31a. The plate has a width of 1.5 m; $\rho_w = 1000 \text{ kg/m}^3$.

Solution

The water pressures at depths A and B are

$$p_A = \rho_w g z_A = (1000 \text{ kg/m}^3)(9.81 \text{ m/s}^2)(2 \text{ m}) = 19.62 \text{ kPa}$$
$$p_B = \rho_w g z_B = (1000 \text{ kg/m}^3)(9.81 \text{ m/s}^2)(5 \text{ m}) = 49.05 \text{ kPa}$$

Since the plate has a constant width, the distributed loading can be viewed in two dimensions as shown in Fig. 9–31b. The intensities of the load at A and B are

$$w_A = bp_A = (1.5 \text{ m})(19.62 \text{ kPa}) = 29.43 \text{ kN/m}$$
$$w_B = bp_B = (1.5 \text{ m})(49.05 \text{ kPa}) = 73.58 \text{ kN/m}$$

From the table on the inside back cover, the magnitude of the resultant force $\mathbf{F}_R$ created by the distributed load is

$$F_R = \text{area of trapezoid}$$
$$= \tfrac{1}{2}(3)(29.4 + 73.6) = 154.5 \text{ kN} \qquad \textit{Ans.}$$

This force acts through the centroid of the area,

$$h = \frac{1}{3}\left(\frac{2(29.43) + 73.58}{29.43 + 73.58}\right)(3) = 1.29 \text{ m} \qquad \textit{Ans.}$$

measured upward from B, Fig. 9–31b.

The same results can be obtained by considering two components of $\mathbf{F}_R$ defined by the triangle and rectangle shown in Fig. 9–31c. Each force acts through its associated centroid and has a magnitude of

$$F_{Re} = (29.43 \text{ kN/m})(3 \text{ m}) = 88.3 \text{ kN}$$
$$F_t = \tfrac{1}{2}(44.15 \text{ kN/m})(3 \text{ m}) = 66.2 \text{ kN}$$

Hence,

$$F_R = F_{Re} + F_t = 88.3 + 66.2 = 154.5 \text{ kN} \qquad \textit{Ans.}$$

The location of $\mathbf{F}_R$ is determined by summing moments about B, Fig. 9–31b and c, i.e.,

$$\curvearrowleft + (M_R)_B = \Sigma M_B; \qquad (154.5)h = 88.3(1.5) + 66.2(1)$$
$$h = 1.29 \text{ m} \qquad \textit{Ans.}$$

E X A M P L E **9.14**

Determine the magnitude of the resultant hydrostatic force acting on the surface of a seawall shaped in the form of a parabola as shown in Fig. 9–32a. The wall is 5 m long; $\rho_w = 1020$ kg/m^3.

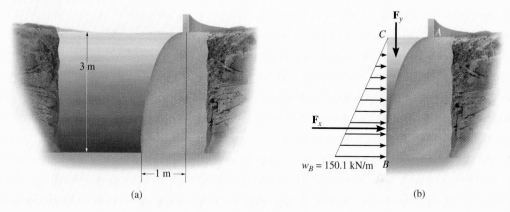

(a)   (b)

Fig. 9–32

Solution

The horizontal and vertical components of the resultant force will be calculated, Fig. 9–32b. Since

$$p_B = \rho_w g z_B = (1020 \text{ kg/m}^3)(9.81 \text{ m/s}^2)(3 \text{ m}) = 30.02 \text{ kPa}$$

then

$$w_B = b p_B = 5 \text{ m}(30.02 \text{ kPa}) = 150.1 \text{ kN/m}$$

Thus,

$$F_x = \tfrac{1}{2}(3 \text{ m})(150.1 \text{ kN/m}) = 225.1 \text{ kN}$$

The area of the parabolic sector ABC can be determined using the table on the inside back cover. Hence, the weight of water within this region is

$$F_y = (\rho_w g b)(\text{area}_{ABC})$$
$$= (1020 \text{ kg/m}^3)(9.81 \text{ m/s}^2)(5 \text{ m})[\tfrac{1}{3}(1 \text{ m})(3 \text{ m})] = 50.0 \text{ kN}$$

The resultant force is therefore

$$F_R = \sqrt{F_x^2 + F_y^2} = \sqrt{(225.1)^2 + (50.0)^2}$$
$$= 231 \text{ kN} \qquad\qquad Ans.$$

EXAMPLE 9.15

Determine the magnitude and location of the resultant force acting on the triangular end plates of the water trough shown in Fig. 9–33a; $\rho_w = 1000 \text{ kg/m}^3$.

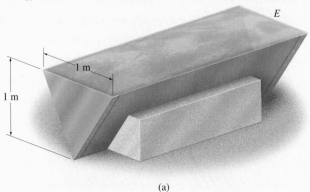

(a)

Solution

The pressure distribution acting on the end plate E is shown in Fig. 9–33b. The magnitude of the resultant force $\mathbf{F}$ is equal to the volume of this loading distribution. We will solve the problem by integration. Choosing the differential volume element shown in the figure, we have

$$dF = dV = p\,dA = \rho_w gz(2x\,dz) = 19\,620zx\,dz$$

The equation of line AB is

$$x = 0.5(1 - z)$$

Hence, substituting and integrating with respect to z from $z = 0$ to $z = 1$ m yields

$$F = V = \int_V dV = \int_0^1 (19\,620)z[0.5(1 - z)]\,dz$$

$$= 9810 \int_0^1 (z - z^2)\,dz = 1635 \text{ N} = 1.64 \text{ kN} \qquad Ans.$$

This resultant passes through the centroid of the volume. Because of symmetry,

$$\overline{x} = 0 \qquad\qquad Ans.$$

Since $\tilde{z} = z$ for the volume element in Fig. 9–33b, then

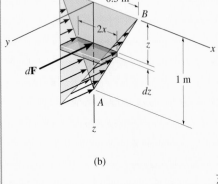

(b)

Fig. 9–33

$$\overline{z} = \frac{\displaystyle\int_V \tilde{z}\,dV}{\displaystyle\int_V dV} = \frac{\displaystyle\int_0^1 z(19\,620)z[0.5(1 - z)]\,dz}{1635} = \frac{9810 \displaystyle\int_0^1 (z^2 - z^3)\,dz}{1635}$$

$$= 0.5 \text{ m} \qquad\qquad Ans.$$

CHAPTER REVIEW

- *Center of Gravity and Centroid.* The *center of gravity* represents a point where the weight of the body can be considered concentrated. The distance $\bar{s}$ to this point can be determined from a balance of moments. This requires that the moment of the weight of all the particles of the body about some point must equal the moment of the entire body about the point, $\bar{s}W = \Sigma \tilde{s}W$. The *centroid* is the location of the geometric center for the body. It is determined in a similar manner, using a moment balance of geometric elements such as line, area, or volume segments. For bodies having a continuous shape, moments are summed (integrated) using differential elements. If the body is a composite of several shapes, each having a known location for its center of gravity or centroid, then the location is determined from a discrete summation using its composite parts.

- *Theorems of Pappus and Guldinus.* These theorems can be used to determine the surface area and volume of a body of revolution. The *surface area* equals the product of the length of the generating curve and the distance traveled by the centroid of the curve needed to generate the area $A = \theta \bar{r} L$. The *volume* of the body equals the product of the generating area and the distance traveled by the centroid of this area needed to generate the volume, $V = \theta \bar{r} A$.

- *Fluid Pressure.* The pressure developed by a liquid at a point on a submerged surface depends upon the depth of the point and the density of the liquid in accordance with Pascal's law, $p = \rho g h = \gamma h$. This pressure will create a *linear distribution* of loading on a flat vertical or inclined surface. If the surface is horizontal, then the loading will be *uniform*. In any case, the resultants of these loadings can be determined by finding the volume or area under the loading curve. The line of action of the resultant force passes through the centroid of the loading diagram.

REVIEW PROBLEMS

*9-1. A circular V-belt has an inner radius of 600 mm and a cross-sectional area as shown. Determine the volume of material required to make the belt.

9-2. A circular V-belt has an inner radius of 600 mm and a cross-sectional area as shown. Determine the surface area of the belt.

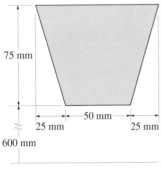

75 mm

50 mm

25 mm   25 mm

600 mm

Probs. 9–1/2

9-3. Locate the centroid $\bar{y}$ of the beam's cross-sectional area.

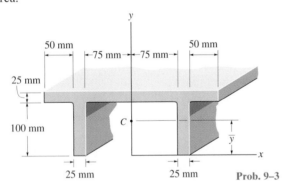

50 mm   50 mm

75 mm 75 mm

25 mm

100 mm

C

$\bar{y}$

25 mm   25 mm   **Prob. 9–3**

9-4. Locate the centroid of the solid.

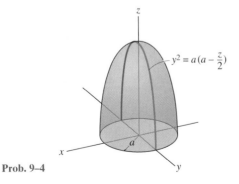

z

$y^2 = a\left(a - \dfrac{z}{2}\right)$

x

a

y

Prob. 9–4

*9-5. Determine the magnitude of the resultant hydrostatic force acting per foot of length on the sea wall; $\gamma_w = 62.4 \ \text{lb/ft}^3$.

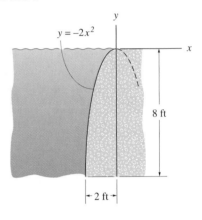

y

$y = -2x^2$

x

8 ft

2 ft

Prob. 9–5

9-6. The tank and compressor have a mass of 15 kg and mass center at G_T, and the motor has a mass of 70 kg and a mass center at G_M. Determine the angle of tilt, θ, of the tank so that the unit will be on the verge of tipping over.

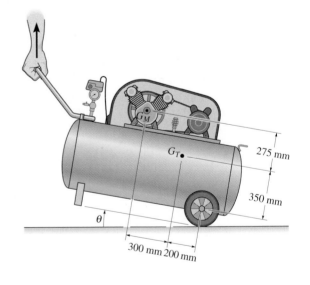

G_M

G_T

275 mm

350 mm

θ

300 mm 200 mm

Prob. 9–6

9-7. The thin-walled channel and stiffener have the cross section shown. If the material has a constant thickness, determine the location $\bar{y}$ of its centroid. The dimensions are indicated to the center of each segment.

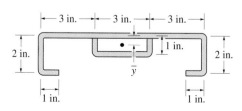

Prob. 9–7

9-8. Locate the center of gravity of the homogeneous rod. The rod has a weight of 2lb/ft. Also, compute the x, y, z components of reaction at the fixed support A.

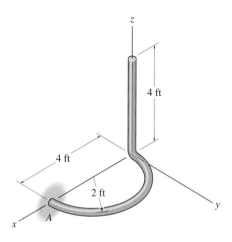

Prob. 9–8

***9-9.** The rectangular bin is filled with coal, which creates a pressure distribution along wall A that varies as shown, i.e., $p = 4z^{1/3}$ lb/ft^2, where z is in feet. Compute the resultant force created by the coal, and its location, measured from the top surface of the coal.

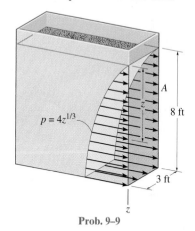

Prob. 9–9

9-10. The load over the plate varies linearly along the sides of the plate such that $p = \frac{2}{3}[x(4 - y)]$ kPa. Determine the resultant force and its position $(\bar{x}, \bar{y})$ on the plate.

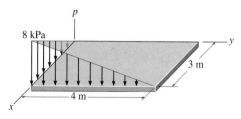

Prob. 9–10

9-11. The pressure loading on the plate is described by the function $p = \{-240/(x + 1) + 340\}$ Pa. Determine the magnitude of the resultant force and coordinates of the point where the line of action of the force intersects the plate.

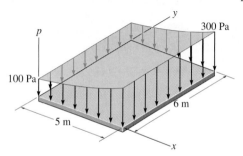

Prob. 9–11

The design of a structural member, such as a beam or column, requires calculation of its cross-sectional moment of inertia. In this chapter, we will discuss how this is done.

Moments of Inertia

- To develop a method for determining the moment of inertia for an area.
- To introduce the product of inertia and show how to determine the maximum and minimum moments of inertia of an area.
- To discuss the mass moment of inertia.

10.1 Definition of Moments of Inertia for Areas

In the last chapter, we determined the centroid for an area by considering the first moment of the area about an axis; that is, for the computation we had to evaluate an integral of the form $\int x \, dA$. An integral of the second moment of an area, such as $\int x^2 \, dA$, is referred to as the *moment of inertia* for the area. The terminology "moment of inertia" as used here is actually a misnomer; however, it has been adopted because of the similarity with integrals of the same form related to mass.

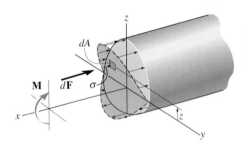

Fig. 10–1

The moment of inertia of an area orginates whenever one relates the normal stress σ (sigma), or force per unit area, acting on the transverse cross section of an elastic beam, to the applied external moment **M**, which causes bending of the beam. From the theory of mechanics of materials, it can be shown that the stress within the beam varies linearly with its distance from an axis passing through the centroid C of the beam's cross-sectional area; i.e., $\sigma = kz$, Fig. 10–1. The magnitude of force acting on the area element dA, shown in the figure, is therefore $dF = \sigma\, dA = kz\, dA$. Since this force is located a distance z from the y axis, the moment of $d\mathbf{F}$ about the y axis is $dM = dFz = kz^2\, dA$. The resulting moment of the entire stress distribution is equal to the applied moment **M**; hence, $M = k\int z^2\, dA$. Here the integral represents the moment of inertia of the area about the y axis. Since integrals of this form often arise in formulas used in mechanics of materials, structural mechanics, fluid mechanics, and machine design, the engineer should become familiar with the methods used for their computation.

Moment of Inertia.   Consider the area A, shown in Fig. 10–2, which lies in the x–y plane. By definition, the moments of inertia of the differential planar area dA about the x and y axes are $dI_x = y^2\, dA$ and $dI_y = x^2\, dA$, respectively. For the entire area the *moments of inertia* are determined by integration; i.e.,

$$I_x = \int_A y^2\, dA$$

$$I_y = \int_A x^2\, dA \tag{10–1}$$

We can also formulate the second moment of dA about the pole O or z axis, Fig. 10–2. This is referred to as the polar moment of inertia, $dJ_O = r^2\, dA$. Here r is the perpendicular distance from the pole (z axis) to the element dA. For the entire area the *polar moment of inertia* is

$$J_O = \int_A r^2\, dA = I_x + I_y \tag{10–2}$$

The relationship between J_O and I_x, I_y is possible since $r^2 = x^2 + y^2$, Fig. 10–2.

From the above formulations it is seen that I_x, I_y, and J_O will *always* be *positive* since they involve the product of distance squared and area. Furthermore, the units for moment of inertia involve length raised to the fourth power, e.g., m^4, mm^4, or ft^4, in^4.

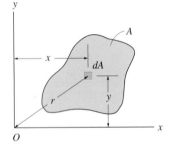

Fig. 10–2

10.2 Parallel-Axis Theorem for an Area

If the moment of inertia for an area is known about an axis passing through its centroid, which is often the case, it is convenient to determine the moment of inertia of the area about a corresponding parallel axis using the *parallel-axis theorem*. To derive this theorem, consider finding the moment of inertia of the shaded area shown in Fig. 10–3 about the x axis. In this case, a differential element dA is located at an arbitrary distance y' from the *centroidal x' axis*, whereas the *fixed distance* between the parallel x and x' axes is defined as d_y. Since the moment of inertia of dA about the x axis is $dI_x = (y' + d_y)^2\, dA$, then for the entire area,

$$I_x = \int_A (y' + d_y)^2\, dA$$

$$= \int_A y'^2\, dA + 2d_y \int_A y'\, dA + d_y^2 \int_A dA$$

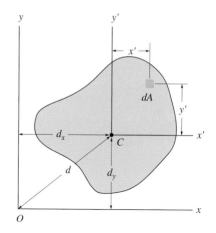

Fig. 10–3

The first integral represents the moment of inertia of the area about the centroidal axis, $\bar{I}_{x'}$. The second integral is zero since the x' axis passes through the area's centroid C; i.e., $\int y'\, dA = \bar{y} \int dA = 0$ since $\bar{y} = 0$. Realizing that the third integral represents the total area A, the final result is therefore

$$\boxed{I_x = \bar{I}_{x'} + Ad_y^2} \qquad (10\text{–}3)$$

A similar expression can be written for I_y; i.e.,

$$\boxed{I_y = \bar{I}_{y'} + Ad_x^2} \qquad (10\text{–}4)$$

And finally, for the polar moment of inertia about an axis perpendicular to the x–y plane and passing through the pole O (z axis), Fig. 10–3, we have

$$\boxed{J_O = \bar{J}_C + Ad^2} \qquad (10\text{–}5)$$

The form of each of these three equations states that *the moment of inertia of an area about an axis is equal to the moment of inertia of the area about a parallel axis passing through the area's centroid plus the product of the area and the square of the perpendicular distance between the axes.*

10.3 Radius of Gyration of an Area

The *radius of gyration* of a planar area has units of length and is a quantity that is often used for the design of columns in structural mechanics. Provided the areas and moments of inertia are *known*, the radii of gyration are determined from the formulas

$$k_x = \sqrt{\frac{I_x}{A}} \qquad k_y = \sqrt{\frac{I_y}{A}} \qquad k_O = \sqrt{\frac{J_O}{A}} \qquad (10\text{–}6)$$

The form of these equations is easily remembered since it is similar to that for finding the moment of inertia of a differential area about an axis. For example, $I_x = k_x^2 A$; whereas for a differential area, $dI_x = y^2\, dA$.

10.4 Moments of Inertia for an Area by Integration

When the boundaries for a planar area are expressed by mathematical functions, Eqs. 10–1 may be integrated to determine the moments of inertia for the area. If the element of area chosen for integration has a differential size in two directions as shown in Fig. 10–2, a double integration must be performed to evaluate the moment of inertia. Most often, however, it is easier to perform only a single integration by choosing an element having a differential size or thickness in only one direction.

PROCEDURE FOR ANALYSIS

- If a single integration is performed to determine the moment of inertia of an area about an axis, it will be necessary to specify the differential element dA.
- Most often this element will be rectangular, such that it will have a finite length and differential width.
- The element should be located so that it intersects the boundary of the area at the *arbitrary point* (x, y). There are two possible ways to orient the element with respect to the axis about which the moment of inertia is to be determined.

Case 1

- The *length* of the element can be oriented *parallel* to the axis. This situation occurs when the rectangular element shown in Fig. 10–4 is used to determine I_y for the area. Direct application of Eq. 10–1, i.e., $I_y = \int x^2\, dA$, can be made in this case since the element has an infinitesimal thickness dx and therefore *all parts* of the element lie at the *same* moment-arm distance x from the y axis.*

Case 2

- The *length* of the element can be oriented *perpendicular* to the axis. Here Eq. 10–1 *does not apply* since all parts of the element will *not* lie at the same moment-arm distance from the axis. For example, if the rectangular element in Fig. 10–4 is used for determining I_x for the area, it will first be necessary to calculate the moment of inertia of the *element* about a horizontal axis passing through the element's centroid and then determine the moment of inertia of the *element* about the x axis by using the parallel-axis theorem. Integration of this result will yield I_x.

*In the case of the element $dA = dx\, dy$, Fig. 10–2, the moment arms y and x are appropriate for the formulation of I_x and I_y (Eq. 10–1) since the *entire* element, because of its infinitesimal size, lies at the specified y and x perpendicular distances from the x and y axes.

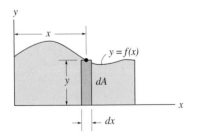

Fig. 10–4

EXAMPLE 10.1

Determine the moment of inertia for the rectangular area shown in Fig. 10–5 with respect to (a) the centroidal x' axis, (b) the axis x_b passing through the base of the rectangle, and (c) the pole or z' axis perpendicular to the $x'-y'$ plane and passing through the centroid C.

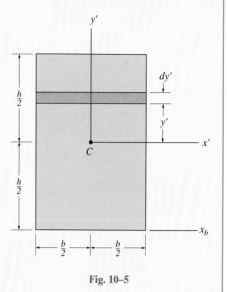

Fig. 10–5

Solution (Case 1)

Part (a). The differential element shown in Fig. 10–5 is chosen for integration. Because of its location and orientation, the *entire element* is at a distance y' from the x' axis. Here it is necessary to integrate from $y' = -h/2$ to $y' = h/2$. Since $dA = b\,dy'$, then

$$\bar{I}_{x'} = \int_A y'^2\,dA = \int_{-h/2}^{h/2} y'^2(b\,dy') = b\int_{-h/2}^{h/2} y'^2\,dy$$

$$= \frac{1}{12}bh^3 \qquad\qquad Ans.$$

Part (b). The moment of inertia about an axis passing through the base of the rectangle can be obtained by using the result of part (a) and applying the parallel-axis theorem, Eq. 10–3.

$$I_{x_b} = \bar{I}_{x'} + Ad_y^2$$

$$= \frac{1}{12}bh^3 + bh\left(\frac{h}{2}\right)^2 = \frac{1}{3}bh^3 \qquad\qquad Ans.$$

Part (c). To obtain the polar moment of inertia about point C, we must first obtain $\bar{I}_{y'}$, which may be found by interchanging the dimensions b and h in the result of part (a), i.e.,

$$\bar{I}_{y'} = \frac{1}{12}hb^3$$

Using Eq. 10–2, the polar moment of inertia about C is therefore

$$\bar{J}_C = \bar{I}_{x'} + \bar{I}_{y'} = \frac{1}{12}bh(h^2 + b^2) \qquad\qquad Ans.$$

E X A M P L E **10.2**

Determine the moment of inertia of the shaded area shown in Fig. 10–6a about the x axis.

Solution I (Case 1)

A differential element of area that is *parallel* to the x axis, as shown in Fig. 10–6a, is chosen for integration. Since the element has a thickness dy and intersects the curve at the *arbitrary point* (x, y), the area is $dA = (100 - x)\, dy$. Furthermore, all parts of the element lie at the same distance y from the x axis. Hence, integrating with respect to y, from $y = 0$ to $y = 200$ mm, yields

$$I_x = \int_A y^2\, dA = \int_A y^2(100 - x)\, dy$$

$$= \int_0^{200} y^2\left(100 - \frac{y^2}{400}\right) dy = 100 \int_0^{200} y^2\, dy - \frac{1}{400}\int_0^{200} y^4\, dy$$

$$= 107(10^6)\ \text{mm}^4 \qquad\qquad Ans.$$

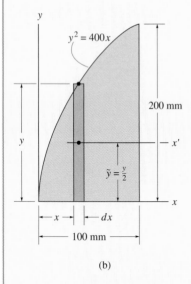

(a)

(b)

Fig. 10–6

Solution II (Case 2)

A differential element *parallel* to the y axis, as shown in Fig. 10–6b, is chosen for integration. It intersects the curve at the *arbitrary point* (x, y). In this case, all parts of the element do *not* lie at the same distance from the x axis, and therefore the parallel-axis theorem must be used to determine the *moment of inertia of the element* with respect to this axis. For a rectangle having a base b and height h, the moment of inertia about its centroidal axis has been determined in part (a) of Example 10.1. There it was found that $\overline{I}_{x'} = \frac{1}{12}bh^3$. For the differential element shown in Fig. 10–6b, $b = dx$ and $h = y$, and thus $d\overline{I}_{x'} = \frac{1}{12}dx\, y^3$. Since the centroid of the element is at $\widetilde{y} = y/2$ from the x axis, the moment of inertia of the element about this axis is

$$dI_x = d\overline{I}_{x'} + dA\, \widetilde{y}^2 = \frac{1}{12}dx\, y^3 + y\, dx\left(\frac{y}{2}\right)^2 = \frac{1}{3}y^3\, dx$$

This result can also be concluded from part (b) of Example 10.1. Integrating with respect to x, from $x = 0$ to $x = 100$ mm, yields

$$I_x = \int dI_x = \int_A \frac{1}{3}y^3\, dx = \int_0^{100} \frac{1}{3}(400x)^{3/2}\, dx$$

$$= 107(10^6)\ \text{mm}^4 \qquad\qquad Ans.$$

EXAMPLE 10.3

Determine the moment of inertia with respect to the x axis of the circular area shown in Fig. 10–7a.

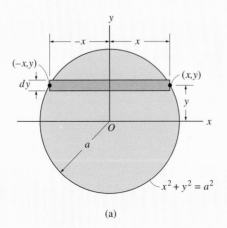

(a)

Solution I (Case 1)

Using the differential element shown in Fig. 10–7a, since $dA = 2x\,dy$, we have

$$I_x = \int_A y^2\,dA = \int_A y^2(2x)\,dy$$

$$= \int_{-a}^{a} y^2(2\sqrt{a^2 - y^2})\,dy = \frac{\pi a^4}{4} \qquad Ans.$$

Solution II (Case 2)

When the differential element is chosen as shown in Fig. 10–7b, the centroid for the element happens to lie on the x axis, and so, applying Eq. 10–3, noting that $d_y = 0$ and for a rectangle $\overline{I}_{x'} = \frac{1}{12}bh^3$, we have

$$dI_x = \frac{1}{12}dx\,(2y)^3$$

$$= \frac{2}{3}y^3\,dx$$

Integrating with respect to x yields

$$I_x = \int_{-a}^{a} \frac{2}{3}(a^2 - x^2)^{3/2}\,dx = \frac{\pi a^4}{4} \qquad Ans.$$

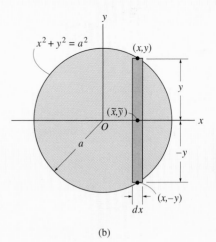

(b)

Fig. 10–7

E X A M P L E 10.4

Determine the moment of inertia of the shaded area shown in Fig. 10–8a about the x axis.

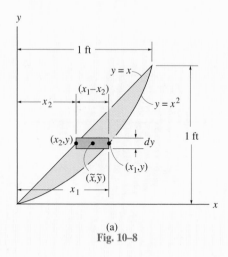

(a)
Fig. 10–8

Solution I *(Case 1)*

The differential element parallel to the x axis is chosen for integration, Fig. 10–8a. The element intersects the curve at the *arbitrary points* (x_2, y) and (x_1, y). Consequently, its area is $dA = (x_1 - x_2)\, dy$. Since all parts of the element lie at the same distance y from the x axis, we have

$$I_x = \int_A y^2 \, dA = \int_0^1 y^2(x_1 - x_2)\, dy = \int_0^1 y^2(\sqrt{y} - y)\, dy$$

$$I_x = \frac{2}{7}y^{7/2} - \frac{1}{4}y^4 \bigg|_0^1 = 0.0357 \text{ ft}^4 \qquad\qquad Ans.$$

Solution II *(Case 2)*

The differential element parallel to the y axis is shown in Fig. 10–8b. It intersects the curves at the *arbitrary points* (x, y_2) and (x, y_1). Since

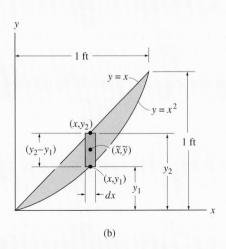

(b)

Fig. 10–8

all parts of its entirety do *not* lie at the same distance from the x axis, we must first use the parallel-axis theorem to find the *element's* moment of inertia about the x axis, using $\bar{I}_{x'} = \frac{1}{12}bh^3$, then integrate this result to determine I_x. Thus,

$$dI_x = d\bar{I}_{x'} + dA\,\tilde{y}^2 = \frac{1}{12}dx\,(y_2 - y_1)^3$$

$$+ \,(y_2 - y_1)\,dx\left(y_1 + \frac{y_2 - y_1}{2}\right)^2$$

$$= \frac{1}{3}(y_2^3 - y_1^3)\,dx = \frac{1}{3}(x^3 - x^6)\,dx$$

$$I_x = \frac{1}{3}\int_0^1 (x^3 - x^6)\,dx = \frac{1}{12}x^4 - \frac{1}{21}x^7 \Big|_0^1 = 0.0357\ \text{ft}^4 \quad \textit{Ans.}$$

By comparison, Solution I requires much less computation. Therefore, if an integral using a particular element appears difficult to evaluate, try solving the problem using an element oriented in the other direction.

10.5 Moments of Inertia for Composite Areas

A composite area consists of a series of connected "simpler" parts or shapes, such as semicircles, rectangles, and triangles. Provided the moment of inertia of each of these parts is known or can be determined about a common axis, then the moment of inertia of the composite area equals the *algebraic sum* of the moments of inertia of all its parts.

Structural members have various cross-sectional shapes, and it is necessary to calculate their moments of inertia in order to determine the stress in these members.

PROCEDURE FOR ANALYSIS

The moment of inertia of a composite area about a reference axis can be determined using the following procedure.

Composite Parts.

- Using a sketch, divide the area into its composite parts and indicate the perpendicular distance from the *centroid* of each part to the reference axis.

Parallel-Axis Theorem.

- The moment of inertia of each part should be determined about its centroidal axis, which is parallel to the reference axis. For the calculation use the table given on the inside back cover.

- If the centroidal axis does not coincide with the reference axis, the parallel-axis theorem, $I = \bar{I} + Ad^2$, should be used to determine the moment of inertia of the part about the reference axis.

Summation.

- The moment of inertia of the entire area about the reference axis is determined by summing the results of its composite parts.

- If a composite part has a "hole," its moment of inertia is found by "subtracting" the moment of inertia for the hole from the moment of inertia of the entire part including the hole.

E X A M P L E **10.5**

Compute the moment of inertia of the composite area shown in Fig. 10–9a about the x axis.

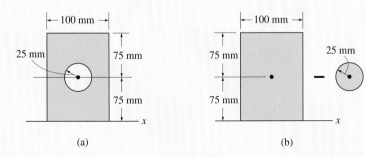

(a)   (b)

Fig. 10–9

Solution

Composite Parts.　The composite area is obtained by *subtracting* the circle from the rectangle as shown in Fig. 10–9b. The centroid of each area is located in the figure.

Parallel-Axis Theorem.　The moments of inertia about the x axis are determined using the parallel-axis theorem and the data in the table on the inside back cover.

Circle

$$I_x = \bar{I}_{x'} + A d_y^2$$

$$= \frac{1}{4}\pi(25)^4 + \pi(25)^2(75)^2 = 11.4(10^6) \text{ mm}^4$$

Rectangle

$$I_x = \bar{I}_{x'} + A d_y^2$$

$$= \frac{1}{12}(100)(150)^3 + (100)(150)(75)^2 = 112.5(10^6) \text{ mm}^4$$

Summation.　The moment of inertia for the composite area is thus

$$I_x = -11.4(10^6) + 112.5(10^6)$$

$$= 101(10^6) \text{ mm}^4 \qquad\qquad Ans.$$

EXAMPLE 10.6

Determine the moments of inertia of the beam's cross-sectional area shown in Fig. 10–10a about the x and y centroidal axes.

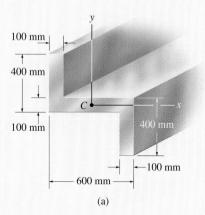

(a)

Solution

Composite Parts. The cross section can be considered as three composite rectangular areas A, B, and D shown in Fig. 10–10b. For the calculation, the centroid of each of these rectangles is located in the figure.

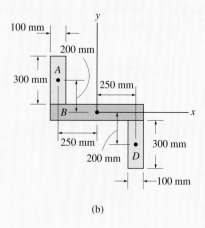

(b)

Fig. 10–10

Parallel-Axis Theorem. From the table on the inside back cover, or Example 10.1, the moment of inertia of a rectangle about its centroidal axis is $\bar{I} = \frac{1}{12} bh^3$. Hence, using the parallel-axis theorem for rectangles A and D, the calculations are as follows:

Rectangle A

$$I_x = \bar{I}_{x'} + Ad_y^2 = \frac{1}{12}(100)(300)^3 + (100)(300)(200)^2$$

$$= 1.425(10^9) \text{ mm}^4$$

$$I_y = \bar{I}_{y'} + Ad_x^2 = \frac{1}{12}(300)(100)^3 + (100)(300)(250)^2$$

$$= 1.90(10^9) \text{ mm}^4$$

Rectangle B

$$I_x = \frac{1}{12}(600)(100)^3 = 0.05(10^9) \text{ mm}^4$$

$$I_y = \frac{1}{12}(100)(600)^3 = 1.80(10^9) \text{ mm}^4$$

Rectangle D

$$I_x = \bar{I}_{x'} + Ad_y^2 = \frac{1}{12}(100)(300)^3 + (100)(300)(200)^2$$

$$= 1.425(10^9) \text{ mm}^4$$

$$I_y = \bar{I}_{y'} + Ad_x^2 = \frac{1}{12}(300)(100)^3 + (100)(300)(250)^2$$

$$= 1.90(10^9) \text{ mm}^4$$

Summation. The moments of inertia for the entire cross section are thus

$$I_x = 1.425(10^9) + 0.05(10^9) + 1.425(10^9)$$
$$= 2.90(10^9) \text{ mm}^4 \qquad \qquad \textit{Ans.}$$
$$I_y = 1.90(10^9) + 1.80(10^9) + 1.90(10^9)$$
$$= 5.60(10^9) \text{ mm}^4 \qquad \qquad \textit{Ans.}$$

*10.6 Product of Inertia for an Area

In general, the moment of inertia for an area is different for every axis about which it is computed. In some applications of structural or mechanical design it is necessary to know the orientation of those axes which give, respectively, the maximum and minimum moments of inertia for the area. The method for determining this is discussed in Sec. 10.7. To use this method, however, one must first compute the product of inertia for the area as well as its moments of inertia for given x, y axes.

The product of inertia for an element of area dA located at point (x, y), Fig. 10–11, is defined as $dI_{xy} = xy\, dA$. Thus, for the entire area A, the *product of inertia* is

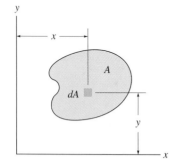

$$I_{xy} = \int_A xy\, dA \qquad (10\text{–}7)$$

Fig. 10–11

If the element of area chosen has a differential size in two directions, as shown in Fig. 10–11, a double integration must be performed to evaluate I_{xy}. Most often, however, it is easier to choose an element having a differential size or thickness in only one direction in which case the evaluation requires only a single integration (see Example 10.7).

Like the moment of inertia, the product of inertia has units of length raised to the fourth power, e.g., m^4, mm^4 or ft^4, in^4. However, since x or y may be a negative quantity, while the element of area is always positive, the product of inertia may be positive, negative, or zero, depending on the location and orientation of the coordinate axes. For example, the product of inertia I_{xy} for an area will be *zero* if either the x or y axis is an axis of *symmetry* for the area. To show this, consider the shaded area in Fig. 10–12, where for every element dA located at point (x, y) there is a corresponding element dA located at $(x, -y)$. Since the products of inertia for these elements are, respectively, $xy\, dA$ and $-xy\, dA$, the algebraic sum or integration of all the elements that are chosen in this way will cancel each other. Consequently, the product of inertia for the total area becomes zero. It also follows from the definition of I_{xy} that the "sign" of this quantity depends on the quadrant where the area is located. As shown in Fig. 10–13, if the area is rotated from one quadrant to another, the sign of I_{xy} will change.

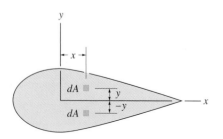

Fig. 10–12

Parallel-Axis Theorem. Consider the shaded area shown in Fig. 10–14, where x' and y' represent a set of axes passing through the *centroid* of the area, and x and y represent a corresponding set of parallel

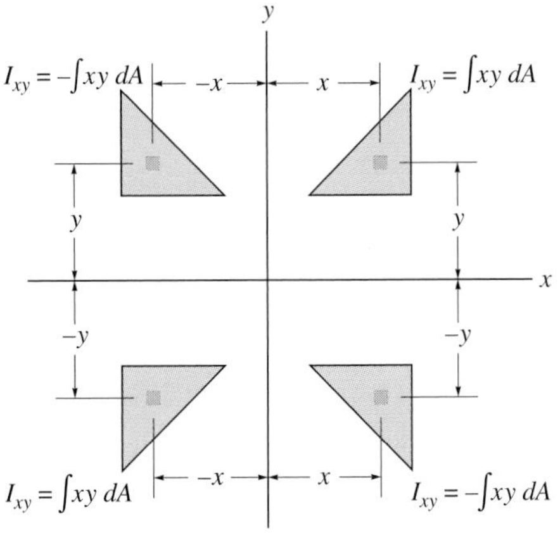

$$I_{xy} = -\int xy\, dA \qquad\qquad I_{xy} = \int xy\, dA$$

$$I_{xy} = \int xy\, dA \qquad\qquad I_{xy} = -\int xy\, dA$$

Fig. 10–13

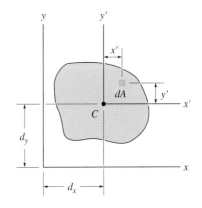

Fig. 10–14

axes. Since the product of inertia of dA with respect to the x and y axes is $dI_{xy} = (x' + d_x)(y' + d_y)\, dA$, then for the entire area,

$$I_{xy} = \int_A (x' + d_x)(y' + d_y)\, dA$$

$$= \int_A x'y'\, dA + d_x \int_A y'\, dA + d_y \int_A x'\, dA + d_x d_y \int_A dA$$

The first term on the right represents the product of inertia of the area with respect to the centroidal axis, $\bar{I}_{x'y'}$. The integrals in the second and third terms are zero since the moments of the area are taken about the centroidal axis. Realizing that the fourth integral represents the total area A, the final result is therefore

$$\boxed{I_{xy} = \bar{I}_{x'y'} + A d_x d_y} \qquad (10\text{–}8)$$

The similarity between this equation and the parallel-axis theorem for moments of inertia should be noted. In particular, it is important that the *algebraic signs* for d_x and d_y be maintained when applying Eq. 10–8. As illustrated in Example 10.8, the parallel-axis theorem finds important application in determining the product of inertia of a *composite area* with respect to a set of x, y axes.

E X A M P L E 10.7

Determine the product of inertia I_{xy} of the triangle shown in Fig. 10–15a.

Solution I

A differential element that has a thickness dx, Fig. 10–15b, has an area $dA = y\,dx$. The product of inertia of the element about the x, y axes is determined using the parallel-axis theorem.

$$dI_{xy} = d\bar{I}_{x'y'} + dA\,\tilde{x}\,\tilde{y}$$

where $(\tilde{x}, \tilde{y})$ locates the *centroid* of the element or the origin of the x', y' axes. Since $d\bar{I}_{x'y'} = 0$, due to symmetry, and $\tilde{x} = x$, $\tilde{y} = y/2$, then

$$dI_{xy} = 0 + (y\,dx)x\left(\frac{y}{2}\right) = \left(\frac{h}{b}x\,dx\right)x\left(\frac{h}{2b}x\right)$$

$$= \frac{h^2}{2b^2}x^3\,dx$$

Integrating with respect to x from $x = 0$ to $x = b$ yields

$$I_{xy} = \frac{h^2}{2b^2}\int_0^b x^3\,dx = \frac{b^2h^2}{8} \qquad \textit{Ans.}$$

Solution II

The differential element that has a thickness dy, Fig. 10–15c, and area $dA = (b - x)\,dy$ can also be used. The *centroid* is located at point $\tilde{x} = x + (b - x)/2 = (b + x)/2$, $\tilde{y} = y$, so the product of inertia of the element becomes

$$dI_{xy} = d\bar{I}_{x'y'} + dA\,\tilde{x}\,\tilde{y}$$

$$= 0 + (b - x)\,dy\left(\frac{b + x}{2}\right)y$$

$$= \left(b - \frac{b}{h}y\right)dy\left[\frac{b + (b/h)y}{2}\right]y = \frac{1}{2}y\left(b^2 - \frac{b^2}{h^2}y^2\right)dy$$

Integrating with respect to y from $y = 0$ to $y = h$ yields

$$I_{xy} = \frac{1}{2}\int_0^h y\left(b^2 - \frac{b^2}{h^2}y^2\right)dy = \frac{b^2h^2}{8} \qquad \textit{Ans.}$$

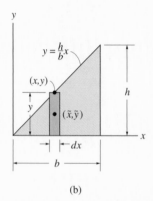

(a)

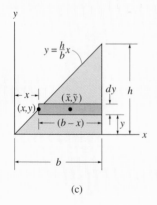

(b)

(c)

Fig. 10–15

E X A M P L E 10.8

Compute the product of inertia of the beam's cross-sectional area, shown in Fig. 10–16a, about the x and y centroidal axes.

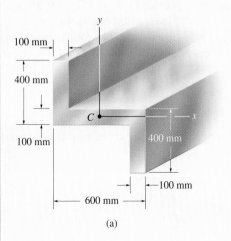

(a)

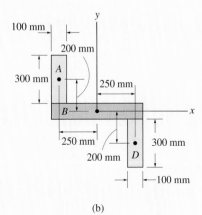

(b)

Fig. 10–16

Solution

As in Example 10.6, the cross section can be considered as three composite rectangular areas A, B, and D, Fig. 10–16b. The coordinates for the centroid of each of these rectangles are shown in the figure. Due to symmetry, the product of inertia of *each rectangle* is *zero* about each set of x', y' axes that passes through the rectangle's centroid. Hence, application of the parallel-axis theorem to each of the rectangles yields

Rectangle A

$$I_{xy} = \bar{I}_{x'y'} + Ad_x d_y$$
$$= 0 + (300)(100)(-250)(200)$$
$$= -1.50(10^9) \text{ mm}^4$$

Rectangle B

$$I_{xy} = \bar{I}_{x'y'} + Ad_x d_y$$
$$= 0 + 0$$
$$= 0$$

Rectangle D

$$I_{xy} = \bar{I}_{x'y'} + Ad_x d_y$$
$$= 0 + (300)(100)(250)(-200)$$
$$= -1.50(10^9) \text{ mm}^4$$

The product of inertia for the entire cross section is therefore

$$I_{xy} = -1.50(10^9) + 0 - 1.50(10^9) = -3.00(10^9) \text{ mm}^4 \quad \textit{Ans.}$$

*10.7 Moments of Inertia for an Area About Inclined Axes

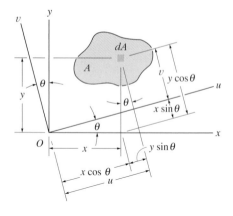

Fig. 10–17

In structural and mechanical design, it is sometimes necessary to calculate the moments and product of inertia I_u, I_v, and I_{uv} for an area with respect to a set of inclined u and v axes when the values for θ, I_x, I_y, and I_{xy} are *known*. To do this we will use *transformation equations* which relate the x, y and u, v coordinates. From Fig. 10–17, these equations are

$$u = x \cos \theta + y \sin \theta$$
$$v = y \cos \theta - x \sin \theta$$

Using these equations, the moments and product of inertia of dA about the u and v axes become

$$dI_u = v^2 \, dA = (y \cos \theta - x \sin \theta)^2 \, dA$$
$$dI_v = u^2 \, dA = (x \cos \theta + y \sin \theta)^2 \, dA$$
$$dI_{uv} = uv \, dA = (x \cos \theta + y \sin \theta)(y \cos \theta - x \sin \theta) \, dA$$

Expanding each expression and integrating, realizing that $I_x = \int y^2 \, dA$, $I_y = \int x^2 \, dA$, and $I_{xy} = \int xy \, dA$, we obtain

$$I_u = I_x \cos^2 \theta + I_y \sin^2 \theta - 2I_{xy} \sin \theta \cos \theta$$
$$I_v = I_x \sin^2 \theta + I_y \cos^2 \theta + 2I_{xy} \sin \theta \cos \theta$$
$$I_{uv} = I_x \sin \theta \cos \theta - I_y \sin \theta \cos \theta + I_{xy}(\cos^2 \theta - \sin^2 \theta)$$

These equations may be simplified by using the trigonometric identities $\sin 2\theta = 2 \sin \theta \cos \theta$ and $\cos 2\theta = \cos^2 \theta - \sin^2 \theta$, in which case

$$I_u = \frac{I_x + I_y}{2} + \frac{I_x - I_y}{2} \cos 2\theta - I_{xy} \sin 2\theta$$
$$I_v = \frac{I_x + I_y}{2} - \frac{I_x - I_y}{2} \cos 2\theta + I_{xy} \sin 2\theta$$
$$I_{uv} = \frac{I_x - I_y}{2} \sin 2\theta + I_{xy} \cos 2\theta$$

(10–9)

If the first and second equations are added together, we can show that the polar moment of inertia about the z axis passing through point O is *independent* of the orientation of the u and v axes; i.e.,

$$J_O = I_u + I_v = I_x + I_y$$

Principal Moments of Inertia. Equations 10–9 show that I_u, I_v, and I_{uv} depend on the angle of inclination, θ, of the u, v axes. We will now determine the orientation of these axes about which the moments of inertia for the area, I_u and I_v, are maximum and minimum. This particular set of axes is called the *principal axes* of the area, and the corresponding moments of inertia with respect to these axes are called the *principal moments of inertia*. In general, there is a set of principal axes for every chosen origin O. For the structural and mechanical design of a member, the origin O is generally located at the cross-sectional area's centroid.

The angle $\theta = \theta_p$, which defines the orientation of the principal axes for the area, may be found by differentiating the first of Eqs. 10–9 with respect to θ and setting the result equal to zero. Thus,

$$\frac{dI_u}{d\theta} = -2\left(\frac{I_x - I_y}{2}\right)\sin 2\theta - 2I_{xy}\cos 2\theta = 0$$

Therefore, at $\theta = \theta_p$,

$$\tan 2\theta_p = \frac{-I_{xy}}{(I_x - I_y)/2} \qquad (10\text{–}10)$$

This equation has two roots, θ_{p_1} and θ_{p_2}, which are 90° apart and so specify the inclination of the principal axes. In order to substitute them into Eq. 10–9, we must first find the sine and cosine of $2\theta_{p_1}$ and $2\theta_{p_2}$. This can be done using the triangles shown in Fig. 10–18, which are based on Eq. 10–10.

For θ_{p_1},

$$\sin 2\theta_{p_1} = -I_{xy}\Big/\sqrt{\left(\frac{I_x - I_y}{2}\right)^2 + I_{xy}^2}$$

$$\cos 2\theta_{p_1} = \left(\frac{I_x - I_y}{2}\right)\Big/\sqrt{\left(\frac{I_x - I_y}{2}\right)^2 + I_{xy}^2}$$

For θ_{p_2},

$$\sin 2\theta_{p_2} = I_{xy}\Big/\sqrt{\left(\frac{I_x - I_y}{2}\right)^2 + I_{xy}^2}$$

$$\cos 2\theta_{p_2} = -\left(\frac{I_x - I_y}{2}\right)\Big/\sqrt{\left(\frac{I_x - I_y}{2}\right)^2 + I_{xy}^2}$$

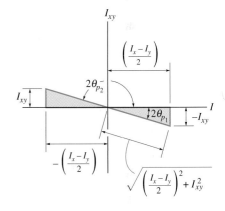

Fig. 10–18

Substituting these two sets of trigonometric relations into the first or second of Eqs. 10–9 and simplifying, we obtain

$$I_{\substack{\max \\ \min}} = \frac{I_x + I_y}{2} \pm \sqrt{\left(\frac{I_x - I_y}{2}\right)^2 + I_{xy}^2} \qquad (10\text{–}11)$$

Depending on the sign chosen, this result gives the maximum or minimum moment of inertia for the area. Furthermore, if the above trigonometric relations for θ_{p_1} and θ_{p_2} are substituted into the third of Eqs. 10–9, it can be shown that $I_{uv} = 0$; that is, the *product of inertia with respect to the principal axes is zero*. Since it was indicated in Sec. 10.6 that the product of inertia is zero with respect to any symmetrical axis, it therefore follows that *any symmetrical axis represents a principal axis of inertia for the area*.

EXAMPLE 10.9

Determine the principal moments of inertia for the beam's cross-sectional area shown in Fig. 10–19a with respect to an axis passing through the centroid.

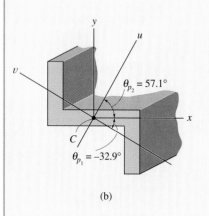

(a)

Solution

The moments and product of inertia of the cross section with respect to the x, y axes have been computed in Examples 10.6 and 10.8. The results are

$$I_x = 2.90(10^9) \text{ mm}^4 \quad I_y = 5.60(10^9) \text{ mm}^4 \quad I_{xy} = -3.00(10^9) \text{ mm}^4$$

Using Eq. 10–10, the angles of inclination of the principal axes u and v are

$$\tan 2\theta_p = \frac{-I_{xy}}{(I_x - I_y)/2} = \frac{3.00(10^9)}{[2.90(10^9) - 5.60(10^9)]/2} = -2.22$$

$$2\theta_{p_1} = -65.8° \quad \text{and} \quad 2\theta_{p_2} = 114.2°$$

Thus, as shown in Fig. 10–19b,

$$\theta_{p_1} = -32.9° \quad \text{and} \quad \theta_{p_2} = 57.1°$$

The principal moments of inertia with respect to the u and v axes are determined from Eq. 10–11. Hence,

$$I_{\substack{max \\ min}} = \frac{I_x + I_y}{2} \pm \sqrt{\left(\frac{I_x - I_y}{2}\right)^2 + I_{xy}^2}$$

$$= \frac{2.90(10^9) + 5.60(10^9)}{2}$$

$$\pm \sqrt{\left[\frac{2.90(10^9) - 5.60(10^9)}{2}\right]^2 + [-3.00(10^9)]^2}$$

$$I_{\substack{max \\ min}} = 4.25(10^9) \pm 3.29(10^9)$$

or

$$I_{max} = 7.54(10^9) \text{ mm}^4 \quad I_{min} = 0.960(10^9) \text{ mm}^4 \quad Ans.$$

Specifically, the maximum moment of inertia, $I_{max} = 7.54(10^9) \text{ mm}^4$, occurs with respect to the selected u axis since by inspection most of the cross-sectional area is farthest away from this axis. Or, stated in another manner, I_{max} occurs about the u axis since it is located within $\pm 45°$ of the y axis, which has the largest value of I ($I_y > I_x$). Also, this may be concluded by substituting the data with $\theta = 57.1°$ into the first of Eqs. 10–9.

(b)

Fig. 10–19

*10.8 Mohr's Circle for Moments of Inertia

Equations 10–9 to 10–11 have a graphical solution that is convenient to use and generally easy to remember. Squaring the first and third of Eqs. 10–9 and adding, it is found that

$$\left(I_u - \frac{I_x + I_y}{2}\right)^2 + I_{uv}^2 = \left(\frac{I_x - I_y}{2}\right)^2 + I_{xy}^2 \qquad (10\text{–}12)$$

In a given problem, I_u and I_{uv} are *variables*, and I_x, I_y, and I_{xy} are *known constants*. Thus, Eq. 10–12 may be written in compact form as

$$(I_u - a)^2 + I_{uv}^2 = R^2$$

When this equation is plotted on a set of axes that represent the respective moment of inertia and the product of inertia, Fig. 10–20, the resulting graph represents a *circle* of radius

$$R = \sqrt{\left(\frac{I_x - I_y}{2}\right)^2 + I_{xy}^2}$$

having its center located at point $(a, 0)$, where $a = (I_x + I_y)/2$. The circle so constructed is called *Mohr's circle*, named after the German engineer Otto Mohr (1835–1918).

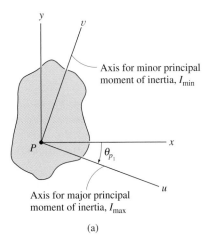

(a)

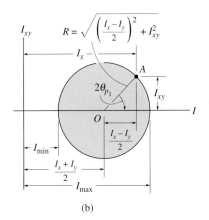

(b)

Fig. 10–20

PROCEDURE FOR ANALYSIS

The main purpose in using Mohr's circle here is to have a convenient means for transforming I_x, I_y, and I_{xy} into the principal moments of inertia. The following procedure provides a method for doing this.

Determine I_x, I_y, and I_{xy}.

- Establish the x, y axes for the area, with the origin located at the point P of interest, and determine I_x, I_y, and I_{xy}, Fig. 10–20a.

Construct the Circle.

- Construct a rectangular coordinate system such that the abscissa represents the moment of inertia I, and the ordinate represents the product of inertia I_{xy}, Fig. 10–20b.

- Determine the center of the circle, O, which is located at a distance $(I_x + I_y)/2$ from the origin, and plot the reference point A having coordinates (I_x, I_{xy}). By definition, I_x is always positive, whereas I_{xy} will be either positive or negative.

- Connect the reference point A with the center of the circle and determine the distance OA by trigonometry. This distance represents the radius of the circle, Fig. 10–20b. Finally, draw the circle.

Principal Moments of Inertia.

- The points where the circle intersects the abscissa give the values of the principal moments of inertia I_{min} and I_{max}. Notice that the *product of inertia will be zero at these points*, Fig. 10–20b.

Principal Axes.

- To find the direction of the major principal axis, determine by trigonometry the angle $2\theta_{p_1}$, *measured from the radius OA to the positive I axis*, Fig. 10–20b. This angle represents *twice* the angle from the x axis of the area in question to the axis of maximum moment of inertia I_{max}, Fig. 10–20a. Both the angle on the circle, $2\theta_{p_1}$, and the angle to the axis on the area, θ_{p_1}, *must be measured in the same sense*, as shown in Fig. 10–20. The axis for minimum moment of inertia I_{min} is perpendicular to the axis for I_{max}.

Using trigonometry, the above procedure may be verified to be in accordance with the equations developed in Sec. 10.7.

EXAMPLE 10.10

Using Mohr's circle, determine the principal moments of inertia for the beam's cross-sectional area shown in Fig. 10–21a, with respect to an axis passing through the centroid.

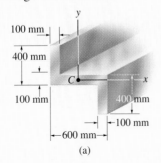

(a)

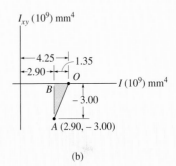

(b)

Solution

Determine I_x, I_y, I_{xy}. The moment of inertia and the product of inertia have been determined in Examples 10.6 and 10.8 with respect to the x, y axes shown in Fig. 10–21a. The results are $I_x = 2.90(10^9)$ mm^4, $I_y = 5.60(10^9)$ mm^4, and $I_{xy} = -3.00(10^9)$ mm^4.

Construct the Circle. The I and I_{xy} axes are shown in Fig. 10–21b. The center of the circle, O, lies at a distance $(I_x + I_y)/2 = (2.90 + 5.60)/2 = 4.25$ from the origin. When the reference point $A(2.90, -3.00)$ is connected to point O, the radius OA is determined from the triangle OBA using the Pythagorean theorem.

$$OA = \sqrt{(1.35)^2 + (-3.00)^2} = 3.29$$

The circle is constructed in Fig. 10–21c.

Principal Moments of Inertia. The circle intersects the I axis at points $(7.54, 0)$ and $(0.960, 0)$. Hence,

$$I_{max} = 7.54(10^9) \text{ mm}^4 \qquad Ans.$$
$$I_{min} = 0.960(10^9) \text{ mm}^4 \qquad Ans.$$

Principal Axes. As shown in Fig. 10–21c, the angle $2\theta_{p_1}$ is determined from the circle by measuring counterclockwise from OA to the direction of the *positive I* axis. Hence,

$$2\theta_{p_1} = 180° - \sin^{-1}\left(\frac{|BA|}{|OA|}\right) = 180° - \sin^{-1}\left(\frac{3.00}{3.29}\right) = 114.2°$$

The principal axis for $I_{max} = 7.54(10^9)$ mm^4 is therefore oriented at an angle $\theta_{p_1} = 57.1°$, measured *counterclockwise*, from the *positive x* axis to the *positive u* axis. The v axis is perpendicular to this axis. The results are shown in Fig. 10–21d.

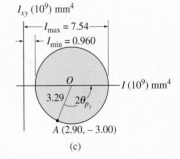

(c)

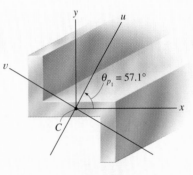

Fig. 10–21

10.9   Mass Moment of Inertia

The mass moment of inertia of a body is a property that measures the resistance of the body to angular acceleration. Since it is used in dynamics to study rotational motion, methods for its calculation will now be discussed.

We define the *mass moment of inertia* as the integral of the "second moment" about an axis of all the elements of mass *dm* which compose the body.* For example, consider the rigid body shown in Fig. 10–22. The body's moment of inertia about the *z* axis is

$$I = \int_m r^2 \, dm \qquad\qquad (10\text{–}13)$$

Here the "moment arm" *r* is the perpendicular distance from the axis to the arbitrary element *dm*. Since the formulation involves *r*, the value of *I* is *unique* for each axis *z* about which it is computed. However, the axis which is generally chosen for analysis passes through the body's mass center *G*. The moment of inertia computed about this axis will be defined as I_G. Realize that because *r* is squared in Eq. 10–13, the mass moment of inertia is always a *positive quantity*. Common units used for its measurement are $kg \cdot m^2$ or $slug \cdot ft^2$.

z

r

dm

Fig. 10–22

---

*Another property of the body which measures the symmetry of the body's mass with respect to a coordinate system is the mass product of inertia. This property most often applies to the three-dimensional motion of a body and is discussed in *Engineering Mechanics: Dynamics* (Chapter 21).

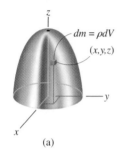

(a)

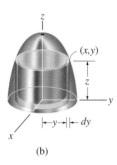

(b)

(c)

Fig. 10–23

If the body consists of material having a variable density, $\rho = \rho(x, y, z)$, the elemental mass dm of the body may be expressed in terms of its density and volume as $dm = \rho\, dV$. Substituting dm into Eq. 10–13, the body's moment of inertia is then computed using *volume elements* for integration; i.e.

$$I = \int_V r^2 \rho\, dV \qquad (10\text{–}14)$$

In the special case of ρ being a *constant*, this term may be factored out of the integral, and the integration is then purely a function of geometry:

$$I = \rho \int_V r^2\, dV \qquad (10\text{–}15)$$

When the elemental volume chosen for integration has differential sizes in all three directions, e.g., $dV = dx\, dy\, dz$, Fig. 10–23a, the moment of inertia of the body must be determined using "triple integration." The integration process can, however, be simplified to a *single integration* provided the chosen elemental volume has a differential size or thickness in only *one direction*. Shell or disk elements are often used for this purpose.

PROCEDURE FOR ANALYSIS

For integration, we will consider only symmetric bodies having surfaces which are generated by revolving a curve about an axis. An example of such a body which is generated about the z axis is shown in Fig. 10–23.

Shell Element

- If a *shell element* having a height z, radius y, and thickness dy is chosen for integration, Fig. 10–23b, then the volume $dV = (2\pi y)(z)\, dy$.
- This element may be used in Eq. 10–14 or 10–15 for determining the moment of inertia I_z of the body about the z axis since the *entire element*, due to its "thinness," lies at the *same* perpendicular distance $r = y$ from the z axis (see Example 10.11).

Disk Element

- If a disk element having a radius y and a thickness dz is chosen for integration, Fig. 10–23c, then the volume $dV = (\pi y^2)\, dz$.
- In this case the element is *finite* in the radial direction, and consequently its parts *do not* all lie at the *same radial distance* r from the z axis. As a result, Eqs. 10–14 or 10–15 *cannot* be used to determine I_z. Instead, to perform the integration using this element, it is first necessary to determine the moment of inertia *of the element* about the z axis and then integrate this result (see Example 10.12).

E X A M P L E   10.11

Determine the mass moment of inertia of the cylinder shown in Fig. 10–24a about the z axis. The density of the material, ρ, is constant.

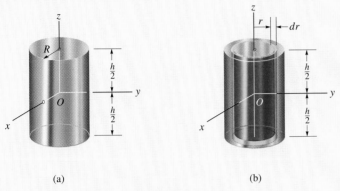

(a)   (b)

Fig. 10–24

Solution

Shell Element.   This problem may be solved using the *shell element* in Fig. 10–24b and single integration. The volume of the element is $dV = (2\pi r)(h)\, dr$, so that its mass is $dm = \rho\, dV = \rho(2\pi h r\, dr)$. Since the *entire element* lies at the same distance r from the z axis, the moment of inertia *of the element* is

$$dI_z = r^2\, dm = \rho 2\pi h r^3\, dr$$

Integrating over the entire region of the cylinder yields

$$I_z = \int_m r^2\, dm = \rho 2\pi h \int_0^R r^3\, dr = \frac{\rho\pi}{2} R^4 h$$

The mass of the cylinder is

$$m = \int_m dm = \rho 2\pi h \int_0^R r\, dr = \rho\pi h R^2$$

so that

$$I_z = \frac{1}{2} mR^2 \qquad\qquad\qquad Ans.$$

E X A M P L E **10.12**

A solid is formed by revolving the shaded area shown in Fig. 10–25a about the y axis. If the density of the material is 5 slug/ft³, determine the mass moment of inertia about the y axis.

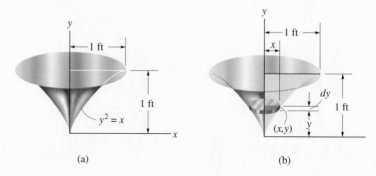

(a)   (b)

Fig. 10–25

Solution

Disk Element. The moment of inertia will be determined using a *disk element*, as shown in Fig. 10–25b. Here the element intersects the curve at the arbitrary point (x, y) and has a mass

$$dm = \rho \, dV = \rho(\pi x^2) \, dy$$

Although all portions of the element are *not* located at the same distance from the y axis, it is still possible to determine the moment of inertia dI_y *of the element* about the y axis. In Example 10.11 it was shown that the moment of inertia of a cylinder about its longitudinal axis is $I = \frac{1}{2}mR^2$, where m and R are the mass and radius of the cylinder. Since the height of the cylinder is not involved in this formula, we can also use it for a disk. Thus, for the disk element in Fig. 10–25b, we have

$$dI_y = \frac{1}{2}(dm)x^2 = \frac{1}{2}[\rho(\pi x^2) \, dy]x^2$$

Substituting $x = y^2$, $\rho = 5$ slug/ft³, and integrating with respect to y, from $y = 0$ to $y = 1$ ft, yields the moment of inertia for the entire solid:

$$I_y = \frac{5\pi}{2} \int_0^1 x^4 \, dy = \frac{5\pi}{2} \int_0^1 y^8 \, dy = 0.873 \text{ slug} \cdot \text{ft}^2 \qquad Ans.$$

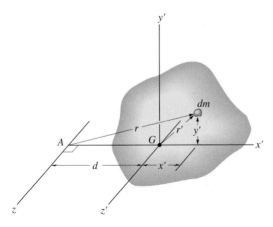

Fig. 10–26

Parallel-Axis Theorem.   If the moment of inertia of the body about an axis passing through the body's mass center is known, then the moment of inertia about any other *parallel axis* may be determined by using the *parallel-axis theorem.* This theorem can be derived by considering the body shown in Fig. 10–26. The z' axis passes through the mass center G, whereas the corresponding *parallel z axis* lies at a constant distance d away. Selecting the differential element of mass dm which is located at point (x', y') and using the Pythagorean theorem, $r^2 = (d + x')^2 + y'^2$, we can express the moment of inertia of the body about the z axis as

$$I = \int_m r^2\, dm = \int_m [(d + x')^2 + y'^2]\, dm$$

$$= \int_m (x'^2 + y'^2)\, dm + 2d \int_m x'\, dm + d^2 \int_m dm$$

Since $r'^2 = x'^2 + y'^2$, the first integral represents I_G. The second integral equals *zero*, since the z' axis passes through the body's mass center, i.e., $\int x'\, dm = \bar{x} \int dm = 0$ since $\bar{x} = 0$. Finally, the third integral represents the total mass m of the body. Hence, the moment of inertia about the z axis can be written as

$$\boxed{I = I_G + md^2} \qquad (10\text{–}16)$$

where

I_G = moment of inertia about the z' axis passing through the mass center G

m = mass of the body

d = perpendicular distance between the parallel axes

Radius of Gyration. Occasionally, the moment of inertia of a body about a specified axis is reported in handbooks using the *radius of gyration, k*. This value has units of length, and when it and the body's mass m are known, the moment of inertia is determined from the equation

$$I = mk^2 \quad \text{or} \quad k = \sqrt{\frac{I}{m}} \qquad (10\text{–}17)$$

Note the *similarity* between the definition of k in this formula and r in the equation $dI = r^2\, dm$, which defines the moment of inertia of an elemental mass dm of the body about an axis.

Composite Bodies. If a body is constructed from a number of simple shapes such as disks, spheres, and rods, the moment of inertia of the body about any axis z can be determined by adding algebraically the moments of inertia of all the composite shapes computed about the z axis. Algebraic addition is necessary since a composite part must be considered as a negative quantity if it has already been included within another part—for example, a "hole" subtracted from a solid plate. The parallel-axis theorem is needed for the calculations if the center of mass of each composite part does not lie on the z axis. For the calculation, then, $I = \Sigma(I_G + md^2)$, where I_G for each of the composite parts is computed by integration or can be determined from a table, such as the one given on the inside back cover of this book.

This flywheel, which operates a metal cutter, has a large moment of inertia about its center. Once it begins rotating it is difficult to stop it and therefore a uniform motion can be effectively transferred to the cutting blade.

EXAMPLE 10.13

If the plate shown in Fig. 10–27a has a density of 8000 kg/m³ and a thickness of 10 mm, determine its mass moment of inertia about an axis directed perpendicular to the page and passing through point O.

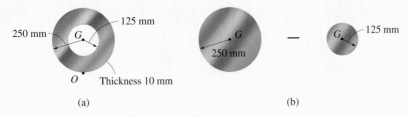

250 mm
125 mm
G
O
Thickness 10 mm

G
250 mm

−

G
125 mm

(a)

(b)

Fig. 10–27

Solution

The plate consists of two composite parts, the 250-mm-radius disk *minus* a 125-mm-radius disk, Fig. 10–27b. The moment of inertia about O can be determined by computing the moment of inertia of each of these parts about O and then *algebraically* adding the results. The computations are performed by using the parallel-axis theorem in conjunction with the data listed in the table on the inside back cover.

Disk. The moment of inertia of a disk about an axis perpendicular to the plane of the disk is $I_G = \frac{1}{2}mr^2$. The mass center of the disk is located at a distance of 0.25 m from point O. Thus,

$$m_d = \rho_d V_d = 8000 \text{ kg/m}^3[\pi(0.25 \text{ m})^2(0.01 \text{ m})] = 15.71 \text{ kg}$$

$$(I_O)_d = \frac{1}{2}m_d r_d^2 + m_d d^2$$

$$= \frac{1}{2}(15.71 \text{ kg})(0.25 \text{ m})^2 + (15.71 \text{ kg})(0.25 \text{ m})^2$$

$$= 1.473 \text{ kg} \cdot \text{m}^2$$

Hole. For the 125-mm-radius disk (hole), we have

$$m_h = \rho_h V_h = 8000 \text{ kg/m}^3[\pi(0.125 \text{ m})^2(0.01 \text{ m})] = 3.93 \text{ kg}$$

$$(I_O)_h = \frac{1}{2}m_h r_h^2 + m_h d^2$$

$$= \frac{1}{2}(3.93 \text{ kg})(0.125 \text{ m})^2 + (3.93 \text{ kg})(0.25 \text{ m})^2$$

$$= 0.276 \text{ kg} \cdot \text{m}^2$$

The moment of inertia of the plate about point O is therefore

$$I_O = (I_O)_d - (I_O)_h$$

$$= 1.473 \text{ kg} \cdot \text{m}^2 - 0.276 \text{ kg} \cdot \text{m}^2$$

$$= 1.20 \text{ kg} \cdot \text{m}^2 \qquad\qquad Ans.$$

E X A M P L E 10.14

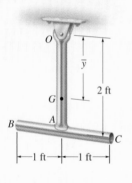

Fig. 10–28

The pendulum in Fig. 10–28 consists of two thin rods each having a weight of 10 lb. Determine the pendulum's mass moment of inertia about an axis passing through (a) the pin at O, and (b) the mass center G of the pendulum.

Solution

Part (a). Using the table on the inside back cover, the moment of inertia of rod OA about an axis perpendicular to the page and passing through the end point O of the rod is $I_O = \frac{1}{3}ml^2$. Hence,

$$(I_{OA})_O = \frac{1}{3}ml^2 = \frac{1}{3}\left(\frac{10 \text{ lb}}{32.2 \text{ ft/s}^2}\right)(2 \text{ ft})^2 = 0.414 \text{ slug} \cdot \text{ft}^2$$

This same value may be computed using $I_G = \frac{1}{12}ml^2$ and the parallel-axis theorem; i.e.,

$$(I_{OA})_O = \frac{1}{12}ml^2 + md^2 = \frac{1}{12}\left(\frac{10 \text{ lb}}{32.2 \text{ ft/s}^2}\right)(2 \text{ ft})^2 + \frac{10 \text{ lb}}{32.2 \text{ ft/s}^2}(1 \text{ ft})^2$$

$$= 0.414 \text{ slug} \cdot \text{ft}^2$$

For rod BC we have

$$(I_{BC})_O = \frac{1}{12}ml^2 + md^2 = \frac{1}{12}\left(\frac{10 \text{ lb}}{32.2 \text{ ft/s}^2}\right)(2 \text{ ft})^2 + \frac{10 \text{ lb}}{32.2 \text{ ft/s}^2}(2 \text{ ft})^2$$

$$= 1.346 \text{ slug} \cdot \text{ft}^2$$

The moment of inertia of the pendulum about O is therefore

$$I_O = 0.414 + 1.346 = 1.76 \text{ slug} \cdot \text{ft}^2 \qquad Ans.$$

Part (b). The mass center G will be located relative to the pin at O. Assuming this distance to be $\bar{y}$, Fig. 10–28, and using the formula for determining the mass center, we have

$$\bar{y} = \frac{\Sigma \tilde{y}m}{\Sigma m} = \frac{1(10/32.2) + 2(10/32.2)}{(10/32.2) + (10/32.2)} = 1.50 \text{ ft}$$

The moment of inertia I_G may be computed in the same manner as I_O, which requires successive applications of the parallel-axis theorem in order to transfer the moments of inertia of rods OA and BC to G. A more direct solution, however, involves applying the parallel-axis theorem using the result for I_O determined above; i.e.,

$$I_O = I_G + md^2; \quad 1.76 \text{ slug} \cdot \text{ft}^2 = I_G + \left(\frac{20 \text{ lb}}{32.2 \text{ ft/s}^2}\right)(1.50 \text{ ft})^2$$

$$I_G = 0.362 \text{ slug} \cdot \text{ft}^2 \qquad Ans.$$

CHAPTER REVIEW

- *Area Moment of Inertia.*   The *area moment of inertia* represents the second moment of the area about an axis, $I = \int r^2 \, dA$. It is frequently used in formulas related to strength and stability of structural members or mechanical elements. If the area shape is irregular, then a differential element must be selected and integration over the entire area must be performed. Tabular values of the moment of inertia of common shapes about their *centroidal axis* are available. To determine the moment of inertia of these shapes about some *other axis*, the parallel-axis theorem must be used, $I = \bar{I} + Ad^2$. If an area is a composite of these shapes, then its moment of inertia is equal to the sum of the moments of inertia of each of its parts.

- *Product of Inertia.*   The *product of inertia* of an area is used to determine the location of an axis about which the moment of inertia for the area is a maximum or minimum. This property is determined from $I_{xy} = \int xy \, dA$, where the integration is performed over the entire area. If the product of inertia for an area is known about its centroidal x', y' axes, then its value can be determined about any x, y axes using the parallel-axis theorem for the product of inertia, $I_{xy} = \bar{I}_{x'y'} + A \, d_x d_y$.

- *Principal Moments of Inertia.*   Provided the moments of inertia I_x and I_y, and the product of inertia I_{xy} are known, then formulas, or Mohr's circle, can be used to determine the maximum and minimum or *principal moments of inertia* for the area, as well as finding the orientation of the principal axes of inertia.

- *Mass Moment of Inertia.*   The *mass moment of inertia* is a property of a body that measures its resistance to a change in its rotation. It is defined as the second moment of the mass elements of the body about an axis, $I = \int r^2 \, dm$. For bodies having axial symmetry, it can be determined by integration, using either disk or shell elements. The mass moment of inertia of a composite body is determined by using tabular values of its composite shapes along with the parallel-axis theorem, $I = \bar{I} + md^2$.

REVIEW PROBLEMS

10-1. Determine the area moment of inertia of the beam's cross-sectional area about the x axis which passes through the centroid C.

10-2. Determine the area moment of inertia of the beam's cross-sectional area about the y axis which passes through the centroid C.

*10-4. Determine the area moments of inertia I_x and I_y of the shaded area.

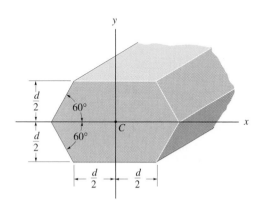

Probs. 10–1/2

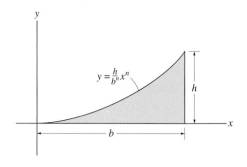

Prob. 10–4

10-3. Determine the mass moment of inertia I_x of the body and express the result in terms of the total mass m of the body. The density is constant.

10-5. Determine the area moments of inertia I_u and I_v and the product of inertia I_{uv} for the semicircular area.

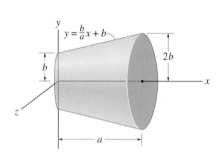

Prob. 10–3

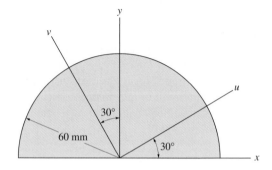

Prob. 10–5

10-6. Determine the area moment of inertia of the shaded area about the y axis.

10-7. Determine the area moment of inertia of the shaded area about the x axis.

10-9. Determine the area moment of inertia of the triangular area about (a) the x axis, and (b) the centroidal x' axis.

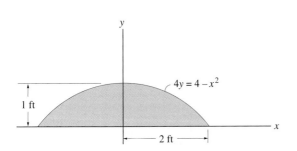

Probs. 10–6/7

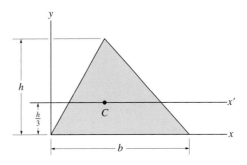

Prob. 10–9

*10-8. Determine the area moment of inertia of the area about the x axis. Then, using the parallel-axis theorem, find the area moment of inertia about the x' axis that passes through the centroid C of the area. $\bar{y} = 120$ mm.

10-10. Determine the product of inertia of the shaded area with respect to the x and y axes.

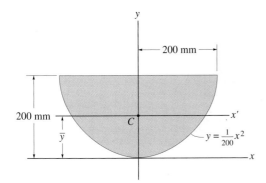

Prob. 10–8

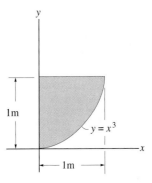

Prob. 10–10

Equilibrium and stability of this articulated crane boom as a function of its position can be analyzed using methods based on work and energy, which are explained in this chapter.

Virtual Work

- To introduce the principle of virtual work and show how it applies to determining the equilibrium configuration of a series of pin-connected members.
- To establish the potential energy function and use the potential-energy method to investigate the type of equilibrium or stability of a rigid body or configuration.

11.1   Definition of Work and Virtual Work

Work of a Force.   In mechanics a force **F** does work only when it undergoes a displacement in the direction of the force. For example, consider the force **F** in Fig. 11–1, which is located on the path s specified by the position vector **r**. If the force moves along the path to a new position $\mathbf{r}' = \mathbf{r} + d\mathbf{r}$, the displacement is $d\mathbf{r}$, and therefore the work dU is a *scalar quantity* defined by the dot product

$$dU = \mathbf{F} \cdot d\mathbf{r}$$

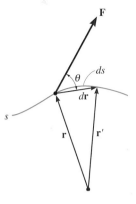

Fig. 11–1

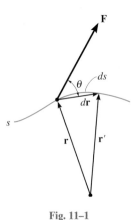

Fig. 11–1

Because $d\mathbf{r}$ is infinitesimal, the magnitude of $d\mathbf{r}$ can be represented by ds, the differential arc segment along the path. If the angle between the tails of $d\mathbf{r}$ and $\mathbf{F}$ is θ, Fig. 11–1, then by definition of the dot product, the above equation may also be written as

$$dU = F\,ds\cos\theta$$

Work expressed by this equation may be interpreted in one of two ways: either as the product of F and the component of displacement in the direction of the force, i.e., $ds\cos\theta$; or as the product of ds and the component of force in the direction of displacement, i.e., $F\cos\theta$. Note that if $0° \le \theta < 90°$, then the force component and the displacement have the *same sense*, so that the work is *positive*; whereas if $90° < \theta \le 180°$, these vectors have an *opposite sense*, and therefore the work is *negative*. Also, $dU = 0$ if the force is *perpendicular* to displacement, since $\cos 90° = 0$, or if the force is applied at a *fixed point*, in which case the displacement $ds = 0$.

The basic unit for work combines the units of force and displacement. In the SI system a *joule* (J) is equivalent to the work done by a force of 1 newton which moves 1 meter in the direction of the force $(1\text{ J} = 1\text{ N}\cdot\text{m})$. In the FPS system, work is defined in units of ft · lb. The moment of a force has the same combination of units; however, the concepts of moment and work are in no way related. A moment is a vector quantity, whereas work is a scalar.

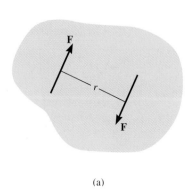

(a)

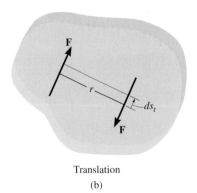

Translation

(b)

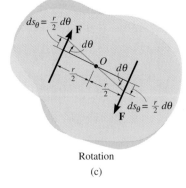

Rotation

(c)

Fig. 11–2

Work of a Couple. The two forces of a couple do work when the couple *rotates* about an axis perpendicular to the plane of the couple. To show this, consider the body in Fig. 11–2a, which is subjected to a couple whose moment has a magnitude $M = Fr$. Any general differential displacement of the body can be considered as a combination of a translation and rotation. When the body *translates* such that the *component of displacement* along the line of action of each force is ds_t, clearly the "positive" work of one force $(F \, ds_t)$ *cancels* the "negative" work of the other $(-F \, ds_t)$, Fig. 11–2b. Consider now a differential *rotation* $d\theta$ of the body about an axis perpendicular to the plane of the couple, which intersects the plane at point O, Fig. 11–2c. (For the derivation, any other point in the plane may also be considered.) As shown, each force undergoes a displacement $ds_\theta = (r/2) \, d\theta$ in the direction of the force; hence, the work of both forces is

$$dU = F\left(\frac{r}{2}d\theta\right) + F\left(\frac{r}{2}d\theta\right) = (Fr) \, d\theta$$

or

$$dU = M \, d\theta$$

The resultant work is *positive* when the sense of $\mathbf{M}$ is the *same* as that of $d\boldsymbol{\theta}$, and negative when they have an opposite sense. As in the case of the moment vector, the *direction and sense* of $d\boldsymbol{\theta}$ are defined by the right-hand rule, where the fingers of the right hand follow the rotation or "curl" and the thumb indicates the direction of $d\boldsymbol{\theta}$. Hence, the line of action of $d\boldsymbol{\theta}$ will be *parallel* to the line of action of $\mathbf{M}$ if movement of the body occurs in the *same plane*. If the body rotates in space, however, the *component* of $d\boldsymbol{\theta}$ in the direction of $\mathbf{M}$ is required. Thus, in general, the work done by a couple is defined by the dot product, $dU = \mathbf{M} \cdot d\boldsymbol{\theta}$.

Virtual Work. The definitions of the work of a force and a couple have been presented in terms of *actual movements* expressed by differential displacements having magnitudes of ds and $d\theta$. Consider now an *imaginary* or *virtual movement*, which indicates a displacement or rotation that is *assumed* and *does not actually exist*. These movements are first-order differential quantities and will be denoted by the symbols δs and $\delta\theta$ (delta s and delta θ), respectively. The *virtual work* done by a force undergoing a virtual displacement δs is

$$\boxed{\delta U = F \cos \theta \, \delta s} \tag{11–1}$$

Similarly, when a couple undergoes a virtual rotation $\delta\theta$ in the plane of the couple forces, the *virtual work* is

$$\boxed{\delta U = M \, \delta\theta} \tag{11–2}$$

11.2 Principle of Virtual Work for a Particle and a Rigid Body

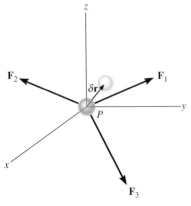

Fig. 11–3

Particle. If the particle in Fig. 11–3 undergoes an imaginary or virtual displacement $\delta\mathbf{r}$, then the virtual work (δU) done by the force system becomes

$$\delta U = \Sigma\mathbf{F} \cdot \delta\mathbf{r}$$
$$= (\Sigma F_x\mathbf{i} + \Sigma F_y\mathbf{j} + \Sigma F_z\mathbf{k}) \cdot (\delta x\mathbf{i} + \delta y\mathbf{j} + \delta z\mathbf{k})$$
$$= \Sigma F_x\,\delta x + \Sigma F_y\,\delta y + \Sigma F_z\,\delta z$$

For equilibrium $\Sigma F_x = 0$, $\Sigma F_y = 0$, $\Sigma F_z = 0$, and so the virtual work must also be zero, i.e.,

$$\delta U = 0$$

In other words, we can write three independent virtual work equations corresponding to the three equations of equilibrium.

For example, consider the free body diagram of the ball which rests on the floor, Fig. 11–4. If we "imagine" the ball to be displaced downwards a virtual amount δy, then the weight does positive virtual work, $W\delta y$, and the normal force does negative virtual work, $-N\delta y$. For equilibrium the total virtual work must be zero, so that $\delta U = W\delta y - N\delta y = (W - N)\delta y = 0$. Since $\delta y \neq 0$, then $N = W$ as required.

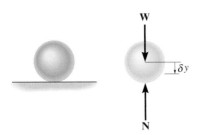

Fig. 11–4

Rigid Body. In a similar manner, we can also write a set of three virtual work equations $(\delta U = 0)$ for a rigid body subjected to a coplanar force system. If these equations involve separate virtual translations in the x and y directions and a virtual rotation about an axis perpendicular to the x–y plane and passing through an arbitrary point O, then it can be shown that they will correspond to the three equilibrium equations, $\Sigma F_x = 0$, $\Sigma F_y = 0$, and $\Sigma M_O = 0$. When writing these equations, it is *not necessary* to include the work done by the *internal forces* acting within the body since a rigid body *does not deform* when subjected to an external loading, and furthermore, when the body moves through a virtual displacement, the internal forces occur in equal but opposite collinear pairs, so that the corresponding work done by each pair of forces *cancels*.

To demonstrate an application, consider the simply supported beam in Fig. 11–5a. When the beam is given a virtual rotation $\delta\theta$ about point B, Fig. 11–5b, the only forces that do work are $\mathbf{P}$ and $\mathbf{A}_y$. Since $\delta y = l\delta\theta$ and $\delta y' = (l/2)\delta\theta$, the virtual work equation for this case is $\delta U = A_y(l\delta\theta) - P(l/2)\delta\theta = (A_y - P/2)l\delta\theta = 0$. Since $\delta\theta \neq 0$, then $A_y = P/2$. Excluding $\delta\theta$, notice that the terms in parentheses actually represent moment equilibrium about point B.

As in the case of a particle, no added advantage is gained by solving rigid-body equilibrium problems using the principle of virtual work. This is because for each application of the virtual-work equation the virtual displacement, common to every term, factors out, leaving an equation that could have been obtained in a more *direct manner* by simply applying the equations of equilibrium.

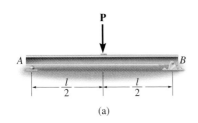

(a)

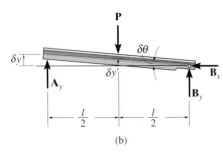

(b)

Fig. 11–5

11.3 Principle of Virtual Work for a System of Connected Rigid Bodies

The method of virtual work is most suitable for solving equilibrium problems that involve a system of several *connected* rigid bodies such as the ones shown in Fig. 11–6. Before we can apply the principle of virtual work to these systems, however, we must first specify the number of degrees of freedom for a system and establish coordinates that define the position of the system.

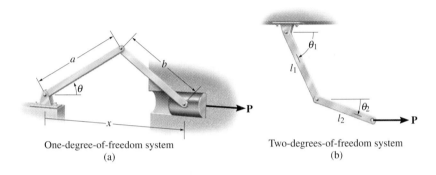

One-degree-of-freedom system
(a)

Two-degrees-of-freedom system
(b)

Fig. 11–6

Degrees of Freedom. A system of connected bodies takes on a unique shape that can be specified provided we know the position of a number of specific points on the system. These positions are defined using *independent coordinates q*, which are measured from fixed reference points. For every coordinate established, the system will have a *degree of freedom* for displacement along the coordinate axis such that it is consistent with the constraining action of the supports. Thus, an n-degree-of-freedom system requires n independent coordinates q_n to specify the location of all its members. For example, the link and sliding-block arrangement shown in Fig. 11–6a is an example of a one-degree-of-freedom system. The independent coordinate $q = \theta$ may be used to specify the location of the two connecting links and the block. The coordinate x could also be used as the independent coordinate. However, since the block is constrained to move within the slot, x is not independent of θ; rather, it can be related to θ using the cosine law, $b^2 = a^2 + x^2 - 2ax \cos \theta$. The double-link arrangement, shown in Fig. 11–6b, is an example of a two-degrees-of-freedom system. To specify the location of each link, the coordinate angles θ_1 and θ_2 must be known since a rotation of one link is independent of a rotation of the other.

During operation the scissors lift has one degree of freedom. Without dismembering the mechanism, the hydraulic force required to provide the lift can be determined *directly* by using the principle of virtual work.

Principle of Virtual Work. The principle of virtual work for a system of rigid bodies whose connections are *frictionless* may be stated as follows: *A system of connected rigid bodies is in equilibrium provided the virtual work done by all the external forces and couples acting on the system is zero for each independent virtual displacement of the system.* Mathematically, this may be expressed as

$$\delta U = 0 \tag{11-3}$$

where δU represents the virtual work of all the external forces (and couples) acting on the system during any independent virtual displacement.

As stated above, if a system has n degrees of freedom it takes n independent coordinates q_n to completely specify the location of the system. Hence, for the system it is possible to write n independent virtual-work equations, one for every virtual displacement taken along each of the independent coordinate axes, while the remaining $n - 1$ independent coordinates are held *fixed*.*

IMPORTANT POINTS

- A force does work when it moves through a displacement in the direction of the force. A couple moment does work when it moves through a collinear rotation. Specifically, positive work is done when the force or couple moment and its displacement have the same sense of direction.

- The principle of virtual work is generally used to determine the equilibrium configuration for a series of multiply-connected members.

- A virtual displacement is imaginary, i.e., does not really happen. It is a differential that is given in the positive direction of the position coordinate.

- Forces or couple moments that do not virtually displace do no virtual work.

*This method of applying the principle of virtual work is sometimes called the *method of virtual displacements* since a virtual displacement is applied, resulting in the calculation of a real force. Although it is not to be used here, realize that we can also apply the principle of virtual work as a method of virtual forces. This method is often used to determine the displacements of points on deformable bodies. See R. C. Hibbeler, *Mechanics of Materials*, 5th edition, Prentice Hall, Inc., 2003.

PROCEDURE FOR ANALYSIS

The equation of virtual work can be used to solve problems involving a system of frictionless connected rigid bodies having a single degree of freedom by using the following procedure.

Free-Body Diagram.

- Draw the free-body diagram of the entire system of connected bodies and define the *independent coordinate q.*
- Sketch the "deflected position" of the system on the free-body diagram when the system undergoes a positive virtual displacement δq.

Virtual Displacements.

- Indicate *position coordinates* s_i, measured from a *fixed point* on the free-body diagram to each of the i number of "active" forces and couples, i.e., those that do work.
- Each coordinate axis should be parallel to the line of action of the "active" force to which it is directed, so that the virtual work along the coordinate axis can be calculated.
- Relate each of the position coordinates s_i to the independent coordinate q; then *differentiate* these expressions in order to express the virtual displacements δs_i in terms of δq.

Virtual Work Equation.

- Write the *virtual-work equation* for the system assuming that, whether possible or not, all the position coordinates s_i undergo *positive* virtual displacements δs_i.
- Using the relations for δs_i, express the work of *each* "active" force and couple in the equation in terms of the *single* independent virtual displacement δq.
- Factor out this common displacement from all the terms and solve for the unknown force, couple, or equilibrium position, q.
- If the system contains n degrees of freedom, n independent coordinates q_n must be specified. Follow the above procedure and let *only one* of the independent coordinates undergo a virtual displacement, while the remaining $n - 1$ coordinates are *held fixed*. In this way, n virtual-work equations can be written, one for each independent coordinate.

EXAMPLE 11.1

Determine the angle θ for equilibrium of the two-member linkage shown in Fig. 11–7a. Each member has a mass of 10 kg.

Solution

Free-Body Diagram. The system has only one degree of freedom since the location of both links may be specified by the single independent coordinate $(q =) \theta$. As shown on the free-body diagram in Fig. 11–7b, when θ undergoes a *positive* (clockwise) virtual rotation $\delta\theta$, only the active forces, **F** and the two 98.1-N weights, do work. (The reactive forces $\mathbf{D}_x$ and $\mathbf{D}_y$ are fixed, and $\mathbf{B}_y$ does not move along its line of action.)

Virtual Displacements. If the origin of coordinates is established at the *fixed* pin support D, the location of **F** and **W** may be specified by the *position coordinates* x_B and y_w, as shown in the figure. In order to determine the work, note that these coordinates are parallel to the lines of action of their associated forces.

Expressing the position coordinates in terms of the independent coordinate θ and taking the derivatives yields

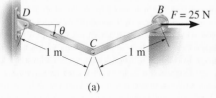

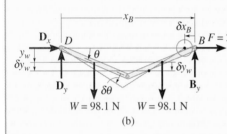

(a)

(b)

Fig. 11–7

$$x_B = 2(1 \cos \theta) \text{ m} \qquad \delta x_B = -2 \sin \theta \, \delta\theta \text{ m} \qquad (1)$$

$$y_w = \tfrac{1}{2}(1 \sin \theta) \text{ m} \qquad \delta y_w = 0.5 \cos \theta \, \delta\theta \text{ m} \qquad (2)$$

It is seen by the *signs* of these equations, and indicated in Fig. 11–7b, that an *increase* in θ (i.e., $\delta\theta$) causes a *decrease* in x_B and an *increase* in y_w.

Virtual-Work Equation. If the virtual displacements δx_B and δy_w were *both positive*, then the forces **W** and **F** would do positive work since the forces and their corresponding displacements would have the same sense. Hence, the virtual-work equation for the displacement $\delta\theta$ is

$$\delta U = 0; \qquad W \, \delta y_w + W \, \delta y_w + F \, \delta x_B = 0 \qquad (3)$$

Substituting Eqs. 1 and 2 into Eq. 3 in order to relate the virtual displacements to the common virtual displacement $\delta\theta$ yields

$$98.1(0.5 \cos \theta \, \delta\theta) + 98.1(0.5 \cos \theta \, \delta\theta) + 25(-2 \sin \theta \, \delta\theta) = 0$$

Notice that the "negative work" done by **F** (force in the opposite sense to displacement) has been *accounted for* in the above equation by the "negative sign" of Eq. 1. Factoring out the *common displacement* $\delta\theta$ and solving for θ, noting that $\delta\theta \neq 0$, yields

$$(98.1 \cos \theta - 50 \sin \theta) \, \delta\theta = 0$$

$$\theta = \tan^{-1}\frac{98.1}{50} = 63.0° \qquad \qquad Ans.$$

If this problem had been solved using the equations of equilibrium, it would have been necessary to dismember the links and apply three scalar equations to *each* link. The principle of virtual work, by means of calculus, has eliminated this task so that the answer is obtained directly.

E X A M P L E   11.2

Determine the angle θ required to maintain equilibrium of the mechanism in Fig. 11–8a. Neglect the weight of the links. The spring is unstretched when $\theta = 0°$, and it maintains a horizontal position due to the roller.

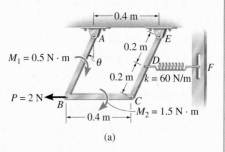

(a)

Solution

Free-Body Diagram. The mechanism has one degree of freedom, and therefore the location of each member may be specified using the independent coordinate θ. When θ undergoes a *positive* virtual displacement $\delta\theta$, as shown on the free-body diagram in Fig. 11–8b, links AB and EC rotate by the same amount since they have the same length, and link BC only translates. Since a couple moment does work *only* when it rotates, the work done by $\mathbf{M}_2$ is zero. The reactive forces at A and E do no work. Why?

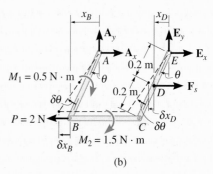

(b)

Fig. 11–8

Virtual Displacements. The position coordinates x_B and x_D are *parallel* to the lines of action of $\mathbf{P}$ and $\mathbf{F}_s$, and these coordinates locate the forces with respect to the *fixed points* A and E. From Fig. 11–8b,

$$x_B = 0.4 \sin\theta \text{ m}$$
$$x_D = 0.2 \sin\theta \text{ m}$$

Thus,

$$\delta x_B = 0.4 \cos\theta \, \delta\theta \text{ m}$$
$$\delta x_D = 0.2 \cos\theta \, \delta\theta \text{ m}$$

Virtual-Work Equation. For *positive* virtual displacements, $\mathbf{F}_s$ is opposite to δx_D and hence does negative work. Thus,

$$\delta U = 0; \qquad M_1 \, \delta\theta + P \, \delta x_B - F_s \, \delta x_D = 0$$

Relating each of the virtual displacements to the *common* virtual displacement $\delta\theta$ yields

$$0.5 \, \delta\theta + 2(0.4 \cos\theta \, \delta\theta) - F_s(0.2 \cos\theta \, \delta\theta) = 0$$
$$(0.5 + 0.8 \cos\theta - 0.2 F_s \cos\theta) \, \delta\theta = 0 \qquad (1)$$

For the arbitrary angle θ, the spring is stretched a distance of $x_D = (0.2 \sin\theta)$ m; and therefore, $F_s = 60 \text{ N/m}(0.2 \sin\theta)$ m = $(12 \sin\theta)$ N. Substituting into Eq. 1 and noting that $\delta\theta \neq 0$, we have

$$0.5 + 0.8 \cos\theta - 0.2(12 \sin\theta) \cos\theta = 0$$

Since $\sin 2\theta = 2 \sin\theta \cos\theta$, then

$$1 = 2.4 \sin 2\theta - 1.6 \cos\theta$$

Solving for θ by trial and error yields

$$\theta = 36.3° \qquad\qquad\qquad \textit{Ans.}$$

EXAMPLE 11.3

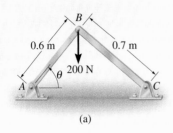

(a)

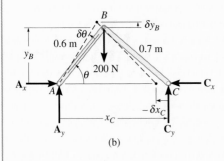

(b)

Fig. 11–9

Determine the horizontal force C_x that the pin at C must exert on BC in order to hold the mechanism shown in Fig. 11–9a in equilibrium when $\theta = 45°$. Neglect the weight of the members.

Solution

Free-Body Diagram. The reaction $\mathbf{C}_x$ can be obtained by *releasing* the pin constraint at C in the x direction and allowing the frame to be displaced in this direction. The system then has only one degree of freedom, defined by the independent coordinate θ, Fig. 11–9b. When θ undergoes a *positive* virtual displacement $\delta\theta$, only $\mathbf{C}_x$ and the 200-N force do work.

Virtual Displacements. Forces $\mathbf{C}_x$ and 200 N are located from the fixed origin A using position coordinates y_B and x_C. From Fig. 11–9b, x_C can be related to θ by the "law of cosines." Hence,

$$(0.7)^2 = (0.6)^2 + x_C^2 - 2(0.6)x_C \cos\theta \tag{1}$$

$$0 = 0 + 2x_C\,\delta x_C - 1.2\,\delta x_C \cos\theta + 1.2 x_C \sin\theta\,\delta\theta$$

$$\delta x_C = \frac{1.2 x_C \sin\theta}{1.2 \cos\theta - 2x_C}\,\delta\theta \tag{2}$$

Also,

$$y_B = 0.6 \sin\theta$$

$$\delta y_B = 0.6 \cos\theta\,\delta\theta \tag{3}$$

Virtual-Work Equation. When y_B and x_C undergo *positive* virtual displacements δy_B and δx_C, $\mathbf{C}_x$ and 200 N do *negative work* since they both act in the opposite sense to $\delta\mathbf{y}_B$ and $\delta\mathbf{x}_C$. Hence,

$$\delta U = 0; \qquad -200\,\delta y_B - C_x\,\delta x_C = 0$$

Substituting Eqs. 2 and 3 into this equation, factoring out $\delta\theta$, and solving for C_x yields

$$-200(0.6 \cos\theta\,\delta\theta) - C_x\frac{1.2 x_C \sin\theta}{1.2 \cos\theta - 2x_C}\,\delta\theta = 0$$

$$C_x = \frac{-120 \cos\theta\,(1.2 \cos\theta - 2x_C)}{1.2 x_C \sin\theta} \tag{4}$$

At the required equilibrium position $\theta = 45°$, the corresponding value of x_C can be found by using Eq. 1, in which case

$$x_C^2 - 1.2 \cos 45°\,x_C - 0.13 = 0$$

Solving for the positive root yields

$$x_C = 0.981 \text{ m}$$

Thus, from Eq. 4,

$$C_x = 114 \text{ N} \qquad\qquad Ans.$$

EXAMPLE 11.4

Determine the equilibrium position of the two-bar linkage shown in Fig. 11–10a. Neglect the weight of the links.

Solution

The system has two degrees of freedom since the *independent coordinates* θ_1 and θ_2 must be known to locate the position of both links. The position coordinate x_B, measured from the fixed point O, is used to specify the location of **P**, Fig. 11–10b and c.

If θ_1 is held *fixed* and θ_2 varies by an amount $\delta\theta_2$, as shown in Fig. 11–10b, the virtual-work equation becomes

$$[\delta U = 0]_{\theta_2}; \qquad P(\delta x_B)_{\theta_2} - M\,\delta\theta_2 = 0 \qquad (1)$$

Here P and M represent the magnitudes of the applied force and couple moment acting on link AB.

When θ_2 is held *fixed* and θ_1 varies by an amount $\delta\theta_1$, as shown in Fig. 11–10c, then AB translates and the virtual-work equation becomes

$$[\delta U = 0]_{\theta_1}; \qquad P(\delta x_B)_{\theta_1} - M\,\delta\theta_1 = 0 \qquad (2)$$

The *position coordinate* x_B may be related to the independent coordinates θ_1 and θ_2 by the equation

$$x_B = l\sin\theta_1 + l\sin\theta_2 \qquad (3)$$

To obtain the variation δx_B in terms of $\delta\theta_2$, it is necessary to take the *partial derivative* of x_B with respect to θ_2 since x_B is a function of both θ_1 and θ_2. Hence,

$$\frac{\partial x_B}{\partial\theta_2} = l\cos\theta_2 \qquad (\delta x_B)_{\theta_2} = l\cos\theta_2\,\delta\theta_2$$

Substituting into Eq. 1, we have

$$(Pl\cos\theta_2 - M)\,\delta\theta_2 = 0$$

Since $\delta\theta_2 \neq 0$, then

$$\theta_2 = \cos^{-1}\left(\frac{M}{Pl}\right) \qquad \textit{Ans.}$$

Using Eq. 3 to obtain the variation of x_B with θ_1 yields

$$\frac{\partial x_B}{\partial\theta_1} = l\cos\theta_1 \qquad (\delta x_B)_{\theta_1} = l\cos\theta_1\,\delta\theta_1$$

Substituting into Eq. 2, we have

$$(Pl\cos\theta_1 - M)\,\delta\theta_1 = 0$$

Since $\delta\theta_1 \neq 0$, then

$$\theta_1 = \cos^{-1}\left(\frac{M}{Pl}\right) \qquad \textit{Ans.}$$

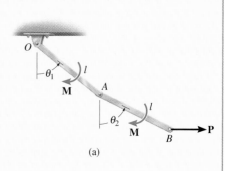

(a)

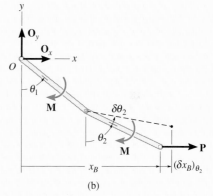

(b)

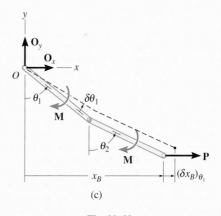

(c)

Fig. 11–10

*11.4 Conservative Forces

The work done by a force when it undergoes a *differential displacement* has been defined as $dU = F \cos \theta \, ds$, Fig. 11–1. If the force is displaced over a path that has a *finite length s*, the work is determined by integrating over the path: i.e.,

$$U = \int_s F \cos \theta \, ds$$

To evaluate the integral, it is necessary to obtain a relationship between F and the component of displacement $ds \cos \theta$. In some instances, however, the work done by a force will be *independent* of its path and, instead, will depend only on the initial and final locations of the force along the path. A force that has this property is called a *conservative force*.

Weight.   Consider the body in Fig. 11–11, which is initially at P'. If the body is moved *down* along the *arbitrary path A* to the second position, then, for a given displacement ds along the path, the displacement component in the direction of **W** has a magnitude of $dy = ds \cos \theta$, as shown. Since both the force and displacement are in the same direction, the work is positive; hence,

$$U = \int_s W \cos \theta \, ds = \int_0^y W \, dy$$

or

$$U = Wy$$

In a similar manner, the work done by the weight when the body moves up a distance y back to P', along the arbitrary path A', is

$$U = -Wy$$

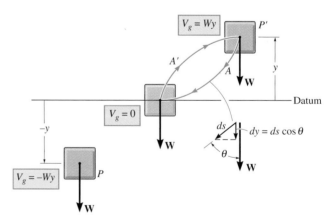

Fig. 11–11

Why is the work negative?

The weight of a body is therefore a conservative force since the work done by the weight depends *only* on the body's *vertical displacement* and is independent of the path along which the body moves.

Elastic Spring. The force developed by an elastic spring ($F_s = ks$) is also a conservative force. If the spring is attached to a body and the body is displaced along *any path*, such that it causes the spring to elongate or compress from a position s_1 to a further position s_2, the work will be negative since the spring exerts a force $\mathbf{F}_s$ *on the body* that is opposite to the body's displacement ds, Fig. 11–12. For either extension or compression, the work is independent of the path and is simply

$$U = \int_{s_1}^{s_2} F_s \, ds = \int_{s_1}^{s_2} (-ks) \, ds$$

$$= -\left(\tfrac{1}{2} k s_2^2 - \tfrac{1}{2} k s_1^2\right)$$

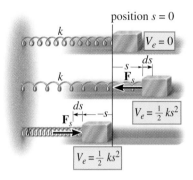

Fig. 11–12

Friction. In contrast to a conservative force, consider the force of *friction* exerted on a sliding body by a fixed surface. The work done by the frictional force depends on the path; the longer the path, the greater the work. Consequently, frictional forces are *nonconservative*, and the work done is dissipated from the body in the form of heat.

*11.5 Potential Energy

When a conservative force acts on a body, it gives the body the capacity to do work. This capacity, measured as *potential energy*, depends on the location of the body.

Gravitational Potential Energy. If a body is located a distance y *above* a fixed horizontal reference or datum, Fig. 11–11, the weight of the body has *positive* gravitational potential energy V_g since $\mathbf{W}$ has the capacity of doing positive work when the body is moved back down to the datum. Likewise, if the body is located a distance y *below* the datum, V_g is *negative* since the weight does negative work when the body is moved back up to the datum. At the datum, $V_g = 0$.

Measuring y as *positive upward*, the gravitational potential energy of the body's weight $\mathbf{W}$ is thus

$$\boxed{V_g = Wy} \qquad (11\text{–}4)$$

Datum

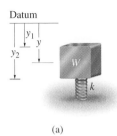

(a)

$F_s = ky_{eq}$

(b)

Fig. 11–13

Elastic Potential Energy. The elastic potential energy V_e that a spring produces on an attached body, when the spring is elongated or compressed from an undeformed position $(s = 0)$ to a final position s, is

$$V_e = \tfrac{1}{2}ks^2 \tag{11–5}$$

Here V_e is *always positive* since in the deformed position the spring has the capacity of doing *positive work* in *returning* the body back to the spring's undeformed position, Fig. 11–12.

Potential Function. In the general case, if a body is subjected to *both* gravitational and elastic forces, the *potential energy or potential function* V of the body can be expressed as the algebraic sum

$$V = V_g + V_e \tag{11–6}$$

where measurement of V depends on the location of the body with respect to a selected datum in accordance with Eqs. 11–4 and 11–5.

In general, if a system of frictionless connected rigid bodies has a *single degree of freedom* such that its position from the datum is defined by the independent coordinate q, then the potential function for the system can be expressed as $V = V(q)$. The work done by all the conservative forces acting on the system in moving it from q_1 to q_2 is measured by the *difference* in V; i.e.,

$$U_{1-2} = V(q_1) - V(q_2) \tag{11–7}$$

For example, the potential function for a system consisting of a block of weight **W** supported by a spring, Fig. 11–13a, can be expressed in terms of its independent coordinate $(q =) \, y$, measured from a fixed datum located at the unstretched length of the spring; we have

$$V = V_g + V_e$$
$$= -Wy + \tfrac{1}{2}ky^2 \tag{11–8}$$

If the block moves from y_1 to a farther downward position y_2, then the work of **W** and $\mathbf{F}_s$ is

$$U_{1-2} = V(y_1) - V(y_2) = -W[y_1 - y_2] + \tfrac{1}{2}ky_1^2 - \tfrac{1}{2}ky_2^2$$

---

*11.6 Potential-Energy Criterion for Equilibrium

System Having One Degree of Freedom. When the displacement of a frictionless connected system is *infinitesimal*, i.e., from q to $q + dq$, Eq. 11–7 becomes

$$dU = V(q) - V(q + dq)$$

or

$$dU = -dV$$

Furthermore, if the system undergoes a *virtual displacement* δq, rather than an actual displacement dq, then $\delta U = -\delta V$. For equilibrium, the principle of virtual work requires that $\delta U = 0$, and therefore, provided the potential function for the system is known, this also requires that $\delta V = 0$. We can also express this requirement as

$$\boxed{\dfrac{dV}{dq} = 0} \qquad (11\text{–}9)$$

Hence, *when a frictionless connected system of rigid bodies is in equilibrium, the first variation or change in V is zero.* This change is determined by taking the *first derivative* of the potential function and setting it equal to zero. For example, using Eq. 11–8 to determine the equilibrium position for the spring and block in Fig. 11–13a, we have

$$\frac{dV}{dy} = -W + ky = 0$$

Hence, the equilibrium position $y = y_{eq}$ is

$$y_{eq} = \frac{W}{k}$$

Of course, the *same result* is obtained by applying $\Sigma F_y = 0$ to the forces acting on the free-body diagram of the block, Fig. 11–13b.

System Having *n* Degrees of Freedom.   When the system of connected bodies has *n* degrees of freedom, the total potential energy stored in the system will be a function of *n* independent coordinates q_n, i.e., $V = V(q_1, q_2, \ldots, q_n)$. In order to apply the equilibrium criterion $\delta V = 0$, it is necessary to determine the change in potential energy δV by using the "chain rule" of differential calculus; i.e.,

$$\delta V = \frac{\partial V}{\partial q_1}\delta q_1 + \frac{\partial V}{\partial q_2}\delta q_2 + \cdots + \frac{\partial V}{\partial q_n}\delta q_n = 0$$

Since the virtual displacements $\delta q_1, \delta q_2, \ldots, \delta q_n$ are independent of one another, the equation is satisfied provided

$$\frac{\partial V}{\partial q_1} = 0, \quad \frac{\partial V}{\partial q_2} = 0, \quad \ldots, \quad \frac{\partial V}{\partial q_n} = 0$$

Hence *it is possible to write n independent equations for a system having n degrees of freedom.*

*11.7 Stability of Equilibrium

A

B

Stable equilibrium   Neutral equilibrium

C

Unstable equilibrium

Fig. 11–14

Once the equilibrium configuration for a body or system of connected bodies is defined, it is sometimes important to investigate the "type" of equilibrium or the stability of the configuration. For example, consider the position of a ball resting at a point on each of the three paths shown in Fig. 11–14. Each situation represents an equilibrium state for the ball. When the ball is at *A*, it is said to be in *stable equilibrium* because if it is given a small displacement up the hill, it will always *return* to its original, lowest, position. At *A*, its total potential energy is a *minimum*. When the ball is at *B*, it is in *neutral equilibrium*. A small displacement to either the left or right of *B* will not alter this condition. The ball *remains* in equilibrium in the displaced position, and therefore its potential energy is *constant*. When the ball is at *C*, it is in *unstable equilibrium*. Here a small displacement will cause the ball's potential energy to be *decreased*, and so it will roll farther *away* from its original, highest position. At *C*, the potential energy of the ball is a *maximum*.

Types of Equilibrium. The example just presented illustrates that one of three types of equilibrium positions can be specified for a body or system of connected bodies.

During high winds and when going around a curve, these sugar-cane trucks can become unstable and tip over since their center of gravity is high off the road when they are fully loaded.

1. *Stable equilibrium* occurs when a small displacement of the system causes the system to return to its original position. In this case the original potential energy of the system is a minimum.

2. *Neutral equilibrium* occurs when a small displacement of the system causes the system to remain in its displaced state. In this case the potential energy of the system remains constant.

3. *Unstable equilibrium* occurs when a small displacement of the system causes the system to move farther away from its original position. In this case the original potential energy of the system is a maximum.

System Having One Degree of Freedom. For *equilibrium* of a system having a single degree of freedom, defined by the independent coordinate q, it has been shown that the first derivative of the potential function for the system must be equal to zero; i.e., $dV/dq = 0$. If the potential function $V = V(q)$ is plotted, Fig. 11–15, the first derivative (equilibrium position) is represented as the slope dV/dq, which is zero when the function is maximum, minimum, or an inflection point.

If the *stability* of a body is to be investigated, it is necessary to determine the *second derivative* of V and evaluate it at the equilibrium position $q = q_{eq}$. As shown in Fig. 11–15a, if $V = V(q)$ is a *minimum*, then

$$\frac{dV}{dq} = 0, \quad \frac{d^2V}{dq^2} > 0 \quad \text{stable equilibrium} \quad (11\text{–}10)$$

If $V = V(q)$ is a *maximum*, Fig. 11–15b, then

$$\frac{dV}{dq} = 0, \quad \frac{d^2V}{dq^2} < 0 \quad \text{unstable equilibrium} \quad (11\text{–}11)$$

If the second derivative is zero, it will be necessary to investigate *higher-order* derivatives to determine the stability. In particular, stable equilibrium will occur if the order of the lowest remaining nonzero derivative is *even* and the sign of this nonzero derivative is positive when it is evaluated at $q = q_{eq}$; otherwise, it is unstable.

If the system is in neutral equilibrium, Fig. 11–15c, it is required that

$$\frac{dV}{dq} = \frac{d^2V}{dq^2} = \frac{d^3V}{dq^3} = \dots = 0 \quad \text{neutral equilibrium} \quad (11\text{–}12)$$

since then V must be constant at and around the "neighborhood" of q_{eq}.

System Having Two Degrees of Freedom. A criterion for investigating stability becomes increasingly complex as the number of degrees of freedom for the system increases. For a system having two degrees of freedom, defined by independent coordinates (q_1, q_2), it may be verified (using the calculus of functions of two variables) that equilibrium and stability occur at a point (q_{1eq}, q_{2eq}) when

$$\frac{\partial V}{\partial q_1} = \frac{\partial V}{\partial q_2} = 0$$

$$\left[\left(\frac{\partial^2 V}{\partial q_1 \, \partial q_2} \right)^2 - \left(\frac{\partial^2 V}{\partial q_1^2} \right) \left(\frac{\partial^2 V}{\partial q_2^2} \right) \right] < 0$$

$$\frac{\partial^2 V}{\partial q_1^2} > 0 \quad \text{or} \quad \frac{\partial^2 V}{\partial q_2^2} > 0$$

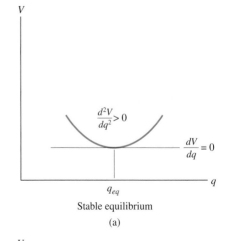

V

$\dfrac{d^2V}{dq^2} > 0$

$\dfrac{dV}{dq} = 0$

q

q_{eq}

Stable equilibrium

(a)

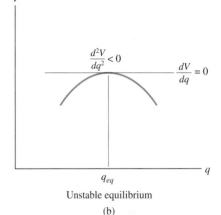

V

$\dfrac{d^2V}{dq^2} < 0$

$\dfrac{dV}{dq} = 0$

q

q_{eq}

Unstable equilibrium

(b)

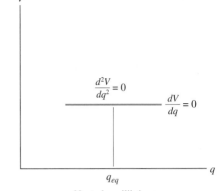

V

$\dfrac{d^2V}{dq^2} = 0$

$\dfrac{dV}{dq} = 0$

q

q_{eq}

Neutral equilibrium

(c)

Fig. 11–15

Both equilibrium and instability occur when

$$\frac{\partial V}{\partial q_1} = \frac{\partial V}{\partial q_2} = 0$$

$$\left[\left(\frac{\partial^2 V}{\partial q_1 \, \partial q_2} \right)^2 - \left(\frac{\partial^2 V}{\partial q_1^2} \right)\left(\frac{\partial^2 V}{\partial q_2^2} \right) \right] < 0$$

$$\frac{\partial^2 V}{\partial q_1^2} < 0 \quad \text{or} \quad \frac{\partial^2 V}{\partial q_2^2} < 0$$

PROCEDURE FOR ANALYSIS

Using potential-energy methods, the equilibrium positions and the stability of a body or a system of connected bodies having a single degree of freedom can be obtained by applying the following procedure.

Potential Function.

- Sketch the system so that it is located at some *arbitrary position* specified by the independent coordinate q.
- Establish a horizontal *datum* through a *fixed point*[*] and express the *gravitational potential energy* V_g in terms of the weight W of each member and its vertical distance y from the datum, $V_g = Wy$.
- Express the elastic potential energy V_e of the system in terms of the stretch or compression, s, of any connecting spring and the spring's stiffness k, $V_e = \frac{1}{2}ks^2$.
- Formulate the potential function $V = V_g + V_e$ and express the *position coordinates* y and s in terms of the independent coordinate q.

Equilibrium Position.

- The equilibrium position is determined by taking the first derivative of V and setting it equal to zero, $\delta V = 0$.

Stability.

- Stability at the equilibrium position is determined by evaluating the second or higher-order derivatives of V.
- If the second derivative is greater than zero, the body is stable, if all derivatives are equal to zero the body is in neutral equilibrium, and if the second derivative is less than zero, the body is unstable.

[*]The location of the datum is *arbitrary* since only the *changes* or differentials of V are required for investigation of the equilibrium position and its stability.

EXAMPLE 11.5

The uniform link shown in Fig. 11–16a has a mass of 10 kg. The spring is unstretched when $\theta = 0°$. Determine the angle θ for equilibrium and investigate the stability at the equilibrium position.

Solution

Potential Function. The datum is established at the top of the link when the *spring is unstretched*, Fig. 11–16b. When the link is located at the arbitrary position θ, the spring increases its potential energy by stretching and the weight decreases its potential energy. Hence,

$$V = V_e + V_g = \frac{1}{2}ks^2 - W\left(s + \frac{l}{2}\cos\theta\right)$$

Since $l = s + l\cos\theta$ or $s = l(1 - \cos\theta)$, then

$$V = \frac{1}{2}kl^2(1 - \cos\theta)^2 - \frac{Wl}{2}(2 - \cos\theta)$$

Equilibrium Position. The first derivative of V gives

$$\frac{dV}{d\theta} = kl^2(1 - \cos\theta)\sin\theta - \frac{Wl}{2}\sin\theta = 0$$

or

$$l\left[kl(1 - \cos\theta) - \frac{W}{2}\right]\sin\theta = 0$$

This equation is satisfied provided

$$\sin\theta = 0 \qquad \theta = 0° \qquad \textit{Ans.}$$

$$\theta = \cos^{-1}\left(1 - \frac{W}{2kl}\right) = \cos^{-1}\left[1 - \frac{10(9.81)}{2(200)(0.6)}\right] = 53.8° \qquad \textit{Ans.}$$

Stability. Determining the second derivative of V gives

$$\frac{d^2V}{d\theta^2} = kl^2(1 - \cos\theta)\cos\theta + kl^2\sin\theta\sin\theta - \frac{Wl}{2}\cos\theta$$

$$= kl^2(\cos\theta - \cos2\theta) - \frac{Wl}{2}\cos\theta$$

Substituting values for the constants, with $\theta = 0°$ and $\theta = 53.8°$, yields

$$\left.\frac{d^2V}{d\theta^2}\right|_{\theta=0°} = 200(0.6)^2(\cos0° - \cos0°) - \frac{10(9.81)(0.6)}{2}\cos0°$$

$$= -29.4 < 0 \qquad \text{(unstable equilibrium at } \theta = 0°) \qquad \textit{Ans.}$$

$$\left.\frac{d^2V}{d\theta^2}\right|_{\theta=53.8°} = 200(0.6)^2(\cos53.8° - \cos107.6°) - \frac{10(9.81)(0.6)}{2}\cos53.8°$$

$$= 46.9 > 0 \qquad \text{(stable equilibrium at } \theta = 53.8°) \qquad \textit{Ans.}$$

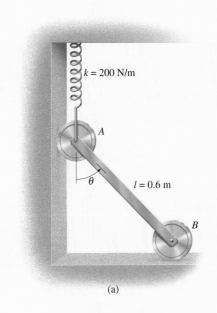

$k = 200$ N/m

A

$l = 0.6$ m

B

(a)

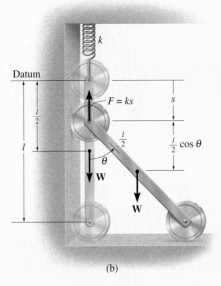

k

Datum

$F = ks$

s

$\frac{l}{2}$

l

$\frac{l}{2}$

$\frac{l}{2}\cos\theta$

θ

W

W

(b)

Fig. 11–16

E X A M P L E **11.6**

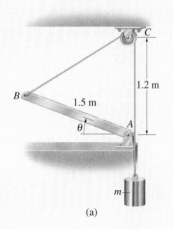

(a)

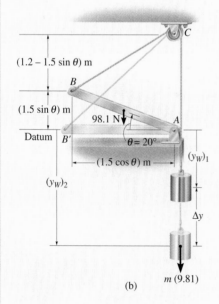

(b)

Fig. 11–17

Determine the mass m of the block required for equilibrium of the uniform 10-kg rod shown in Fig. 11–17a when $\theta = 20°$. Investigate the stability at the equilibrium position.

Solution

Potential Function. The datum is established through point A, Fig. 11–17b. When $\theta = 0°$, the block is assumed to be suspended $(y_W)_1$ below the datum. Hence, in the position θ,

$$V = V_e + V_g = 98.1\left(\frac{1.5 \sin \theta}{2}\right) - m(9.81)(\Delta y) \qquad (1)$$

The distance $\Delta y = (y_W)_2 - (y_W)_1$ may be related to the independent coordinate θ by measuring the difference in cord lengths $B'C$ and BC. Since

$$B'C = \sqrt{(1.5)^2 + (1.2)^2} = 1.92$$
$$BC = \sqrt{(1.5 \cos \theta)^2 + (1.2 - 1.5 \sin \theta)^2} = \sqrt{3.69 - 3.60 \sin \theta}$$

then

$$\Delta y = B'C - BC = 1.92 - \sqrt{3.69 - 3.60 \sin \theta}$$

Substituting the above result into Eq. 1 yields

$$V = 98.1\left(\frac{1.5 \sin \theta}{2}\right) - m(9.81)(1.92 - \sqrt{3.69 - 3.60 \sin \theta}) \qquad (2)$$

Equilibrium Position.

$$\frac{dV}{d\theta} = 73.6 \cos \theta - \left[\frac{m(9.81)}{2}\right]\left(\frac{3.60 \cos \theta}{\sqrt{3.69 - 3.60 \sin \theta}}\right) = 0$$

$$\left.\frac{dV}{d\theta}\right|_{\theta=20°} = 69.14 - 10.58m = 0$$

$$m = \frac{69.14}{10.58} = 6.53 \text{ kg} \qquad\qquad Ans.$$

Stability. Taking the second derivative of Eq. 2, we obtain

$$\frac{d^2V}{d\theta^2} = -73.6 \sin \theta - \left[\frac{m(9.81)}{2}\right]\left(\frac{-1}{2}\right)\frac{-(3.60 \cos \theta)^2}{(3.69 - 3.60 \sin \theta)^{3/2}}$$

$$- \left[\frac{m(9.81)}{2}\right]\left(\frac{-3.60 \sin \theta}{\sqrt{3.69 - 3.60 \sin \theta}}\right)$$

For the equilibrium position $\theta = 20°$, with $m = 6.53$ kg, then

$$\frac{d^2V}{d\theta^2} = -47.6 < 0 \qquad \text{(unstable equilibrium at } \theta = 20°) \quad Ans.$$

EXAMPLE 11.7

The homogeneous block having a mass m rests on the top surface of the cylinder, Fig. 11–18a. Show that this is a condition of unstable equilibrium if $h > 2R$.

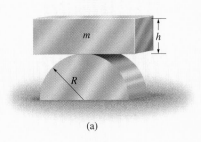

(a)

Solution

Potential Function. The datum is established at the base of the cylinder, Fig. 11–18b. If the block is displaced by an amount θ from the equilibrium position, the potential function may be written in the form

$$V = V_e + V_g$$
$$= 0 + mgy$$

From Fig. 11–18b,

$$y = \left(R + \frac{h}{2}\right)\cos\theta + R\theta\sin\theta$$

Thus,

$$V = mg\left[\left(R + \frac{h}{2}\right)\cos\theta + R\theta\sin\theta\right]$$

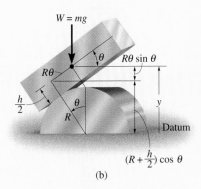

(b)

Fig. 11–18

Equilibrium Position.

$$\frac{dV}{d\theta} = mg\left[-\left(R + \frac{h}{2}\right)\sin\theta + R\sin\theta + R\theta\cos\theta\right] = 0$$

$$= mg\left(-\frac{h}{2}\sin\theta + R\theta\cos\theta\right) = 0$$

Obviously, $\theta = 0°$ is the equilibrium position that satisfies this equation.

Stability. Taking the second derivative of V yields

$$\frac{d^2V}{d\theta^2} = mg\left(-\frac{h}{2}\cos\theta + R\cos\theta - R\theta\sin\theta\right)$$

At $\theta = 0°$,

$$\left.\frac{d^2V}{d\theta^2}\right|_{\theta=0°} = -mg\left(\frac{h}{2} - R\right)$$

Since all the constants are positive, the block is in unstable equilibrium if $h > 2R$, for then $d^2V/d\theta^2 < 0$.

CHAPTER REVIEW

- *Principle of Virtual Work.* The forces on a body will do *virtual work* when the body undergoes an *imaginary* differential displacement or rotation. For equilibrium, the sum of the virtual work done by all the forces acting on the body must be equal to zero for any virtual displacement. This is referred to as the *principle of virtual work*, and it is useful for finding the equilibrium configuration for a mechanism or a reactive force acting on a series of connected members. If this system has one degree of freedom, then its position can be specified by one independent coordinate q. To apply the principle of virtual work, it is first necessary to use *position coordinates* to locate all the forces and moments on the mechanism that will do work when the mechanism undergoes a virtual movement δq. The coordinates are related to the independent coordinate q and then these expressions are differentiated in order to relate the *virtual* coordinate displacements to δq. Finally, the equation of virtual work is written for the mechanism in terms of the common displacement δq, and then it is set equal to zero. By factoring δq out of the equation, it is then possible to determine either the unknown force or couple moment, or the equilibrium position q.

- *Potential Energy Criterion for Equilibrium.* When a system is subjected only to conservative forces, such as weight or spring forces, then the equilibrium configuration can be determined using the *potential energy function V* for the system. This function is established by expressing the weight and spring potential energy for the system in terms of the independent coordinate q. Once it is formulated, its first derivative is set equal to zero, $dV/dq = 0$. The solution yields the equilibrium position q_{eq} for the system. The stability of the system can be investigated by taking the second derivative of V. If this is evaluated at q_{eq} and $d^2V/dq^2 > 0$, then *stable equilibrium* occurs. If $d^2V/dq^2 < 0$, then *unstable*

REVIEW PROBLEMS

11-1. The uniform links AB and BC each weigh 2 lb and the cylinder weighs 20 lb. Determine the horizontal force **P** required to hold the mechanism in the position when $\theta = 45°$. The spring has an unstretched length of 6 in.

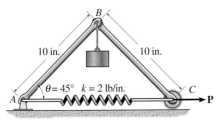

Prob. 11–1

11-2. The spring attached to the mechanism has an unstretched length when $\theta = 90°$. Determine the position θ for equilibrium and investigate the stability of the mechanism at this position. Disk A is pin-connected to the frame at B and has a weight of 20 lb. Neglect the weight of the bars.

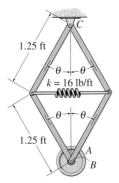

Prob. 11–2

***11-3.** The toggle joint is subjected to the load **P.** Determine the compressive force F it creates on the cylinder at A as a function of θ.

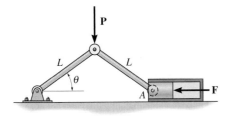

Prob. 11–3

11-4. The uniform beam AB weighs 100 lb. If both springs DE and BC are unstretched when $\theta = 90°$, determine the angle θ for equilibrium using the principle of potential energy. Investigate the stability at the equilibrium position. Both springs always act in the horizontal position because of the roller guides at C and E.

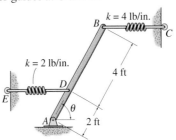

Prob. 11–4

11-5. The uniform bar AB weighs 10 lb. If the attached spring is unstretched when $\theta = 90°$, use the method of virtual work and determine the angle θ for equilibrium. Note that the spring always remains in the vertical position due to the roller guide.

11-6. Solve Prob. 11-5 using the principle of potential energy. Investigate the stability of the bar when it is in the equilibrium position.

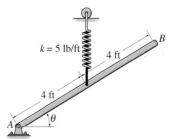

Probs. 11–5/6

***11-7.** The punch press consists of the ram R, connecting rod AB, and a flywheel. If a torque of $M = 50\ \text{N} \cdot \text{m}$ is applied to the flywheel, determine the force **F** applied at the ram to hold the rod in the position $\theta = 60°$.

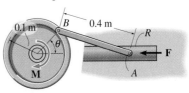

Prob. 11–7

Mathematical Expressions

Quadratic Formula

If $ax^2 + bx + c = 0$, then $x = \dfrac{-b \pm \sqrt{b^2 - 4ac}}{2a}$

Hyperbolic Functions

$\sinh x = \dfrac{e^x - e^{-x}}{2}$, $\cosh x = \dfrac{e^x + e^{-x}}{2}$, $\tanh x = \dfrac{\sinh x}{\cosh x}$

Trigonometric Identities

$\sin \theta = \dfrac{A}{C}$, $\csc \theta = \dfrac{C}{A}$

$\cos \theta = \dfrac{B}{C}$, $\sec \theta = \dfrac{C}{B}$

$\tan \theta = \dfrac{A}{B}$, $\cot \theta = \dfrac{B}{A}$

$\sin^2 \theta + \cos^2 \theta = 1$

$\sin(\theta \pm \phi) = \sin \theta \cos \phi \pm \cos \theta \sin \phi$

$\sin 2\theta = 2 \sin \theta \cos \theta$

$\cos(\theta \pm \phi) = \cos \theta \cos \phi \mp \sin \theta \sin \phi$

$\cos 2\theta = \cos^2 \theta - \sin^2 \theta$

$\cos \theta = \pm\sqrt{\dfrac{1 + \cos 2\theta}{2}}$, $\sin \theta = \pm\sqrt{\dfrac{1 - \cos 2\theta}{2}}$

$\tan \theta = \dfrac{\sin \theta}{\cos \theta}$

$1 + \tan^2 \theta = \sec^2 \theta$ $\qquad$ $1 + \cot^2 \theta = \csc^2 \theta$

Power-Series Expansions

$\sin x = x - \dfrac{x^3}{3!} + \cdots$, $\quad \cos x = 1 - \dfrac{x^2}{2!} + \cdots$

$\sinh x = x + \dfrac{x^3}{3!} + \cdots$, $\quad \cosh x = 1 + \dfrac{x^2}{2!} + \cdots$

Derivatives

$\dfrac{d}{dx}(u^n) = nu^{n-1}\dfrac{du}{dx}$ $\qquad$ $\dfrac{d}{dx}(\sin u) = \cos u \dfrac{du}{dx}$

$\dfrac{d}{dx}(uv) = u\dfrac{dv}{dx} + v\dfrac{du}{dx}$ $\qquad$ $\dfrac{d}{dx}(\cos u) = -\sin u\dfrac{du}{dx}$

$\dfrac{d}{dx}\left(\dfrac{u}{v}\right) = \dfrac{v\dfrac{du}{dx} - u\dfrac{dv}{dx}}{v^2}$ $\qquad$ $\dfrac{d}{dx}(\tan u) = \sec^2 u\dfrac{du}{dx}$

$\dfrac{d}{dx}(\cot u) = -\csc^2 u\dfrac{du}{dx}$ $\qquad$ $\dfrac{d}{dx}(\sinh u) = \cosh u\dfrac{du}{dx}$

$\dfrac{d}{dx}(\sec u) = \tan u \sec u\dfrac{du}{dx}$ $\dfrac{d}{dx}(\cosh u) = \sinh u\dfrac{du}{dx}$

$\dfrac{d}{dx}(\csc u) = -\csc u \cot u\dfrac{du}{dx}$

Integrals

$$\int x^n \, dx = \frac{x^{n+1}}{n+1} + C, \, n \neq -1$$

$$\int \frac{dx}{a+bx} = \frac{1}{b} \ln(a+bx) + C$$

$$\int \frac{dx}{a+bx^2} = \frac{1}{2\sqrt{-ba}} \ln\left[\frac{a+x\sqrt{-ab}}{a-x\sqrt{-ab}}\right] + C, \quad ab < 0$$

$$\int \frac{x \, dx}{a+bx^2} = \frac{1}{2b} \ln(bx^2+a) + C$$

$$\int \frac{x^2 \, dx}{a+bx^2} = \frac{x}{b} - \frac{a}{b\sqrt{ab}} \tan^{-1}\frac{x\sqrt{ab}}{a} + C, \, ab > 0$$

$$\int \sqrt{a+bx} \, dx = \frac{2}{3b} \sqrt{(a+bx)^3} + C$$

$$\int x\sqrt{a+bx} \, dx = \frac{-2(2a-3bx)\sqrt{(a+bx)^3}}{15b^2} + C$$

$$\int x^2\sqrt{a+bx} \, dx =$$

$$\frac{2(8a^2 - 12abx + 15b^2x^2)\sqrt{(a+bx)^3}}{105b^3} + C$$

$$\int \sqrt{a^2-x^2} \, dx = \frac{1}{2}\left[x\sqrt{a^2-x^2} + a^2 \sin^{-1}\frac{x}{a}\right] + C, \quad a > 0$$

$$\int x\sqrt{a^2-x^2} \, dx = -\frac{1}{3}\sqrt{(a^2-x^2)^3} + C$$

$$\int x^2\sqrt{a^2-x^2} \, dx = -\frac{x}{4}\sqrt{(a^2-x^2)^3}$$

$$+ \frac{a^2}{8}\left(x\sqrt{a^2-x^2} + a^2 \sin^{-1}\frac{x}{a}\right) + C, \, a > 0$$

$$\int \sqrt{x^2 \pm a^2} \, dx =$$

$$\frac{1}{2}\left[x\sqrt{x^2 \pm a^2} \pm a^2 \ln(x + \sqrt{x^2 \pm a^2})\right] + C$$

$$\int x\sqrt{x^2 \pm a^2} \, dx = \frac{1}{3}\sqrt{(x^2 \pm a^2)^3} + C$$

$$\int x^2\sqrt{x^2 \pm a^2} \, dx = \frac{x}{4}\sqrt{(x^2 \pm a^2)^3}$$

$$\mp \frac{a^2}{8} x \sqrt{x^2 \pm a^2} - \frac{a^4}{8} \ln(x + \sqrt{x^2 \pm a^2}) + C$$

$$\int \frac{dx}{\sqrt{a+bx}} = \frac{2\sqrt{a+bx}}{b} + C$$

$$\int \frac{x \, dx}{\sqrt{x^2 \pm a^2}} = \sqrt{x^2 \pm a^2} + C$$

$$\int \frac{dx}{\sqrt{a+bx+cx^2}} = \frac{1}{\sqrt{c}} \ln\left[\sqrt{a+bx+cx^2} + \right.$$

$$\left. x\sqrt{c} + \frac{b}{2\sqrt{c}}\right] + C, \, c > 0$$

$$= \frac{1}{\sqrt{-c}} \sin^{-1}\left(\frac{-2cx-b}{\sqrt{b^2-4ac}}\right) + C, \, c < 0$$

$$\int \sin x \, dx = -\cos x + C$$

$$\int \cos x \, dx = \sin x + C$$

$$\int x \cos(ax) \, dx = \frac{1}{a^2} \cos(ax) + \frac{x}{a} \sin(ax) + C$$

$$\int x^2 \cos(ax) \, dx = \frac{2x}{a^2} \cos(ax) + \frac{a^2x^2-2}{a^3} \sin(ax) + C$$

$$\int e^{ax} \, dx = \frac{1}{a} e^{ax} + C$$

$$\int x e^{ax} \, dx = \frac{e^{ax}}{a^2}(ax-1) + C$$

$$\int \sinh x \, dx = \cosh x + C$$

$$\int \cosh x \, dx = \sinh x + C$$

Numerical and Computer Analysis

Occasionally the application of the laws of mechanics will lead to a system of equations for which a closed-form solution is difficult or impossible to obtain. When confronted with this situation, engineers will often use a numerical method which in most cases can be programmed on a microcomputer or "programmable" pocket calculator. Here we will briefly present a computer program for solving a set of linear algebraic equations and three numerical methods which can be used to solve an algebraic or transcendental equation, evaluate a definite integral, and solve an ordinary differential equation. Application of each method will be explained by example, and an associated computer program written in Microsoft BASIC, which is designed to run on most personal computers, is provided.* A text on numerical analysis should be consulted for further discussion regarding a check of the accuracy of each method and the inherent errors that can develop from the methods.

B.1 Linear Algebraic Equations

Application of the equations of static equilibrium or the equations of motion sometimes requires solving a set of linear algebraic equations. The computer program listed in Fig. B–1 can be used for this purpose. It is based on the method of a Gaussian elimination and can solve at

---

*Similar types of programs can be written or purchased for programmable pocket calculators.

```
 1 PRINT"Linear system of equations":PRINT        20 PRINT"Unknowns"                  39 NEXT I
 2 DIM A(10,11)                                    21 FOR I = 1 TO N                   40 FOR I = M+1 TO N
 3 INPUT"Input number of equations : ",N           22 PRINT "X(";I;")=";A(I,N+1)       41 FC=A(I,M)/A(M,M)
 4 PRINT                                            23 NEXT I                           42 FOR J = M+1 TO N+1
 5 PRINT"A  coefficients"                           24 END                              43 A(I,J)=A(I,J)-FC*A(M,J)
 6 FOR I = 1 TO N                                   25 REM Subroutine Guassian          44 NEXT J
 7 FOR J = 1 TO N                                   26 FOR M=1 TO N                      45 NEXT I
 8 PRINT "A(";I;",";J;                              27 NP=M                             46 NEXT M
 9 INPUT")=",A(I,J)                                 28 BG=ABS(A(M,M))                   47 A(N,N+1)=A(N,N+1)/A(N,N)
10 NEXT J                                           29 FOR I = M TO N                   48 FOR I = N-1 TO 1 STEP -1
11 NEXT I                                           30 IF ABS(A(I,M))<=BG THEN 33       49 SM=0
12 PRINT                                            31 BG=ABS(A(I,M))                   50 FOR J=I+1 TO N
13 PRINT"B  coefficients"                           32 NP=I                             51 SM=SM+A(I,J)*A(J,N+1)
14 FOR I = 1 TO N                                   33 NEXT I                           52 NEXT J
15 PRINT "B(";I;                                    34 IF NP=M THEN 40                  53 A(I,N+1)=(A(I,N+1)-SM)/A(I,I)
16 INPUT")=",A(I,N+1)                               35 FOR I = M TO N+1                 54 NEXT I
17 NEXT I                                           36 TE=A(M,I)                        55 RETURN
18 GOSUB 25                                         37 A(M,I)=A(NP,I)
19 PRINT                                            38 A(NP,I)=TE
```

Fig. B–1

most 10 equations with 10 unknowns. To do so, the equations should first be written in the following general format:

$$A_{11}x_1 + A_{12}x_2 + \cdots + A_{1n}x_n = B_1$$
$$A_{21}x_1 + A_{22}x_2 + \cdots + A_{2n}x_n = B_2$$
$$\vdots$$
$$A_{n1}x_1 + A_{n2}x_2 + \cdots + A_{nn}x_n = B_n$$

The "A" and "B" coefficients are "called" for when running the program. The output presents the unknowns $x_1, \ldots, x_n$.

E X A M P L E   **B.1**

Solve the two equations

$$3x_1 + x_2 = 4$$
$$2x_1 - x_2 = 10$$

Solution

When the program begins to run, it first calls for the number of equations (2); then the A coefficients in the sequence $A_{11} = 3$, $A_{12} = 1$, $A_{21} = 2$, $A_{22} = -1$; and finally the B coefficients $B_1 = 4$, $B_2 = 10$. The output appears as

Unknowns

$$X(1) = 2.8 \qquad \text{Ans.}$$
$$X(2) = -4.4 \qquad \text{Ans.}$$

B.2 Simpson's Rule

Simpson's rule is a numerical method that can be used to determine the area under a curve given as a graph or as an explicit function $y = f(x)$. Likewise, it can be used to compute the value of a definite integral which involves the function $y = f(x)$. To do so, the area must be subdivided into an *even number* of strips or intervals having a width h. The curve between three consecutive ordinates is approximated by a parabola, and the entire area or definite integral is then determined from the formula

$$\int_{x_0}^{x_n} f(x)\, dx \simeq \frac{h}{3}[y_0 + 4(y_1 + y_3 + \cdots + y_{n-1})$$

$$+ 2(y_2 + y_4 + \cdots + y_{n-2}) + y_n] \quad \text{(B–1)}$$

The computer program for this equation is given in Fig. B–2. For its use, we must first specify the function (on line 6 of the program). The upper and lower limits of the integral and the number of intervals are called for when the program is executed. The value of the integral is then given as the output.

```
1 PRINT"Simpson's rule":PRINT
2 PRINT" To execute this program :":PRINT
3 PRINT"   1- Modify right-hand side of the equation given below,
4 PRINT"      then press RETURN key"
5 PRINT"   2- Type  RUN 6":PRINT:EDIT 6
6 DEF FNF(X)=LOG(X)
7 PRINT:INPUT" Enter Lower Limit = ",A
8 INPUT" Enter Upper Limit = ",B
9 INPUT" Enter Number (even) of Intervals = ",N%
10 H=(B-A)/N%:AR=FNF(A):X=A+H
11 FOR J%=2 TO N%
12 K=2*(2-J%+2*INT(J%/2))
13 AR=AR+K*FNF(X)
14 X=X+H:NEXT J%
15 AR=H*(AR+FNF(B))/3
16 PRINT" Integral = ",AR
17 END
```

Fig. B–2

E X A M P L E **B.2**

Evaluate the definite integral

$$\int_2^5 \ln x \, dx$$

Solution

The interval $x_0 = 2$ to $x_6 = 5$ will be divided into six equal parts ($n = 6$), each having a width $h = (5 - 2)/6 = 0.5$. We then compute $y = f(x) = \ln x$ at each point of subdivision.

n	x_n	y_n
0	2	0.693
1	2.5	0.916
2	3	1.099
3	3.5	1.253
4	4	1.386
5	4.5	1.504
6	5	1.609

Thus, Eq. B–1 becomes

$$\int_2^5 \ln x \, dx \simeq \frac{0.5}{3}[0.693 + 4(0.916 + 1.253 + 1.504)$$

$$+ 2(1.099 + 1.386) + 1.609]$$

$$\simeq 3.66 \qquad\qquad\qquad Ans.$$

This answer is equivalent to the exact answer to three significant figures. Obviously, accuracy to a greater number of significant figures can be improved by selecting a smaller interval h (or larger n).

Using the computer program, we first specify the function $\ln x$, line 6 in Fig. B–2. During execution, the program input requires the upper and lower limits 2 and 5, and the number of intervals $n = 6$. The output appears as

$$\text{Integral} = 3.66082 \qquad\qquad Ans.$$

B.3 The Secant Method

The secant method is used to find the real roots of an algebraic or transcendental equation $f(x) = 0$. The method derives its name from the fact that the formula used is established from the slope of the secant line to the graph $y = f(x)$. This slope is $[f(x_n) - f(x_{n-1})]/(x_n - x_{n-1})$, and the secant formula is

$$x_{n+1} = x_n - f(x_n)\left[\frac{x_n - x_{n-1}}{f(x_n) - f(x_{n-1})}\right] \tag{B-2}$$

For application it is necessary to provide two initial guesses, x_0 and x_1, and thereby evaluate x_2 from Eq. B–2 ($n = 1$). One then proceeds to reapply Eq. B–2 with x_1 and the calculated value of x_2 and obtain x_3 ($n = 2$), etc., until the value $x_{n+1} \simeq x_n$. One can see this will occur if x_n is approaching the root of the function $f(x) = 0$, since the correction term on the right of Eq. B–2 will tend toward zero. In particular, the larger the slope, the smaller the correction to x_n, and the faster the root will be found. On the other hand, if the slope is very small in the neighborhood of the root, the method leads to large corrections for x_n, and convergence to the root is slow and may even lead to a failure to find it. In such cases other numerical techniques must be used for solution.

A computer program based on Eq. B–2 is listed in Fig. B–3. We must first specify the function on line 7 of the program. When the program is executed, two initial guesses, x_0 and x_1, must be entered in order to approximate the solution. The output specifies the value of the root. If it cannot be determined, this is so stated.

```
1 PRINT"Secant method":PRINT
2 PRINT" To execute this program :":PRINT
3 PRINT"    1) Modify right hand side of the equation given below,"
4 PRINT"       then press RETURN key."
5 PRINT"    2) Type  RUN 7"
6 PRINT:EDIT 7
7 DEF FNF(X)=.5*SIN(X)-2*COS(X)+1.3
8 INPUT"Enter point #1 =",X
9 INPUT"Enter point #2 =",X1
10 IF X=X1 THEN 14
11 EP=.00001:TL=2E-20
12 FP=(FNF(X1)-FNF(X))/(X1-X)
13 IF ABS(FP)>TL THEN 15
14 PRINT"Root can not be found.":END
15 DX=FNF(X1)/FP
16 IF ABS(DX)>EP THEN 19
17 PRINT "Root = ";X1;"      Function evaluated at this root = ";FNF(X1)
18 END
19 X=X1:X1=X1-DX
20 GOTO 12
```

Fig. B–3

E X A M P L E B.3

Determine the root of the equation

$$f(x) = 0.5 \sin x - 2 \cos x + 1.30 = 0$$

Solution

Guesses of the initial roots will be $x_0 = 45°$ and $x_1 = 30°$. Applying Eq. B–2,

$$x_2 = 30° - (-0.1821)\frac{(30° - 45°)}{(-0.1821 - 0.2393)} = 36.48°$$

Using this value in Eq. B–2, along with $x_1 = 30°$, we have

$$x_3 = 36.48° - (-0.0108)\frac{36.48° - 30°}{(-0.0108 + 0.1821)} = 36.89°$$

Repeating the process with this value and $x_2 = 36.48°$ yields

$$x_4 = 36.89° - (0.0005)\left[\frac{36.89° - 36.48°}{(0.0005 + 0.0108)}\right] = 36.87°$$

Thus $x = 36.9°$ is appropriate to three significant figures.

If the problem is solved using the computer program, first we specify the function, line 7 in Fig. B–3. During execution, the first and second guesses must be entered in radians. Choosing these to be 0.8 rad and 0.5 rad, the result appears as

Root = 0.6435022

Function evaluated at this root = 1.66893E−06

This result converted from radians to degrees is therefore

$$x = 36.9° \qquad\qquad \textit{Ans.}$$

Review for the Fundamentals of Engineering Examination

The Fundamentals of Engineering (FE) exam is given semiannually by the National Council of Engineering Examiners (NCEE) and is one of the requirements for obtaining a Professional Engineering License. A portion of this exam contains problems in statics, and this appendix provides a review of the subject matter most often asked on this exam. Before solving any of the problems, you should review the sections indicated in each chapter in order to become familiar with the boldfaced definitions and the procedures used to solve the various types of problems. Also, review the example problems in these sections.

The following problems are arranged in the same sequence as the topics in each chapter. Besides helping as a preparation for the FE exam, these problems also provide additional examples for general practice of the subject matter. Solutions to *all the problems* are given at the back of this appendix.

Chapter 2—Review All Sections

C-1. Two forces act on the hook. Determine the magnitude of the resultant force.

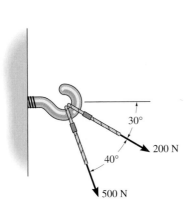

Prob. C–1

C-2. The force $F = 450$ lb acts on the frame. Resolve this force into components acting along members AB and AC, and determine the magnitude of each component.

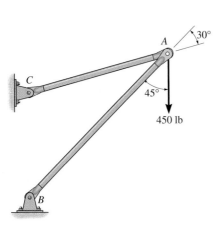

Prob. C–2

C-3. Determine the magnitude and direction of the resultant force.

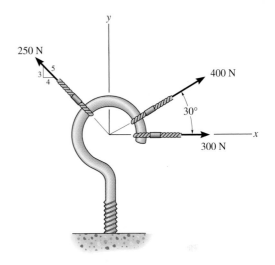

Prob. C–3

C-4. If $\mathbf{F} = \{30\mathbf{i} + 50\mathbf{j} - 45\mathbf{k}\}$ N, determine the magnitude and coordinate direction angles of the force.

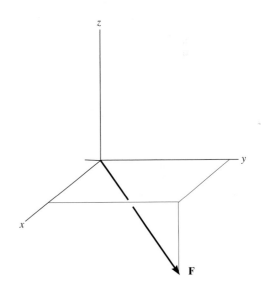

Prob. C–4

C-5. The force has a component of 20 N directed along the $-y$ axis as shown. Represent the force **F** as a Cartesian vector.

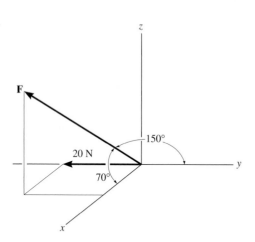

Prob. C–5

C-7. The cables supporting the antenna are subjected to the forces shown. Represent each force as a Cartesian vector.

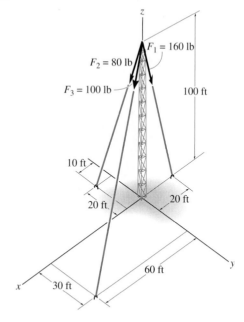

Prob. C–7

C-6. The force acts on the beam as shown. Determine its coordinate direction angles.

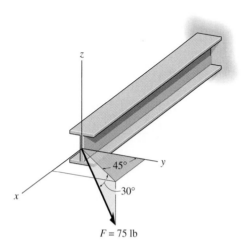

Prob. C–6

C-8. Determine the angle θ between the two cords.

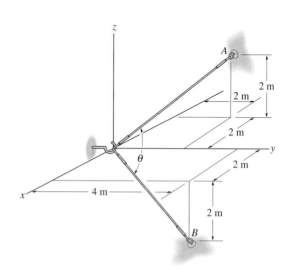

Prob. C–8

C-9. Determine the component the of projection of the force **F** along the pipe *AB*.

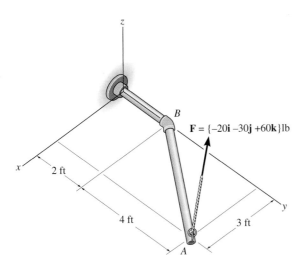

$$\mathbf{F} = \{-20\mathbf{i} - 30\mathbf{j} + 60\mathbf{k}\}\text{lb}$$

Prob. C–9

Chapter 3—Review Sections 3.1–3.3

C-10. The crate at *D* has a weight of 550 lb. Determine the force in each supporting cable.

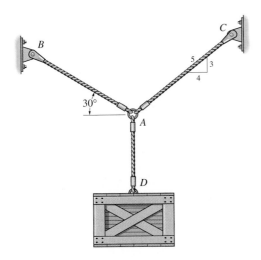

Prob. C–10

C-11. The beam has a weight of 700 lb. Determine the shortest cable *ABC* that can be used to lift it if the maximum force the cable can sustain is 1500 lb.

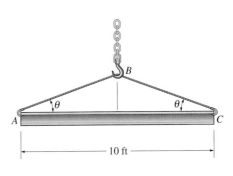

Prob. C–11

C-12. The block has a mass of 5 kg and rests on the smooth plane. Determine the unstretched length of the spring.

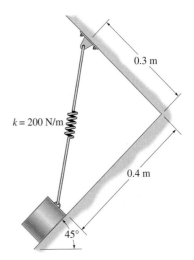

Prob. C–12

C-13. The post can be removed by a vertical force of 400 lb. Determine the force P that must be applied to the cord in order to pull the post out of the ground.

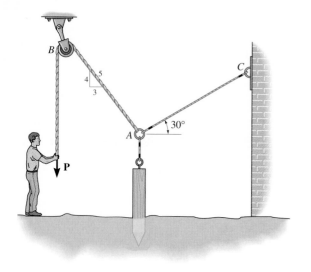

Prob. C–13

C-15. Determine the moment of the force about point O. Neglect the thickness of the member.

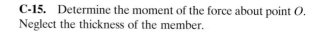

Prob. C–15

Chapter 4—Review All Sections

C-14. Determine the moment of the force about point O.

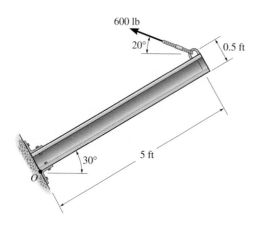

Prob. C–14

C-16. Determine the moment of the force about point O.

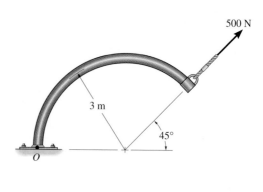

Prob. C–16

C-17. Determine the moment of the force about point A. Express the result as a Cartesian vector.

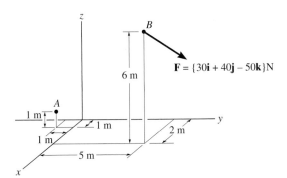

Prob. C–17

C-19. Determine the resultant couple moment acting on the beam.

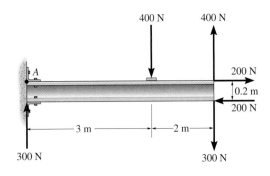

Prob. C–19

C-18. Determine the moment of the force about point A. Express the result as a Cartesian vector.

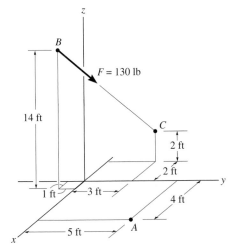

Prob. C–18

C-20. Determine the resultant couple moment acting on the triangular plate.

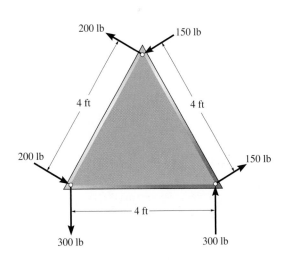

Prob. C–20

C-21. Replace the loading shown by an equivalent resultant force and couple-moment system at point A.

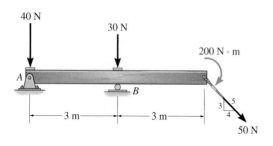

Prob. C–21

C-23. Replace the loading shown by an equivalent single resultant force and specify where the force acts, measured from point O.

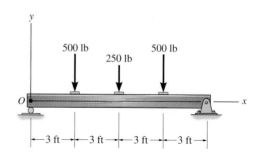

Prob. C–23

C-22. Replace the loading shown by an equivalent resultant force and couple-moment system at point A.

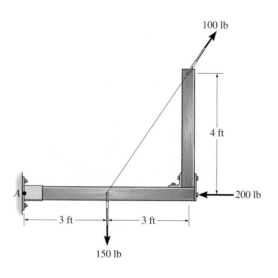

Prob. C–22

C-24. Replace the loading shown by an equivalent single resultant force and specify the x and y coordinates of its line of action.

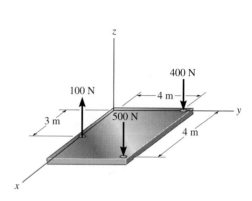

Prob. C–24

C-25. Replace the loading shown by an equivalent single resultant force and specify the x and y coordinates of its line of action.

C-27. Determine the resultant force and specify where it acts on the beam measured from A.

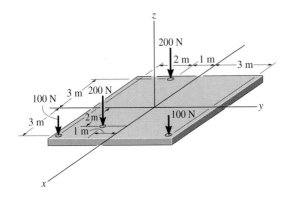

Prob. C–25

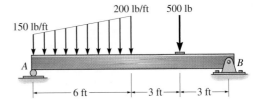

Prob. C–27

C-26. Determine the resultant force and specify where it acts on the beam measured from A.

C-28. Determine the resultant force and specify where it acts on the beam measured from A.

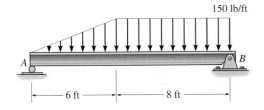

Prob. C–26

Prob. C–28

Chapter 5—Review Sections 5.1–5.6

C-29. Determine the horizontal and vertical components of reaction at the supports. Neglect the thickness of the beam.

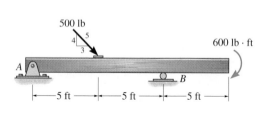

Prob. C–29

C-30. Determine the horizontal and vertical components of reaction at the supports.

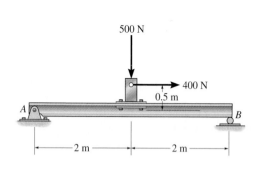

Prob. C–30

C-31. Determine the components of reaction at the fixed support A. Neglect the thickness of the beam.

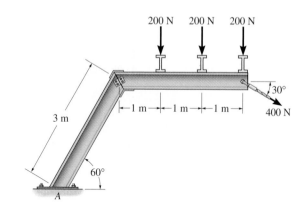

Prob. C–31

C-32. Determine the tension in the cable and the horizontal and vertical components of reaction at the pin A. Neglect the size of the pulley.

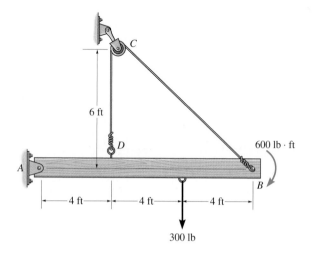

Prob. C–32

C-33. The uniform plate has a weight of 500 lb. Determine the tension in each of the supporting cables.

C-35. Determine the force in members AE and DC. State if the members are in tension or compression.

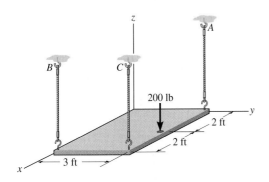

Prob. C–33

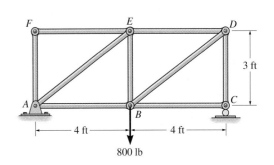

Prob. C–35

Chapter 6—Review Sections 6.1–6.4, 6.6

C-34. Determine the force in each member of the truss. State if the members are in tension or compression.

C-36. Determine the force in members BC, CF, and FE. State if the members are in tension or compression.

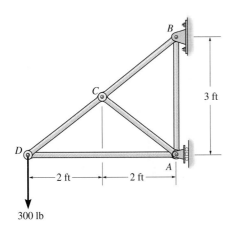

Prob. C–34

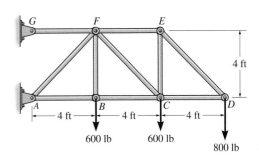

Prob. C–36

C-37. Determine the force in members *GF, FC*, and *CD*. State if the members are in tension or compression.

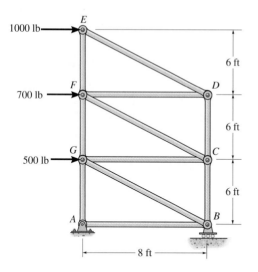

Prob. C–37

C-39. Determine the horizontal and vertical components of reaction at pin *C*.

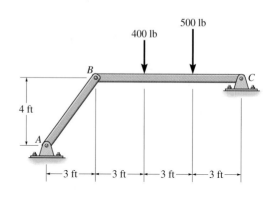

Prob. C–39

C-38. Determine the force *P* needed to hold the 60-lb weight in equilibrium.

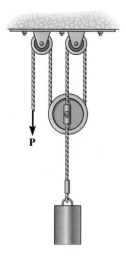

Prob. C–38

C-40. Determine the horizontal and vertical components of reaction at pin *C*.

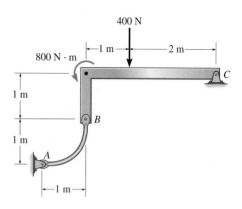

Prob. C–40

C-41. Determine the normal force that the 100-lb plate A exerts on the 30-lb plate B.

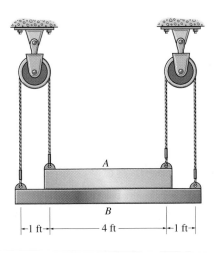

Prob. C–41

C-42. Determine the force P needed to lift the load. Also, determine the proper placement x of the hook for equilibrium. Neglect the weight of the beam.

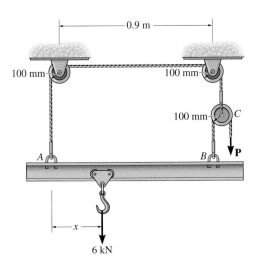

Prob. C–42

Chapter 7—Review Section 7.1

C-43. Determine the internal normal force, shear force, and moment acting in the beam at point B.

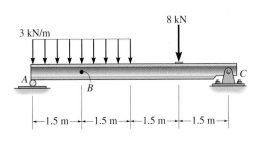

Prob. C–43

C-44. Determine the internal normal force, shear force, and moment acting in the beam at point B, which is located just to the left of the 800-lb force.

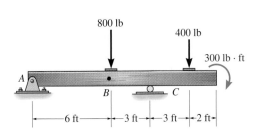

Prob. C–44

C-45. Determine the internal normal force, shear force, and moment acting in the beam at point B.

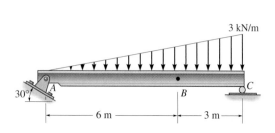

Prob. C–45

C-47. Determine the vertical force P needed to rotate the 200-lb spool. The coefficient of static friction at all contacting surfaces is $\mu_s = 0.4$.

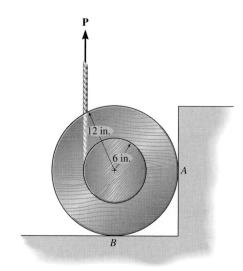

Prob. C–47

Chapter 8—Review Sections 8.1–8.2

C-46. Determine the force P needed to move the 100-lb block. The coefficient of static friction is $\mu_s = 0.3$, and the coefficient of kinetic friction is $\mu_k = 0.25$. Neglect tipping.

C-48. Block A has a weight of 30 lb and block B weighs 50 lb. If the coefficient of static friction is $\mu_s = 0.4$ between all contacting surfaces, determine the frictional force at each surface.

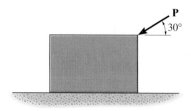

Prob. C–46

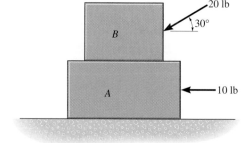

Prob. C–48

C-49. Determine the force P necessary to move the 250-lb crate which has a center of gravity at G. The coefficient of static friction at the floor is $\mu_s = 0.4$.

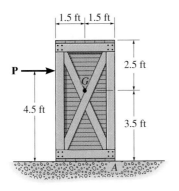

Prob. C–49

C-50. The filing cabinet A has a mass of 60 kg and center of mass at G. It rests on a 10-kg plank. Determine the smallest force P needed to move it. The coefficient of static friction between the cabinet A and the plank B is $\mu_s = 0.4$, and between the plank and the floor $\mu_s = 0.3$.

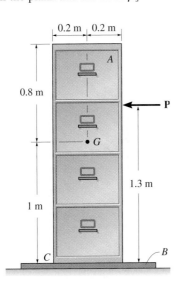

Prob. C–50

Chapter 9—Review Sections 9.1–9.3
(Integration is covered in the mathematics portion of the exam.)

C-51. Determine the location $(\bar{x}, \bar{y})$ of the centroid of the area.

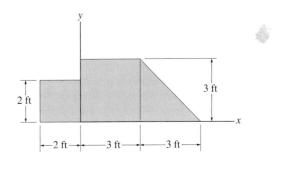

Prob. C–51

C-52. Determine the location $(\bar{x}, \bar{y})$ of the centroid of the area.

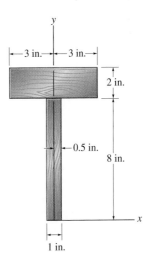

Prob. C–52

Chapter 10—Review Sections 10.1–10.5
(Integration is covered in the mathematics portion of the exam.)
C-53. Determine the moment of inertia of the cross-sectional area of the channel with respect to the y axis.

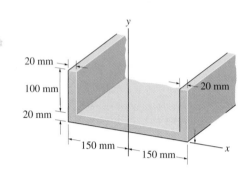

Prob. C–53

C-54. Determine the moment of inertia of the area with respect to the x axis.

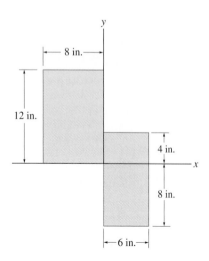

Prob. C–54

C-55. Determine the moment of inertia of the cross-sectional area of the T-beam with respect to the x' axis passing through the centroid of the cross section.

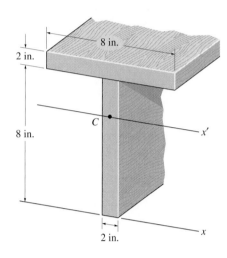

Prob. C–55

Partial Solutions and Answers

C-1. $F_R = \sqrt{200^2 + 500^2 - 2(200)(500)\cos 140°}$
$= 666 \text{ N}$ *Ans.*

C-2. $\dfrac{F_{AB}}{\sin 105°} = \dfrac{450}{\sin 30°}$
$= 869 \text{ lb}$ *Ans.*

$\dfrac{F_{AC}}{\sin 45°} = \dfrac{450}{\sin 30°}$
$F_{AC} = 636 \text{ lb}$ *Ans.*

C-3. $F_{Rx} = 300 + 400\cos 30° - 250\left(\dfrac{4}{5}\right) = 446.4 \text{ N}$

$F_{Ry} = 400\sin 30° + 250\left(\dfrac{3}{5}\right) = 350 \text{ N}$

$F_R = \sqrt{(446.4)^2 + 350^2} = 567 \text{ N}$ *Ans.*
$\theta = \tan^{-1}\dfrac{350}{446.4} = 38.1°$ ⦨ *Ans.*

C-4. $F = \sqrt{30^2 + 50^2 + (-45)^2} = 73.7 \text{ N}$ *Ans.*
$\alpha = \cos^{-1}\left(\dfrac{30}{73.7}\right) = 66.0°$ *Ans.*

$\beta = \cos^{-1}\left(\dfrac{50}{73.7}\right) = 47.2°$ *Ans.*

$\gamma = \cos^{-1}\left(\dfrac{-45}{73.7}\right) = 128°$ *Ans.*

C-5. $F_y = -20$
$\dfrac{F_y}{|F|} = \cos\beta$
$|F| = \left|\dfrac{-20}{\cos 150°}\right| = 23.09 \text{ N}$

$\cos\gamma = \sqrt{1 - \cos^2 70° - \cos^2 150°}$
$\gamma = 68.61° \text{ (From Fig. } \gamma < 90°)$
$\mathbf{F} = 23.09\cos 70°\mathbf{i} + 23.09\cos 150°\mathbf{j}$
$\qquad + 23.09\cos 68.61°\mathbf{k}$
$= \{7.90\mathbf{i} - 20\mathbf{j} + 8.42\mathbf{k}\} \text{ N}$ *Ans.*

C-6. $F_x = 75\cos 30°\sin 45° = 45.93$
$F_y = 75\cos 30°\cos 45° = 45.93$
$F_z = -75\sin 30° = -37.5$
$\alpha = \cos^{-1}\left(\dfrac{45.93}{75}\right) = 52.2°$ *Ans.*

$\beta = \cos^{-1}\left(\dfrac{45.93}{75}\right) = 52.2°$ *Ans.*

$\gamma = \cos^{-1}\left(\dfrac{-37.5}{75}\right) = 120°$ *Ans.*

C-7. $\mathbf{F}_1 = 160 \text{ lb}\left(-\dfrac{20}{102.0}\mathbf{i} - \dfrac{100}{102.0}\mathbf{k}\right)$
$= \{-31.4\mathbf{i} - 157\mathbf{k}\} \text{ lb}$ *Ans.*

$\mathbf{F}_2 = 80 \text{ lb}\left(\dfrac{10}{102.5}\mathbf{i} - \dfrac{20}{102.5}\mathbf{j} - \dfrac{100}{102.5}\mathbf{k}\right)$
$= \{7.81\mathbf{i} - 15.6\mathbf{j} - 78.1\mathbf{k}\} \text{ lb}$ *Ans.*

$\mathbf{F}_3 = 100 \text{ lb}\left(\dfrac{60}{120.4}\mathbf{i} + \dfrac{30}{120.4}\mathbf{j} - \dfrac{100}{120.4}\mathbf{k}\right)$
$= \{49.8\mathbf{i} + 24.9\mathbf{j} - 83.0\mathbf{k}\} \text{ lb}$ *Ans.*

C-8. $\mathbf{r}_{OA} = \{-2\mathbf{i} + 2\mathbf{j} + 2\mathbf{k}\} \text{ m}$
$\mathbf{r}_{OB} = \{2\mathbf{i} + 4\mathbf{j} - 2\mathbf{k}\} \text{ m}$

$\cos\theta = \dfrac{\mathbf{r}_{OA}\cdot\mathbf{r}_{OB}}{|r_{OA}||r_{OB}|}$

$\dfrac{(-2\mathbf{i} + 2\mathbf{j} + 2\mathbf{k})\cdot(2\mathbf{i} + 4\mathbf{j} - 2\mathbf{k})}{\sqrt{12}\sqrt{24}} = 0$

$\theta = 90°$ *Ans.*

C-9. $|F_{AB}| = \mathbf{F}\cdot\mathbf{u}_{AB}$
$= (-20\mathbf{i} - 30\mathbf{j} + 60\mathbf{k})\cdot\left(-\dfrac{3}{5}\mathbf{i} - \dfrac{4}{5}\mathbf{j}\right) = 36 \text{ lb}$ *Ans.*

C-10. $\xrightarrow{+}\Sigma F_x = 0; \dfrac{4}{5}F_{AC} - F_{AB}\cos 30° = 0$

$+\uparrow\Sigma F_y = 0; \dfrac{3}{5}F_{AC} + F_{AB}\sin 30° - 550 = 0$

$F_{AB} = 478 \text{ lb}$ *Ans.*, $F_{AC} = 518 \text{ lb}$ *Ans.*

C-11. $+\uparrow\Sigma F_y = 0; -2(1500)\sin\theta + 700 = 0$
$\theta = 13.5°$
$L_{ABC} = 2\left(\dfrac{5 \text{ ft}}{\cos 13.5°}\right) = 10.3 \text{ ft}$

C-12. $+\nearrow\Sigma F_x = 0; \dfrac{4}{5}(F_{sp}) - 5(9.81)\sin 45° = 0$

$F_{sp} = 43.35 \text{ N}$
$F_{sp} = k(l - l_0); 43.35 = 200(0.5 - l_0)$
$l_0 = 0.283 \text{ m}$ *Ans.*

C-13. At A:
$\xleftarrow{+}\Sigma F_x = 0; \dfrac{3}{5}P - T_{AC}\cos 30° = 0$

$+\uparrow\Sigma F_y = 0; \dfrac{4}{5}P + T_{AC}\sin 30° - 400 = 0$

$P = 349 \text{ lb}$ *Ans.*, $T_{AC} = 242 \text{ lb}$ *Ans.*

C-14. $\downarrow+M_O = 600\sin 50° (5) + 600\cos 50° (0.5)$
$= 2.49 \text{ kip}\cdot\text{ft}$ *Ans.*

C-15. $\mathord{\curvearrowright} + M_O = 50 \sin 60° (0.1 + 0.2 \cos 45° + 0.1)$
$- 50 \cos 60° (0.2 \sin 45°)$
$= 11.2 \text{ N} \cdot \text{m}$ *Ans.*

C-16. $\mathord{\curvearrowleft} + M_O = 500 \sin 45° (3 + 3 \cos 45°)$
$- 500 \cos 45° (3 \sin 45°)$
$= 1.06 \text{ kN} \cdot \text{m}$ *Ans.*

C-17. $\mathbf{M}_A = \mathbf{r}_{AB} \times \mathbf{F} = \begin{vmatrix} \mathbf{i} & \mathbf{j} & \mathbf{k} \\ 1 & 6 & 5 \\ 30 & 40 & -50 \end{vmatrix}$
$= \{-500\mathbf{i} + 200\mathbf{j} - 140\mathbf{k}\} \text{ N} \cdot \text{m}$ *Ans.*

C-18. $\mathbf{F} = 130 \text{ lb} \left(-\dfrac{3}{13}\mathbf{i} + \dfrac{4}{13}\mathbf{j} - \dfrac{12}{13}\mathbf{k} \right)$
$= \{-30\mathbf{i} + 40\mathbf{j} - 120\mathbf{k}\} \text{ lb}$

$\mathbf{M}_A = \mathbf{r}_{AB} \times \mathbf{F} = \begin{vmatrix} \mathbf{i} & \mathbf{j} & \mathbf{k} \\ -3 & -6 & 14 \\ -30 & 40 & -120 \end{vmatrix}$
$= \{160\mathbf{i} - 780\mathbf{j} - 300\mathbf{k}\} \text{ lb} \cdot \text{ft}$ *Ans.*

C-19. $\mathord{\curvearrowright} + M_{C_R} = \Sigma M_A = 400(3) - 400(5) + 300(5)$
$+ 200(0.2) = 740 \text{ N} \cdot \text{m}$ *Ans.*

Also,
$\mathord{\curvearrowright} + M_{C_R} = 300(5) - 400(2) + 200(0.2)$
$= 740 \text{ N} \cdot \text{m}$ *Ans.*

C-20. $\mathord{\curvearrowleft} + M_{C_R} = 300(4) + 200(4) + 150(4)$
$= 2600 \text{ lb} \cdot \text{ft}$ *Ans.*

C-21. $\overset{+}{\rightarrow} F_{Rx} = \Sigma F_x;$ $F_{Rx} = \dfrac{4}{5}(50) = 40 \text{ N}$

$+ \downarrow F_{Ry} = \Sigma F_y;$ $F_{Ry} = 40 + 30 + \dfrac{3}{5}(50)$
$= 100 \text{ N}$
$F_R = \sqrt{(40)^2 + (100)^2} = 108 \text{ N}$ *Ans.*
$\theta = \tan^{-1}\left(\dfrac{100}{40} \right) = 68.2° \, \searrow$ *Ans.*

$+ \downarrow M_{A_R} = \Sigma M_A;\ M_{A_R} = 30(3) + \dfrac{3}{5}(50)(6) + 200$
$= 470 \text{ N} \cdot \text{m}$ *Ans.*

C-22. $\overset{+}{\leftarrow} F_{Rx} = \Sigma F_x;\ F_{Rx} = 200 - \dfrac{3}{5}(100) = 140 \text{ lb}$

$+ \downarrow F_{Ry} = \Sigma F_y;\ F_{Ry} = 150 - \dfrac{4}{5}(100) = 70 \text{ lb}$
$F_R = \sqrt{140^2 + 70^2} = 157 \text{ lb}$ *Ans.*
$\theta = \tan^{-1}\left(\dfrac{70}{140} \right) = 26.6° \, \nearrow$ *Ans.*

$+ \downarrow M_{A_R} = \Sigma M_A;\ M_{A_R} = \dfrac{3}{5}(100)(4) - \dfrac{4}{5}(100)(6) + 150(3)$
$M_{R_A} = 210 \text{ lb} \cdot \text{ft}$ *Ans.*

C-23. $+ \downarrow F_R = \Sigma F_y;$ $F_R = 500 + 250 + 500$
$= 1250 \text{ lb}$ *Ans.*
$+ \downarrow F_R x = \Sigma M_O;$ $1250(x) = 500(3) + 250(6) + 500(9)$
$x = 6 \text{ ft}$ *Ans.*

C-24. $+ \downarrow F_R = \Sigma F_z;$ $F_R = 400 + 500 - 100$
$= 800 \text{ N}$ *Ans.*
$M_{Rx} = \Sigma M_x;$ $-800y = -400(4) - 500(4)$
$y = 4.50 \text{ m}$ *Ans.*
$M_{Ry} = \Sigma M_y;$ $800x = 500(4) - 100(3)$
$x = 2.125 \text{ m}$ *Ans.*

C-25. $+ \downarrow F_R = \Sigma F_y;$ $F_R = 200 + 200 + 100 + 100$
$= 600 \text{ N}$ *Ans.*
$M_{Rx} = \Sigma M_x;$ $-600y = 200(1) + 200(1)$
$+ 100(3) - 100(3)$
$y = -0.667 \text{ m}$ *Ans.*
$M_{Ry} = \Sigma M_y;$ $600x = 100(3) + 100(3)$
$+ 200(2) - 200(3)$
$x = 0.667 \text{ m}$ *Ans.*

C-26. $F_R = \dfrac{1}{2}(6)(150) + 8(150) = 1650 \text{ lb}$ *Ans.*
$+ \downarrow M_{A_R} = \Sigma M_A;$
$1650\, d = \left[\dfrac{1}{2}(6)(150) \right](4) + [8(150)](10)$
$d = 8.36 \text{ ft}$ *Ans.*

C-27. $F_R = \displaystyle\int w(x)\, dx = \int_0^4 2.5x^3\, dx = 160 \text{ N}$ *Ans.*
$+ \downarrow M_{A_R} = \Sigma M_A;$

$x = \dfrac{\displaystyle\int xw(x)\, dx}{\displaystyle\int w(x)\, dx} = \dfrac{\displaystyle\int_0^4 2.5x^4\, dx}{160} = 3.20 \text{ m}$ *Ans.*

C-28. $+ \downarrow F_R = \Sigma F_y;$ $F_R = \dfrac{1}{2}(50)(6) + 150(6) + 500$
$= 1550 \text{ lb}$ *Ans.*
$+ \downarrow M_{A_R} = \Sigma M_A;$
$1550\, d = \left[\dfrac{1}{2}(50)(6) \right](4) + [150(6)](3) + 500(9)$
$d = 5.03 \text{ ft}$ *Ans.*

C-29. $\overset{+}{\rightarrow} \Sigma F_x = 0;\ -A_x + 500\left(\dfrac{3}{5} \right) = 0$
$A_x = 300 \text{ lb}$ *Ans.*
$+ \mathord{\curvearrowleft}\Sigma M_A = 0;\ B_y(10) - 500\left(\dfrac{4}{5} \right)(5) - 600 = 0$
$B_y = 260 \text{ lb}$ *Ans.*

$+\uparrow\Sigma F_y = 0; \quad A_y + 260 - 500\left(\dfrac{4}{5}\right) = 0$

$$A_y = 140 \text{ lb} \quad \textbf{\textit{Ans.}}$$

C-30. $\xrightarrow{+}\Sigma F_x = 0; -A_x + 400 = 0; A_x = 400 \text{ N} \quad \textbf{\textit{Ans.}}$

$\downarrow+\Sigma M_A = 0; \quad B_y(4) - 400(0.5) - 500(2) = 0$

$$B_y = 300 \text{ N} \quad \textbf{\textit{Ans.}}$$

$+\uparrow\Sigma F_y = 0; \quad A_y + 300 - 500 = 0$

$$A_y = 200 \text{ N} \quad \textbf{\textit{Ans.}}$$

C-31. $\xrightarrow{+}\Sigma F_x = 0; -A_x + 400 \cos 30° = 0$

$$A_x = 346 \text{ N} \quad \textbf{\textit{Ans.}}$$

$+\uparrow\Sigma F_y = 0; A_y - 200 - 200 - 200 - 400 \sin 30° = 0$

$$A_y = 800 \text{ N} \quad \textbf{\textit{Ans.}}$$

$\downarrow+\Sigma M_A = 0; M_A - 200(2.5) - 200(3.5) - 200(4.5)$

$-400 \sin 30°(4.5) - 400 \cos 30°(3 \sin 60°) = 0$

$$M_A = 3.90 \text{ kN}\cdot\text{m} \quad \textbf{\textit{Ans.}}$$

C-32. $+\uparrow\Sigma M_A = 0; T(4) + \dfrac{3}{5}T(12) - 300(8) - 600 = 0$

$$T = 267.9 = 268 \text{ lb} \quad \textbf{\textit{Ans.}}$$

$\xrightarrow{+}\Sigma F_x = 0; \quad A_x - \left(\dfrac{4}{5}\right)(267.9) = 0$

$$A_x = 214 \text{ lb} \quad \textbf{\textit{Ans.}}$$

$+\uparrow\Sigma F_y = 0; A_y + 267.9 + \left(\dfrac{3}{5}\right)(267.9) - 300 = 0$

$$A_y = -129 \text{ lb} \quad \textbf{\textit{Ans.}}$$

C-33. $\Sigma F_z = 0; T_A + T_B + T_C - 200 - 500 = 0$

$\Sigma M_x = 0; T_A(3) + T_C(3) - 500(1.5) - 200(3) = 0$

$\Sigma M_y = 0; -T_B(4) - T_C(4) + 500(2) + 200(2) = 0$

$T_A = 350 \text{ lb}, T_B = 250 \text{ lb}, T_C = 100 \text{ lb} \quad \textbf{\textit{Ans.}}$

C-34. Joint D:

$+\uparrow\Sigma F_y = 0; \dfrac{3}{5}F_{CD} - 300 = 0; F_{CD} = 500 \text{ lb (T)} \quad \textbf{\textit{Ans.}}$

$\xrightarrow{+}\Sigma F_x = 0; -F_{AD} + \dfrac{4}{5}(500) = 0;$

$$F_{AD} = 400 \text{ lb (C)}\textbf{\textit{Ans.}}$$

Joint C:

$+\searrow\Sigma F_y = 0; F_{CA} = 0 \quad \textbf{\textit{Ans.}}$

$+\nearrow\Sigma F_x = 0; F_{CB} - 500 = 0;$

$$F_{CB} = 500 \text{ lb (T)}\textbf{\textit{Ans.}}$$

Joint A:

$+\uparrow\Sigma F_y = 0; F_{AB} = 0 \quad \textbf{\textit{Ans.}}$

C-35. $Ax = 0, Ay = Cy = 400 \text{ lb}$

Joint A:

$+\uparrow\Sigma F_y = 0; -\dfrac{3}{5}F_{AE} + 400 = 0; F_{AE} = 667 \text{ lb (C)}\textbf{\textit{Ans.}}$

Joint C:

$+\uparrow\Sigma F_y = 0; -F_{DC} + 400 = 0; F_{DC} = 400 \text{ lb (C)} \quad \textbf{\textit{Ans.}}$

C-36. Section truss through *FE*, *FC*, *BC*. Use the right segment.

$+\uparrow\Sigma F_y = 0; \quad F_{CF} \sin 45° - 600 - 800 = 0$

$$F_{CE} = 1980 \text{ lb (T)} \quad \textbf{\textit{Ans.}}$$

$+\uparrow\Sigma M_C = 0; F_{FE}(4) - 800(4) = 0$

$$F_{FE} = 800 \text{ lb (T)} \quad \textbf{\textit{Ans.}}$$

$\downarrow+\Sigma M_F = 0; F_{BC}(4) - 600(4) - 800(8) = 0$

$$F_{BC} = 2200 \text{ lb (C)} \quad \textbf{\textit{Ans.}}$$

C-37. Section truss through *GF*, *FC*, *DC*. Use the top segment.

$+\uparrow\Sigma M_C = 0; F_{GF}(8) - 700(6) - 1000(12) = 0$

$$F_{GF} = 2025 \text{ lb (T)} \quad \textbf{\textit{Ans.}}$$

$\xrightarrow{+}\Sigma F_x = 0; -\dfrac{4}{5}F_{FC} + 700 + 1000 = 0$

$$F_{FC} = 2125 \text{ lb (C)} \quad \textbf{\textit{Ans.}}$$

$\downarrow+\Sigma M_F = 0; F_{CD}(8) - 1000(6) = 0$

$$F_{CD} = 750 \text{ lb (C)} \quad \textbf{\textit{Ans.}}$$

C-38. $+\uparrow\Sigma F_y = 0; \quad 3P - 60 = 0$

$$P = 20 \text{ lb} \quad \textbf{\textit{Ans.}}$$

C-39. $+\uparrow\Sigma M_C = 0; -\left(\dfrac{4}{5}\right)(F_{AB})(9) + 400(6) + 500(3) = 0$

$$F_{AB} = 541.67 \text{ lb}$$

$\xrightarrow{+}\Sigma F_x = 0; -C_x + \dfrac{3}{5}(541.67) = 0$

$$C_x = 325 \text{ lb} \quad \textbf{\textit{Ans.}}$$

$+\uparrow\Sigma F_y = 0; C_y + \dfrac{4}{5}(541.67) - 400 - 500 = 0$

$$C_y = 467 \text{ lb} \quad \textbf{\textit{Ans.}}$$

C-40. $+\uparrow\Sigma M_C = 0; F_{AB} \cos 45°(1) - F_{AB} \sin 45°(3) + 800$

$+ 400(2) = 0$

$$F_{AB} = 1131.37 \text{ N}$$

$\xrightarrow{+}\Sigma F_x = 0; -C_x + 1131.37 \cos 45° = 0$

$$C_x = 800 \text{ N} \quad \textbf{\textit{Ans.}}$$

$+\uparrow\Sigma F_y = 0; -C_y + 1131.37 \sin 45° - 400 = 0$

$$C_y = 400 \text{ N} \quad \textbf{\textit{Ans.}}$$

C-41. Plate A:

$+\uparrow\Sigma F_y = 0; 2T + N_{AB} - 100 = 0$

Plate B:

$+\uparrow\Sigma F_y = 0; 2T - N_{AB} - 30 = 0$

$T = 32.5 \text{ lb}, N_{AB} = 35 \text{ lb} \quad \textbf{\textit{Ans.}}$

C-42. Pulley C:

$+\uparrow\Sigma F_y = 0; T - 2P = 0; T = 2P$

Beam:

$+\uparrow\Sigma F_y = 0; 2P + P - 6 = 0$

$$P = 2 \text{ kN} \quad \textbf{\textit{Ans.}}$$

$+\uparrow\Sigma M_A = 0; 2(1) - 6(x) = 0$

$$x = 0.333 \text{ m} \quad \textbf{\textit{Ans.}}$$

C-43. $A_y = 8.75$ kN. Use segment AB:

$\xrightarrow{+} \Sigma F_x = 0;$ $N_B = 0$ *Ans.*

$+\uparrow \Sigma F_y = 0;$ $8.75 - 3(1.5) - V_B = 0$

               $V_B = 4.25$ kN *Ans.*

$+\curvearrowleft \Sigma M_B = 0;$ $M_B + 3(1.5)(0.75) - 8.75(1.5) = 0$

               $M_B = 9.75$ kN $\cdot$ m *Ans.*

C-44. $A_y = 0$, $A_y = 100$ lb. Use segment AB.

$\xrightarrow{+} \Sigma F_x = 0;$ $N_B = 0$ *Ans.*

$+\uparrow \Sigma F_y = 0;$ $100 - V_B = 0$

               $V_B = 100$ lb *Ans.*

$+\curvearrowleft \Sigma M_B = 0;$ $M_B - 100(6) = 0$

               $M_B = 600$ lb $\cdot$ ft *Ans.*

C-45. $A_x = 0$, $A_y = 4.5$ kN, $w_B = 2$ kN/m. Use segment AB.

$\xrightarrow{+} \Sigma F_x = 0;$ $N_B = 0$ *Ans.*

$+\uparrow \Sigma F_y = 0;$ $4.5 - \dfrac{1}{2}(6)(2) + V_B = 0$

               $V_B = 1.5$ kN *Ans.*

$+\curvearrowleft \Sigma M_B = 0;$ $M_B + \left[\dfrac{1}{2}(6)(2)\right](2) - 4.5(6) = 0$

               $M_B = 15$ kN $\cdot$ m *Ans.*

C-46. $+\uparrow \Sigma F_y = 0;$ $N_b - P \sin 30° - 100 = 0$

$\xrightarrow{+} \Sigma F_x = 0;$ $-P \cos 30° + 0.3 N_b = 0$

               $P = 41.9$ lb *Ans.*

C-47. $\xrightarrow{+} \Sigma F_x = 0;$ $0.4N_B - N_A = 0$

$+\curvearrowleft \Sigma M_B = 0;$ $0.4N_A(12) + N_A(12) - P(6) = 0$

$+\curvearrowleft \Sigma F_y = 0;$ $P + 0.4N_A + N_B - 200 = 0$

               $P = 98.2$ lb *Ans.*

C-48. Block B:

$+\uparrow \Sigma F_y = 0;$ $N_B - 20 \sin 30° - 50 = 0$

               $N_B = 60$ lb

$\xrightarrow{+} \Sigma F_x = 0;$ $F_B - 20 \cos 30° = 0$

               $F_B = 17.3$ lb $(<0.4(60$ lb$))$ *Ans.*

Blocks A and B:

$+\uparrow \Sigma F_y = 0;$ $N_A - 30 - 50 - 20 \sin 30° = 0$

               $N_A = 90$ lb

$\xrightarrow{+} \Sigma F_x = 0;$ $F_A - 20 \cos 30° - 10 = 0$

               $F_A = 27.3$ lb $(<0.4(90$ lb$))$ *Ans.*

C-49. If slipping occurs:

$+\uparrow \Sigma F_y = 0;$ $N_C - 250$ lb $= 0$

               $N_C = 250$ lb

$\xrightarrow{+} \Sigma F_x = 0;$ $P - 0.4(250) = 0$

               $P = 100$ lb

If tipping occurs:

$\curvearrowleft + \Sigma M_A = 0;$ $-P(4.5) + 250(1.5) = 0$

               $P = 83.3$ lb *Ans.*

C-50. P for A to slip on B:

$+\uparrow \Sigma F_y = 0;$ $N_A - 60(9.81) = 0$

               $N_A = 588.6$ N

$\xrightarrow{+} \Sigma F_x = 0;$ $0.4(588.6) - P = 0$

               $P = 235$ N

P for B to slip:

$+\uparrow \Sigma F_y = 0;$ $N_B - 60(9.81) - 10(9.81) = 0$

               $N_B = 686.7$ N

$\xrightarrow{+} \Sigma F_x = 0;$ $0.3(686.7) - P = 0$

               $P = 206$ N

P to tip A:

$\curvearrowleft + \Sigma M_C = 0;$ $P(1.3) - 60(9.81)(0.2) = 0$

               $P = 90.6$ N *Ans.*

C-51. $\bar{x} = \dfrac{\Sigma \tilde{x} A}{\Sigma A} =$

$$\frac{(-1)(2)(2)+1.5(3)(3)+4\left(\dfrac{1}{2}\right)(3)(3)}{2(2)+3(3)+\dfrac{1}{2}(3)(3)} = 1.57 \text{ ft} \ \textit{Ans.}$$

$\bar{y} = \dfrac{\Sigma \tilde{y} A}{\Sigma A} =$

$$\frac{1(2)(2)+1.5(3)(3)+1\left(\dfrac{1}{2}\right)(3)(3)}{2(2)+3(3)+\dfrac{1}{2}(3)(3)} = 1.26 \text{ ft} \quad \textit{Ans.}$$

C-52. $\bar{x} = 0$ (symmetry) *Ans.*

$$\bar{y} = \frac{\Sigma \tilde{y} A}{\Sigma A} = \frac{4(1(8)) + 9(6)(2)}{1(8) + 6(2)} = 7 \text{ in.} \ \textit{Ans.}$$

C-53. $I_y = \dfrac{1}{12}(120)(300)^3 - \dfrac{1}{12}(100)(260)^3$

     $= 124 \,(10^6) \text{ mm}^4$ *Ans.*

C-54. $I = \Sigma(\bar{I} + Ad^2) = \left[\dfrac{1}{12}(8)(12)^3 + (8)(12)(6)^2\right]$

         $+ \left[\dfrac{1}{12}(6)(12)^3 + (6)(12)(-2)^2\right] = 5760 \text{ in}^4$ *Ans.*

C-55. $\bar{x} = \dfrac{\Sigma \tilde{x} A}{\Sigma A} = \dfrac{4(8)(2) + 9(2)(8)}{8(2) + 2(8)} = 6.5$ in.

$\bar{I}_{x'} = \Sigma(\bar{I} + Ad^2) = \left[\dfrac{1}{12}(2)(8)^3 + (8)(2)(6.5 - 4)^2\right]$

         $+ \left[\dfrac{1}{12}(8)(2)^3 + 2(8)(9 - 6.5)^2\right] = 291 \text{ in}^4$ *Ans.*

Answers to Selected Problems

Chapter 2

2–1. $F_3 = 428$ lb, $\alpha = 88.3°$, $\beta = 20.6°$, $\gamma = 69.5°$

2–2. $F_3 = 250$ lb, $\alpha = 87.0°$, $\beta = 143°$, $\gamma = 53.1°$

2–3. $F_{BA} = 215$ lb, $\theta = 52.7°$

2–5. $\phi = \dfrac{\theta}{2}$, $F_R = 2F \cos\left(\dfrac{\theta}{2}\right)$

2–6. $\theta = 74.0°$, $\phi = 33.9°$

2–7. Proj $F_{AB} = 70.5$ N, Proj $F_{AC} = 65.1$ N

2–9. $\theta = 60°$, $P = 40$ lb, $T = 69.3$ lb

Chapter 3

3–1. $F_A = 34.6$ lb, $F_B = 57.3$ lb

3–2. $F = 40.8$ lb

3–4. Romeo can climb up the rope.
Romeo and Juliet can climb down.

3–5. $F_1 = 8.26$ kN, $F_2 = 3.84$ kN, $F_3 = 12.2$ kN

3–6. $\theta = 90°$, $F_{AC} = 160$ lb, $\theta = 120°$, $F_{AB} = 160$ lb

3–8. $W = 240$ lb

3–9. $F_{CD} = 625$ lb, $F_{CA} = F_{CB} = 198$ lb

3–10. $F_1 = 0$, $F_2 = 311$ lb, $F_3 = 238$ lb

Chapter 4

4–1. $\alpha = 70.8°$, $\beta = 39.8°$, $\gamma = 56.7°$ or
$\alpha = 109°$, $\beta = 140°$, $\gamma = 123°$

4–2. $\mathbf{M}_O = \{298\mathbf{i} + 15.1\mathbf{j} - 200\mathbf{k}\}$ lb $\cdot$ in

4–3. $P = 23.8$ lb

4–5. $\mathbf{M}_{a-a} = 59.7$ N $\cdot$ m

4–6. $\mathbf{M}_{CR} = \{63.6\mathbf{i} - 170\mathbf{j} + 264\mathbf{k}\}$ N $\cdot$ m

4–7. $\mathbf{F}_R = \{14.3\mathbf{i} + 21.4\mathbf{j} - 42.9\mathbf{k}\}$ lb,
$\mathbf{M}_A = \{-1.93\mathbf{i} + 0.429\mathbf{j} - 0.429\mathbf{k}\}$ kip $\cdot$ ft

4–9. $\mathbf{M}_O = \{1.06\mathbf{i} + 1.06\mathbf{j} - 4.03\mathbf{k}\}$ N $\cdot$ m,
$\alpha = 75.7°$, $\beta = 75.7°$, $\gamma = 160°$

4–10. $\mathbf{F}_R = \{-70\mathbf{i} + 140\mathbf{j} - 408\mathbf{k}\}$ N,
$\mathbf{M}_{RP} = \{-26\mathbf{i} + 357\mathbf{j} + 127\mathbf{k}\}$ N $\cdot$ m

Chapter 5

5–1. $F_{CD} = 1.02$ kN, $A_z = -208$ N, $B_z = -139$ N,
$A_y = 573$ N, $B_y = 382$ N

5–3. $F = 354$ N

5–4. $N_A = 8.00$ kN, $B_x = 5.20$ kN, $B_y = 5.00$ kN

5–5. $N_B = 400$ N, $F_A = 721$ N

5–7. $N_B = 957$ N, $A_y = 743$ N, $A_x = 0$

5–8. $A_x = 0$, $A_y = 0$, $A_z = B_z = C_z = 5.33$ lb

5–9. $A_x = 0$, $A_y = -200$ N, $A_z = 150$ N,
$(M_A)_x = -100$ N $\cdot$ m, $(M_A)_y = 0$,
$(M_A)_z = -500$ N $\cdot$ m

Chapter 6

6–2. $\theta = 16.1°$

6–3. $A_x = 1.40$ kN, $A_y = 250$ N,
$C_x = 500$ N, $C_y = 1.70$ kN

6–4. $\theta = 21.7°$

6–5. $B_x = B_y = 220$ N, $A_x = 300$ N, $A_y = 80.4$ N

6–6. $A_x = 117$ N, $A_y = 397$ N,
$B_x = 97.4$ N, $B_y = 97.4$ N

6–7. $P = \dfrac{kL}{2 \tan \theta \sin \theta}(2 - \csc \theta)$

6–9. $F_{AD} = 2.47$ kip (T), $F_{AC} = F_{AB} = 1.22$ kip (C)

Chapter 7

7–1. $a = 0.366L$

7–3. For $0 \le x < 3$ m: $V = 1.50$ kN,
$M = 1.50x$ kN $\cdot$ m, For $3 < x \le 6$m:
$V = -4.50$ kN, $M = \{27.0 - 4.50x\}$ kN $\cdot$ m

7–5. $N_C = 0$, $V_C = 9.00$ kN,
$M_C = -62.5$ kN $\cdot$ m,
$N_B = 0$, $V_B = 27.5$ kN,
$M_B = -184.5$ kN $\cdot$ m

7–6. $T_{max} = 76.7$ lb

Chapter 8

8–1. $s = 0.750$ m
8–2. **a)** $W = 6.97$ kN, **b)** $W = 15.3$ kN
8–4. The cam cannot support the broom.
8–5. $P = 60$ lb for two cartons.
$P' = 90$ lb for three cartons.
8–6. $M = 2.50$ kip $\cdot$ ft

Chapter 9

9–2. $A = 1.25$ m^2
9–3. $\bar{y} = 87.5$ mm
9–4. $\bar{x} = \bar{y} = 0, \ \bar{z} = \dfrac{2}{3}a$
9–6. $\theta = 37.8°$
9–7. $\bar{y} = 0.600$ in.
9–8. $\bar{x} = 1.22$ ft, $\bar{y} = 0.778$ ft, $\bar{z} = 0.778$ ft,
$M_{Ax} = 16.0$ lb $\cdot$ ft, $M_{Ay} = 57.1$ lb $\cdot$ ft,
$M_{Az} = 0, \ A_x = 0, \ A_y = 0, \ A_z = 20.6$ lb
9–10. $F_R = 24.0$ kN, $\bar{x} = 2.00$ m, $\bar{y} = 1.33$ m
9–11. $F_R = 7.62$ kN, $\bar{x} = 2.74$ m, $\bar{y} = 3.00$ m

Chapter 10

10–1. $I_y = 0.0954d^4$
10–2. $I_y = 0.187d^4$
10–3. $I_x = \frac{93}{70}mb^2$
10–5. $I_u = 5.09(10^6)$ mm^4, $I_v = 5.09(10^6)$ mm^4,
$I_{uv} = 0$
10–6. $I_y = 2.13$ ft^4
10–7. $I_x = 0.610$ ft^4
10–9. **a)** $I_x = \dfrac{bh^3}{12}$, **b)** $\bar{I}_{x'} = \dfrac{bh^3}{36}$
10–10. $I_{xy} = 0.1875$ m^4

Chapter 11

11–1. $P = 5.28$ lb
11–2. $\theta = 37.8°$, stable
11–4. Stable at $\theta = 90°$,
Unstable at $\theta = 9.47°$
11–5. $\theta = 90°, \theta = 30°$
11–6. Unstable at $\theta = 90°$,
Stable at $\theta = 30°$,

INDEX

Active forces, on a particle, 61
Actual movement vs. imaginary movement, 371
Addition:
    algebraic, 19
    of Cartesian vectors, 38
    of collinear vectors, 18
    scalar, 19
    of system of coplanar forces, 27–33
        Cartesian vector notation, 28
        scalar notation, 27
    vector, 18–19
Addition of system of coplanar forces, 27–33
    Cartesian vector notation, 28
    coplanar force resultants, 29–30
    of system of coplanar forces, coplanar force resultants, 29–30
Algebraic addition, 19
Algebraic manipulations of an equation, 10, 14–15
Algebraic scalars, 27
Analysis:
    frictional, screws, 276–278
    general procedure for, 14
    scalar, moment of a force about a specified axis, 96–97
    structural, 177–219
    vector, moment of a force about a specified axis, 97–99
Angle:
    of kinetic friction, 262
    lead, 276
Archimedes, 4
Area moments of inertia, 333–334, 365
Area(s):
    centroid, 298
    moment of inertia for, 333–334
    parallel-axis theorem for, 335
    radius of gyration for, 335–336
Arrowhead, graphical use of, 27
Axial force, 223
Axis:
    moment, 79–80, 98
    specified, moment of a force about, 96–101
    scalar analysis, 96–97
        vector analysis, 97–99
        of a wrench, 117

Base units, 6–7, 9
Beam(s), 232
    cantilevered, 232
    floor, 178
    procedure for analysis, 233
    resistance to shear, 232
    sign convention, 232, 233
    simply supported, 232
Bending moment, 223, 256

Bending-moment diagram, 232
    for a beam, procedure for analysis, 233
Body:
    center of gravity for, 297
    center of mass for, 297
    centroid for, 298–299
        area, 298
        line, 298–299
        symmetry, 299
        volume, 298
    deformable-body mechanics, 3
    rigid bodies, 5, 15
        principle of virtual work for, 372
    rigid-body equilbrium, 131–175
        conditions for, 131–133
        constraints for a rigid body, 164–166
        equations of equilibrium, 145–153, 163
        equilibrium in two dimensions, 133–144
        three-force members, 154–156
        two-force members, 154–156
    rigid-body mechanics, 3
        dynamics, 3
        statics, 3
    translation of, 136, 139
    uniform, 137
    volume, 317, 329
Bridge truss, design of, 217

Cable(s), 61, 244–255
    flexible, 244–245, 256
    inextensible, 245, 256
    perfectly flexible, 245
    subjected to concentrated loads, 245–247
        examples, 246–247
    subjected to distributed load, 248–251
        examples, 250–251
    subjected to its own weight, 252–255
        example, 254–255
Cantilevered beam, 232
Cart lift, design of, 217
Cartesian unit vectors, 28
Cartesian vector formulation:
    cross product, 85–86
    moment of a force (vector formulation), 88–89
Cartesian vector notation, 28–29, 31, 32
Cartesian vectors, 28, 31, 34–37, 55
    addition of, 38
        examples, 39–42
Cartesian unit vectors, 35
    concurrent force systems, 38

PRINCIPLES OF DYNAMICS

Although each of these planes is rather large, from a distance their motion
can be modeled as if each plane were a particle.

CHAPTER

12

Kinematics of a Particle

CHAPTER OBJECTIVES

- To introduce the concepts of position, displacement, velocity, and acceleration.
- To study particle motion along a straight line and represent this motion graphically.
- To investigate particle motion along a curved path using different coordinate systems.
- To present an analysis of dependent motion of two particles.
- To examine the principles of relative motion of two particles using translating axes.

12.1   Introduction

Mechanics is a branch of the physical sciences that is concerned with the state of rest or motion of bodies subjected to the action of forces. The mechanics of rigid bodies is divided into two areas: statics and dynamics. *Statics* is concerned with the equilibrium of a body that is either at rest or moves with constant velocity. The foregoing treatment is concerned with *dynamics* which deals with the accelerated motion of a body. Here the subject of dynamics will be presented in two parts: *kinematics*, which treats only the geometric aspects of the motion, and *kinetics*, which is the analysis of the forces causing the motion. To develop these principles, the dynamics of a particle will be discussed first, followed by topics in rigid-body dynamics in two and then three dimensions.

Historically, the principles of dynamics developed when it was possible to make an accurate measurement of time. Galileo Galilei (1564–1642) was one of the first major contributors to this field. His work consisted of experiments using pendulums and falling bodies. The most significant contributions in dynamics, however, were made by Isaac Newton (1642–1727), who is noted for his formulation of the three fundamental laws of motion and the law of universal gravitational attraction. Shortly after these laws were postulated, important techniques for their application were developed by Euler, D'Alembert, Lagrange, and others.

There are many problems in engineering whose solutions require application of the principles of dynamics. Typically the structural design of any vehicle, such as an automobile or airplane, requires consideration of the motion to which it is subjected. This is also true for many mechanical devices, such as motors, pumps, movable tools, industrial manipulators, and machinery. Furthermore, predictions of the motions of artificial satellites, projectiles, and spacecraft are based on the theory of dynamics. With further advances in technology, there will be an even greater need for knowing how to apply the principles of this subject.

Problem Solving. Dynamics is considered to be more involved than statics since both the forces applied to a body and its motion must be taken into account. Also, many applications require using calculus, rather than just algebra and trigonometry. In any case, the most effective way of learning the principles of dynamics is *to solve problems*. To be successful at this, it is necessary to present the work in a logical and orderly manner as suggested by the following sequence of steps:

1. Read the problem carefully and try to correlate the actual physical situation with the theory studied.

2. Draw any necessary diagrams and tabulate the problem data.

3. Establish a coordinate system and apply the relevant principles, generally in mathematical form.

4. Solve the necessary equations algebraically as far as practical; then, use a consistent set of units and complete the solution numerically. Report the answer with no more significant figures than the accuracy of the given data.

5. Study the answer using technical judgment and common sense to determine whether or not it seems reasonable.

6. Once the solution has been completed, review the problem. Try to think of other ways of obtaining the same solution.

In applying this general procedure, do the work as neatly as possible. Being neat generally stimulates clear and orderly thinking, and vice versa.

12.2 Rectilinear Kinematics: Continuous Motion

We will begin our study of dynamics by discussing the kinematics of a particle that moves along a rectilinear or straight line path. Recall that a *particle* has a mass but negligible size and shape. Therefore we must limit application to those objects that have dimensions that are of no consequence in the analysis of the motion. In most problems, one is interested in bodies of finite size, such as rockets, projectiles, or vehicles. Such objects may be considered as particles, provided motion of the body is characterized by motion of its mass center and any rotation of the body is neglected.

Rectilinear Kinematics. The kinematics of a particle is characterized by specifying, at any given instant, the particle's position, velocity, and acceleration.

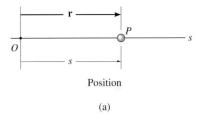

Position

(a)

Position. The straight-line path of a particle will be defined using a single coordinate axis s, Fig. 12–1*a*. The origin O on the path is a fixed point, and from this point the *position vector* **r** is used to specify the location of the particle P at any given instant. Notice that **r** is *always* along the s axis, and so its direction never changes. What will change is its magnitude and its sense or arrowhead direction. For analytical work it is therefore convenient to represent **r** by an *algebraic scalar s*, representing the *position coordinate* of the particle, Fig. 12–1*a*. The magnitude of s (and **r**) is the distance from O to P, usually measured in meters (m) or feet (ft), and the sense (or arrowhead direction of **r**) is defined by the algebraic sign on s. Although the choice is arbitrary, in this case s is positive since the coordinate axis is positive to the right of the origin. Likewise, it is negative if the particle is located to the left of O.

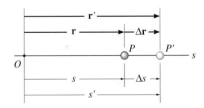

Displacement

(b)

Fig. 12–1

Displacement. The *displacement* of the particle is defined as the *change* in its *position*. For example, if the particle moves from P to P', Fig. 12–1*b*, the displacement is $\Delta\mathbf{r} = \mathbf{r}' - \mathbf{r}$. Using algebraic scalars to represent $\Delta\mathbf{r}$, we also have

$$\Delta s = s' - s$$

Here Δs is *positive* since the particle's final position is to the *right* of its initial position, i.e., $s' > s$. Likewise, if the final position were to the *left* of its initial position, Δs would be *negative*.

Since the displacement of a particle is a *vector quantity*, it should be distinguished from the distance the particle travels. Specifically, the *distance traveled* is a *positive scalar* which represents the total length of path over which the particle travels.

Velocity.   If the particle moves through a displacement $\Delta\mathbf{r}$ from P to P' during the time interval Δt, Fig. 12–1b, the *average velocity* of the particle during this time interval is

$$\mathbf{v}_{\text{avg}} = \frac{\Delta\mathbf{r}}{\Delta t}$$

If we take smaller and smaller values of Δt, the magnitude of $\Delta\mathbf{r}$ becomes smaller and smaller. Consequently, the *instantaneous velocity* is defined as $\mathbf{v} = \lim_{\Delta t \to 0}(\Delta\mathbf{r}/\Delta t)$, or

$$\mathbf{v} = \frac{d\mathbf{r}}{dt}$$

Representing $\mathbf{v}$ as an algebraic scalar, Fig. 12–1c, we can also write

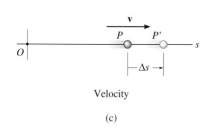

Velocity

(c)

$(\xrightarrow{+})$

$$\boxed{v = \frac{ds}{dt}}$$

$(12\text{–}1)$

Since Δt or dt is always positive, the sign used to define the *sense* of the velocity is the same as that of Δs or ds. For example, if the particle is moving to the *right*, Fig. 12–1c, the velocity is *positive;* whereas if it is moving to the *left*, the velocity is *negative.* (This is emphasized here by the arrow written at the left of Eq. 12–1.) The *magnitude* of the velocity is known as the *speed*, and it is generally expressed in units of m/s or ft/s.

   Occasionally, the term "average speed" is used. The *average speed* is always a positive scalar and is defined as the total distance traveled by a particle, s_T, divided by the elapsed time Δt; i.e.,

$$(v_{\text{sp}})_{\text{avg}} = \frac{s_T}{\Delta t}$$

For example the particle in Fig. 12–1d travels along the path of length s_T in time Δt, so its average speed is $(v_{\text{sp}})_{\text{avg}} = s_T/\Delta t$, but its average velocity is $v_{\text{avg}} = -\Delta s/\Delta t$.

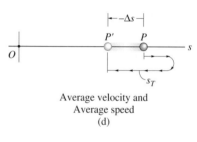

Average velocity and
Average speed
(d)

Fig. 12–1

Acceleration. Provided the velocity of the particle is known at the two points P and P', the *average acceleration* of the particle during the time interval Δt is defined as

$$\mathbf{a}_{avg} = \frac{\Delta \mathbf{v}}{\Delta t}$$

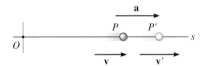

Acceleration

(e)

Here $\Delta \mathbf{v}$ represents the difference in the velocity during the time interval Δt, i.e., $\Delta \mathbf{v} = \mathbf{v}' - \mathbf{v}$, Fig. 12–1e.

The *instantaneous acceleration* at time t is found by taking smaller and smaller values of Δt and corresponding smaller and smaller values of $\Delta \mathbf{v}$, so that $\mathbf{a} = \lim_{\Delta t \to 0} (\Delta \mathbf{v}/\Delta t)$ or, using algebraic scalars,

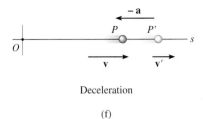

Deceleration

(f)

$(\stackrel{+}{\rightarrow})$
$$a = \frac{dv}{dt}$$
(12–2)

Substituting Eq. 12–1 into this result, we can also write

$(\stackrel{+}{\rightarrow})$
$$a = \frac{d^2 s}{dt^2}$$

Both the average and instantaneous acceleration can be either positive or negative. In particular, when the particle is *slowing down*, or its speed is decreasing, it is said to be *decelerating*. In this case, v' in Fig. 12–1f is *less* than v, and so $\Delta v = v' - v$ will be negative. Consequently, a will also be negative, and therefore it will act to the *left*, in the opposite *sense* to v. Also, note that when the *velocity* is *constant*, the *acceleration is zero* since $\Delta v = v - v = 0$. Units commonly used to express the magnitude of acceleration are m/s² or ft/s².

A differential relation involving the displacement, velocity, and acceleration along the path may be obtained by eliminating the time differential dt between Eqs. 12–1 and 12–2. Realize that although we can then establish another equation, by doing so it will *not* be independent of Eqs. 12–1 and 12–2. Show that

$(\stackrel{+}{\rightarrow})$
$$a\,ds = v\,dv$$
(12–3)

Constant Acceleration, $a = a_c$. When the acceleration is constant, each of the three kinematic equations $a_c = dv/dt$, $v = ds/dt$, and $a_c \, ds = v \, dv$ may be integrated to obtain formulas that relate a_c, v, s, and t.

Velocity as a Function of Time. Integrate $a_c = dv/dt$, assuming that initially $v = v_0$ when $t = 0$.

$$\int_{v_0}^{v} dv = \int_{0}^{t} a_c \, dt$$

$(\overset{+}{\rightarrow})$

$$\boxed{v = v_0 + a_c t}$$
Constant Acceleration

(12–4)

Position as a Function of Time. Integrate $v = ds/dt = v_0 + a_c t$, assuming that initially $s = s_0$ when $t = 0$.

$$\int_{s_0}^{s} ds = \int_{0}^{t} (v_0 + a_c t) \, dt$$

$(\overset{+}{\rightarrow})$

$$\boxed{s = s_0 + v_0 t + \tfrac{1}{2} a_c t^2}$$
Constant Acceleration

(12–5)

Velocity as a Function of Position. Either solve for t in Eq. 12–4 and substitute into Eq. 12–5, or integrate $v \, dv = a_c \, ds$, assuming that initially $v = v_0$ at $s = s_0$.

$$\int_{v_0}^{v} v \, dv = \int_{s_0}^{s} a_c \, ds$$

$(\overset{+}{\rightarrow})$

$$\boxed{v^2 = v_0^2 + 2a_c(s - s_0)}$$
Constant Acceleration

(12–6)

This equation is not independent of Eqs. 12–4 and 12–5 since it can be obtained by eliminating t between these equations.

The magnitudes and signs of s_0, v_0, and a_c, used in the above three equations are determined from the chosen origin and positive direction of the s axis as indicated by the arrow written at the left of each equation. Also, it is important to remember that these equations are useful *only when the acceleration is constant and when* $t = 0$, $s = s_0$, $v = v_0$. A common example of constant accelerated motion occurs when a body falls freely toward the earth. If air resistance is neglected and the distance of fall is short, then the *downward* acceleration of the body when it is close to the earth is constant and approximately 9.81 m/s² or 32.2 ft/s². The proof of this is given in Example 13.2.

IMPORTANT POINTS

- Dynamics is concerned with bodies that have accelerated motion.
- Kinematics is a study of the geometry of the motion.
- Kinetics is a study of the forces that cause the motion.
- Rectilinear kinematics refers to straight-line motion.
- Speed refers to the magnitude of velocity.
- Average speed is the total distance traveled divided by the total time. This is different from the average velocity which is the displacement divided by the time.
- The acceleration, $a = dv/dt$, is negative when the particle is slowing down or decelerating.
- A particle can have an acceleration and yet have zero velocity.
- The relationship $a \, ds = v \, dv$ is derived from $a = dv/dt$ and $v = ds/dt$, by eliminating dt.

PROCEDURE FOR ANALYSIS

The equations of rectilinear kinematics should be applied using the following procedure.

Coordinate System

- Establish a position coordinate s along the path and specify its *fixed origin* and positive direction.
- Since motion is along a straight line, the particle's position, velocity, and acceleration can be represented as algebraic scalars. For analytical work the sense of s, v, and a is then determined from their *algebraic signs*.
- The positive sense for each scalar can be indicated by an arrow shown alongside each kinematic equation as it is applied.

Kinematic Equations

- If a relationship is known between any *two* of the four variables a, v, s and t, then a third variable can be obtained by using one of the kinematic equations, $a = dv/dt$, $v = ds/dt$ or $a \, ds = v \, dv$, which relates all three variables.*
- Whenever integration is performed, it is important that the position and velocity be known at a given instant in order to evaluate either the constant of integration if an indefinite integral is used, or the limits of integration if a definite integral is used.
- Remember that Eqs. 12–4 through 12–6 have only a limited use. Never apply these equations unless it is absolutely certain that the *acceleration is constant.*

*Some standard differentiation and integration formulas are given in Appendix A.

During the time this rocket undergoes rectilinear motion, its altitude as a function of time can be measured and expressed as $s = s(t)$. Its velocity can then be found using $v = ds/dt$, and its acceleration can be determined from $a = dv/dt$.

E X A M P L E 12.1

The car in Fig. 12–2 moves in a straight line such that for a short time its velocity is defined by $v = (3t^2 + 2t)$ ft/s, where t is in seconds. Determine its position and acceleration when $t = 3$ s. When $t = 0$, $s = 0$.

Fig. 12–2

Solution

Coordinate System. The position coordinate extends from the fixed origin O to the car, positive to the right.

Position. Since $v = f(t)$, the car's position can be determined from $v = ds/dt$, since this equation relates v, s, and t. Noting that $s = 0$ when $t = 0$, we have*

$(\xrightarrow{+})$
$$v = \frac{ds}{dt} = (3t^2 + 2t)$$

$$\int_0^s ds = \int_0^t (3t^2 + 2t)\, dt$$

$$s \Big|_0^s = t^3 + t^2 \Big|_0^t$$

$$s = t^3 + t^2$$

When $t = 3$ s,

$$s = (3)^3 + (3)^2 = 36 \text{ ft} \qquad\qquad Ans.$$

Acceleration. Knowing $v = f(t)$, the acceleration is determined from $a = dv/dt$, since this equation relates a, v, and t.

$(\xrightarrow{+})$
$$a = \frac{dv}{dt} = \frac{d}{dt}(3t^2 + 2t)$$

$$= 6t + 2$$

When $t = 3$ s,

$$a = 6(3) + 2 = 20 \text{ ft/s}^2 \rightarrow \qquad\qquad Ans.$$

The formulas for constant acceleration *cannot* be used to solve this problem. Why?

*The *same result* can be obtained by evaluating a constant of integration C rather than using definite limits on the integral. For example, integrating $ds = (3t^2 + 2t)\, dt$ yields $s = t^3 + t^2 + C$. Using the condition that at $t = 0$, $s = 0$, then $C = 0$.

EXAMPLE 12.2

A small projectile is fired vertically *downward* into a fluid medium with an initial velocity of 60 m/s. Due to the resistance of the fluid the projectile experiences a deceleration equal to $a = (-0.4v^3)$ m/s², where v is in m/s. *Determine the projectile's velocity and position 4 s after it is fired.

Solution

Coordinate System. Since the motion is downward, the position coordinate is positive downward, with origin located at O, Fig. 12–3.

Velocity. Here $a = f(v)$ and so we must determine the velocity as a function of time using $a = dv/dt$, since this equation relates v, a, and t. (Why not use $v = v_0 + a_c t$?) Separating the variables and integrating, with $v_0 = 60$ m/s when $t = 0$, yields

$(+\downarrow)$
$$a = \frac{dv}{dt} = -0.4v^3$$

$$\int_{60 \text{ m/s}}^{v} \frac{dv}{-0.4v^3} = \int_{0}^{t} dt$$

$$\frac{1}{-0.4}\left(\frac{1}{-2}\right)\frac{1}{v^2}\Big|_{60}^{v} = t - 0$$

$$\frac{1}{0.8}\left[\frac{1}{v^2} - \frac{1}{(60)^2}\right] = t$$

$$v = \left\{\left[\frac{1}{(60)^2} + 0.8t\right]^{-1/2}\right\} \text{ m/s}$$

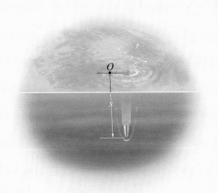

Fig. 12–3

Here the positive root is taken, since the projectile is moving downward. When $t = 4$ s,

$$v = 0.559 \text{ m/s} \downarrow \qquad\qquad Ans.$$

Position. Knowing $v = f(t)$, we can obtain the projectile's position from $v = ds/dt$, since this equation relates s, v, and t. Using the initial condition $s = 0$, when $t = 0$, we have

$(+\downarrow)$
$$v = \frac{ds}{dt} = \left[\frac{1}{(60)^2} + 0.8t\right]^{-1/2}$$

$$\int_{0}^{s} ds = \int_{0}^{t}\left[\frac{1}{(60)^2} + 0.8t\right]^{-1/2} dt$$

$$s = \frac{2}{0.8}\left[\frac{1}{(60)^2} + 0.8t\right]^{1/2}\Big|_{0}^{t}$$

$$s = \frac{1}{0.4}\left\{\left[\frac{1}{(60)^2} + 0.8t\right]^{1/2} - \frac{1}{60}\right\} \text{ m}$$

When $t = 4$ s,

$$s = 4.43 \text{ m} \qquad\qquad Ans.$$

*Note that to be dimensionally homogeneous, the constant 0.4 has units of s/m².

EXAMPLE 12.3

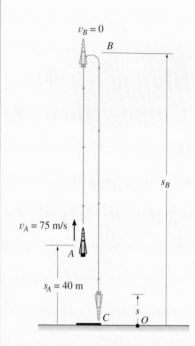

Fig. 12–4

During a test a rocket is traveling upward at 75 m/s, and when it is 40 m from the ground its engine fails. Determine the maximum height s_B reached by the rocket and its speed just before it hits the ground. While in motion the rocket is subjected to a constant downward acceleration of 9.81 m/s² due to gravity. Neglect the effect of air resistance.

Solution

Coordinate System. The origin O for the position coordinate s is taken at ground level with positive upward, Fig. 12–4.

Maximum Height. Since the rocket is traveling *upward*, $v_A = +75$ m/s when $t = 0$. At the maximum height $s = s_B$ the velocity $v_B = 0$. For the entire motion, the acceleration is $a_c = -9.81$ m/s² (negative since it acts in the *opposite* sense to positive velocity or positive displacement). Since a_c is *constant* the rocket's position may be related to its velocity at the two points A and B on the path by using Eq. 12–6, namely,

$$(+\uparrow) \qquad v_B^2 = v_A^2 + 2a_c(s_B - s_A)$$
$$0 = (75 \text{ m/s})^2 + 2(-9.81 \text{ m/s}^2)(s_B - 40 \text{ m})$$
$$s_B = 327 \text{ m} \qquad\qquad Ans.$$

Velocity. To obtain the velocity of the rocket just before it hits the ground, we can apply Eq. 12–6 between points B and C, Fig. 12–4.

$$(+\uparrow) \qquad v_C^2 = v_B^2 + 2a_c(s_C - s_B)$$
$$= 0 + 2(-9.81 \text{ m/s}^2)(0 - 327 \text{ m})$$
$$v_C = -80.1 \text{ m/s} = 80.1 \text{ m/s}\downarrow \qquad Ans.$$

The negative root was chosen since the rocket is moving downward. Similarly, Eq. 12–6 may also be applied between points A and C, i.e.,

$$(+\uparrow) \qquad v_C^2 = v_A^2 + 2a_c(s_C - s_A)$$
$$= (75 \text{ m/s})^2 + 2(-9.81 \text{ m/s}^2)(0 - 40 \text{ m})$$
$$v_C = -80.1 \text{ m/s} = 80.1 \text{ m/s} \downarrow$$

Note: It should be realized that the rocket is subjected to a *deceleration* from A to B of 9.81 m/s², and then from B to C it is *accelerated* at this rate. Furthermore, even though the rocket momentarily comes to *rest* at B ($v_B = 0$) the acceleration at B is 9.81 m/s² downward!

E X A M P L E **12.4**

A metallic particle is subjected to the influence of a magnetic field as it travels downward through a fluid that extends from plate A to plate B, Fig. 12–5. If the particle is released from rest at the midpoint C, $s = 100$ mm, and the acceleration is $a = (4s)$ m/s^2, where s is in meters, determine the velocity of the particle when it reaches plate B, $s = 200$ mm, and the time it needs to travel from C to B.

Solution

Coordinate System. As shown in Fig. 12–5, s is taken positive downward, measured from plate A.

Velocity. Since $a = f(s)$, the velocity as a function of position can be obtained by using $v\, dv = a\, ds$. Why not use the formulas for constant acceleration? Realizing that $v = 0$ at $s = 100$ mm $= 0.1$ m, we have

$(+\downarrow)$
$$v\, dv = a\, ds$$

$$\int_0^v v\, dv = \int_{0.1}^s 4s\, ds$$

$$\tfrac{1}{2}v^2 \Big|_0^v = \frac{4}{2}s^2 \Big|_{0.1}^s$$

$$v = 2(s^2 - 0.01)^{1/2} \qquad (1)$$

At $s = 200$ mm $= 0.2$ m,

$$v_B = 0.346 \text{ m/s} = 346 \text{ mm/s} \downarrow \qquad \text{Ans.}$$

The positive root is chosen since the particle is traveling downward, i.e., in the $+s$ direction.

Time. The time for the particle to travel from C to B can be obtained using $v = ds/dt$ and Eq. 1, where $s = 0.1$ m when $t = 0$. From Appendix A,

$(+\downarrow)$
$$ds = v\, dt$$

$$= 2(s^2 - 0.01)^{1/2}\, dt$$

$$\int_{0.1}^s \frac{ds}{(s^2 - 0.01)^{1/2}} = \int_0^t 2\, dt$$

$$\ln(\sqrt{s^2 - 0.01} + s)\Big|_{0.1}^s = 2t \Big|_0^t$$

$$\ln(\sqrt{s^2 - 0.01} + s) + 2.33 = 2t$$

At $s = 200$ mm $= 0.2$ m,

$$t = \frac{\ln(\sqrt{(0.2)^2 - 0.01} + 0.2) + 2.33}{2} = 0.658 \text{ s} \qquad \text{Ans.}$$

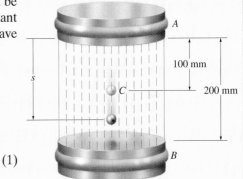

Fig. 12–5

EXAMPLE **12.5**

A particle moves along a horizontal path with a velocity of $v = (3t^2 - 6t)$ m/s, where t is the time in seconds. If it is initially located at the origin O, determine the distance traveled in 3.5 s, and the particle's average velocity and average speed during the time interval.

Solution

Coordinate System. Here we will assume positive motion to the right, measured from the origin O, Fig. 12–6a.

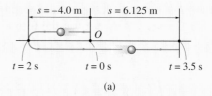

(a)

Fig. 12–6

Distance Traveled. Since $v = f(t)$, the position as a function of time may be found by integrating $v = ds/dt$ with $t = 0$, $s = 0$.

$(\overset{+}{\rightarrow})$
$$ds = v \, dt$$
$$= (3t^2 - 6t) \, dt$$

$$\int_0^s ds = 3 \int_0^t t^2 \, dt - 6 \int_0^t t \, dt$$

$$s = (t^3 - 3t^2) \text{ m} \qquad (1)$$

In order to determine the distance traveled in 3.5 s, it is necessary to investigate the path of motion. The graph of the velocity function, Fig. 12–6b, reveals that for $0 \le t < 2$ s the velocity is *negative*, which means the particle is traveling to the *left*, and for $t > 2$ s the velocity is *positive*, and hence the particle is traveling to the *right*. Also, $v = 0$ at $t = 2$ s. The particle's position when $t = 0$, $t = 2$ s, and $t = 3.5$ s can be determined from Eq. 1. This yields

$$s|_{t=0} = 0 \qquad s|_{t=2\,s} = -4.0 \text{ m} \qquad s|_{t=3.5\,s} = 6.125 \text{ m}$$

The path is shown in Fig. 12–6a. Hence, the distance traveled in 3.5 s is

$$s_T = 4.0 + 4.0 + 6.125 = 14.125 \text{ m} = 14.1 \text{ m} \qquad \textit{Ans.}$$

Velocity. The *displacement* from $t = 0$ to $t = 3.5$ s is

$$\Delta s = s|_{t=3.5\,s} - s|_{t=0} = 6.12 - 0 = 6.12 \text{ m}$$

and so the average velocity is

$$v_{avg} = \frac{\Delta s}{\Delta t} = \frac{6.12}{3.5 - 0} = 1.75 \text{ m/s} \rightarrow \qquad \textit{Ans.}$$

The average speed is defined in terms of the *distance traveled* s_T. This positive scalar is

$$(v_{sp})_{avg} = \frac{s_T}{\Delta t} = \frac{14.125}{3.5 - 0} = 4.04 \text{ m/s} \qquad \textit{Ans.}$$

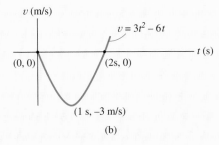

(b)

Fig. 12–6

12.3 Rectilinear Kinematics: Erratic Motion

When a particle's motion during a time period is erratic, it may be difficult to obtain a continuous mathematical function to describe its position, velocity, or acceleration. Instead, the motion may best be described graphically using a series of curves that can be generated experimentally from computer output. If the resulting graph describes the relationship between any two of the variables, a, v, s, t, a graph describing the relationship between the other variables can be established by using the kinematic equations $a = dv/dt$, $v = ds/dt$, $a\,ds = v\,dv$. Several situations occur frequently.

Given the s–t Graph, Construct the v–t Graph. If the position of a particle can be *determined experimentally* during a time period t, the s–t graph for the particle can be plotted, Fig. 12–7a. To determine the particle's velocity as a function of time, i.e., the v–t graph, we must use $v = ds/dt$ since this equation relates v, s, and t. Therefore, the velocity at any instant is determined by measuring the *slope* of the s–t graph, i.e.,

$$\frac{ds}{dt} = v$$

slope of
s–t graph = velocity

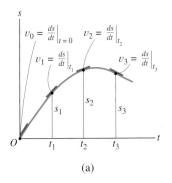

(a)

For example, measurement of the slopes v_0, v_1, v_2, v_3 at the intermediate points $(0, 0)$, (t_1, s_1), (t_2, s_2), (t_3, s_3) on the s–t graph, Fig. 12–7a, gives the corresponding points on the v–t graph shown in Fig. 12–7b.

It may also be possible to establish the v–t graph *mathematically*, provided the segments of the s–t graph can be expressed in the form of equations $s = f(t)$. Corresponding equations describing the segments of the v–t graph are then determined by time *differentiation*, since $v = ds/dt$.

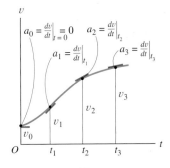

(b)

Fig. 12–7

Given the v–t Graph, Construct the a–t Graph. When the particle's v–t graph is known, as in Fig. 12–8a, the acceleration as a function of time, i.e., the a–t graph, can be determined using $a = dv/dt$. (Why?) Hence, the acceleration at any instant is determined by measuring the slope of the v–t graph, i.e.,

$$\frac{dv}{dt} = a$$

slope of
v–t graph = acceleration

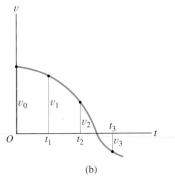

(a)

For example, measurement of the slopes a_0, a_1, a_2, a_3 at the intermediate points $(0, v_0)$, (t_1, v_1), (t_2, v_2), (t_3, v_3) on the v–t graph, Fig. 12–8a, yields the corresponding points on the a–t graph shown in Fig. 12–8b.

Any segments of the a–t graph can also be determined *mathematically*, provided the equations of the corresponding segments of the v–t graph are known, $v = g(t)$. This is done by simply taking the time *derivative* of $v = g(t)$, since $a = dv/dt$.

Since differentiation reduces a polynomial of degree n to that of degree $n - 1$, then if the s–t graph is parabolic (a second-degree curve), the v–t graph will be a sloping line (a first-degree curve), and the a–t graph will be a constant or a horizontal line (a zero-degree curve).

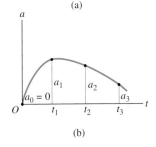

(b)

Fig. 12–8

E X A M P L E   12.6

A bicycle moves along a straight road such that its position is described by the graph shown in Fig. 12–9a. Construct the v–t and a–t graphs for $0 \le t \le 30$ s.

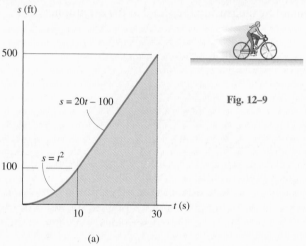

Fig. 12–9

(a)

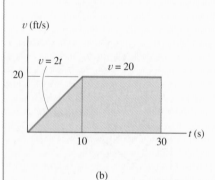

(b)

(c)

Fig. 12–9

Solution

v–t Graph. Since $v = ds/dt$, the v–t graph can be determined by differentiating the equations defining the s–t graph, Fig. 12–9a. We have

$$0 \le t < 10 \text{ s}; \qquad s = t^2 \qquad v = \frac{ds}{dt} = 2t$$

$$10 \text{ s} < t \le 30 \text{ s}; \qquad s = 20t - 100 \qquad v = \frac{ds}{dt} = 20$$

The results are plotted in Fig. 12–9b. We can also obtain specific values of v by measuring the *slope* of the s–t graph at a given instant. For example, at $t = 20$ s, the slope of the s–t graph is determined from the straight line from 10 s to 30 s, i.e.,

$$t = 20 \text{ s}; \qquad v = \frac{\Delta s}{\Delta t} = \frac{500 - 100}{30 - 10} = 20 \text{ ft/s}$$

a–t Graph. Since $a = dv/dt$, the a–t graph can be determined by differentiating the equations defining the lines of the v–t graph. This yields

$$0 \le t < 10 \text{ s}; \qquad v = 2t \qquad a = \frac{dv}{dt} = 2$$

$$10 < t \le 30 \text{ s}; \qquad v = 20 \qquad a = \frac{dv}{dt} = 0$$

The results are plotted in Fig. 12–9c. Show that $a = 2$ ft/s² when $t = 5$ s by measuring the slope of the v–t graph.

Given the *a–t* Graph, Construct the *v–t* Graph.   If the *a–t* graph is given, Fig. 12–10*a*, the *v–t* graph may be constructed using $a = dv/dt$, written in integrated form as

$$\Delta v = \int a\, dt$$

$$\frac{\text{change in}}{\text{velocity}} = \frac{\text{area under}}{a\text{–}t \text{ graph}}$$

Hence, to construct the *v–t* graph, we begin by first knowing the particle's initial velocity v_0 and then add to this small increments of area (Δv) determined from the *a–t* graph. In this manner, successive points, $v_1 = v_0 + \Delta v$, etc., for the *v–t* graph are determined, Fig. 12–10*b*. Notices that an algebraic addition of the area increments is necessary, since areas lying above the *t* axis correspond to an increase in *v* ("positive" area), whereas those lying below the axis indicate a decrease in *v* ("negative" area).

   If segments of the *a–t* graph can be described by a series of equations, then each of these equations may be *integrated* to yield equations describing the corresponding segments of the *v–t* graph. Hence, if the *a–t* graph is linear (a first-degree curve), integration will yield a *v–t* graph that is parabolic (a second-degree curve), etc.

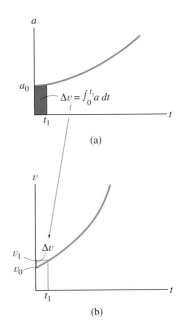

(a)

(b)

Fig. 12–10

Given the *v–t* Graph, Construct the *s–t* Graph.   When the *v–t* graph is given, Fig. 12–11*a*, it is possible to determine the *s–t* graph using $v = ds/dt$, written in integrated form

$$\Delta s = \int v\, dt$$

$$\text{displacement} = \text{area under} \\ v\text{–}t \text{ graph}$$

In the same manner as stated above, we begin by knowing the particle's initial position s_0 and add (algebraically) to this small area increments Δs determined from the *v–t* graph, Fig. 12–11*b*.

   If it is possible to describe segments of the *v–t* graph by a series of equations, then each of these equations may be *integrated* to yield equations that describe corresponding segments of the *s–t* graph.

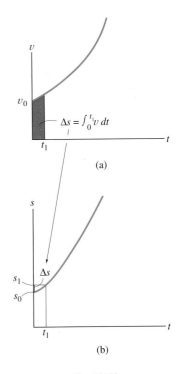

(a)

(b)

Fig. 12–11

E X A M P L E 12.7

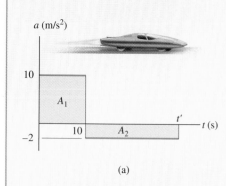

(a)

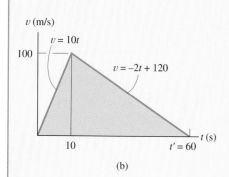

(b)

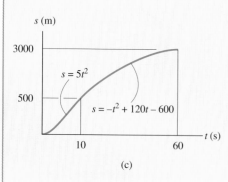

(c)

Fig. 12–12

The test car in Fig. 12–12a starts from rest and travels along a straight track such that it accelerates at a constant rate for 10 s and then decelerates at a constant rate. Draw the v–t and s–t graphs and determine the time t' needed to stop the car. How far has the car traveled?

Solution

v–t Graph. Since $dv = a\,dt$, the v–t graph is determined by integrating the straight-line segments of the a–t graph. Using the *initial condition* $v = 0$ when $t = 0$, we have

$$0 \le t < 10\text{ s}; \quad a = 10; \quad \int_0^v dv = \int_0^t 10\,dt, \quad v = 10t$$

When $t = 10$ s, $v = 10(10) = 100$ m/s. Using this as the *initial condition* for the next time period, we have

$$10\text{ s} < t \le t'; \quad a = -2; \quad \int_{100}^v dv = \int_{10}^t -2\,dt, \quad v = -2t + 120$$

When $t = t'$ we require $v = 0$. This yields, Fig. 12–12b,

$$t' = 60\text{ s} \qquad \qquad Ans.$$

A more direct solution for t' is possible by realizing that the area under the a–t graph is equal to the change in the car's velocity. We require $\Delta v = 0 = A_1 + A_2$, Fig. 12–12a. Thus

$$0 = 10\text{ m/s}^2(10\text{ s}) + (-2\text{ m/s}^2)(t' - 10\text{ s}) = 0$$

$$t' = 60\text{ s} \qquad \qquad Ans.$$

s–t Graph. Since $ds = v\,dt$, integrating the equations of the v–t graph yields the corresponding equations of the s–t graph. Using the *initial conditions* $s = 0$ when $t = 0$, we have

$$0 \le t \le 10\text{ s}; \quad v = 10t; \quad \int_0^s ds = \int_0^t 10t\,dt, \quad s = 5t^2$$

When $t = 10$ s, $s = 5(10)^2 = 500$ m. Using this *initial condition*,

$$10\text{ s} \le t \le 60\text{ s}; \quad v = -2t + 120; \quad \int_{500}^s ds = \int_{10}^t (-2t + 120)\,dt$$

$$s - 500 = -t^2 + 120t - [-(10)^2 + 120(10)]$$

$$s = -t^2 + 120t - 600$$

When $t' = 60$ s, the position is

$$s = -(60)^2 + 120(60) - 600 = 3000\text{ m} \qquad Ans.$$

The s–t graph is shown in Fig. 12–12c. Note that a direct solution for s is possible when $t' = 60$ s, since the *triangular area* under the v–t graph would yield the displacement $\Delta s = s - 0$ from $t = 0$ to $t' = 60$ s. Hence,

$$\Delta s = \tfrac{1}{2}(60)(100) = 3000\text{ m} \qquad Ans.$$

Given the a–s Graph, Construct the v–s Graph. In some cases an a–s graph for the particle can be constructed, so that points on the v–s graph can be determined by using $v\,dv = a\,ds$. Integrating this equation between the limits $v = v_0$ at $s = s_0$ and $v = v_1$ at $s = s_1$, we have,

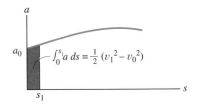

$$\tfrac{1}{2}(v_1^2 - v_0^2) = \int_{s_0}^{s_1} a\,ds$$

area under
a–s graph

Thus, the initial small segment of area under the a–s graph, $\int_{s_0}^{s_1} a\,ds$, shown colored in Fig. 12–13a, equals one-half the difference in the squares of the speed, $\tfrac{1}{2}(v_1^2 - v_0^2)$. Therefore, if the area is determined and the initial value of v_0 at $s_0 = 0$ is known, then $v_1 = (2\int_{s_0}^{s_1} a\,ds + v_0^2)^{1/2}$, Fig. 12–13b. Successive points on the v–s graph can be constructed in this manner starting from the initial velocity v_0.

Another way to construct the v–s graph is to first determine the equations which define the segments of the a–s graph. Then the corresponding equations defining the segments of the v–s graph can be obtained directly from integration, using $v\,dv = a\,ds$.

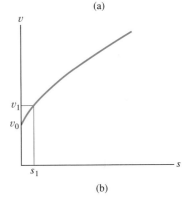

(a)

(b)

Fig. 12–13

Given the v–s Graph, Construct the a–s Graph. If the v–s graph is known, the acceleration a at any position s can be determined using $a\,ds = v\,dv$, written as

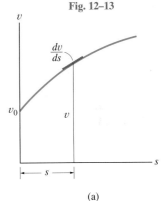

(a)

$$a = v\left(\frac{dv}{ds}\right)$$

acceleration = velocity times
slope of
v–s graph

Thus, at any point (s, v) in Fig. 12–14a, the slope dv/ds of the v–s graph is measured. Then since v and dv/ds are known, the value of a can be calculated, Fig. 12–14b.

We can also determine the segments describing the a–s graph analytically, provided the equations of the corresponding segments of the v–s graph are known. As above, this requires integration using $a\,ds = v\,dv$.

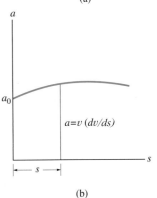

(b)

Fig. 12–14

EXAMPLE 12.8

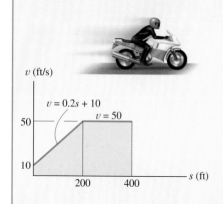

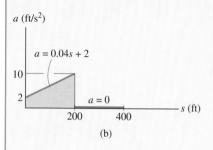

(b)

Fig. 12–15

The v–s graph describing the motion of a motorcycle is shown in Fig. 12–15a. Construct the a–s graph of the motion and determine the time needed for the motorcycle to reach the position $s = 400$ ft.

Solution

a–s Graph. Since the equations for segments of the v–s graph are given, the a–s graph can be determined using $a\,ds = v\,dv$.

$$0 \le s < 200 \text{ ft}; \quad v = 0.2s + 10$$

$$a = v\frac{dv}{ds} = (0.2s + 10)\frac{d}{ds}(0.2s + 10) = 0.04s + 2$$

$$200 \text{ ft} < s \le 400 \text{ ft}; \quad v = 50;$$

$$a = v\frac{dv}{ds} = (50)\frac{d}{ds}(50) = 0$$

The results are plotted in Fig. 12–15b.

Time. The time can be obtained using the v–s graph and $v = ds/dt$, because this equation relates v, s, and t. For the first segment of motion, $s = 0$ at $t = 0$, so

$$0 \le s < 200 \text{ ft}; \quad v = 0.2s + 10; \quad dt = \frac{ds}{v} = \frac{ds}{0.2s + 10}$$

$$\int_0^t dt = \int_0^s \frac{ds}{0.2s + 10}$$

$$t = 5\ln(0.2s + 10) - 5\ln 10$$

At $s = 200$ ft, $t = 5\ln[0.2(200) + 10] - 5\ln 10 = 8.05$ s. Therefore, for the second segment of motion,

$$200 \text{ ft} < s \le 400 \text{ ft}; \quad v = 50; \quad dt = \frac{ds}{v} = \frac{ds}{50}$$

$$\int_{8.05}^t dt = \int_{200}^s \frac{ds}{50}$$

$$t - 8.05 = \frac{s}{50} - 4$$

$$t = \frac{s}{50} + 4.05$$

Therefore, at $s = 400$ ft,

$$t = \frac{400}{50} + 4.05 = 12.0 \text{ s} \qquad \qquad \textit{Ans.}$$

12.4 General Curvilinear Motion

Curvilinear motion occurs when the particle moves along a curved path. Since this path is often described in three dimensions, vector analysis will be used to formulate the particle's position, velocity, and acceleration.* In this section the general aspects of curvilinear motion are discussed, and in subsequent sections three types of coordinate systems often used to analyze this motion will be introduced.

Position. Consider a particle located at point P on a space curve defined by the path function s, Fig. 12–16a. The position of the particle, measured from a fixed point O, will be designated by the *position vector* $\mathbf{r} = \mathbf{r}(t)$. This vector is a function of time since, in general, both its magnitude and direction change as the particle moves along the curve.

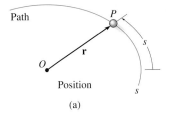

Position

(a)

Displacement. Suppose that during a small time interval Δt the particle moves a distance Δs along the curve to a new position P', defined by $\mathbf{r}' = \mathbf{r} + \Delta \mathbf{r}$, Fig. 12–16b. The *displacement* $\Delta \mathbf{r}$ represents the change in the particle's position and is determined by vector subtraction; i.e., $\Delta \mathbf{r} = \mathbf{r}' - \mathbf{r}$.

Velocity. During the time Δt, the *average velocity* of the particle is defined as

$$\mathbf{v}_{\text{avg}} = \frac{\Delta \mathbf{r}}{\Delta t}$$

The *instantaneous velocity* is determined from this equation by letting $\Delta t \to 0$, and consequently the direction of $\Delta \mathbf{r}$ *approaches* the *tangent* to the curve at point P. Hence, $\mathbf{v} = \lim_{\Delta t \to 0} (\Delta \mathbf{r}/\Delta t)$ or

$$\boxed{\mathbf{v} = \frac{d\mathbf{r}}{dt}} \qquad (12\text{–}7)$$

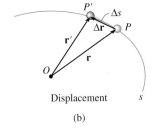

Displacement

(b)

Since $d\mathbf{r}$ will be tangent to the curve at P, the *direction* of $\mathbf{v}$ is also *tangent to the curve*, Fig. 12–16c. The *magnitude* of $\mathbf{v}$, which is called the *speed*, may be obtained by noting that the magnitude of the displacement $\Delta \mathbf{r}$ is the length of the straight line segment from P to P', Fig. 12–16b. Realizing that this length, Δr, approaches the arc length Δs as $\Delta t \to 0$, we have $v = \lim_{\Delta t \to 0} (\Delta r/\Delta t) = \lim_{\Delta t \to 0} (\Delta s/\Delta t)$, or

$$\boxed{v = \frac{ds}{dt}} \qquad (12\text{–}8)$$

Thus, the *speed* can be obtained by differentiating the path function s with respect to time.

*A summary of some of the important concepts of vector analysis is given in Appendix C.

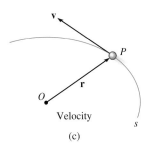

Velocity

(c)

Fig. 12–16

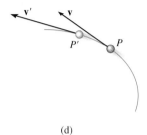

(d)

Acceleration.   If the particle has a velocity **v** at time t and a velocity $\mathbf{v}' = \mathbf{v} + \Delta\mathbf{v}$ at $t + \Delta t$, Fig. 12–16d, then the *average acceleration* of the particle during the time interval Δt is

$$\mathbf{a}_{\text{avg}} = \frac{\Delta\mathbf{v}}{\Delta t}$$

where $\Delta\mathbf{v} = \mathbf{v}' - \mathbf{v}$. To study this time rate of change, the two velocity vectors in Fig. 12–16d are plotted in Fig. 12–16e such that their tails are located at the fixed point O' and their arrowheads touch points on the curve. This curve is called a *hodograph*, and when constructed, it describes the locus of points for the arrowhead of the velocity vector in the same manner as the *path s* describes the locus of points for the arrowhead of the position vector, Fig. 12–16a.

To obtain the *instantaneous acceleration*, let $\Delta t \to 0$ in the above equation. In the limit $\Delta\mathbf{v}$ will approach the *tangent to the hodograph*, and so $\mathbf{a} = \lim_{\Delta t \to 0} (\Delta\mathbf{v}/\Delta t)$, or

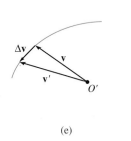

(e)

$$\mathbf{a} = \frac{d\mathbf{v}}{dt} \tag{12–9}$$

Substituting Eq. 12–7 into this result, we can also write

$$\mathbf{a} = \frac{d^2\mathbf{r}}{dt^2}$$

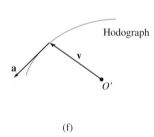

(f)

By definition of the derivative, **a** acts *tangent to the hodograph*, Fig. 12–16f, and therefore, *in general*, **a** *is not tangent to the path of motion*, Fig. 12–16g. To clarify this point, realize that $\Delta\mathbf{v}$ and consequently **a** must account for the change made in *both* the magnitude *and* direction of the velocity **v** as the particle moves from P to P', Fig. 12–16d. Just a magnitude change increases (or decreases) the "length" of **v**, and this in itself would allow **a** to remain tangent to the path. However, in order for the particle to follow the path, the directional change always "swings" the velocity vector toward the "inside" or "concave side" of the path, and therefore **a** *cannot* remain tangent to the path. In summary, **v** is always tangent to the *path* and **a** is always tangent to the *hodograph*.

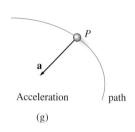

(g)

Fig. 12–16

12.5   Curvilinear Motion: Rectangular Components

Occasionally the motion of a particle can best be described along a path that is represented using a fixed *x, y, z* frame of reference.

Position.   If at a given instant the particle *P* is at point (*x, y, z*) on the curved path *s*, Fig. 12–17*a*, its location is then defined by the *position vector*

$$\mathbf{r} = x\mathbf{i} + y\mathbf{j} + z\mathbf{k} \qquad (12\text{–}10)$$

Because of the particle motion and the shape of the path, the *x, y, z* components of **r** are generally all functions of time; i.e., $x = x(t)$, $y = y(t)$, $z = z(t)$, so that $\mathbf{r} = \mathbf{r}(t)$.

In accordance with the discussion in Appendix C, the *magnitude* of **r** is *always positive* and defined from Eq. C–3 as

$$r = \sqrt{x^2 + y^2 + z^2}$$

The *direction* of **r** is specified by the components of the unit vector $\mathbf{u}_r = \mathbf{r}/r$.

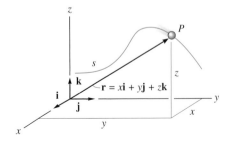

Position

(a)

Velocity.   The first time derivative of **r** yields the velocity **v** of the particle. Hence,

$$\mathbf{v} = \frac{d\mathbf{r}}{dt} = \frac{d}{dt}(x\mathbf{i}) + \frac{d}{dt}(y\mathbf{j}) + \frac{d}{dt}(z\mathbf{k})$$

When taking this derivative, it is necessary to account for changes in *both* the magnitude and direction of each of the vector's components. The derivative of the **i** component of **v** is therefore

$$\frac{d}{dt}(x\mathbf{i}) = \frac{dx}{dt}\mathbf{i} + x\frac{d\mathbf{i}}{dt}$$

The second term on the right side is zero, since the *x, y, z* reference frame is *fixed*, and therefore the *direction* (and the *magnitude*) of **i** does not change with time. Differentiation of the **j** and **k** components may be carried out in a similar manner, which yields the final result,

$$\mathbf{v} = \frac{d\mathbf{r}}{dt} = v_x\mathbf{i} + v_y\mathbf{j} + v_z\mathbf{k} \qquad (12\text{–}11)$$

where

$$v_x = \dot{x} \quad v_y = \dot{y} \quad v_z = \dot{z} \qquad (12\text{–}12)$$

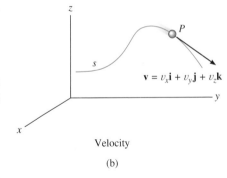

Velocity

(b)

Fig. 12–17

The "dot" notation $\dot{x}$, $\dot{y}$, $\dot{z}$ represents the first time derivatives of the parametric equations $x = x(t)$, $y = y(t)$, $z = z(t)$, respectively.

The velocity has a *magnitude* defined as the positive value of

$$v = \sqrt{v_x^2 + v_y^2 + v_z^2}$$

and a *direction* that is specified by the components of the unit vector $\mathbf{u}_v = \mathbf{v}/v$. This direction is *always tangent to the path*, as shown in Fig. 12–17b.

Acceleration.   The acceleration of the particle is obtained by taking the first time derivative of Eq. 12–11 (or the second time derivative of Eq. 12–10). Using dots to represent the derivatives of the components, we have

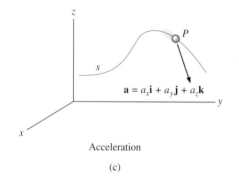

$\mathbf{a} = a_x\mathbf{i} + a_y\mathbf{j} + a_z\mathbf{k}$

Acceleration

(c)

Fig. 12–17

$$\mathbf{a} = \frac{d\mathbf{v}}{dt} = a_x\mathbf{i} + a_y\mathbf{j} + a_z\mathbf{k} \qquad (12\text{–}13)$$

where

$$
\begin{aligned}
a_x &= \dot{v}_x = \ddot{x} \\
a_y &= \dot{v}_y = \ddot{y} \\
a_z &= \dot{v}_z = \ddot{z}
\end{aligned}
\qquad (12\text{–}14)
$$

Here a_x, a_y, a_z represent, respectively, the first time derivatives of the functions $v_x = v_x(t)$, $v_y = v_y(t)$, $v_z = v_z(t)$, or the second time derivatives of the functions $x = x(t)$, $y = y(t)$, $z = z(t)$.

The acceleration has a *magnitude* defined by the positive value of

$$a = \sqrt{a_x^2 + a_y^2 + a_z^2}$$

and a *direction* specified by the components of the unit vector $\mathbf{u}_a = \mathbf{a}/a$. Since $\mathbf{a}$ represents the time rate of *change* in velocity, in general $\mathbf{a}$ will *not* be tangent to the path, Fig. 12–17c.

IMPORTANT POINTS

- Curvilinear motion can cause changes in *both* the magnitude and direction of the position, velocity, and acceleration vectors.
- The velocity vector is always directed *tangent* to the path.
- In general, the acceleration vector is *not* tangent to the path, but rather, it is tangent to the hodograph.
- If the motion is described using rectangular coordinates, then the components along each of the axes do not change direction, only their magnitude and sense (algebraic sign) will change.
- By considering the component motions, the direction of motion of the particle is automatically taken into account.

PROCEDURE FOR ANALYSIS

Coordinate System

- A rectangular coordinate system can be used to solve problems for which the motion can conveniently be expressed in terms of its x, y, z components.

Kinematic Quantities

- Since *rectilinear motion* occurs along *each* coordinate axis, the motion of each component is found using $v = ds/dt$ and $a = dv/dt$; or in cases where the motion is not expressed as a function of time, the equation $a\, ds = v\, dv$ can be used.
- Once the x, y, z components of $\mathbf{v}$ and $\mathbf{a}$ have been determined, the magnitudes of these vectors are found from the Pythagorean theorem, Eq. C–3, and their directions from the components of their unit vectors, Eqs. C–4 and C–5.

As the airplane takes off, its path of motion can be established by knowing its horizontal position $x = x(t)$, and its vertical position or altitude $y = y(t)$, both of which can be found from navigation equipment. By plotting the results from these equations the path can be shown, and by taking the time derivatives, the velocity and acceleration of the plane at any instant can be determined.

EXAMPLE 12.9

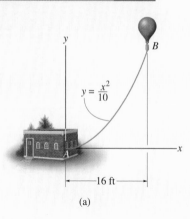

(a)

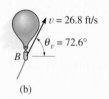

(b)

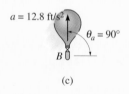

(c)

Fig. 12–18

At any instant the horizontal position of the weather balloon in Fig. 12–18a is defined by $x = (8t)$ ft, where t is in seconds. If the equation of the path is $y = x^2/10$, determine (a) the distance of the balloon from the station at A when $t = 2$ s, (b) the magnitude and direction of the velocity when $t = 2$ s, and (c) the magnitude and direction of the acceleration when $t = 2$ s.

Solution

Position. When $t = 2$ s, $x = 8(2)$ ft $= 16$ ft, and so

$$y = (16)^2/10 = 25.6 \text{ ft}$$

The straight-line distance from A to B is therefore

$$r = \sqrt{(16)^2 + (25.6)^2} = 30.2 \text{ ft} \qquad Ans.$$

Velocity. Using Eqs. 12–12 and application of the chain rule of calculus the components of velocity when $t = 2$ s are

$$v_x = \dot{x} = \frac{d}{dt}(8t) = 8 \text{ ft/s} \rightarrow$$

$$v_y = \dot{y} = \frac{d}{dt}(x^2/10) = 2x\dot{x}/10 = 2(16)(8)/10 = 25.6 \text{ ft/s} \uparrow$$

When $t = 2$ s, the magnitude of velocity is therefore

$$v = \sqrt{(8)^2 + (25.6)^2} = 26.8 \text{ ft/s} \qquad Ans.$$

The direction is tangent to the path, Fig. 12–18b, where

$$\theta_v = \tan^{-1}\frac{v_y}{v_x} = \tan^{-1}\frac{25.6}{8} = 72.6° \qquad Ans.$$

Acceleration. The components of acceleration are determined from Eqs. 12–14 and application of the chain rule, noting that $\ddot{x} = d^2(8t)/dt^2 = 0$. We have

$$a_x = \dot{v}_x = 0$$

$$a_y = \dot{v}_y = \frac{d}{dt}(2x\dot{x}/10) = 2(\dot{x})\dot{x}/10 + 2x(\ddot{x})/10$$

$$= 2(8)^2/10 + 2(16)(0)/10 = 12.8 \text{ ft/s}^2 \uparrow$$

Thus

$$a = \sqrt{(0)^2 + (12.8)^2} = 12.8 \text{ ft/s}^2 \qquad Ans.$$

The direction of **a**, as shown in Fig. 12–18c, is

$$\theta_a = \tan^{-1}\frac{12.8}{0} = 90° \qquad Ans.$$

Note: It is also possible to obtain v_y and a_y by first expressing $y = f(t) = (8t)^2/10 = 6.4t^2$ and then taking successive time derivatives.

E X A M P L E 12.10

The motion of a box B moving along the spiral conveyor shown in Fig. 12–19 is defined by the position vector $\mathbf{r} = \{0.5 \sin(2t)\mathbf{i} + 0.5 \cos(2t)\mathbf{j} - 0.2t\mathbf{k}\}$ m, where t is in seconds and the arguments for sine and cosine are in radians (π rad $= 180°$). Determine the location of the box when $t = 0.75$ s and the magnitudes of its velocity and acceleration at this instant.

Solution

Position. Evaluating $\mathbf{r}$ when $t = 0.75$ s yields

$$\mathbf{r}|_{t=0.75\ \text{s}} = \{0.5 \sin(1.5\ \text{rad})\mathbf{i} + 0.5 \cos(1.5\ \text{rad})\mathbf{j} - 0.2(0.75)\mathbf{k}\}\ \text{m}$$
$$= \{0.499\mathbf{i} + 0.0354\mathbf{j} - 0.150\mathbf{k}\}\ \text{m} \qquad \textit{Ans.}$$

The distance of the box from the origin O is

$$r = \sqrt{(0.499)^2 + (0.0354)^2 + (-0.150)^2} = 0.522\ \text{m} \qquad \textit{Ans.}$$

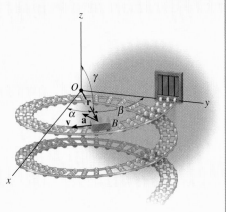

Fig. 12–19

The direction of $\mathbf{r}$ is obtained from the components of the unit vector,

$$\mathbf{u}_r = \frac{\mathbf{r}}{r} = \frac{0.499}{0.522}\mathbf{i} + \frac{0.0354}{0.522}\mathbf{j} - \frac{0.150}{0.522}\mathbf{k}$$
$$= 0.955\mathbf{i} + 0.0678\mathbf{j} - 0.287\mathbf{k}$$

Hence, the coordinate direction angles α, β, γ, Fig. 12–19, are

$$\alpha = \cos^{-1}(0.955) = 17.2° \qquad \textit{Ans.}$$
$$\beta = \cos^{-1}(0.0678) = 86.1° \qquad \textit{Ans.}$$
$$\gamma = \cos^{-1}(-0.287) = 107° \qquad \textit{Ans.}$$

Velocity. The velocity is defined by

$$\mathbf{v} = \frac{d\mathbf{r}}{dt} = \frac{d}{dt}[0.5 \sin(2t)\mathbf{i} + 0.5 \cos(2t)\mathbf{j} - 0.2t\mathbf{k}]$$
$$= \{1 \cos(2t)\mathbf{i} - 1 \sin(2t)\mathbf{j} - 0.2\mathbf{k}\}\ \text{m/s}$$

Hence, when $t = 0.75$ s the magnitude of velocity, or the speed, is

$$v = \sqrt{v_x^2 + v_y^2 + v_z^2}$$
$$= \sqrt{[1 \cos(1.5\ \text{rad})]^2 + [-1 \sin(1.5\ \text{rad})]^2 + (-0.2)^2}$$
$$= 1.02\ \text{m/s} \qquad \textit{Ans.}$$

The velocity is tangent to the path as shown in Fig. 12–19. Its coordinate direction angles can be determined from $\mathbf{u}_v = \mathbf{v}/v$.

Acceleration. The acceleration $\mathbf{a}$ of the box, which is shown in Fig. 12–19, is *not* tangent to the path. Show that

$$\mathbf{a} = \frac{d\mathbf{v}}{dt} = \{-2 \sin(2t)\mathbf{i} - 2 \cos(2t)\mathbf{j}\}\ \text{m/s}^2$$

At $t = 0.75$s, $a = 2$ m/s^2 \qquad \textit{Ans.}

12.6 Motion of a Projectile

The free-flight motion of a projectile is often studied in terms of its rectangular components, since the projectile's acceleration *always* acts in the vertical direction. To illustrate the kinematic analysis, consider a projectile launched at point (x_0, y_0), as shown in Fig. 12–20. The path is defined in the x–y plane such that the initial velocity is $\mathbf{v}_0$, having components $(\mathbf{v}_0)_x$ and $(\mathbf{v}_0)_y$. When air resistance is neglected, the only force acting on the projectile is its weight, which causes the projectile to have a *constant downward acceleration* of approximately $a_c = g = 9.81 \text{ m/s}^2$ or $g = 32.2 \text{ ft/s}^2$.*

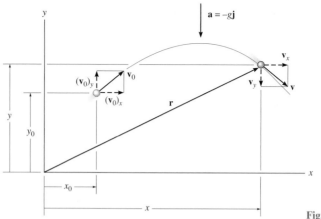

Fig. 12–20

Horizontal Motion. Since $a_x = 0$, application of the constant acceleration equations, 12–4 to 12–6, yields

$$(\xrightarrow{+})\, v = v_0 + a_c t; \qquad\qquad v_x = (v_0)_x$$
$$(\xrightarrow{+})\, x = x_0 + v_0 t + \tfrac{1}{2} a_c t^2; \qquad x = x_0 + (v_0)_x t$$
$$(\xrightarrow{+})\, v^2 = v_0^2 + 2a_c(s - s_0); \qquad v_x = (v_0)_x$$

The first and last equations indicate that *the horizontal component of velocity always remains constant during the motion.*

Vertical Motion. Since the positive y axis is directed upward, then $a_y = -g$. Applying Eqs. 12–4 to 12–6, we get

$$(+\uparrow)\, v = v_0 + a_c t; \qquad\qquad v_y = (v_0)_y - gt$$
$$(+\uparrow)\, y = y_0 + v_0 t + \tfrac{1}{2} a_c t^2; \qquad y = y_0 + (v_0)_y t - \tfrac{1}{2} g t^2$$
$$(+\uparrow)\, v^2 = v_0^2 + 2a_c(y - y_0); \qquad v_y^2 = (v_0)_y^2 - 2g(y - y_0)$$

Recall that the last equation can be formulated on the basis of eliminating the time t between the first two equations, and therefore *only two of the above three equations are independent of one another.*

*This assumes that the earth's gravitational field does not vary with altitude.

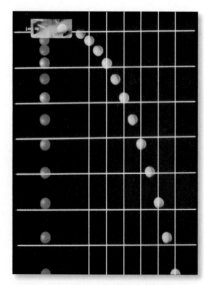

Each picture in this sequence is taken after the same time interval. The red ball falls from rest, whereas the yellow ball is given a horizontal velocity when released. Notice both balls are subjected to the same downward acceleration since they remain at the same elevation at any instant. This acceleration causes the difference in elevation to increase between successive photos. Also, note the horizontal distance between successive photos of the yellow ball is constant since the velocity in the horizontal direction remains constant.

To summarize, problems involving the motion of a projectile can have at most three unknowns since only three independent equations can be written; that is, *one* equation in the *horizontal direction* and *two* in the *vertical direction*. Once $\mathbf{v}_x$ and $\mathbf{v}_y$ are obtained, the resultant velocity $\mathbf{v}$, which is *always tangent* to the path, is defined by the *vector sum* as shown in Fig. 12–20.

PROCEDURE FOR ANALYSIS

Free-flight projectile motion problems can be solved using the following procedure.

Coordinate System

- Establish the fixed x, y coordinate axes and sketch the trajectory of the particle. Between any *two points* on the path specify the given problem data and the *three unknowns*. In all cases the acceleration of gravity acts downward. The particle's initial and final velocities should be represented in terms of their x and y components.

- Remember that positive and negative position, velocity, and acceleration components always act in accordance with their associated coordinate directions.

Kinematic Equations

- Depending upon the known data and what is to be determined, a choice should be made as to which three of the following four equations should be applied between the two points on the path to obtain the most direct solution to the problem.

Horizontal Motion

- The *velocity* in the horizontal or x direction is *constant*, i.e., $(v_x) = (v_0)_x$, and

$$x = x_0 + (v_0)_x t$$

Vertical Motion

- In the vertical or y direction *only two* of the following three equations can be used for solution.

$$v_y = (v_0)_y + a_c t$$
$$y = y_0 + (v_0)_y t + \frac{1}{2} a_c t^2$$
$$v_y^2 = (v_0)_y^2 + 2a_c(y - y_0)$$

- For example, if the particle's final velocity v_y is not needed, then the first and third of these equations (for y) will not be useful.

Gravel falling off the end of this conveyor belt follows a path that can be predicted using the equations of constant acceleration. In this way the location of the accumulated pile can be determined. Rectilinear coordinates are used for the analysis since the acceleration is only in the vertical direction.

EXAMPLE **12.11**

A sack slides off the ramp, shown in Fig. 12–21, with a horizontal velocity of 12 m/s. If the height of the ramp is 6 m from the floor, determine the time needed for the sack to strike the floor and the range R where sacks begin to pile up.

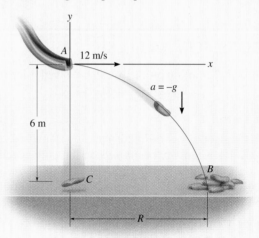

Fig. 12–21

Solution

Coordinate System. The origin of coordinates is established at the beginning of the path, point A, Fig. 12–21. The initial velocity of a sack has components $(v_A)_x = 12$ m/s and $(v_A)_y = 0$. Also, between points A and B the acceleration is $a_y = -9.81$ m/s². Since $(v_B)_x = (v_A)_x = 12$ m/s, the three unknowns are $(v_B)_y$, R, and the time of flight t_{AB}. Here we do not need to determine $(v_B)_y$.

Vertical Motion. The vertical distance from A to B is known, and therefore we can obtain a direct solution for t_{AB} by using the equation

$(+\uparrow)$ $\qquad\qquad y = y_0 + (v_0)_y t_{AB} + \frac{1}{2}a_c t^2_{AB}$

$$-6 \text{ m} = 0 + 0 + \tfrac{1}{2}(-9.81 \text{ m/s}^2)t^2_{AB}$$

$$t_{AB} = 1.11 \text{ s} \qquad\qquad\qquad Ans.$$

This calculation also indicates that if a sack were released *from rest* at A, it would take the same amount of time to strike the floor at C, Fig. 12–21.

Horizontal Motion. Since t has been calculated, R is determined as follows:

$(\xrightarrow{+})$ $\qquad\qquad\qquad\qquad x = x_0 + (v_0)_x t_{AB}$

$$R = 0 + 12 \text{ m/s} (1.11 \text{ s})$$

$$R = 13.3 \text{ m} \qquad\qquad\qquad Ans.$$

EXAMPLE 12.12

The chipping machine is designed to eject wood chips at $v_O = 25$ ft/s as shown in Fig. 12–22. If the tube is oriented at $30°$ from the horizontal, determine how high, h, the chips strike the pile if they land on the pile 20 ft from the tube.

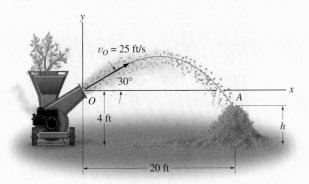

Fig. 12–22

Solution

Coordinate System. When the motion is analyzed between points O and A, the three unknowns are represented as the height h, time of flight t_{OA}, and vertical component of velocity $(v_A)_y$. (Note that $(v_A)_x = (v_O)_x$.) With the origin of coordinates at O, Fig. 12–22, the initial velocity of a chip has components of

$$(v_O)_x = (25 \cos 30°) \text{ ft/s} = 21.65 \text{ ft/s} \rightarrow$$
$$(v_O)_y = (25 \sin 30°) \text{ ft/s} = 12.5 \text{ ft/s} \uparrow$$

Also, $(v_A)_x = (v_O)_x = 21.65$ ft/s and $a_y = -32.2$ ft/s². Since we do not need to determine $(v_A)_y$, we have

Horizontal Motion

$(\xrightarrow{+})$
$$x_A = x_O + (v_O)_x t_{OA}$$
$$20 \text{ ft} = 0 + (21.65 \text{ ft/s}) t_{OA}$$
$$t_{OA} = 0.9238 \text{ s}$$

Vertical Motion. Relating t_{OA} to the initial and final elevations of a chip, we have

$(+\uparrow)$ $\quad y_A = y_O + (v_O)_y t_{OA} + \frac{1}{2} a_c t_{OA}^2$
$$(h - 4 \text{ ft}) = 0 + (12.5 \text{ ft/s})(0.9238 \text{ s}) + \frac{1}{2}(-32.2 \text{ ft/s}^2)(0.9238 \text{ s})^2$$
$$h = 1.81 \text{ ft} \qquad\qquad Ans.$$

EXAMPLE 12.13

The track for this racing event was designed so that riders jump off the slope at 30°, from a height of 1 m. During a race it was observed that the rider shown in Fig. 12–23a remained in mid air for 1.5 s. Determine the speed at which he was traveling off the slope, the horizontal distance he travels before striking the ground, and the maximum height he attains. Neglect the size of the bike and rider.

(a)

Fig. 12–23

(b)

Solution

Coordinate System. As shown in Fig. 12–23b, the origin of the coordinates is established at A. Between the end points of the path AB the three unknowns are the initial speed v_A, range R, and the vertical component of velocity v_B.

Vertical Motion. Since the time of flight and the vertical distance between the ends of the path are known, we can determine v_A.

$(+\uparrow)$ $(s_B)_y = (s_A)_y + (v_A)_y t_{AB} + \frac{1}{2}a_c t^2_{AB}$

$-1 = 0 + v_A \sin 30° (1.5) + \frac{1}{2}(-9.81)(1.5)^2$

$v_A = 13.38 \text{ m/s} = 13.4 \text{ m/s}$ *Ans.*

Horizontal Motion. The range R can now be determined.

$(\xrightarrow{+})$ $(s_B)_x = (s_A)_x + (v_A)_x t_{AB}$

$R = 0 + 13.38 \cos 30°(1.5)$

$= 17.4 \text{ m}$ *Ans.*

In order to find the maximum height h we will consider the path AC, Fig. 12–23b. Here the three unknowns become the time of flight t_{AC}, the horizontal distance from A to C, and the height h. At the maximum height $(v_C)_y = 0$, and since v_A is known, we can determine h directly without considering t_{AC} using the following equation.

$(v_C)^2_y = (v_A)^2_y + 2a_c[(s_C)_y - (s_A)_y]$

$(0)^2 = (13.38 \sin 30°)^2 + 2(-9.81)[(h-1) - 0]$

$h = 3.28 \text{ m}$ *Ans.*

Show that the bike will strike the ground at B with a velocity having components of

$(v_B)_x = 11.6 \text{ m/s} \rightarrow, \quad (v_B)_y = 8.02 \text{ m/s} \downarrow$

12.7 Curvilinear Motion: Normal and Tangential Components

When the path along which a particle is moving is *known*, it is often convenient to describe the motion using n and t coordinates which act normal and tangent to the path, respectively, and at the instant considered have their *origin located at the particle.*

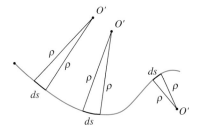

Position

(a)

Planar Motion. Consider the particle P shown in Fig. 12–24a, which is moving in a plane along a fixed curve, such that at a given instant it is at position s, measured from point O. We will now consider a coordinate system that has its origin at a *fixed point* on the curve, and at the instant considered this origin happens to *coincide* with the location of the particle. The t axis is *tangent* to the curve at P and is positive in the direction of *increasing s.* We will designate this positive direction with the unit vector $\mathbf{u}_t$. A unique choice for the *normal axis* can be made by noting that geometrically the curve is constructed from a series of differential arc segments ds, Fig. 12–24b. Each segment ds is formed from the arc of an associated circle having a *radius of curvature* ρ (rho) and *center of curvature* O'. The normal axis n is perpendicular to the t axis and is directed from P *toward* the center of curvature O', Fig. 12–24a. This positive direction, which is *always* on the concave side of the curve, will be designated by the unit vector $\mathbf{u}_n$. The plane which contains the n and t axes is referred to as the *osculating plane*, and in this case it is fixed in the plane of motion.*

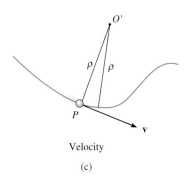

Radius of curvature

(b)

Velocity. Since the particle is moving, s is a function of time. As indicated in Sec. 12.4, the particle's velocity $\mathbf{v}$ has a *direction* that is *always tangent to the path*, Fig. 12–24c, and a *magnitude* that is determined by taking the time derivative of the path function $s = s(t)$, i.e., $v = ds/dt$ (Eq. 12–8). Hence

$$\mathbf{v} = v\mathbf{u}_t \qquad (12\text{–}15)$$

where

$$v = \dot{s} \qquad (12\text{–}16)$$

Velocity

(c)

Fig. 12–24

---

*The osculating plane may also be defined as that plane which has the greatest contact with the curve at a point. It is the limiting position of a plane contacting both the point and the arc segment ds. As noted above, the osculating plane is always coincident with a plane curve; however, each point on a three-dimensional curve has a unique osculating plane.

Acceleration. The acceleration of the particle is the time rate of change of the velocity. Thus,

$$\mathbf{a} = \dot{\mathbf{v}} = \dot{v}\mathbf{u}_t + v\dot{\mathbf{u}}_t \tag{12–17}$$

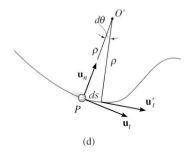

(d)

In order to determine the time derivative $\dot{\mathbf{u}}_t$, note that as the particle moves along the arc ds in time dt, $\mathbf{u}_t$ preserves its magnitude of unity; however, its *direction* changes, and becomes $\mathbf{u}_t'$, Fig. 12–24d. As shown in Fig. 12–24e, we require $\mathbf{u}_t' = \mathbf{u}_t + d\mathbf{u}_t$. Here $d\mathbf{u}_t$ stretches between the arrowheads of $\mathbf{u}_t$ and $\mathbf{u}_t'$, which lie on an infinitesimal arc of radius $u_t = 1$. Hence, $d\mathbf{u}_t$ has a *magnitude* of $du_t = (1) \, d\theta$, and its *direction* is defined by $\mathbf{u}_n$. Consequently, $d\mathbf{u}_t = d\theta\mathbf{u}_n$, and therefore the time derivative becomes $\dot{\mathbf{u}}_t = \dot{\theta}\mathbf{u}_n$. Since $ds = \rho \, d\theta$, Fig. 12–24d, then $\dot{\theta} = \dot{s}/\rho$, and therefore

$$\dot{\mathbf{u}}_t = \dot{\theta}\mathbf{u}_n = \frac{\dot{s}}{\rho}\mathbf{u}_n = \frac{v}{\rho}\mathbf{u}_n$$

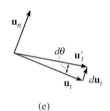

(e)

Substituting into Eq. 12–17, $\mathbf{a}$ can be written as the sum of its two components,

$$\boxed{\mathbf{a} = a_t\mathbf{u}_t + a_n\mathbf{u}_n} \tag{12–18}$$

where

$$\boxed{a_t = \dot{v}} \quad \text{or} \quad \boxed{a_t ds = v \, dv} \tag{12–19}$$

and

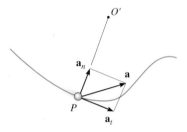

Acceleration

(f)

Fig. 12–24

$$\boxed{a_n = \frac{v^2}{\rho}} \tag{12–20}$$

These two mutually perpendicular components are shown in Fig. 12–24f, in which case the *magnitude* of acceleration is the positive value of

$$a = \sqrt{a_t^2 + a_n^2} \tag{12–21}$$

To summarize these concepts, consider the following two special cases of motion.

1. If the particle moves along a straight line, then $\rho \to \infty$ and from Eq. 12–20, $a_n = 0$. Thus $a = a_t = \dot{v}$, and we can conclude that the *tangential component of acceleration represents the time rate of change in the magnitude of the velocity.*

2. If the particle moves along a curve with a constant speed, then $a_t = \dot{v} = 0$ and $a = a_n = v^2/\rho$. Therefore, the *normal component of acceleration represents the time rate of change in the direction of the velocity.* Since $\mathbf{a}_n$ *always* acts towards the center of curvature, this component is sometimes referred to as the *centripetal acceleration.*

As a result of these interpretations, a particle moving along the curved path in Fig. 12–25 will have accelerations directed as shown.

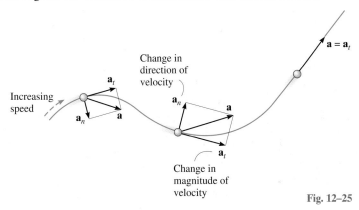

Fig. 12–25

Three-Dimensional Motion. If the particle is moving along a space curve, Fig. 12–26, then at a given instant the *t* axis is uniquely specified; however, an infinite number of straight lines can be constructed normal to the tangent axis at *P*. As in the case of planar motion, we will choose the positive *n* axis directed from *P* toward the path's center of curvature *O'*. This axis is referred to as the *principal normal* to the curve at *P*. With the *n* and *t* axes so defined, Eqs. 12–15 to 12–21 can be used to determine **v** and **a**. Since $\mathbf{u}_t$ and $\mathbf{u}_n$ are always perpendicular to one another and lie in the osculating plane, for spatial motion a third unit vector, $\mathbf{u}_b$, defines a *binormal axis b* which is perpendicular to $\mathbf{u}_t$ and $\mathbf{u}_n$, Fig. 12–26.

Since the three unit vectors are related to one another by the vector cross product, e.g., $\mathbf{u}_b = \mathbf{u}_t \times \mathbf{u}_n$, Fig. 12–26, it may be possible to use this relation to establish the direction of one of the axes, if the directions of the other two are known. For example, no motion occurs in the $\mathbf{u}_b$ direction, and so if this direction and $\mathbf{u}_t$ are known, then $\mathbf{u}_n$ can be determined, where in this case $\mathbf{u}_n = \mathbf{u}_b \times \mathbf{u}_t$, Fig. 12–26. Remember, though, that $\mathbf{u}_n$ is always on the concave side of the curve.

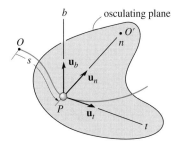

Fig. 12–26

Motorists traveling along this clover-leaf interchange experience a normal acceleration due to the change in direction of their velocity. A tangential component of acceleration occurs when the cars' speed is increased or decreased.

PROCEDURE FOR ANALYSIS

Coordinate System

- Provided the *path* of the particle is *known*, we can establish a set of n and t coordinates having a *fixed origin* which is coincident with the particle at the instant considered.
- The positive tangent axis acts in the direction of motion and the positive normal axis is directed toward the path's center of curvature.
- The n and t axes are particularly advantageous for studying the velocity and acceleration of the particle, because the t and n components of **a** are expressed by Eqs. 12–19 and 12–20, respectively.

Velocity

- The particle's *velocity* is always tangent to the path.
- The magnitude of velocity is found from the time derivative of the path function.

$$v = \dot{s}$$

Tangential Acceleration

- The tangential component of acceleration is the result of the time rate of change in the magnitude of velocity. This component acts in the positive s direction if the particle's speed is increasing or in the opposite direction if the speed is decreasing.
- The relations between a_t, v, t and s are the same as for rectilinear motion, namely,

$$a_t = \dot{v} \quad a_t\,ds = v\,dv$$

- If a_t is constant, $a_t = (a_t)_c$, the above equations, when integrated, yield

$$s = s_0 + v_0 t + \tfrac{1}{2}(a_t)_c t^2$$
$$v = v_0 + (a_t)_c t$$
$$v^2 = v_0^2 + 2(a_t)_c(s - s_0)$$

Normal Acceleration

- The normal component of acceleration is the result of the time rate of change in the direction of the particle's velocity. This component is *always* directed toward the center of curvature of the path, i.e., along the positive n axis.
- The magnitude of this component is determined from

$$a_n = \frac{v^2}{\rho}$$

- If the path is expressed as $y = f(x)$, the radius of curvature ρ at any point on the path is determined from the equation

$$\rho = \frac{[1 + (dy/dx)^2]^{3/2}}{|d^2y/dx^2|}$$

The derivation of this result is given in any standard calculus text.

E X A M P L E 12.14

When the skier reaches point A along the parabolic path in Fig. 12–27a, he has a speed of 6 m/s which is increasing at 2 m/s². Determine the direction of his velocity and the direction and magnitude of his acceleration at this instant. Neglect the size of the skier in the calculation.

Solution

Coordinate System.   Although the path has been expressed in terms of its x and y coordinates, we can still establish the origin of the n, t axes at the fixed point A on the path and determine the components of $\mathbf{v}$ and $\mathbf{a}$ along these axes, Fig. 12–27a.

Velocity.   By definition, the velocity is always directed tangent to the path. Since $y = \frac{1}{20}x^2$, $dy/dx = \frac{1}{10}x$, then $dy/dx|_{x=10} = 1$. Hence, at A, $\mathbf{v}$ makes an angle of $\theta = \tan^{-1} 1 = 45°$ with the x axis, Fig. 12–27. Therefore,

$$v_A = 6 \text{ m/s} \qquad 45° \; \nearrow \; \mathbf{v}_A \qquad \qquad Ans.$$

Acceleration.   The acceleration is determined from $\mathbf{a} = \dot{v}\mathbf{u}_t + (v^2/\rho)\mathbf{u}_n$. However, it is first necessary to determine the radius of curvature of the path at A (10 m, 5 m). Since $d^2y/dx^2 = \frac{1}{10}$, then

$$\rho = \frac{[1 + (dy/dx)^2]^{3/2}}{|d^2y/dx^2|} = \frac{[1 + (\frac{1}{10}x)^2]^{3/2}}{|\frac{1}{10}|}\bigg|_{x=10 \text{ m}} = 28.28 \text{ m}$$

The acceleration becomes

$$\mathbf{a}_A = \dot{v}\mathbf{u}_t + \frac{v^2}{\rho}\mathbf{u}_n$$

$$= 2\mathbf{u}_t + \frac{(6 \text{ m/s})^2}{28.28 \text{ m}}\mathbf{u}_n$$

$$= \{2\mathbf{u}_t + 1.273\mathbf{u}_n\} \text{ m/s}^2$$

As shown in Fig. 12–27b,

$$a = \sqrt{(2)^2 + (1.273)^2} = 2.37 \text{ m/s}^2$$

$$\phi = \tan^{-1}\frac{2}{1.273} = 57.5°$$

Thus, $57.5° - 45° = 12.5°$ so that,

$$a = 2.37 \text{ m/s}^2 \qquad 12.5° \; \nearrow \; \mathbf{a}_A \qquad \qquad Ans.$$

   Note: By using n, t coordinates, we were able to readily solve this problem since the n and t components account for the *separate* changes in the magnitude and direction of $\mathbf{v}$.

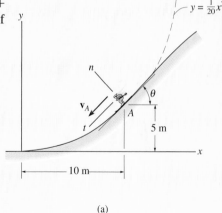

(a)

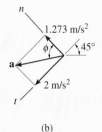

(b)

Fig. 12–27

EXAMPLE 12.15

A race car C travels around the horizontal circular track that has a radius of 300 ft, Fig. 12–28. If the car increases its speed at a constant rate of 7 ft/s², starting from rest, determine the time needed for it to reach an acceleration of 8 ft/s². What is its speed at this instant?

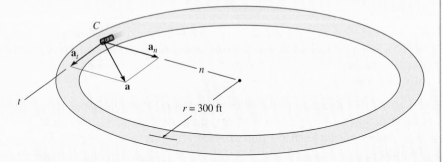

Fig. 12–28

Solution

Coordinate System. The origin of the n and t axes is coincident with the car at the instant considered. The t axis is in the direction of motion, and the positive n axis is directed toward the center of the circle. This coordinate system is selected since the path is known.

Acceleration. The magnitude of acceleration can be related to its components using $a = \sqrt{a_t^2 + a_n^2}$. Here $a_t = 7$ ft/s². Since $a_n = v^2/\rho$, the velocity as a function of time is

$$v = v_0 + (a_t)_c t$$
$$v = 0 + 7t$$

Thus

$$a_n = \frac{v^2}{\rho} = \frac{(7t)^2}{300} = 0.163t^2 \text{ ft/s}^2$$

The time needed for the acceleration to reach 8 ft/s² is therefore

$$a = \sqrt{a_t^2 + a_n^2}$$
$$8 = \sqrt{(7)^2 + (0.163t^2)^2}$$

Solving for the positive value of t yields

$$0.163t^2 = \sqrt{(8)^2 - (7)^2}$$
$$t = 4.87 \text{ s} \qquad \qquad Ans.$$

Velocity. The speed at time $t = 4.87$ s is

$$v = 7t = 7(4.87) = 34.1 \text{ ft/s} \qquad \qquad Ans.$$

EXAMPLE 12.16

The boxes in Fig. 12–29a travel along the industrial conveyor. If a box as in Fig. 12–29b starts from rest at A and increases its speed such that $a_t = (0.2t) \text{ m/s}^2$, where t is in seconds, determine the magnitude of its acceleration when it arrives at point B.

Solution

Coordinate System. The position of the box at any instant is defined from the fixed point A using the position or path coordinate s, Fig. 12–29b. The acceleration is to be determined at B, so the origin of the n, t axes is at this point.

Acceleration. To determine the acceleration components $a_t = \dot{v}$ and $a_n = v^2/\rho$, it is first necessary to formulate v and $\dot{v}$ so that they may be evaluated at B. Since $v_A = 0$ when $t = 0$, then

$$a_t = \dot{v} = 0.2t \qquad (1)$$

$$\int_0^v dv = \int_0^t 0.2t \, dt$$

$$v = 0.1t^2 \qquad (2)$$

The time needed for the box to reach point B can be determined by realizing that the position of B is $s_B = 3 + 2\pi(2)/4 = 6.142 \text{ m}$, Fig. 12–29b, and since $s_A = 0$ when $t = 0$ we have

$$v = \frac{ds}{dt} = 0.1t^2$$

$$\int_0^{6.142} ds = \int_0^{t_B} 0.1t^2 \, dt$$

$$6.142 = 0.0333t_B^3$$

$$t_B = 5.690 \text{ s}$$

Substituting into Eqs. 1 and 2 yields

$$(a_B)_t = \dot{v}_B = 0.2(5.690) = 1.138 \text{ m/s}^2$$
$$v_B = 0.1(5.69)^2 = 3.238 \text{ m/s}$$

At B, $\rho_B = 2 \text{ m}$, so that

$$(a_B)_n = \frac{v_B^2}{\rho_B} = \frac{(3.238 \text{ m/s})^2}{2 \text{ m}} = 5.242 \text{ m/s}^2$$

The magnitude of $\mathbf{a}_B$, Fig. 12–29c, is therefore

$$a_B = \sqrt{(1.138)^2 + (5.242)^2} = 5.36 \text{ m/s}^2 \qquad \textit{Ans.}$$

(a)

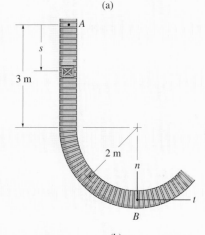

(b)

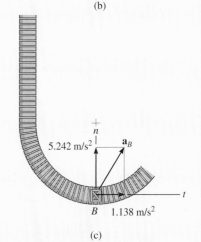

(c)

Fig. 12–29

12.8 Curvilinear Motion: Cylindrical Components

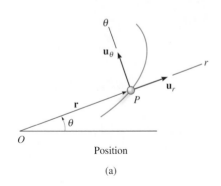

Position

(a)

Fig. 12–30

In some engineering problems it is often convenient to express the path of motion in terms of cylindrical coordinates, r, θ, z. If motion is restricted to the plane, the polar coordinates r and θ are used.

Polar Coordinates. We can specify the location of particle P shown in Fig. 12–30a using both the *radial coordinate* r, which extends outward from the fixed origin O to the particle, and a *transverse coordinate* θ, which is the counterclockwise angle between a fixed reference line and the r axis. The angle is generally measured in degrees or radians, where $1 \text{ rad} = 180°/\pi$. The positive directions of the r and θ coordinates are defined by the unit vectors $\mathbf{u}_r$ and $\mathbf{u}_\theta$, respectively. Here $\mathbf{u}_r$ or the radial direction $+r$ extends from P along increasing r, when θ is held fixed, and $\mathbf{u}_\theta$ or $+\theta$ extends from P in a direction that occurs when r is held fixed and θ is increased. Note that these directions are perpendicular to one another.

Position.   At any instant the position of the particle, Fig. 12–30a, is defined by the position vector

$$\mathbf{r} = r\mathbf{u}_r \qquad (12\text{–}22)$$

Velocity.   The instantaneous velocity $\mathbf{v}$ is obtained by taking the time derivative of $\mathbf{r}$. Using a dot to represent time differentiation, we have

$$\mathbf{v} = \dot{\mathbf{r}} = \dot{r}\mathbf{u}_r + r\dot{\mathbf{u}}_r$$

To evaluate $\dot{\mathbf{u}}_r$, notice that $\mathbf{u}_r$ changes only its direction with respect to time, since by definition the magnitude of this vector is always one unit. Hence, during the time Δt, a change Δr will not cause a change in the direction of $\mathbf{u}_r$; however, a change $\Delta\theta$ will cause $\mathbf{u}_r$ to become $\mathbf{u}_r'$, where $\mathbf{u}_r' = \mathbf{u}_r + \Delta\mathbf{u}_r$, Fig. 12–30b. The time change in $\mathbf{u}_r$ is then $\Delta\mathbf{u}_r$. For small angles $\Delta\theta$ this vector has a magnitude $\Delta u_r \approx 1(\Delta\theta)$ and acts in the $\mathbf{u}_\theta$ direction. Therefore, $\Delta\mathbf{u}_r = \Delta\theta\mathbf{u}_\theta$, and so

$$\dot{\mathbf{u}}_r = \lim_{\Delta t \to 0} \frac{\Delta\mathbf{u}_r}{\Delta t} = \left(\lim_{\Delta t \to 0} \frac{\Delta\theta}{\Delta t} \right) \mathbf{u}_\theta$$

$$\dot{\mathbf{u}}_r = \dot{\theta}\mathbf{u}_\theta \qquad (12\text{–}23)$$

Substituting into the above equation for $\mathbf{v}$, the velocity can be written in component form as

$$\boxed{\mathbf{v} = v_r\mathbf{u}_r + v_\theta\mathbf{u}_\theta} \qquad (12\text{–}24)$$

where

$$\boxed{\begin{aligned} v_r &= \dot{r} \\ v_\theta &= r\dot{\theta} \end{aligned}} \qquad (12\text{–}25)$$

These components are shown graphically in Fig. 12–30c. The *radial component* $\mathbf{v}_r$ is a measure of the rate of increase or decrease in the length of the radial coordinate, i.e., $\dot{r}$; whereas the *transverse component* $\mathbf{v}_\theta$ can be interpreted as the rate of motion along the circumference of a circle having a radius r. In particular, the term $\dot{\theta} = d\theta/dt$ is called the *angular velocity*, since it indicates the time rate of change of the angle θ. Common units used for this measurement are rad/s.

Since $\mathbf{v}_r$ and $\mathbf{v}_\theta$ are mutually perpendicular, the *magnitude* of velocity or speed is simply the positive value of

$$v = \sqrt{(\dot{r})^2 + (r\dot{\theta})^2} \qquad (12\text{–}26)$$

and the *direction* of $\mathbf{v}$ is, of course, tangent to the path at P, Fig. 12–30c.

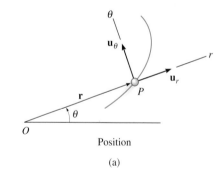

Position

(a)

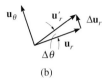

(b)

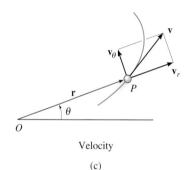

Velocity

(c)

Fig. 12–30

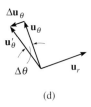

(d)

Acceleration. Taking the time derivatives of Eq. 12–24, using Eqs. 12–25, we obtain the particle's instantaneous acceleration,

$$\mathbf{a} = \dot{\mathbf{v}} = \ddot{r}\mathbf{u}_r + \dot{r}\dot{\mathbf{u}}_r + \dot{r}\dot{\theta}\mathbf{u}_\theta + r\ddot{\theta}\mathbf{u}_\theta + r\dot{\theta}\dot{\mathbf{u}}_\theta$$

To evaluate the term involving $\dot{\mathbf{u}}_\theta$, it is necessary only to find the change made in the direction of $\mathbf{u}_\theta$ since its magnitude is always unity. During the time Δt, a change Δr will not change the direction of $\mathbf{u}_\theta$, although a change $\Delta\theta$ will cause $\mathbf{u}_\theta$ to become $\mathbf{u}'_\theta$, where $\mathbf{u}'_\theta = \mathbf{u}_\theta + \Delta\mathbf{u}_\theta$, Fig. 12–30d. The time change in $\mathbf{u}_\theta$ is thus $\Delta\mathbf{u}_\theta$. For small angles this vector has a magnitude $\Delta u_\theta \approx 1(\Delta\theta)$ and acts in the $-\mathbf{u}_r$, direction; i.e., $\Delta\mathbf{u}_\theta = -\Delta\theta\mathbf{u}_r$. Thus,

$$\dot{\mathbf{u}}_\theta = \lim_{\Delta t \to 0} \frac{\Delta\mathbf{u}_\theta}{\Delta t} = -\left(\lim_{\Delta t \to 0} \frac{\Delta\theta}{\Delta t}\right)\mathbf{u}_r$$

$$\dot{\mathbf{u}}_\theta = -\dot{\theta}\mathbf{u}_r \qquad (12\text{–}27)$$

Substituting this result and Eq. 12–23 into the above equation for $\mathbf{a}$, we can write the acceleration in component form as

$$\boxed{\mathbf{a} = a_r\mathbf{u}_r + a_\theta\mathbf{u}_\theta} \qquad (12\text{–}28)$$

where

$$\boxed{\begin{array}{l} a_r = \ddot{r} - r\dot{\theta}^2 \\ a_\theta = r\ddot{\theta} + 2\dot{r}\dot{\theta} \end{array}} \qquad (12\text{–}29)$$

The term $\ddot{\theta} = d^2\theta/dt^2 = d/dt(d\theta/dt)$ is called the *angular acceleration* since it measures the change made in the angular velocity during an instant of time. Units for this measurement are rad/s².

Since $\mathbf{a}_r$, and $\mathbf{a}_\theta$ are always perpendicular, the *magnitude* of acceleration is simply the positive value of

$$a = \sqrt{(\ddot{r} - r\dot{\theta}^2)^2 + (r\ddot{\theta} + 2\dot{r}\dot{\theta})^2} \qquad (12\text{–}30)$$

The *direction* is determined from the vector addition of its two components. In general, $\mathbf{a}$ will *not* be tangent to the path, Fig. 12–30e.

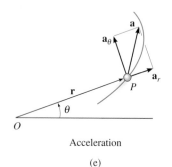

Acceleration

(e)

Fig. 12–30

Cylindrical Coordinates. If the particle P moves along a space curve as shown in Fig. 12–31, then its location may be specified by the three *cylindrical coordinates, $r\,\theta$, z*. The z coordinate is identical to that used for rectangular coordinates. Since the unit vector defining its direction, $\mathbf{u}_z$, is constant, the time derivatives of this vector are zero, and therefore the position, velocity, and acceleration of the particle can be written in terms of its cylindrical coordinates as follows:

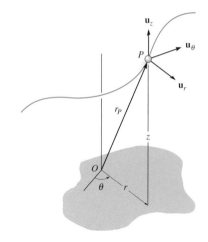

$$\mathbf{r}_p = r\mathbf{u}_r + z\mathbf{u}_z$$
$$\mathbf{v} = \dot{r}\mathbf{u}_r + r\dot{\theta}\mathbf{u}_\theta + \dot{z}\mathbf{u}_z$$
$$\mathbf{a} = (\ddot{r} - r\dot{\theta}^2)\mathbf{u}_r + (r\ddot{\theta} + 2\dot{r}\dot{\theta})\mathbf{u}_\theta + \ddot{z}\mathbf{u}_z$$

Fig. 12–31

Time Derivatives. The equations of kinematics require that we obtain the time derivatives $\dot{r}$, $\ddot{r}$, $\dot{\theta}$, and $\ddot{\theta}$ in order to evaluate the r and θ components of $\mathbf{v}$ and $\mathbf{a}$. Two types of problems generally occur:

1. If the coordinates are specified as time parametric equations, $r = r(t)$ and $\theta = \theta(t)$, then the time derivatives can be found directly. For example, consider

$$r = 4t^2 \qquad \theta = (8t^3 + 6)$$
$$\dot{r} = 8t \qquad \dot{\theta} = 24t^2$$
$$\ddot{r} = 8 \qquad \ddot{\theta} = 48t$$

2. If the time-parametric equations are not given, then it will be necessary to specify the path $r = f(\theta)$ and find the *relationship* between the time derivatives using the chain rule of calculus. Consider the following examples.

$$r = 5\theta^2$$
$$\dot{r} = 10\theta\dot{\theta}$$
$$\ddot{r} = 10[(\dot{\theta})\dot{\theta} + \theta(\ddot{\theta})]$$
$$= 10\dot{\theta}^2 + 10\theta\ddot{\theta}$$

The spiral motion of this boy can be followed by using cylindrical components. Here the radial coordinate r is constant, the transverse coordinate θ will increase with time as the boy rotates about the vertical, and his altitude z will decrease with time.

or

$$r^2 = 6\theta^3$$

$$2r\dot{r} = 18\theta^2\dot{\theta}$$

$$2[(\dot{r})\dot{r} + r(\ddot{r})] = 18[(2\theta\dot{\theta})\dot{\theta} + \theta^2(\ddot{\theta})]$$

$$\dot{r}^2 + r\ddot{r} = 9(2\theta\dot{\theta}^2 + \theta^2\ddot{\theta})$$

If two of the *four* time derivatives $\dot{r}$, $\ddot{r}$, $\dot{\theta}$, and $\ddot{\theta}$ are *known*, then the other two can be obtained from the equations for first and second time derivatives of $r = f(\theta)$. See Example 12.19. In some problems, however, two of these time derivatives may *not* be known; instead the magnitude of the particle's velocity or acceleration may be specified. If this is the case, Eqs. 12–26 and 12–30 $[v^2 = \dot{r}^2 + (\dot{r}\theta)^2$ and $a^2 = (\ddot{r} - r\dot{\theta}^2)^2 + (r\ddot{\theta} + 2\dot{r}\dot{\theta})^2]$ may be used to obtain the necessary relationships involving $\dot{r}$, $\ddot{r}$, $\dot{\theta}$, and $\ddot{\theta}$. See Example 12.20.

PROCEDURE FOR ANALYSIS

Coordinate System

- Polar coordinates are a suitable choice for solving problems for which data regarding the angular motion of the radial coordinate r is given to describe the particle's motion. Also, some paths of motion can conveniently be described in terms of these coordinates.
- To use polar coordinates, the origin is established at a fixed point, and the radial line r is directed to the particle.
- The transverse coordinate θ is measured from a fixed reference line to the radial line.

Velocity and Acceleration

- Once r and the four time derivatives $\dot{r}$, $\ddot{r}$, $\dot{\theta}$, and $\ddot{\theta}$ have been evaluated at the instant considered, their values can be substituted into Eqs. 12–25 and 12–29 to obtain the radial and transverse components of **v** and **a.**
- If it is necessary to take the time derivatives of $r = f(\theta)$, it is very important to use the chain rule of calculus.
- Motion in three dimensions requires a simple extension of the above procedure to include $\dot{z}$ and $\ddot{z}$.

Besides the examples which follow, further examples involving the calculation of a_r and a_θ can be found in the "kinematics" sections of Examples 13.10 through 13.12.

E X A M P L E 12.17

The amusement park ride shown in Fig. 12–32a consists of a chair that is rotating in a horizontal circular path of radius r such that the arm OB has an angular velocity $\dot{\theta}$ and angular acceleration $\ddot{\theta}$. Determine the radial and transverse components of velocity and acceleration of the passenger. Neglect his size in the calculation.

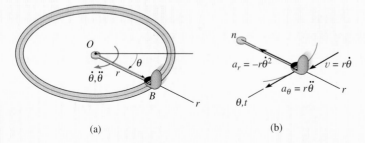

(a)   (b)

Fig. 12–32

Solution

Coordinate System.   Since the angular motion of the arm is reported, polar coordinates are chosen for the solution, Fig. 12–32a. Here θ is not related to r, since the radius is constant for all θ.

Velocity and Acceleration.   Equations 12–25 and 12–29 will be used for the solution, and so it is first necessary to specify the first and second time derivatives of r and θ. Since r is *constant*, we have

$$r = r \qquad \dot{r} = 0 \qquad \ddot{r} = 0$$

Thus

$$v_r = \dot{r} = 0 \qquad\qquad\qquad\qquad Ans.$$
$$v_\theta = r\dot{\theta} \qquad\qquad\qquad\qquad Ans.$$
$$a_r = \ddot{r} - r\dot{\theta}^2 = -r\dot{\theta}^2 \qquad\qquad Ans.$$
$$a_\theta = r\ddot{\theta} + 2\dot{r}\dot{\theta} = r\ddot{\theta} \qquad\qquad Ans.$$

These results are shown in Fig. 12–32b. Also shown are the n, t axes, which in this special case of circular motion happen to be *colinear* with the r and θ axes, respectively. In particular note that $v = v_\theta = v_t = r\dot{\theta}$. Also,

$$-a_r = a_n = \frac{v^2}{\rho} = \frac{(r\dot{\theta})^2}{r} = r\dot{\theta}^2$$

$$a_\theta = a_t = \frac{dv}{dt} = \frac{d}{dt}(r\dot{\theta}) = \frac{dr}{dt}\dot{\theta} + r\frac{d\dot{\theta}}{dt} = 0 + r\ddot{\theta}$$

E X A M P L E 12.18

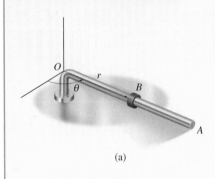

(a)

The rod OA in Fig. 12–33a is rotating in the horizontal plane such that $\theta = (t^3)$ rad. At the same time, the collar B is sliding outward along OA so that $r = (100t^2)$ mm. If in both cases t is in seconds, determine the velocity and acceleration of the collar when $t = 1$ s.

Solution

Coordinate System. Since time-parametric equations of the path are given, it is not necessary to relate r to θ.

Velocity and Acceleration. Determining the time derivatives and evaluating when $t = 1$ s, we have

$$r = 100t^2 \Big|_{t=1\,\text{s}} = 100 \text{ mm} \qquad \theta = t^3 \Big|_{t=1\,\text{s}} = 1 \text{ rad} = 57.3°$$

$$\dot{r} = 200t \Big|_{t=1\,\text{s}} = 200 \text{ mm/s} \qquad \dot{\theta} = 3t^2 \Big|_{t=1\,\text{s}} = 3 \text{ rad/s}$$

$$\ddot{r} = 200 \Big|_{t=1\,\text{s}} = 200 \text{ mm/s}^2 \qquad \ddot{\theta} = 6t \Big|_{t=1\,\text{s}} = 6 \text{ rad/s}^2.$$

As shown in Fig. 12–33b,

$$\mathbf{v} = \dot{r}\mathbf{u}_r + r\dot{\theta}\mathbf{u}_\theta$$
$$= 200\mathbf{u}_r + 100(3)\mathbf{u}_\theta$$
$$= \{200\mathbf{u}_r + 300\mathbf{u}_\theta\} \text{ mm/s}$$

The magnitude of $\mathbf{v}$ is

$$v = \sqrt{(200)^2 + (300)^2} = 361 \text{ mm/s} \qquad \qquad Ans.$$

$$\delta = \tan^{-1}\left(\frac{300}{200}\right) = 56.3° \qquad \delta + 57.3° = 114° \qquad Ans.$$

(b)

As shown in Fig. 12–33c,

$$\mathbf{a} = (\ddot{r} - r\dot{\theta}^2)\mathbf{u}_r + (r\ddot{\theta} + 2\dot{r}\dot{\theta})\mathbf{u}_\theta$$
$$= [200 - 100(3)^2]\mathbf{u}_r + [100(6) + 2(200)3]\mathbf{u}_\theta$$
$$= \{-700\mathbf{u}_r + 1800\mathbf{u}_\theta\} \text{ mm/s}^2$$

The magnitude of $\mathbf{a}$ is

$$a = \sqrt{(700)^2 + (1800)^2} = 1930 \text{ mm/s}^2 \qquad \qquad Ans.$$

$$\phi = \tan^{-1}\left(\frac{1800}{700}\right) = 68.7° \qquad (180° - \phi) + 57.3° = 169° \qquad Ans.$$

(c)

Fig. 12–33

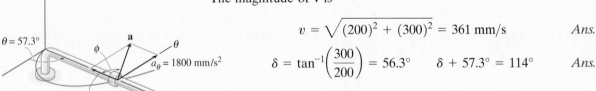

EXAMPLE 12.19

The searchlight in Fig. 12–34a casts a spot of light along the face of a wall that is located 100 m from the searchlight. Determine the magnitudes of the velocity and acceleration at which the spot appears to travel across the wall at the instant $\theta = 45°$. The searchlight is rotating at a constant rate of $\dot{\theta} = 4$ rad/s.

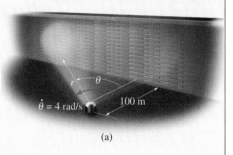

(a)

Solution

Coordinate System. Polar coordinates will be used to solve this problem since the angular rate of the searchlight is given. To find the necessary time derivatives it is first necessary to relate r to θ. From Fig. 12–34a, this relation is

$$r = 100/\cos\theta = 100\sec\theta$$

Velocity and Acceleration. Using the chain rule of calculus, noting that $d(\sec\theta) = \sec\theta\tan\theta\,d\theta$, and $d(\tan\theta) = \sec^2\theta\,d\theta$, we have

$$\dot{r} = 100(\sec\theta\tan\theta)\dot{\theta}$$
$$\ddot{r} = 100(\sec\theta\tan\theta)\dot{\theta}(\tan\theta)\dot{\theta} + 100\sec\theta(\sec^2\theta)\dot{\theta}(\dot{\theta})$$
$$+ 100\sec\theta\tan\theta(\ddot{\theta})$$
$$= 100\sec\theta\tan^2\theta(\dot{\theta})^2 + 100\sec^3\theta(\dot{\theta})^2 + 100(\sec\theta\tan\theta)\ddot{\theta}$$

Since $\dot{\theta} = 4$ rad/s = constant, then $\ddot{\theta} = 0$, and the above equations, when $\theta = 45°$, become

$$r = 100\sec 45° = 141.4$$
$$\dot{r} = 400\sec 45°\tan 45° = 565.7$$
$$\ddot{r} = 1600(\sec 45°\tan^2 45° + \sec^3 45°) = 6788.2$$

As shown in Fig. 12–34b,

$$\mathbf{v} = \dot{r}\mathbf{u}_r + r\dot{\theta}\mathbf{u}_\theta$$
$$= 565.7\mathbf{u}_r + 141.4(4)\mathbf{u}_\theta$$
$$= \{565.7\mathbf{u}_r + 565.7\mathbf{u}_\theta\}\text{ m/s}$$
$$v = \sqrt{v_r^2 + v_\theta^2} = \sqrt{(565.7)^2 + (565.7)^2}$$
$$= 800\text{ m/s} \qquad\qquad Ans.$$

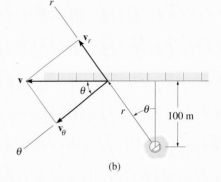

(b)

As shown in Fig. 12–34c,

$$\mathbf{a} = (\ddot{r} - r\dot{\theta}^2)\mathbf{u}_r + (r\ddot{\theta} + 2\dot{r}\dot{\theta})\mathbf{u}_\theta$$
$$= [6788.2 - 141.4(4)^2]\mathbf{u}_r + [141.4(0) + 2(565.7)4]\mathbf{u}_\theta$$
$$= \{4525.5\mathbf{u}_r + 4525.5\mathbf{u}_\theta\}\text{ m/s}^2$$
$$a = \sqrt{a_r^2 + a_\theta^2} = \sqrt{(4525.5)^2 + (4525.5)^2}$$
$$= 6400\text{ m/s}^2 \qquad\qquad Ans.$$

Note: It is also possible to find a without having to calculate $\ddot{r}$ (or a_r). As shown in Fig. 12–34d, since $a_\theta = 4525.5$ m/s², then by vector resolution, $a = 4525.5/\cos 45° = 6400$ m/s².

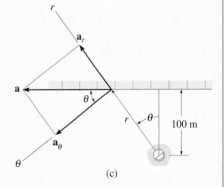

(c)

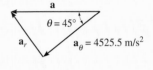

(d)

Fig. 12–34

E X A M P L E 12.20

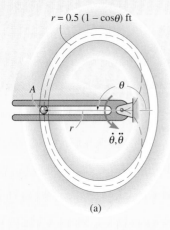

$r = 0.5 (1 - \cos\theta)$ ft

(a)

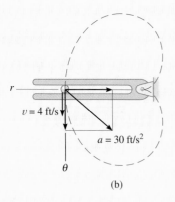

(b)

Fig. 12–35

Due to the rotation of the forked rod, the ball A in Fig. 12–35a travels around the slotted path, a portion of which is in the shape of a cardioid, $r = 0.5(1 - \cos\theta)$ ft, where θ is in radians. If the ball's velocity is $v = 4$ ft/s and its acceleration is $a = 30$ ft/s^2 at the instant $\theta = 180°$, determine the angular velocity $\dot{\theta}$ and angular acceleration $\ddot{\theta}$ of the fork.

Solution

Coordinate System. This path is most unusual, and mathematically it is best expressed using polar coordinates, as done here, rather than rectangular coordinates. Also, $\dot{\theta}$ and $\ddot{\theta}$ must be determined so r, θ coordinates are an obvious choice.

Velocity and Acceleration. Determining the time derivatives of r using the chain rule of calculus yields

$$r = 0.5(1 - \cos\theta)$$
$$\dot{r} = 0.5(\sin\theta)\dot{\theta}$$
$$\ddot{r} = 0.5(\cos\theta)\dot{\theta}(\dot{\theta}) + 0.5(\sin\theta)\ddot{\theta}$$

Evaluating these results at $\theta = 180°$, we have

$$r = 1 \text{ ft} \qquad \dot{r} = 0 \qquad \ddot{r} = -0.5\dot{\theta}^2$$

Since $v = 4$ ft/s, using Eq. 12–26 to determine $\dot{\theta}$ yields

$$v = \sqrt{(\dot{r})^2 + (r\dot{\theta})^2}$$
$$4 = \sqrt{(0)^2 + (1\dot{\theta})^2}$$
$$\dot{\theta} = 4 \text{ rad/s} \qquad\qquad Ans.$$

In a similar manner, $\ddot{\theta}$ can be found using Eq. 12–30.

$$a = \sqrt{(\ddot{r} - r\dot{\theta}^2)^2 + (r\ddot{\theta} + 2\dot{r}\dot{\theta})^2}$$
$$30 = \sqrt{[-0.5(4)^2 - 1(4)^2]^2 + [1\ddot{\theta} + 2(0)(4)]^2}$$
$$(30)^2 = (-24)^2 + \ddot{\theta}^2$$
$$\ddot{\theta} = 18 \text{ rad/s}^2 \qquad\qquad Ans.$$

Vectors **a** and **v** are shown in Fig. 12–35b.

12.9 Absolute Dependent Motion Analysis of Two Particles

In some types of problems the motion of one particle will *depend* on the corresponding motion of another particle. This dependency commonly occurs if the particles are interconnected by inextensible cords which are wrapped around pulleys. For example, the movement of block A downward along the inclined plane in Fig. 12–36 will cause a corresponding movement of block B up the other incline. We can show this mathematically by first specifying the location of the blocks using *position coordinates* s_A and s_B. Note that each of the coordinate axes is (1) referenced from a *fixed* point (O) or *fixed* datum line, (2) measured along each inclined plane in the direction of motion of block A and block B, and (3) has a positive sense from C to A and D to B. If the total cord length is l_T, the position coordinates are related by the equation

$$s_A + l_{CD} + s_B = l_T$$

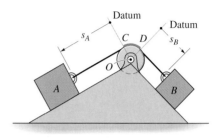

Fig. 12–36

Here l_{CD} is the length of the cord passing over arc CD. Taking the time derivative of this expression, realizing that l_{CD} and l_T *remain constant*, while s_A and s_B measure the lengths of the changing segments of the cord, we have

$$\frac{ds_A}{dt} + \frac{ds_B}{dt} = 0 \qquad \text{or} \qquad v_B = -v_A$$

The negative sign indicates that when block A has a velocity downward, i.e., in the direction of positive s_A, it causes a corresponding upward velocity of block B; i.e., B moves in the negative s_B direction.

In a similar manner, time differentiation of the velocities yields the relation between the accelerations, i.e.,

$$a_B = -a_A$$

A more complicated example involving dependent motion of two blocks is shown in Fig. 12–37a. In this case, the position of block A is specified by s_A, and the position of the *end* of the cord from which block B is suspended is defined by s_B. Here we have chosen coordinate axes which are (1) referenced from fixed points or datums, (2) measured in the direction of motion of each block, and (3) positive to the right (s_A) and positive downward (s_B). During the motion, the red colored segments of the cord in Fig. 12–37a *remain constant*. If l represents the total length of cord minus these segments, then the position coordinates can be related by the equation

$$2s_B + h + s_A = l$$

Since l and h are constant during the motion, the two time derivatives yield

$$2v_B = -v_A \qquad 2a_B = -a_A$$

Hence, when B moves downward ($+s_B$), A moves to the left ($-s_A$) with two times the motion.

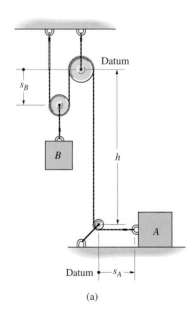

(a)

Fig. 12–37

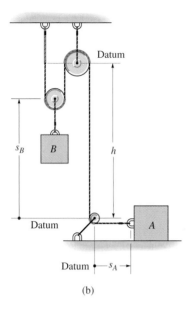

(b)

Fig. 12–37

The motion of the traveling block on this oil rig depends upon the motion of the cable connected to the winch which operates it. It is important to be able to relate these motions in order to determine the power requirements of the winch and the force in the cable caused by accelerated motion.

This example can also be worked by defining the position of block B from the center of the bottom pulley (a fixed point), Fig. 12–37b. In this case

$$2(h - s_B) + h + s_A = l$$

Time differentiation yields

$$2v_B = v_A \quad 2a_B = a_A$$

Here the signs are the same. Why?

PROCEDURE FOR ANALYSIS

The above method of relating the dependent motion of one particle to that of another can be performed using algebraic scalars or position coordinates provided each particle moves along a rectilinear path. When this is the case, only the magnitudes of the velocity and acceleration of the particles will change, not their line of direction. The following procedure is required.

Position-Coordinate Equation

- Establish position coordinates which have their origin located at a *fixed* point or datum.

- The coordinates are directed along the path of motion and extend to a point having the same motion as each of the particles.

- It is *not necessary* that the *origin* be the *same* for each of the coordinates; however, it is *important* that each coordinate axis selected be directed along the *path of motion* of the particle.

- Using geometry or trigonometry, relate the coordinates to the total length of the cord, l_T, or to that portion of cord, l, which *excludes* the segments that do not change length as the particles move—such as arc segments wrapped over pulleys.

- If a problem involves a *system* of two or more cords wrapped around pulleys, then the position of a point on one cord must be related to the position of a point on another cord using the above procedure. Separate equations are written for a fixed length of each cord of the system and the positions of the two particles are then related by these equations (see Examples 12.22 and 12.23).

Time Derivatives

- Two successive time derivatives of the position-coordinate equations yield the required velocity and acceleration equations which relate the motions of the particles.

- The signs of the terms in these equations will be consistent with those that specify the positive and negative sense of the position coordinates.

E X A M P L E 12.21

Determine the speed of block A in Fig. 12–38 if block B has an upward speed of 6 ft/s.

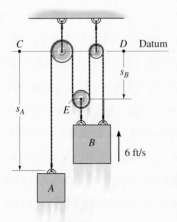

Fig. 12–38

Solution

Position-Coordinate Equation. There is *one cord* in this system having segments which are changing length. Position coordinates s_A and s_B will be used since each is measured from a fixed point (C or D) and extends along each block's *path of motion*. In particular, s_B is directed to point E since motion of B and E is the *same*.

The red colored segments of the cord in Fig. 12–38 remain at a constant length and do not have to be considered as the blocks move. The remaining length of cord, l, is also constant and is related to the changing position coordinates s_A and s_B by the equation

$$s_A + 3s_B = l$$

Time Derivative. Taking the time derivative yields

$$v_A + 3v_B = 0$$

so that when $v_B = -6$ ft/s (upward),

$$v_A = 18 \text{ ft/s} \downarrow \qquad\qquad Ans.$$

EXAMPLE 12.22

Determine the speed of block A in Fig. 12–39 if block B has an upward speed of 6 ft/s.

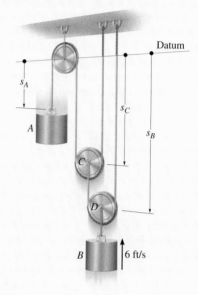

Fig. 12–39

Solution

Position-Coordinate Equation.   As shown, the positions of blocks A and B are defined using coordinates s_A and s_B. Since the system has *two* cords which change length, it will be necessary to use a third coordinate, s_C, in order to relate s_A to s_B. In other words, the length of one of the cords can be expressed in terms of s_A and s_C, and the length of the other cord can be expressed in terms of s_B and s_C.

The red colored segments of the cords in Fig. 12–39 do not have to be considered in the analysis. Why? For the remaining cord lengths, say l_1 and l_2, we have

$$s_A + 2s_C = l_1 \qquad s_B + (s_B - s_C) = l_2$$

Eliminating s_C yields an equation defining the positions of both blocks, i.e.,

$$s_A + 4s_B = 2l_2 + l_1$$

Time Derivative.   The time derivative gives

$$v_A + 4v_B = 0$$

so that when $v_B = -6$ ft/s (upward),

$$v_A = +24 \text{ ft/s} = 24 \text{ ft/s} \downarrow \qquad\qquad Ans.$$

EXAMPLE 12.23

Determine the speed with which block B rises in Fig. 12–40 if the end of the cord at A is pulled down with a speed of 2 m/s.

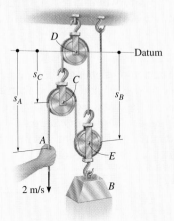

Fig. 12–40

Solution

Position-Coordinate Equation. The position of point A is defined by s_A, and the position of block B is specified by s_B since point E on the pulley will have the *same motion* as the block. Both coordinates are measured from a horizontal datum passing through the *fixed* pin at pulley D. Since the system consists of *two* cords, the coordinates s_A and s_B cannot be related directly. Instead, by establishing a third position coordinate, s_C, we can now express the length of one of the cords in terms of s_B and s_C, and the length of the other cord in terms of s_A, s_B, and s_C.

Excluding the red colored segments of the cords in Fig. 12–40, the remaining constant cord lengths l_1 and l_2 (along with the hook and link dimensions) can be expressed as

$$s_C + s_B = l_1$$
$$(s_A - s_C) + (s_B - s_C) + s_B = l_2$$

Eliminating s_C yields

$$s_A + 4s_B = l_2 + 2l_1$$

As required, this equation relates the position s_B of block B to the position s_A of point A.

Time Derivative. The time derivative gives

$$v_A + 4v_B = 0$$

so that when $v_A = 2$ m/s (downward),

$$v_B = -0.5 \text{ m/s} = 0.5 \text{ m/s} \uparrow \qquad\qquad \textit{Ans.}$$

E X A M P L E **12.24**

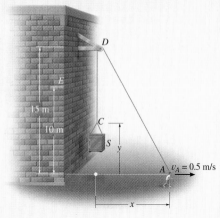

Fig. 12–41

A man at A is hoisting a safe S as shown in Fig. 12–41 by walking to the right with a constant velocity $v_A = 0.5$ m/s. Determine the velocity and acceleration of the safe when it reaches the elevation at E. The rope is 30 m long and passes over a small pulley at D.

Solution

Position-Coordinate Equation. This problem is unlike the previous examples since rope segment DA changes both direction and magnitude. However, the ends of the rope, which define the positions of S and A, are specified by means of the x and y coordinates measured from a fixed point and *directed along the paths of motion* of the ends of the rope.

The x and y coordinates may be related since the rope has a fixed length $l = 30$ m, which at all times is equal to the length of segment DA plus CD. Using the Pythagorean theorem to determine l_{DA}, we have $l_{DA} = \sqrt{(15)^2 + x^2}$; also, $l_{CD} = 15 - y$. Hence,

$$l = l_{DA} + l_{CD}$$

$$30 = \sqrt{(15)^2 + x^2} + (15 - y)$$

$$y = \sqrt{225 + x^2} - 15 \qquad (1)$$

Time Derivatives. Taking the time derivative, using the chain rule, where $v_S = dy/dt$ and $v_A = dx/dt$, yields

$$v_S = \frac{dy}{dt} = \left[\frac{1}{2}\frac{2x}{\sqrt{225 + x^2}}\right]\frac{dx}{dt}$$

$$= \frac{x}{\sqrt{225 + x^2}}v_A. \qquad (2)$$

At $y = 10$ m, x is determined from Eq. 1, i.e., $x = 20$ m. Hence, from Eq. 2 with $v_A = 0.5$ m/s,

$$v_S = \frac{20}{\sqrt{225 + (20)^2}}(0.5) = 0.4 \text{ m/s} = 400 \text{ mm/s}\uparrow \qquad Ans.$$

The acceleration is determined by taking the time derivative of Eq. 2. Since v_A is constant, then $a_A = dv_A/dt = 0$, and we have

$$a_S = \frac{d^2y}{dt^2} = \left[\frac{-x(dx/dt)}{(225 + x^2)^{3/2}}\right]xv_A + \left[\frac{1}{\sqrt{225 + x^2}}\right]\left(\frac{dx}{dt}\right)v_A + \left[\frac{1}{\sqrt{225 + x^2}}\right]x\frac{dv_A}{dt} = \frac{225v_A^2}{(225 + x^2)^{3/2}}$$

At $x = 20$ m, with $v_A = 0.5$ m/s, the acceleration becomes

$$a_S = \frac{225(0.5 \text{ m/s})^2}{[225 + (20 \text{ m})^2]^{3/2}} = 0.00360 \text{ m/s}^2 = 3.60 \text{ mm/s}^2\uparrow \qquad Ans.$$

Note that the constant velocity at A causes the other end C of the rope to have an acceleration since $\mathbf{v}_A$ causes segment DA to change its direction as well as its length.

12.10 Relative-Motion Analysis of Two Particles Using Translating Axes

Throughout this chapter the absolute motion of a particle has been determined using a single fixed reference frame for measurement. There are many cases, however, where the path of motion for a particle is complicated, so that it may be feasible to analyze the motion in parts by using two or more frames of reference. For example, the motion of a particle located at the tip of an airplane propeller, while the plane is in flight, is more easily described if one observes first the motion of the airplane from a fixed reference and then superimposes (vectorially) the circular motion of the particle measured from a reference attached to the airplane. Any type of coordinates—rectangular, cylindrical, etc.—may be chosen to describe these two different motions.

In this section only *translating frames of reference* will be considered for the analysis. Relative-motion analysis of particles using rotating frames of reference will be treated in Secs. 16.8 and 20.4, since such an analysis depends on prior knowledge of the kinematics of line segments.

Position. Consider particles A and B, which move along the arbitrary paths aa and bb, respectively, as shown in Fig. 12–42a. The *absolute position* of each particle, $\mathbf{r}_A$ and $\mathbf{r}_B$, is measured from the common origin O of the *fixed* x, y, z reference frame. The origin of a second frame of reference x', y', z' is attached to and moves with particle A. The axes of this frame are *only permitted to translate* relative to the fixed frame. The *relative position* of "B with respect to A" is designated by a *relative-position vector* $\mathbf{r}_{B/A}$. Using vector addition, the three vectors shown in Fig. 12–42a can be related by the equation*

$$\boxed{\mathbf{r}_B = \mathbf{r}_A + \mathbf{r}_{B/A}} \qquad (12\text{–}33)$$

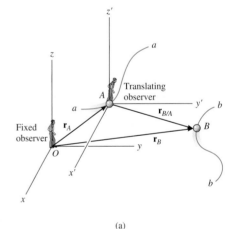

Fig. 12–42

Velocity. An equation that relates the velocities of the particles can be determined by taking the time derivative of Eq. 12–33, i.e.,

$$\boxed{\mathbf{v}_B = \mathbf{v}_A + \mathbf{v}_{B/A}} \qquad (12\text{–}34)$$

Here $\mathbf{v}_B = d\mathbf{r}_B/dt$ and $\mathbf{v}_A = d\mathbf{r}_A/dt$ refer to *absolute velocities*, since they are observed from the fixed frame; whereas the *relative velocity* $\mathbf{v}_{B/A} = d\mathbf{r}_{B/A}/dt$ is observed from the translating frame. It is important

---

*An easy way to remember the setup of this equation, and others like it, is to note the "cancellation" of the subscript A between the two terms, i.e., $\mathbf{r}_B = \mathbf{r}_A + \mathbf{r}_{B/A}$.

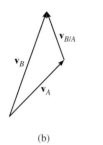

(b)

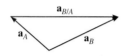

(c)

Fig. 12–42

to note that since the x', y', z' axes translate, the *components* of $\mathbf{r}_{B/A}$ will *not* change direction and therefore the time derivative of this vector's components will only have to account for the change in the vector's magnitude. Equation 12–34 therefore states that the velocity of B is equal to the velocity of A plus (vectorially) the relative velocity of "B with respect to A," as measured by the *translating observer* fixed in the x', y', z' reference, Fig. 12–42b.

Acceleration. The time derivative of Eq. 12–34 yields a similar vector relationship between the *absolute* and *relative accelerations* of particles A and B.

$$\mathbf{a}_B = \mathbf{a}_A + \mathbf{a}_{B/A} \qquad (12\text{–}35)$$

Here $\mathbf{a}_{B/A}$ is the acceleration of B as seen by the observer located at A and translating with the x', y', z' reference frame. The vector addition is shown in Fig. 12–42c.

The pilots of these jet planes flying close to one another must be aware of their relative positions and velocities at all times in order to avoid a collision.

PROCEDURE FOR ANALYSIS

- When applying the relative-position equation, $\mathbf{r}_B = \mathbf{r}_A + \mathbf{r}_{B/A}$, it is first necessary to specify the locations of the fixed x, y, z, and translating x', y', z' axes.
- Usually, the origin A of the translating axes is located at a point having a *known position*, $\mathbf{r}_A$, Fig. 12–42a.
- A graphical representation of the vector addition $\mathbf{r}_B = \mathbf{r}_A + \mathbf{r}_{B/A}$ can be shown, and both the known and unknown quantities labeled on this sketch.
- Since vector addition forms a triangle, there can be at most *two unknowns*, represented by the magnitudes and/or directions of the vector quantities.
- These unknowns can be solved for either graphically, using trigonometry (law of sines, law of cosines), or by resolving each of the three vectors $\mathbf{r}_B$, $\mathbf{r}_A$, and $\mathbf{r}_{B/A}$ into rectangular or Cartesian components, thereby generating a set of scalar equations.
- The relative-motion equations $\mathbf{v}_B = \mathbf{v}_A + \mathbf{v}_{B/A}$ and $\mathbf{a}_B = \mathbf{a}_A + \mathbf{a}_{B/A}$ are applied in the same manner as explained above, except in this case the origin O of the fixed x, y, z axes does not have to be specified, Figs. 12–42b and 12–42c.

EXAMPLE 12.25

A train, traveling at a constant speed of 60 mi/h, crosses over a road as shown in Fig. 12–43a. If automobile A is traveling at 45 mi/h along the road, determine the magnitude and direction of the relative velocity of the train with respect to the automobile.

Solution I

Vector Analysis.   The relative velocity $\mathbf{v}_{T/A}$ is measured from the translating x', y' axes attached to the automobile, Fig. 12–43a. It is determined from $\mathbf{v}_T = \mathbf{v}_A + \mathbf{v}_{T/A}$. Since $\mathbf{v}_T$ and $\mathbf{v}_A$ are known in *both* magnitude and direction, the unknowns become the x and y components of $\mathbf{v}_{T/A}$. Using the x, y axes in Fig. 12–43a and a Cartesian vector analysis, we have

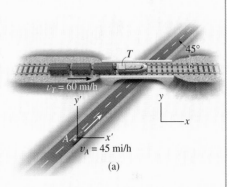

(a)

$$\mathbf{v}_T = \mathbf{v}_A + \mathbf{v}_{T/A}$$
$$60\mathbf{i} = (45\cos45°\,\mathbf{i} + 45\sin45°\,\mathbf{j}) + \mathbf{v}_{T/A}$$
$$\mathbf{v}_{T/A} = \{28.2\mathbf{i} - 31.8\mathbf{j}\} \text{ mi/h} \qquad Ans.$$

The magnitude of $\mathbf{v}_{T/A}$ is thus

$$v_{T/A} = \sqrt{(28.2)^2 + (-31.8)^2} = 42.5 \text{ mi/h} \qquad Ans.$$

From the direction of each component, Fig. 12–43b, the direction of $\mathbf{v}_{T/A}$ defined from the x axis is

$$\tan\theta = \frac{(v_{T/A})_y}{(v_{T/A})_x} = \frac{31.8}{28.2}$$
$$\theta = 48.5° \qquad Ans.$$

Note that the vector addition shown in Fig. 12–43b indicates the correct sense for $\mathbf{v}_{T/A}$. This figure anticipates the answer and can be used to check it.

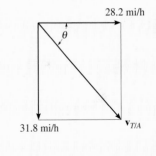

(b)

Solution II

Scalar Analysis.   The unknown components of $\mathbf{v}_{T/A}$ can also be determined by applying a scalar analysis. We will assume these components act in the *positive* x and y directions. Thus,

$$\mathbf{v}_T = \mathbf{v}_A + \mathbf{v}_{T/A}$$

$$\begin{bmatrix} 60 \text{ mi/h} \\ \rightarrow \end{bmatrix} = \begin{bmatrix} 45 \text{ mi/h} \\ \nearrow^{45°} \end{bmatrix} + \begin{bmatrix} (v_{T/A})_x \\ \rightarrow \end{bmatrix} + \begin{bmatrix} (v_{T/A})_y \\ \uparrow \end{bmatrix}$$

Resolving each vector into its x and y components yields

$(\xrightarrow{+})$ $60 = 45\cos45° + (v_{T/A})_x + 0$

$(+\uparrow)$ $0 = 45\sin45° + 0 + (v_{T/A})_y$

Solving, we obtain the previous results,

$$(v_{T/A})_x = 28.2 \text{ mi/h} = 28.2 \text{ mi/h} \rightarrow$$
$$(v_{T/A})_y = -31.8 \text{ mi/h} = 31.8 \text{ mi/h} \downarrow \qquad Ans.$$

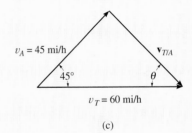

(c)

Fig. 12–43

E X A M P L E 12.26

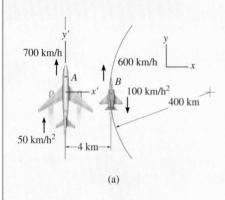

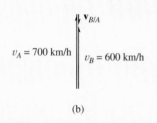

(a)

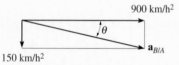

(b)

900 km/h²

θ

$\mathbf{a}_{B/A}$

150 km/h²

(c)

Fig. 12–44

Plane *A* in Fig. 12–44*a* is flying along a straight-line path, whereas plane *B* is flying along a circular path having a radius of curvature of $\rho_B = 400$ km. Determine the velocity and acceleration of *B* as measured by the pilot of *A*.

Solution

Velocity. The *x*, *y* axes are located at an arbitrary fixed point. Since the motion relative to plane *A* is to be determined, the *translating frame of reference x′, y′* is attached to it, Fig. 12–44*a*. Applying the relative-velocity equation in scalar form since the velocity vectors of both planes are parallel at the instant shown, we have

$(+\uparrow)$
$$v_B = v_A + v_{B/A}$$
$$600 = 700 + v_{B/A}$$
$$v_{B/A} = -100 \text{ km/h} = 100 \text{ km/h} \downarrow \qquad Ans.$$

The vector addition is shown in Fig. 12–44*b*.

Acceleration. Plane *B* has both tangential and normal components of acceleration, since it is flying along a *curved path*. From Eq. 12–20, the magnitude of the normal component is

$$(a_B)_n = \frac{v_B^2}{\rho} = \frac{(600 \text{ km/h})^2}{400 \text{ km}} = 900 \text{ km/h}^2$$

Applying the relative-acceleration equation, we have

$$\mathbf{a}_B = \mathbf{a}_A + \mathbf{a}_{B/A}$$
$$900\mathbf{i} - 100\mathbf{j} = 50\mathbf{j} + \mathbf{a}_{B/A}$$

Thus,

$$\mathbf{a}_{B/A} = \{900\mathbf{i} - 150\mathbf{j}\} \text{ km/h}^2$$

From Fig. 12–44*c*, the magnitude and direction of $\mathbf{a}_{B/A}$ are therefore

$$a_{B/A} = 912 \text{ km/h}^2 \quad \theta = \tan^{-1}\frac{150}{900} = 9.46° \quad Ans.$$

Notice that the solution to this problem is possible using a translating frame of reference, since the pilot in plane *A* is "translating." Observation of plane *A* with respect to the pilot of plane *B*, however, must be obtained using a *rotating* set of axes attached to plane *B*. (This assumes, of course, that the pilot of *B* is fixed in the rotating frame, so he does not turn his eyes to follow the motion of *A*.) The analysis for this case is given in Example 16.21.

EXAMPLE 12.27

At the instant shown in Fig. 12–45 cars A and B are traveling with speeds of 18 m/s and 12 m/s, respectively. Also at this instant, A has a decrease in speed of 2 m/s^2, and B has an increase in speed of 3 m/s^2. Determine the velocity and acceleration of B with respect to A.

Solution

Velocity. The fixed x, y axes are established at a point on the ground and the translating x', y' axes are attached to car A, Fig. 12–45a. Why? The relative velocity is determined from $\mathbf{v}_B = \mathbf{v}_A + \mathbf{v}_{B/A}$. What are the two unknowns? Using a Cartesian vector analysis, we have

$$\mathbf{v}_B = \mathbf{v}_A + \mathbf{v}_{B/A}$$

$$-12\mathbf{j} = (-18 \cos 60°\mathbf{i} - 18 \sin 60°\mathbf{j}) + \mathbf{v}_{B/A}$$

$$\mathbf{v}_{B/A} = \{9\mathbf{i} + 3.588\mathbf{j}\} \text{ m/s}$$

Thus,

$$v_{B/A} = \sqrt{(9)^2 + (3.588)^2} = 9.69 \text{ m/s} \qquad Ans.$$

Noting that $\mathbf{v}_{B/A}$ has $+\mathbf{i}$ and $+\mathbf{j}$ components, Fig. 12–45b, its direction is

$$\tan \theta = \frac{(v_{B/A})_y}{(v_{B/A})_x} = \frac{3.588}{9}$$

$$\theta = 21.7° \quad \measuredangle \qquad Ans.$$

Acceleration. Car B has both tangential and normal components of acceleration. Why? The magnitude of the normal component is

$$(a_B)_n = \frac{v_B^2}{\rho} = \frac{(12 \text{ m/s})^2}{100 \text{ m}} = 1.440 \text{ m/s}^2$$

Applying the equation for relative acceleration yields

$$\mathbf{a}_B = \mathbf{a}_A + \mathbf{a}_{B/A}$$

$$(-1.440\mathbf{i} - 3\mathbf{j}) = (2 \cos 60°\mathbf{i} + 2 \sin 60°\mathbf{j}) + \mathbf{a}_{B/A}$$

$$\mathbf{a}_{B/A} = \{-2.440\mathbf{i} - 4.732\mathbf{j}\} \text{ m/s}^2$$

Here $\mathbf{a}_{B/A}$ has $-\mathbf{i}$ and $-\mathbf{j}$ components. Thus, from Fig. 12–45c,

$$a_{B/A} = \sqrt{(2.440)^2 + (4.732)^2} = 5.32 \text{ m/s}^2 \qquad Ans.$$

$$\tan \phi = \frac{(a_{B/A})_y}{(a_{B/A})_x} = \frac{4.732}{2.440}$$

$$\phi = 62.7° \quad \swarrow \qquad Ans.$$

It is possible to obtain the relative acceleration of $\mathbf{a}_{A/B}$ using this method? Refer to the comment made at the end of Example 12.26.

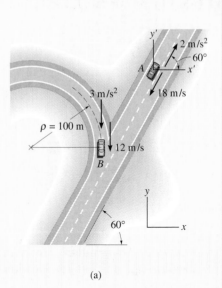

(a)

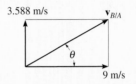

(b)

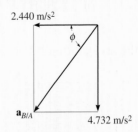

(c)

Fig. 12–45

DESIGN PROJECT

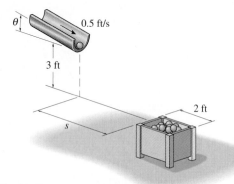

12–1D. DESIGN OF A MARBLE-SORTING DEVICE

Marbles roll off the production chute at 0.5 ft/s. Determine the range for the angle $0 \leq \theta \leq 30°$ for a selected position s for the placement of the hopper relative to the end of the chute. Submit a drawing of the device that shows the path the marbles take.

Fig. 12–1D

CHAPTER REVIEW

- *Rectilinear Kinematics.* Rectilinear kinematics refers to motion along a straight line. A position coordinate s specifies the location of the particle on the line, and the displacement Δs is the change in this position.

  The average velocity is a vector quantity, defined as the displacement divided by the time interval.

  $$\mathbf{v}_{avg} = \frac{\Delta \mathbf{r}}{\Delta t}$$

  This is different than the average speed, which is a scalar and is the total distance traveled divided by the time of travel.

  $$(v_{sp})_{avg} = \frac{s_T}{\Delta t}$$

  The time, position, instantaneous velocity, and instantaneous acceleration are related by the differential equations

  $$v = ds/dt \qquad a = dv/dt \qquad a\,ds = v\,dv$$

  If the acceleration is known to be constant, then integration of these equations yields

  $$v = v_0 + a_c t$$
  $$s = s_0 + v_0 t + \tfrac{1}{2}a_c t^2$$
  $$v^2 = v_0^2 + 2a_c(s - s_0)$$

- *Graphical Solutions.* If the motion is erratic, then it can be described by a graph. If one of these graphs is given, then the others can be established using the differential relations, $v = ds/dt$, $a = dv/dt$, or $a\,ds = v\,dv$. For example, if the v–t graph is known, then values of the s–t graph are determined from $\Delta s = \int v\,dt = $ area increments under the v–t graph. Values of the a–t graph are determined from $a = dv/dt = $ slope of v–t graph.

- *Curvilinear Motion, x, y, z.* For this case, motion along the path is resolved into rectilinear motion along the x, y, z axes. The equation of the path is used to relate the motion along each axis.

- *Projectile Motion.* Free flight motion of a projectile follows a parabolic path. It has a constant velocity in the horizontal direction and constant acceleration of $g = 9.81$ m/s^2 or 32.2 ft/s^2 in the vertical direction. Any two of the three equations for constant acceleration apply in the vertical direction, and in the horizontal direction only $x = x_0 + (v_0)_x t$ applies.

- *Curvilinear Motion n, t.* If normal and tangential axes are used for the analysis, then **v** is always in the positive t direction. The acceleration has two components. The tangential component, **a**$_t$, accounts for the change in the magnitude of the velocity; a slowing down is in the negative t direction, and a speeding up is in the positive t direction. The normal component **a**$_n$ accounts for the change in the direction of the velocity. This component is always in the positive n direction.

- *Curvilinear Motion, r, θ, z.* If the path of motion is expressed in polar coordinates, then the velocity and acceleration components can be written as

$$v_r = \dot{r} \qquad a_r = \ddot{r} - r\dot{\theta}^2$$
$$v_\theta = r\dot{\theta} \qquad a_\theta = r\ddot{\theta} + 2\dot{r}\dot{\theta}$$

To apply these equations, it is necessary to determine $r, \dot{r}, \ddot{r}, \dot{\theta}, \ddot{\theta}$ at the instant considered. If the path $r = f(\theta)$ is given, then the chain rule of calculus must be used to obtain the time derivatives. Once the data is substituted into the equations, then the algebraic sign of the results will indicate the direction of the components of v or a along each axis.

- *Absolute Dependent Motion of Two Particles.* The dependent motion of blocks that are suspended from pulleys and cables can be related by the geometry of the system. This is done by first establishing position coordinates, measured from a fixed origin to each block so that they are directed along the line of motion of the blocks. Using geometry and/or trigonometry, the coordinates are then related to the cable length in order to formulate a position coordinate equation. The first time derivative of this equation gives a relationship between the velocities of the blocks, and a second time derivative gives the relationship between their accelerations.

- *Relative Motion Analysis Using Translating Axes.* If two particles A and B undergo independent motions, then these motions can be related to their relative motion. Using a translating set of axes attached to one of the particles (A), the velocity and acceleration equations become

$$\mathbf{v}_B = \mathbf{v}_A + \mathbf{v}_{B/A}$$

$$\mathbf{a}_B = \mathbf{a}_A + \mathbf{a}_{B/A}$$

For planar motion, each of these equations produces two scalar equations, one in the x, and the other in the y direction. For solution, the vectors can be expressed in Cartesian form or the x and y scalar components can be written directly.

The design of conveyors for a bottling plant requires knowledge of the forces that act on them and the ability to predict the motion of the bottles they transport.

Kinetics of a Particle: Force and Acceleration

- To state Newton's Laws of Motion and Gravitational Attraction and to define mass and weight.
- To analyze the accelerated motion of a particle using the equation of motion with different coordinate systems.
- To investigate central-force motion and apply it to problems in space mechanics.

13.1   Newton's Laws of Motion

Many of the earlier notions about dynamics were dispelled after 1590 when Galileo performed experiments to study the motions of pendulums and falling bodies. The conclusions drawn from these experiments gave some insight as to the effects of forces acting on bodies in motion. The general laws of motion of a body subjected to forces were not known, however, until 1687, when Isaac Newton first presented three basic laws governing the motion of a particle. In a slightly reworded form, Newton's three laws of motion can be stated as follows:

First Law: A particle originally at rest, or moving in a straight line with a constant velocity, will remain in this state provided the particle is not subjected to an unbalanced force.

Second Law: A particle acted upon by an unbalanced force **F** *experiences an acceleration* **a** *that has the same direction as the force and a magnitude that is directly proportional to the force.**

Third Law: The mutual forces of action and reaction between two particles are equal, opposite, and collinear.

*Stated another way, the unbalanced force acting on the particle is proportional to the time rate of change of the particle's linear momentum. See footnote † on next page.

The first and third laws were used extensively in developing the concepts of statics. Although these laws are also considered in dynamics, Newton's second law of motion forms the basis for most of this study, since this law relates the accelerated motion of a particle to the forces that act on it.

Measurements of force and acceleration can be recorded in a laboratory so that in accordance with the second law, if a known unbalanced force $\mathbf{F}$ is applied to a particle, the acceleration $\mathbf{a}$ of the particle may be measured. Since the force and acceleration are directly proportional, the constant of proportionality, m, may be determined from the ratio $m = F/a$.* The positive scalar m is called the *mass* of the particle. Being constant during any acceleration, m provides a quantitative measure of the resistance of the particle to a change in its velocity.

If the mass of the particle is m, Newton's second law of motion may be written in mathematical form as

$$\mathbf{F} = m\mathbf{a}$$

This equation, which is referred to as the *equation of motion*, is one of the most important formulations in mechanics.† As previously stated, its validity is based solely on *experimental evidence*. In 1905, however, Albert Einstein developed the theory of relativity and placed limitations on the use of Newton's second law for describing general particle motion. Through experiments it was proven that *time* is not an absolute quantity as assumed by Newton; as a result, the equation of motion fails to predict the exact behavior of a particle, especially when the particle's speed approaches the speed of light (0.3 Gm/s). Developments of the theory of quantum mechanics by Erwin Schrödinger and others indicate further that conclusions drawn from using this equation are also invalid when particles are the size of an atom and move close to one another. For the most part, however, these requirements regarding particle speed and size are not encountered in engineering problems, so their effects will not be considered in this book.

---

*Recall that the units of force in the SI system and mass in the FPS system are derived from this equation, where N = kg·m/s² and slug = lb·s²/ft (see Sec. 1.3 of *Statics*). If, however, the units of force, mass, length, and time were *all* selected arbitrarily, then it is necessary to write $F = kma$, where k (a dimensionless constant) would have to be determined experimentally in order to preserve the equality.

†Since m is constant, we can also write $\mathbf{F} = d(m\mathbf{v})/dt$, where $m\mathbf{v}$ is the particle's linear momentum.

Newton's Law of Gravitational Attraction. Shortly after formulating his three laws of motion, Newton postulated a law governing the mutual attraction between any two particles. In mathematical form this law can be expressed as

$$F = G\frac{m_1 m_2}{r^2} \tag{13–1}$$

where

> F = force of attraction between the two particles
>
> G = universal constant of gravitation; according to experimental evidence $G = 66.73(10^{-12})$ m^3/(kg·s^2)
>
> m_1, m_2 = mass of each of the two particles
>
> r = distance between the centers of the two particles

Any two particles or bodies have a mutually attractive gravitational force acting between them. In the case of a particle located at or near the surface of the earth, however, the only gravitational force having any sizable magnitude is that between the earth and the particle. This force is termed the "weight" and, for our purpose, it will be the only gravitational force considered.

Mass and Weight. *Mass* is a property of matter by which we can compare the response of one body with that of another. As indicated above, this property manifests itself as a gravitational attraction between two bodies and provides a quantitative measure of the resistance of matter to a change in velocity. It is an *absolute* quantity since the measurement of mass can be made at any location. The weight of a body, however, is *not absolute* since it is measured in a gravitational field, and hence its magnitude depends on where the measurement is made. From Eq. 13–1, we can develop a general expression for finding the weight W of a particle having a mass $m_1 = m$. Let m_2 be the mass of the earth and r the distance between the earth's center and the particle. Then, if $g = Gm_2/r^2$, we have

$$W = mg$$

By comparison with $F = ma$, we term g the acceleration due to gravity. For most engineering calculations g is measured at a point on the surface of the earth at sea level, and at a latitude of 45°, considered the "standard location."

The mass and weight of a body are measured differently in the SI and FPS systems of units, and the method of defining these units should be thoroughly understood.

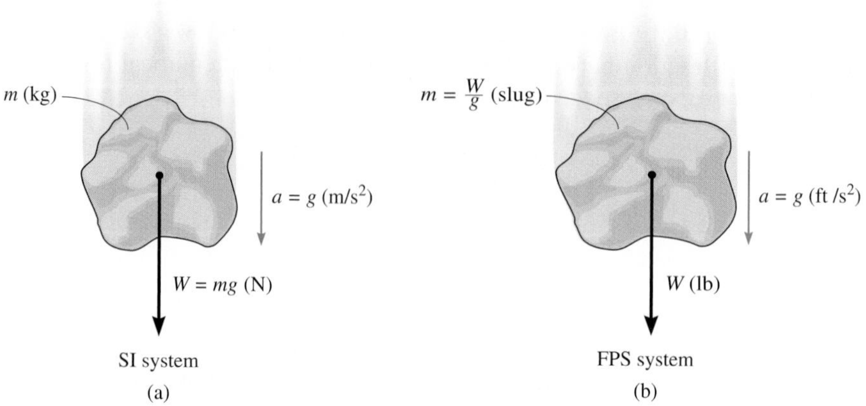

m (kg)

$m = \dfrac{W}{g}$ (slug)

$a = g$ (m/s^2)

$a = g$ (ft /s^2)

$W = mg$ (N)

W (lb)

SI system

FPS system

(a)

(b)

Fig. 13–1

SI System of Units. In the SI system the mass of the body is specified in kilograms, and the weight must be calculated using the equation of motion, $F = ma$. Hence, if a body has a mass m (kg) and is located at a point where the acceleration due to gravity is g (m/s^2), then the weight is expressed in *newtons* as $W = mg$ (N), Fig. 13–1a. In particular, if the body is located at the "standard location," the acceleration due to gravity is $g = 9.80665$ m/s^2. For calculations, the value $g = 9.81$ m/s^2 will be used, so that

$$W = mg \text{ (N)} \qquad (g = 9.81 \text{ m/s}^2) \qquad (13\text{–}2)$$

Therefore, a body of mass 1 kg has a weight of 9.81 N; a 2-kg body weighs 19.62 N; and so on.

FPS System of Units. In the FPS system the weight of the body is specified in pounds, and the mass must be calculated from $F = ma$. Hence, if a body has a weight W (lb) and is located at a point where the acceleration due to gravity is g (ft/s^2) then the mass is expressed in *slugs* as $m = W/g$ (slug), Fig. 13–1b. Since the acceleration of gravity at the standard location is approximately 32.2 ft/s^2($= 9.81$ m/s^2), the mass of the body measured in slugs is

$$m = \frac{W}{g} \text{ (slug)} \quad (g = 32.2 \text{ ft/s}^2) \qquad (13\text{–}3)$$

Therefore, a body weighing 32.2 lb has a mass of 1 slug; a 64.4-lb body has a mass of 2 slugs; and so on.

13.2 The Equation of Motion

When more than one force acts on a particle, the resultant force is determined by a vector summation of all the forces; i.e., $\mathbf{F}_R = \Sigma\mathbf{F}$. For this more general case, the equation of motion may be written as

$$\Sigma\mathbf{F} = m\mathbf{a} \qquad (13\text{--}4)$$

To illustrate application of this equation, consider the particle P shown in Fig. 13–2a, which has a mass m and is subjected to the action of two forces, $\mathbf{F}_1$ and $\mathbf{F}_2$. We can graphically account for the magnitude and direction of each force acting on the particle by drawing the particle's *free-body diagram*, Fig. 13–2b. Since the *resultant* of these forces *produces* the vector $m\mathbf{a}$, its magnitude and direction can be represented graphically on the *kinetic diagram*, shown in Fig. 13–2c.* The equal sign written between the diagrams symbolizes the *graphical* equivalency between the free-body diagram and the kinetic diagram; i.e., $\Sigma\mathbf{F} = m\mathbf{a}$.† In particular, note that if $\mathbf{F}_R = \Sigma\mathbf{F} = \mathbf{0}$, then the acceleration is also zero, so that the particle will either remain at *rest* or move along a straight-line path with *constant velocity*. Such are the conditions of *static equilibrium*, Newton's first law of motion.

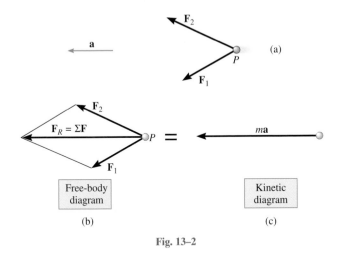

Free-body diagram
(b)

Kinetic diagram
(c)

Fig. 13–2

*Recall the free-body diagram considers the particle to be free of its surroundings and shows all the forces acting on the particle. The kinetic diagram pertains to the particle's motion as caused by the forces.

† The equation of motion can also be rewritten in the form $\Sigma\mathbf{F} - m\mathbf{a} = \mathbf{0}$. The vector $-m\mathbf{a}$ is referred to as the *inertia force vector*. If it is treated in the same way as a "force vector," then the state of "equilibrium" created is referred to as *dynamic equilibrium*. This method for application is often referred to as the *D'Alembert principle*, named after the French mathematician Jean le Rond d'Alembert.

Inertial Frame of Reference. Whenever the equation of motion is applied, it is required that measurements of the acceleration be made from a *Newtonian* or *inertial frame of reference. Such a coordinate system does not rotate and is either fixed or translates in a given direction with a constant velocity (zero acceleration).* This definition ensures that the particle's *acceleration* measured by observers in two different inertial frames of reference will always be the *same.* For example, consider the particle P moving with an absolute acceleration $\mathbf{a}_P$ along a straight path as shown in Fig. 13–3. If the observer is *fixed* in the inertial x, y frame of reference, this acceleration, $\mathbf{a}_P$, will be measured by the observer regardless of the direction and magnitude of the velocity $\mathbf{v}_O$ of the frame of reference. On the other hand, if the observer is *fixed* in the noninertial x', y' frame of reference, Fig. 13–3, the observer will not measure the particle's acceleration as $\mathbf{a}_P$. Instead, if the frame is *accelerating* at $\mathbf{a}_{O'}$ the particle will appear to have an acceleration of $\mathbf{a}_{P/O'} = \mathbf{a}_P - \mathbf{a}_{O'}$. Also, if the frame is *rotating*, as indicated by the curl, then the particle will appear to move along a *curved path*, in which case it will appear to have other components of acceleration (see Sec. 16.8). In any case, the measured acceleration from this observer cannot be used in Newton's law of motion to determine the forces acting on the particle.

When studying the motions of rockets and satellites, it is justifiable to consider the inertial reference frame as fixed to the stars, whereas dynamics problems concerned with motions on or near the surface of the earth may be solved by using an inertial frame which is assumed fixed to the earth. Even though the earth both rotates about its own axis and revolves about the sun, the accelerations created by these rotations are relatively small and can be neglected in most computations.

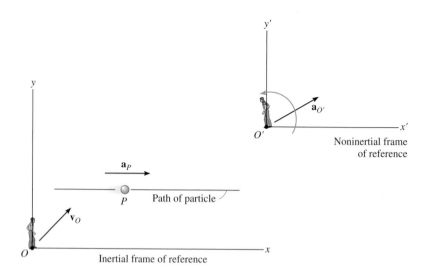

Fig. 13–3

Fig. 1

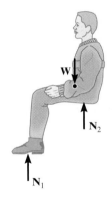

At rest or constant velocity

We are all familiar with the sensation one feels when sitting in a car that is subjected to a forward acceleration. Often people think this is caused by a "force" which acts on them and tends to push them back in their seats; however, this is not the case. Instead, this sensation occurs due to their inertia or the resistance of their mass to a change in velocity.

Consider the passenger in Fig. 1 who is strapped to the seat of a rocket sled. Provided the sled is at rest or is moving with constant velocity, then no force is exerted on his back as shown on his free-body diagram.

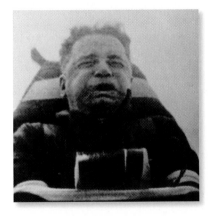

Fig. 2

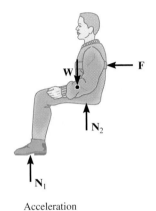

Acceleration

When the thrust of the rocket engine causes the sled to accelerate, then the seat upon which he is sitting exerts a force **F** on him which pushes him forward, Fig. 2. In the photo, notice that the inertia of his head resists this change in motion (acceleration), and so his head moves back against the seat and his face, which is nonrigid, tends to distort.

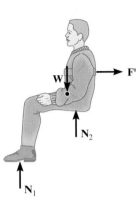

Fig. 3

Deceleration

Upon deceleration, Fig. 3, the force of the seatbelt **F′** tends to pull him to a stop, and so his head leaves contact with the back of the seat and his face distorts forward, again due to his inertia or tendency to continue to move forward. No force is pulling him forward, although this is the sensation he receives.

13.3 Equation of Motion for a System of Particles

The equation of motion will now be extended to include a system of n particles isolated within an enclosed region in space, as shown in Fig. 13–4a. In particular, there is no restriction in the way the particles are connected, and as a result the following analysis will apply equally well to the motion of a solid, liquid, or gas system. At the instant considered, the arbitrary ith particle, having a mass m_i, is subjected to a system of internal forces and a resultant external force. The *resultant internal force*, represented symbolically as $\mathbf{f}_i$, is determined from the forces which the other particles exert on the ith particle. Usually these forces are developed by direct contact, although the summation extends over all n particles within the dashed boundary. The *resultant external force* $\mathbf{F}_i$ represents, for example, the effect of gravitational, electrical, magnetic, or contact forces between the ith particle and adjacent bodies or particles *not* included within the system.

The free-body and kinetic diagrams for the ith particle are shown in Fig. 13–4b. Applying the equation of motion yields

$$\Sigma \mathbf{F} = m\mathbf{a}; \qquad\qquad \mathbf{F}_i + \mathbf{f}_i = m_i \mathbf{a}_i$$

When the equation of motion is applied to each of the other particles of the system, similar equations will result. If all these equations are added together *vectorially*, we obtain

$$\Sigma \mathbf{F}_i + \Sigma \mathbf{f}_i = \Sigma m_i \mathbf{a}_i$$

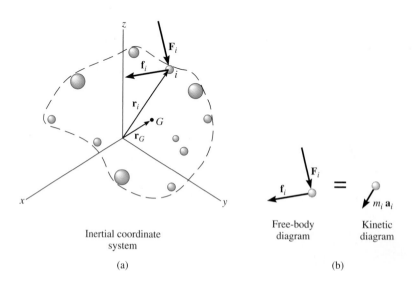

Inertial coordinate
system

(a)

Free-body
diagram

Kinetic
diagram

(b)

Fig. 13–4

The summation of the internal forces, if carried out, will equal zero, since internal forces between particles all occur in equal but opposite collinear pairs. Consequently, only the sum of the external forces will remain, and therefore the equation of motion, written for the system of particles, becomes

$$\Sigma \mathbf{F}_i = \Sigma m_i \mathbf{a}_i \qquad (13\text{--}5)$$

If $\mathbf{r}_G$ is a position vector which locates the *center of mass* G of the particles, Fig. 13–4a, then by definition of the center of mass, $m\mathbf{r}_G = \Sigma m_i \mathbf{r}_i$, where $m = \Sigma m_i$ is the total mass of all the particles. Differentiating this equation twice with respect to time, assuming that no mass is entering or leaving the system, yields

$$m\mathbf{a}_G = \Sigma m_i \mathbf{a}_i$$

Substituting this result into Eq. 13–5, we obtain

$$\boxed{\Sigma \mathbf{F} = m\mathbf{a}_G} \qquad (13\text{--}6)$$

Hence, the sum of the external forces acting on the system of particles is equal to the total mass of the particles times the acceleration of its center of mass G. Since in reality all particles must have a finite size to possess mass, Eq. 13–6 justifies application of the equation of motion to a *body* that is represented as a single particle.

IMPORTANT POINTS

- The equation of motion is based on experimental evidence and is valid only when applied from an inertial frame of reference.
- The equation of motion states that the *unbalanced force* on a particle causes it to accelerate.
- An inertial frame of reference does not rotate, rather it has axes that either translate with constant velocity or are at rest.
- Mass is a property of matter that provides a quantitative measure of its resistance to a change in velocity. It is an absolute quantity.
- Weight is a force that is caused by the earth's gravitation. It is not absolute; rather it depends on the altitude of the mass from the earth's surface.

13.4 Equations of Motion: Rectangular Coordinates

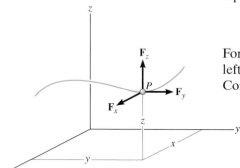

Fig. 13–5

When a particle is moving relative to an inertial x, y, z frame of reference, the forces acting on the particle, as well as its acceleration, may be expressed in terms of their $\mathbf{i}$, $\mathbf{j}$, $\mathbf{k}$ components, Fig. 13–5. Applying the equation of motion, we have

$$\Sigma \mathbf{F} = m\mathbf{a}$$

$$\Sigma F_x \mathbf{i} + \Sigma F_y \mathbf{j} + \Sigma F_z \mathbf{k} = m(a_x \mathbf{i} + a_y \mathbf{j} + a_z \mathbf{k})$$

For this equation to be satisfied, the respective $\mathbf{i}$, $\mathbf{j}$, $\mathbf{k}$ components on the left side must equal the corresponding components on the right side. Consequently, we may write the following three scalar equations:

$$
\begin{aligned}
\Sigma F_x &= ma_x \\
\Sigma F_y &= ma_y \\
\Sigma F_z &= ma_z
\end{aligned}
\tag{13–7}
$$

In particular, if the particle is constrained to move only in the x–y plane, then the first two of these equations are used to specify the motion.

PROCEDURE FOR ANALYSIS

The equations of motion are used to solve problems which require a relationship between the forces acting on a particle and the accelerated motion they cause.

Free-Body Diagram

- Select the inertial coordinate system. Most often, rectangular or x, y, z coordinates are chosen to analyze problems for which the particle has *rectilinear motion*.

- Once the coordinates are established, draw the particle's free-body diagram. Drawing this diagram is *very important* since it provides a graphical representation that accounts for *all the forces* ($\Sigma \mathbf{F}$) which act on the particle, and thereby makes it possible to resolve these forces into their x, y, z components.

- The direction and sense of the particle's acceleration $\mathbf{a}$ should also be established. If the senses of its components are unknown, for mathematical convenience assume that they are in the *same direction* as the *positive* inertial coordinate axes.

- The acceleration may be represented as the $m\mathbf{a}$ vector on the kinetic diagram.*

- Identify the unknowns in the problem.

*It is a convention in this text always to use the kinetic diagram as a graphical aid when developing the proofs and theory. The particle's acceleration or its components will be shown as blue colored vectors near the free-body diagram in the examples.

Equations of Motion

- If the forces can be resolved directly from the free-body diagram, apply the equations of motion in their scalar component form.

- If the geometry of the problem appears complicated, which often occurs in three dimensions, Cartesian vector analysis can be used for the solution.

- *Friction.* If a moving particle contacts a rough surface, it may be necessary to use the *frictional equation*, which relates the coefficient of kinetic friction μ_k to the magnitudes of the frictional and normal forces $\mathbf{F}_f$ and $\mathbf{N}$ acting at the surfaces of contact, i.e., $F_f = \mu_k N$. Remember that $\mathbf{F}_f$ always acts on the free-body diagram such that it opposes the motion of the particle relative to the surface it contacts. If the particle is on the verge of relative motion then the coefficient of static friction should be used.

- *Spring.* If the particle is connected to an *elastic spring* having negligible mass, the spring force F_s can be related to the deformation of the spring by the equation $F_s = ks$. Here k is the spring's stiffness measured as a force per unit length, and s is the stretch or compression defined as the difference between the deformed length l and the undeformed length l_0, i.e., $s = l - l_0$.

Kinematics

- If the velocity or position of the particle is to be found, it will be necessary to apply the proper kinematic equations once the particle's acceleration is determined from $\Sigma \mathbf{F} = m\mathbf{a}$.

- If *acceleration is a function of time*, use $a = dv/dt$ and $v = ds/dt$ which, when integrated, yield the particle's velocity and position.

- If *acceleration is a function of displacement*, integrate $a\, ds = v\, dv$ to obtain the velocity as a function of position.

- If *acceleration is constant*, use $v = v_0 + a_c t$, $s = s_0 + v_0 t + \frac{1}{2} a_c t^2$, $v^2 = v_0^2 + 2a_c(s - s_0)$ to determine the velocity or position of the particle.

- If the problem involves the dependent motion of several particles, use the method outlined in Sec. 12.9 to relate their accelerations.

- In all cases, make sure the positive inertial coordinate directions used for writing the kinematic equations are the same as those used for writing the equations of motion; otherwise, simultaneous solution of the equations will result in errors.

- If the solution for an unknown vector component yields a negative scalar, it indicates that the component acts in the direction opposite to that which was assumed.

EXAMPLE 13.1

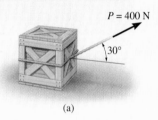

(a)

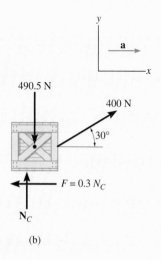

(b)

Fig. 13–6

The 50-kg crate shown in Fig. 13–6a rests on a horizontal plane for which the coefficient of kinetic friction is $\mu_k = 0.3$. If the crate is subjected to a 400-N towing force as shown, determine the velocity of the crate in 3 s starting from rest.

Solution

Using the equations of motion, we can relate the crate's acceleration to the force causing the motion. The crate's velocity can then be determined using kinematics.

Free-Body Diagram. The weight of the crate is $W = mg = 50 \text{ kg} (9.81 \text{ m/s}^2) = 490.5 \text{ N}$. As shown in Fig. 13–6b, the frictional force has a magnitude $F = \mu_k N_C$ and acts to the left, since it opposes the motion of the crate. The acceleration **a** is assumed to act horizontally, in the positive x direction. There are two unknowns, namely N_C and a. (We can also use the alternative procedure of drawing the crate's free-body *and* kinetic diagrams, Fig. 13–6c, prior to applying the equations of motion.)

Equations of Motion. Using the data shown on the free-body diagram, we have

$$\xrightarrow{+} \Sigma F_x = ma_x; \qquad 400 \cos 30° - 0.3N_C = 50a \qquad (1)$$

$$+\uparrow \Sigma F_y = ma_y; \quad N_C - 490.5 + 400 \sin 30° = 0 \qquad (2)$$

Solving Eq. 2 for N_C, substituting the result into Eq. 1, and solving for a yields

$$N_C = 290.5 \text{ N}$$
$$a = 5.19 \text{ m/s}^2$$

Kinematics. Note that the acceleration is *constant*, since the applied force **P** is constant. Since the initial velocity is zero, the velocity of the crate in 3 s is

$$(\xrightarrow{+}) \qquad\qquad v = v_0 + a_c t$$
$$= 0 + 5.19(3)$$
$$= 15.6 \text{ m/s} \rightarrow \qquad\qquad Ans.$$

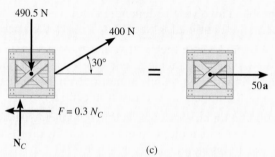

(c)

EXAMPLE 13.2

A 10-kg projectile is fired vertically upward from the ground, with an initial velocity of 50 m/s, Fig. 13–7a. Determine the maximum height to which it will travel if (a) atmospheric resistance is neglected; and (b) atmospheric resistance is measured as $F_D = (0.01v^2)$ N, where v is the speed at any instant, measured in m/s.

(a)

Solution

In both cases the known force on the projectile can be related to its acceleration using the equation of motion. Kinematics can then be used to relate the projectile's acceleration to its position.

Part (a) Free-Body Diagram. As shown in Fig. 13–7b, the projectile's weight is $W = mg = 10(9.81) = 98.1$ N. We will assume the unknown acceleration **a** acts upward in the *positive z* direction.

Equation of Motion

$$+\uparrow \Sigma F_z = ma_z; \qquad -98.1 = 10a, \qquad a = -9.81 \text{ m/s}^2$$

The result indicates that the projectile, like every object having free-flight motion near the earth's surface, is subjected to a *constant* downward acceleration of 9.81 m/s².

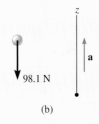

(b)

Kinematics. Initially, $z_0 = 0$ and $v_0 = 50$ m/s, and at the maximum height $z = h$, $v = 0$. Since the acceleration is *constant*, then

$$(+\uparrow) \qquad v^2 = v_0^2 + 2a_c(z - z_0)$$
$$0 = (50)^2 + 2(-9.81)(h - 0)$$
$$h = 127 \text{ m} \qquad\qquad Ans.$$

Part (b) Free-Body Diagram. Since the force $F_D = (0.01v^2)$ N tends to retard the upward motion of the projectile, it acts downward as shown on the free-body diagram, Fig. 13–7c.

Equation of Motion

$$+\uparrow \Sigma F_z = ma_z; \qquad -0.01v^2 - 98.1 = 10a, \qquad a = -0.01v^2 - 9.81$$

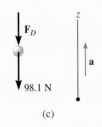

(c)

Fig. 13–7

Kinematics. Here the acceleration is *not constant* since F_D depends on the velocity. Since $a = f(v)$, we can relate a to position using

$$(+\uparrow)\, a\, dz = v\, dv; \qquad (-0.001v^2 - 9.81)\, dz = v\, dv$$

Separating the variables and integrating, realizing that initially $z_0 = 0$, $v_0 = 50$ m/s (positive upward), and at $z = h$, $v = 0$, we have

$$\int_0^h dz = -\int_{50}^0 \frac{v\, dv}{0.001v^2 + 9.81} = -500 \ln(v^2 + 9810)\Big|_{50}^0$$

$$h = 114 \text{ m} \qquad\qquad Ans.$$

The answer indicates a lower elevation than that obtained in part (a) due to atmospheric resistance.

*Note that if the projectile were fired downward, with z positive downward, the equation of motion would then be $-0.01v^2 + 98.1 = 10a$.

E X A M P L E 13.3

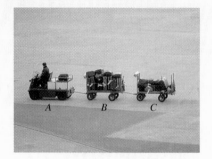

The baggage truck *A* shown in the photo has a weight of 900 lb and tows a 550-lb cart *B* and a 325-lb cart *C*. For a short time the driving frictional force developed at the wheels of the truck is $F_A = (40t)$ lb, where *t* is in seconds. If the truck starts from rest, determine its speed in 2 seconds. Also, what is the horizontal force acting on the coupling between the truck and cart *B* at this instant? Neglect the size of the truck and carts.

(a)

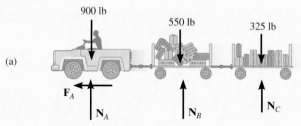

Solution

Free-Body Diagram. As shown in Fig. 13–8*a*, it is the frictional driving force that gives both the truck and carts an acceleration. Here we have considered all three vehicles.

Equation of Motion. Only motion in the horizontal direction has to be considered.

$$\xleftarrow{+} \Sigma F_x = ma_x; \qquad 40t = \left(\frac{900 + 550 + 325}{32.2}\right)a$$

$$a = 0.7256t$$

Kinematics. Since the acceleration is a function of time, the velocity of the truck is obtained using $a = dv/dt$ with the initial condition that $v_0 = 0$ at $t = 0$. We have

$$\int_0^v dv = \int_0^2 0.7256t \, dt; \qquad v = 0.3628t^2 \Big|_0^2 = 1.45 \text{ ft/s} \qquad \textit{Ans.}$$

Free-Body Diagram. In order to determine the force between the truck and cart *B*, we can consider a free-body diagram of the truck so that we can "expose" the coupling force **T** as external to the free-body diagram, Fig. 13–8*b*.

Equation of Motion. When $t = 2$ s, then

$$\xleftarrow{+} \Sigma F_x = ma_x; \qquad 40(2) - T = \left(\frac{900}{32.2}\right)[0.7256(2)]$$

$$T = 39.4 \text{ lb} \qquad \textit{Ans.}$$

Try and obtain this same result by considering a free-body diagram of carts *B* and *C*.

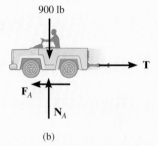

(b)

Fig. 13–8

EXAMPLE 13.4

A smooth 2-kg collar C, shown in Fig. 13–9a, is attached to a spring having a stiffness $k = 3$ N/m and an unstretched length of 0.75 m. If the collar is released from rest at A, determine its acceleration and the normal force of the rod on the collar at the instant $y = 1$ m.

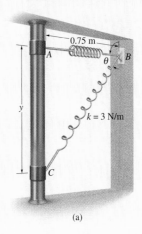

(a)

Solution

Free-Body Diagram. The free-body diagram of the collar when it is located at the arbitrary position y is shown in Fig. 13–9b. Note that the weight is $W = 2(9.81) = 19.62$ N. Furthermore, the collar is *assumed* to be accelerating so that "$\mathbf{a}$" acts downward in the *positive* y direction. There are four unknowns, namely, N_C, F_s, a, and θ.

Equations of Motion

$$\xrightarrow{+} \Sigma F_x = ma_x; \qquad -N_C + F_s \cos\theta = 0 \qquad (1)$$

$$+\downarrow \Sigma F_y = ma_y; \qquad 19.62 - F_s \sin\theta = 2a \qquad (2)$$

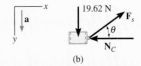

(b)

Fig. 13–9

From Eq. 2 it is seen that the acceleration depends on the magnitude and direction of the spring force. Solution for N_C and a is possible once F_s and θ are known.

The magnitude of the spring force is a function of the stretch s of the spring; i.e., $F_s = ks$. Here the unstretched length is $AB = 0.75$ m, Fig. 13–9a; therefore, $s = CB - AB = \sqrt{y^2 + (0.75)^2} - 0.75$. Since $k = 3$ N/m, then

$$F_s = ks = 3(\sqrt{y^2 + (0.75)^2} - 0.75) \qquad (3)$$

From Fig. 13–9a, the angle θ is related to y by trigonometry.

$$\tan\theta = \frac{y}{0.75} \qquad (4)$$

Substituting $y = 1$ m into Eqs. 3 and 4 yields $F_s = 1.50$ N and $\theta = 53.1°$. Substituting these results into Eqs. 1 and 2, we obtain

$$N_C = 0.900 \text{ N} \qquad\qquad Ans.$$

$$a = 9.21 \text{ m/s}^2 \downarrow \qquad\qquad Ans.$$

E X A M P L E 13.5

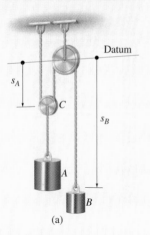

s_A

Datum

C

s_B

A

B

(a)

T T

$2T$

(b)

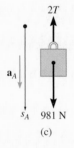

$2T$

$\mathbf{a}_A$

s_A 981 N

(c)

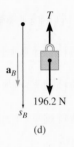

T

$\mathbf{a}_B$

196.2 N

s_B

(d)

Fig. 13–10

The 100-kg block A shown in Fig. 13–10a is released from rest. If the masses of the pulleys and the cord are neglected, determine the speed of the 20-kg block B in 2 s.

Solution

Free-Body Diagrams. Since the mass of the pulleys is *neglected*, then for pulley C, $ma = 0$ and we can apply $\Sigma F_y = 0$ as shown in Fig. 13–10b. The free-body diagrams for blocks A and B are shown in Fig. 13–10c and d, respectively. One can see that for A to remain static requires $T = 490.5$ N, whereas for B to remain static requires $T = 196.2$ N. Hence A will move down while B moves up. Here we will *assume* both blocks accelerate downward, in the direction of $+s_A$ and $+s_B$. The three unknowns are T, a_A, and a_B.

Equations of Motion.

Block A (Fig. 13–10c):

$$+\downarrow \Sigma F_y = ma_y; \qquad 981 - 2T = 100a_A \qquad (1)$$

Block B (Fig. 13–10d):

$$+\downarrow \Sigma F_y = ma_y; \qquad 196.2 - T = 20a_B \qquad (2)$$

Kinematics. The necessary third equation is obtained by relating a_A to a_B using a dependent motion analysis. Using the technique developed in Sec. 12.9, the coordinates s_A and s_B measure the positions of A and B from the fixed datum, Fig. 13–10a. It is seen that

$$2s_A + s_B = l$$

where l is constant and represents the total vertical length of cord. Differentiating this expression twice with respect to time yields

$$2a_A = -a_B \qquad (3)$$

Notice that in writing Eqs. 1 to 3, the *positive direction was always assumed downward*. It is very important to be *consistent* in this assumption since we are seeking a simultaneous solution of equations. The solution yields

$$T = 327.0 \text{ N}$$
$$a_A = 3.27 \text{ m/s}^2$$
$$a_B = -6.54 \text{ m/s}^2$$

Hence when block A accelerates *downward*, block B accelerates *upward*. Since a_B is constant, the velocity of block B in 2 s is thus

$$(+\downarrow) \qquad\qquad v = v_0 + a_B t$$
$$= 0 + (-6.54)(2)$$
$$= -13.1 \text{ m/s} \qquad\qquad Ans.$$

The negative sign indicates that block B is moving upward.

13.5 Equations of Motion: Normal and Tangential Coordinates

When a particle moves over a curved path which is known, the equation of motion for the particle may be written in the tangential, normal, and binormal directions. We have

$$\Sigma \mathbf{F} = m\mathbf{a}$$

$$\Sigma F_t \mathbf{u}_t + \Sigma F_n \mathbf{u}_n + \Sigma F_b \mathbf{u}_b = m\mathbf{a}_t + m\mathbf{a}_n$$

Here ΣF_t, ΣF_n, ΣF_b represent the sums of all the force components acting on the particle in the tangential, normal, and binormal directions, respectively, Fig. 13–11. Note that there is no motion of the particle in the binormal direction, since the particle is constrained to move along the path. The above equation is satisfied provided

$$\begin{aligned} \Sigma F_t &= ma_t \\ \Sigma F_n &= ma_n \\ \Sigma F_b &= 0 \end{aligned}$$ (13–8)

Recall that $a_t (= dv/dt)$ represents the time rate of change in the magnitude of velocity. Consequently, if $\Sigma \mathbf{F}_t$ acts in the direction of motion, the particle's speed will increase, whereas if it acts in the opposite direction, the particle will slow down. Likewise, $a_n (= v^2/\rho)$ represents the time rate of change in the velocity's direction. Since this vector *always* acts in the positive n direction, i.e., toward the path's center of curvature, then $\Sigma \mathbf{F}_n$, which causes $\mathbf{a}_n$, also acts in this direction. For example, when the particle is constrained to travel in a circular path with a constant speed, there is a normal force exerted on the particle by the constraint in order to change the direction of the particle's velocity ($\mathbf{a}_n$). Since this force is always directed toward the center of the path, it is often referred to as the *centripetal force*.

The centrifuge is used to subject a passenger to very large normal accelerations caused by high rotations. Realize that these accelerations are caused by the unbalanced normal force exerted on the passenger by the seat of the centrifuge.

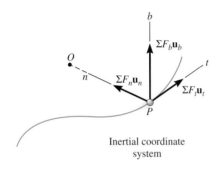

Inertial coordinate system

Fig. 13–11

PROCEDURE FOR ANALYSIS

When a problem involves the motion of a particle along a *known curved path*, normal and tangential coordinates should be considered for the analysis since the acceleration components can be readily formulated. The method for applying the equations of motion, which relate the forces to the acceleration, has been outlined in the procedure given in Sec. 13.4. Specifically, for *t, n, b* coordinates it may be stated as follows:

Free-Body Diagram

- Establish the inertial *t, n, b* coordinate system at the particle and draw the particle's free-body diagram.

- The particle's normal acceleration $\mathbf{a}_n$ *always* acts in the positive *n* direction.

- If the tangential acceleration $\mathbf{a}_t$ is unknown, assume it acts in the positive *t* direction.

- Identify the unknowns in the problem.

Equations of Motion

- Apply the equations of motion, Eqs. 13–8.

Kinematics

- Formulate the tangential and normal components of acceleration; i.e., $a_t = dv/dt$ or $a_t = v\, dv/ds$ and $a_n = v^2/\rho$.

- If the path is defined as $y = f(x)$, the radius of curvature at the point where the particle is located can be obtained from $\rho = [1 + (dy/dx)^2]^{3/2}/|d^2y/dx^2|$.

E X A M P L E   13.6

Determine the banking angle θ for the race track so that the wheels of the racing cars shown in Fig. 13–12a will not have to depend upon friction to prevent any car from sliding up or down the track. Assume the cars have negligible size a mass m, and travel around the curve of radius ρ with a speed v.

(a)

Solution

Before looking at the following solution, give some thought as to why it should be solved using t, n, b coordinates.

Free-Body Diagram.   As shown in Fig. 13–12b, and as stated in the problem, no frictional force acts on the car. Here $\mathbf{N}_C$ represents the *resultant* of the ground on all four wheels. Since a_n can be calculated, the unknowns are N_C and θ.

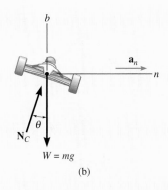

(b)

Fig. 13–12

Equations of Motion.   Using the n, b axes shown,

$$\xrightarrow{\pm} \Sigma F_n = ma_n; \qquad N_C \sin \theta = m \frac{v^2}{\rho} \qquad\qquad (1)$$

$$+\uparrow \Sigma F_b = 0; \qquad N_C \cos \theta - mg = 0 \qquad\qquad (2)$$

Eliminating N_C and m from these equations by dividing Eq. 1 by Eq. 2, we obtain

$$\tan \theta = \frac{v^2}{g\rho}$$

$$\theta = \tan^{-1}\left(\frac{v^2}{g\rho}\right) \qquad\qquad Ans.$$

Notice that the result is independent of the mass of the car. Also, a force summation in the tangential direction is of no consequence to the solution. If it were considered, then $a_t = dv/dt = 0$, since the car moves with *constant speed*. A further analysis of this problem is discussed in Prob. 21–48.

E X A M P L E **13.7**

The 3-kg disk D is attached to the end of a cord as shown in Fig. 13–13a. The other end of the cord is attached to a ball-and-socket joint located at the center of a platform. If the platform is rotating rapidly, and the disk is placed on it and released from rest as shown, determine the time it takes for the disk to reach a speed great enough to break the cord. The maximum tension the cord can sustain is 100 N, and the coefficient of kinetic friction between the disk and the platform is $\mu_k = 0.1$.

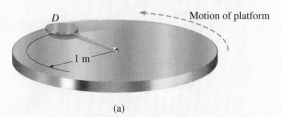

(a)

Solution

Free-Body Diagram. The frictional force has a magnitude $F = \mu_k N_D = 0.1 N_D$ and a sense of direction that opposes the *relative motion* of the disk with respect to the platform. It is this force that gives the disk a tangential component of acceleration causing v to increase, thereby causing T to increase until it reaches 100 N. The weight of the disk is $W = 3(9.81) = 29.43$ N. Since a_n can be related to v, the unknowns are N_D, a_t, and v.

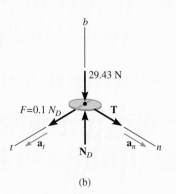

(b)

Fig. 13–13

Equations of Motion.

$$\Sigma F_n = ma_n; \qquad T = 3\left(\frac{v^2}{1}\right) \qquad (1)$$

$$\Sigma F_t = ma_t; \qquad 0.1 N_D = 3a_t \qquad (2)$$

$$\Sigma F_b = 0; \qquad N_D - 29.43 = 0 \qquad (3)$$

Setting $T = 100$ N, Eq. 1 can be solved for the critical speed v_{cr} of the disk needed to break the cord. Solving all the equations, we obtain

$$N_D = 29.43 \text{ N}$$
$$a_t = 0.981 \text{ m/s}^2$$
$$v_{cr} = 5.77 \text{ m/s}$$

Kinematics. Since a_t is *constant*, the time needed to break the cord is

$$v_{cr} = v_0 + a_t t$$
$$5.77 = 0 + (0.981)t$$
$$t = 5.89 \text{ s} \qquad \qquad Ans.$$

EXAMPLE 13.8

Design of the ski jump shown in the photo requires knowing the type of forces that will be exerted on the skier and his approximate trajectory. If in this case the jump can be approximated by the parabola shown in Fig. 13–14a, determine the normal force on the 150-lb skier the instant he arrives at the end of the jump, point A, where his velocity is 65 ft/s. Also, what is his acceleration at this point?

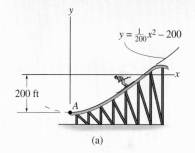

Solution

Why consider using n, t coordinates to solve this problem?

Free-Body Diagram. The free-body diagram for the skier when he is at A is shown in Fig. 13–14b. Since the path is *curved*, there are two components of acceleration, $\mathbf{a}_n$ and $\mathbf{a}_t$. Since a_n can be calculated, the unknowns are a_t and N_A.

Equations of Motion.

$$+\uparrow \Sigma F_n = ma_n; \qquad N_A - 150 = \frac{150}{32.2}\left(\frac{(65)^2}{\rho}\right) \qquad (1)$$

$$\xleftarrow{+} \Sigma F_t = ma_t; \qquad 0 = \frac{150}{32.2}a_t \qquad (2)$$

The radius of curvature ρ for the path must be determined at point $A(0,-200\text{ ft})$. Here $y = \frac{1}{200}x^2 - 200$, $dy/dx = \frac{1}{100}x$, $d^2y/dx^2 = \frac{1}{100}$, so that at $x = 0$,

$$\rho = \frac{[1 + (dy/dx)^2]^{3/2}}{|d^2y/dx^2|}\bigg|_{x=0} = \frac{[1 + (0)^2]^{3/2}}{|\frac{1}{100}|} = 100\text{ ft}$$

Substituting into Eq. 1 and solving for N_A, we have

$$N_A = 347\text{ lb} \qquad\qquad Ans.$$

Kinematics. From Eq. 2,

$$a_t = 0$$

Thus,

$$a_n = \frac{v^2}{\rho} = \frac{(65)^2}{100} = 42.2\text{ ft/s}^2$$

$$a_A = a_n = 42.2\text{ ft/s}^2 \uparrow \qquad\qquad Ans.$$

Show that when the skier is in mid-air his acceleration is 32.2 ft/s².

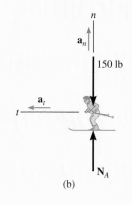

Fig. 13–14

E X A M P L E **13.9**

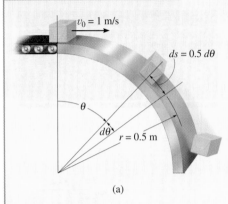

(a)

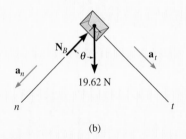

(b)

Fig. 13–15

Packages, each having a mass of 2 kg, are delivered from a conveyor to a smooth circular ramp with a velocity of $v_0 = 1$ m/s as shown in Fig. 13–15a. If the effective radius of the ramp is 0.5 m, determine the angle $\theta = \theta_{max}$ at which each package begins to leave the surface.

Solution

Free-Body Diagram.   The free-body diagram for a package, when it is located at the *general position* θ, is shown in Fig. 13–15b. The package must have a tangential acceleration $\mathbf{a}_t$, since its *speed* is always *increasing* as it slides downward. The weight is $W = 2(9.81) = 19.62$ N. Specify the three unknowns.

Equations of Motion.

$$+\swarrow \Sigma F_n = ma_n; \qquad -N_B + 19.62 \cos \theta = 2\frac{v^2}{0.5} \qquad (1)$$

$$+\searrow \Sigma F_t = ma_t; \qquad 19.62 \sin \theta = 2a_t \qquad (2)$$

At the instant $\theta = \theta_{max}$, the package leaves the surface of the ramp so that $N_B = 0$. Therefore, there are three unknowns, v, a_t, and θ.

Kinematics.   The third equation for the solution is obtained by noting that the magnitude of tangential acceleration a_t may be related to the speed of the package v and the angle θ. Since $a_t\, ds = v\, dv$ and $ds = r\, d\theta = 0.5\, d\theta$, Fig. 13–15a, we have

$$a_t = \frac{v\, dv}{0.5\, d\theta} \qquad (3)$$

To solve, substitute Eq. 3 into Eq. 2 and separate the variables. This gives

$$v\, dv = 4.905 \sin \theta\, d\theta$$

Integrate both sides, realizing that when $\theta = 0°$, $v_0 = 1$ m/s.

$$\int_1^v v\, dv = 4.905 \int_{0°}^{\theta} \sin \theta\, d\theta$$

$$\frac{v^2}{2}\bigg|_1^v = -4.905 \cos \theta \bigg|_{0°}^{\theta}$$

$$v^2 = 9.81(1 - \cos \theta) + 1$$

Substituting into Eq. 1 with $N_B = 0$ and solving for $\cos \theta_{max}$ yields

$$19.62 \cos \theta_{max} = \frac{2}{0.5}[9.81(1 - \cos \theta_{max}) + 1]$$

$$\cos \theta_{max} = \frac{43.24}{58.86}$$

$$\theta_{max} = 42.7° \hspace{4cm} Ans.$$

13.6 Equations of Motion: Cylindrical Coordinates

When all the forces acting on a particle are resolved into cylindrical components, i.e., along the unit-vector directions $\mathbf{u}_r$, $\mathbf{u}_\theta$, $\mathbf{u}_z$, Fig. 13–16, the equation of motion may be expressed as

$$\Sigma \mathbf{F} = m\mathbf{a}$$

$$\Sigma F_r\mathbf{u}_r + \Sigma F_\theta\mathbf{u}_\theta + \Sigma F_z\mathbf{u}_z = ma_r\mathbf{u}_r + ma_\theta\mathbf{u}_\theta + ma_z\mathbf{u}_z$$

To satisfy this equation, the respective $\mathbf{u}_r$, $\mathbf{u}_\theta$, $\mathbf{u}_z$ components on the left side must equal the corresponding components on the right side. Consequently, we may write the following three scalar equations of motion:

$$\begin{aligned} \Sigma F_r &= ma_r \\ \Sigma F_\theta &= ma_\theta \\ \Sigma F_z &= ma_z \end{aligned} \qquad (13\text{–}9)$$

If the particle is constrained to move only in the $r{-}\theta$ plane, then only the first two of Eqs. 13–9 are used to specify the motion.

Tangential and Normal Forces. The most straightforward type of problem involving cylindrical coordinates requires the determination of the resultant force components ΣF_r, ΣF_θ, ΣF_z causing a particle to move with a *known* acceleration. If, however, the particle's accelerated motion is not completely specified at the given instant, then some information regarding the directions or magnitudes of the forces acting on the particle must be known or computed in order to solve Eqs. 13–9. For example, the force $\mathbf{P}$ causes the particle in Fig. 13–17a to move along a path $r = f(\theta)$. The *normal force* $\mathbf{N}$ which the path exerts on the particle is always *perpendicular to the tangent of the path*, whereas the frictional force $\mathbf{F}$ always acts along the tangent in the opposite direction of motion. The *directions* of $\mathbf{N}$ and $\mathbf{F}$ can be specified relative to the radial coordinate by using the angle ψ (psi), Fig. 13–17b, which is defined between the *extended* radial line and the tangent to the curve.

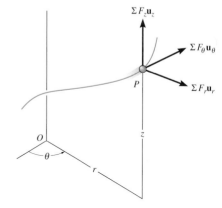

Inertial coordinate system

Fig. 13–16

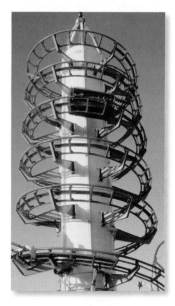

As the car of weight W descends the spiral track, the resultant normal force which the track exerts on the car can be represented by its three cylindrical components. $-N_r$ creates a radial acceleration, $-a_r$, N_θ creates a transverse acceleration a_θ, and the difference, $W - N_z$, creates an azimuthal acceleration $-a_z$.

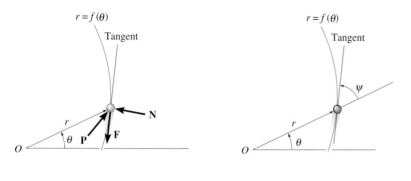

(a)

Fig. 13–17

(b)

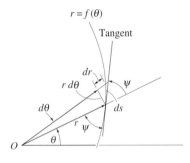

$r = f(\theta)$

Tangent

(c)

Fig. 13–17

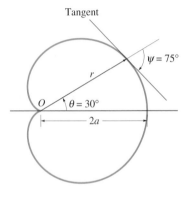

Tangent

$\psi = 75°$

r

O $\theta = 30°$

$2a$

Fig. 13–18

This angle can be obtained by noting that when the particle is displaced a distance ds along the path, Fig. 13–17c, the component of displacement in the radial direction is dr and the component of displacement in the transverse direction is $r\,d\theta$. Since these two components are mutually perpendicular, the angle ψ can be determined from $\tan\psi = r\,d\theta/dr$, or

$$\tan\psi = \frac{r}{dr/d\theta} \tag{13–10}$$

If ψ is calculated as a positive quantity, it is measured from the *extended radial line* to the tangent in a counterclockwise sense or in the positive direction of θ. If it is negative, it is measured in the opposite direction to positive θ. For example, consider the cardioid $r = a(1 + \cos\theta)$, shown in Fig. 13–18. Because $dr/d\theta = -a\sin\theta$, then when $\theta = 30°$, $\tan\psi = a(1 + \cos 30°)/(-a\sin 30°) = -3.732$, or $\psi = -75°$, measured clockwise, as shown in the figure.

PROCEDURE FOR ANALYSIS

Cylindrical or polar coordinates are a suitable choice for the analysis of a problem for which data regarding the angular motion of the radial line r are given, or in cases where the path can be conveniently expressed in terms of these coordinates. Once these coordinates have been established, the equations of motion can be applied in order to relate the forces acting on the particle to its acceleration components. The method for doing this has been outlined in the procedure for analysis given in Sec. 13.4. The following is a summary of this procedure.

Free-Body Diagram

- Establish the r, θ, z inertial coordinate system and draw the particle's free-body diagram.

- Assume that $\mathbf{a}_r$, $\mathbf{a}_\theta$, $\mathbf{a}_z$ act in the *positive directions* of r, θ, z if they are unknown.

- Identify all the unknowns in the problem.

Equations of Motion

- Apply the equations of motion, Eqs. 13–9.

Kinematics

- Use the methods of Sec. 12.8 to determine r and the time derivatives $\dot{r}$, $\ddot{r}$, $\dot{\theta}$, $\ddot{\theta}$, $\ddot{z}$, and then evaluate the acceleration components $a_r = \ddot{r} - r\dot{\theta}^2$, $a_\theta = r\ddot{\theta} + 2\dot{r}\dot{\theta}$, $a_z = \ddot{z}$.

- If any of the acceleration components is computed as a negative quantity, it indicates that it acts in its negative coordinate direction.

- When taking the time derivatives of $r = f(\theta)$, it is very important to use the chain rule of calculus.

E X A M P L E 13.10

The 2-lb block in Fig. 13–19a moves on a smooth horizontal track, such that its path is specified in polar coordinates by the parametric equations $r = (10t^2)$ ft and $\theta = (0.5t)$ rad, where t is in seconds. Determine the magnitude of the tangential force $\mathbf{F}$ causing the motion at the instant $t = 1$ s.

Solution

Free-Body Diagram. As shown on the block's free-body diagram, Fig. 13–19b, the normal force of the track on the block, $\mathbf{N}$, and the tangential force $\mathbf{F}$ are located at an angle ψ from the r and θ axes. This angle can be obtained from Eq. 13–10. To do so, we must first express the path as $r = f(\theta)$ by eliminating the parameter t between r and θ. This yields $r = 40\theta^2$. Also, when $t = 1$ s, $\theta = 0.5(1\text{ s}) = 0.5$ rad. Thus,

$$\tan \psi = \frac{r}{dr/d\theta} = \frac{40\theta^2}{40(2\theta)}\bigg|_{\theta = 0.5 \text{ rad}} = 0.25$$

$$\psi = 14.04°$$

Because ψ is a positive quantity, it is measured counterclockwise from the r axis to the tangent (the same direction as θ) as shown in Fig. 13–19b. There are presently four unknowns: F, N, a_r and a_θ.

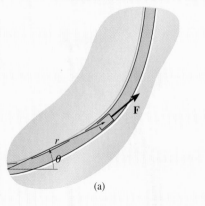

(a)

Equations of Motion.

$$+\downarrow \Sigma F_r = ma_r; \quad F\cos 14.04° - N\sin 14.04° = \frac{2}{32.2}a_r \quad (1)$$

$$\uparrow + \Sigma F_\theta = ma_\theta; \quad F\sin 14.04° + N\cos 14.04° = \frac{2}{32.2}a_\theta \quad (2)$$

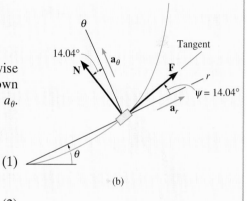

(b)

Fig. 13–19

Kinematics. Since the motion is specified, the coordinates and the required time derivatives can be calculated and evaluated at $t = 1$ s.

$$r = 10t^2\bigg|_{t=1\text{ s}} = 10 \text{ ft} \qquad \theta = 0.5t\bigg|_{t=1\text{ s}} = 0.5 \text{ rad}$$

$$\dot{r} = 20t\bigg|_{t=1\text{ s}} = 20 \text{ ft/s} \quad \dot{\theta} = 0.5 \text{ rad/s}$$

$$\ddot{r} = 20 \text{ ft/s}^2 \qquad\qquad \ddot{\theta} = 0$$

$$a_r = \ddot{r} - r\dot{\theta}^2 = 20 - 10(0.5)^2 = 17.5 \text{ ft/s}^2$$

$$a_\theta = r\ddot{\theta} + 2\dot{r}\dot{\theta} = 10(0) + 2(20)(0.5) = 20 \text{ ft/s}^2$$

Substituting into Eqs. 1 and 2 and solving, we get

$$F = 1.36 \text{ lb} \qquad\qquad Ans.$$
$$N = 0.942 \text{ lb}$$

EXAMPLE 13.11

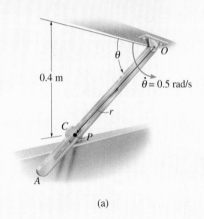

(a)

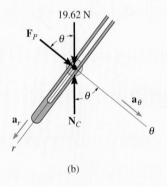

(b)

Fig. 13-20

The smooth 2-kg cylinder C in Fig. 13–20a has a peg P through its center which passes through the slot in arm OA. If the arm rotates in the *vertical plane* at a constant rate $\dot{\theta} = 0.5$ rad/s, determine the force that the arm exerts on the peg at the instant $\theta = 60°$.

Solution

Why is it a good idea to use polar coordinates to solve this problem?

Free-Body Diagram. The free-body diagram for the cylinder is shown in Fig. 13–20b. The force on the peg, $\mathbf{F}_P$, acts perpendicular to the slot in the arm. As usual, $\mathbf{a}_r$ and $\mathbf{a}_\theta$ are assumed to act in the directions of *positive r* and θ, respectively. Identify the four unknowns.

Equations of Motion. Using the data in Fig. 13–20b, we have

$$+\searrow \Sigma F_r = ma_r; \qquad 19.62 \sin\theta - N_C \sin\theta = 2a_r \qquad (1)$$

$$+\swarrow \Sigma F_\theta = ma_\theta; \qquad 19.62 \cos\theta + F_P - N_C \cos\theta = 2a_\theta \qquad (2)$$

Kinematics. From Fig. 13–20a, r can be related to θ by the equation

$$r = \frac{0.4}{\sin\theta} = 0.4 \csc\theta$$

Since $d(\csc\theta) = -(\csc\theta \cot\theta)\,d\theta$ and $d(\cot\theta) = -(\csc^2\theta)\,d\theta$, then r and the necessary time derivatives become

$$\dot{\theta} = 0.5 \qquad r = 0.4 \csc\theta$$
$$\ddot{\theta} = 0 \qquad \dot{r} = -0.4(\csc\theta \cot\theta)\dot{\theta}$$
$$= -0.2 \csc\theta \cot\theta$$
$$\ddot{r} = -0.2(-\csc\theta \cot\theta)(\dot{\theta})\cot\theta - 0.2\csc\theta(-\csc^2\theta)\dot{\theta}$$
$$= 0.1 \csc\theta(\cot^2\theta + \csc^2\theta)$$

Evaluating these formulas at $\theta = 60°$, we get

$$\dot{\theta} = 0.5 \qquad r = 0.462$$
$$\ddot{\theta} = 0 \qquad \dot{r} = -0.133$$
$$\ddot{r} = 0.192$$
$$a_r = \ddot{r} - r\dot{\theta}^2 = 0.192 - 0.462(0.5)^2 = 0.0770$$
$$a_\theta = r\ddot{\theta} + 2\dot{r}\dot{\theta} = 0 + 2(-0.133)(0.5) = -0.133$$

Substituting these results into Eqs. 1 and 2 with $\theta = 60°$ and solving yields

$$N_C = 19.4 \text{ N} \qquad F_P = -0.356 \text{ N} \qquad Ans.$$

The negative sign indicates that $\mathbf{F}_P$ acts opposite to that shown in Fig. 13–20b.

EXAMPLE 13.12

A can C, having a mass of 0.5 kg, moves along a grooved horizontal slot shown in Fig. 13–21a. The slot is in the form of a spiral, which is defined by the equation $r = (0.1\theta)$ m, where θ is in radians. If the arm OA is rotating at a constant rate $\dot{\theta} = 4$ rad/s in the horizontal plane, determine the force it exerts on the can at the instant $\theta = \pi$ rad. Neglect friction and the size of the can.

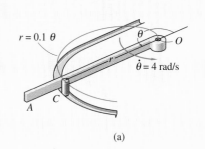

(a)

Solution

Free-Body Diagram. The driving force $\mathbf{F}_C$ acts perpendicular to the arm OA, whereas the normal force of the wall of the slot on the can, $\mathbf{N}_C$, acts perpendicular to the tangent to the curve at $\theta = \pi$ rad, Fig. 13–21b. As usual, $\mathbf{a}_r$ and $\mathbf{a}_\theta$ are assumed to act in the *positive directions* of r and θ, respectively. Since the path is specified, the angle ψ which the extended radial line r makes with the tangent, Fig. 13–21c, can be determined from Eq. 13–10. We have $r = 0.1\theta$, so that $dr/d\theta = 0.1$, and therefore

$$\tan \psi = \frac{r}{dr/d\theta} = \frac{0.1\theta}{0.1} = \theta$$

When $\theta = \pi$, $\psi = \tan^{-1}\pi = 72.3°$, so that $\phi = 90° - \psi = 17.7°$, as shown in Fig. 13–21c. Identify the four unknowns in Fig. 13–21b.

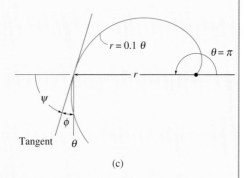

(b)

Equations of Motion. Using $\phi = 17.7°$ and the data shown in Fig. 13–21b, we have

$$\xleftarrow{+} \Sigma F_r = ma_r; \qquad N_C \cos 17.7° = 0.5a_r \qquad (1)$$

$$+\downarrow \Sigma F_\theta = ma_\theta; \qquad F_C - N_C \sin 17.7° = 0.5a_\theta \qquad (2)$$

Kinematics. The time derivatives of r and θ are

$$\dot{\theta} = 4 \text{ rad/s} \qquad r = 0.1\theta$$
$$\ddot{\theta} = 0 \qquad \dot{r} = 0.1\dot{\theta} = 0.1(4) = 0.4 \text{ m/s}$$
$$\ddot{r} = 0.1\ddot{\theta} = 0$$

At the instant $\theta = \pi$ rad,

$$a_r = \ddot{r} - r\dot{\theta}^2 = 0 - 0.1(\pi)(4)^2 = -5.03 \text{ m/s}^2$$
$$a_\theta = r\ddot{\theta} + 2\dot{r}\dot{\theta} = 0 + 2(0.4)(4) = 3.20 \text{ m/s}^2$$

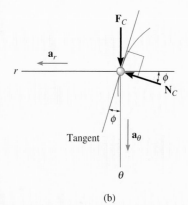

(c)

Fig. 13–21

Substituting these results into Eqs. 1 and 2 and solving yields

$$N_C = -2.64 \text{ N}$$
$$F_C = 0.800 \text{ N} \qquad\qquad Ans.$$

What does the negative sign for N_C indicate?

*13.7 Central-Force Motion and Space Mechanics

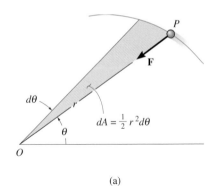

(a)

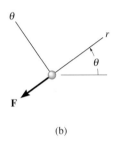

(b)

Fig. 13–22

If a particle is moving only under the influence of a force having a line of action which is always directed toward a fixed point, the motion is called *central-force motion*. This type of motion is commonly caused by electrostatic and gravitational forces.

In order to determine the motion, we will consider the particle P shown in Fig. 13–22a, which has a mass m and is acted upon only by the central force **F**. The free-body diagram for the particle is shown in Fig. 13–22b. Using polar coordinates (r, θ), the equations of motion, Eqs. 13–9, become

$$-F = m\left[\frac{d^2r}{dt^2} - r\left(\frac{d\theta}{dt}\right)^2\right]$$

$$0 = m\left(r\frac{d^2\theta}{dt^2} + 2\frac{dr}{dt}\frac{d\theta}{dt}\right) \tag{13–11}$$

The second of these equations may be written in the form

$$\frac{1}{r}\left[\frac{d}{dt}\left(r^2\frac{d\theta}{dt}\right)\right] = 0$$

so that integrating yields

$$r^2\frac{d\theta}{dt} = h \tag{13–12}$$

Here h is a constant of integration. From Fig. 13–22a notice that the shaded area described by the radius r, as r moves through an angle $d\theta$, is $dA = \frac{1}{2}r^2 d\theta$. If the *areal velocity* is defined as

$$\frac{dA}{dt} = \frac{1}{2}r^2\frac{d\theta}{dt} = \frac{h}{2} \tag{13–13}$$

then, it is seen that the areal velocity for a particle subjected to central-force motion is *constant*. In other words, the particle will sweep out equal segments of area per unit of time as it travels along the path. To obtain the *path of motion*, $r = f(\theta)$, the independent variable t must be eliminated from Eqs. 13–11. Using the chain rule of calculus and Eq. 13–12, the time derivatives of Eqs. 13–11 may be replaced by

$$\frac{dr}{dt} = \frac{dr}{d\theta}\frac{d\theta}{dt} = \frac{h}{r^2}\frac{dr}{d\theta}$$

$$\frac{d^2r}{dt^2} = \frac{d}{dt}\left(\frac{h}{r^2}\frac{dr}{d\theta}\right) = \frac{d}{d\theta}\left(\frac{h}{r^2}\frac{dr}{d\theta}\right)\frac{d\theta}{dt} = \left[\frac{d}{d\theta}\left(\frac{h}{r^2}\frac{dr}{d\theta}\right)\right]\frac{h}{r^2}$$

Substituting a new dependent variable (xi) $\xi = 1/r$ into the second equation, we have

$$\frac{d^2r}{dt^2} = -h^2\xi^2\frac{d^2\xi}{d\theta^2}$$

Also, the square of Eq. 13–12 becomes

$$\left(\frac{d\theta}{dt}\right)^2 = h^2\xi^4$$

Substituting these last two equations into the first of Eqs. 13–11 yields

$$-h^2\xi^2\frac{d^2\xi}{d\theta^2} - h^2\xi^3 = -\frac{F}{m}$$

or

$$\frac{d^2\xi}{d\theta^2} + \xi = \frac{F}{mh^2\xi^2} \tag{13-14}$$

This satellite is subjected to a central force and as such its orbital motion can be closely predicted using the equations developed in this section.

This differential equation defines the path over which the particle travels when it is subjected to the central force* $\mathbf{F}$.

For application, the force of gravitational attraction will be considered. Some common examples of central-force systems which depend on gravitation include the motion of the moon and artificial satellites about the earth, and the motion of the planets about the sun. As a typical problem in space mechanics, consider the trajectory of a space satellite or space vehicle launched into free-flight orbit with an initial velocity $\mathbf{v}_0$, Fig. 13–23. It will be assumed that this velocity is initially *parallel* to the tangent at the surface of the earth, as shown in the figure.† Just after the satellite is released into free flight, the only force acting on it is the gravitational force of the earth. (Gravitational attractions involving other bodies such as the moon or sun will be neglected, since for orbits close to the earth their effect is small in comparison with the earth's gravitation.) According to Newton's law of gravitation, force $\mathbf{F}$ will always act between the mass centers of the earth and the satellite, Fig. 13–23. From Eq. 13–1, this force of attraction has a magnitude of

$$F = G\frac{M_em}{r^2}$$

where M_e and m represent the mass of the earth and the satellite, respectively, G is the gravitational constant, and r is the distance between

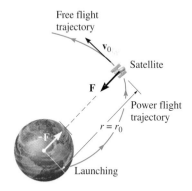

Free flight trajectory

$\mathbf{v}_0$

Satellite

$\mathbf{F}$

Power flight trajectory

$r = r_0$

Launching

Fig. 13–23

---

*In the derivation, $\mathbf{F}$ is considered positive when it is directed toward point O. If $\mathbf{F}$ is oppositely directed, the right side of Eq. 13–14 should be negative.

†The case where $\mathbf{v}_0$ acts at some initial angle θ to the tangent is best described using the conservation of angular momentum.

the mass centers. Setting $\xi = 1/r$ in the foregoing equation and substituting the result into Eq. 13–14, we obtain

$$\frac{d^2\xi}{d\theta^2} + \xi = \frac{GM_e}{h^2} \qquad (13\text{–}15)$$

This second-order ordinary differential equation has constant coefficients and is nonhomogeneous. The solution is represented as the sum of the complementary and particular solutions. The complementary solution is obtained when the term on the right is equal to zero. It is

$$\xi_c = C \cos(\theta - \phi)$$

where C and ϕ are constants of integration. The particular solution is

$$\xi_p = \frac{GM_e}{h^2}$$

Thus, the complete solution to Eq. 13–15 is

$$\xi = \xi_c + \xi_p$$
$$= \frac{1}{r} = C \cos(\theta - \phi) + \frac{GM_e}{h^2} \qquad (13\text{–}16)$$

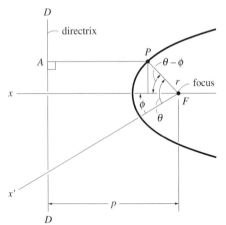

Fig. 13–24

The validity of this result may be checked by substitution into Eq. 13–15.

Equation 13–16 represents the *free-flight trajectory* of the satellite. It is the equation of a conic section expressed in terms of polar coordinates. As shown in Fig. 13–24, a *conic section* is defined as the locus of point P, which moves in a plane in such a way that the ratio of its distance from a fixed point F to its distance from a fixed line is constant. The fixed point is called the *focus*, and the fixed line DD is called the *directrix*. The constant ratio is called the *eccentricity* of the conic section and is denoted by e. Thus,

$$e = \frac{FP}{PA}$$

which may be written in the form

$$FP = r = e(PA) = e[p - r \cos(\theta - \phi)]$$

or

$$\frac{1}{r} = \frac{1}{p} \cos(\theta - \phi) + \frac{1}{ep}$$

Comparing this equation with Eq. 13–16, it is seen that the eccentricity of the conic section for the trajectory is

$$e = \frac{Ch^2}{GM_e} \qquad (13\text{–}17)$$

and the fixed distance from the focus to the directrix is

$$p = \frac{1}{C}$$ (13–18)

Provided the polar angle θ is measured from the x axis (an axis of symmetry since it is perpendicular to the directrix), the angle ϕ is zero, Fig. 13–24, and therefore Eq. 13–16 reduces to

$$\frac{1}{r} = C \cos \theta + \frac{GM_e}{h^2}$$ (13–19)

The constants h and C are determined from the data obtained for the position and velocity of the satellite at the end of the *power-flight trajectory*. For example, if the initial height or distance to the space vehicle is r_0 (measured from the center of the earth) and its initial speed is v_0 at the beginning of its free flight, Fig. 13–25, then the constant h may be obtained from Eq. 13–12. When $\theta = \phi = 0°$, the velocity $\mathbf{v}_0$ has no radial component; therefore, from Eq. 12–25, $v_0 = r_0(d\theta/dt)$, so that

$$h = r_0^2 \frac{d\theta}{dt}$$

or

$$\boxed{h = r_0 v_0}$$ (13–20)

To determine C, use Eq. 13–19 with $\theta = 0°$, $r = r_0$, and substitute Eq. 13–20 for h:

$$\boxed{C = \frac{1}{r_0}\left(1 - \frac{GM_e}{r_0 v_0^2}\right)}$$ (13–21)

The equation for the free-flight trajectory therefore becomes

$$\boxed{\frac{1}{r} = \frac{1}{r_0}\left(1 - \frac{GM_e}{r_0 v_0^2}\right)\cos \theta + \frac{GM_e}{r_0^2 v_0^2}}$$ (13–22)

The type of path taken by the satellite is determined from the value of the eccentricity of the conic section as given by Eq. 13–17. If

$$\boxed{\begin{array}{ll} e = 0 & \text{free-flight trajectory is a circle} \\ e = 1 & \text{free-flight trajectory is a parabola} \\ e < 1 & \text{free-flight trajectory is an ellipse} \\ e > 1 & \text{free-flight trajectory is a hyperbola} \end{array}}$$ (13–23)

Each of these trajectories is shown in Fig. 13–25. From the curves it is seen that when the satellite follows a parabolic path, it is "on the border" of never returning to its initial starting point. The initial launch velocity, $\mathbf{v}_0$, required for the satellite to follow a parabolic path is called the *escape velocity*. The speed, v_e, can be determined by using the second of Eqs. 13–23 with Eqs. 13–17, 13–20, and 13–21. It is left as an exercise to show that

$$v_e = \sqrt{\frac{2GM_e}{r_0}} \qquad (13\text{–}24)$$

The speed v_c required to launch a satellite into a *circular orbit* can be found using the first of Eqs. 13–23. Since e is related to h and C, Eq. 13–17, C must be zero to satisfy this equation (from Eq. 13–20, h cannot be zero); and therefore, using Eq. 13–21, we have

$$v_c = \sqrt{\frac{GM_e}{r_0}} \qquad (13\text{–}25)$$

Provided r_0 represents a minimum height for launching, in which frictional resistance from the atmosphere is neglected, speeds at launch which are less than v_c will cause the satellite to reenter the earth's atmosphere and either burn up or crash, Fig. 13–25.

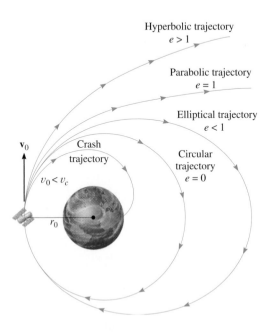

Fig. 13–25

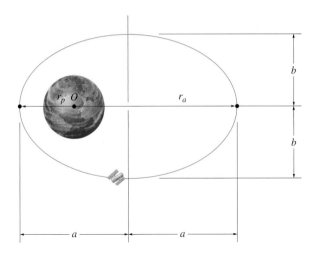

Fig. 13–26

All the trajectories attained by planets and most satellites are elliptical, Fig. 13–26. For a satellite's orbit about the earth, the *minimum distance* from the orbit to the center of the earth O (which is located at one of the foci of the ellipse) is r_p and can be found using Eq. 13–22 with $\theta = 0°$. Therefore;

$$r_p = r_0 \tag{13–26}$$

This minimum distance is called the *perigee* of the orbit. The *apogee* or maximum distance r_a can be found using Eq. 13–22 with $\theta = 180°$.* Thus,

$$r_a = \frac{r_0}{(2GM_e/r_0v_0^2) - 1} \tag{13–27}$$

With reference to Fig. 13–26, the semimajor axis a of the ellipse is

$$a = \frac{r_p + r_a}{2} \tag{13–28}$$

Using analytical geometry, it can be shown that the minor axis b is determined from the equation

$$b = \sqrt{r_p r_a} \tag{13–29}$$

---

*Actually, the terminology perigee and apogee pertains only to orbits about the *earth*. If any other heavenly body is located at the focus of an elliptical orbit, the minimum and maximum distances are referred to respectively as the *periapsis* and *apoapsis* of the orbit.

Furthermore, by direct integration, the area of an ellipse is

$$A = \pi ab = \frac{\pi}{2}(r_p + r_a)\sqrt{r_p r_a} \qquad (13\text{–}30)$$

The areal velocity has been defined by Eq. 13–13, $dA/dt = h/2$. Integrating yields $A = hT/2$, where T is the *period* of time required to make one orbital revolution. From Eq. 13–30, the period is

$$\boxed{T = \frac{\pi}{h}(r_p + r_a)\sqrt{r_p r_a}} \qquad (13\text{–}31)$$

In addition to predicting the orbital trajectory of earth satellites, the theory developed in this section is valid, to a surprisingly close approximation, at predicting the actual motion of the planets traveling around the sun. In this case the mass of the sun, M_s, should be substituted for M_e when the appropriate formulas are used.

The fact that the planets do indeed follow elliptic orbits about the sun was discovered by the German astronomer Johannes Kepler in the early seventeenth century. His discovery was made *before* Newton had developed the laws of motion and the law of gravitation, and so at the time it provided important proof as to the validity of these laws. Kepler's laws, developed after 20 years of planetary observation, are summarized as follows:

1. Every planet moves in its orbit such that the line joining it to the sun sweeps over equal areas in equal intervals of time, whatever the line's length.

2. The orbit of every planet is an ellipse with the sun placed at one of its foci.

3. The square of the period of any planet is directly proportional to the cube of the minor axis of its orbit.

A mathematical statement of the first and second laws is given by Eqs. 13–13 and 13–22, respectively. The third law can be shown from Eq. 13–31 using Eqs. 13–19, 13–28, and 13–29.

EXAMPLE 13.13

A satellite is launched 600 km from the surface of the earth, with an initial velocity of 30 Mm/h acting parallel to the tangent at the surface of the earth, Fig. 13–27. Assuming that the radius of the earth is 6378 km and that its mass is $5.976(10^{24})$ kg, determine (a) the eccentricity of the orbital path, and (b) the velocity of the satellite at apogee.

Solution

Part (a). The eccentricity of the orbit is obtained using Eq. 13–17. The constants h and C are first determined from Eqs. 13–20 and 13–21. Since

$$r_p = r_0 = 6378 \text{ km} + 600 \text{ km} = 6.978(10^6) \text{ m}$$

$$v_0 = 30 \text{ Mm/h} = 8333.3 \text{ m/s}$$

then

$$h = r_p v_0 = 6.978(10^6)(8333.3) = 58.15(10^9) \text{ m}^2/\text{s}$$

$$C = \frac{1}{r_p}\left(1 - \frac{GM_e}{r_p v_0^2}\right)$$

$$= \frac{1}{6.978(10^6)}\left\{1 - \frac{66.73(10^{-12})[5.976(10^{24})]}{6.978(10^6)(8333.3)^2}\right\} = 25.4(10^{-9}) \text{ m}^{-1}$$

Hence,

$$e = \frac{Ch^2}{GM_e} = \frac{2.54(10^{-8})[58.15(10^9)]^2}{66.73(10^{-12})[5.976(10^{24})]} = 0.215 < 1 \qquad Ans.$$

From Eq. 13–23, observe that the orbit is an *ellipse*.

Part (b). If the satellite were launched at the apogee A shown in Fig. 13–27, with a velocity $\mathbf{v}_A$, the same orbit would be maintained provided that

$$h = r_p v_0 = r_a v_A = 58.15(10^9) \text{ m}^2/\text{s}$$

Using Eq. 13–27, we have

$$r_a = \frac{r_p}{\dfrac{2GM_e}{r_p v_0^2} - 1} = \frac{6.978(10^6)}{\left\{\dfrac{2[66.73(10^{-12})][5.976(10^{24})]}{6.978(10^6)(8333.3)^2} - 1\right\}} = 10.804(10^6)$$

Thus,

$$v_A = \frac{58.15(10^9)}{10.804(10^6)} = 5382.2 \text{ m/s} = 19.4 \text{ Mm/h} \qquad Ans.$$

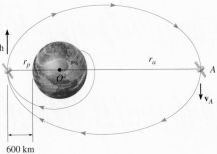

$v_0 = 30 \text{ Mm/h}$

r_p O r_a A

$\mathbf{v}_A$

600 km

Fig. 13–27

DESIGN PROJECTS

13–1D. DESIGN OF A RAMP CATAPULT

The block B has a mass of 20 kg and is to be catapulted from the table. Design a catapulting mechanism that can be attached to the table and to the container of the block, using cables and pulleys. Neglect the mass of the container, assume the operator can exert a constant tension of 120 N on a single cable during operation, and that the maximum movement of his arm is 0.5 m. The coefficient of kinetic friction between the table and container is $\mu_k = 0.2$. Submit a drawing of your design, and calculate the maximum range R to where the block will strike the ground. Compare your value with that of others in the class.

13–2D. DESIGN OF A WATER-BALLOON LAUNCHER

Design a method for launching a 0.25-lb water balloon. Hold a contest with other students to see who can launch the balloon the farthest or hit a target. Materials should consist of a single rubber band of specified length and stiffness, and if necessary no more than three pieces of wood of specified size. Submit a report to show your calculations of where the balloon is predicted to strike the ground from the point at which it was launched. Compare this with the actual value R and discuss why the two distances may be different.

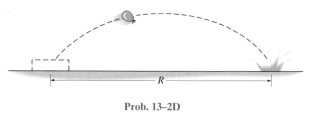

Prob. 13–2D

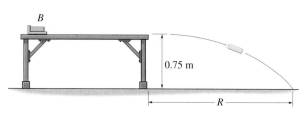

Prob. 13–1D

CHAPTER REVIEW

- **Kinetics.**   Kinetics is the study of the relationship between forces and the acceleration they cause. This relationship is based on Newton's second law of motion, expressed mathematically as $\Sigma \mathbf{F} = m\mathbf{a}$. Here the mass m is the proportionality constant between the resultant force $\Sigma \mathbf{F}$ acting on the particle and the acceleration $\mathbf{a}$ caused by this resultant. The mass represents the quantity of matter contained within the particle. It measures the resistance to a change in its motion.

  Before applying the equation of motion, it is important to draw the particle's free-body diagram first in order to account for all of the forces that act on the particle. Graphically, this diagram can be set equal to the kinetic diagram, which shows the result of the forces, that is, the $m\mathbf{a}$ vector.

- **Inertial Coordinate Systems.**   When applying the equation of motion, it is important to make measurements of the acceleration from an inertial coordinate system. This system has axes that do not rotate but are either fixed or translate with a constant velocity. Various types of inertial coordinate systems can be used to apply $\Sigma \mathbf{F} = m\mathbf{a}$ in component form. Rectangular x, y, z axes are used to describe rectilinear motion along each of the axes. Normal and tangential n, t axes are often used when the path is known. Recall that $\mathbf{a}_n$ is always directed in the $+n$ direction. It indicates the change in the velocity direction. And $\mathbf{a}_t$ is tangent to the path. It indicates the change in the velocity magnitude. Finally, cylindrical coordinates are useful when angular motion of the radial coordinate r is specified or when the path can conveniently be described with these coordinates. For some problems, the direction of the forces on the free-body diagram will require finding the angle ψ between the extended radial coordinate and the tangent to the curve. This angle can be determined using

$$\tan \psi = \frac{r}{dr/d\theta}$$

- **Central-Force Motion.**   When a single force acts upon a particle, such as the free-flight trajectory of a satellite in a gravitational field, then the motion is referred to as central-force motion. The orbit depends upon the eccentricity e, and as a result, the trajectory can either be circular, parabolic, elliptical, or hyperbolic.

In order to properly design the loop of this roller coaster it is necessary to ensure that the cars have enough energy to be able to make the loop without leaving the tracks.

Kinetics of a Particle: Work and Energy

- To develop the principle of work and energy and apply it to solve problems that involve force, velocity, and displacement.
- To study problems that involve power and efficiency.
- To introduce the concept of a conservative force and apply the theorem of conservation of energy to solve kinetic problems.

14.1 The Work of a Force

In mechanics a force **F** does *work* on a particle only when the particle undergoes a *displacement in the direction of the force*. For example, consider the force **F** acting on the particle in Fig. 14–1. If the particle moves along the path *s* from position **r** to a new position **r′**, the displacement is then $d\mathbf{r} = \mathbf{r}' - \mathbf{r}$. The magnitude of $d\mathbf{r}$ is represented by ds, the differential segment along the path. If the angle between the tails of $d\mathbf{r}$ and **F** is θ, Fig. 14–1, then the work dU which is done by **F** is a

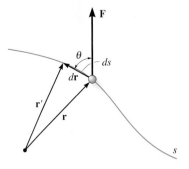

Fig. 14–1

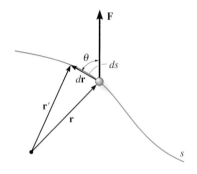

Fig. 14–1

scalar quantity, defined by

$$dU = F \, ds \, \cos\theta$$

By definition of the dot product (see Eq. C–14) this equation may also be written as

$$dU = \mathbf{F} \cdot d\mathbf{r}$$

This result may be interpreted in one of two ways: either as the product of F and the component of displacement in the direction of the force, i.e., $ds \cos\theta$, or as the product of ds and the component of force in the direction of displacement, i.e., $F \cos\theta$. Note that if $0° \le \theta < 90°$, then the force component and the displacement have the *same sense* so that the work is *positive*; whereas if $90° < \theta \le 180°$, these vectors have an *opposite sense*, and therefore the work is *negative*. Also, $dU = 0$ if the force is *perpendicular* to displacement, since $\cos 90° = 0$, or if the force is applied at a *fixed point*, in which case the displacement is zero.

The basic unit for work in the SI system is called a joule (J). This unit combines the units of force and displacement. Specifically, 1 *joule* of work is done when a force of 1 newton moves 1 meter along its line of action $(1\,J = 1\,N \cdot m)$. The moment of a force has this same combination of units $(N \cdot m)$; however, the concepts of moment and work are in no way related. A moment is a vector quantity, whereas work is a scalar. In the FPS system work is generally defined by writing the units as $ft \cdot lb$, which is distinguished from the units for a moment, written as $lb \cdot ft$.

Work of a Variable Force. If the particle undergoes a finite displacement along its path from r_1 to r_2 or s_1 to s_2, Fig. 14–2a, the work is determined by integration. If **F** is expressed as a function of position, $F = F(s)$, we have

$$U_{1-2} = \int_{r_1}^{r_2} \mathbf{F} \cdot d\mathbf{r} = \int_{s_1}^{s_2} F \cos\theta \, ds \qquad (14\text{–}1)$$

If the working component of the force, $F \cos\theta$, is plotted versus s, Fig. 14–2b, the integral in this equation can be interpreted as the *area under the curve* from position s_1 to position s_2.

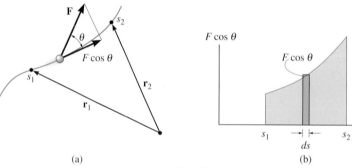

(a)

(b)

Fig. 14–2

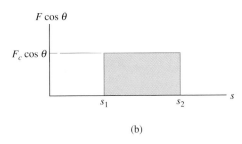

Fig. 14–3

Work of a Constant Force Moving Along a Straight Line.

If the force F_c has a constant magnitude and acts at a constant angle θ from its straight-line path, Fig. 14–3a, then the component of F_c in the direction of displacement is $F_c \cos\theta$. The work done by F_c when the particle is displaced from s_1 to s_2 is determined from Eq. 14–1, in which case

$$U_{1-2} = F_c \cos\theta \int_{s_1}^{s_2} ds$$

or

$$\boxed{U_{1-2} = F_c \cos\theta(s_2 - s_1)} \qquad (14\text{–}2)$$

Here the work of F_c represents the *area of the rectangle* in Fig. 14–3b.

Work of a Weight.

Consider a particle which moves up along the path s shown in Fig. 14–4 from position s_1 to position s_2. At an intermediate point, the displacement $d\mathbf{r} = dx\mathbf{i} + dy\mathbf{j} + dz\mathbf{k}$. Since $\mathbf{W} = -W\mathbf{j}$, applying Eq. 14–1 yields

$$U_{1-2} = \int \mathbf{F} \cdot d\mathbf{r} = \int_{\mathbf{r}_1}^{\mathbf{r}_2} (-W\mathbf{j}) \cdot (dx\mathbf{i} + dy\mathbf{j} + dz\mathbf{k})$$

$$= \int_{y_1}^{y_2} -W\,dy = -W(y_2 - y_1)$$

or

$$\boxed{U_{1-2} = -W\Delta y} \qquad (14\text{–}3)$$

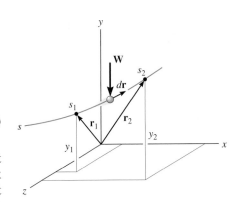

Fig. 14–4

Thus, the work done is equal to the magnitude of the particle's weight times its vertical displacement. In the case shown in Fig. 14–4 the work is *negative*, since W is downward and Δy is upward. Note, however, that if the particle is displaced *downward* $(-\Delta y)$, the work of the weight is *positive*. Why?

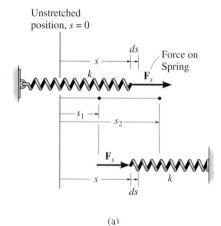

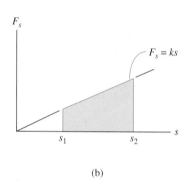

(a)

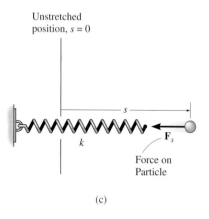

(b)

(c)

Fig. 14–5

Work of a Spring Force.

The magnitude of force developed in a linear elastic spring when the spring is displaced a distance s from its unstretched position is $F_s = ks$, where k is the spring stiffness. If the spring is elongated or compressed from a position s_1 to a further position s_2, Fig. 14–5a, the work done *on the spring* by F_s is *positive*, since in each case the force and displacement are in the *same direction*. We require

$$U_{1-2} = \int_{s_1}^{s_2} F_s \, ds = \int_{s_1}^{s_2} ks \, ds$$

$$= \tfrac{1}{2}ks_2^2 - \tfrac{1}{2}ks_1^2$$

This equation represents the trapezoidal area under the line $F_s = ks$, Fig. 14–5b.

If a particle (or body) is attached to a spring, then the force F_s exerted on the particle is *opposite* to that exerted on the spring, Fig. 14–5c. Consequently, the force will do *negative work* on the particle when the particle is moving so as to further elongate (or compress) the spring. Hence, the above equation becomes

$$U_{1-2} = -(\tfrac{1}{2}ks_2^2 - \tfrac{1}{2}ks_1^2) \qquad (14\text{–}4)$$

When this equation is used, a mistake in sign can be eliminated if one simply notes the direction of the spring force acting on the particle and compares it with the direction of displacement of the particle—if both are in the *same direction, positive work* results; if they are *opposite* to one another, the *work is negative*.

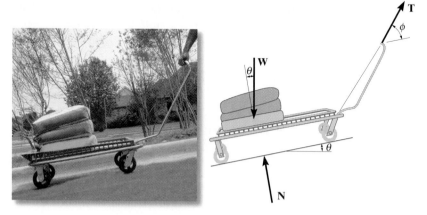

The forces acting on the cart as it is pulled a distance s up the incline, are shown on its free-body diagram. The constant towing force **T** does positive work of $U_T = (T \cos\phi)s$, the weight does negative work of $U_W = -(W \sin\theta)s$, and the normal force **N** does no work since there is no displacement of this force along its line of action.

E X A M P L E 14.1

The 10-kg block shown in Fig. 14–6a rests on the smooth incline. If the spring is originally stretched 0.5 m, determine the total work done by all the forces acting on the block when a horizontal force $P = 400$ N pushes the block up the plane $s = 2$ m.

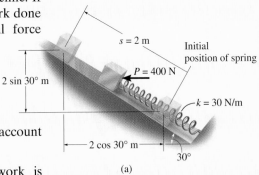

(a)

Solution

First the free-body diagram of the block is drawn in order to account for all the forces that act on the block, Fig. 14–6b.

Horizontal Force P. Since this force is *constant*, the work is determined using Eq. 14–2. The result can be calculated as the force times the component of displacement in the direction of the force; i.e.,

$$U_P = 400 \text{ N } (2 \text{ m cos } 30°) = 692.8 \text{ J}$$

or the displacement times the component of force in the direction of displacement, i.e.,

$$U_P = 400 \text{ N cos } 30°(2 \text{ m}) = 692.8 \text{ J}$$

(b)

Fig. 14–6

Spring Force $\mathbf{F_s}$. In the initial position the spring is stretched $s_1 = 0.5$ m and in the final position it is stretched $s_2 = 0.5 + 2 = 2.5$ m. We require the work to be negative since the force and displacement are in opposite directions. The work of $\mathbf{F}_s$ is thus

$$U_s = -[\tfrac{1}{2}(30 \text{ N/M})(2.5 \text{ m})^2 - \tfrac{1}{2}(30 \text{ N/m})(0.5 \text{ m})^2] = -90 \text{ J}$$

Weight W. Since the weight acts in the opposite direction to its vertical displacement, the work is negative; i.e.,

$$U_W = -98.1 \text{ N } (2 \text{ m sin } 30°) = -98.1 \text{ J}$$

Note that it is also possible to consider the component of weight in the direction of displacement; i.e.,

$$U_W = -(98.1 \sin 30° \text{ N})2 \text{ m} = -98.1 \text{ J}$$

Normal Force $\mathbf{N_B}$. This force does *no work* since it is *always* perpendicular to the displacement.

Total Work. The work of all the forces when the block is displaced 2 m is thus

$$U_T = 692.8 - 90 - 98.1 = 505 \text{ J} \qquad \textit{Ans.}$$

14.2 Principle of Work and Energy

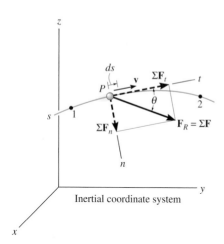

Inertial coordinate system

Fig. 14–7

Consider a particle P in Fig. 14–7, which at the instant considered is located on the path as measured from an inertial coordinate system. If the particle has a mass m and is subjected to a system of external forces represented by the resultant $F_R = \Sigma F$, then the equation of motion for the particle in the tangential direction is $\Sigma F_t = ma_t$. Applying the kinematic equation $a_t = v \, dv/ds$ and integrating both sides, assuming initially that the particle has a position $s = s_1$ and a speed $v = v_1$, and later at $s = s_2, v = v_2$, yields

$$\Sigma \int_{s_1}^{s_2} F_t \, ds = \int_{v_1}^{v_2} mv \, dv$$

$$\Sigma \int_{s_1}^{s_2} F_t \, ds = \tfrac{1}{2}mv_2^2 - \tfrac{1}{2}mv_1^2 \tag{14–5}$$

From Fig. 14–7, $\Sigma F_t = \Sigma F \cos\theta$ and since work is defined from Eq. 14–1, the final result may be written as

$$\Sigma U_{1-2} = \tfrac{1}{2}mv_2^2 - \tfrac{1}{2}mv_1^2 \tag{14–6}$$

This equation represents the *principle of work and energy* for the particle. The term on the left is the sum of the work done by *all* the forces acting on the particle as the particle moves from point 1 to point 2. The two terms on the right side, which are of the form $T = \tfrac{1}{2}mv^2$, define the particle's final and initial *kinetic energy*, respectively. These terms are always *positive* scalars. Furthermore, Eq. 14–6 must be dimensionally homogeneous so that the kinetic energy has the same units as work, e.g., joules (J) or ft·lb.

When Eq. 14–6 is applied, it is often symbolized in the form

$$\boxed{T_1 + \Sigma U_{1-2} = T_2} \tag{14–7}$$

which states that the particle's initial kinetic energy plus the work done by all the forces acting on the particle as it moves from its initial to its final position is equal to the particle's final kinetic energy.

As noted from the derivation, the principle of work and energy represents an integrated form of $\Sigma F_t = ma_t$, obtained by using the kinematic equation $a_t = v \, dv/ds$. As a result, this principle will provide a convenient *substitution* for $\Sigma F_t = ma_t$ when solving those types of kinetic problems which involve force, velocity, and displacement, since these variables are involved in the terms of Eq. 14–7. For example, if a particle's initial speed is known and the work of all the forces acting on the particle can be determined, then Eq. 14–7 provides a *direct means* of obtaining the final speed v_2 of the particle after it undergoes a specified

If an oncoming car strikes these crash barrels, the car's kinetic energy will be transformed into work, which causes the barrels, and to some extent the car, to be deformed. By knowing the amount of energy absorbed by each barrel it is possible to design a crash cushion such as this.

displacement. If instead v_2 is determined by means of the equation of motion, a two-step process is necessary; i.e., apply $\Sigma F_t = ma_t$ to obtain a_t, then integrate $a_t = v\, dv/ds$ to obtain v_2. Note that the principle of work and energy cannot be used, for example, to determine forces directed *normal* to the path of motion, since these forces do no work on the particle. Instead $\Sigma F_n = ma_n$ must be applied. For curved paths, however, the magnitude of the normal force is a function of speed. Hence, it may be easier to obtain this speed using the principle of work and energy, and then substitute this quantity into the equation of motion $\Sigma F_n = mv^2/\rho$ to obtain the normal force.

PROCEDURE FOR ANALYSIS

The principle of work and energy is used to solve kinetic problems that involve *velocity, force*, and *displacement*, since these terms are involved in the equation. For application it is suggested that the following procedure be used.

Work (Free-Body Diagram)

- Establish the inertial coordinate system and draw a free-body diagram of the particle in order to account for all the forces that do work on the particle as it moves along its path.

Principle of Work and Energy

- Apply the principle of work and energy, $T_1 + \Sigma U_{1-2} = T_2$.
- The kinetic energy at the initial and final points is always positive, since it involves the speed squared $\left(T = \frac{1}{2}mv^2\right)$.
- A force does work when it moves through a displacement in the direction of the force.
- Work is *positive* when the force component is in the *same direction* as its displacement, otherwise it is negative.
- Forces that are functions of displacement must be integrated to obtain the work. Graphically, the work is equal to the area under the force-displacement curve.
- The work of a weight is the product of the weight magnitude and the vertical displacement, $U_W = \pm Wy$. It is positive when the weight moves downwards.
- The work of a spring is of the form $U_s = \frac{1}{2}ks^2$, where k is the spring stiffness and s is the stretch or compression of the spring.

Numerical application of this procedure is illustrated in the examples following Sec. 14.3.

14.3 Principle of Work and Energy for a System of Particles

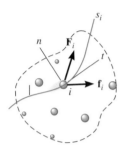

Inertial coordinate system

Fig. 14–8

The principle of work and energy can be extended to include a system of n particles isolated within an enclosed region of space as shown in Fig. 14–8. Here the arbitrary ith particle, having a mass m_i, is subjected to a resultant external force F_i and a resultant internal force f_i which each of the other particles exerts on the ith particle. Using Eq. 14–5, which applies in the tangential direction, the principle of work and energy written for the ith particle is thus

$$\tfrac{1}{2}m_i v_{i1}^2 + \int_{s_{i1}}^{s_{i2}} (F_i)_t \, ds + \int_{s_{i1}}^{s_{i2}} (f_i)_t \, ds = \tfrac{1}{2}m_i v_{i2}^2.$$

Similar equations result if the principle of work and energy is applied to each of the other particles of the system. Since both work and kinetic energy are scalars, the results may be added together algebraically, so that

$$\Sigma \tfrac{1}{2}m_i v_{i1}^2 + \Sigma \int_{s_{i1}}^{s_{i2}} (F_i)_t \, ds + \Sigma \int_{s_{i1}}^{s_{i2}} (f_i)_t \, ds = \Sigma \tfrac{1}{2}m_i v_{i2}^2$$

We can write this equation symbolically as

$$\Sigma T_1 + \Sigma U_{1-2} = \Sigma T_2 \qquad (14\text{–}8)$$

This equation states that the system's initial kinetic energy (ΣT_1) plus the work done by all the external and internal forces acting on the particles of the system (ΣU_{1-2}) is equal to the system's final kinetic energy (ΣT_2). To maintain this balance of energy, strict accountability of the work done by all the forces must be made. In this regard, note that although the internal forces on adjacent particles occur in equal but opposite collinear pairs, the total work done by each of these forces will, in general, *not cancel* out since the paths over which corresponding particles travel will be *different*. There are, however, two important exceptions to this rule which often occur in practice. If the particles are contained within the boundary of a *translating rigid body*, the internal forces all undergo the same displacement, and therefore the internal work will be zero. Also, particles connected by inextensible cables make up a system that has internal forces which are displaced by an equal amount. In this case, adjacent particles exert equal but opposite internal forces that have components which undergo the same displacement, and therefore the work of these forces cancels. On the other hand, note that if the body is assumed to be *nonrigid*, the particles of the body are displaced along *different paths*, and some of the energy due to force interactions would be given off and lost as heat or stored in the body if permanent deformations occur. We will discuss these effects briefly at the end of this section and in Sec. 15.4. Throughout this text, however, the principle of work and energy will be applied to the solution of problems only when direct accountability of these energy losses does not have to be made.

The procedure for analysis outlined in Sec. 14.2 provides a method for applying Eq. 14–8; however, only one equation applies for the entire system. If the particles are connected by cords, other equations can generally be obtained by using the kinematic principles outlined in Sec. 12.9 in order to relate the particle's speeds. See Example 14.6.

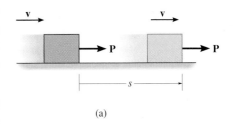

(a)

Work of Friction Caused by Sliding.

A special class of problems will now be investigated which requires a careful application of Eq. 14–8. These problems all involve cases where a body is sliding over the surface of another body in the presence of friction. Consider, for example, a block which is translating a distance s over a rough surface as shown in Fig. 14–9a. If the applied force **P** just balances the *resultant* frictional force $\mu_k N$, Fig. 14–9b, then due to equilibrium a constant velocity **v** is maintained, and one would expect Eq. 14–8 to be applied as follows:

$$\tfrac{1}{2}mv^2 + Ps - \mu_k Ns = \tfrac{1}{2}mv^2$$

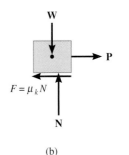

(b)

Indeed this equation is satisfied if $P = \mu_k N$; however, as one realizes from experience, the sliding motion will *generate heat*, a form of energy which seems not to be accounted for in the work-energy equation. In order to explain this paradox and thereby more closely represent the nature of friction, we should actually model the block so that the surfaces of contact are *deformable* (nonrigid).* Recall that the rough portions at the bottom of the block act as "teeth," and when the block slides these teeth *deform slightly* and either break off or vibrate due to interlocking effects and pull away from "teeth" at the contacting surface, Fig. 14–9c. As a result, frictional forces that act on the block at these points are displaced slightly, due to the localized deformations, and then they are replaced by other frictional forces as other points of contact are made. At any instant, the *resultant* **F** of all these frictional forces remains essentially constant, i.e., $\mu_k N$; however, due to the many *localized deformations*, the actual displacement s' of $\mu_k N$ is *not* the same displacement s as the applied force **P**. Instead, s' will be *less* than s $(s' < s)$, and therefore the *external work* done by the resultant frictional force will be $\mu_k Ns'$ and not $\mu_k Ns$. The remaining amount of work, $\mu_k N(s - s')$, manifests itself as an increase in *internal energy*, which in fact causes the block's temperature to rise.

(c)

Fig. 14–9

In summary then, Eq. 14–8 can be applied to problems involving sliding friction; however, it should be fully realized that the work of the resultant frictional force is not represented by $\mu_k Ns$; instead, this term represents *both* the external work of friction $(\mu_k Ns')$ and internal work $[\mu_k N(s - s')]$ which is converted into various forms of internal energy, such as heat.

*See Chapter 8 of *Engineering Mechanics: Statics.*

See B. A. Sherwood and W. H. Bernard, "Work and Heat Transfer in the Presence of Sliding Friction," *Am. J. Phys.* 52, 1001 (1984).

EXAMPLE **14.2**

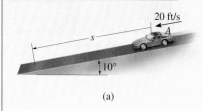

(a)

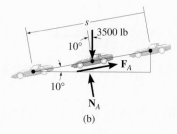

(b)

Fig. 14–10

The 3500-lb automobile shown in Fig. 14–10a is traveling down the 10° inclined road at a speed of 20 ft/s. If the driver jams on the brakes, causing his wheels to lock, determine how far s his tires skid on the road. The coefficient of kinetic friction between the wheels and the road is $\mu_k = 0.5$.

Solution I

This problem can be solved using the principle of work and energy, since it involves force, velocity, and displacement.

Work (Free-Body Diagram). As shown in Fig. 14–10b, the normal force N_A does no work since it never undergoes displacement along its line of action. The weight, 3500 lb, is displaced $s \sin 10°$ and does positive work. Why? The frictional force F_A does both external and internal work when it is *thought* to undergo a displacement s. This work is negative since it is in the opposite direction to displacement. Applying the equation of equilibrium normal to the road, we have

$$+ \nwarrow \Sigma F_n = 0; \qquad N_A - 3500 \cos 10° \text{ lb} = 0 \qquad N_A = 3446.8 \text{ lb}$$

Thus,

$$F_A = 0.5 N_A = 1723.4 \text{ lb}$$

Principle of Work and Energy.

$$T_1 + \Sigma U_{1-2} = T_2$$

$$\frac{1}{2}\left(\frac{3500 \text{ lb}}{32.2 \text{ ft/s}^2}\right)(20 \text{ ft/s})^2 + \{3500 \text{ lb}(s \sin 10°) - (1723.4 \text{ lb})s\} = 0$$

Solving for s yields

$$s = 19.5 \text{ ft} \qquad\qquad Ans.$$

Solution II

If this problem is solved by using the equation of motion, *two steps* are involved. First, from the free-body diagram, Fig. 14–10b, the equation of motion is applied along the incline. This yields

$$+ \swarrow \Sigma F_s = ma_s; \qquad 3500 \sin 10° \text{ lb} - 1723.4 \text{ lb} = \frac{3500 \text{ lb}}{32.2 \text{ ft/s}^2}a$$

$$a = -10.3 \text{ ft/s}^2$$

Then, using the integrated form of $a\,ds = v\,dv$ (kinematics), since a is constant, we have

$$(+ \swarrow) \quad v^2 = v_0^2 + 2a_c(s - s_0);$$

$$(0)^2 = (20 \text{ ft/s})^2 + 2(-10.3 \text{ ft/s}^2)(s - 0)$$

$$s = 19.5 \text{ ft} \qquad\qquad Ans.$$

EXAMPLE 14.3

For a short time the crane in Fig. 14–11a lifts the 2.50-Mg beam with a force of $F = (28 + 3s^2)$ kN. Determine the speed of the beam when it has risen $s = 3$ m. Also, how much time does it take to attain this height starting from rest?

Solution

We can solve part of this problem using the principle of work and energy since it involves force, velocity, and displacement. Kinematics must be used to determine the time.

Work (Free-Body Diagram). As shown on the free-body diagram, Fig. 14–11b, the towing force **F** does positive work, which must be determined by integration since this force is a variable. Also, the weight is constant and will do negative work since the displacement is upwards.

(a)

Principles of Work and Energy.

$$T_1 + \Sigma U_{1-2} = T_2$$

$$0 + \int_0^s (28 + 3s^2)(10^3)ds - (2.50)(10^3)(9.81)s = \tfrac{1}{2}(2.50)(10^3)v^2$$

$$28(10^3)s + (10^3)s^3 - 24.525(10^3)s = 1.25(10^3)v^2$$

$$v = (2.78s + 0.8s^3)^{\frac{1}{2}}$$

When $s = 3$ m,

$$v = 5.47 \text{ m/s} \qquad\qquad Ans.$$

Kinematics. Since we were able to express the velocity as a function of displacement, the time can be determined using $v = ds/dt$. In this case,

$$(2.78s + 0.8s^3)^{\frac{1}{2}} = \frac{ds}{dt}$$

$$t = \int_0^3 \frac{ds}{(2.78s + 0.8s^3)^{\frac{1}{2}}}$$

The integration can be performed numerically using a pocket calculator. The result is

$$t = 1.79 \text{ s} \qquad\qquad Ans.$$

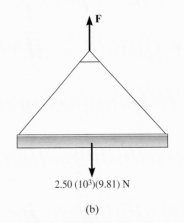

2.50 $(10^3)(9.81)$ N

(b)

Fig. 14–11

E X A M P L E   **14.4**

The platform P, shown in Fig. 14–12a, has negligible mass and is tied down so that the 0.4-m-long cords keep a 1-m long spring compressed 0.6 m when *nothing* is on the platform. If a 2-kg block is placed on the platform and released from rest after the platform is pushed down 0.1 m, Fig. 14–12b, determine the maximum height h the block rises in the air, measured from the ground.

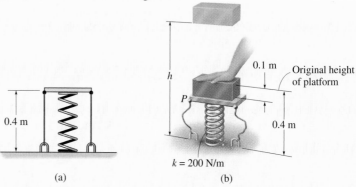

(a)   (b)

Fig. 14–12

Solution

Work (Free-Body Diagram).   Since the block is released from rest and later reaches its maximum height, the initial and final velocities are zero. The free-body diagram of the block when it is still in contact with the platform is shown in Fig. 14–12c. Note that the weight does negative work and the spring force does positive work. Why? In particular, the *initial compression* in the spring is $s_1 = 0.6$ m + 0.1 m = 0.7 m. Due to the cords, the spring's *final compression* is $s_2 = 0.6$ m (after the block leaves the platform). The bottom of the block rises from a height of $(0.4$ m $- 0.1$ m$) = 0.3$ m to a final height h.

(c)

Principle of Work and Energy.

$$T_1 + \Sigma U_{1-2} = T_2$$

$$\tfrac{1}{2}mv_1^2 + \{-(\tfrac{1}{2}ks_2^2 - \tfrac{1}{2}ks_1^2) - W\Delta y\} = \tfrac{1}{2}mv_2^2$$

Note that here $s_1 = 0.7$ m $> s_2 = 0.6$ m and so the work of the spring as determined from Eq. 14–4 will indeed be positive once the calculation is made. Thus,

$$0 + \{-[\tfrac{1}{2}(200 \text{ N/m})(0.6 \text{ m})^2 - \tfrac{1}{2}(200 \text{ N/m})(0.7 \text{ m})^2]$$
$$- (19.62 \text{ N})[h - (0.3 \text{ m})]\} = 0$$

Solving yields

$$h = 0.963 \text{ m} \qquad\qquad Ans.$$

E X A M P L E 14.5

Packages having a mass of 2 kg are delivered from a conveyor to a smooth circular ramp with a velocity of $v_0 = 1$ m/s as shown in Fig. 14–13a. If the radius of the ramp is 0.5 m, determine the angle $\theta = \theta_{max}$ at which each package begins to leave the surface.

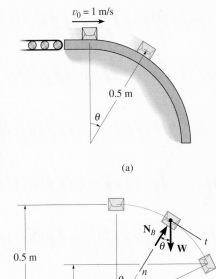

(a)

Solution

Work (Free-Body Diagram).   The free-body diagram of the block is shown at the intermediate location θ. The weight $W = 2(9.81)$ = 19.62 N does positive work during the displacement. If a package is assumed to leave the surface when $\theta = \theta_{max}$ then the weight moves through a vertical displacement of $[0.5 - 0.5 \cos \theta_{max}]$ m, as shown in the figure.

Principle of Work and Energy.

$$T_1 + \Sigma U_{1-2} = T_2$$

$$\tfrac{1}{2}(2 \text{ kg})(1 \text{ m/s})^2 + \{19.62 \text{ N}(0.5 - 0.5 \cos \theta_{max})\text{m}\} = \tfrac{1}{2}(2 \text{ kg})v_2^2$$

$$v_2^2 = 9.81(1 - \cos \theta_{max}) + 1 \qquad (1)$$

(b)

Fig. 14–13

Equation of Motion.   There are two unknowns in Eq. 1, θ_{max} and v_2. A second equation relating these two variables may be obtained by applying the equation of motion in the *normal direction* to the forces on the free-body diagram. (The principle of work and energy has replaced application of $\Sigma F_t = ma_t$ as noted in the derivation.) Thus,

$$+ \swarrow \Sigma F_n = ma_n; \qquad -N_B + 19.62 \text{ N} \cos \theta = (2 \text{ kg})\left(\frac{v^2}{0.5 \text{ m}}\right)$$

When the package leaves the ramp at $\theta = \theta_{max}$, $N_B = 0$ and $v = v_2$; hence, this equation becomes

$$\cos \theta_{max} = \frac{v_2^2}{4.905} \qquad (2)$$

Eliminating the unknown v_2^2 between Eqs. 1 and 2 gives

$$4.905 \cos \theta_{max} = 9.81(1 - \cos \theta_{max}) + 1$$

Solving, we have

$$\cos \theta_{max} = 0.735$$
$$\theta_{max} = 42.7° \qquad \qquad Ans.$$

This problem has also been solved in Example 13.9. If the two methods of solution are compared, it will be apparent that a work-energy approach yields a more direct solution.

E X A M P L E 14.6

The blocks A and B shown in Fig. 14–14a have a mass of 10 kg and 100 kg, respectively. Determine the distance B travels from the point where it is released from rest to the point where its speed becomes 2 m/s.

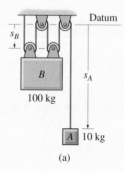

(a)

Solution

This problem may be solved by considering the blocks separately and applying the principle of work and energy to each block. However, the work of the (unknown) cable tension can be eliminated from the analysis by considering blocks A and B together as a *system*. The solution will require simultaneous solution of the equations of work and energy *and* kinematics. To be consistent with our sign convention, we will assume both blocks move in the positive *downward* direction.

Work (Free-Body Diagram). As shown on the free-body diagram of the system, Fig. 14–14b, the cable force **T** and reactions **R**$_1$ and **R**$_2$ do

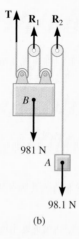

(b)

Fig. 14–14

no work, since these forces represent the reactions at the supports and consequently do not move while the blocks are being displaced. The weights both do positive work since, as stated above, they are both assumed to move downward.

Principle of Work and Energy.   Realizing the blocks are released from rest, we have

$$\Sigma T_1 + \Sigma U_{1\text{-}2} = \Sigma T_2$$

$$\{\tfrac{1}{2}m_A(v_A)_1^2 + \tfrac{1}{2}m_B(v_B)_1^2\} + \{W_A\,\Delta s_A + W_B\,\Delta s_B\} =$$
$$\{\tfrac{1}{2}m_A(v_A)_2^2 + \tfrac{1}{2}m_B(v_B)_2^2\}$$

$$\{0 + 0\} + \{98.1\text{ N}(\Delta s_A) + 981\text{ N}(\Delta s_B)\} =$$
$$\{\tfrac{1}{2}(10\text{ kg})(v_A)_2^2 + \tfrac{1}{2}(100\text{ kg})(2\text{ m/s})^2\} \qquad (1)$$

Kinematics.   Using the methods of kinematics discussed in Sec. 12.9, it may be seen from Fig. 14–14a that at any given instant the total length *l* of all the vertical segments of cable may be expressed in terms of the position coordinates s_A and s_B as

$$s_A + 4s_B = l$$

Hence, a change in position yields the displacement equation

$$\Delta s_A + 4\,\Delta s_B = 0$$

$$\Delta s_A = -4\,\Delta s_B \qquad (2)$$

As required, both of these displacements are positive downward. Taking the time derivative yields

$$v_A = -4v_B = -4(2\text{ m/s}) = -8\text{ m/s}$$

Retaining the negative sign in Eq. 2 and substituting into Eq. 1 yields

$$\Delta s_B = 0.883\text{ m}\!\downarrow \qquad\qquad \textit{Ans.}$$

14.4 Power and Efficiency

Power. *Power* is defined as the amount of work performed per unit of time. Hence, the *power* generated by a machine or engine that performs an amount of work dU within the time interval dt is

$$P = \frac{dU}{dt} \tag{14–9}$$

Provided the work dU is expressed by $dU = \mathbf{F} \cdot d\mathbf{r}$, then it is also possible to write

$$P = \frac{dU}{dt} = \frac{\mathbf{F} \cdot d\mathbf{r}}{dt} = \mathbf{F} \cdot \frac{d\mathbf{r}}{dt}$$

or

$$P = \mathbf{F} \cdot \mathbf{v} \tag{14–10}$$

Hence, power is a *scalar*, where in the formulation $\mathbf{v}$ represents the velocity of the point which is acted upon by the force $\mathbf{F}$.

The basic units of power used in the SI and FPS systems are the watt (W) and horsepower (hp), respectively. These units are defined as

$$1\text{ W} = 1\text{ J/s} = 1\text{ N} \cdot \text{m/s}$$
$$1\text{ hp} = 550\text{ ft} \cdot \text{lb/s}$$

For conversion between the two systems of units, 1 hp = 746 W.

The term "power" provides a useful basis for determining the type of motor or machine which is required to do a certain amount of work in a given time. For example, two pumps may each be able to empty a reservoir if given enough time; however, the pump having the larger power will complete the job sooner.

The power output of this locomotive comes from the driving frictional force **F** developed at its wheels. It is this force that overcomes the frictional resistance of the cars in tow and is able to lift the weight of the train up the grade.

Efficiency. The *mechanical efficiency* of a machine is defined as the ratio of the output of useful power produced by the machine to the input of power supplied to the machine. Hence,

$$\epsilon = \frac{\text{power output}}{\text{power input}} \tag{14–11}$$

If energy applied to the machine occurs during the *same time interval* at which it is removed, then the efficiency may also be expressed in terms of the ratio of output energy to input energy; i.e.

$$\epsilon = \frac{\text{energy output}}{\text{energy input}} \qquad\qquad (14\text{–}12)$$

Since machines consist of a series of moving parts, frictional forces will always be developed within the machine, and as a result, extra energy or power is needed to overcome these forces. Consequently, *the efficiency of a machine is always less than 1.*

The power requirements of this elevator depend upon the vertical force **F** that acts on the elevator and causes it to move upwards. If the velocity of the elevator is **v**, then the power output is $P = \mathbf{F} \cdot \mathbf{v}$.

PROCEDURE FOR ANALYSIS

The power supplied to a body can be computed using the following procedure.

- First determine the external force **F** acting on the body which causes the motion. This force is usually developed by a machine or engine placed either within or external to the body.

- If the body is accelerating, it may be necessary to draw its free-body diagram and apply the equation of motion ($\Sigma\mathbf{F} = m\mathbf{a}$) to determine **F**.

- Once **F** and the velocity **v** of the point where **F** is applied have been found, the power is determined by multiplying the force magnitude by the component of velocity acting in the direction of **F**, (i.e., $P = \mathbf{F} \cdot \mathbf{v} = Fv \cos\theta$).

- In some problems the power may be found by calculating the work done by **F** per unit of time ($P_{\text{avg}} = \Delta U/\Delta t$, or $P = dU/dt$).

E X A M P L E 14.7

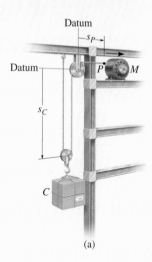

(a)

(b)

Fig. 14–15

The motor M of the hoist shown in Fig. 14–15a operates with an efficiency of $\epsilon = 0.85$. Determine the power that must be supplied to the motor to lift the 75-lb crate C at the instant point P on the cable has an acceleration of 4 ft/s² and a velocity of 2 ft/s. Neglect the mass of the pulley and cable.

Solution

In order to compute the power output of the motor, it is first necessary to determine the tension in the cable since this force is developed by the motor.

From the free-body diagram, Fig. 14–15b, we have

$$+\downarrow \quad \Sigma F_y = ma_y; \qquad -2T + 75 \text{ lb} = \frac{75 \text{ lb}}{32.2 \text{ ft/s}^2} a_c \qquad (1)$$

The acceleration of the crate can be obtained by using kinematics to relate it to the known acceleration of point P, Fig. 14–15a. Using the methods of Sec. 12.9, the coordinates s_C and s_P in Fig. 14–15a can be related to a constant portion of cable length l which is changing in the vertical and horizontal directions. We have $2s_C + s_P = l$. Taking the second time derivative of this equation yields

$$2a_C = -a_P \qquad (2)$$

Since $a_P = +4$ ft/s², then $a_C = (-4 \text{ ft/s}^2)/2 = -2$ ft/s². What does the negative sign indicate? Substituting this result into Eq. 1 and *retaining* the negative sign since the acceleration in *both* Eqs. 1 and 2 is considered positive downward, we have

$$-2T + 75 \text{ lb} = \frac{75 \text{ lb}}{32.2 \text{ ft/s}^2}(-2 \text{ ft/s}^2)$$

$$T = 39.8 \text{ lb}$$

The power output, measured in units of horsepower, required to draw the cable in at a rate of 2 ft/s is therefore

$$P = \mathbf{T} \cdot \mathbf{v} = (39.8 \text{ lb})(2 \text{ ft/s})[1 \text{ hp}/(550 \text{ ft} \cdot \text{lb/s})]$$

$$= 0.145 \text{ hp}$$

This *power output* requires that the motor provide a *power input* of

$$\text{power input} = \frac{1}{\epsilon}(\text{power output})$$

$$= \frac{1}{0.85}(0.145 \text{ hp}) = 0.170 \text{ hp} \qquad \qquad Ans.$$

Since the velocity of the crate is constantly changing, notice that this power requirement is *instantaneous*.

E X A M P L E **14.8**

The sports car shown in Fig. 14–16*a* has a mass of 2 Mg and is traveling at a speed of 25 m/s, when the brakes to all the wheels are applied. If the coefficient of kinetic friction is $\mu_k = 0.35$, determine the power developed by the friction force when the car skids. Then find the car's speed after it has slid 10 m.

19.62 kN

F_C

N_C

(a)   (b)

Fig. 14–16

Solution
As shown on the free-body diagram, Fig. 14–16*b*, the normal force $\mathbf{N}_C$ and frictional force $\mathbf{F}_C$ represent the *resultant forces* of all four wheels.
Applying the equation of equilibrium in the *y* direction to determine N_C, we have

$$+\uparrow \Sigma F_y = 0\,; \qquad N_C = 19.62 \text{ kN}$$

The kinetic frictional force is therefore

$$F_C = 0.35(19.62 \text{ kN}) = 6.867 \text{ kN}$$

The velocity of the car can be determined when $s = 10$ m by applying the principle of work and energy. Why?

$$T_1 + \Sigma U_{1-2} = T_2$$
$$\tfrac{1}{2}(2000 \text{ kg})(25 \text{ m/s})^2 - 6.867(10^3)\text{N}(10 \text{ m}) = \tfrac{1}{2}(2000 \text{ kg})v^2$$
$$v = 23.59 \text{ m/s}$$

The power of the frictional force at this instant is therefore

$$P = |\mathbf{F}_C \cdot \mathbf{v}| = 6.867(10^3)\text{N}(25 \text{ m/s}) = 172 \text{ kW} \qquad \textit{Ans.}$$

14.5 Conservative Forces and Potential Energy

Conservative Force. When the work done by a force in moving a particle from one point to another is *independent of the path* followed by the particle, then this force is called a *conservative force*. The weight of a particle and the force of an elastic spring are two examples of conservative forces often encountered in mechanics. The work done by the weight of a particle is *independent of the path* since it depends only on the particle's *vertical displacement*. The work done by a spring force *acting on a particle* is *independent of the path* of the particle, since it depends only on the extension or compression *s* of the spring.

In contrast to a conservative force, consider the force of friction exerted *on a sliding object* by a fixed surface. The work done by the frictional force *depends on the path*—the longer the path, the greater the work. Consequently, *frictional forces are nonconservative*. The work is dissipated from the body in the form of heat.

Potential Energy. Energy may be defined as the capacity for doing work. When energy comes from the *motion* of the particle, it is referred to as *kinetic energy*. When it comes from the *position* of the particle, measured from a fixed datum or reference plane, it is called potential energy. Thus, *potential energy* is a measure of the amount of work a conservative force will do when it moves from a given position to the datum. In mechanics, the potential energy due to gravity (weight) or an elastic spring is important.

Gravitational Potential Energy. If a particle is located a distance *y* *above* an arbitrarily selected datum, as shown in Fig. 14–17, the particle's weight **W** has positive *gravitational potential energy*, V_g, since **W** has the capacity of doing positive work when the particle is moved back down to the datum. Likewise, if the particle is located a distance *y below* the datum, V_g is negative since the weight does negative work when the particle is moved back up to the datum. At the datum $V_g = 0$.

In general, if *y* is *positive upward*, the gravitational potential energy of the particle of weight *W* is*

$$V_g = Wy$$

(14–13)

---

*Here the weight is assumed to be *constant*. This assumption is suitable for small differences in elevation Δy. If the elevation change is significant, however, a variation of weight with elevation must be taken into account.

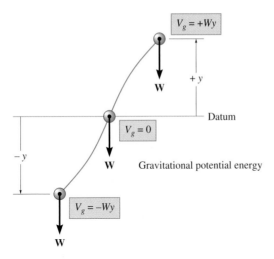

$V_g = +Wy$

$+y$

W

Datum

$V_g = 0$

$-y$

W

Gravitational potential energy

$V_g = -Wy$

W

Fig. 14–17

Elastic Potential Energy. When an elastic spring is elongated or compressed a distance s from its unstretched position, the elastic potential energy V_e due to the spring's configuration can be expressed as

$$V_e = +\tfrac{1}{2}ks^2 \qquad (14\text{–}14)$$

Here V_e is *always positive* since, in the deformed position, the force of the spring has the *capacity* for always doing positive work on the particle when the spring is returned to its unstretched position, Fig. 14–18.

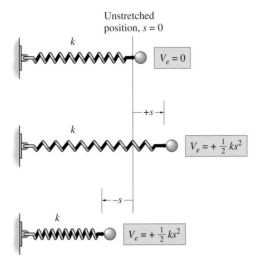

Unstretched position, $s = 0$

k

$V_e = 0$

$+s$

k

$V_e = +\tfrac{1}{2}ks^2$

$-s$

k

$V_e = +\tfrac{1}{2}ks^2$

Elastic potential energy

Fig. 14–18

Potential Function. In the general case, if a particle is subjected to both gravitational and elastic forces, the particle's potential energy can be expressed as a *potential function*, which is the algebraic sum

$$V = V_g + V_e \qquad (14\text{–}15)$$

Measurement of V depends on the location of the particle with respect to a selected datum in accordance with Eqs. 14–13 and 14–14.

If the particle is located at an arbitrary point (x, y, z) in space, this potential function is then $V = V(x, y, z)$. The work done by a conservative force in moving the particle from point (x_1, y_1, z_1) to point (x_2, y_2, z_2) is measured by the *difference* of this function, i.e.,

$$U_{1\text{–}2} = V_1 - V_2 \qquad (14\text{–}16)$$

For example, the potential function for a particle of weight W suspended from a spring can be expressed in terms of its position, s, measured from a datum located at the unstretched length of the spring, Fig. 14–19. We have

$$V = V_g + V_e$$
$$= -Ws + \tfrac{1}{2}ks^2$$

If the particle moves from s_1 to a lower position s_2, then applying Eq. 14–16 it can be seen that the work of $\mathbf{W}$ and $\mathbf{F}_s$ is

$$U_{1\text{–}2} = V_1 - V_2 = (-Ws_1 + \tfrac{1}{2}ks_1^2) - (-Ws_2 + \tfrac{1}{2}ks_2^2)$$
$$= W(s_2 - s_1) - (\tfrac{1}{2}ks_2^2 - \tfrac{1}{2}ks_1^2)$$

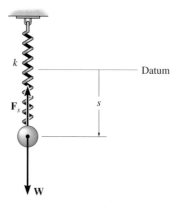

Datum

k

$\mathbf{F}_s$

s

$\mathbf{W}$

Fig. 14–19

When the displacement along the path is infinitesimal, i.e., from point (x, y, z) to $(x + dx, y + dy, z + dz)$, Eq. 14–16 becomes

$$dU = V(x, y, z) - V(x + dx, y + dy, z + dz)$$
$$= -dV(x, y, z) \qquad (14\text{–}17)$$

Provided both the force and displacement are defined using rectangular coordinates, then the work can also be expressed as

$$dU = \mathbf{F} \cdot d\mathbf{r} = (F_x\mathbf{i} + F_y\mathbf{j} + F_z\mathbf{k}) \cdot (dx\mathbf{i} + dy\mathbf{j} + dz\mathbf{k})$$
$$= F_x dx + F_y dy + F_z\, dz$$

Substituting this result into Eq. 14–17 and expressing the differential $dV(x, y, z)$ in terms of its partial derivatives yields

$$F_x dx + F_y dy + F_z\, dz = -\left(\frac{\partial V}{\partial x}dx + \frac{\partial V}{\partial y}dy + \frac{\partial V}{\partial z}dz\right)$$

Since changes in x, y, and z are all independent of one another, this equation is satisfied provided

$$F_x = -\frac{\partial V}{\partial x}, \qquad F_y = -\frac{\partial V}{\partial y}, \qquad F_z = -\frac{\partial V}{\partial z} \qquad (14\text{–}18)$$

Thus,

$$\mathbf{F} = -\frac{\partial V}{\partial x}\mathbf{i} - \frac{\partial V}{\partial y}\mathbf{j} - \frac{\partial V}{\partial z}\mathbf{k}$$

$$= -\left(\frac{\partial}{\partial x}\mathbf{i} + \frac{\partial}{\partial y}\mathbf{j} + \frac{\partial}{\partial z}\mathbf{k}\right)V$$

or

$$\mathbf{F} = -\nabla V \qquad (14\text{–}19)$$

where ∇ (del) represents the vector operator $\nabla = (\partial/\partial x)\mathbf{i} + (\partial/\partial y)\mathbf{j} + (\partial/\partial z)\mathbf{k}$.

Equation 14–19 relates a force $\mathbf{F}$ to its potential function V and thereby provides a mathematical criterion for proving that $\mathbf{F}$ is conservative. For example, the gravitational potential function for a weight located a distance y above a datum is $V_g = Wy$. To prove that $\mathbf{W}$ is conservative, it is necessary to show that it satisfies Eq. 14–19 (or Eq. 14–18), in which case

$$F_y = -\frac{\partial V}{\partial y}; \qquad F = -\frac{\partial}{\partial y}(Wy) = -W$$

The negative sign indicates that $\mathbf{W}$ acts downward, opposite to positive y, which is upward.

14.6 Conservation of Energy

The weight of the sacks resting on this platform causes potential energy to be stored in the supporting springs. As each sack is removed, the platform will *rise* slightly since some of the potential energy within the springs will be transferred into an increase in gravitational potential energy of the remaining sacks. Such a device is useful for removing the sacks without having to bend over to pick them up as they are unloaded.

When a particle is acted upon by a system of *both* conservative and nonconservative forces, the portion of the work done by the *conservative forces* can be written in terms of the difference in their potential energies using Eq. 14–16, i.e., $(\Sigma U_{1-2})_{\text{cons.}} = V_1 - V_2$. As a result, the principle of work and energy can be written as

$$T_1 + V_1 + (\Sigma U_{1-2})_{\text{noncons.}} = T_2 + V_2 \qquad (14\text{–}20)$$

Here $(\Sigma U_{1-2})_{\text{noncons.}}$ represents the work of the nonconservative forces acting on the particle. If *only conservative forces* are applied to the body, this term is zero and then we have

$$\boxed{T_1 + V_1 = T_2 + V_2} \qquad (14\text{–}21)$$

This equation is referred to as the *conservation of mechanical energy* or simply the *conservation of energy*. It states that during the motion the sum of the particle's kinetic and potential energies remains *constant*. For this to occur, kinetic energy must be transformed into potential energy, and vice versa. For example, if a ball of weight **W** is dropped from a height h above the ground (datum), Fig. 14–20, the potential energy of the ball is maximum before it is dropped, at which time its kinetic energy is zero. The total mechanical energy of the ball in its initial position is thus

$$E = T_1 + V_1 = 0 + Wh = Wh$$

When the ball has fallen a distance $h/2$, its speed can be determined by using $v^2 = v_0^2 + 2a_c(y - y_0)$, which yields $v = \sqrt{2g(h/2)} = \sqrt{gh}$. The energy of the ball at the mid-height position is therefore

$$E = T_2 + V_2 = \frac{1}{2}\frac{W}{g}(\sqrt{gh})^2 + W\frac{h}{2} = Wh$$

Just before the ball strikes the ground, its potential energy is zero and its speed is $v = \sqrt{2gh}$. Here, again, the total energy of the ball is

$$E = T_3 + V_3 = \frac{1}{2}\frac{W}{g}(\sqrt{2gh})^2 + 0 = Wh$$

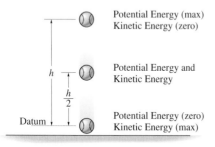

Potential Energy (max)
Kinetic Energy (zero)

Potential Energy and Kinetic Energy

Potential Energy (zero)
Kinetic Energy (max)

Datum

h

$\dfrac{h}{2}$

Fig. 14–20

Note that when the ball comes in contact with the ground, it deforms somewhat, and provided the ground is hard enough, the ball will rebound off the surface, reaching a new height h', which will be less than the height h from which it was first released. Neglecting air friction, the difference in height accounts for an energy loss, $E_l = W(h - h')$, which occurs during the collision. Portions of this loss produce noise, localized deformation of the ball and ground, and heat.

System of Particles. If a system of particles is *subjected only to conservative forces*, then an equation similar to Eq. 14–21 can be written for the particles. Applying the ideas of the preceding discussion, Eq. 14–8 ($\Sigma T_1 + \Sigma U_{1-2} = \Sigma T_2$) becomes

$$\Sigma T_1 + \Sigma V_1 = \Sigma T_2 + \Sigma V_2 \qquad (14\text{–}22)$$

Here, the sum of the system's initial kinetic and potential energies is equal to the sum of the system's final kinetic and potential energies. In other words, $\Sigma T + \Sigma V = \text{const.}$

It is important to remember that only problems involving conservative force systems (weights and springs) may be solved by using the conservation of energy theorem. As stated previously, friction or other drag-resistant forces, which depend upon velocity or acceleration, are nonconservative. A portion of the work done by such forces is transformed into thermal energy, and consequently this energy dissipates into the surroundings and may not be recovered.

PROCEDURE FOR ANALYSIS

The conservation of energy equation is used to solve problems involving *velocity, displacement*, and *conservative force systems*. It is generally *easier to apply* than the principle of work and energy because the energy equation just requires specifying the particle's kinetic and potential energies at only *two points* along the path, rather than determining the work when the particle moves through a *displacement*. For application it is suggested that the following procedure be used.

Potential Energy
- Draw two diagrams showing the particle located at its initial and final points along the path.
- If the particle is subjected to a vertical displacement, establish the fixed horizontal datum from which to measure the particle's gravitational potential energy V_g.
- Data pertaining to the elevation y of the particle from the datum and the extension or compression s of any connecting springs can be determined from the geometry associated with the two diagrams.
- Recall $V_g = Wy$, where y is positive upward from the datum and negative downward from the datum; also $V_e = \frac{1}{2}ks^2$, which is *always positive.*

Conservation of Energy
- Apply the equation $T_1 + V_1 = T_2 + V_2$.
- When determining the kinetic energy, $T = \frac{1}{2}mv^2$, the particle's speed v must be measured from an inertial reference frame.

E X A M P L E 14.9

The gantry structure in the photo is used to test the response of an airplane during a crash. As shown in Fig. 14–21a, the plane, having a mass of 8 Mg, is hoisted back until $\theta = 60°$, and then the pull-back cable AC is released when the plane is at rest. Determine the speed of the plane just before crashing into the ground, $\theta = 15°$. Also, what is the maximum tension developed in the supporting cable during the motion? Neglect the effect of lift caused by the wings during the motion and the size of the airplane.

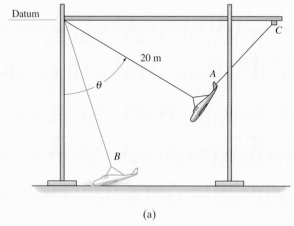

(a)

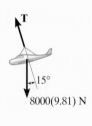

(b)

Fig. 14–21

Solution

Since the force of the cable does *no work* on the plane, it must be obtained using the equation of motion. First, however, we must determine the plane's speed at B.

Potential Energy. For convenience, the datum has been established at the top of the gantry.

Conservation of Energy.

$$T_A + V_A = T_B + V_B$$

$$0 - 8000 \text{ kg } (9.81 \text{ m/s}^2)(20 \cos 60° \text{ m}) =$$

$$\tfrac{1}{2}(8000 \text{ kg})v_B^2 - 8000 \text{ kg } (9.81 \text{ m/s}^2)(20 \cos 15° \text{ m})$$

$$v_B = 13.5 \text{ m/s} \qquad Ans.$$

Equation of Motion. Using the data tabulated on the free-body diagram when the plane is at B, Fig. 14–21b, we have

$$+\nwarrow \ \Sigma F_n = ma_n;$$

$$T - 8000 \ (9.81)\text{N} \cos 15° = (8000 \text{ kg})\frac{(13.5 \text{ m/s})^2}{20 \text{ m}}$$

$$T = 149 \text{ kN} \qquad Ans.$$

E X A M P L E **14.10**

The ram R shown in Fig. 14–22a has a mass of 100 kg and is released from rest 0.75 m from the top of a spring, A, that has a stiffness $k_A = 12$ kN/m. If a second spring B, having a stiffness $k_B = 15$ kN/m, is "nested" in A, determine the maximum displacement of A needed to stop the downward motion of the ram. The unstretched length of each spring is indicated in the figure. Neglect the mass of the springs.

Solution

Potential Energy. We will *assume* that the ram compresses *both* springs at the instant it comes to rest. The datum is located through the center of gravity of the ram at its initial position, Fig. 14–22b. When the kinetic energy is reduced to zero $(v_2 = 0)$ A is compressed a distance s_A and B compresses $s_B = s_A - 0.1$ m.

Conservation of Energy.

$$T_1 + V_1 = T_2 + V_2$$

$$0 + 0 = 0 + \{\tfrac{1}{2}k_A s_A^2 + \tfrac{1}{2}k_B(s_A - 0.1)^2 - Wh\}$$

$$0 + 0 = 0 + \{\tfrac{1}{2}(12\ 000\ \text{N/m})s_A^2 + \tfrac{1}{2}(15\ 000\ \text{N/m})(s_A - 0.1\text{m})^2$$
$$- 981\ \text{N}(0.75\ \text{m} + s_A)\}$$

Rearranging the terms,

$$13\ 500s_A^2 - 2481s_A - 660.75 = 0$$

Using the quadratic formula and solving for the positive root,* we have

$$s_A = 0.331\ \text{m} \qquad\qquad Ans.$$

Since $s_B = 0.331$ m $- 0.1$ m $= 0.231$ m, which is positive, the assumption that *both* springs are compressed by the ram is correct.

*The second root, $s_A = -0.148$ m, does not represent the physical situation. Since positive s is measured downward, the negative sign indicates that spring A would have to be "extended" by an amount of 0.148 m to stop the ram.

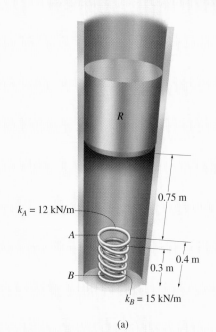

$k_A = 12$ kN/m

A

B

$k_B = 15$ kN/m

0.75 m

0.4 m

0.3 m

(a)

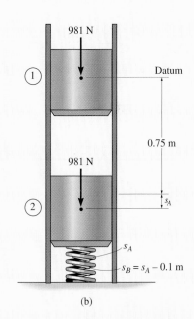

981 N

Datum

981 N

0.75 m

s_A

s_A

$s_B = s_A - 0.1$ m

(b)

Fig. 14–22

EXAMPLE 14.11

A smooth 2-kg collar C, shown in Fig. 14–23a, fits loosely on the vertical shaft. If the spring is unstretched when the collar is in the position A, determine the speed at which the collar is moving when $y = 1$ m, if (a) it is released from rest at A, and (b) it is released at A with an *upward* velocity $v_A = 2$ m/s.

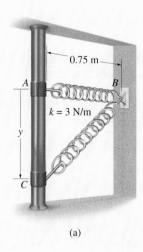

0.75 m

A B

$k = 3$ N/m

y

C

(a)

Solution

Part (a)

Potential Energy.   For convenience, the datum is established through AB, Fig. 14–23b. When the collar is at C, the gravitational potential energy is $-(mg)y$, since the collar is *below* the datum, and the elastic potential energy is $\frac{1}{2}ks_{CB}^2$. Here $s_{CB} = 0.5$ m, which represents the *stretch* in the spring as shown in the figure.

Conservation of Energy.

$$T_A + V_A = T_C + V_C$$

$$0 + 0 = \tfrac{1}{2}mv_C^2 + \{\tfrac{1}{2}ks_{CB}^2 - mgy\}$$

$$0 + 0 = \{\tfrac{1}{2}(2\text{ kg})v_C^2\} + \{\tfrac{1}{2}(3\text{ N/m})(0.5\text{ m})^2 - 2(9.81)\text{ N}(1\text{ m})\}$$

$$v_C = 4.39\text{ m/s} \downarrow \qquad\qquad\qquad Ans.$$

This problem can also be solved by using the equation of motion or the principle of work and energy. Note that in *both* of these methods the variation of the magnitude and direction of the spring force must be taken into account (see Example 13.4). Here, however, the above method of solution is clearly advantageous since the calculations depend *only* on data calculated at the initial and final points of the path.

Part (b)

Conservation of Energy. If $v_A = 2$ m/s, using the data in Fig. 14–23b, we have

$$T_A + V_A = T_C + V_C$$

$$\tfrac{1}{2}mv_A^2 + 0 = \tfrac{1}{2}mv_C^2 + \{\tfrac{1}{2}ks_{CB}^2 - mgy\}$$

$$\tfrac{1}{2}(2 \text{ kg})(2 \text{ m/s})^2 + 0 = \tfrac{1}{2}(2 \text{ kg})v_C^2 + \{\tfrac{1}{2}(3 \text{ N/m})(0.5 \text{ m})^2$$

$$- 2(9.81) \text{ N}(1 \text{ m})\}$$

$$v_C = 4.82 \text{ m/s} \downarrow \qquad\qquad Ans.$$

Note that the kinetic energy of the collar depends only on the *magnitude* of velocity, and therefore it is immaterial if the collar is moving up or down at 2 m/s when released at A.

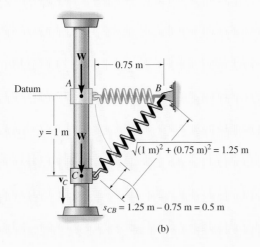

(b)

Fig. 14–23

DESIGN PROJECTS

14–1D.   DESIGN OF A CAR BUMPER

The body of an automobile is to be protected by a spring-loaded bumper, which is attached to the automobile's frame. Design the bumper so that it will stop a 3500-lb car traveling freely at 5 mi/h and not deform the springs more than 3 in. Submit a sketch of your design showing the placement of the springs and their stiffness. Plot the load-deflection diagram for the bumper during a direct collision with a rigid wall, and also plot the deceleration of the car as a function of the springs' displacement.

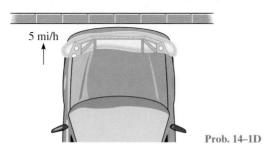

Prob. 14–1D

14–2D.   DESIGN OF AN ELEVATOR HOIST

It is required that an elevator and its contents, having a maximum weight of 500 lb, be lifted $y = 20$ ft, starting from rest and then stopping after 6 seconds. A single motor and cable-winding drum can be mounted anywhere and used for the operation. During any lift or descent the acceleration should not exceed $10 \ \text{ft/s}^2$. Design a cable-and-pulley system for the elevator, and estimate the material cost if the cable is $1.30/ft and pulleys are $3.50 each. Submit a drawing of your design, and include plots of the power output required of the motor and the elevator's speed versus the height y traveled.

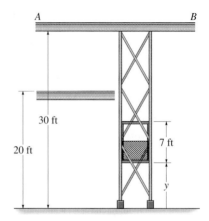

Prob. 14–2D

CHAPTER REVIEW

- *Work of a Force.* A force does work when it undergoes a displacement along its line of action. If the force varies with the displacement, then $U = \int F \, ds$. Graphically, this represents the area under the $F-s$ diagram. If the force is constant, then for a displacement Δs in the direction of the force, $U = F \Delta s$.

  A typical example of this case is the work of weight, $U = W \Delta y$.

  Here, Δy is the vertical displacement. A spring force, $F = ks$, depends upon the elongation or compression s of the spring. The work is determined by integration to be $U = \frac{1}{2} ks^2$.

- *The Principle of Work and Energy.* If the equation of motion in the tangential direction, $\Sigma F_t = ma_t$, is combined with the kinematic equation, $a_t \, ds = v \, dv$, we obtain the principle of work and energy.

$$T_1 + \Sigma U_{1-2} = T_2$$

  Here, the initial kinetic energy of the particle $(T_1 = \frac{1}{2} mv_1{}^2)$, plus the work done by all the forces that act on the particle as it moves from its initial position to its final position (ΣU_{1-2}), is equal to the final kinetic energy of the particle $(T_1 = \frac{1}{2} mv_2{}^2)$.

  The principle of work and energy is useful for solving problems that involve force, velocity, and displacement. For application, the free-body diagram of the particle should be drawn in order to identify the forces that do work.

- *Power and Efficiency.* Power is the time-rate of doing work. It is defined by $P = dU/dt$, or $P = \mathbf{F} \cdot \mathbf{v}$. For application, the force $\mathbf{F}$ creating the power and its velocity $\mathbf{v}$ must be specified. Efficiency represents the ratio of power output to power input. Due to frictional losses, it is always less than one.

- *Conservation of Energy.* A conservative force is one that does work which is independent of its path. Two examples are the weight of a particle and the spring force. Friction is a nonconservative force since the work depends upon the length of the path. The longer the path is, the more work is done. The work done by a conservative force depends upon its position relative to a datum. When this work is referenced from a datum, it is called potential energy. For weight, it is $V_g = Wy$, and for a spring it is $V_e = \frac{1}{2} kx^2$.

  Mechanical energy consists of kinetic energy T and gravitational and elastic potential energies V. According to the conservation of energy, this sum is constant and has the same value at any two positions on the path. That is,

$$T_1 + V_1 = T_2 + V_2$$

  If motion of the particle is caused only by gravitational and spring forces, then this equation can be used to solve problems involving displacement and velocity.

The velocities of the vehicles involved in this accident can be estimated using the principles of impulse and momentum.

Kinetics of a Particle: Impulse and Momentum

- To develop the principle of linear impulse and momentum for a particle.
- To study the conservation of linear momentum for particles.
- To analyze the mechanics of impact.
- To introduce the concept of angular impulse and momentum.
- To solve problems involving steady fluid streams and propulsion with variable mass.

15.1 Principle of Linear Impulse and Momentum

In this section we will integrate the equation of motion with respect to time and thereby obtain the principle of impulse and momentum. It will then be shown that the resulting equation will be useful for solving problems involving force, velocity, and time.

The equation of motion for a particle of mass m can be written as

$$\Sigma \mathbf{F} = m\mathbf{a} = m\frac{d\mathbf{v}}{dt} \qquad (15\text{--}1)$$

where $\mathbf{a}$ and $\mathbf{v}$ are both measured from an inertial frame of reference. Rearranging the terms and integrating between the limits $\mathbf{v} = \mathbf{v}_1$ at $t = t_1$ and $\mathbf{v} = \mathbf{v}_2$ at $t = t_2$, we have

$$\Sigma \int_{t_1}^{t_2} \mathbf{F}\, dt = m \int_{\mathbf{v}_1}^{\mathbf{v}_2} d\mathbf{v}$$

The impulse tool is used to remove the dent in the automobile fender. To do so its end is first screwed into a hole drilled in the fender, then the weight is gripped and jerked upwards, striking the stop ring. The impulse developed is transferred along the shaft of the tool and pulls suddenly on the dent.

or

$$\Sigma \int_{t_1}^{t_2} \mathbf{F} \, dt = m\mathbf{v}_2 - m\mathbf{v}_1 \qquad (15\text{–}2)$$

This equation is referred to as the *principle of linear impulse and momentum.* From the derivation it can be seen that it is simply a time integration of the equation of motion. It provides a *direct means* of obtaining the particle's final velocity $\mathbf{v}_2$ after a specified time period when the particle's initial velocity is known and the forces acting on the particle are either constant or can be expressed as functions of time. By comparison, if $\mathbf{v}_2$ was determined using the equation of motion, a two-step process would be necessary; i.e., apply $\Sigma\mathbf{F} = m\mathbf{a}$ to obtain $\mathbf{a}$, then integrate $\mathbf{a} = d\mathbf{v}/dt$ to obtain $\mathbf{v}_2$.

Linear Momentum. Each of the two vectors of the form $\mathbf{L} = m\mathbf{v}$ in Eq. 15–2 is referred to as the particle's *linear momentum.* Since m is a positive scalar, the linear-momentum vector has the same direction as $\mathbf{v}$, and its magnitude mv has units of mass–velocity, e.g., $kg \cdot m/s$, or $slug \cdot ft/s$.

Linear Impulse. The integral $\mathbf{I} = \int \mathbf{F} \, dt$ in Eq. 15–2 is referred to as the *linear impulse.* This term is a vector quantity which measures the effect of a force during the time the force acts. Since time is a positive scalar, the impulse acts in the same direction as the force, and its magnitude has units of force–time, e.g., $N \cdot s$ or $lb \cdot s$.* If the force is expressed as a function of time, the impulse may be determined by direct evaluation of the integral. In particular, the magnitude of the impulse $\mathbf{I} = \int_{t_1}^{t_2} \mathbf{F} \, dt$ can be represented experimentally by the shaded area under the curve of force versus time, Fig. 15–1. When the force is constant in both magnitude and direction, the resulting impulse becomes $\mathbf{I} = \int_{t_1}^{t_2} \mathbf{F}_c \, dt = \mathbf{F}_c(t_2 - t_1)$, which represents the shaded rectangular area shown in Fig. 15–2.

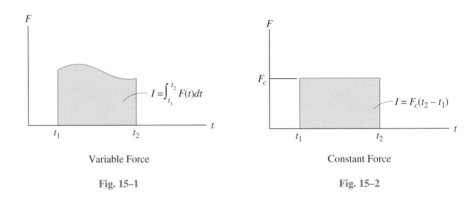

Variable Force

Fig. 15–1

Constant Force

Fig. 15–2

*Although the units for impulse and momentum are defined differently, one can show that Eq. 15–2 is dimensionally homogeneous.

Principle of Linear Impulse and Momentum. For problem solving, Eq. 15–2 will be rewritten in the form

$$mv_1 + \Sigma \int_{t_1}^{t_2} F \, dt = mv_2 \qquad (15\text{–}3)$$

Initial
momentum
diagram

+

which states that the initial momentum of the particle at t_1 plus the sum of all the impulses applied to the particle from t_1 to t_2 is equivalent to the final momentum of the particle at t_2. These three terms are illustrated graphically on the *impulse and momentum diagrams* shown in Fig. 15–3. The two *momentum diagrams* are simply outlined shapes of the particle which indicate the direction and magnitude of the particle's initial and final momenta, mv_1 and mv_2, respectively, Fig. 15–3. Similar to the free-body diagram, the *impulse diagram* is an outlined shape of the particle showing all the impulses that act on the particle when it is located at some intermediate point along its path. In general, whenever the magnitude or direction of a force *varies* with time, the impulse is represented on the impulse diagram as $\int_{t_1}^{t_2} F \, dt$. If the force is *constant*, the impulse applied to the particle is $F_c(t_2 - t_1)$, and it acts in the same direction as F_c.

$\Sigma \int_{t_1}^{t_2} F \, dt$

Impulse
diagram

‖

Scalar Equations. If each of the vectors in Eq. 15–3 is resolved into its x, y, z components, we can write symbolically the following three scalar equations:

mv_2

Final
momentum
diagram

$$m(v_x)_1 + \Sigma \int_{t_1}^{t_2} F_x \, dt = m(v_x)_2$$

$$m(v_y)_1 + \Sigma \int_{t_1}^{t_2} F_y \, dt = m(v_y)_2 \qquad (15\text{–}4)$$

$$m(v_z)_1 + \Sigma \int_{t_1}^{t_2} F_z \, dt = m(v_z)_2$$

Fig. 15–3

These equations represent the principle of linear impulse and momentum for the particle in the x, y, z directions, respectively.

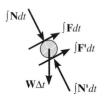

As the wheels of the pitching machine rotate, they apply frictional impulses to the ball, thereby giving it a linear momentum. These impulses are shown on the impulse diagram. Here both the frictional and normal impulses vary with time. By comparison, the weight impulse is constant and is very small since the time Δt the ball is in contact with the wheels is very small.

PROCEDURE FOR ANALYSIS

The principle of linear impulse and momentum is used to solve problems involving *force, time*, and *velocity*, since these terms are involved in the formulation. For application it is suggested that the following procedure be used.*

Free-Body Diagram

- Establish the *x, y, z* inertial frame of reference and draw the particle's free-body diagram in order to account for all the forces that produce impulses on the particle.

- The direction and sense of the particle's initial and final velocities should be established.

- If a vector is unknown, assume that the sense of its components is in the direction of the positive inertial coordinate(s).

- As an alternative procedure, draw the impulse and momentum diagrams for the particle as discussed in reference to Fig. 15–3.

Principle of Impulse and Momentum

- In accordance with the established coordinate system apply the principle of linear impulse and momentum, $m\mathbf{v}_1 + \Sigma \int_{t_1}^{t_2} \mathbf{F}\, dt = m\mathbf{v}_2$. If motion occurs in the *x–y* plane, the two scalar component equations can be formulated by either resolving the vector components of $\mathbf{F}$ from the free-body diagram, or by using the data on the impulse and momentum diagrams.

- Realize that every force acting on the particle's free-body diagram will create an impulse, even though some of these forces will do no work.

- Forces that are functions of time must be integrated to obtain the impulse. Graphically, the impulse is equal to the area under the force–time curve.

- If the problem involves the dependent motion of several particles, use the method outlined in Sec. 12.9 to relate their velocities. Make sure the positive coordinate directions used for writing these kinematic equations are the *same* as those used for writing the equations of impulse and momentum.

*This procedure will be followed when developing the proofs and theory in the text.

E X A M P L E 15.1

The 100-kg stone shown in Fig. 15–4a is originally at rest on the smooth horizontal surface. If a towing force of 200 N, acting at an angle of 45°, is applied to the stone for 10 s, determine the final velocity and the normal force which the surface exerts on the stone during the time interval.

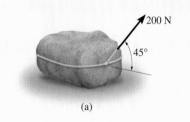

(a)

Solution

This problem can be solved using the principle of impulse and momentum since it involves force, velocity, and time.

Free-Body Diagram.   See Fig. 15–4b. Since all the forces acting are *constant*, the impulses are simply the product of the force magnitude and 10 s $[\mathbf{I} = \mathbf{F}_c(t_2 - t_1)]$. Note the alternative procedure of drawing the stone's impulse and momentum diagrams, Fig. 15–4c.

Principle of Impulse and Momentum.   Resolving the vectors in Fig. 15–4b along the x, y axes and applying Eqs. 15–4 yields

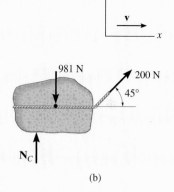

(b)

$(\overset{+}{\rightarrow})$
$$m(v_x)_1 + \Sigma \int_{t_1}^{t_2} F_x \, dt = m(v_x)_2$$
$$0 + 200 \text{ N}(10 \text{ s}) \cos 45° = (100 \text{ kg})v_2$$
$$v_2 = 14.1 \text{ m/s} \qquad \qquad Ans.$$

$(+\uparrow)$
$$m(v_y)_1 + \Sigma \int_{t_1}^{t_2} F_y \, dt = m(v_y)_2$$
$$0 + N_C(10 \text{ s}) - 981 \text{ N}(10 \text{ s}) + 200 \text{ N}(10 \text{ s}) \sin 45° = 0$$
$$N_C = 840 \text{ N} \qquad \qquad Ans.$$

Since no motion occurs in the y direction, direct application of the equilibrium equation $\Sigma F_y = 0$ gives the same result for N_C.

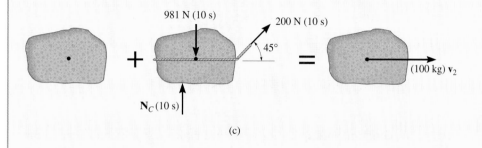

(c)

Fig. 15–4

E X A M P L E   15.2

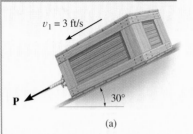

$v_1 = 3$ ft/s

P   30°

(a)

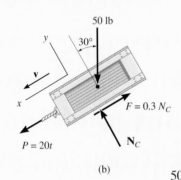

50 lb

y 30°

v

x $F = 0.3 N_C$

$P = 20t$ **N_C**

(b)

Fig. 15–5

The 50-lb crate shown in Fig. 15–5a is acted upon by a force having a variable magnitude $P = (20t)$ lb, where t is in seconds. Determine the crate's velocity 2 s after **P** has been applied. The initial velocity is $v_1 = 3$ ft/s down the plane, and the coefficient of kinetic friction between the crate and the plane is $\mu_k = 0.3$.

Solution

Free-Body Diagram. See Fig. 15–5b. Since the magnitude of force $P = 20t$ varies with time, the impulse it creates must be determined by integrating over the 2-s time interval. The weight, normal force, and frictional force (which acts opposite to the direction of motion) are all *constant*, so that the impulse created by each of these forces is simply the magnitude of the force times 2 s.

Principle of Impulse and Momentum. Applying Eqs. 15–4 in the x direction, we have

$$(+ \swarrow) \qquad\qquad m(v_x)_1 + \Sigma \int_{t_1}^{t_2} F_x \, dt = m(v_x)_2$$

$$\frac{50 \text{ lb}}{32.2 \text{ ft/s}^2}(3 \text{ ft/s}) + \int_0^2 20t \, dt - 0.3N_C(2 \text{ s}) + (50 \text{ lb})(2 \text{ s}) \sin 30° = \frac{50 \text{ lb}}{32.2 \text{ ft/s}^2}v_2$$

$$4.66 + 40 - 0.6N_C + 50 = 1.55v_2$$

The equation of equilibrium can be applied in the y direction. Why?

$$+\nwarrow \Sigma F_y = 0; \qquad\qquad N_C - 50 \cos 30° \text{ lb} = 0$$

Solving,

$$N_C = 43.3 \text{ lb}$$
$$v_2 = 44.2 \text{ ft/s} \checkmark \qquad\qquad Ans.$$

Note: We can also solve this problem using the equation of motion. From Fig. 15–5b,

$$+\swarrow \Sigma F_x = ma_x; \qquad 20t - 0.3(43.3) + 50 \sin 30° = \frac{50}{32.2}a$$

$$a = 12.88t + 7.734$$

Using kinematics

$$+\swarrow dv = a \, dt; \qquad \int_3^v dv = \int_0^2 (12.88t + 7.734) \, dt$$

$$v = 44.2 \text{ ft/s} \qquad\qquad Ans.$$

By comparison, application of the principle of impulse and momentum eliminates the need for using kinematics $(a = dv/dt)$ and thereby yields an easier method for solution.

E X A M P L E 15.3

Blocks A and B shown in Fig. 15–6a have a mass of 3 kg and 5 kg, respectively. If the system is released from rest, determine the velocity of block B in 6 s. Neglect the mass of the pulleys and cord.

Solution

Free-Body Diagram. See Fig. 15–6b. Since the weight of each block is constant, the cord tensions will also be constant. Furthermore, since the mass of pulley D is neglected, the cord tension $T_A = 2T_B$. Note that the blocks are both assumed to be traveling downward in the positive coordinate directions, s_A and s_B.

Principle of Impulse and Momentum.

Block A:

$(+\downarrow)$ $\qquad m(v_A)_1 + \Sigma \int_{t_1}^{t_2} F_y \, dt = m(v_A)_2$

$\qquad 0 - 2T_B(6\text{ s}) + 3(9.81)\text{ N}(6\text{ s}) = (3\text{ kg})(v_A)_2 \qquad (1)$

Block B:

$(+\downarrow)$ $\qquad m(v_B)_1 + \Sigma \int_{t_1}^{t_2} F_y \, dt = m(v_B)_2$

$\qquad 0 + 5(9.81)\text{ N}(6\text{ s}) - T_B(6\text{ s}) = (5\text{ kg})(v_B)_2 \qquad (2)$

Kinematics. Since the blocks are subjected to dependent motion, the velocity of A may be related to that of B by using the kinematic analysis discussed in Sec. 12.9. A horizontal datum is established through the fixed point at C, Fig. 15–6a, and the position coordinates, s_A and s_B, are related to the constant total length l of the vertical segments of the cord by the equation

$$2s_A + s_B = l$$

Taking the time derivative yields

$$2v_A = -v_B \qquad (3)$$

As indicated by the negative sign, when B moves downward A moves upward. *Substituting this result into Eq. 1 and solving Eqs. 1 and 2 yields

$$(v_B)_2 = 35.8 \text{ m/s} \downarrow \qquad \textit{Ans.}$$

$$T_B = 19.2 \text{ N}$$

*Realize that the *positive* (downward) direction for $\mathbf{v}_A$ and $\mathbf{v}_B$ is *consistent* in Figs. 15–6a and 15–6b and in Eqs. 1 to 3. Why is this important?

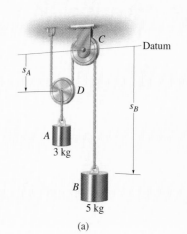

(a)

$T_A = 2T_B$

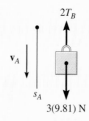

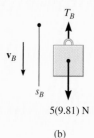

(b)

Fig. 15–6

15.2 Principle of Linear Impulse and Momentum for a System of Particles

The principle of linear impulse and momentum for a system of particles moving relative to an inertial reference, Fig. 15–7, is obtained from the equation of motion applied to all the particles in the system, i.e.,

$$\Sigma \mathbf{F}_i = \Sigma m_i \frac{d\mathbf{v}_i}{dt} \tag{15-5}$$

The term on the left side represents only the sum of the *external forces* acting on the system of particles. Recall that the internal forces $\mathbf{f}_i$ acting between particles do not appear with this summation, since by Newton's third law they occur in equal but opposite collinear pairs and therefore cancel out. Multiplying both sides of Eq. 15–5 by dt and integrating between the limits $t = t_1$, $\mathbf{v}_i = (\mathbf{v}_i)_1$ and $t = t_2$, $\mathbf{v}_i = (\mathbf{v}_i)_2$ yields

$$\Sigma m_i(\mathbf{v}_i)_1 + \Sigma \int_{t_1}^{t_2} \mathbf{F}_i \, dt = \Sigma m_i(\mathbf{v}_i)_2 \tag{15-6}$$

This equation states that the initial linear momenta of the system plus the impulses of all the *external forces* acting on the system from t_1 to t_2 are equal to the system's final linear momenta.

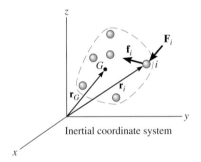

Inertial coordinate system

Fig. 15–7

Since the location of the mass center G of the system is determined from $m\mathbf{r}_G = \Sigma m_i \mathbf{r}_i$, where $m = \Sigma m_i$ is the total mass of all the particles, Fig. 15–7, then taking the time derivatives, we have

$$m\mathbf{v}_G = \Sigma m_i \mathbf{v}_i$$

which states that the total linear momentum of the system of particles is equivalent to the linear momentum of a "fictitious" aggregate particle of mass $m = \Sigma m_i$ moving with the velocity of the mass center of the system. Substituting into Eq. 15–6 yields

$$m(\mathbf{v}_G)_1 + \Sigma \int_{t_1}^{t_2} \mathbf{F}_i \, dt = m(\mathbf{v}_G)_2 \qquad (15\text{–}7)$$

Here the initial linear momentum of the aggregate particle plus the external impulses acting on the system of particles from t_1 to t_2 is equal to the aggregate particle's final linear momentum. Since in reality all particles must have nonzero size to possess mass, the above equation justifies application of the principle of linear impulse and momentum to a rigid body represented as a single particle.

15.3 Conservation of Linear Momentum for a System of Particles

When the sum of the *external impulses* acting on a system of particles is *zero*, Eq. 15–6 reduces to a simplified form, namely,

$$\Sigma m_i(\mathbf{v}_i)_1 = \Sigma m_i(\mathbf{v}_i)_2 \qquad (15\text{–}8)$$

This equation is referred to as the *conservation of linear momentum*. It states that the total linear momentum for a system of particles remains constant during the time period t_1 to t_2. Substituting $m\mathbf{v}_G = \Sigma m_i\mathbf{v}_i$ into Eq. 15–8, we can also write

$$(\mathbf{v}_G)_1 = (\mathbf{v}_G)_2 \qquad (15\text{–}9)$$

which indicates that the velocity $\mathbf{v}_G$ of the mass center for the system of particles does not change when no external impulses are applied to the system.

The conservation of linear momentum is often applied when particles collide or interact. For application, a careful study of the free-body diagram for the *entire* system of particles should be made in order to identify the forces which create either external or internal impulses and thereby determine in what direction(s) linear momentum is conserved. As stated earlier, the *internal impulses* for the system will always cancel out, since they occur in equal but opposite collinear pairs. If the time period over which the motion is studied is *very short*, some of the external impulses may also be neglected or considered approximately equal to zero. The forces causing these negligible impulses are called *nonimpulsive forces*. By comparison, forces which are very large and act for a very short period of time produce a significant change in momentum and are called *impulsive forces*. They, of course, cannot be neglected in the impulse-momentum analysis.

The hammer in the top photo applies an impulsive force to the stake. During this extremely short time of contact the weight of the stake can be considered nonimpulsive, and provided the stake is driven into soft ground, the impulse of the ground acting on the stake can also be considered nonimpulsive. By contrast, if the stake is used in a concrete chipper to break concrete, then two impulsive forces act on the stake: one at its top due to the chipper and the other on its bottom due to the rigidity of the concrete.

Impulsive forces normally occur due to an explosion or the striking of one body against another, whereas nonimpulsive forces may include the weight of a body, the force imparted by a slightly deformed spring having a relatively small stiffness, or for that matter, any force that is very small compared to other larger (impulsive) forces. When making this distinction between impulsive and nonimpulsive forces, it is important to realize that this only applies during the time t_1 to t_2. To illustrate, consider the effect of striking a tennis ball with a racket as shown in the photo. During the *very short* time of interaction, the force of the racket on the ball is impulsive since it changes the ball's momentum drastically. By comparison, the ball's weight will have a negligible effect on the change in momentum, and therefore it is nonimpulsive. Consequently, it can be neglected from an impulse-momentum analysis during this time. If an impulse-momentum analysis is considered during the much longer time of flight after the racket-ball interaction, then the impulse of the ball's weight is important since it, along with air resistance, causes the change in the momentum of the ball.

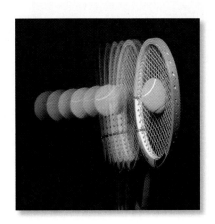

PROCEDURE FOR ANALYSIS

Generally, the principle of linear impulse and momentum or the conservation of linear momentum is applied to a *system of particles* in order to determine the final velocities of the particles *just after* the time period considered. By applying these equations to the entire system, the internal impulses acting within the system, which may be unknown, are *eliminated* from the analysis. For application it is suggested that the following procedure be used.

Free-Body Diagram.

- Establish the *x, y, z* inertial frame of reference and draw the free-body diagram for each particle of the system in order to identify the internal and external forces.

- The conservation of linear momentum applies to the system in a given direction when no external forces or if nonimpulsive forces act on the system in that direction.

- Establish the direction and sense of the particles' initial and final velocities. If the sense is unknown, assume it is along a positive inertial coordinate axis.

- As an alternative procedure, draw the impulse and momentum diagrams for each particle of the system.

Momentum Equations.

- Apply the principle of linear impulse and momentum or the conservation of linear momentum in the appropriate directions.

- If it is necessary to determine the *internal impulse* $\int F\, dt$ acting on only one particle of a system, then the particle must be *isolated* (free-body diagram), and the principle of linear impulse and momentum must be applied *to the particle.*

- After the impulse is calculated, and provided the time Δt for which the impulse acts is known, then the *average impulsive force* F_{avg} can be determined from $F_{avg} = \int F\, dt / \Delta t$.

EXAMPLE 15.4

The 15-Mg boxcar A is coasting at 1.5 m/s on the horizontal track when it encounters a 12-Mg tank car B coasting at 0.75 m/s toward it as shown in Fig. 15–8a. If the cars meet and couple together, determine (a) the speed of both cars just after the coupling, and (b) the average force between them if the coupling takes place in 0.8 s.

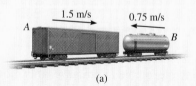

(a)

Solution

Part (a) Free-Body Diagram.* Here we have considered *both* cars as a single system, Fig. 15–8b. By inspection, momentum is conserved in the x direction since the coupling force $\mathbf{F}$ is *internal* to the system and will therefore cancel out. It is assumed both cars, when coupled, move at $\mathbf{v}_2$ in the positive x direction.

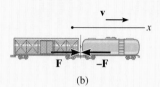

(b)

Conservation of Linear Momentum.

$$(\xrightarrow{+}) \qquad m_A(v_A)_1 + m_B(v_B)_1 = (m_A + m_B)v_2$$

$$(15\,000 \text{ kg})(1.5 \text{ m/s}) - 12\,000 \text{ kg}(0.75 \text{ m/s}) = (27\,000 \text{ kg})v_2$$

$$v_2 = 0.5 \text{ m/s} \rightarrow \qquad\qquad Ans.$$

Part (b). The average (impulsive) coupling force, $\mathbf{F}_{avg}$, can be determined by applying the principle of linear momentum to *either one* of the cars.

Free-Body Diagram. As shown in Fig. 15–8c, by isolating the boxcar the coupling force is *external* to the car.

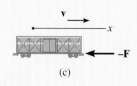

(c)

Fig. 15–8

Principle of Impulse and Momentum. Since $\int F \, dt = F_{avg}\Delta t = F_{avg}(0.8)$, we have

$$(\xrightarrow{+}) \qquad m_A(v_A)_1 + \Sigma \int F \, dt = m_A v_2$$

$$(15\,000 \text{ kg})(1.5 \text{ m/s}) - F_{avg}(0.8 \text{ s}) = (15\,000 \text{ kg})(0.5 \text{ m/s})$$

$$F_{avg} = 18.8 \text{ kN} \qquad\qquad Ans.$$

Solution was possible here since the boxcar's final velocity was obtained in Part (a). Try solving for F_{avg} by applying the principle of impulse and momentum to the tank car.

---

*Only horizontal forces are shown on the free-body diagram.

E X A M P L E 15.5

The 1200-lb cannon shown in Fig. 15–9a fires an 8-lb projectile with a muzzle velocity of 1500 ft/s relative to the ground. If firing takes place in 0.03 s, determine (a) the recoil velocity of the cannon just after firing, and (b) the average impulsive force acting on the projectile. The cannon support is fixed to the ground, and the horizontal recoil of the cannon is absorbed by two springs.

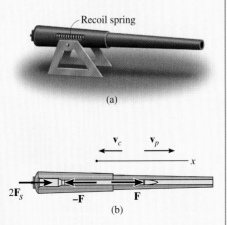

Recoil spring

(a)

Solution

Part (a) Free-Body Diagram.* As shown in Fig. 15–9b, we have considered the projectile and cannon as a single system, since the impulsive forces, **F**, between the cannon and projectile are *internal* to the system and will therefore cancel from the analysis. Furthermore, during the time $\Delta t = 0.03$ s, the two recoil springs which are attached to the support each exert a *nonimpulsive force* $\mathbf{F}_s$ on the cannon. This is because Δt is very short, so that during this time the cannon only moves through a very small distances.† Consequently, $F_s = ks \approx 0$, where k is the spring's stiffness. Hence it may be concluded that momentum for the system is conserved in the *horizontal direction*. Here we will assume that the cannon moves to the left, while the projectile moves to the right after firing.

$2\mathbf{F}_s$ $-\mathbf{F}$ $\mathbf{F}$

$\mathbf{v}_c$ $\mathbf{v}_p$ x

(b)

Conservation of Linear Momentum.

$(\overset{+}{\rightarrow})$ $m_c(v_c)_1 + m_p(v_p)_1 = -m_c(v_c)_2 + m_p(v_p)_2$

$$0 + 0 = -\frac{1200\ \text{lb}}{32.2\ \text{ft/s}^2}(v_c)_2 + \frac{8\ \text{lb}}{32.2\ \text{ft/s}^2}(1500\ \text{ft/s})$$

$$(v_c)_2 = 10\ \text{ft/s} \leftarrow \qquad \textit{Ans.}$$

Part (b). The average impulsive force exerted by the cannon on the projectile can be determined by applying the principle of linear impulse and momentum to the projectile (or to the cannon). Why?

Principle of Impulse and Momentum. Using the data in Fig. 15–9c, noting that $\int F\,dt = F_{\text{avg}}\,\Delta t = F_{\text{avg}}(0.03)$, we have

$(\overset{+}{\rightarrow})$ $m(v_p)_1 + \Sigma \int F\,dt = m(v_p)_2$

$$0 + F_{\text{avg}}(0.03\ \text{s}) = \frac{8\ \text{lb}}{32.2\ \text{ft/s}^2}(1500\ \text{ft/s})$$

$$F_{\text{avg}} = 12.4(10^3)\ \text{lb} = 12.4\ \text{kip} \qquad \textit{Ans.}$$

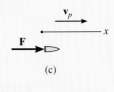

$\mathbf{v}_p$ x

$\mathbf{F}$

(c)

Fig. 15–9

*Only horizontal forces are shown on the free-body diagram.

†If the cannon is firmly fixed to its support (no springs), the reactive force of the support on the cannon must be considered as an external impulse to the system, since the support would allow no movement of the cannon. This, of course, assumes the earth's movement is neglected.

E X A M P L E 15.6

The 350-Mg tugboat T shown in Fig. 15–10a is used to pull the 50-Mg barge B with a rope R. If the barge is initially at rest and the tugboat is coasting freely with a velocity of $(v_T)_1 = 3$ m/s while the rope is *slack*, determine the velocity of the tugboat *directly after* the rope becomes taut. Assume the rope does not stretch. Neglect the frictional effects of the water.

(a)   (b)

Solution

Free-Body Diagram. *   As shown in Fig. 15–10b, we have considered the entire system (tugboat and barge). Hence, the impulsive force created between the tugboat and the barge is *internal* to the system, and therefore momentum of the system is conserved during the instant of towing.

The alternative procedure of drawing the system's impulse and momentum diagrams is shown in Fig. 15–10c.

Conservation of Momentum.   Noting that $(v_B)_2 = (v_T)_2$, we have

$$(\xleftarrow{+})\qquad m_T(v_T)_1 + m_B(v_B)_1 = m_T(v_T)_2 + m_B(v_B)_2$$
$$350(10^3)\ \text{kg}(3\ \text{m/s}) + 0 = 350(10^3)\ \text{kg}(v_T)_2 + 50(10^3)\ \text{kg}(v_T)_2$$

Solving,

$$(v_T)_2 = 2.62\ \text{m/s} \leftarrow \qquad\qquad Ans.$$

This value represents the tugboat's velocity *just after* the towing impulse. Use this result and show that the towing *impulse* is 131 kN · s.

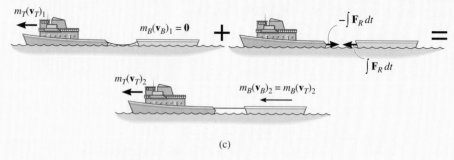

(c)

Fig. 15–10

*Only horizontal forces are shown on the free-body diagram.

EXAMPLE 15.7

An 800-kg rigid pile P shown in Fig. 15–11a is driven into the ground using a 300-kg hammer H. The hammer falls from rest at a height $y_0 = 0.5$ m and strikes the top of the pile. Determine the impulse which the hammer imparts on the pile if the pile is surrounded entirely by loose sand so that after striking, the hammer does *not* rebound off the pile.

Solution

Conservation of Energy.   The velocity at which the hammer strikes the pile can be determined using the conservation of energy equation applied to the hammer. With the datum at the top of the pile, Fig. 15–11a, we have

$$T_0 + V_0 = T_1 + V_1$$

$$\tfrac{1}{2}m_H(v_H)_0^2 + W_H y_0 = \tfrac{1}{2}m_H(v_H)_1^2 + W_H y_1$$

$$0 + 300(9.81) \text{ N}(0.5 \text{ m}) = \frac{1}{2}(300 \text{ kg})(v_H)_1^2 + 0$$

$$(v_H)_1 = 3.13 \text{ m/s}$$

Free-Body Diagram.   From the physical aspects of the problem, the free-body diagram of the hammer and pile, Fig. 15–11b, indicates that during the *short time* occurring *just before* to *just after* the *collision*, the weights of the hammer and pile and the resistance force $\mathbf{F}_s$ of the sand are all *nonimpulsive*. The impulsive force $\mathbf{R}$ is internal to the system and therefore cancels. Consequently, momentum is conserved in the vertical direction during this short time.

Conservation of Momentum.   Since the hammer does not rebound off the pile just after collision, then $(v_H)_2 = (v_P)_2 = v_2$.

$$(+\downarrow) \qquad m_H(v_H)_1 + m_P(v_P)_1 = m_H v_2 + m_P v_2$$

$$(300 \text{ kg})(3.13 \text{ m/s}) + 0 = (300 \text{ kg})v_2 + (800 \text{ kg})v_2$$

$$v_2 = 0.854 \text{ m/s}$$

Principle of Impulse and Momentum.   The impulse which the pile imparts to the hammer can now be determined since $\mathbf{v}_2$ is known. From the free-body diagram for the hammer, Fig. 15–11c, we have

$$(+\downarrow) \qquad m_H(v_H)_1 + \Sigma \int_{t_1}^{t_2} F_y\, dt = m_H v_2$$

$$(300 \text{ kg})(3.13 \text{ m/s}) - \int R\, dt = (300 \text{ kg})(0.854 \text{ m/s})$$

$$\int R\, dt = 683 \text{ N} \cdot \text{s} \qquad\qquad \textit{Ans.}$$

The equal but opposite impulse acts on the pile. Try finding this impulse by applying the principle of impulse and momentum to the pile.

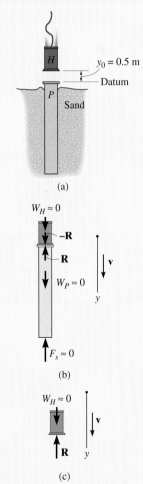

(a)

(b)

(c)

Fig. 15–11

E X A M P L E 15.8

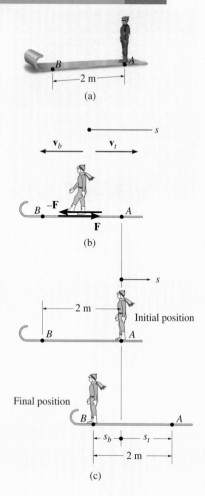

Fig. 15–12

A boy having a mass of 40 kg stands on the back of a 15-kg toboggan which is originally at rest, Fig. 15–12a. If he walks to the front B and stops, determine the distance the toboggan moves. Neglect friction between the bottom of the toboggan and the ground (ice).

Solution I

Free-Body Diagram. The unknown frictional force of the boy's shoes on the bottom of the toboggan can be *excluded* from the analysis if the toboggan and boy on it are considered as a single system. In this way the frictional force **F** becomes *internal* and the conservation of momentum applies, Fig. 15–12b.

Conservation of Momentum. Since both the initial and final momenta of the system are zero (because the initial and final velocities are zero), the system's momentum must also be zero when the boy is at some intermediate point between A and B. Thus

$$(\stackrel{+}{\rightarrow}) \qquad -m_b v_b + m_t v_t = 0 \qquad (1)$$

Here the two unknowns v_b and v_t represent the velocities of the boy moving to the left and the toboggan moving to the right. Both are measured from a *fixed inertial reference* on the ground.

At any instant the *position* of point A on the toboggan and the *position* of the boy must be determined by integration. Since $v = ds/dt$, then $-m_b ds_b + m_t ds_t = 0$. Assuming the initial position of point A to be at the origin, Fig. 15–12c, then at the final position we have $-m_b s_b + m_t s_t = 0$. Since $s_b + s_t = 2$ m, or $s_b = (2 - s_t)$ then

$$-m_b(2 - s_t) + m_t s_t = 0 \qquad (2)$$

$$s_t = \frac{2m_b}{m_b + m_t} = \frac{2(40)}{40 + 15} = 1.45 \text{ m} \qquad Ans.$$

Solution II

The problem may also be solved by considering the relative motion of the boy with respect to the toboggan, $\mathbf{v}_{b/t}$. This velocity is related to the velocities of the boy and toboggan by the equation $\mathbf{v}_b = \mathbf{v}_t + \mathbf{v}_{b/t}$, Eq. 12–34. Since positive motion is assumed to be to the right in Eq. 1, $\mathbf{v}_b$ and $\mathbf{v}_{b/t}$ are negative, because the boy's motion is to the left. Hence, in scalar form, $-v_b = v_t - v_{b/t}$, and Eq. 1 then becomes $m_b(v_t - v_{b/t}) + m_t v_t = 0$. Integrating gives

$$m_b(s_t - s_{b/t}) + m_t s_t = 0$$

Realizing $s_{b/t} = 2$ m, we obtain Eq. 2.

15.4 Impact

Impact occurs when two bodies collide with each other during a very *short* period of time, causing relatively large (impulsive) forces to be exerted between the bodies. The striking of a hammer on a nail, or a golf club on a ball, are common examples of impact loadings.

In general, there are two types of impact. *Central impact* occurs when the direction of motion of the mass centers of the two colliding particles is along a line passing through the mass centers of the particles. This line is called the *line of impact*, Fig. 15–13a. When the motion of one or both of the particles is at an angle with the line of impact, Fig. 15–13b, the impact is said to be *oblique impact*.

Central Impact. To illustrate the method for analyzing the mechanics of impact, consider the case involving the central impact of two *smooth* particles A and B shown in Fig. 15–14.

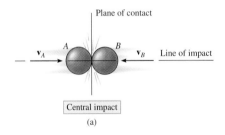

Central impact

(a)

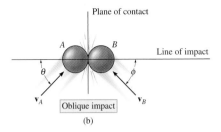

Oblique impact

(b)

Fig. 15–13

- The particles have the initial momenta shown in Fig. 15–14a. Provided $(v_A)_1 > (v_B)_1$, collision will eventually occur.

- During the collision the particles must be thought of as *deformable* or nonrigid. The particles will undergo a *period of deformation* such that they exert an equal but opposite deformation impulse $\int P\, dt$ on each other, Fig. 15–14b.

- Only at the instant of *maximum deformation* will both particles move with a common velocity **v**, since their relative motion is zero, Fig. 15–14c.

- Afterward a *period of restitution* occurs, in which case the particles will either return to their original shape or remain permanently deformed. The equal but opposite *restitution impulse* $\int R\, dt$ pushes the particles apart from one another, Fig. 15–14d. In reality, the physical properties of any two bodies are such that the deformation impulse is *always greater* than that of restitution, i.e., $\int P\, dt > \int R\, dt$.

- Just after separation the particles will have the final momenta shown in Fig. 15–14e, where $(v_B)_2 > (v_A)_2$.

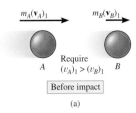

Require
$(v_A)_1 > (v_B)_1$

Before impact

(a)

Fig. 15–14

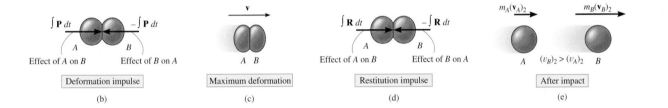

Effect of A on B   Effect of B on A

Deformation impulse

(b)

A   B

Maximum deformation

(c)

Effect of A on B   Effect of B on A

Restitution impulse

(d)

$(v_B)_2 > (v_A)_2$

After impact

(e)

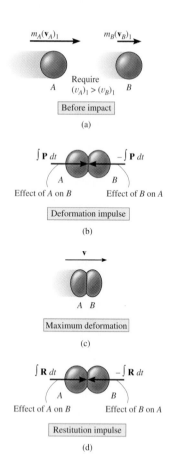

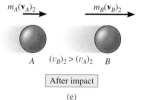

Fig. 15–14

In most problems the initial velocities of the particles will be *known*, and it will be necessary to determine their final velocities $(v_A)_2$ and $(v_B)_2$. In this regard, *momentum* for the *system of particles* is *conserved* since during collision the internal impulses of deformation and restitution *cancel*. Hence, referring to Fig. 15–14*a* and Fig. 15–14*e* we require

$$(\xrightarrow{+}) \qquad m_A(v_A)_1 + m_B(v_B)_1 = m_A(v_A)_2 + m_B(v_B)_2 \qquad (15\text{–}10)$$

In order to obtain a second equation necessary to solve for $(v_A)_2$ and $(v_B)_2$, we must apply the principle of impulse and momentum to *each particle*. For example, during the deformation phase for particle A, Figs. 15–14*a*, 15–14*b*, and 15–14*c*, we have

$$(\xrightarrow{+}) \qquad m_A(v_A)_1 - \int P\, dt = m_A v$$

For the restitution phase, Figs. 15–14*c*, 15–14*d*, and 15–14*e*,

$$(\xrightarrow{+}) \qquad m_A v - \int R\, dt = m_A(v_A)_2$$

The ratio of the restitution impulse to the deformation impulse is called the *coefficient of restitution, e.* From the above equations, this value for particle A is

$$e = \frac{\displaystyle\int R\, dt}{\displaystyle\int P\, dt} = \frac{v - (v_A)_2}{(v_A)_1 - v}$$

In a similar manner, we can establish e by considering particle B, Fig. 15–14. This yields

$$e = \frac{\displaystyle\int R\, dt}{\displaystyle\int P\, dt} = \frac{(v_B)_2 - v}{v - (v_B)_1}$$

If the unknown v is eliminated from the above two equations, the coefficient of restitution can be expressed in terms of the particles' initial and final velocities as

$$(\xrightarrow{+}) \qquad \boxed{e = \frac{(v_B)_2 - (v_A)_2}{(v_A)_1 - (v_B)_1}} \qquad (15\text{–}11)$$

   Provided a value for e is specified, Eqs. 15–10 and 15–11 may be solved simultaneously to obtain $(v_A)_2$ and $(v_B)_2$. In doing so, however, it is important to carefully establish a sign convention for defining the positive direction for both $\mathbf{v}_A$ and $\mathbf{v}_B$ and then use it *consistently* when writing *both* equations. As noted from the application shown, and indicated symbolically by the arrow in parentheses, we have defined the positive direction to the right when referring to the motions of both A and B. Consequently, if a negative value results from the solution of either $(v_A)_2$ or $(v_B)_2$, it indicates motion is to the left.

Coefficient of Restitution.   With reference to Figs. 15–14*a* and 15–14*e*, it is seen that Eq. 15–11 states that e is equal to the ratio of the relative velocity of the particles' separation *just after impact*, $(v_B)_2 - (v_A)_2$, to the relative velocity of the particles' approach *just before impact*, $(v_A)_1 - (v_B)_1$. By measuring these relative velocities experimentally, it has been found that e varies appreciably with impact velocity as well as with the size and shape of the colliding bodies. For these reasons the coefficient of restitution is reliable only when used with data which closely approximate the conditions which were known to exist when measurements of it were made. In general e has a value between zero and one, and one should be aware of the physical meaning of these two limits.

The quality of a manufactured tennis ball is measured by the height of its bounce, which can be related to its coefficient of restitution. Using the mechanics of oblique impact, engineers can design a separation device to maintain quality control of tennis balls after production.

Elastic Impact $(e = 1)$:   If the collision between the two particles is *perfectly elastic*, the deformation impulse ($\int \mathbf{P}\, dt$) is equal and opposite to the restitution impulse ($\int \mathbf{R}\, dt$). Although in reality this can never be achieved, $e = 1$ for an elastic collision.

Plastic Impact $(e = 0)$:   The impact is said to be *inelastic or plastic* when $e = 0$. In this case there is no restitution impulse given to the particles ($\int \mathbf{R}\, dt = \mathbf{0}$), so that after collision both particles couple or stick *together* and move with a common velocity.

   From the above derivation it should be evident that the principle of work and energy cannot be used for the analysis of impact problems since it is not possible to know how the *internal forces* of deformation and restitution vary or displace during the collision. By knowing the particle's velocities before and after collision, however, the energy loss during collision can be calculated on the basis of the difference in the particle's kinetic energy. This energy loss, $\Sigma U_{1-2} = \Sigma T_2 - \Sigma T_1$, occurs because some of the initial kinetic energy of the particle is transformed into thermal energy as well as creating sound and localized deformation of the material when the collision occurs. In particular, if the impact is *perfectly elastic*, no energy is lost in the collision; whereas if the collision is *plastic*, the energy lost during collision is a maximum.

PROCEDURE FOR ANALYSIS (CENTRAL IMPACT)

In most cases the *final velocities* of two smooth particles are to be determined *just after* they are subjected to direct central impact. Provided the coefficient of restitution, the mass of each particle, and each particle's initial velocity *just before* impact are known, the solution to the problem can be obtained using the following two equations:

• The conservation of momentum applies to the system of particles, $\Sigma mv_1 = \Sigma mv_2$.

• The coefficient of restitution, $e = [(v_B)_2 - (v_A)_2]/[(v_A)_1 - (v_B)_1]$, relates the relative velocities of the particles along the line of impact, just before and just after collision.

When applying these two equations, the sense of an unknown velocity can be assumed. If the solution yields a negative magnitude, the velocity acts in the opposite sense.

Oblique Impact. When oblique impact occurs between two smooth particles, the particles move away from each other with velocities having unknown directions as well as unknown magnitudes. Provided the initial velocities are known, four unknowns are present in the problem. As shown in Fig. 15–15a, these unknowns may be represented either as $(v_A)_2, (v_B)_2, \theta_2$, and ϕ_2, or as the x and y components of the final velocities.

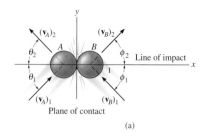

Plane of contact

(a)

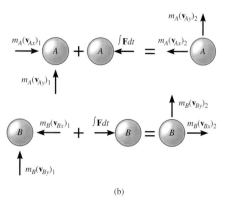

(b)

Fig. 15–15

PROCEDURE FOR ANALYSIS (OBLIQUE IMPACT)

If the y axis is established within the plane of contact and the x axis along the line of impact, the impulsive forces of deformation and restitution act *only in the x direction*, Fig. 15–15b. Resolving the velocity or momentum vectors into components along the x and y axes, Fig. 15–15b, it is possible to write four independent scalar equations in order to determine $(v_{Ax})_2, (v_{Ay})_2, (v_{Bx})_2$, and $(v_{By})_2$.

• Momentum of the system is conserved *along the line of impact*, x axis, so that $\Sigma m(v_x)_1 = \Sigma m(v_x)_2$.

• The coefficient of restitution, $e = [(v_{Bx})_2 - (v_{Ax})_2]/[(v_{Ax})_1 - (v_{Bx})_1]$, relates the relative-velocity *components* of the particles *along the line of impact* (x axis).

• Momentum of particle A is conserved along the y axis, perpendicular to the line of impact, since no impulse acts on particle A in this direction.

• Momentum of particle B is conserved along the y axis, perpendicular to the line of impact, since no impulse acts on particle B in this direction.

Application of these four equations is illustrated numerically in Example 15.11.

EXAMPLE 15.9

The bag A, having a weight of 6 lb, is released from rest at the position $\theta = 0°$, as shown in Fig. 15–16a. After falling to $\theta = 90°$, it strikes an 18-lb box B. If the coefficient of restitution between the bag and box is $e = 0.5$, determine the velocities of the bag and box just after impact and the loss of energy during collision.

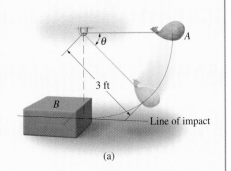
(a)

Solution

This problem involves central impact. Why? Before analyzing the mechanics of the impact, however, it is first necessary to obtain the velocity of the bag *just before* it strikes the box.

Conservation of Energy. With the datum at $\theta = 0°$, Fig. 15–16b, we have

$$T_0 + V_0 = T_1 + V_1$$

$$0 + 0 = \frac{1}{2}\left(\frac{6 \text{ lb}}{32.2 \text{ ft/s}^2}\right)(v_A)_1^2 - 6 \text{ lb}(3 \text{ ft}); \quad (v_A)_1 = 13.9 \text{ ft/s}$$

Conservation of Momentum. After impact we will assume A and B travel to the left. Applying the conservation of momentum to the system, we have

$$(\xrightarrow{+}) \quad m_B(v_B)_1 + m_A(v_A)_1 = m_B(v_B)_2 + m_A(v_A)_2$$

$$0 + \frac{6 \text{ lb}}{32.2 \text{ ft/s}^2}(13.9 \text{ ft/s}) = \frac{18 \text{ lb}}{32.2 \text{ ft/s}^2}(v_B)_2 + \frac{6 \text{ lb}}{32.2 \text{ ft/s}^2}(v_A)_2$$

$$(v_A)_2 = 13.9 - 3(v_B)_2 \qquad (1)$$

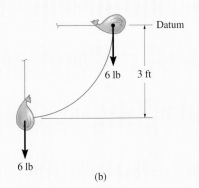
(b)

Coefficient of Restitution. Realizing that for separation to occur after collision $(v_B)_2 > (v_A)_2$, Fig. 15–16c, we have

$$(\xrightarrow{+}) \quad e = \frac{(v_B)_2 - (v_A)_2}{(v_A)_1 - (v_B)_1}; \quad 0.5 = \frac{(v_B)_2 - (v_A)_2}{13.9 \text{ ft/s} - 0}$$

$$(v_A)_2 = (v_B)_2 - 6.95 \qquad (2)$$

Solving Eqs. 1 and 2 simultaneously yields

$$(v_A)_2 = -1.74 \text{ ft/s} = 1.74 \text{ ft/s} \rightarrow \quad \text{and} \quad (v_B)_2 = 5.21 \text{ ft/s} \leftarrow \quad \textit{Ans.}$$

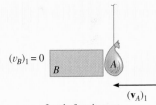
Just before impact

Loss of Energy. Applying the principle of work and energy to the bag and box just before and just after collision, we have

$$\Sigma U_{1-2} = T_2 - T_1;$$

$$\Sigma U_{1-2} = \left[\frac{1}{2}\left(\frac{18 \text{ lb}}{32.2 \text{ ft/s}^2}\right)(5.21 \text{ ft/s})^2 + \frac{1}{2}\left(\frac{6 \text{ lb}}{32.2 \text{ ft/s}^2}\right)(1.74 \text{ ft/s})^2\right] -$$

$$\left[\frac{1}{2}\left(\frac{6 \text{ lb}}{32.2 \text{ ft/s}^2}\right)(13.9 \text{ ft/s})^2\right]$$

$$\Sigma U_{1-2} = -10.1 \text{ ft} \cdot \text{lb} \qquad \textit{Ans.}$$

Why is there an energy loss?

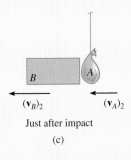
Just after impact
(c)

Fig. 15–16

EXAMPLE 15.10

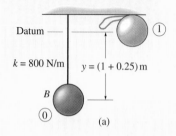

Datum

$k = 800$ N/m $y = (1 + 0.25)$ m

B

(a)

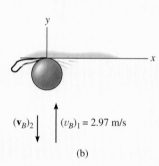

y

x

$(\mathbf{v}_B)_2$ $(v_B)_1 = 2.97$ m/s

(b)

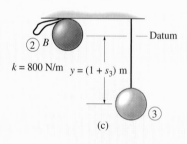

B

Datum

$k = 800$ N/m $y = (1 + s_3)$ m

(c)

Fig. 15–17

The ball B shown in Fig. 15–17a has a mass of 1.5 kg and is suspended from the ceiling by a 1-m-long elastic cord. If the cord is *stretched* downward 0.25 m and the ball is released from rest, determine how far the cord stretches after the ball rebounds from the ceiling. The stiffness of the cord is $k = 800$ N/m, and the coefficient of restitution between the ball and ceiling is $e = 0.8$. The ball makes a central impact with the ceiling.

Solution

First we must obtain the velocity of the ball *just before* it strikes the ceiling using energy methods, then consider the impulse and momentum between the ball and ceiling, and finally again use energy methods to determine the stretch in the cord.

Conservation of Energy. With the datum located as shown in Fig. 15–17a, realizing that initially $y = y_0 = (1 + 0.25)$ m $= 1.25$ m, we have

$$T_0 + V_0 = T_1 + V_1$$

$$\tfrac{1}{2}m(v_B)_0^2 - W_B y_0 + \tfrac{1}{2}ks^2 = \tfrac{1}{2}m(v_B)_1^2 + 0$$

$$0 - 1.5(9.81) \text{ N}(1.25 \text{ m}) + \tfrac{1}{2}(800 \text{ N/m})(0.25 \text{ m})^2 = \tfrac{1}{2}(1.5 \text{ kg})(v_B)_1^2$$

$$(v_B)_1 = 2.97 \text{ m/s} \uparrow$$

The interaction of the ball with the ceiling will now be considered using the principles of impact.* Since an unknown portion of the mass of the ceiling is involved in the impact, the conservation of momentum for the ball-ceiling system will not be written. The "velocity" of this portion of ceiling is zero since it (or the earth) are assumed to remain at rest *both* before and after impact.

Coefficient of Restitution. Fig. 15–17b.

$$(+\uparrow) \quad e = \frac{(v_B)_2 - (v_A)_2}{(v_A)_1 - (v_B)_1}; \quad 0.8 = \frac{(v_B)_2 - 0}{0 - 2.97 \text{ m/s}}$$

$$(v_B)_2 = -2.37 \text{ m/s} = 2.37 \text{ m/s} \downarrow$$

Conservation of Energy. The maximum stretch s_3 in the cord may be determined by again applying the conservation of energy equation to the ball just after collision. Assuming that $y = y_3 = (1 + s_3)$ m, Fig. 15–17c, then

$$T_2 + V_2 = T_3 + V_3$$

$$\tfrac{1}{2}m(v_B)_2^2 + 0 = \tfrac{1}{2}m(v_B)_3^2 - W_B y_3 + \tfrac{1}{2}ks_3^2$$

$$\tfrac{1}{2}(1.5 \text{ kg})(2.37 \text{ m/s})^2 = 0 - 9.81(1.5) \text{ N}(1 \text{ m} + s_3) + \tfrac{1}{2}(800 \text{ N/m})s_3^2$$

$$400s_3^2 - 14.72s_3 - 18.94 = 0$$

Solving this quadratic equation for the positive root yields

$$s_3 = 0.237 \text{ m} = 237 \text{ mm} \qquad \text{Ans.}$$

*The weight of the ball is considered a nonimpulsive force.

EXAMPLE 15.11

Two smooth disks A and B, having a mass of 1 kg and 2 kg, respectively, collide with the velocities shown in Fig. 15–18a. If the coefficient of restitution for the disks is $e = 0.75$, determine the x and y components of the final velocity of each disk just after collision.

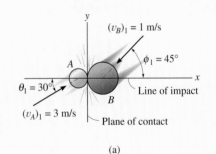

(a)

Solution

The problem involves *oblique impact*. Why? In order to seek a solution, we have established the x and y axes along the line of impact and the plane of contact, respectively, Fig. 15–18a.

Resolving each of the initial velocities into x and y components, we have

$$(v_{Ax})_1 = 3 \cos 30° = 2.60 \text{ m/s} \qquad (v_{Ay})_1 = 3 \sin 30° = 1.50 \text{ m/s}$$
$$(v_{Bx})_1 = -1 \cos 45° = -0.707 \text{ m/s} \quad (v_{By})_1 = -1 \sin 45° = -0.707 \text{ m/s}$$

The four unknown velocity components after collision are assumed to act in the positive directions, Fig. 15–18b. Since the impact occurs only in the x direction (line of impact), the conservation of momentum for *both* disks can be applied in this direction. Why?

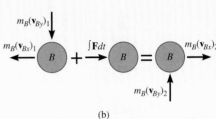

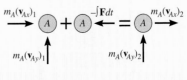

(b)

Conservation of "x" Momentum. In reference to the momentum diagrams, we have

$(\xrightarrow{+})$
$$m_A(v_{Ax})_1 + m_B(v_{Bx})_1 = m_A(v_{Ax})_2 + m_B(v_{Bx})_2$$
$$1 \text{ kg}(2.60 \text{ m/s}) + 2 \text{ kg}(-0.707 \text{ m/s}) = 1 \text{ kg}(v_{Ax})_2 + 2 \text{ kg}(v_{Bx})_2$$
$$(v_{Ax})_2 + 2(v_{Bx})_2 = 1.18 \qquad (1)$$

Coefficient of Restitution (x). Both disks are *assumed* to have components of velocity in the $+x$ direction after collision, Fig. 15–18b.

$(\xrightarrow{+})$
$$e = \frac{(v_{Bx})_2 - (v_{Ax})_2}{(v_{Ax})_1 - (v_{Bx})_1}; \quad 0.75 = \frac{(v_{Bx})_2 - (v_{Ax})_2}{2.60 \text{ m/s} - (-0.707 \text{ m/s})}$$
$$(v_{Bx})_2 - (v_{Ax})_2 = 2.48 \qquad (2)$$

Solving Eqs. 1 and 2 for $(v_{Ax})_2$ and $(v_{Bx})_2$ yields

$$(v_{Ax})_2 = -1.26 \text{ m/s} = 1.26 \text{ m/s} \leftarrow \qquad (v_{Bx})_2 = 1.22 \text{ m/s} \rightarrow \qquad Ans.$$

Conservation of "y" Momentum. The momentum of *each disk* is *conserved* in the y direction (plane of contact), since the disks are smooth and therefore *no* external impulse acts in this direction. From Fig. 15–18b,

$(+\uparrow) \; m_A(v_{Ay})_1 = m_A(v_{Ay})_2; \qquad (v_{Ay})_2 = 1.50 \text{ m/s}\uparrow \qquad\qquad Ans.$
$(+\uparrow) \; m_B(v_{By})_1 = m_B(v_{By})_2; \; (v_{By})_2 = -0.707 \text{ m/s} = 0.707 \text{ m/s}\downarrow \; Ans.$

Show that when the velocity components are summed, one obtains the results shown in Fig. 15–18c.

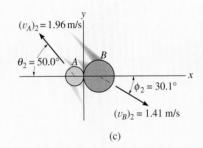

(c)

Fig. 15–18

15.5 Angular Momentum

The *angular momentum* of a particle about point O is defined as the "moment" of the particle's linear momentum about O. Since this concept is analogous to finding the moment of a force about a point, the angular momentum, $\mathbf{H}_O$, is sometimes referred to as the *moment of momentum*.

Scalar Formulation. If a particle is moving along a curve lying in the x-y plane, Fig. 15–19, the angular momentum at any instant can be determined about point O (actually the z axis) by using a scalar formulation. The *magnitude* of $\mathbf{H}_O$ is

$$(H_O)_z = (d)(mv) \tag{15–12}$$

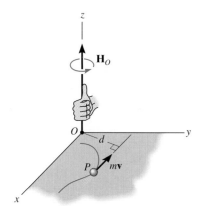

Fig. 15–19

Here d is the moment arm or perpendicular distance from O to the line of action of $m\mathbf{v}$. Common units for $(H_O)_z$ are $\text{kg} \cdot \text{m}^2/\text{s}$ or $\text{slug} \cdot \text{ft}^2/\text{s}$. The *direction* of $\mathbf{H}_O$ is defined by the right-hand rule. As shown, the curl of the fingers of the right hand indicates the sense of rotation of $m\mathbf{v}$ about O, so that in this case the thumb (or $\mathbf{H}_O$) is directed perpendicular to the x-y plane along the $+z$ axis.

Vector Formulation. If the particle is moving along a space curve, Fig. 15–20, the vector cross product can be used to determine the *angular momentum* about O. In this case

$$\mathbf{H}_O = \mathbf{r} \times m\mathbf{v} \tag{15–13}$$

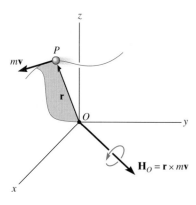

Fig. 15–20

Here $\mathbf{r}$ denotes a position vector drawn from point O to the particle P. As shown in the figure, $\mathbf{H}_O$ is *perpendicular* to the shaded plane containing $\mathbf{r}$ and $m\mathbf{v}$.

In order to evaluate the cross product, $\mathbf{r}$ and $m\mathbf{v}$ should be expressed in terms of their Cartesian components, so that the angular momentum is determined by evaluating the determinant:

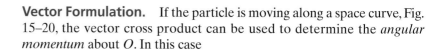

$$\mathbf{H}_O = \begin{vmatrix} \mathbf{i} & \mathbf{j} & \mathbf{k} \\ r_x & r_y & r_z \\ mv_x & mv_y & mv_z \end{vmatrix} \tag{15–14}$$

15.6 Relation Between Moment of a Force and Angular Momentum

The moments about point O of all the forces acting on the particle in Fig. 15–21a may be related to the particle's angular momentum by using the equation of motion. If the mass of the particle is constant, we may write

$$\Sigma \mathbf{F} = m\dot{\mathbf{v}}$$

The moments of the forces about point O can be obtained by performing a cross-product multiplication of each side of this equation by the position vector $\mathbf{r}$, which is measured in the x, y, z inertial frame of reference. We have

$$\Sigma \mathbf{M}_O = \mathbf{r} \times \Sigma \mathbf{F} = \mathbf{r} \times m\dot{\mathbf{v}}$$

From Appendix C, the derivative of $\mathbf{r} \times m\mathbf{v}$ can be written as

$$\dot{\mathbf{H}}_O = \frac{d}{dt}(\mathbf{r} \times m\mathbf{v}) = \dot{\mathbf{r}} \times m\mathbf{v} + \mathbf{r} \times m\dot{\mathbf{v}}$$

The first term on the right side, $\dot{\mathbf{r}} \times m\mathbf{v} = m(\dot{\mathbf{r}} \times \dot{\mathbf{r}}) = \mathbf{0}$, since the cross product of a vector with itself is zero. Hence, the above equation becomes

$$\boxed{\Sigma \mathbf{M}_O = \dot{\mathbf{H}}_O} \qquad (15\text{–}15)$$

This equation states that *the resultant moment about point O of all the forces acting on the particle is equal to the time rate of change of the particle's angular momentum about point O*. This result is similar to Eq. 15–1, i.e.,

$$\boxed{\Sigma \mathbf{F} = \dot{\mathbf{L}}} \qquad (15\text{–}16)$$

Here $\mathbf{L} = m\mathbf{v}$, so *the resultant force acting on the particle is equal to the time rate of change of the particle's linear momentum.*

From the derivations, it is seen that Eqs. 15–15 and 15–16 are actually another way of stating Newton's second law of motion. In other sections of this book it will be shown that these equations have many practical applications when extended and applied to the solution of problems involving either a system of particles or a rigid body.

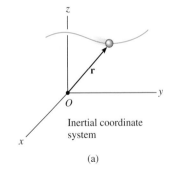

Inertial coordinate system

(a)

Fig. 15–21

System of Particles.   An equation having the same form as Eq. 15–15 may be derived for the system of particles shown in Fig. 15–21b. The forces acting on the arbitrary ith particle of the system consist of a resultant *external force* $\mathbf{F}_i$ and a resultant *internal force* $\mathbf{f}_i$. Expressing the moments of these forces about point O, using the form of Eq. 15–15, we have

$$(\mathbf{r}_i \times \mathbf{F}_i) + (\mathbf{r}_i \times \mathbf{f}_i) = (\dot{\mathbf{H}}_i)_O$$

Here $\mathbf{r}_i$ is the position vector drawn from the origin O of an inertial frame of reference to the ith particle, and $(\dot{\mathbf{H}}_i)_O$ is the time rate of change in the angular momentum of the ith particle about O. Similar equations can be written for each of the other particles of the system. When the results are summed vectorially, the result is

$$\Sigma(\mathbf{r}_i \times \mathbf{F}_i) + \Sigma(\mathbf{r}_i \times \mathbf{f}_i) = \Sigma(\dot{\mathbf{H}}_i)_O$$

The second term is zero since the internal forces occur in equal but opposite collinear pairs, and hence the moment of each pair about point O is zero. Dropping the index notation, the above equation can be written in a simplified form as

$$\Sigma\mathbf{M}_O = \dot{\mathbf{H}}_O \tag{15–17}$$

which states that *the sum of the moments about point O of all the external forces acting on a system of particles is equal to the time rate of change of the total angular momentum of the system about point O.* Although O has been chosen here as the origin of coordinates, it actually can represent any *fixed point* in the inertial frame of reference.

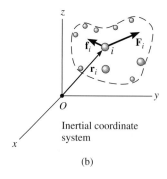

Inertial coordinate system

(b)

Fig. 15–21

E X A M P L E **15.12**

The box shown in Fig. 15–22a has a mass m and is traveling down the smooth circular ramp such that when it is at the angle θ it has a speed v. Determine its angular momentum about point O at this instant and the rate of increase in its speed, i.e., a_t.

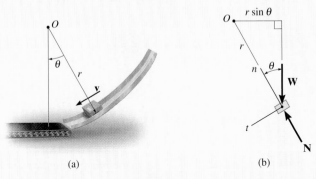

(a)   (b)

Fig. 15–22

Solution

Since $\mathbf{v}$ is tangent to the path, applying Eq. 15–12 the angular momentum is

$$H_O = rmv \downarrow \qquad\qquad Ans.$$

The rate of increase in its speed (dv/dt) can be found by applying Eq. 15–15. From the free-body diagram of the block, Fig. 15–22b, it is seen that only the weight $W = mg$ contributes a moment about point O. We have

$$\curvearrowright + \Sigma M_O = \dot{H}_O; \qquad mg(r \sin \theta) = \frac{d}{dt}(rmv)$$

Since r and m are constant,

$$mgr \sin \theta = rm \frac{dv}{dt}$$

$$\frac{dv}{dt} = g \sin \theta \qquad\qquad Ans.$$

This same result can, of course, be obtained from the equation of motion applied in the tangential direction, Fig. 15–22b, i.e.,

$$+\swarrow \Sigma F_t = ma_t; \qquad mg \sin \theta = m\left(\frac{dv}{dt}\right)$$

$$\frac{dv}{dt} = g \sin \theta \qquad\qquad Ans.$$

15.7 Angular Impulse and Momentum Principles

Principle of Angular Impulse and Momentum. If Eq. 15–15 is rewritten in the form $\Sigma \mathbf{M}_O \, dt = d\mathbf{H}_O$ and integrated, we have, assuming that at time $t = t_1$, $\mathbf{H}_O = (\mathbf{H}_O)_1$ and at time $t = t_2$, $\mathbf{H}_O = (\mathbf{H}_O)_2$,

$$\Sigma \int_{t_1}^{t_2} \mathbf{M}_O \, dt = (\mathbf{H}_O)_2 - (\mathbf{H}_O)_1$$

or

$$(\mathbf{H}_O)_1 + \Sigma \int_{t_1}^{t_2} \mathbf{M}_O \, dt = (\mathbf{H}_O)_2 \qquad (15\text{–}18)$$

This equation is referred to as the *principle of angular impulse and momentum*. The initial and final angular momenta $(\mathbf{H}_O)_1$ and $(\mathbf{H}_O)_2$ are defined as the moment of the linear momentum of the particle $(\mathbf{H}_O = \mathbf{r} \times m\mathbf{v})$ at the instants t_1 and t_2, respectively. The second term on the left side, $\Sigma \int \mathbf{M}_O \, dt$, is called the *angular impulse*. It is determined by integrating, with respect to time, the moments of all the forces acting on the particle over the time period t_1 to t_2. Since the moment of a force about point O is $\mathbf{M}_O = \mathbf{r} \times \mathbf{F}$, the angular impulse may be expressed in vector form as

$$\text{angular impulse} = \int_{t_1}^{t_2} \mathbf{M}_O \, dt = \int_{t_1}^{t_2} (\mathbf{r} \times \mathbf{F}) \, dt \qquad (15\text{–}19)$$

Here $\mathbf{r}$ is a position vector which extends from point O to any point on the line of action of $\mathbf{F}$.

In a similar manner, using Eq. 15–18, the principle of angular impulse and momentum for a system of particles may be written as

$$\Sigma (\mathbf{H}_O)_1 + \Sigma \int_{t_1}^{t_2} \mathbf{M}_O \, dt = \Sigma (\mathbf{H}_O)_2 \qquad (15\text{–}20)$$

Here the first and third terms represent the angular momenta of the system of particles $[\Sigma \mathbf{H}_O = \Sigma (\mathbf{r}_i \times m\mathbf{v}_i)]$ at the instants t_1 and t_2. The second term is the sum of the angular impulses given to all the particles from t_1 to t_2. Recall that these impulses are created only by the moments of the external forces acting on the system where, for the ith particle, $\mathbf{M}_O = \mathbf{r}_i \times \mathbf{F}_i$.

Vector Formulation. Using impulse and momentum principles, it is therefore possible to write two equations which define the particle's motion, namely, Eqs. 15–3 and 15–18, restated as

$$m\mathbf{v}_1 + \Sigma \int_{t_1}^{t_2} \mathbf{F} \, dt = m\mathbf{v}_2$$

$$(\mathbf{H}_O)_1 + \Sigma \int_{t_1}^{t_2} \mathbf{M}_O \, dt = (\mathbf{H}_O)_2$$

(15–21)

Scalar Formulation. In general, the above equations may be expressed in x, y, z component form, yielding a total of six independent scalar equations. If the particle is confined to move in the x–y plane, three independent scalar equations may be written to express the motion, namely,

$$m(v_x)_1 + \Sigma \int_{t_1}^{t_2} F_x \, dt = m(v_x)_2$$

$$m(v_y)_1 + \Sigma \int_{t_1}^{t_2} F_y \, dt = m(v_y)_2$$

$$(H_O)_1 + \Sigma \int_{t_1}^{t_2} M_O \, dt = (H_O)_2$$

(15–22)

The first two of these equations represent the principle of linear impulse and momentum in the x and y directions, which has been discussed in Sec. 15.1, and the third equation represents the principle of angular impulse and momentum about the z axis.

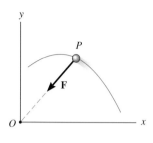

Fig. 15–23

Provided air resistance is neglected, the passengers on this amusement-park ride are subjected to a conservation of angular momentum about the axis of rotation. As shown on the free-body diagram, the line of action of the normal force **N** of the seat on the passenger passes through the axis, and the passenger's weight **W** is parallel to it. No angular impulse acts around the z axis.

Conservation of Angular Momentum. When the angular impulses acting on a particle are all zero during the time t_1 to t_2, Eq. 15–18 reduces to the following simplified form:

$$(\mathbf{H}_O)_1 = (\mathbf{H}_O)_2 \qquad (15\text{–}23)$$

This equation is known as the *conservation of angular momentum*. It states that from t_1 to t_2 the particle's angular momentum remains constant. Obviously, if no external impulse is applied to the particle, both linear and angular momentum will be conserved. In some cases, however, the particle's angular momentum will be conserved and linear momentum may not. An example of this occurs when the particle is subjected *only* to a *central force* (see Sec. 13.7). As shown in Fig. 15–23, the impulsive central force **F** is always directed toward point O as the particle moves along the path. Hence, the angular impulse (moment) created by **F** about the z axis passing through point O is always zero, and therefore angular momentum of the particle is conserved about this axis.

From Eq. 15–20, we can also write the conservation of angular momentum for a system of particles, namely,

$$\Sigma(\mathbf{H}_O)_1 = \Sigma(\mathbf{H}_O)_2 \qquad (15\text{–}24)$$

In this case the summation must include the angular momenta of all particles in the system.

PROCEDURE FOR ANALYSIS

When applying the principles of angular impulse and momentum, or the conservation of angular momentum, it is suggested that the following procedure be used.

Free-Body Diagram.

- Draw the particle's free-body diagram in order to determine any axis about which angular momentum may be conserved. For this to occur, the moments of the forces (or impulses) must be parallel or pass through the axis so as to create zero moment throughout the time period t_1 to t_2.

- The direction and sense of the particle's initial and final velocities should also be established.

- An alternative procedure would be to draw the impulse and momentum diagrams for the particle.

Momentum Equations.

- Apply the principle of angular impulse and momentum, $(\mathbf{H}_o)_1$ $\Sigma \int_{t_1}^{t_2} \mathbf{M}_O \, dt = (\mathbf{H}_O)_2$, or if appropriate, the conservation of angular momentum, $(\mathbf{H}_O)_1 = (\mathbf{H}_O)_2$.

E X A M P L E 15.13

The 5-kg block of negligible size rests on the smooth horizontal plane, Fig. 15–24a. It is attached at A to a slender rod of negligible mass. The rod is attached to a ball-and-socket joint at B. If a moment $M = (3t)$ N·m, where t is in seconds, is applied to the rod and a horizontal force $P = 10$ N is applied to the block, determine the speed of the block in 4 s starting from rest.

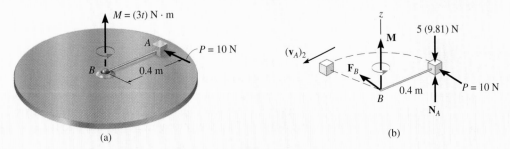

(a)   (b)

Fig. 15–24

Solution

Free-Body Diagram. If we consider the system of both the rod and block, Fig. 15–24b, then the resultant force reaction $\mathbf{F}_B$ at the ball-and-socket can be eliminated from the analysis by applying the principle of angular impulse and momentum about the z axis. If this is done, the angular impulses created by the weight and normal reaction $\mathbf{N}_A$ are also eliminated, since they act parallel to the z axis and therefore create zero moment about this axis.

Principle of Angular Impulse and Momentum.

$$(H_z)_1 + \Sigma \int_{t_1}^{t_2} M_z \, dt = (H_z)_2$$

$$(H_z)_1 + \int_{t_1}^{t_2} M \, dt + r_{BA} P(\Delta t) = (H_z)_2$$

$$0 + \int_0^4 3t \, dt + (0.4 \text{ m})(10 \text{ N})(4 \text{ s}) = 5 \text{ kg}(v_A)_2 (0.4 \text{ m})$$

$$24 + 16 = 2(v_A)_2$$

$$(v_A)_2 = 20 \text{ m/s} \qquad\qquad Ans.$$

EXAMPLE 15.14

(a)

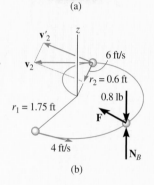

(b)

Fig. 15–25

The 0.8-lb ball B, shown in Fig. 15–25a, is attached to a cord which passes through a hole at A in a smooth table. When the ball is $r_1 = 1.75$ ft from the hole, it is rotating around in a circle such that its speed is $v_1 = 4$ ft/s. By applying a force $\mathbf{F}$ the cord is pulled downward through the hole with a constant speed $v_c = 6$ ft/s. Determine (a) the speed of the ball at the instant it is $r_2 = 0.6$ ft from the hole, and (b) the amount of work done by $\mathbf{F}$ in shortening the radial distance from r_1 to r_2. Neglect the size of the ball.

Solution

Part (a) Free-Body Diagram. As the ball moves from r_1 to r_2, Fig. 15–25b, the cord force $\mathbf{F}$ on the ball always passes through the z axis, and the weight and $\mathbf{N}_B$ are parallel to it. Hence the moments, or angular impulses created by these forces, are all *zero* about this axis. Therefore, the conservation of angular momentum applies about the z axis.

Conservation of Angular Momentum. The ball's velocity $\mathbf{v}_2$ is resolved into two components. The radial component, 6 ft/s, is known; however, it produces zero angular momentum about the z axis. Thus,

$$\mathbf{H}_1 = \mathbf{H}_2$$
$$r_1 m_B v_1 = r_2 m_B v_2'$$
$$1.75 \text{ ft}\left(\frac{0.8 \text{ lb}}{32.2 \text{ ft/s}^2}\right)4 = 0.6 \text{ ft}\left(\frac{0.8 \text{ lb}}{32.2 \text{ ft/s}^2}\right)v_2'$$
$$v_2' = 11.67 \text{ ft/s}$$

The speed of the ball is thus

$$v_2 = \sqrt{(11.67)^2 + (6)^2}$$
$$= 13.1 \text{ ft/s}$$

Part (b). The only force that does work on the ball is $\mathbf{F}$. (The normal force and weight do not move vertically.) The initial and final kinetic energies of the ball can be determined so that from the principle of work and energy we have

$$T_1 + \Sigma U_{1-2} = T_2$$
$$\frac{1}{2}\left(\frac{0.8 \text{ lb}}{32.2 \text{ ft/s}^2}\right)(4 \text{ ft/s})^2 + U_F = \frac{1}{2}\left(\frac{0.8 \text{ lb}}{32.2 \text{ ft/s}^2}\right)(13.1 \text{ ft/s})^2$$

$$U_F = 1.94 \text{ ft} \cdot \text{lb} \qquad Ans.$$

EXAMPLE 15.15

The 2-kg disk shown in Fig. 15–26a rests on a smooth horizontal surface and is attached to an elastic cord that has a stiffness $k_c = 20$ N/m and is initially unstretched. If the disk is given a velocity $(v_D)_1 = 1.5$ m/s, perpendicular to the cord, determine the rate at which the cord is being stretched and the speed of the disk at the instant the cord is stretched 0.2 m.

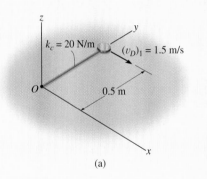

(a)

Solution

Free-Body Diagram. After the disk has been launched, it slides along the path shown in Fig. 15–26b. By inspection, angular momentum about point O (or the z axis) is *conserved*, since none of the forces produce an angular impulse about this axis. Also, when the distance is 0.7 m, only the component $(\mathbf{v}'_D)_2$ produces angular momentum of the disk about O.

Conservation of Angular Momentum. The component $(\mathbf{v}'_D)_2$ can be obtained by applying the conservation of angular momentum about O (the z axis), i.e.,

$$(\mathbf{H}_O)_1 = (\mathbf{H}_O)_2$$
$$r_1 m_D (v_D)_1 = r_2 m_D (v'_D)_2$$
$$(\curvearrowleft +) \qquad 0.5 \text{ m}(2 \text{ kg})(1.5 \text{ m/s}) = 0.7 \text{ m}(2 \text{ kg})(v'_D)_2$$
$$(v'_D)_2 = 1.07 \text{ m/s}$$

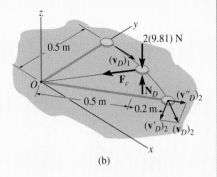

(b)

Fig. 15–26

Conservation of Energy. The speed of the disk may be obtained by applying the conservation of energy equation at the point where the disk was launched and at the point where the cord is stretched 0.2 m.

$$T_1 + V_1 = T_2 + V_2$$
$$\tfrac{1}{2}(2 \text{ kg})(1.5 \text{ m/s})^2 + 0 = \tfrac{1}{2}(2 \text{ kg})(v_D)_2^2 + \tfrac{1}{2}(20 \text{ N/m})(0.2 \text{ m})^2$$

Thus,

$$(v_D)_2 = 1.36 \text{ m/s} \qquad \qquad Ans.$$

Having determined $(v_D)_2$ and its component $(v'_D)_2$, the rate of stretch of the cord $(v''_D)_2$ is determined from the Pythagorean theorem,

$$(v''_D)_2 = \sqrt{(v_D)_2^2 - (v'_D)_2^2}$$
$$= \sqrt{(1.36)^2 - (1.07)^2}$$
$$= 0.838 \text{ m/s} \qquad \qquad Ans.$$

*15.8 Steady Fluid Streams

Knowledge of the forces developed by steadily moving fluid streams is of importance in the design and analysis of turbines, pumps, blades, and fans. To illustrate how the principle of impulse and momentum may be used to determine these forces, consider the diversion of a steady stream of fluid (liquid or gas) by a fixed pipe, Fig. 15–27a. The fluid enters the pipe with a velocity $\mathbf{v}_A$ and exits with a velocity $\mathbf{v}_B$. The impulse and momentum diagrams for the fluid stream are shown in Fig. 15–27b. The force $\Sigma\mathbf{F}$, shown on the impulse diagram, represents the resultant of all the external forces acting on the fluid stream. It is this loading which gives the fluid stream an impulse whereby the original momentum of the fluid is changed in both its magnitude and direction. Since the flow is steady, $\Sigma\mathbf{F}$ will be *constant* during the time interval dt. During this time the fluid stream is in motion, and as a result a small amount of fluid, having a mass dm, is about to enter the pipe with a velocity $\mathbf{v}_A$ at time t. If this element of mass and the mass of fluid in the pipe are considered as a "closed system," then at time $t + dt$ a corresponding element of mass dm must leave the pipe with a velocity $\mathbf{v}_B$. Also, the fluid stream *within* the pipe section has a mass m and an *average velocity* $\mathbf{v}$ which is constant during the time interval dt. Applying the principle of linear impulse and momentum to the fluid stream, we have

$$dm\, \mathbf{v}_A + m\mathbf{v} + \Sigma\mathbf{F}\, dt = dm\, \mathbf{v}_B + m\mathbf{v}$$

Fig. 15–27

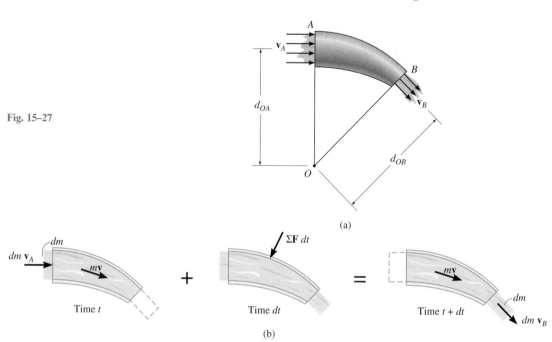

(a)

(b)

Force Resultant. Solving for the resultant force yields

$$\Sigma \mathbf{F} = \frac{dm}{dt}(\mathbf{v}_B - \mathbf{v}_A) \qquad (15\text{–}25)$$

Provided the motion of the fluid can be represented in the *x–y* plane, it is usually convenient to express this vector equation in the form of two scalar component equations, i.e.,

$$\Sigma F_x = \frac{dm}{dt}(v_{Bx} - v_{Ax})$$
$$\Sigma F_y = \frac{dm}{dt}(v_{By} - v_{Ay})$$
$$(15\text{–}26)$$

The term dm/dt is called the *mass flow* and indicates the constant amount of fluid which flows either into or out of the pipe per unit of time. If the cross-sectional areas and densities of the fluid at the entrance A and exit B are A_A, ρ_A and A_B, ρ_B, respectively, Fig. 15–27c, then *continuity of mass* requires that $dm = \rho\, dV = \rho_A(ds_A A_A) = \rho_B(ds_B A_B)$. Hence, during the time dt, since $v_A = ds_A/dt$ and $v_B = ds_B/dt$, we have

$$\frac{dm}{dt} = \rho_A v_A A_A = \rho_B v_B A_B = \rho_A Q_A = \rho_B Q_B \qquad (15\text{–}27)$$

Here $Q = vA$ is the volumetric *flow rate*, which measures the volume of fluid flowing per unit of time.

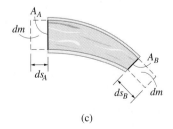

(c)

The conveyor belt must supply frictional forces to the gravel that falls upon it in order to change the momentum of the gravel stream, so that it begins to travel along the belt.

The air on one side of this fan is essentially at rest, and as it passes through the blades its momentum is increased. To change the momentum of the air flow in this manner, the blades must exert a horizontal thrust on the air stream. As the blades turn faster, the equal but opposite thrust of the air on the blades could overcome the rolling resistance of the wheels on the ground and begin to move the frame of the fan.

Moment Resultant. In some cases it is necessary to obtain the support reactions on the fluid-carrying device. If Eq. 15–25 does not provide enough information to do this, the principle of angular impulse and momentum must be used. The formulation of this principle applied to fluid streams can be obtained from Eq. 15–17, $\Sigma \mathbf{M}_O = \dot{\mathbf{H}}_O$, which states that the moment of all the external forces acting on the system about point O is equal to the time rate of change of angular momentum about O. In the case of the pipe shown in Fig. 15–27a, the flow is steady in the x–y plane; hence we have

$(\curvearrowleft +)$

$$\Sigma M_O = \frac{dm}{dt}(d_{OB}v_B - d_{OA}v_A) \qquad (15\text{–}28)$$

where the moment arms d_{OB} and d_{OA} are directed from O to the *geometric center* or *centroid* of the openings at A and B.

PROCEDURE FOR ANALYSIS

Problems involving steady flow can be solved using the following procedure.

Kinematic Diagram.

- If the device is *moving*, a *kinematic diagram* may be helpful for determining the entrance and exit velocities of the fluid flowing onto the device, since a *relative-motion analysis* of velocity will be involved.
- The measurement of velocities v_A and v_B must be made by an observer fixed in an inertial frame of reference.
- Once the velocity of the fluid flowing onto the device is determined, the mass flow is calculated using Eq. 15–27.

Free-Body Diagram.

- Draw a free-body diagram of the device which is directing the fluid in order to establish the forces $\Sigma \mathbf{F}$ that act on it. These external forces will include the support reactions, the weight of the device and the fluid contained within it, and the static pressure forces of the fluid at the entrance and exit sections of the device.*

Equations of Steady Flow.

- Apply the equations of steady flow, Eqs. 15–26 and 15–28, using the appropriate components of velocity and force shown on the kinematic and free-body diagrams.

*In the SI system pressure is measured using the *pascal* (Pa), where 1 Pa = 1 N/m².

E X A M P L E 15.16

Determine the components of reaction which the fixed pipe joint at A exerts on the elbow in Fig. 15–28a, if water flowing through the pipe is subjected to a static gauge pressure of 100 kPa at A. The discharge at B is $Q_B = 0.2$ m³/s. Water has a density $\rho_w = 1000$ kg/m³, and the water-filled elbow has a mass of 20 kg and center of mass at G.

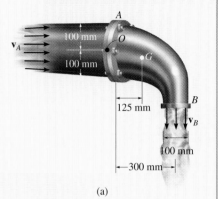

(a)

Solution

Using a fixed inertial coordinate system, the velocity of flow at A and B and the mass flow rate can be obtained from Eq. 15–27. Since the density of water is constant, $Q_B = Q_A = Q$. Hence,

$$\frac{dm}{dt} = \rho_w Q = (1000 \text{ kg/m}^3)(0.2 \text{ m}^3/\text{s}) = 200 \text{ kg/s}$$

$$v_B = \frac{Q}{A_B} = \frac{0.2 \text{ m}^3/\text{s}}{\pi(0.05 \text{ m})^2} = 25.46 \text{ m/s}\downarrow$$

$$v_A = \frac{Q}{A_A} = \frac{0.2 \text{ m}^3/\text{s}}{\pi(0.1 \text{ m})^2} = 6.37 \text{ m/s} \rightarrow$$

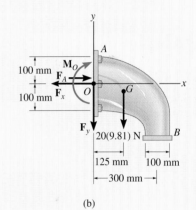

(b)

Fig. 15–28

Free-Body Diagram. As shown on the free-body diagram, Fig. 15–28b, the *fixed* connection at A exerts a resultant couple moment $\mathbf{M}_O$ and force components $\mathbf{F}_x$ and $\mathbf{F}_y$ on the elbow. Due to the static pressure of water in the pipe, the pressure force acting on the fluid at A is $F_A = p_A A_A$. Since 1 kPa $= 1000$ N/m²,

$$F_A = p_A A_A = [100(10^3) \text{ N/m}^2][\pi(0.1 \text{ m})^2] = 3141.6 \text{ N}$$

There is no static pressure acting at B, since the water is discharged at atmospheric pressure; i.e., the pressure measured by a gauge at B is equal to zero, $p_B = 0$.

Equations of Steady Flow.

$$\xrightarrow{+} \Sigma F_x = \frac{dm}{dt}(v_{Bx} - v_{Ax}); \quad -F_x + 3141.6 \text{ N} = 200 \text{ kg/s}(0 - 6.37 \text{ m/s})$$

$$F_x = 4.41 \text{ kN} \qquad \qquad Ans.$$

$$+\uparrow \Sigma F_y = \frac{dm}{dt}(v_{By} - v_{Ay}); \quad -F_y - 20(9.81) \text{ N} = 200 \text{ kg/s}(-25.46 \text{ m/s} - 0)$$

$$F_y = 4.90 \text{ kN} \qquad \qquad Ans.$$

If moments are summed about point O, Fig. 15–28b, then $\mathbf{F}_x$, $\mathbf{F}_y$, and the static pressure $\mathbf{F}_A$ are eliminated, as well as the moment of momentum of the water entering at A, Fig. 15–28a. Hence,

$$\curvearrowleft + \Sigma M_O = \frac{dm}{dt}(d_{OB}v_B - d_{OA}v_A)$$

$$M_O + 20(9.81) \text{ N}(0.125 \text{ m}) = 200 \text{ kg/s}[(0.3 \text{ m})(25.46 \text{ m/s}) - 0]$$

$$M_O = 1.50 \text{ kN} \cdot \text{m} \qquad \qquad Ans.$$

E X A M P L E 15.17

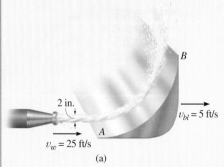

2 in.

$v_{bl} = 5$ ft/s

A

$v_w = 25$ ft/s

(a)

A 2-in.-diameter water jet having a velocity of 25 ft/s impinges upon a single moving blade, Fig. 15–29a. If the blade is moving at 5 ft/s away from the jet, determine the horizontal and vertical components of force which the blade is exerting on the water. What power does the water generate on the blade? Water has a specific weight of $\gamma_w = 62.4$ lb/ft^3.

Solution

Kinematic Diagram. From a fixed inertial coordinate system, Fig. 15–29b, the rate at which water enters the blade is

$$\mathbf{v}_A = \{25\mathbf{i}\} \text{ ft/s}$$

The *relative-flow velocity* of the water onto the blade is $\mathbf{v}_{w/bl} = \mathbf{v}_w - \mathbf{v}_{bl} = 25\mathbf{i} - 5\mathbf{i} = \{20\mathbf{i}\}$ ft/s. Since the blade is moving with a velocity of $\mathbf{v}_{bl} = \{5\mathbf{i}\}$ ft/s, the velocity of flow at B measured from x, y is the vector sum, shown in Fig. 15–29b. Here,

$$\mathbf{v}_B = \mathbf{v}_{bl} + \mathbf{v}_{w/bl}$$
$$= \{5\mathbf{i} + 20\mathbf{j}\} \text{ ft/s}$$

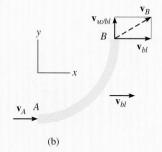

(b)

Thus, the mass flow of water *onto* the blade that undergoes a momentum change is

$$\frac{dm}{dt} = \rho_w(v_{w/bl})A_A = \frac{62.4}{32.2}(20)\left[\pi\left(\frac{1}{12}\right)^2\right] = 0.846 \text{ slug/s}$$

Free-Body Diagram. The free-body diagram of a section of water acting on the blade is shown in Fig. 15–29c. The weight of the water will be neglected in the calculation, since this force will be small compared to the reactive components $\mathbf{F}_x$ and $\mathbf{F}_y$.

Equations of Steady Flow.

$$\Sigma\mathbf{F} = \frac{dm}{dt}(\mathbf{v}_B - \mathbf{v}_A)$$

$$-F_x\mathbf{i} + F_y\mathbf{j} = 0.846(5\mathbf{i} + 20\mathbf{j} - 25\mathbf{i})$$

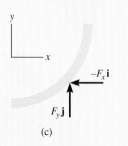

$-F_x\mathbf{i}$

$F_y\mathbf{j}$

(c)

Fig. 15–29

Equating the respective $\mathbf{i}$ and $\mathbf{j}$ components gives

$$F_x = 0.846(20) = 16.9 \text{ lb} \leftarrow \qquad\qquad Ans.$$
$$F_y = 0.846(20) = 16.9 \text{ lb}\uparrow \qquad\qquad Ans.$$

The water exerts equal but opposite forces on the blade.

Since the water force which causes the blade to move forward horizontally with a velocity of 5 ft/s is $F_x = 16.9$ lb, then from Eq. 14–10 the power is

$$P = \mathbf{F} \cdot \mathbf{v}; \qquad P = \frac{16.9 \text{ lb}(5 \text{ ft/s})}{550 \text{ hp}/(\text{ft} \cdot \text{lb/s})} = 0.154 \text{ hp} \qquad Ans.$$

*15.9 Propulsion with Variable Mass

In the previous section we considered the case in which a *constant* amount of mass dm enters and leaves a "*closed system*." There are, however, two other important cases involving mass flow, which are represented by a system that is either gaining or losing mass. In this section we will discuss each of these cases separately.

A System That Loses Mass. Consider a device such as a rocket which at an instant of time has a mass m and is moving forward with a velocity $\mathbf{v}$, Fig. 15–30a. At this same instant the device is expelling an amount of mass m_e with a mass flow velocity $\mathbf{v}_e$. For the analysis, the "*closed system*" includes *both the mass m of the device and the expelled mass m_e.* The impulse and momentum diagrams for the system are shown in Fig. 15–30b. During the time dt, the velocity of the device is increased from $\mathbf{v}$ to $\mathbf{v} + d\mathbf{v}$ since an amount of mass dm_e has been ejected and thereby gained in the exhaust. This increase in forward velocity, however, does not change the velocity $\mathbf{v}_e$ of the expelled mass, since this mass moves at a constant speed once it has been ejected. The impulses are created by $\Sigma\mathbf{F}_s$, which represents the resultant of all the external forces that *act on the system* in the direction of motion. This force resultant *does not include* the force which causes the device to move forward, since this force (called a *thrust*) is *internal to the system*; that is, the thrust acts with equal magnitude but opposite direction on the mass m of the device and the expelled exhaust mass m_e.* Applying the principle of impulse and momentum to the system, Fig. 15–30b, we have

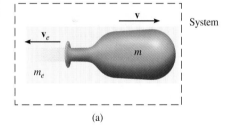

(a)

$$(\xrightarrow{+}) \quad mv - m_e v_e + \Sigma F_s \, dt = (m - dm_e)(v + dv) - (m_e + dm_e)v_e$$

or

$$\Sigma F_s \, dt = -v \, dm_e + m \, dv - dm_e \, dv - v_e \, dm_e$$

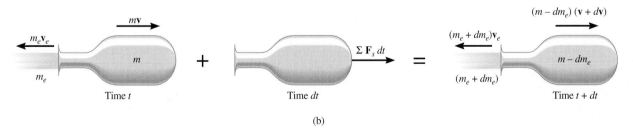

(b)

Fig. 15–30

---

*$\Sigma\mathbf{F}_s$ represents the external resultant force *acting on the system*, which is different from $\Sigma\mathbf{F}$, the resultant force acting only on the device.

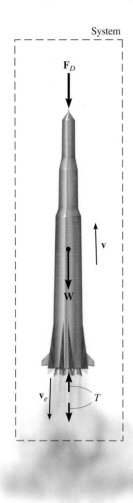

System

Fig. 15–31

Without loss of accuracy, the third term on the right side may be neglected since it is a "second-order" differential. Dividing by dt gives

$$\Sigma F_s = m\frac{dv}{dt} - (v + v_e)\frac{dm_e}{dt}$$

The relative velocity of the device as seen by an observer moving with the particles of the ejected mass is $v_{D/e} = (v + v_e)$, and so the final result can be written as

$$\Sigma F_s = m\frac{dv}{dt} - v_{D/e}\frac{dm_e}{dt} \qquad (15\text{–}29)$$

Here the term dm_e/dt represents the rate at which mass is being ejected.

To illustrate an application of Eq. 15–29, consider the rocket shown in Fig. 15–31, which has a weight $\mathbf{W}$ and is moving upward against an atmospheric drag force $\mathbf{F}_D$. The system to be considered consists of the mass of the rocket and the mass of ejected gas m_e. Applying Eq. 15–29 to this system gives

$$(+\uparrow) \qquad -F_D - W = \frac{W}{g}\frac{dv}{dt} - v_{D/e}\frac{dm_e}{dt}$$

The last term of this equation represents the *thrust* $\mathbf{T}$ which the engine exhaust exerts on the rocket, Fig. 15–31. Recognizing that $dv/dt = a$, we may therefore write

$$(+\uparrow) \qquad T - F_D - W = \frac{W}{g}a$$

If a free-body diagram of the rocket is drawn, it becomes obvious that this equation represents an application of $\Sigma\mathbf{F} = m\mathbf{a}$ for the rocket.

A System That Gains Mass. A device such as a scoop or a shovel may gain mass as it moves forward. For example, the device shown in Fig. 15–32a has a mass m and is moving forward with a velocity $\mathbf{v}$. At this instant, the device is collecting a particle stream of mass m_i. The flow velocity $\mathbf{v}_i$ of this injected mass is constant and independent of the velocity $\mathbf{v}$ such that $v > v_i$. The system to be considered includes both the mass of the device and the mass of the injected particles. The impulse and momentum diagrams for this system are shown in Fig. 15–32b. Along with an increase in mass dm_i gained by the device, there is an assumed increase in velocity $d\mathbf{v}$ during the time interval dt. This increase is caused by the impulse created by $\Sigma\mathbf{F}_s$, the resultant of all the external forces *acting on the system* in the direction of motion. The force summation does not

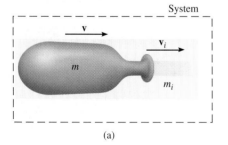

System

(a)

Fig. 15–32

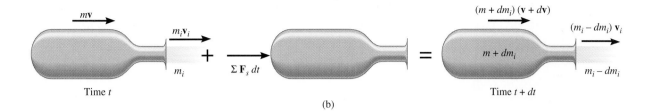

Time t

(b)

Time $t + dt$

include the retarding force of the injected mass acting on the device. Why? Applying the principle of impulse and momentum to the system, we have

$$(\overset{+}{\rightarrow}) \qquad mv + m_i v_i + \Sigma F_s \, dt = (m + dm_i)(v + dv) + (m_i - dm_i)v_i$$

Using the same procedure as in the previous case, we may write this equation as

$$\Sigma F_s = m\frac{dv}{dt} + (v - v_i)\frac{dm_i}{dt}$$

Since the relative velocity of the device as seen by an observer moving with the particles of the injected mass is $v_{D/i} = (v - v_i)$, the final result can be written as

$$\Sigma F_s = m\frac{dv}{dt} + v_{D/i}\frac{dm_i}{dt} \qquad (15\text{--}30)$$

where dm_i/dt is the rate of mass injected into the device. The last term in this equation represents the magnitude of force **R**, which the injected mass *exerts on the device*. Since $dv/dt = a$, Eq. 15–30 becomes

$$\Sigma F_s - R = ma$$

This is the application of $\Sigma \mathbf{F} = m\mathbf{a}$, Fig. 15–32c.

As in the case of steady flow, problems which are solved using Eqs. 15–29 and 15–30 should be accompanied by the necessary free-body diagram. With this diagram one can then determine ΣF_s *for the system* and isolate the force exerted on the device by the particle stream.

The scraper box behind this tractor represents a device that gains mass. If the tractor maintains a constant velocity v, then $dv/dt = 0$ and, because the soil is originally at rest, $v_{D/i} = v$. By Eq. 15–30, the horizontal towing force on the scraper box is then $T = 0 + v(dm/dt)$, where dm/dt is the rate of soil accumulation in the box.

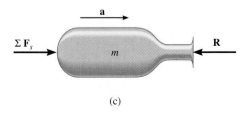

(c)

E X A M P L E 15.18

The initial combined mass of a rocket and its fuel is m_0. A total mass m_f of fuel is consumed at a constant rate of $dm_e/dt = c$ and expelled at a constant speed of u relative to the rocket. Determine the maximum velocity of the rocket, i.e., at the instant the fuel runs out. Neglect the change in the rocket's weight with altitude and the drag resistance of the air. The rocket is fired vertically from rest.

Solution

Since the rocket is losing mass as it moves upward, Eq. 15–29 can be used for the solution. The only *external force* acting on the *system* consisting of the rocket and a portion of the expelled mass is the weight **W**, Fig. 15–33. Hence,

$$+\uparrow \Sigma F_s = m\frac{dv}{dt} - v_{D/e}\frac{dm_e}{dt}; \qquad -W = m\frac{dv}{dt} - uc \qquad (1)$$

The rocket's velocity is obtained by integrating this equation.

At any given instant t during the flight, the mass of the rocket can be expressed as $m = m_0 - (dm_e/dt)t = m_0 - ct$. Since $W = mg$, Eq. 1 becomes

$$-(m_0 - ct)g = (m_0 - ct)\frac{dv}{dt} - uc$$

Separating the variables and integrating, realizing that $v = 0$ at $t = 0$, we have

$$\int_0^v dv = \int_0^t \left(\frac{uc}{m_0 - ct} - g\right)dt$$

$$v = -u\ln(m_0 - ct) - gt\Big|_0^t = u\ln\left(\frac{m_0}{m_0 - ct}\right) - gt \qquad (2)$$

Note that lift off requires the first term on the right to be greater than the second during the initial phase of motion. The time t' needed to consume all the fuel is given by

$$m_f = \left(\frac{dm_e}{dt}\right)t' = ct'$$

Hence,

$$t' = m_f/c$$

Substituting into Eq. 2 yields

$$v_{\max} = u\ln\left(\frac{m_0}{m_0 - m_f}\right) - \frac{gm_f}{c} \qquad Ans.$$

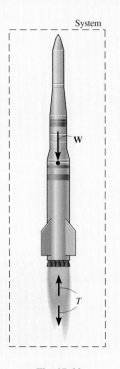

System

W

T

Fig. 15–33

EXAMPLE 15.19

A chain of length l, Fig. 15–34a, has a mass m. Determine the magnitude of force **F** required to (a) raise the chain with a constant speed v_c, starting from rest when $y = 0$; and (b) lower the chain with a constant speed v_c, starting from rest when $y = l$.

Solution

Part (a).   As the chain is raised, all the suspended links are given a sudden impulse downward by each added link which is lifted off the ground. Thus, the *suspended portion* of the chain may be considered as a device which is *gaining mass*. The system to be considered is the length of chain y which is suspended by **F** at any instant, including the next link which is about to be added but is still at rest, Fig. 15–34b. The forces acting on this system *exclude* the internal forces **P** and $-$**P**, which act between the added link and the suspended portion of the chain. Hence, $\Sigma F_s = F - mg(y/l)$.

To apply Eq. 15–30, it is also necessary to find the rate at which mass is being added to the system. The velocity $\mathbf{v}_c$ of the chain is equivalent to $\mathbf{v}_{D/i}$. Why? Since v_c is constant, $dv_c/dt = 0$ and $dy/dt = v_c$. Integrating, using the initial condition that $y = 0$ at $t = 0$, gives $y = v_c t$. Thus, the mass of the system at any instant is $m_s = m(y/l) = m(v_c t/l)$, and therefore the *rate* at which mass is *added* to the suspended chain is

$$\frac{dm_i}{dt} = m\left(\frac{v_c}{l}\right)$$

Applying Eq. 15–30 to the system, using this data, we have

$$+\uparrow \Sigma F_s = m\frac{dv_c}{dt} + v_{D/i}\frac{dm_i}{dt}$$

$$F - mg\left(\frac{y}{l}\right) = 0 + v_c m\left(\frac{v_c}{l}\right)$$

Hence,

$$F = (m/l)(gy + v_c^2) \qquad \textit{Ans.}$$

Part (b).   When the chain is being lowered, the links which are expelled (given zero velocity) *do not* impart an impulse to the *remaining* suspended links. Why? Thus, the system in Part (a) cannot be considered. Instead, the equation of motion will be used to obtain the solution. At time t the portion of chain still off the floor is y. The free-body diagram for a suspended portion of the chain is shown in Fig. 15–34c. Thus,

$$+\uparrow\Sigma F = ma; \qquad F - mg\left(\frac{y}{l}\right) = 0$$

$$F = mg\left(\frac{y}{l}\right) \qquad \textit{Ans.}$$

(a)

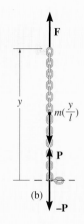

(b)

(c)

Fig. 15–34

DESIGN PROJECT

15–1D.   DESIGN OF A CRANBERRY SELECTOR

The quality of a cranberry depends upon its firmness, which in turn is related to its bounce. Through experiment, it is found that berries that bounce to a height of $2.5 \leq h' \leq 3.25$ ft, when released from rest at a height of $h = 4$ ft, are appropriate for processing. Using this information, determine the berry's range of the allowable coefficient of restitution, and then design a manner in which good and bad berries can be separated. Submit a drawing of your design, and show calculations as to how the selection and collection of berries is made from your established geometry.

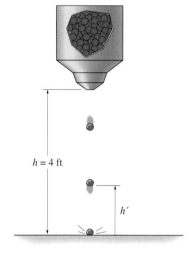

$h = 4$ ft

h'

Prob. 15–1D

CHAPTER REVIEW

- *Impulse.*   An impulse that acts on the particle is defined by

$$\mathbf{I} = \int \mathbf{F}\, dt$$

Graphically this represents the area under the *F–t* diagram. If the force is constant, then the impulse becomes

$$\mathbf{I} = \mathbf{F}_c(t_2 - t_1)$$

- *Principle of Impulse and Momentum.*   When the equation of motion, $\Sigma\mathbf{F} = m\mathbf{a}$, and the kinematic equation, $a = dv/dt$, are combined, we obtain the principle of impulse and momentum.

$$m\mathbf{v}_1 + \Sigma \int_{t_1}^{t_2} \mathbf{F}\, dt = m\mathbf{v}_2$$

Here, the initial momentum of the particle, $m\mathbf{v}_1$, plus all of the impulses that are applied to the particle during the time t_1 to t_2, $\Sigma \int \mathbf{F}\, dt$, equals the final momentum $m\mathbf{v}_2$ of the particle. This is a vector equation that can be resolved into components and is used to solve problems that involve force, velocity, and time. For application, the free-body diagram should be drawn in order to account for all the impulses that act on the particle.

- *Conservation of Linear Momentum.*   If the principle of impulse and momentum is applied to a system of particles, then the collisions between the particles produce internal impulses that are equal, opposite, and collinear, and therefore cancel from the equation. Furthermore, if an external impulse is small, that is, the force is small and the time is short, then the impulse can be classified as nonimpulsive and can be neglected. Consequently, momentum for the system of particles is conserved, and so

$$\Sigma(m\mathbf{v}_i)_1 = \Sigma(m\mathbf{v}_i)_2$$

This equation is useful for finding the final velocity of a particle when internal impulses are exerted between two particles. If the internal impulse is to be determined, then one of the particles is isolated and the principle of impulse and momentum is applied to this particle.

- *Impact.*   When two particles collide (*A* and *B*), the internal impulse between them is equal, opposite, and collinear. Consequently, the conservation of momentum for this system applies along the line of impact. If the final velocities are unknown, a second equation is needed for solution. Here, we use the coefficient of restitution, *e*. This experimentally determined coefficient depends upon the physical properties of the colliding particles. It can be expressed as the ratio of the relative velocity after collision to the relative velocity before collision

$$e = \frac{(v_B)_2 - (v_A)_2}{(v_A)_1 - (v_B)_1}$$

If the collision is elastic, no energy is lost and $e = 1$. For a plastic collision $e = 0$.

If the impact is oblique, then conservation of momentum for the system and the coefficient of restitution equation apply along the line of impact. Also, conservation of momentum for each particle applies perpendicular to this line, because no impulse acts on the particles in this direction.

- *Principle of Angular Impulse and Momentum.* The moment of the linear momentum about an axis (z) is called the angular momentum. Its magnitude is

$$(H_O)_z = (d)(mv)$$

In three dimensions, the cross product is used.

$$\mathbf{H}_O = \mathbf{r} \times m\mathbf{v}$$

The principle of angular impulse and momentum is derived from taking moments of the equation of motion about an inertial axis, using $\mathbf{a} = d\mathbf{v}/dt$. The result is

$$(\mathbf{H}_O)_1 + \Sigma \int_{t_1}^{t_2} \mathbf{M}_O \, dt = (\mathbf{H}_O)_2$$

This equation is often used to eliminate unknown impulses by summing the moments about an axis through which the lines of action of these impulses produce no moment. For this reason, a free-body diagram should accompany the solution.

- *Steady Fluid Streams.* Impulse and momentum methods are often used to determine the forces that a device exerts on the mass flow of a fluid—liquid or gas. To do so, a free-body diagram of the fluid mass in contact with the device is drawn in order to identify these forces. Also, the velocity of the fluid as it flows into and out of of the device is calculated. The equations of steady flow involve summing the forces and the moments to determine these reactions. These equations are

$$\Sigma F_x = \frac{dm}{dt}(v_{Bx} - v_{Ax})$$

$$\Sigma F_y = \frac{dm}{dt}(v_{By} - v_{Ay})$$

$$\Sigma M_O = \frac{dm}{dt}(d_{OB}v_B - d_{OA}v_A)$$

- *Propulsion with Variable Mass.* Some devices, such as a rocket, lose mass as they are propelled forward. Others gain mass, such as a shovel. We can account for this mass loss or gain by applying the principle of impulse and momentum to the device. From this equation, the force exerted on the device by the mass flow can then be determined. For a mass loss, the equation is

$$\Sigma F_s = m\frac{dv}{dt} - v_{D/e}\frac{dm_e}{dt}$$

And for a mass gain, it is

$$\Sigma F_s = m\frac{dv}{dt} + v_{D/i}\frac{dm_i}{dt}$$

Kinematics and Kinetics of a Particle

The topics and problems pressented in Chapters 12 through 15 have all been *categorized* in order to provide a *clear focus* for learning the various problem-solving principles involved. In engineering practice, however, it is most important to be able to *identify* an appropriate method for the solution of a particular problem. In this regard, one must fully understand the limitations and use of the equations of dynamics, and be able to recognize which equations and principles to use for the problem's soluton. For these reasons, we will now summarize the equations and principles of particle dynamics and provide the opportunity for applying them to a variety of problems.

Kinematics. Problems in kinematics require a study only of the geometry of motion, and do not account for the forces causing the motion. When the equations of kinematics are applied, one should clearly establish a fixed origin and select an appropriate coordinate system used to define the position of the particle. Once the positive direction of each coordinate axis is established, then the directions of the components of position, velocity, and acceleration can be determined from the algebraic sign of their numerical quantities.

Rectilinear Motion.

Variable Acceleration. If a mathematical (or graphical) relationship is established between *any two* of the *four* variables s, v, a, and t, then a third variable can be determined by solving one of the following equations which relates all three variables.

$$v = \frac{ds}{dt} \qquad a = \frac{dv}{dt} \qquad a\,ds = v\,dv$$

Constant Acceleration. Be *absolutely certain* that the acceleration is constant when using the following equations:

$$s = s_0 + v_0 t + \tfrac{1}{2}a_c t^2 \qquad v = v_0 + a_c t \qquad v^2 = v_0^2 + 2a_c(s - s_0)$$

Curvilinear Motion.

x, y, z Coordinates. These coordinates are often used when the motion can be resolved into horizontal and vertical components. They are also useful for studying projectile motion since the acceleration of the projectile is *always* downward.

$$v_x = \dot{x} \qquad a_x = \dot{v}_x$$
$$v_y = \dot{y} \qquad a_y = \dot{v}_y$$
$$v_z = \dot{z} \qquad a_z = \dot{v}_z$$

n, t, b Coordinates. These coordinates are particularly advantageous for studying the particle's *acceleration* along a known path. This is because the t and n components of **a** represent the separate changes in the magnitude and direction of the velocity, respectively, and these components can be readily formulated.

$$v = \dot{s}$$
$$a_t = \dot{v} = v\frac{dv}{ds}$$
$$a_n = \frac{v^2}{\rho}$$

where

$$\rho = \left| \frac{[1 + (dy/dx)^2]^{3/2}}{d^2y/dx^2} \right|$$

when the path $y = f(x)$ is given.

r, θ, z Coordinates. These coordinates are used when data regarding the angular motion of the radial coordinate r is given to describe the particle's motion. Also, some paths of motion can conveniently be described using these coordinates.

$$v_r = \dot{r} \qquad a_r = \ddot{r} - r\dot{\theta}^2$$
$$v_\theta = r\dot{\theta} \qquad a_\theta = r\ddot{\theta} + 2\dot{r}\dot{\theta}$$
$$v_z = \dot{z} \qquad a_z = \ddot{z}$$

Relative Motion. If the origin of a *translating* coordinate system is established at particle A, then for particle B,

$$\mathbf{r}_B = \mathbf{r}_A + \mathbf{r}_{B/A}$$
$$\mathbf{v}_B = \mathbf{v}_A + \mathbf{v}_{B/A}$$
$$\mathbf{a}_B = \mathbf{a}_A + \mathbf{a}_{B/A}$$

Here the relative motion is measured by an observer fixed in the translating coordinate system.

Kinetics. Problems in kinetics involve the analysis of forces which cause the motion. When applying the equations of kinetics, it is absolutely necessary that measurements of the motion be made from an *inertial coordinate system*, i.e., one that does not rotate and is either fixed or translates with constant velocity. If a problem requires *simultaneous solution* of the equations of kinetics and kinematics, then it is important that the coordinate systems selected for writing each of the equations define the *positive directions* of the axes in the *same* manner.

Equations of Motion. These equations are used to solve for the particle's acceleration or the forces causing the motion. If they are used to determine a particle's position, velocity, or time of motion, then kinematics will also have to be considered in the solution. Before applying the equations of motion, *always draw a free-body diagram* to identify all the forces acting on the particle. Also, establish the direction of the particle's acceleration or its components. (A kinetic diagram may accompany the solution in order to graphically account for the $m\mathbf{a}$ vector.)

$$\Sigma F_x = ma_x \qquad \Sigma F_n = ma_n \qquad \Sigma F_r = ma_r$$
$$\Sigma F_y = ma_y \qquad \Sigma F_t = ma_t \qquad \Sigma F_\theta = ma_\theta$$
$$\Sigma F_z = ma_z \qquad \Sigma F_b = 0 \qquad \Sigma F_z = ma_z$$

Work and Energy. The equation of work and energy represents an integrated form of the tangential equation of motion, $\Sigma F_t = ma_t$, combined with kinematics ($a_t\, ds = v\, dv$). *It is used to solve problems involving force, velocity, and displacement.* Before applying this equation, *always draw a free-body diagram* in order to identify the forces which do work on the particle.

$$T_1 + \Sigma U_{1-2} = T_2$$

where

$$T = \tfrac{1}{2}mv^2 \qquad \text{(kinetic energy)}$$

$$U_F = \int_{s_1}^{s_2} F \cos\theta\, ds \qquad \text{(work of a variable force)}$$

$$U_{F_c} = F_c \cos\theta\,(s_2 - s_1) \qquad \text{(work of a constant force)}$$

$$U_W = -W\,\Delta y \qquad \text{(work of a weight)}$$

$$U_s = -(\tfrac{1}{2}ks_2^2 - \tfrac{1}{2}ks_1^2) \qquad \text{(work of an elastic spring)}$$

If the forces acting on the particle are *conservative forces*, i.e., those that *do not* cause a dissipation of energy, such as friction, then apply the conservation of energy equation. This equation is easier to use than the equation of work and energy since it applies only at *two points* on the path and *does not* require calculation of the work done by a force as the particle moves along the path.

$$T_1 + V_1 = T_2 + V_2$$

where

$$V_g = Wy \qquad \text{(gravitational potential energy)}$$

$$V_e = \tfrac{1}{2} k s^2 \qquad \text{(elastic potential energy)}$$

If the *power* developed by a force is to be determined, use

$$P = \frac{dU}{dt} = \mathbf{F} \cdot \mathbf{v}$$

where $\mathbf{v}$ is the velocity of a particle acted upon by the force $\mathbf{F}$.

Impulse and Momentum. The equation of *linear impulse and momentum* is an integrated form of the equation of motion, $\Sigma \mathbf{F} = m\mathbf{a}$, combined with kinematics ($\mathbf{a} = d\mathbf{v}/dt$). *It is used to solve problems involving force, velocity, and time.* Before applying this equation, one should *always draw the free-body diagram*, in order to identify all the forces that cause impulses on the particle. From the diagram the impulsive and nonimpulsive forces should be identified. Recall that the nonimpulsive forces can be neglected in the analysis during the time of impact. Also, establish the direction of the particle's velocity just before and just after the impulses are applied. As an alternative procedure, impulse and momentum diagrams may accompany the solution in order to graphically account for the terms in the equation.

$$m\mathbf{v}_1 + \Sigma \int_{t_1}^{t_2} \mathbf{F}\, dt = m\mathbf{v}_2$$

If several particles are involved in the problem, consider applying the *conservation of momentum* to the system in order to eliminate the internal impulses from the analysis. This can be done in a specified direction, provided no external impulses act on the particles in that direction.

$$\Sigma m\mathbf{v}_1 = \Sigma m\mathbf{v}_2$$

If the problem involves impact and the coefficient of restitution e is given, then apply the following equation.

$$e = \frac{(v_B)_2 - (v_A)_2}{(v_A)_1 - (v_B)_1} \qquad \text{(along line of impact)}$$

Remember that during impact the principle of work and energy cannot be used, since the particles deform and therefore the work due to the internal forces will be unknown. The principle of work and energy can be used, however, to determine the energy loss during the collision once the particle's initial and final velocities are determined.

The *principle of angular impulse and momentum* and the *conservation of angular momentum* may be applied about an axis in order to *eliminate* some of the unknown impulses acting on the particle during the time period when its motion is studied. Investigation of the particle's free-body diagram (or the impulse diagram) will aid in choosing the axis for application.

$$(\mathbf{H}_O)_1 + \Sigma \int_{t_1}^{t_2} \mathbf{M}_O \, dt = (\mathbf{H}_O)_2$$

$$(\mathbf{H}_O)_1 = (\mathbf{H}_O)_2$$

The following problems provide an opportunity for applying the above concepts. They are presented in *random order* so that practice may be gained in identifying the various types of problems and developing the skills necessary for their solution.

REVIEW PROBLEMS

R1-1. A sports car can accelerate at 6 m/s² and decelerate at 8 m/s². If the maximum speed it can attain is 60 m/s, determine the shortest time it takes to travel 900 m starting from rest and then stopping when s = 900 m.

R1-2. A 2-kg particle rests on a smooth horizontal plane and is acted upon by forces $F_x = 0$ and $F_y = 3$ N. If $x = 0$, $y = 0$, $v_x = 6$ m/s, and $v_y = 2$ m/s when $t = 0$, determine the equation $y = f(x)$ which describes the path.

R1-3. Determine the velocity of each block 2 s after the blocks are released from rest. Neglect the mass of the pulleys and cord.

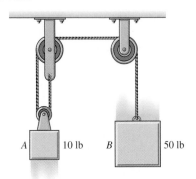

A   10 lb   B   50 lb

Prob. R1–3

*R1-4. To test the manufactured properties of 2-lb steel balls, each ball is released from rest as shown and strikes a 45° inclined surface. If the coefficient of restitution is to be e = 0.8, determine the distance s to where the ball must strike the horizontal plane at A. At what speed does the ball strike A?

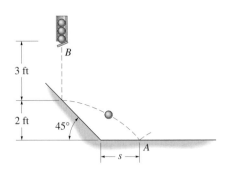

3 ft

2 ft

45°

B

A

s

Prob. R1–4

R1-5. The 90-lb force is required to drag the 200-lb block 60 ft up the *rough* inclined plane at constant velocity. If the force is removed when the block reaches point B, and the block is then released from rest, determine the block's velocity when it slides back down the plane and reaches point A.

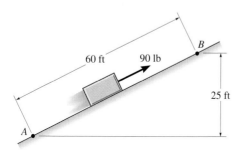

Prob. R1–5

R1-6. The motor at C pulls in the cable with an acceleration $a_c = (3t^2)$ m/s², where t is in seconds. The motor at D draws in its cable at $a_D = 5$ m/s². If both motors start at the same instant from rest when $d = 3$ m, determine (a) the time needed for $d = 0$, and (b) the relative velocity of block A with respect to block B when this occurs.

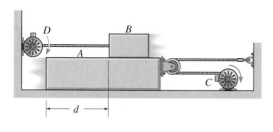

Prob. R1–6

R1-7. A spring having a stiffness of 5 kN/m is compressed 400 mm. The stored energy in the spring is used to drive a machine which requires 80 W of power. Determine how long the spring can supply energy at the required rate.

*R1-8.** The baggage truck A has a mass of 800 kg and is used to pull each of the 300-kg cars. Determine the tension in the couplings at B and C if the tractive force F on the truck is $F = 480$ N. What is the speed of the truck when $t = 2$ s, starting from the rest? The car wheels are free to roll. Neglect the mass of the wheels.

Prob. R1–8

R1-9. The baggage truck A has a mass of 800 kg and is used to pull each of the 300-kg cars. If the tractive force F on the truck is $F = 480$ N, determine the initial acceleration of the truck. What is the acceleration of the truck if the coupling at C suddenly fails? The car wheels are free to roll. Neglect the mass of the wheels.

Prob. R1–9

R1-10. A 2-kg particle rests on a smooth horizontal plane and is acted upon by forces $F_x = (8x)$ N, where x is in meters, and $F_y = 0$. If $x = 0$, $y = 0$, $v_x = 4$ m/s, and $v_y = 6$ m/s when $t = 0$, determine the equation $y = f(x)$ which describes the path.

R1-11. Determine the speed of block B if the end of the cable at C is pulled downward with a speed of 10 ft/s. What is the relative velocity of the block with respect to C.

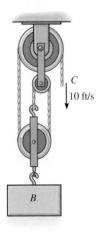

Prob. R1–11

*■**R1-12.** Packages having a mass of 2.5 kg ride on the surface of the conveyor belt. If the belt starts from rest and with constant acceleration increases to a speed of 0.75 m/s in 2 s, determine the maximum angle of tilt, θ, so that none of the packages slip on the inclined surface AB of the belt. The coefficient of static friction between the belt and each package is $\mu_s = 0.3$. At what angle ϕ do the packages first begin to slip off the surface of the belt if the belt is moving at a constant speed of 0.75 m/s?

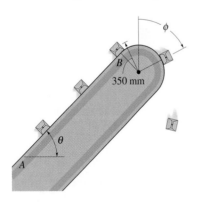

Prob. R1–12

R1-13. A projectile, initially at the origin, moves along a straight-line path through a fluid medium such that its velocity is $v = 1800(1 - e^{-0.3t})$ mm/s, where t is in seconds. Determine the displacement of the projectile during the first 3 s.

R1-14. The speed of a train during the first minute of its motion has been recorded as follows:

t (s)	0	20	40	60
v (m/s)	0	16	21	24

Plot the v–t graph, approximating the curve as straight line segments between the given points. Determine the total distance traveled.

R1-15. A train car, having a mass of 25 Mg, travels up a $10°$ incline with a constant speed of 80 km/h. Determine the power required to overcome the force of gravity.

***R1-16.** The slotted arm AB drives the pin C through the spiral groove described by the equation $r = (1.5\theta)$ ft, where θ is in radians. If the arm starts from rest when $\theta = 60°$ and is driven at an angular rate of $\dot{\theta} = (4t)$ rad/s, where t is in seconds, determine the radial and transverse components of velocity and acceleration of the pin when $t = 1$ s.

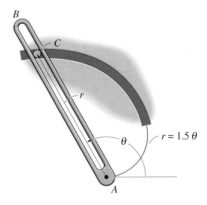

Prob. R1–16

R1-17. The chain has a mass of 3 kg/m. If the coefficient of kinetic friction between the chain and the plane is $\mu_k = 0.2$, determine the velocity at which the end A will pass point B when the chain is released from rest.

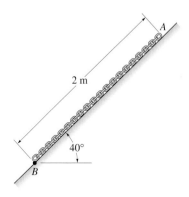

Prob. R1–17

R1-19. The collar of negligible size has a mass of 0.25 kg and is attached to a spring having an unstretched length of 100 mm. If the collar is released from rest at A and travels along the smooth guide, determine its speed just before it strikes B.

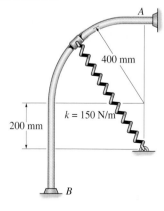

Prob. R1–19

R1-18. The 6-lb ball is fired from a tube by a spring having a stiffness $k = 20$ lb/in. Determine how far the spring must be compressed to fire the ball from the compressed position to a height of 8 ft, at which point it has a velocity of 6 ft/s.

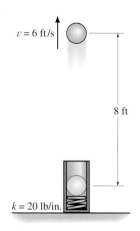

Prob. R1–18

***R1-20.** A crate has a weight of 1500 lb. If it is pulled along the ground at a constant speed for a distance of 20 ft, and the towing cable makes an angle of 15° with the horizontal, determine the tension in the cable and the work done by the towing force. The coefficient of kinetic friction between the crate and the ground is $\mu_k = 0.55$.

R1-21. Disk A weighs 2 lb and is sliding on a smooth horizontal plane with a velocity of 3 ft/s. Disk B weighs 11 lb and is initially at rest. If after the impact A has a velocity of 1 ft/s directed along the positive x axis, determine the velocity of B after impact. How much kinetic energy is lost in the collision?

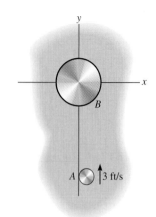

Prob. R1–21

R1-22. A particle is moving along a circular path of 2-m radius such that its position as a function of time is given by $\theta = (5t^2)$ rad, where t is in seconds. Determine the magnitude of the particle's acceleration when $\theta = 30°$. The particle starts from rest when $\theta = 0°$.

R1-23. If the end of the cable at A is pulled down with a speed of 2 m/s, determine the speed at which block B rises.

R1-26. The 20-lb block B rests on the surface of a table for which the coefficient of kinetic friction is $\mu_k = 0.1$. Determine the speed of the 10-lb block A after it has moved downward 2 ft from rest. Neglect the mass of the pulleys and cords.

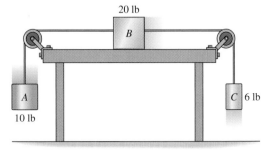

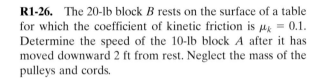

Prob. R1–26

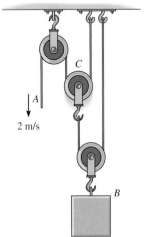

Prob. R1–23

*R1-24. A rifle has a mass of 2.5 kg. If it is loosely gripped and a 1.5-g bullet is fired from it with a horizontal muzzle velocity of 1400 m/s, determine the recoil velocity of the rifle just after firing.

R1-25. The drinking fountain is designed such that the nozzle is located from the edge of the basin as shown. Determine the maximum and minimum speed at which water can be ejected from the nozzle so that it does not splash over the sides of the basin at B and C.

R1-27. The 5-lb ball, attached to the cord, is struck by the boy. Determine the smallest speed he must impart to the ball so that it will swing around in a vertical circle, without causing the cord to become slack.

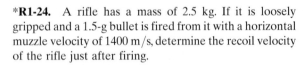

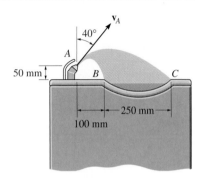

Prob. R1–25

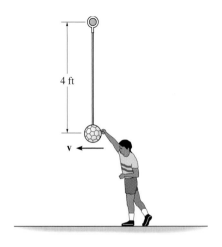

Prob. R1–27

*R1-28. Two winding drum D is drawing in the cable at an accelerated rate of $5\,\text{m/s}^2$. Determine the cable tension if the suspended crate has a mass of 800 kg.

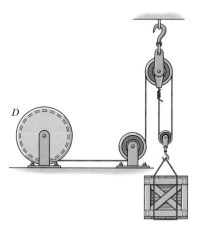

Prob. R1–28

R1-29. The particle P travels with a constant speed of 300 mm/s along the curve. Determine its acceleration when it is located at point (200 mm, 100 mm).

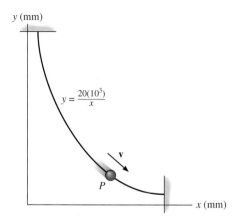

y (mm)

$$y = \frac{20(10^3)}{x}$$

x (mm)

Prob. R1–29

R1-30. The block has a mass of 0.5 kg and moves within the smooth vertical slot. If the block starts from rest when the *attached* spring is in the unstretched position at A, determine the *constant* vertical force F which must be applied to the cord so that the block attains a speed $v_B = 2.5$ m/s when it reaches B; $s_B = 0.15$ m. Neglect the mass of the cord and pulley.

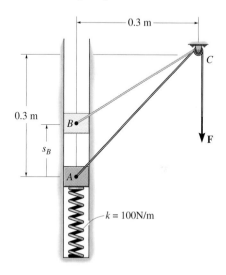

0.3 m

0.3 m

B

s_B

A

C

F

$k = 100\text{N/m}$

Prob. R1–30

R1-31. The rocket sled has a mass of 4 Mg and travels from rest along the smooth horizontal track such that it maintains a constant power output of 450 kW. Neglect the loss of fuel mass and air resistance, and determine how far it must travel to reach a speed of $v = 60$ m/s.

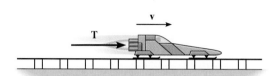

v

T

Prob. R1–31

*R1-32. The spool, which has a mass of 4 kg, slides along the rotating rod. At the instant shown, the angular rate of rotation of the rod is $\dot{\theta} = 6$ rad/s and this rotation is increasing at $\ddot{\theta} = 2$ rad/s². At this same instant, the spool has a velocity of 3 m/s and an acceleration of 1 m/s², both measured relative to the rod and directed away from the center O when $r = 0.5$ m. Determine the radial frictional force and the normal force, both exerted by the rod on the spool at this instant.

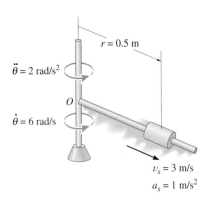

Prob. R1–32

R1-33. A skier starts from rest at A (30 ft, 0) and descends the smooth slope, which may be approximated by a parabola. If she has a weight of 120 lb, determine the normal force she exerts on the ground at the instant she arrives at point B.

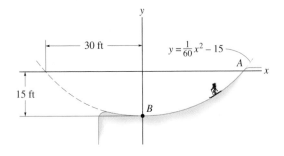

Prob. R1–33

R1-34. The small 2-lb collar starting from rest at A slides down along the smooth rod. During the motion, the collar is acted upon by a force $\mathbf{F} = \{10\mathbf{i} + 6y\mathbf{j} + 2z\mathbf{k}\}$ lb, where x, y, z are in feet. Determine the collar's speed when it strikes the wall at B.

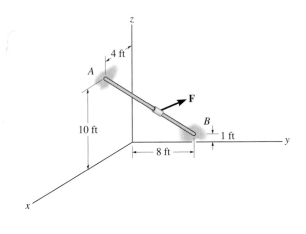

Prob. R1–34

R1-35. A ball having a mass of 200 g is released from rest at a height of 400 mm above a very large fixed metal surface. If the ball rebounds to a height of 325 mm above the surface, determine the coefficient of restitution between the ball and the surface.

*R1-36. Packages having a mass of 6 kg slide down a smooth chute and land horizontally with a speed of 3 m/s on the surface of a conveyor belt. If the coefficient of kinetic friction between the belt and a package is $\mu_k = 0.2$, determine the time needed to bring the package to rest on the belt if the belt is moving in the same direction as the package with a speed $v = 1$ m/s.

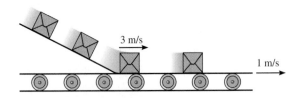

Prob. R1–36

R1-37. The blocks A and B weigh 10 and 30 lb, respectively. They are connected together by a light cord and ride in the frictionless grooves. Determine the speed of each block after block A moves 6 ft up along the plane. The blocks are released from rest.

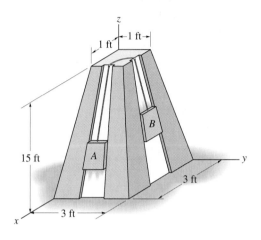

Prob. R1–37

R1-38. The motor M pulls in its attached rope with an acceleration $a_p = 6$ m/s². Determine the towing force exerted by M on the rope in order to move the 50-kg crate up the inclined plane. The coefficient of kinetic friction between the crate and the plane is $\mu_k = 0.3$. Neglect the mass of the pulleys and rope.

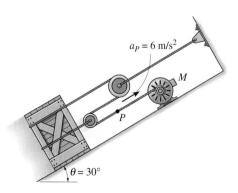

Prob. R1–38

R1-39. If a particle has an initial velocity $v_0 = 12$ ft/s to the right, and a constant acceleration of 2 ft/s² to the left, determine the particle's displacement in 10 s. Originally $s_0 = 0$.

*****R1-40.** A 3-lb block, initially at rest at point A, slides along the smooth parabolic surface. Determine the normal force acting on the block when it reaches B. Neglect the size of the block.

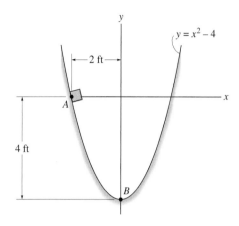

Prob. R1–40

R1-41. At a given instant the 10-lb block A is moving downward with a speed of 6 ft/s. Determine its speed 2 s later. Block B has a weight of 4 lb, and the coefficient of kinetic friction between it and the horizontal plane is $\mu_k = 0.2$. Neglect the mass of the pulleys and cord.

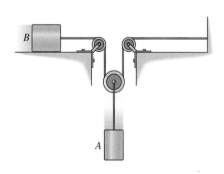

Prob. R1–41

R1-42. A freight train starts from rest and travels with a constant acceleration of $0.5 \, \text{ft/s}^2$. After a time t' it maintains a constant speed so that when $t = 160 \, \text{s}$ it has traveled 2000 ft. Determine the time t' and draw the v–t graph for the motion.

R1-43. The crate, having a weight of 50 lb, is hoisted by the pulley system and motor M. If the crate starts from rest and, by constant acceleration, attains a speed of 12 ft/s after rising 10 ft, determine the power that must be supplied to the motor at the instant $s = 10 \, \text{ft}$. The motor has an efficiency $\epsilon = 0.74$.

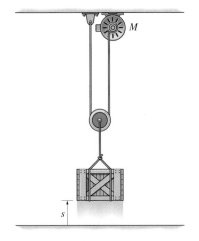

Prob. R1–43

*R1-44. An automobile is traveling with a *constant speed* along a horizontal circular curve that has a radius $\rho = 750 \, \text{ft}$. If the magnitude of acceleration is $a = 8 \, \text{ft/s}^2$, determine the speed at which the automobile is traveling.

R1-45. Block B rests on a smooth surface. If the coefficients of friction between A and B are $\mu_s = 0.4$ and $\mu_k = 0.3$, determine the acceleration of each block if (a) $F = 6 \, \text{lb}$, and (b) $F = 50 \, \text{lb}$.

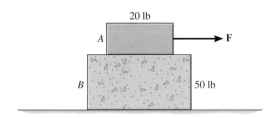

Prob. R1–45

R1-46. The 100-kg crate is subjected to the action of two forces, $F_1 = 800 \, \text{N}$ and $F_2 = 1.5 \, \text{kN}$, as shown. If it is originally at rest, determine the distance it slides in order to attain a speed of 6 m/s. The coefficient of kinetic friction between the crate and the surface is $\mu_k = 0.2$.

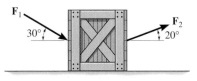

Prob. R1–46

R1-47. A 20-kg block is originally at rest on a horizontal surface for which the coefficient of static friction is $\mu_s = 0.6$ and the coefficient of kinetic friction is $\mu_k = 0.5$. If a horizontal force F is applied such that it varies with time as shown, determine the speed of the block in 10 s. *Hint:* First determine the time needed to overcome friction and start the block moving.

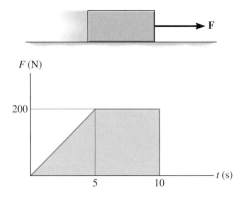

Prob. R1–47

*R1-48. Two smooth billiard balls A and B have an equal mass of $m = 200$ g. If A strikes B with a velocity of $(v_A)_1 = 2$ m/s as shown, determine their final velocities just after collision. Ball B is originally at rest and the coefficient of restitution is $e = 0.75$.

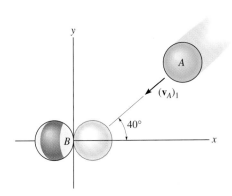

Prob. R1–48

R1-49. If a 150-lb crate is released from rest at A, determine its speed after it slides 30 ft down the plane. The coefficient of kinetic friction between the crate and plane is $\mu_k = 0.3$.

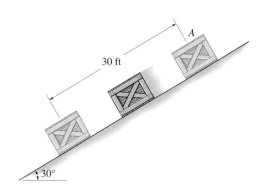

Prob. R1–49

R1-50. Determine the tension developed in the two cords and the acceleration of each block. Neglect the mass of the pulleys and cords. *Hint:* Since the system consists of *two* cords, relate the motion of block A to C, and of block B to C. Then, by elimination, relate the motion of A to B.

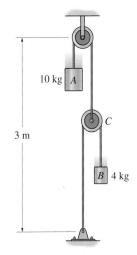

Prob. R1–50

R1-51. The bottle rests at a distance of 3 ft from the center of the horizontal platform. If the coefficient of static friction between the bottle and the platform is $\mu_s = 0.3$, determine the maximum speed that the bottle can attain before slipping. Assume the angular motion of the platform is slowly increasing.

*R1-52. Work Prob. R1–51 assuming that the platform starts rotating from rest so that the speed of the bottle is increased at 2 ft/s^2.

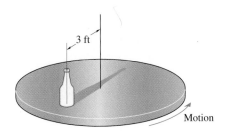

Probs. R1–51/R1–52

The wind turbine rotates about a fixed axis with variable angular motion.

Planar Kinematics of a Rigid Body

- To classify the various types of rigid-body planar motion.
- To investigate rigid-body translation and show how to analyze motion about a fixed axis.
- To study planar motion using an absolute motion analysis.
- To provide a relative motion analysis of velocity and acceleration using a translating frame of reference.
- To show how to find the instantaneous center of zero velocity and determine the velocity of a point on a body using this method.
- To provide a relative motion analysis of velocity and acceleration using a rotating frame of reference.

16.1 Rigid-Body Motion

In this chapter, the planar kinematics of a rigid body will be discussed. This study is important for the design of gears, cams, and mechanisms used for many mechanical operations. Furthermore, once the kinematics of a rigid body is thoroughly understood, then it will be possible to apply the equations of motion, which relate the forces on the body to the body's motion.

When all the particles of a rigid body move along paths which are equidistant from a fixed plane, the body is said to undergo *planar motion*. There are three types of rigid body planar motion; in order of increasing complexity, they are

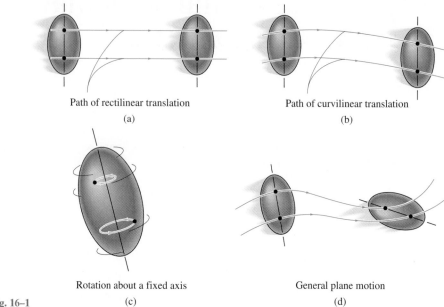

Path of rectilinear translation

(a)

Path of curvilinear translation

(b)

Rotation about a fixed axis

(c)

General plane motion

(d)

Fig. 16–1

1. *Translation.* This type of motion occurs if every line segment on the body remains parallel to its original direction during the motion. When the paths of motion for any two particles of the body are along equidistant straight lines the motion is called *rectilinear translation*, Fig. 16–1a. However, if the paths of motion are along curved lines which are equidistant the motion is called *curvilinear translation*, Fig. 16–1b.

2. *Rotation about a fixed axis.* When a rigid body rotates about a fixed axis, all the particles of the body, except those which lie on the axis of rotation, move along circular paths, Fig. 16–1c.

3. *General plane motion.* When a body is subjected to general plane motion, it undergoes a combination of translation *and* rotation, Fig. 16–1d. The translation occurs within a reference plane, and the rotation occurs about an axis perpendicular to the reference plane.

In the following sections we will consider each of these motions in detail. Examples of bodies undergoing these motions are shown in Fig. 16–2.

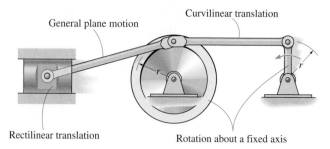

General plane motion

Curvilinear translation

Rectilinear translation

Rotation about a fixed axis

Fig. 16–2

16.2 Translation

Consider a rigid body which is subjected to either rectilinear or curvilinear translation in the x–y plane, Fig. 16–3.

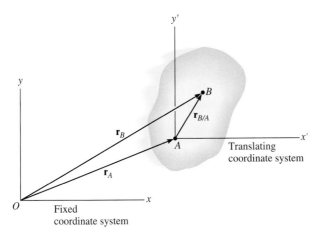

Fixed
coordinate system

Fig. 16–3

Position. The locations of points A and B in the body are defined from the fixed x, y reference frame by using *position vectors* $\mathbf{r}_A$ and $\mathbf{r}_B$. The translating x', y' coordinate system is *fixed in the body* and has its origin located at A, hereafter referred to as the *base point*. The position of B with respect to A is denoted by the *relative-position vector* $\mathbf{r}_{B/A}$ ("$\mathbf{r}$ of B with respect to A"). By vector addition,

$$\mathbf{r}_B = \mathbf{r}_A + \mathbf{r}_{B/A}$$

Velocity. A relationship between the instantaneous velocities of A and B is obtained by taking the time derivative of the position equation, which yields $\mathbf{v}_B = \mathbf{v}_A + d\mathbf{r}_{B/A}/dt$. Here $\mathbf{v}_A$ and $\mathbf{v}_B$ denote *absolute velocities* since these vectors are measured from the x, y axes. The term $d\mathbf{r}_{B/A}/dt = \mathbf{0}$, since the *magnitude* of $\mathbf{r}_{B/A}$ is *constant* by definition of a rigid body, and because the body is translating the *direction* of $\mathbf{r}_{B/A}$ is *constant*. Therefore,

$$\mathbf{v}_B = \mathbf{v}_A$$

Acceleration. Taking the time derivative of the velocity equation yields a similar relationship between the instantaneous accelerations of A and B:

$$\mathbf{a}_B = \mathbf{a}_A$$

The above two equations indicate that *all points in a rigid body subjected to either rectilinear or curvilinear translation move with the same velocity and acceleration*. As a result, the kinematics of particle motion, discussed in Chapter 12, may also be used to specify the kinematics of points located in a translating rigid body.

Passengers on this amusement ride are subjected to curvilinear translation since the vehicle moves in a circular path yet it always remains in an upright position.

16.3 Rotation About a Fixed Axis

When a body is rotating about a fixed axis, any point P located in the body travels along a *circular path*. To study this motion it is first necessary to discuss the angular motion of the body about the axis.

Angular Motion. A point is without dimension, and so it has no angular motion. *Only lines or bodies undergo angular motion.* For example, consider the body shown in Fig. 16–4a and the angular motion of a radial line r located within the shaded plane and directed from point O on the axis of rotation to point P.

Angular Position. At the instant shown, the *angular position* of r is defined by the angle θ, measured between a *fixed* reference line and r.

Angular Displacement. The change in the angular position, which can be measured as a differential $d\theta$, is called the *angular displacement*.* This vector has a *magnitude* of $d\theta$, measured in degrees, radians, or revolutions, where 1 rev = 2π rad. Since motion is about a *fixed axis*, the direction of $d\theta$ is *always* along the axis. Specifically, the *direction* is determined by the right-hand rule; that is, the fingers of the right hand are curled with the sense of rotation, so that in this case the thumb, or $d\theta$, points upward, Fig. 16–4a. In two dimensions, as shown by the top view of the shaded plane, Fig. 16–4b, both θ and $d\theta$ are directed counterclockwise, and so the thumb points outward from the page.

Angular Velocity. The time rate of change in the angular position is called the *angular velocity* $\boldsymbol{\omega}$ (omega). Since $d\boldsymbol{\theta}$ occurs during an instant of time dt, then,

$$(\curvearrowright+) \qquad \boxed{\omega = \frac{d\theta}{dt}} \qquad (16\text{–}1)$$

This vector has a *magnitude* which is often measured in rad/s. It is expressed here in scalar form since its *direction* is always along the axis of rotation, i.e., in the same direction as $d\boldsymbol{\theta}$, Fig. 16–4a. When indicating the angular motion in the shaded plane, Fig. 16–4b, we can refer to the sense of rotation as clockwise or counterclockwise. Here we have *arbitrarily* chosen counterclockwise rotations as *positive* and indicated this by the curl shown in parentheses next to Eq. 16–1. Realize, however, that the directional sense of $\boldsymbol{\omega}$ is actually outward from the page.

(a)

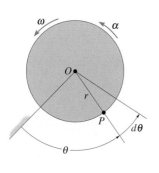

(b)

Fig. 16–4

---

*It is shown in Sec. 20.1 that finite rotations or finite angular displacements are *not* vector quantities, although differential rotations $d\boldsymbol{\theta}$ are vectors.

Angular Acceleration. The *angular acceleration* α (alpha) measures the time rate of change of the angular velocity. The *magnitude* of this vector may be written as

$(\downarrow+)$
$$\alpha = \frac{d\omega}{dt}$$
(16–2)

Using Eq. 16–1, it is possible to express α as

$(\downarrow+)$
$$\alpha = \frac{d^2\theta}{dt^2}$$
(16–3)

The line of action of $\boldsymbol{\alpha}$ is the same as that for $\boldsymbol{\omega}$, Fig. 16–4a; however, its sense of *direction* depends on whether $\boldsymbol{\omega}$ is increasing or decreasing. In particular, if $\boldsymbol{\omega}$ is decreasing, then $\boldsymbol{\alpha}$ is called an *angular deceleration* and it therefore has a sense of direction which is opposite to $\boldsymbol{\omega}$.

By eliminating dt from Eqs. 16–1 and 16–2, we obtain a differential relation between the angular acceleration, angular velocity, and angular displacement, namely,

$(\downarrow+)$
$$\alpha \, d\theta = \omega \, d\omega$$
(16–4)

The similarity between the differential relations for angular motion and those developed for rectilinear motion of a particle ($v = ds/dt$, $a = dv/dt$, and $a \, ds = v \, dv$) should be apparent.

Constant Angular Acceleration. If the angular acceleration of the body is constant, $\boldsymbol{\alpha} = \boldsymbol{\alpha}_c$, then Eqs. 16–1, 16–2, and 16–4, when integrated, yield a set of formulas which relate the body's angular velocity, angular position, and time. These equations are similar to Eqs. 12–4 to 12–6 used for rectilinear motion. The results are

$(\downarrow+)$
$(\downarrow+)$
$(\downarrow+)$

$$\omega = \omega_0 + \alpha_c t \qquad (16\text{–}5)$$
$$\theta = \theta_0 + \omega_0 t + \tfrac{1}{2}\alpha_c t^2 \qquad (16\text{–}6)$$
$$\omega^2 = \omega_0^2 + 2\alpha_c(\theta - \theta_0) \qquad (16\text{–}7)$$
Constant Angular Acceleration

Here θ_0 and ω_0 are the initial values of the body's angular position and angular velocity, respectively.

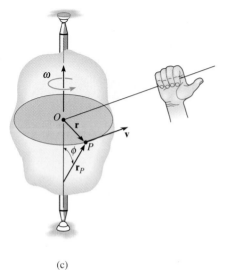

(c)

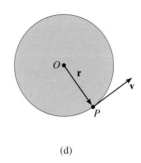

(d)

Fig. 16–4

Motion of Point P. As the rigid body in Fig. 16–4c rotates, point P travels along a *circular path* of radius r and center at point O. This path is contained within the shaded plane shown in top view, Fig. 16–4d.

Position. The position of P is defined by the position vector $\mathbf{r}$, which extends from O to P.

Velocity. The velocity of P has a magnitude which can be found from its polar coordinate components $v_r = \dot{r}$ and $v_\theta = r\dot{\theta}$, Eqs. 12–25. Since r is constant, the radial component $v_r = \dot{r} = 0$, and so $v = v_\theta = r\dot{\theta}$. Because $\omega = \dot{\theta}$, Eq. 16–1, the velocity is

$$v = \omega r \qquad (16\text{–}8)$$

As shown in Figs. 16–4c and 16–4d, the *direction* of $\mathbf{v}$ is *tangent* to the circular path.

Both the magnitude and direction of $\mathbf{v}$ can also be accounted for by using the cross product of $\boldsymbol{\omega}$ and $\mathbf{r}_P$ (see Appendix C). Here, $\mathbf{r}_P$ is directed from *any point* on the axis of rotation to point P, Fig. 16–4c. We have

$$\mathbf{v} = \boldsymbol{\omega} \times \mathbf{r}_P \qquad (16\text{–}9)$$

The order of the vectors in this formulation is important, since the cross product is not commutative, i.e., $\boldsymbol{\omega} \times \mathbf{r}_P \neq \mathbf{r}_P \times \boldsymbol{\omega}$. In this regard, notice in Fig. 16–4c how the correct direction of $\mathbf{v}$ is established by the right-hand rule. The fingers of the right hand are curled from $\boldsymbol{\omega}$ toward $\mathbf{r}_P$ ($\boldsymbol{\omega}$ "cross" $\mathbf{r}_P$). The thumb indicates the correct direction of $\mathbf{v}$, which is tangent to the path in the direction of motion. From Eq. C–8, the magnitude of $\mathbf{v}$ in Eq. 16–9 is $v = \omega r_P \sin\phi$, and since $r = r_P \sin\phi$, Fig. 16–4c, then $v = \omega r$, which agrees with Eq. 16–8. As a special case, the position vector $\mathbf{r}$ can be chosen for $\mathbf{r}_P$. Here $\mathbf{r}$ lies in the plane of motion and again the velocity of point P is

$$\mathbf{v} = \boldsymbol{\omega} \times \mathbf{r} \qquad (16\text{–}10)$$

Acceleration. The acceleration of P can be expressed in terms of its normal and tangential components.* Since $a_t = dv/dt$ and $a_n = v^2/\rho$, where $\rho = r$, $v = \omega r$, and $\alpha = d\omega/dt$, we have

$$a_t = \alpha r \qquad (16\text{--}11)$$

$$a_n = \omega^2 r \qquad (16\text{--}12)$$

The *tangential component of acceleration*, Figs. 16–4e and 16–4f, represents the time rate of change in the velocity's magnitude. If the speed of P is increasing, then $\mathbf{a}_t$ acts in the same direction as $\mathbf{v}$; if the speed is decreasing, $\mathbf{a}_t$ acts in the opposite direction of $\mathbf{v}$; and finally, if the speed is constant, $\mathbf{a}_t$ is zero.

The *normal component of acceleration* represents the time rate of change in the velocity's direction. The *direction* of $\mathbf{a}_n$ is always toward O, the center of the circular path, Figs. 16–4e and 16–4f.

Like the velocity, the acceleration of point P may be expressed in terms of the vector cross product. Taking the time derivative of Eq. 16–9 we have

$$\mathbf{a} = \frac{d\mathbf{v}}{dt} = \frac{d\boldsymbol{\omega}}{dt} \times \mathbf{r}_P + \boldsymbol{\omega} \times \frac{d\mathbf{r}_P}{dt}$$

Recalling that $\boldsymbol{\alpha} = d\boldsymbol{\omega}/dt$, and using Eq. 16–9 ($d\mathbf{r}_P/dt = \mathbf{v} = \boldsymbol{\omega} \times \mathbf{r}_P$), yields

$$\mathbf{a} = \boldsymbol{\alpha} \times \mathbf{r}_P + \boldsymbol{\omega} \times (\boldsymbol{\omega} \times \mathbf{r}_P) \qquad (16\text{--}13)$$

From the definition of the cross product, the first term on the right has a magnitude $a_t = \alpha r_P \sin \phi = \alpha r$, and by the right-hand rule, $\boldsymbol{\alpha} \times \mathbf{r}_P$ is in the direction of $\mathbf{a}_t$, Fig. 16–4e. Likewise, the second term has a magnitude $a_n = \omega^2 r_P \sin \phi = \omega^2 r$, and applying the right-hand rule twice, first to determine the result $\mathbf{v}_P = \boldsymbol{\omega} \times \mathbf{r}_P$ then $\boldsymbol{\omega} \times \mathbf{v}_P$, it can be seen that this result is in the same direction as $\mathbf{a}_n$, shown in Fig. 16–4e. Noting that this is also the *same* direction as $-\mathbf{r}$, which lies in the plane of motion, we can express $\mathbf{a}_n$ in a much simpler form as $\mathbf{a}_n = -\omega^2\mathbf{r}$. Hence, Eq. 16–12 can be identified by its two components as

$$\boxed{\begin{aligned} \mathbf{a} &= \mathbf{a}_t + \mathbf{a}_n \\ &= \boldsymbol{\alpha} \times \mathbf{r} - \omega^2 \mathbf{r} \end{aligned}} \qquad (16\text{--}14)$$

Since $\mathbf{a}_t$ and $\mathbf{a}_n$ are perpendicular to one another, if needed the magnitude of acceleration can be determined from the Pythagorean theorem; namely, $a = \sqrt{a_n^2 + a_t^2}$, Fig. 16–4f.

*Polar coordinates can also be used. Since $a_r = \ddot{r} - r\dot{\theta}^2$ and $a_\theta = r\ddot{\theta} + 2\dot{r}\dot{\theta}$, substituting $\dot{r} = \ddot{r} = 0$, $\dot{\theta} = \omega$, $\ddot{\theta} = \alpha$, we obtain Eqs. 16–11 and 16–12.

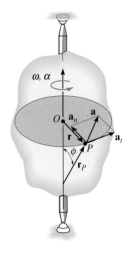

(e)

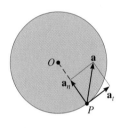

(f)

Fig. 16–4

The many gears used in the operation of a crane all rotate about fixed axes. Engineers must be able to relate their angular motions in order to properly design this gear system.

IMPORTANT POINTS

- A body can undergo two types of translation. During rectilinear translation all points follow parallel straight-line paths, and during curvilinear translation the points follow curved paths that are the same shape and are equidistant from one another.

- All the points on a translating body move with the same velocity and acceleration.

- Points located on a body that rotates about a fixed axis follow circular paths.

- The relationship $\alpha \, d\theta = \omega \, d\omega$ is derived from $\alpha = d\omega/dt$ and $\omega = d\theta/dt$ by eliminating dt.

- Once the angular motions ω and α are known, the velocity and acceleration of any point on the body can be determined.

- The velocity always acts tangent to the path of motion.

- The acceleration has two components. The tangential acceleration measures the rate of change in the magnitude of the velocity and can be determined using $a_t = \alpha r$. The normal acceleration measures the rate of change in the direction of the velocity and can be determined from $a_n = \omega^2 r$.

PROCEDURE FOR ANALYSIS

The velocity and acceleration of a point located on a rigid body that is rotating about a fixed axis can be determined using the following procedure.

Angular Motion.

- Establish the positive sense of direction along the axis of rotation and show it alongside each kinematic equation as it is applied.
- If a relationship is known between any *two* of the four variables α, ω, θ, and t, then a third variable can be obtained by using one of the following kinematic equations which relates all three variables.

$$\omega = \frac{d\theta}{dt} \quad \alpha = \frac{d\omega}{dt} \quad \alpha\, d\theta = \omega\, d\omega$$

- If the body's angular acceleration is *constant*, then the following equations can be used:

$$\omega = \omega_0 + \alpha_c t$$
$$\theta = \theta_0 + \omega_0 t + \tfrac{1}{2}\alpha_c t^2$$
$$\omega^2 = \omega_0^2 + 2\alpha_c(\theta - \theta_0)$$

- Once the solution is obtained, the sense of θ, ω, and α is determined from the algebraic signs of their numerical quantities.

Motion of P.

- In most cases the velocity of P and its two components of acceleration can be determined from the scalar equations

$$v = \omega r$$
$$a_t = \alpha r$$
$$a_n = \omega^2 r$$

- If the geometry of the problem is difficult to visualize, the following vector equations should be used:

$$\mathbf{v} = \boldsymbol{\omega} \times \mathbf{r}_P = \boldsymbol{\omega} \times \mathbf{r}$$
$$\mathbf{a}_t = \boldsymbol{\alpha} \times \mathbf{r}_P = \boldsymbol{\alpha} \times \mathbf{r}$$
$$\mathbf{a}_n = \boldsymbol{\omega} \times (\boldsymbol{\omega} \times \mathbf{r}_P) = -\omega^2 \mathbf{r}$$

Here $\mathbf{r}_P$ is directed from any point on the axis of rotation to point P, whereas $\mathbf{r}$ lies in the plane of motion of P. Either of these vectors, along with $\boldsymbol{\omega}$ and $\boldsymbol{\alpha}$, should be expressed in terms of its $\mathbf{i}$, $\mathbf{j}$, $\mathbf{k}$ components, and, if necessary, the cross products determined by using a determinant expansion (see Eq. C–12).

E X A M P L E   **16.1**

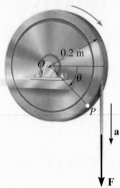

Fig. 16–5

A cord is wrapped around a wheel which is initially at rest as shown in Fig. 16–5. If a force is applied to the cord and gives it an acceleration $a = (4t)$ m/s^2, where t is in seconds, determine as a function of time (a) the angular velocity of the wheel, and (b) the angular position of line OP in radians.

Solution

Part (a).   The wheel is subjected to rotation about a fixed axis passing through point O. Thus, point P on the wheel has motion about a circular path, and the acceleration of this point has *both* tangential and normal components. The tangential component is $(a_P)_t = (4t)$ m/s^2, since the cord is wrapped around the wheel and moves *tangent* to it. Hence the angular acceleration of the wheel is

$$(\curvearrowright +) \qquad\qquad (a_P)_t = \alpha r$$
$$(4t) \text{ m/s}^2 = \alpha(0.2 \text{ m})$$
$$\alpha = 20t \text{ rad/s}^2 \;\downarrow$$

Using this result, the wheel's angular velocity ω can now be determined from $\alpha = d\omega/dt$, since this equation relates α, t, and ω. Integrating, with the initial condition that $\omega = 0$ at $t = 0$, yields

$$(\curvearrowright +) \qquad\qquad \alpha = \frac{d\omega}{dt} = (20t) \text{ rad/s}^2$$
$$\int_0^\omega d\omega = \int_0^t 20t \, dt$$
$$\omega = 10t^2 \text{ rad/s} \;\downarrow \qquad\qquad Ans.$$

Why not use Eq. 16–5 ($\omega = \omega_0 + \alpha_c t$) to obtain this result?

Part (b).   Using this result, the angular position θ of OP can be found from $\omega = d\theta/dt$, since this equation relates θ, ω, and t. Integrating, with the initial condition $\theta = 0$ at $t = 0$, we have

$$(\curvearrowright +) \qquad\qquad \frac{d\theta}{dt} = \omega = (10t^2) \text{ rad/s}$$
$$\int_0^\theta d\theta = \int_0^t 10t^2 \, dt$$
$$\theta = 3.33t^3 \text{ rad} \qquad\qquad Ans.$$

EXAMPLE 16.2

The motor shown in the photo is used to turn a wheel and attached blower contained within the housing. The details of the design are shown in Fig. 16–6a. If the pulley A connected to the motor begins rotating from rest with an angular acceleration of $\alpha_A = 2 \text{ rad/s}^2$, determine the magnitudes of the velocity and acceleration of point P on the wheel, after the wheel B has turned one revolution. Assume the transmission belt does not slip on the pulley and wheel.

Solution

Angular Motion. First we will convert the one revolution to radians. Since there are 2π rad in one revolution, then

$$\theta_B = 1 \text{ rev}\left(\frac{2\pi \text{ rad}}{1 \text{ rev}}\right) = 6.283 \text{ rad}$$

We can find the angular velocity of pulley A provided we first find its angular displacement. Since the belt does not slip, an equivalent length of belt s must be unraveled from both the pulley and wheel at all times. Thus,

$$s = \theta_A r_A = \theta_B r_B; \qquad \theta_A(0.15 \text{ m}) = 6.283(0.4 \text{ m})$$
$$\theta_A = 16.76 \text{ rad}$$

Since α_A is constant, the angular velocity of pulley A is therefore

$$(\uparrow+) \qquad \omega^2 = \omega_0^2 + 2\alpha_c(\theta - \theta_0)$$
$$\omega_A^2 = 0 + 2(2 \text{ rad/s}^2)(16.76 \text{ rad} - 0)$$
$$\omega_A = 8.188 \text{ rad/s}$$

The belt has the same speed and tangential component of acceleration as it passes over the pulley and wheel. Thus,

$$v = \omega_A r_A = \omega_B r_B; \qquad 8.188 \text{ rad/s}(0.15 \text{ m}) = \omega_B(0.4 \text{ m})$$
$$\omega_B = 3.070 \text{ rad/s}$$

$$a_t = \alpha_A r_A = \alpha_B r_B; \qquad 2 \text{ rad/s}^2(0.15 \text{ m}) = \alpha_B(0.4 \text{ m})$$
$$\alpha_B = 0.750 \text{ rad/s}^2$$

Motion of P. As shown on the kinematic diagram in Fig. 16–6b, we have

$$v_P = \omega_B r_B = 3.070 \text{ rad/s}(0.4 \text{ m}) = 1.23 \text{ m/s} \qquad \textit{Ans.}$$
$$(a_P)_t = \alpha_B r_B = 0.750 \text{ rad/s}^2(0.4 \text{ m}) = 0.3 \text{ m/s}^2$$
$$(a_P)_n = \omega_B^2 r_B = (3.070 \text{ rad/s})^2(0.4 \text{ m}) = 3.77 \text{ m/s}^2$$

Thus

$$a_P = \sqrt{(0.3)^2 + (3.77)^2} = 3.78 \text{ m/s}^2 \qquad \textit{Ans.}$$

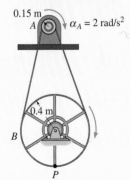

(a)

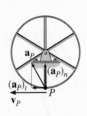

(b)

Fig. 16–6

*16.4 Absolute Motion Analysis

A body subjected to *general plane motion* undergoes a *simultaneous* translation and rotation. If the body is represented by a thin slab, the slab translates in the plane and rotates about an axis perpendicular to the plane. The motion can be completely specified by knowing *both* the angular rotation of a line fixed in the body and the motion of a point on the body. One way to define these motions is to use a rectilinear position coordinate s to locate the point along its path and an angular position coordinate θ to specify the orientation of the line. The two coordinates are then related using the geometry of the problem. By *direct application* of the time-differential equations $v = ds/dt$, $a = dv/dt$, $\omega = d\theta/dt$, and $\alpha = d\omega/dt$, the *motion* of the point and the *angular motion* of the line can then be related. In some cases, this procedure may also be used to relate the motions of one body to those of a connected body, or to study the motion of a body subjected to rotation about a fixed axis.

PROCEDURE FOR ANALYSIS

The velocity and acceleration of a point P undergoing rectilinear motion can be related to the angular velocity and angular acceleration of a line contained within a body using the following procedure.

Position Coordinate Equation.

- Locate point P using a position coordinate s, which is measured from a *fixed origin* and is *directed along the straight-line path of motion* of point P.
- Measure from a fixed reference line the angular position θ of a line lying in the body.
- From the dimensions of the body, relate s to θ, $s = f(\theta)$, using geometry and/or trigonometry.

Time Derivatives.

- Take the first derivative of $s = f(\theta)$ with respect to time to get a relationship between v and ω.
- Take the second time derivative to get a relationship between a and α.
- In each case the chain rule of calculus must be used when taking the derivatives of the position coordinate equation.

The dumping bin on the truck rotates about a fixed axis passing through the pin at A. It is operated by the extension of the hydraulic cylinder BC. The angular position of the bin can be specified using the angular position coordinate θ, and the position of point C on the bin is specified using the rectilinear coordinate s. Since a and b are fixed lengths, then the coordinates can be related by the cosine law, $s = \sqrt{a^2 + b^2 - 2ab \cos \theta}$. Using the chain rule, the time derivative of this equation relates the speed at which the hydraulic cylinder extends, to the angular velocity of the bin, i.e.,

$$v = \frac{1}{2}(a^2 + b^2 - 2ab \cos \theta)^{-\frac{1}{2}}(2ab \sin \theta)\,\omega.$$

E X A M P L E 16.3

The end of rod R shown in Fig. 16–7 maintains contact with the cam by means of a spring. If the cam rotates about an axis through point O with an angular acceleration $\boldsymbol{\alpha}$ and angular velocity $\boldsymbol{\omega}$, determine the velocity and acceleration of the rod when the cam is in the arbitrary position θ.

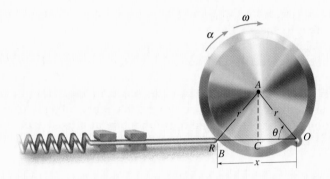

Fig. 16–7

Solution

Position Coordinate Equation. Coordinates θ and x are chosen in order to relate the *rotational motion* of the line segment OA on the cam to the *rectilinear motion* of the rod. These coordinates are measured from the *fixed point* O and may be related to each other using trigonometry. Since $OC = CB = r \cos \theta$, Fig. 16–7, then

$$x = 2r \cos \theta$$

Time Derivatives. Using the chain rule of calculus, we have

$$\frac{dx}{dt} = -2r(\sin \theta)\frac{d\theta}{dt}$$

$$v = -2r\omega \sin \theta \qquad\qquad Ans.$$

$$\frac{dv}{dt} = -2r\left(\frac{d\omega}{dt}\right)\sin \theta - 2r\omega(\cos \theta)\frac{d\theta}{dt}$$

$$a = -2r(\alpha \sin \theta + \omega^2 \cos \theta) \qquad\qquad Ans.$$

The negative signs indicate that v and a are opposite to the direction of positive x.

E X A M P L E 16.4

At a given instant, the cylinder of radius r, shown in Fig. 16–8, has an angular velocity $\boldsymbol{\omega}$ and angular acceleration $\boldsymbol{\alpha}$. Determine the velocity and acceleration of its center G if the cylinder rolls without slipping.

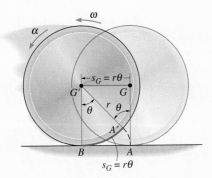

Fig. 16–8

Solution

Position Coordinate Equation. By inspection, point G moves *horizontally* to the left from G to G' as the cylinder rolls, Fig. 16–8. Consequently its new location G' will be specified by the *horizontal* position coordinate s_G, which is measured from the original position (G) of the cylinder's center. Notice also, that as the cylinder rolls (without slipping), points on its surface contact the ground such that the arc length $A'B$ of contact must be equal to the distance s_G. Consequently, the motion requires the radial line GA to rotate θ to the position $G'A'$. Since the arc $A'B = r\theta$, then G travels a distance

$$s_G = r\theta$$

Time Derivatives. Taking successive time derivatives of this equation, realizing that r is constant, $\omega = d\theta/dt$, and $\alpha = d\omega/dt$, gives the necessary relationships:

$$s_G = r\theta$$
$$v_G = r\omega \qquad\qquad\qquad\qquad \textit{Ans.}$$
$$a_G = r\alpha \qquad\qquad\qquad\qquad \textit{Ans.}$$

Remember that these relationships are valid only if the cylinder (disk, wheel, ball, etc.) rolls *without* slipping.

EXAMPLE 16.5

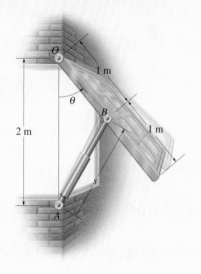

Fig. 16–9

The large window in Fig. 16–9 is opened using a hydraulic cylinder AB. If the cylinder extends at a constant rate of 0.5 m/s, determine the angular velocity and angular acceleration of the window at the instant $\theta = 30°$.

Solution

Position Coordinate Equation. The angular motion of the window can be obtained using the coordinate θ, whereas the extension or motion *along the hydraulic cylinder* is defined using a coordinate s, which measures the length from the fixed point A to the moving point B. These coordinates can be related using the law of cosines, namely,

$$s^2 = (2 \text{ m})^2 + (1 \text{ m})^2 - 2(2 \text{ m})(1 \text{ m}) \cos \theta$$

$$s^2 = 5 - 4 \cos \theta \qquad (1)$$

When $\theta = 30°$,

$$s = 1.239 \text{ m}$$

Time Derivatives. Taking the time derivatives of Eq. 1, we have

$$2s \frac{ds}{dt} = 0 - 4(-\sin \theta) \frac{d\theta}{dt}$$

$$s(v_s) = 2(\sin \theta)\omega \qquad (2)$$

Since $v_s = 0.5$ m/s, then at $\theta = 30°$,

$$(1.239 \text{ m})(0.5 \text{ m/s}) = 2 \sin 30°\omega$$

$$\omega = 0.620 \text{ rad/s} \qquad Ans.$$

Taking the time derivative of Eq. 2 yields

$$\frac{ds}{dt} v_s + s \frac{dv_s}{dt} = 2(\cos \theta) \frac{d\theta}{dt}\omega + 2(\sin \theta) \frac{d\omega}{dt}$$

$$v_s^2 + sa_s = 2(\cos \theta)\omega^2 + 2(\sin \theta)\alpha$$

Since $a_s = dv_s/dt = 0$, then

$$(0.5 \text{ m/s})^2 + 0 = 2 \cos 30°(0.620 \text{ rad/s})^2 + 2 \sin 30°\alpha$$

$$\alpha = -0.415 \text{ rad/s}^2 \qquad Ans.$$

Because the result is negative, it indicates the window has an angular deceleration.

16.5 Relative-Motion Analysis: Velocity

The general plane motion of a rigid body can be described as a *combination* of translation and rotation. To view these "component" motions *separately* we will use a *relative-motion analysis* involving two sets of coordinate axes. The *x, y* coordinate system is fixed and measures the *absolute* position of two points *A* and *B* on the body, Fig. 16–10a. The origin of the *x', y'* coordinate system will be attached to the selected "base point" *A*, which generally has a *known* motion. The axes of this coordinate system do not rotate with the body; rather they will only be allowed to *translate* with respect to the fixed frame.

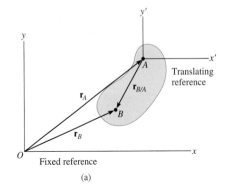

Translating reference

Fixed reference

(a)

Fig. 16–10

Position. The position vector $\mathbf{r}_A$ in Fig. 16–10a specifies the location of the "base point" *A*, and the relative-position vector $\mathbf{r}_{B/A}$ locates point *B* with respect to point *A*. By vector addition, the *position* of *B* is then

$$\mathbf{r}_B = \mathbf{r}_A + \mathbf{r}_{B/A}$$

Displacement. During an instant of time *dt*, points *A* and *B* undergo displacements $d\mathbf{r}_A$ and $d\mathbf{r}_B$ as shown in Fig. 16–10b. If we consider the general plane motion by its component parts then the *entire body* first *translates* by an amount $d\mathbf{r}_A$ so that *A*, the base point, moves to its *final position* and point *B* moves to *B'*, Fig. 16–10c. The body is then *rotated* about *A* by an amount $d\theta$ so that *B'* undergoes a *relative displacement* $d\mathbf{r}_{B/A}$ and thus moves to its final position *B*. Due to the rotation about *A*, $dr_{B/A} = r_{B/A}\, d\theta$, and the displacement of *B* is

$$d\mathbf{r}_B = d\mathbf{r}_A + d\mathbf{r}_{B/A}$$

due to rotation about *A*

due to translation of *A*

due to translation and rotation

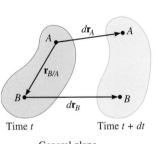

Time *t*   Time *t* + *dt*

General plane motion

(b)

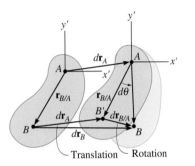

Translation   Rotation

(c)

As the slider block A moves horizontally to the left with a velocity $\mathbf{v}_A$, it causes the link CB to rotate counterclockwise, such that $\mathbf{v}_B$ is directed tangent to its circular path, i.e., upward to the left. The connecting rod AB is subjected to general plane motion, and at the instant shown it has an angular velocity $\boldsymbol{\omega}$.

Velocity. To determine the relationship between the velocities of points A and B, it is necessary to take the time derivative of the position equation, or simply divide the displacement equation by dt. This yields

$$\frac{d\mathbf{r}_B}{dt} = \frac{d\mathbf{r}_A}{dt} + \frac{d\mathbf{r}_{B/A}}{dt}$$

The terms $d\mathbf{r}_B/dt = \mathbf{v}_B$ and $d\mathbf{r}_A/dt = \mathbf{v}_A$ are measured from the fixed x, y axes and represent the *absolute velocities* of points A and B, respectively. The magnitude of the third term is $r_{B/A} d\theta/dt = r_{B/A}\dot{\theta} = r_{B/A}\omega$, where ω is the angular velocity of the body at the instant considered. We will denote this term as the *relative velocity* $\mathbf{v}_{B/A}$, since it represents the velocity of B with respect to A as measured by an observer fixed to the translating x', y' axes. Since the body is rigid, realize that this observer only sees point B move along a *circular arc* that has a radius of curvature $r_{B/A}$. In other words, *the body appears to move as if it were rotating with an angular velocity* $\boldsymbol{\omega}$ *about the z' axis passing through A*. Consequently, $\mathbf{v}_{B/A}$ has a magnitude of $v_{B/A} = \omega r_{B/A}$ and a *direction* which is perpendicular to $\mathbf{r}_{B/A}$. We therefore have

$$\boxed{\mathbf{v}_B = \mathbf{v}_A + \mathbf{v}_{B/A}} \tag{16–15}$$

where

$$\mathbf{v}_B = \text{velocity of point } B$$
$$\mathbf{v}_A = \text{velocity of the base point } A$$
$$\mathbf{v}_{B/A} = \text{relative velocity of "}B \text{ with respect to } A\text{"}$$

This relative motion is *circular*, the *magnitude* is $v_{B/A} = \omega r_{B/A}$ and the *direction* is perpendicular to $\mathbf{r}_{B/A}$.

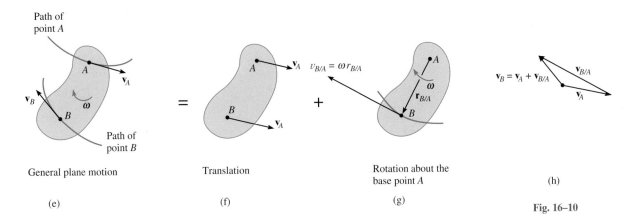

General plane motion

(e)

Translation

(f)

Rotation about the base point A

(g)

(h)

Fig. 16–10

Each of the three terms in Eq. 16–15 is represented graphically on the *kinematic diagrams* in Figs. 16–10e, 16–10f, and 16–10g. Here it is seen that the velocity of B, Fig. 16–10e, is determined by considering the entire body to translate with a velocity of $\mathbf{v}_A$, Fig. 16–10f, and rotate about A with an angular velocity $\boldsymbol{\omega}$, Fig. 16–10g. Vector addition of these two effects, applied to B, yields $\mathbf{v}_B$, as shown in Fig. 16–10h.

Since the relative velocity $\mathbf{v}_{B/A}$ represents the effect of *circular motion*, about A, this term can be expressed by the cross product $\mathbf{v}_{B/A} = \boldsymbol{\omega} \times \mathbf{r}_{B/A}$, Eq. 16–9. Hence, for application, we can also write Eq. 16–15 as

$$\mathbf{v}_B = \mathbf{v}_A + \boldsymbol{\omega} \times \mathbf{r}_{B/A} \qquad (16\text{–}16)$$

where

$\mathbf{v}_B$ = velocity of B

$\mathbf{v}_A$ = velocity of the base point A

$\boldsymbol{\omega}$ = angular velocity of the body

$\mathbf{r}_{B/A}$ = relative-position vector drawn from A to B

The velocity equation 16–15 or 16–16 may be used in a practical manner to study the general plane motion of a rigid body which is either pin-connected to or in contact with other moving bodies. When applying this equation, points A and B should generally be selected as points on the body which are pin-connected to other bodies, or as points in contact with adjacent bodies which have a *known motion*. For example, both points A and B on link AB, Fig. 16–11a, have circular paths of motion since the wheel and link CB move in circular paths. The *directions* of $\mathbf{v}_A$ and $\mathbf{v}_B$ can therefore be established since they are always *tangent* to their paths of motion, Fig. 16–11b. In the case of the wheel in Fig. 16–12, which rolls without slipping, point A can be selected at the ground. Here A (momentarily) has zero velocity since the ground does not move. Furthermore, the center of the wheel, B, moves along a horizontal path so that $\mathbf{v}_B$ is horizontal.

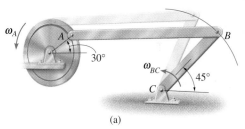

(a)

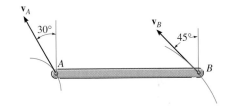

(b)

Fig. 16–11

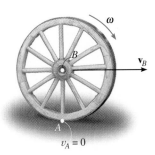

Fig. 16–12

PROCEDURE FOR ANALYSIS

The relative velocity equation can be applied either by using Cartesian vector analysis, or by writing the x and y scalar component equations directly. For application, it is suggested that the following procedure be used.

VECTOR ANALYSIS

Kinematic Diagram.

- Establish the directions of the fixed x, y coordinates and draw a kinematic diagram of the body. Indicate on it the velocities $\mathbf{v}_A$, $\mathbf{v}_B$ of points A and B, the angular velocity $\boldsymbol{\omega}$, and the relative-position vector $\mathbf{r}_{B/A}$.

- If the magnitudes of $\mathbf{v}_A$, $\mathbf{v}_B$, or $\boldsymbol{\omega}$ are unknown, the sense of direction of these vectors may be assumed.

Velocity Equation.

- To apply $\mathbf{v}_B = \mathbf{v}_A + \boldsymbol{\omega} \times \mathbf{r}_{B/A}$, express the vectors in Cartesian vector form and substitute them into the equation. Evaluate the cross product and then equate the respective $\mathbf{i}$ and $\mathbf{j}$ components to obtain two scalar equations.

- If the solution yields a *negative* answer for an *unknown* magnitude, it indicates the sense of direction of the vector is opposite to that shown on the kinematic diagram.

SCALAR ANALYSIS

Kinematic Diagram.

- If the velocity equation is to be applied in scalar form, then the magnitude and direction of the relative velocity $\mathbf{v}_{B/A}$ must be established. Draw a kinematic diagram such as shown in Fig. 16–10g, which shows the relative motion. Since the body is considered to be "pinned" momentarily at the base point A, the magnitude is $v_{B/A} = \omega r_{B/A}$. The sense of direction of $\mathbf{v}_{B/A}$ is established from the diagram, such that $\mathbf{v}_{B/A}$ acts perpendicular to $\mathbf{r}_{B/A}$ in accordance with the rotational motion $\boldsymbol{\omega}$ of the body.*

Velocity Equation.

- Write Eq. 16–15 in symbolic form, $\mathbf{v}_B = \mathbf{v}_A + \mathbf{v}_{B/A}$, and underneath each of the terms represent the vectors graphically by showing their magnitudes and directions. The scalar equations are determined from the x and y components of these vectors.

*The notation $\mathbf{v}_B = \mathbf{v}_A + \mathbf{v}_{B/A(\text{pin})}$ may be helpful in recalling that A is "pinned."

EXAMPLE 16.6

The link shown in Fig. 16–13a is guided by two blocks at A and B, which move in the fixed slots. If the velocity of A is 2 m/s downward, determine the velocity of B at the instant $\theta = 45°$.

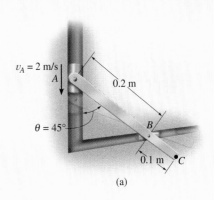

(a)

Solution (Vector Analysis)

Kinematic Diagram. Since points A and B are restricted to move along the fixed slots and $\mathbf{v}_A$ is directed downward, the velocity $\mathbf{v}_B$ must be directed horizontally to the right, Fig. 16–13b. This motion causes the link to rotate counterclockwise; that is, by the right-hand rule the angular velocity $\boldsymbol{\omega}$ is directed outward, perpendicular to the plane of motion. Knowing the magnitude and direction of $\mathbf{v}_A$ and the lines of action of $\mathbf{v}_B$ and $\boldsymbol{\omega}$, it is possible to apply the velocity equation $\mathbf{v}_B = \mathbf{v}_A + \boldsymbol{\omega} \times \mathbf{r}_{B/A}$ to points A and B in order to solve for the two unknown magnitudes v_B and ω. Since $\mathbf{r}_{B/A}$ is needed, it is also shown in Fig. 16–13b.

Velocity Equation. Expressing each of the vectors in Fig. 16–13b in terms of their $\mathbf{i}$, $\mathbf{j}$, $\mathbf{k}$ components and applying Eq. 16–16 to A, the base point, and B, we have

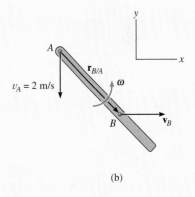

(b)

$$\mathbf{v}_B = \mathbf{v}_A + \boldsymbol{\omega} \times \mathbf{r}_{B/A}$$
$$v_B\mathbf{i} = -2\mathbf{j} + [\omega\mathbf{k} \times (0.2 \sin 45°\mathbf{i} - 0.2 \cos 45°\mathbf{j})]$$
$$v_B\mathbf{i} = -2\mathbf{j} + 0.2\omega \sin 45°\mathbf{j} + 0.2\omega \cos 45°\mathbf{i}$$

Equating the $\mathbf{i}$ and $\mathbf{j}$ components gives

$$v_B = 0.2\omega \cos 45° \qquad 0 = -2 + 0.2\omega \sin 45°$$

Fig. 16–13

Thus,

$$\omega = 14.1 \text{ rad/s}\curvearrowleft$$
$$v_B = 2 \text{ m/s} \rightarrow \qquad\qquad \textit{Ans.}$$

Since both results are *positive*, the *directions* of $\mathbf{v}_B$ and $\boldsymbol{\omega}$ are indeed *correct* as shown in Fig. 16–13b. It should be emphasized that these results are *valid only* at the instant $\theta = 45°$. A recalculation for $\theta = 44°$ yields $v_B = 2.07$ m/s and $\omega = 14.4$ rad/s; whereas when $\theta = 46°$, $v_B = 1.93$ m/s and $\omega = 13.9$ rad/s, etc.

Now that once the velocity of a point (A) on the link and the angular velocity are *known*, the velocity of any other point on the link can be determined. As an exercise, see if you can apply Eq. 16–16 to points A and C or to points B and C and show that when when $\theta = 45°$, $v_C = 3.16$ m/s, directed at $\theta = 18.4°$ up from the horizontal.

E X A M P L E 16.7

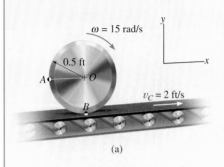

(a)

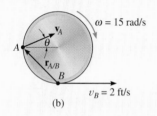

(b)

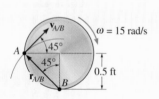

Relative motion
(c)

Fig. 16–14

The cylinder shown in Fig. 16–14a rolls without slipping on the surface of a conveyor belt which is moving at 2 ft/s. Determine the velocity of point A. The cylinder has a clockwise angular velocity $\omega = 15$ rad/s at the instant shown.

Solution I (Vector Analysis)

Kinematic Diagram. Since no slipping occurs, point B on the cylinder has the same velocity as the conveyor, Fig. 16–14b. Also, the angular velocity of the cylinder is known, so we can apply the velocity equation to B, the base point, and A to determine $\mathbf{v}_A$.

Velocity Equation.

$$\mathbf{v}_A = \mathbf{v}_B + \boldsymbol{\omega} \times \mathbf{r}_{A/B}$$
$$(v_A)_x\mathbf{i} + (v_A)_y\mathbf{j} = 2\mathbf{i} + (-15\mathbf{k}) \times (-0.5\mathbf{i} + 0.5\mathbf{j})$$
$$(v_A)_x\mathbf{i} + (v_A)_y\mathbf{j} = 2\mathbf{i} + 7.50\mathbf{j} + 7.50\mathbf{i}$$

so that

$$(v_A)_x = 2 + 7.50 = 9.50 \text{ ft/s} \qquad (1)$$
$$(v_A)_y = 7.50 \text{ ft/s} \qquad (2)$$

Thus,

$$v_A = \sqrt{(9.50)^2 + (7.50)^2} = 12.1 \text{ ft/s} \qquad \textit{Ans.}$$

$$\theta = \tan^{-1}\frac{7.50}{9.50} = 38.3° \qquad \textit{Ans.}$$

Solution II (Scalar Analysis)

As an alternative procedure, the scalar components of $\mathbf{v}_A = \mathbf{v}_B + \mathbf{v}_{A/B}$ can be obtained directly. From the kinematic diagram showing the relative "circular" motion $\mathbf{v}_{A/B}$, Fig. 16–14c, we have

$$v_{A/B} = \omega r_{A/B} = (15 \text{ rad/s})\left(\frac{0.5 \text{ ft}}{\cos 45°}\right) = 10.6 \text{ ft/s} \angle^{45°}$$

Thus

$$\mathbf{v}_A = \mathbf{v}_B + \mathbf{v}_{A/B}$$

$$\begin{bmatrix} (v_A)_x \\ \rightarrow \end{bmatrix} + \begin{bmatrix} (v_A)_y \\ \uparrow \end{bmatrix} = \begin{bmatrix} 2 \text{ ft/s} \\ \rightarrow \end{bmatrix} + \begin{bmatrix} 10.6 \text{ ft/s} \\ \angle^{45°} \end{bmatrix}$$

Equating the x and y components gives the same results as before, namely,

$$(\xrightarrow{+}) \qquad (v_A)_x = 2 + 10.6 \cos 45° = 9.50 \text{ ft/s}$$
$$(+\uparrow) \qquad (v_A)_y = 0 + 10.6 \sin 45° = 7.50 \text{ ft/s}$$

EXAMPLE 16.8

The collar C in Fig. 16–15a is moving downward with a velocity of 2 m/s. Determine the angular velocities of CB and AB at this instant.

Solution I (Vector Analysis)

Kinematic Diagram. The downward motion of C causes B to move to the right. Also, CB and AB rotate counterclockwise. To solve, we will write the appropriate kinematic equation for each link.

Velocity Equation.

Link CB (general plane motion): See Fig. 16–15b.

$$\mathbf{v}_B = \mathbf{v}_C + \boldsymbol{\omega}_{CB} \times \mathbf{r}_{B/C}$$
$$v_B\mathbf{i} = -2\mathbf{j} + \omega_{CB}\mathbf{k} \times (0.2\mathbf{i} - 0.2\mathbf{j})$$
$$v_B\mathbf{i} = -2\mathbf{j} + 0.2\omega_{CB}\mathbf{j} + 0.2\omega_{CB}\mathbf{i}$$

$$v_B = 0.2\omega_{CB} \qquad\qquad (1)$$
$$0 = -2 + 0.2\omega_{CB} \qquad\qquad (2)$$
$$\omega_{CB} = 10 \text{ rad/s} \curvearrowleft \qquad\qquad Ans.$$
$$v_B = 2 \text{ m/s} \rightarrow$$

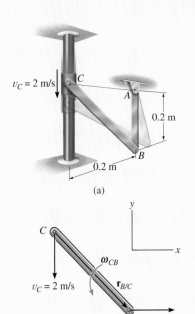
(a)

Link AB (rotation about a fixed axis): See Fig. 16–15c.

$$\mathbf{v}_B = \boldsymbol{\omega}_{AB} \times \mathbf{r}_B$$
$$2\mathbf{i} = \omega_{AB}\mathbf{k} \times (-0.2\mathbf{j})$$
$$2 = 0.2\omega_{AB}$$
$$\omega_{AB} = 10 \text{ rad/s} \curvearrowleft \qquad\qquad Ans.$$

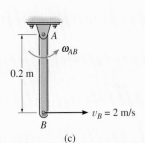
(c)

Solution II (Scalar Analysis)

The scalar component equations of $\mathbf{v}_B = \mathbf{v}_C + \mathbf{v}_{B/C}$ can be obtained directly. The kinematic diagram in Fig. 16–15d shows the relative "circular" motion $\mathbf{v}_{B/C}$. We have

$$\mathbf{v}_B = \mathbf{v}_C + \mathbf{v}_{B/C}$$

$$\begin{bmatrix} v_B \\ \rightarrow \end{bmatrix} = \begin{bmatrix} 2 \text{ m/s} \\ \downarrow \end{bmatrix} + \begin{bmatrix} \omega_{CB}(0.2\sqrt{2} \text{ m}) \\ \measuredangle\,^{45°} \end{bmatrix}$$

Resolving these vectors in the x and y directions yields

$(\xrightarrow{+}) \qquad v_B = 0 + \omega_{CB}(0.2\sqrt{2}\cos 45°)$
$(+\uparrow) \qquad 0 = -2 + \omega_{CB}(0.2\sqrt{2}\sin 45°)$

which is the same as Eqs. 1 and 2.

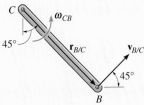
Relative motion
(d)

Fig. 16–15

EXAMPLE 16.9

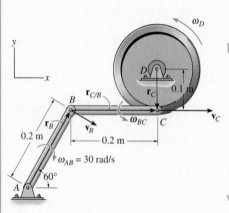

(a)

Fig. 16–16

The bar AB of the linkage shown in Fig. 16–16a has a clockwise angular velocity of 30 rad/s when $\theta = 60°$. Determine the angular velocities of member BC and the wheel at this instant.

Solution (Vector Analysis)

Kinematic Diagram.   By inspection, the velocities of points B and C are defined by the rotation of link AB and the wheel about their fixed axes. The position vectors and the angular velocity of each member are shown on the kinematic diagram in Fig. 16–16b. To solve, we will write the appropriate kinematic equation for each member.

Velocity Equation.

Link AB (rotation about a fixed axis):

$$\mathbf{v}_B = \boldsymbol{\omega}_{AB} \times \mathbf{r}_B$$
$$= (-30\mathbf{k}) \times (0.2 \cos 60°\mathbf{i} + 0.2 \sin 60°\mathbf{j})$$
$$= \{5.20\mathbf{i} - 3.0\mathbf{j}\} \text{ m/s}$$

Link BC (general plane motion):

$$\mathbf{v}_C = \mathbf{v}_B + \boldsymbol{\omega}_{BC} \times \mathbf{r}_{C/B}$$
$$v_C\mathbf{i} = 5.20\mathbf{i} - 3.0\mathbf{j} + (\omega_{BC}\mathbf{k}) \times (0.2\mathbf{i})$$
$$v_C\mathbf{i} = 5.20\mathbf{i} + (0.2\omega_{BC} - 3.0)\mathbf{j}$$
$$v_C = 5.20 \text{ m/s}$$
$$0 = 0.2\omega_{BC} - 3.0$$
$$\omega_{BC} = 15 \text{ rad/s} \curvearrowleft \qquad \qquad Ans.$$

Wheel (rotation about a fixed axis):

$$\mathbf{v}_C = \boldsymbol{\omega}_D \times \mathbf{r}_C$$
$$5.20\mathbf{i} = (\omega_D\mathbf{k}) \times (-0.1\mathbf{j})$$
$$5.20 = 0.1\omega_D$$
$$\omega_D = 52 \text{ rad/s} \curvearrowleft \qquad \qquad Ans.$$

Note that, by inspection, Fig. 16–16a, $v_B = (0.2)(30) = 6$ m/s, $\measuredangle^{30°}$ and $\mathbf{v}_C$ is directed to the right. As an exercise, use this information and try to obtain ω_{BC} by applying $\mathbf{v}_C = \mathbf{v}_B + \mathbf{v}_{C/B}$ using scalar components.

(b)

16.6 Instantaneous Center of Zero Velocity

The velocity of any point B located on a rigid body can be obtained in a very direct way if one chooses the base point A to be a point that has *zero velocity* at the instant considered. In this case, $\mathbf{v}_A = \mathbf{0}$, and therefore the velocity equation, $\mathbf{v}_B = \mathbf{v}_A + \boldsymbol{\omega} \times \mathbf{r}_{B/A}$, becomes $\mathbf{v}_B = \boldsymbol{\omega} \times \mathbf{r}_{B/A}$. For a body having general plane motion, point A so chosen is called the *instantaneous center of zero velocity (IC)*, and it lies on the *instantaneous axis of zero velocity*. This axis is always perpendicular to the plane of motion, and the intersection of the axis with this plane defines the location of the IC. Since point A is coincident with the IC, then $\mathbf{v}_B = \boldsymbol{\omega} \times \mathbf{r}_{B/IC}$ and so point B moves momentarily about the IC in a *circular path;* in other words, the body appears to rotate about the instantaneous axis. The *magnitude* of $\mathbf{v}_B$ is simply $v_B = \omega r_{B/IC}$, where ω is the angular velocity of the body. Due to the circular motion, the *direction* of $\mathbf{v}_B$ must always be *perpendicular* to $\mathbf{r}_{B/IC}$.

For example, consider the wheel in Fig. 16–17a. If it rolls *without slipping*, then the point of *contact* with the ground has *zero velocity*. Hence this point represents the IC for the wheel, Fig. 16–17b. If it is imagined that the wheel is momentarily pinned at this point, the velocities of points B, C, O, and so on, can be found using $v = \omega r$. Here the radial distances $r_{B/IC}$, $r_{C/IC}$, and $r_{O/IC}$, shown in Fig. 16–17b, must be determined from the geometry of the wheel.

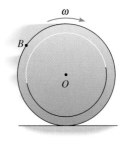

(a)

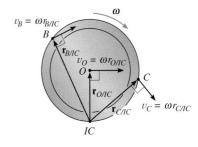

(b)

Fig. 16–17

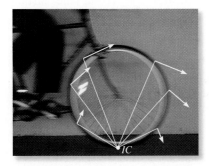

The IC for this bicycle wheel is at the ground. There the spokes are somewhat visible, whereas at the top of the wheel they become blurred. Note also how points on the side portions of the wheel move as shown by their velocities.

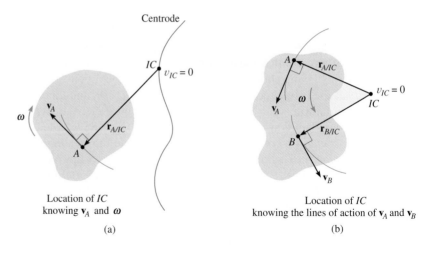

Location of IC
knowing $\mathbf{v}_A$ and $\boldsymbol{\omega}$

(a)

Location of IC
knowing the lines of action of $\mathbf{v}_A$ and $\mathbf{v}_B$

(b)

Fig. 16–18

Location of the IC. To locate the IC we can use the fact that the *velocity* of a point on the body is *always perpendicular* to the *relative-position vector* extending from the IC to the point. Several possibilities exist:

- *Given the velocity $\mathbf{v}_A$ of a point A on the body, and the angular velocity $\boldsymbol{\omega}$ of the body*, Fig. 16–18a. In this case, the IC is located along the line drawn perpendicular to $\mathbf{v}_A$ at A, such that the distance from A to the IC is $r_{A/IC} = v_A/\omega$. Note that the IC lies up and to the right of A since $\mathbf{v}_A$ must cause a clockwise angular velocity $\boldsymbol{\omega}$ about the IC.

- *Given the lines of action of two nonparallel velocities $\mathbf{v}_A$ and $\mathbf{v}_B$*, Fig. 16–18b. Construct at points A and B line segments that are perpendicular to $\mathbf{v}_A$ and $\mathbf{v}_B$. Extending these perpendiculars to their *point of intersection* as shown locates the IC at the instant considered.

- *Given the magnitude and direction of two parallel velocities $\mathbf{v}_A$ and $\mathbf{v}_B$*. Here the location of the IC is determined by proportional triangles. Examples are shown in Fig. 16–18c and d. In both cases $r_{A/IC} = v_A/\omega$ and $r_{B/IC} = v_B/\omega$. If d is a known distance between points A and B, then in Fig. 16–18c, $r_{A/IC} + r_{B/IC} = d$ and in Fig. 16–18d, $r_{B/IC} - r_{A/IC} = d$. As a special case, note that if the body is *translating*, $\mathbf{v}_A = \mathbf{v}_B$, then the IC would be located at infinity, in which case $r_{A/IC} = r_{B/IC} \rightarrow \infty$. This being the case, $\omega = (v_A/r_{A/IC}) = (v_B/r_{B/IC}) \rightarrow 0$, as expected.

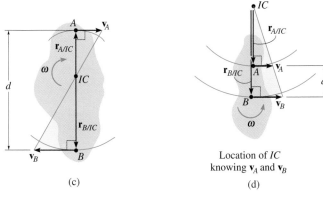

(c)

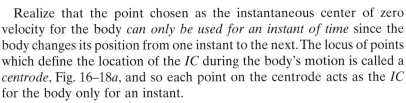

Location of *IC*
knowing $\mathbf{v}_A$ and $\mathbf{v}_B$

(d)

Fig. 16–18

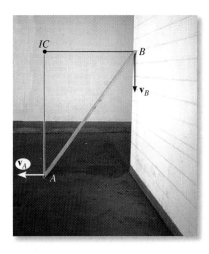

Realize that the point chosen as the instantaneous center of zero velocity for the body *can only be used for an instant of time* since the body changes its position from one instant to the next. The locus of points which define the location of the *IC* during the body's motion is called a *centrode*, Fig. 16–18a, and so each point on the centrode acts as the *IC* for the body only for an instant.

Although the *IC* may be conveniently used to determine the velocity of any point in a body, it generally *does not have zero acceleration* and therefore it *should not* be used for finding the accelerations of points in a body.

As the board slides downward to the left it is subjected to general plane motion. Since the directions of the velocities of its ends *A* and *B* are known, the *IC* is located as shown. At this instant the board will momentarily rotate about this point. Draw the board in several other positions, establish the *IC* for each case and sketch the centrode.

PROCEDURE FOR ANALYSIS

The velocity of a point on a body which is subjected to general plane motion can be determined with reference to its instantaneous center of zero velocity provided the location of the *IC* is first established using one of the three methods described above.

- As shown on the kinematic diagram in Fig. 16–19, the body is imagined as "extended and pinned" at the *IC* such that, at the instant considered, it rotates about this pin with its angular velocity $\boldsymbol{\omega}$.

- The *magnitude* of velocity for each of the arbitrary points *A*, *B*, and *C* on the body can be determined by using the equation $v = \omega r$, where r is the radial distance from the *IC* to each point.

- The line of action of each velocity vector $\mathbf{v}$ is *perpendicular* to its associated radial line $\mathbf{r}$, and the velocity has a *sense of direction* which tends to move the point in a manner consistent with the angular rotation $\boldsymbol{\omega}$ of the radial line, Fig. 16–19.

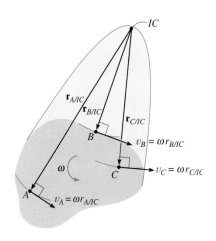

Fig. 16–19

E X A M P L E 16.10

Show how to determine the location of the instantaneous center of zero velocity for (a) member *BC* shown in Fig. 16–20*a*; and (b) the link *CB* shown in Fig. 16–20*b*.

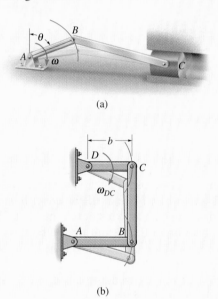

(a)

(b)

Fig. 16–20

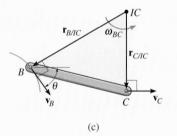

(c)

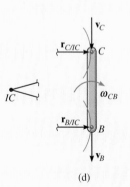

(d)

Solution

Part (a). As shown in Fig. 16–20*a*, point *B* has a velocity $\mathbf{v}_B$, which is caused by the clockwise rotation of link *AB*. Point *B* moves in a circular path such that $\mathbf{v}_B$ is perpendicular to *AB*, and so it acts at an angle θ from the horizontal as shown in Fig. 16–20*c*. The motion of point *B* causes the piston to move forward *horizontally* with a velocity $\mathbf{v}_C$. When lines are drawn perpendicular to $\mathbf{v}_B$ and $\mathbf{v}_C$, Fig. 16–20*c*, they intersect at the *IC*.

Part (b). Points *B* and *C* follow circular paths of motion since rods *AB* and *DC* are each subjected to rotation about a fixed axis, Fig. 16–20*b*. Since the velocity is always tangent to the path, at the instant considered, $\mathbf{v}_C$ on rod *DC* and $\mathbf{v}_B$ on rod *AB* are both directed vertically downward, along the axis of link *CB*, Fig. 16–20*d*. Radial lines drawn perpendicular to these two velocities form parallel lines which intersect at "infinity;" i.e., $r_{C/IC} \to \infty$ and $r_{B/IC} \to \infty$. Thus, $\omega_{CB} = (v_C/r_{C/IC}) = \to 0$. As a result, rod *CB* momentarily *translates*. An instant later, however, *CB* will move to a tilted position, causing the instantaneous center to move to some finite location.

E X A M P L E 16.11

Block D shown in Fig. 16–21a moves with a speed of 3 m/s. Determine the angular velocities of links BD and AB, at the instant shown.

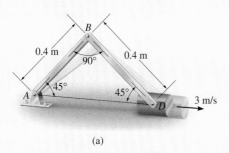

(a)

Fig. 16–21

Solution

As D moves to the right, it causes arm AB to rotate clockwise about point A. Hence, $\mathbf{v}_B$ is directed perpendicular to AB. The instantaneous center of zero velocity for BD is located at the intersection of the line segments drawn perpendicular to $\mathbf{v}_B$ and $\mathbf{v}_D$, Fig. 16–21b. From the geometry,

$$r_{B/IC} = 0.4 \tan 45°\text{m} = 0.4 \text{ m}$$

$$r_{D/IC} = \frac{0.4 \text{ m}}{\cos 45°} = 0.566 \text{ m}$$

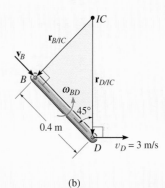

(b)

Since the magnitude of $\mathbf{v}_D$ is known, the angular velocity of link BD is

$$\omega_{BD} = \frac{v_D}{r_{D/IC}} = \frac{3 \text{ m/s}}{0.566 \text{ m}} = 5.30 \text{ rad/s} \nwarrow \qquad Ans.$$

The velocity of B is therefore

$$v_B = \omega_{BD}(r_{B/IC}) = 5.30 \text{ rad/s}(0.4 \text{ m}) = 2.12 \text{ m/s} \searrow^{45°}$$

From Fig. 16–21c, the angular velocity of AB is

$$\omega_{AB} = \frac{v_B}{r_{B/A}} = \frac{2.12 \text{ m/s}}{0.4 \text{ m}} = 5.30 \text{ rad/s} \downarrow \qquad Ans.$$

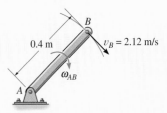

(c)

EXAMPLE 16.12

The cylinder shown in Fig. 16–22a rolls without slipping between the two moving plates E and D. Determine the angular velocity of the cylinder and the velocity of its center C at the instant shown.

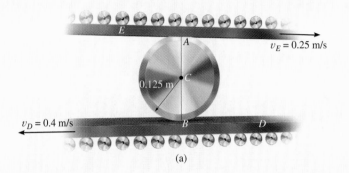

(a)

Fig. 16–22

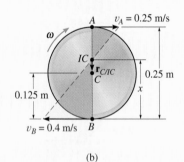

(b)

Solution

Since no slipping occurs, the contact points A and B on the cylinder have the same velocities as the plates E and D, respectively. Furthermore, the velocities $\mathbf{v}_A$ and $\mathbf{v}_B$ are *parallel*, so that by the proportionality of right triangles the IC is located at a point on line AB, Fig. 16–22b. Assuming this point to be a distance x from B, we have

$$v_B = \omega x; \qquad\qquad 0.4 \text{ m/s} = \omega x$$

$$v_A = \omega(0.25 \text{ m} - x); \qquad 0.25 \text{ m/s} = \omega(0.25 \text{ m} - x)$$

Dividing one equation into the other eliminates ω and yields

$$0.4(0.25 - x) = 0.25x$$

$$x = \frac{0.1}{0.65} = 0.154 \text{ m}$$

Hence, the angular velocity of the cylinder is

$$\omega = \frac{v_B}{x} = \frac{0.4 \text{ m/s}}{0.154 \text{ m}} = 2.60 \text{ rad/s} \downarrow \qquad\qquad Ans.$$

The velocity of point C is therefore

$$v_C = \omega r_{C/IC} = 2.60 \text{ rad/s}(0.154 \text{ m} - 0.125 \text{ m})$$
$$= 0.0750 \text{ m/s} \leftarrow \qquad\qquad Ans.$$

16.7 Relative-Motion Analysis: Acceleration

An equation that relates the accelerations of two points on a rigid body subjected to general plane motion may be determined by differentiating the velocity equation $\mathbf{v}_B = \mathbf{v}_A + \mathbf{v}_{B/A}$ with respect to time. This yields

$$\frac{d\mathbf{v}_B}{dt} = \frac{d\mathbf{v}_A}{dt} + \frac{d\mathbf{v}_{B/A}}{dt}$$

The terms $d\mathbf{v}_B/dt = \mathbf{a}_B$ and $d\mathbf{v}_A/dt = \mathbf{a}_A$ are measured from a set of *fixed x, y axes* and represent the *absolute accelerations* of points B and A. The last term represents the acceleration of B with respect to A as measured by an observer fixed to translating x', y' axes which have their origin at the base point A. In Sec. 16.5 it was shown that to this observer point B appears to move along a *circular arc* that has a radius of curvature $r_{B/A}$. Consequently, $\mathbf{a}_{B/A}$ can be expressed in terms of its tangential and normal components of motion; i.e., $\mathbf{a}_{B/A} = (\mathbf{a}_{B/A})_t + (\mathbf{a}_{B/A})_n$, where $(a_{B/A})_t = \alpha r_{B/A}$ and $(a_{B/A})_n = \omega^2 r_{B/A}$. Hence, the relative-acceleration equation can be written in the form

$$\mathbf{a}_B = \mathbf{a}_A + (\mathbf{a}_{B/A})_t + (\mathbf{a}_{B/A})_n \qquad (16\text{–}17)$$

where

$\mathbf{a}_B$ = acceleration of point B

$\mathbf{a}_A$ = acceleration of point A

$(\mathbf{a}_{B/A})_t$ = relative tangential acceleration component of "B with respect to A." The *magnitude* is $(a_{B/A})_t = \alpha r_{B/A}$, and the *direction* is perpendicular to $\mathbf{r}_{B/A}$.

$(\mathbf{a}_{B/A})_n$ = relative normal acceleration component of "B with respect to A." The *magnitude* is $(a_{B/A})_n = \omega^2 r_{B/A}$, and the *direction* is always from B towards A.

Each of the four terms in Eq. 16–17 is represented graphically on the *kinematic diagrams* shown in Fig. 16–23. Here it is seen that at a given instant the acceleration of B, Fig. 16–23a, is determined by considering the body to translate with an acceleration $\mathbf{a}_A$, Fig. 16–23b, and simultaneously rotate about the base point A with an instantaneous angular velocity $\boldsymbol{\omega}$ and angular acceleration $\boldsymbol{\alpha}$, Fig. 16–23c. Vector addition of these two effects, applied to B, yields $\mathbf{a}_B$, as shown in Fig. 16–23d. It should be noted from Fig. 16–23a that since points A and B move along *curved paths*, the accelerations of these points will have *both tangential and normal components*. (Recall that the acceleration of a point is *tangent to the path only* when the path is *rectilinear* or when it is an inflection point on a curve.)

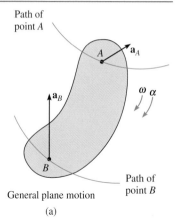

Path of point A

Path of point B

General plane motion

(a)

$\parallel$

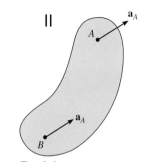

Translation

(b)

$+$

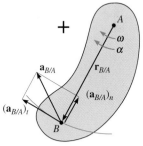

Rotation about the base point A

(c)

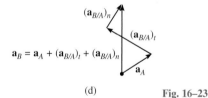

$\mathbf{a}_B = \mathbf{a}_A + (\mathbf{a}_{B/A})_t + (\mathbf{a}_{B/A})_n$

(d)

Fig. 16–23

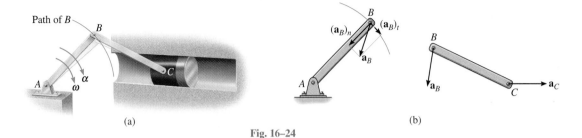

(a)

(b)

Fig. 16–24

Since the relative-acceleration components represent the effect of *circular motion* observed from translating axes having their origin at the base point A, these terms can be expressed as $(\mathbf{a}_{B/A})_t = \boldsymbol{\alpha} \times \mathbf{r}_{B/A}$ and $(\mathbf{a}_{B/A})_n = -\omega^2 \mathbf{r}_{B/A}$, Eq. 16–14. Hence, Eq. 16–17 becomes

$$\mathbf{a}_B = \mathbf{a}_A + \boldsymbol{\alpha} \times \mathbf{r}_{B/A} - \omega^2 \mathbf{r}_{B/A} \qquad (16\text{–}18)$$

where

$\mathbf{a}_B$ = acceleration of point B

$\mathbf{a}_A$ = acceleration of the base point A

$\boldsymbol{\alpha}$ = angular acceleration of the body

$\boldsymbol{\omega}$ = angular velocity of the body

$\mathbf{r}_{B/A}$ = relative-position vector drawn from A to B

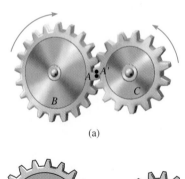

(a)

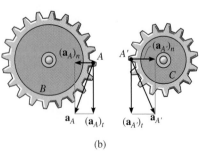

(b)

Fig. 16–25

If Eq. 16–17 or 16–18 is applied in a practical manner to study the accelerated motion of a rigid body which is pin connected to two other bodies, it should be realized that points which are *coincident at the pin* move with the *same acceleration*, since the path of motion over which they travel is the *same*. For example, point B lying on either rod AB or BC of the crank mechanism shown in Fig. 16–24a has the same acceleration, since the rods are pin connected at B. Here the motion of B is along a *curved path*, so that $\mathbf{a}_B$ can be expressed in terms of its tangential and normal components. At the other end of rod BC point C moves along a *straight-lined path*, which is defined by the piston. Hence, $\mathbf{a}_C$ is horizontal, Fig. 16–24b.

If two bodies contact one another *without slipping*, and the *points in contact* move along *different paths*, the *tangential components* of acceleration of the points will be the *same*; however, the *normal components* will *not* be the same. For example, consider the two meshed gears in Fig. 16–25a. Point A is located on gear B and a coincident point A' is located on gear C. Due to the rotational motion, $(\mathbf{a}_A)_t = (\mathbf{a}_{A'})_t$; however, since both points follow different curved paths, $(\mathbf{a}_A)_n \neq (\mathbf{a}_{A'})_n$ and therefore $\mathbf{a}_A \neq \mathbf{a}_{A'}$, Fig. 16–25b.

PROCEDURE FOR ANALYSIS

The relative acceleration equation can be applied between any two points A and B on a body either by using a Cartesian vector analysis, or by writing the x and y scalar component equations directly.

Velocity Analysis.

- Determine the angular velocity $\boldsymbol{\omega}$ of the body by using a velocity analysis as discussed in Sec. 16.5 or 16.6. Also, determine the velocities $\mathbf{v}_A$ and $\mathbf{v}_B$ of points A and B if these points move along curved paths.

VECTOR ANALYSIS

Kinematic Diagram.

- Establish the directions of the fixed x, y coordinates and draw the kinematic diagram of the body. Indicate on it $\mathbf{a}_A$, $\mathbf{a}_B$, $\boldsymbol{\omega}$, $\boldsymbol{\alpha}$, and $\mathbf{r}_{B/A}$.

- If points A and B move along *curved paths*, then their accelerations should be indicated in terms of their tangential and normal components, i.e., $\mathbf{a}_A = (\mathbf{a}_A)_t + (\mathbf{a}_A)_n$ and $\mathbf{a}_B = (\mathbf{a}_B)_t + (\mathbf{a}_B)_n$.

Acceleration Equation.

- To apply $\mathbf{a}_B = \mathbf{a}_A + \boldsymbol{\alpha} \times \mathbf{r}_{B/A} - \omega^2 \mathbf{r}_{B/A}$ express the vectors in Cartesian vector form and substitute them into the equation. Evaluate the cross product and then equate the respective $\mathbf{i}$ and $\mathbf{j}$ components to obtain two scalar equations.

- If the solution yields a *negative* answer for an *unknown* magnitude, it indicates that the sense of direction of the vector is opposite to that shown on the kinematic diagram.

SCALAR ANALYSIS

Kinematic Diagram.

- If the equation $\mathbf{a}_B = \mathbf{a}_A + (\mathbf{a}_{B/A})_t + (\mathbf{a}_{B/A})_n$ is applied, then the magnitudes and directions of the relative-acceleration components $(\mathbf{a}_{B/A})_t$ and $(\mathbf{a}_{B/A})_n$ must be established. To do this draw a kinematic diagram such as shown in Fig. 16–23c. Since the body is considered to be momentarily "pinned" at the base point A, the *magnitudes* are $(a_{B/A})_t = \alpha r_{B/A}$ and $(a_{B/A})_n = \omega^2 r_{B/A}$. Their *sense of direction* is established from the diagram such that $(\mathbf{a}_{B/A})_t$ acts perpendicular to $\mathbf{r}_{B/A}$, in accordance with the rotational motion $\boldsymbol{\alpha}$ of the body, and $(\mathbf{a}_{B/A})_n$ is directed from B towards A.*

Acceleration Equation.

- Represent the vectors in $\mathbf{a}_B = \mathbf{a}_A + (\mathbf{a}_{B/A})_t + (\mathbf{a}_{B/A})_n$ graphically by showing their magnitudes and directions underneath each term. The scalar equations are determined from the x and y components of these vectors.

*The notation $\mathbf{a}_B = \mathbf{a}_A + (\mathbf{a}_{B/A(\text{pin})})_t + (\mathbf{a}_{B/A(\text{pin})})_n$ may be helpful in recalling that A is assumed to be pinned.

The mechanism for a window is shown. Here CA rotates about a fixed axis through C, and AB undergoes general plane motion. Since point A moves along a curved path it has two components of acceleration, whereas point B moves along a straight track and the direction of its acceleration is specified.

E X A M P L E 16.13

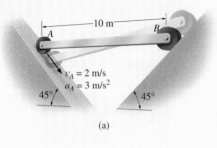

$v_A = 2$ m/s
$a_A = 3$ m/s^2

45°   45°

(a)

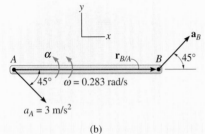

(b)

The rod AB shown in Fig. 16–26a is confined to move along the inclined planes at A and B. If point A has an acceleration of 3 m/s^2 and a velocity of 2 m/s, both directed down the plane at the instant the rod becomes horizontal, determine the angular acceleration of the rod at this instant.

Solution I (Vector Analysis)

We will apply the acceleration equation to points A and B on the rod. To do so it is first necessary to determine the angular velocity of the rod. Show that it is $\omega = 0.283$ rad/s $\uparrow$ using either the velocity equation or the method of instantaneous centers.

Kinematic Diagram. Since points A and B both move along straight-line paths, they have *no* components of acceleration normal to the paths. There are two unknowns in Fig. 16–26b, namely, a_B and α.

Acceleration Equation. Applying Eq. 16–18 to points A and B on the rod, and expressing each of the vectors in Cartesian vector form, we have

$$\mathbf{a}_B = \mathbf{a}_A + \boldsymbol{\alpha} \times \mathbf{r}_{B/A} - \omega^2 \mathbf{r}_{B/A}$$

$$a_B \cos 45°\mathbf{i} + a_B \sin 45°\mathbf{j} = 3 \cos 45°\mathbf{i} - 3 \sin 45°\mathbf{j} + (\alpha\mathbf{k}) \times (10\mathbf{i}) - (0.283)^2(10\mathbf{i})$$

Carrying out the cross product and equating the **i** and **j** components yields

$$a_B \cos 45° = 3 \cos 45° - (0.283)^2(10) \qquad (1)$$

$$a_B \sin 45° = -3 \sin 45° + \alpha(10) \qquad (2)$$

Solving, we have

$$a_B = 1.87 \text{ m/s}^2 \angle^{45°}$$

$$\alpha = 0.344 \text{ rad/s}^2 \uparrow \qquad \qquad Ans.$$

Solution II (Scalar Analysis)

As an alternative procedure, the scalar component equations 1 and 2 can be obtained directly. From the kinematic diagram, showing the relative-acceleration components $(\mathbf{a}_{B/A})_t$ and $(\mathbf{a}_{B/A})_n$, Fig. 16–26c, we have

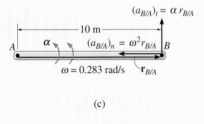

(c)

Fig. 16–26

$$\mathbf{a}_B = \mathbf{a}_A + (\mathbf{a}_{B/A})_t + (\mathbf{a}_{B/A})_n$$

$$\begin{bmatrix} a_B \\ \angle 45° \end{bmatrix} = \begin{bmatrix} 3 \text{ m/s}^2 \\ 45° \end{bmatrix} + \begin{bmatrix} \alpha(10 \text{ m}) \\ \uparrow \end{bmatrix} + \begin{bmatrix} (0.283 \text{ rad/s})^2(10 \text{ m}) \\ \leftarrow \end{bmatrix}$$

Equating the x and y components yields Eqs. 1 and 2, and the solution proceeds as before.

EXAMPLE 16.14

At a given instant, the cylinder of radius r, shown in Fig. 16–27a, has an angular velocity $\boldsymbol{\omega}$ and angular acceleration $\boldsymbol{\alpha}$. Determine the velocity and acceleration of its center G if it rolls without slipping.

(a)

Solution (Vector Analysis)

As the cylinder rolls, point G moves along a straight line, and point A, located on the rim of the cylinder, moves along a curved path called a *cycloid*, Fig. 16–27b. We will apply the velocity and acceleration equations to these two points.

Velocity Analysis. Since no slipping occurs, at the instant A contacts the ground, $\mathbf{v}_A = \mathbf{0}$. Thus, from the kinematic diagram in Fig. 16–27c we have

$$\mathbf{v}_G = \mathbf{v}_A + \boldsymbol{\omega} \times \mathbf{r}_{G/A}$$
$$v_G\mathbf{i} = \mathbf{0} + (-\omega\mathbf{k}) \times (r\mathbf{j})$$
$$v_G = \omega r \qquad (1) \quad Ans.$$

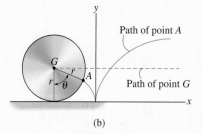

(b)

This same result can also be obtained directly by noting that point A represents the instantaneous center of zero velocity.

Kinematic Diagram. The acceleration of point G is horizontal since it moves along a *straight-line path*. *Just before* point A touches the ground, its velocity is directed downward along the y axis, Fig. 16–27b, and just after contact, its velocity is directed *upward*. For this reason, point A begins to accelerate upward when it leaves the ground at A, Fig. 16–27d. The magnitudes of $\mathbf{a}_A$ and $\mathbf{a}_G$ are unknown.

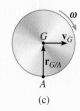

(c)

Acceleration Equation.

$$\mathbf{a}_G = \mathbf{a}_A + \boldsymbol{\alpha} \times \mathbf{r}_{G/A} - \omega^2 \mathbf{r}_{G/A}$$
$$a_G\mathbf{i} = a_A\mathbf{j} + (-\alpha\mathbf{k}) \times (r\mathbf{j}) - \omega^2(r\mathbf{j})$$

Evaluating the cross product and equating the $\mathbf{i}$ and $\mathbf{j}$ components yields

$$a_G = \alpha r \qquad (2) \quad Ans.$$
$$a_A = \omega^2 r \qquad (3)$$

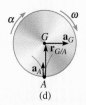

(d)

Fig. 16–27

These important results, that $v_G = \omega r$ and $a_G = \alpha r$, were also obtained in Example 16.4. They apply to any circular object, such as a ball, pulley, disk, etc., that rolls *without* slipping. Also, the fact that $a_A = \omega^2 r$ indicates that the instantaneous center of zero velocity, point A, *is not* a point of zero acceleration.

EXAMPLE 16.15

The ball rolls without slipping and has the angular motion shown in Fig. 16–28a. Determine the accelerations of point B and point A at this instant.

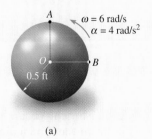

(a)

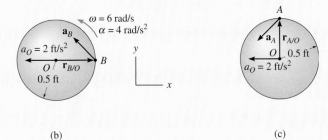

(b)   (c)

Fig. 16–28

Solution (Vector Analysis)

Kinematic Diagram. Using the results of the previous example, the center of the ball has an acceleration of $a_O = \alpha r = (4 \text{ rad/s}^2)(0.5 \text{ ft}) = 2 \text{ ft/s}^2$. We will apply the acceleration equation to points O and B and points O and A.

Acceleration Equation.

For point B, Fig. 16–28b,

$$\mathbf{a}_B = \mathbf{a}_O + \boldsymbol{\alpha} \times \mathbf{r}_{B/O} - \omega^2 \mathbf{r}_{B/O}$$
$$\mathbf{a}_B = -2\mathbf{i} + (4\mathbf{k}) \times (0.5\mathbf{i}) - (6)^2(0.5\mathbf{i})$$
$$\mathbf{a}_B = \{-20\mathbf{i} + 2\mathbf{j}\} \text{ ft/s}^2 \qquad\qquad Ans.$$

For point A, Fig. 16–28c,

$$\mathbf{a}_A = \mathbf{a}_O + \boldsymbol{\alpha} \times \mathbf{r}_{A/O} - \omega^2 \mathbf{r}_{A/O}$$
$$\mathbf{a}_A = -2\mathbf{i} + (4\mathbf{k}) \times (0.5\mathbf{j}) - (6)^2(0.5\mathbf{j})$$
$$\mathbf{a}_A = \{-4\mathbf{i} - 18\mathbf{j}\} \text{ ft/s}^2 \qquad\qquad Ans.$$

EXAMPLE 16.16

The spool shown in Fig. 16–29a unravels from the cord, such that at the instant shown it has an angular velocity of 3 rad/s and an angular acceleration of 4 rad/s². Determine the acceleration of point B.

Solution I (Vector Analysis)

The spool "appears" to be rolling downward without slipping at point A. Therefore, we can use the results of Example 16.14 to determine the acceleration of point G, i.e.,

$$a_G = \alpha r = 4 \text{ rad/s}^2 (0.5 \text{ ft}) = 2 \text{ ft/s}^2$$

We will apply the acceleration equation at points G and B.

Kinematic Diagram. Point B moves along a *curved path* having an *unknown* radius of curvature.* Its acceleration will be represented by its unknown x and y components as shown in Fig. 16–29b.

Acceleration Equation.

$$\mathbf{a}_B = \mathbf{a}_G + \boldsymbol{\alpha} \times \mathbf{r}_{B/G} - \omega^2 \mathbf{r}_{B/G}$$

$$(a_B)_x \mathbf{i} + (a_B)_y \mathbf{j} = -2\mathbf{j} + (-4\mathbf{k}) \times (0.75\mathbf{j}) - (3)^2(0.75\mathbf{j})$$

Equating the **i** and **j** terms, the component equations are

$$(a_B)_x = 4(0.75) = 3 \text{ ft/s}^2 \rightarrow \quad (1)$$

$$(a_B)_y = -2 - 6.75 = -8.75 \text{ ft/s}^2 = 8.75 \text{ ft/s}^2 \downarrow \quad (2)$$

The magnitude and direction of $\mathbf{a}_B$ are therefore

$$a_B = \sqrt{(3)^2 + (8.75)^2} = 9.25 \text{ ft/s}^2 \qquad \textit{Ans.}$$

$$\theta = \tan^{-1}\frac{8.75}{3} = 71.1° \quad \textit{Ans.}$$

Solution II (Scalar Analysis)

This problem may be solved by writing the scalar component equations directly. The kinematic diagram in Fig. 16–29c shows the relative-acceleration components $(\mathbf{a}_{B/G})_t$ and $(\mathbf{a}_{B/G})_n$. Thus,

$$\mathbf{a}_B = \mathbf{a}_G + (\mathbf{a}_{B/G})_t + (\mathbf{a}_{B/G})_n$$

$$\begin{bmatrix} (a_B)_x \\ \rightarrow \end{bmatrix} + \begin{bmatrix} (a_B)_y \\ \uparrow \end{bmatrix}$$

$$= \begin{bmatrix} 2 \text{ ft/s}^2 \\ \downarrow \end{bmatrix} + \begin{bmatrix} 4 \text{ rad/s}^2(0.75 \text{ ft}) \\ \rightarrow \end{bmatrix} + \begin{bmatrix} (3 \text{ rad/s})^2(0.75 \text{ ft}) \\ \downarrow \end{bmatrix}$$

The x and y components yield Eqs. 1 and 2 above.

*Realize that the path's radius of curvature ρ is *not* equal to the radius of the spool since the spool is *not* rotating about point G. Furthermore, ρ is *not* defined as the distance from A (IC) to B, since the location of the IC depends only on the velocity of a point and *not* the geometry of its path.

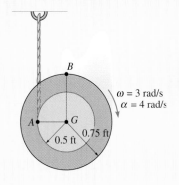

(a)

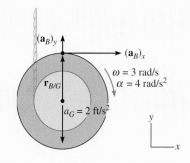

(b)

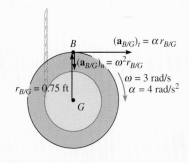

(c)

Fig. 16–29

EXAMPLE 16.17

The collar C in Fig. 16–30a is moving downward with an acceleration of 1 m/s². At the instant shown, it has a speed of 2 m/s which gives links CB and AB an angular velocity $\omega_{AB} = \omega_{CB} = 10$ rad/s. (See Example 16.8.) Determine the angular accelerations of CB and AB at this instant.

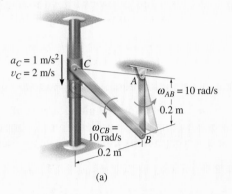

(a)

Solution (Vector Analysis)

Kinematic Diagram. The kinematic diagrams of *both* links AB and CB are shown in Fig. 16–30b. To solve, we will apply the appropriate kinematic equation to each link.

Acceleration Equation.

Link AB (rotation about a fixed axis):

$$\mathbf{a}_B = \boldsymbol{\alpha}_{AB} \times \mathbf{r}_B - \omega_{AB}^2 \mathbf{r}_B$$
$$\mathbf{a}_B = (\alpha_{AB}\mathbf{k}) \times (-0.2\mathbf{j}) - (10)^2(-0.2\mathbf{j})$$
$$\mathbf{a}_B = 0.2\alpha_{AB}\mathbf{i} + 20\mathbf{j}$$

Note that $\mathbf{a}_B$ has two components since it moves along a *curved path.*

Link BC (general plane motion): Using the result for $\mathbf{a}_B$ and applying Eq. 16–18, we have

$$\mathbf{a}_B = \mathbf{a}_C + \boldsymbol{\alpha}_{CB} \times \mathbf{r}_{B/C} - \omega_{CB}^2 \mathbf{r}_{B/C}$$
$$0.2\alpha_{AB}\mathbf{i} + 20\mathbf{j} = -1\mathbf{j} + (\alpha_{CB}\mathbf{k}) \times (0.2\mathbf{i} - 0.2\mathbf{j}) - (10)^2(0.2\mathbf{i} - 0.2\mathbf{j})$$
$$0.2\alpha_{AB}\mathbf{i} + 20\mathbf{j} = -1\mathbf{j} + 0.2\alpha_{CB}\mathbf{j} + 0.2\alpha_{CB}\mathbf{i} - 20\mathbf{i} + 20\mathbf{j}$$

Thus,

$$0.2\alpha_{AB} = 0.2\alpha_{CB} - 20$$
$$20 = -1 + 0.2\alpha_{CB} + 20$$

Solving,

$$\alpha_{CB} = 5 \text{ rad/s}^2 \text{\reflectbox{$\circlearrowright$}} \qquad\qquad Ans.$$
$$\alpha_{AB} = -95 \text{ rad/s}^2 = 95 \text{ rad/s}^2 \downcurvearrowright \qquad\qquad Ans.$$

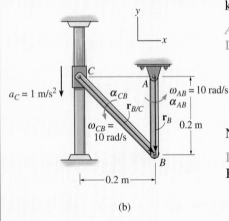

(b)

Fig. 16–30

E X A M P L E 16.18

The crankshaft AB of an engine turns with a clockwise angular acceleration of 20 rad/s², Fig. 16–31a. Determine the acceleration of the piston at the instant AB is in the position shown. At this instant $\omega_{AB} = 10$ rad/s and $\omega_{BC} = 2.43$ rad/s.

Solution (Vector Analysis)

Kinematic Diagram. The kinematic diagrams for both AB and BC are shown in Fig. 16–31b. Here $\mathbf{a}_C$ is vertical since C moves along a straight-line path.

Acceleration Equation. Expressing each of the position vectors in Cartesian vector form

$$\mathbf{r}_B = \{-0.25 \sin 45°\mathbf{i} + 0.25 \cos 45°\mathbf{j}\} \text{ ft} = \{-0.177\mathbf{i} + 0.177\mathbf{j}\} \text{ ft}$$
$$\mathbf{r}_{C/B} = \{0.75 \sin 13.6°\mathbf{i} + 0.75 \cos 13.6°\mathbf{j}\} \text{ ft} = \{0.176\mathbf{i} + 0.729\mathbf{j}\} \text{ ft}$$

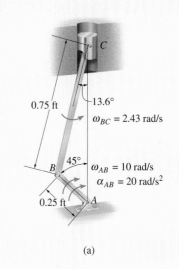

(a)

Crankshaft AB (rotation about a fixed axis):

$$\mathbf{a}_B = \boldsymbol{\alpha}_{AB} \times \mathbf{r}_B - \omega_{AB}^2 \mathbf{r}_B$$
$$= (-20\mathbf{k}) \times (-0.177\mathbf{i} + 0.177\mathbf{j}) - (10)^2(-0.177\mathbf{i} + 0.177\mathbf{j})$$
$$= \{21.21\mathbf{i} - 14.14\mathbf{j}\} \text{ ft/s}^2$$

Connecting Rod BC (general plane motion): Using the result for $\mathbf{a}_B$ and noting that $\mathbf{a}_C$ is in the vertical direction, we have

$$\mathbf{a}_C = \mathbf{a}_B + \boldsymbol{\alpha}_{BC} \times \mathbf{r}_{C/B} - \omega_{BC}^2 \mathbf{r}_{C/B}$$
$$a_C\mathbf{j} = 21.21\mathbf{i} - 14.14\mathbf{j} + (\alpha_{BC}\mathbf{k}) \times (0.176\mathbf{i} + 0.729\mathbf{j}) - (2.43)^2(0.176\mathbf{i} + 0.729\mathbf{j})$$
$$a_C\mathbf{j} = 21.21\mathbf{i} - 14.14\mathbf{j} + 0.176\alpha_{BC}\mathbf{j} - 0.729\alpha_{BC}\mathbf{i} - 1.04\mathbf{i} - 4.30\mathbf{j}$$
$$0 = 20.17 - 0.729\alpha_{BC}$$
$$a_C = 0.176\alpha_{BC} - 18.45$$

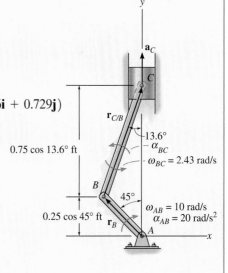

Solving yields

$$\alpha_{BC} = 27.7 \text{ rad/s}^2 \;\curvearrowleft$$
$$a_C = -13.6 \text{ ft/s}^2 \qquad\qquad Ans.$$

Since the piston is moving upward, the negative sign for a_C indicates that the piston is decelerating, i.e., $\mathbf{a}_C = \{-13.6\mathbf{j}\}$ ft/s². This causes the speed of the piston to decrease until AB becomes vertical, at which time the piston is momentarily at rest.

(b)

Fig. 16–31

16.8 Relative-Motion Analysis Using Rotating Axes

In the previous sections the relative-motion analysis for velocity and acceleration was described using a translating coordinate system. This type of analysis is useful for determining the motion of points on the *same* rigid body, or the motion of points located on several pin-connected rigid bodies. In some problems, however, rigid bodies (mechanisms) are constructed such that *sliding* will occur at their connections. The kinematic analysis for such cases is best performed if the motion is analyzed using a coordinate system which both *translates* and *rotates*. Furthermore, this frame of reference is useful for analyzing the motions of two points on a mechanism which are *not* located in the *same* rigid body and for specifying the kinematics of particle motion when the particle is moving along a rotating path.

In the following analysis two equations are developed which relate the velocity and acceleration of two points, one of which is the origin of a moving frame of reference subjected to both a translation and a rotation in the plane.* Due to the generality in the derivation which follows, these two points may represent either two particles moving independently of one another or two points located on the same (or different) rigid bodies.

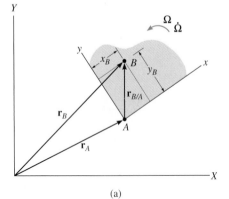

(a)

Fig. 16–32

Position.   Consider the two points A and B shown in Fig. 16–32a. Their location is specified by the position vectors $\mathbf{r}_A$ and $\mathbf{r}_B$, which are measured from the fixed X, Y, Z coordinate system. As shown in the figure, the "base point" A represents the origin of the x, y, z coordinate system, which is assumed to be both translating and rotating with respect to the X, Y, Z system. The position of B with respect to A is specified by the relative-position vector $\mathbf{r}_{B/A}$. The components of this vector may be expressed either in terms of unit vectors along the X, Y axes, i.e., $\mathbf{I}$ and $\mathbf{J}$, or by unit vectors along the x, y axes, i.e., $\mathbf{i}$ and $\mathbf{j}$. For the development which follows, $\mathbf{r}_{B/A}$ will be measured relative to the moving x, y frame of reference. Thus, if B has coordinates (x_B, y_B), Fig. 16–32a, then

$$\mathbf{r}_{B/A} = x_B\mathbf{i} + y_B\mathbf{j}$$

Using vector addition, the three position vectors in Fig. 16–32a are related by the equation

$$\mathbf{r}_B = \mathbf{r}_A + \mathbf{r}_{B/A} \tag{16–19}$$

At the instant considered, point A has a velocity $\mathbf{v}_A$ and an acceleration $\mathbf{a}_A$, while the angular velocity and angular acceleration of the x, y axes are Ω (omega) and $\dot{\Omega} = d\Omega/dt$, respectively. All these vectors are measured from the X, Y, Z frame of reference, although they may be

*The more general, three-dimensional motion of the points is developed in Sec. 20.4.

expressed in terms of either **I, J, K** or **i, j, k** components. Since planar motion is specified, by the right-hand rule Ω and $\dot{\Omega}$ are always directed *perpendicular* to the reference plane of motion, whereas $\mathbf{v}_A$ and $\mathbf{a}_A$ lie in this plane.

Velocity. The velocity of point B is determined by taking the time derivative of Eq. 16–19, which yields

$$\mathbf{v}_B = \mathbf{v}_A + \frac{d\mathbf{r}_{B/A}}{dt} \qquad (16\text{--}20)$$

The last term in this equation is evaluated as follows:

$$\frac{d\mathbf{r}_{B/A}}{dt} = \frac{d}{dt}(x_B\mathbf{i} + y_B\mathbf{j})$$

$$= \frac{dx_B}{dt}\mathbf{i} + x_B\frac{d\mathbf{i}}{dt} + \frac{dy_B}{dt}\mathbf{j} + y_B\frac{d\mathbf{j}}{dt}$$

$$= \left(\frac{dx_B}{dt}\mathbf{i} + \frac{dy_B}{dt}\mathbf{j}\right) + \left(x_B\frac{d\mathbf{i}}{dt} + y_B\frac{d\mathbf{j}}{dt}\right) \qquad (16\text{--}21)$$

The two terms in the first set of parentheses represent the components of velocity of point B as measured by an observer attached to the moving x, y, z coordinate system. These terms will be denoted by vector $(\mathbf{v}_{B/A})_{xyz}$. In the second set of parentheses the instantaneous time rate of change of the unit vectors $\mathbf{i}$ and $\mathbf{j}$ is measured by an observer located in the fixed X, Y, Z coordinate system. These changes, $d\mathbf{i}$ and $d\mathbf{j}$, are due *only* to the instantaneous *rotation* $d\theta$ of the x, y, z axes, causing $\mathbf{i}$ to become $\mathbf{i}' = \mathbf{i} + d\mathbf{i}$ and $\mathbf{j}$ to become $\mathbf{j}' = \mathbf{j} + d\mathbf{j}$, Fig. 16–32b. As shown, the *magnitudes* of both $d\mathbf{i}$ and $d\mathbf{j}$ equal 1 $(d\theta)$, since $i = i' = j = j' = 1$. The *direction* of $d\mathbf{i}$ is defined by $+\mathbf{j}$, since $d\mathbf{i}$ is tangent to the path described by the arrowhead of $\mathbf{i}$ in the limit as $\Delta t \rightarrow dt$. Likewise, $d\mathbf{j}$ acts in the $-\mathbf{i}$ direction, Fig. 16–32b. Hence,

$$\frac{d\mathbf{i}}{dt} = \frac{d\theta}{dt}(\mathbf{j}) = \Omega\mathbf{j} \qquad \frac{d\mathbf{j}}{dt} = \frac{d\theta}{dt}(-\mathbf{i}) = -\Omega\mathbf{i}$$

Viewing the axes in three dimensions, Fig. 16–32c, and nothing that $\Omega = \Omega\mathbf{k}$, we can express the above derivatives in terms of the cross product as

$$\frac{d\mathbf{i}}{dt} = \Omega \times \mathbf{i} \qquad \frac{d\mathbf{j}}{dt} = \Omega \times \mathbf{j} \qquad (16\text{--}22)$$

Substituting these results into Eq. 16–21 and using the distributive property of the vector cross product, we obtain

$$\frac{d\mathbf{r}_{B/A}}{dt} = (\mathbf{v}_{B/A})_{xyz} + \Omega \times (x_B\mathbf{i} + y_B\mathbf{j}) = (\mathbf{v}_{B/A})_{xyz} + \Omega \times \mathbf{r}_{B/A} \quad (16\text{--}23)$$

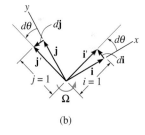

(b)

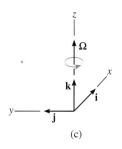

(c)

Fig. 16–32

Hence, Eq. 16–20 becomes

$$\mathbf{v}_B = \mathbf{v}_A + \boldsymbol{\Omega} \times \mathbf{r}_{B/A} + (\mathbf{v}_{B/A})_{xyz} \qquad (16\text{–}24)$$

where

$\mathbf{v}_B$ = velocity of B, measured from the X, Y, Z reference

$\mathbf{v}_A$ = velocity of the origin A of the x, y, z reference, measured from the X, Y, Z reference

$(\mathbf{v}_{B/A})_{xyz}$ = relative velocity of "B with respect to A," as measured by an observer attached to the *rotating* x, y, z reference

$\boldsymbol{\Omega}$ = angular velocity of the x, y, z reference, measured from the X, Y, Z reference

$\mathbf{r}_{B/A}$ = relative position of "B with respect to A"

Comparing Eq. 16–24 with Eq. 16–16 ($\mathbf{v}_B = \mathbf{v}_A + \boldsymbol{\Omega} \times \mathbf{r}_{B/A}$), which is valid for a translating frame of reference, it can be seen that the only difference between the equations is represented by the term $(\mathbf{v}_{B/A})_{xyz}$.

When applying Eq. 16–24 it is often useful to understand what each of the terms represents. In order of appearance, they are as follows:

$\mathbf{v}_B$ $\left\{\begin{array}{l}\text{absolute velocity of } B\end{array}\right.$ $\left.\begin{array}{l}\text{motion of } B \text{ observed} \\ \text{from the } X, Y, Z \text{ frame}\end{array}\right\}$

(equals)

$\mathbf{v}_A$ $\left\{\begin{array}{l}\text{absolute velocity of origin} \\ \text{of } x, y, z \text{ frame}\end{array}\right.$

(plus)

$\left.\begin{array}{l}\text{motion of } x, y, z \text{ frame} \\ \text{observed from the } X, Y, Z \\ \text{frame}\end{array}\right\}$

$\boldsymbol{\Omega} \times \mathbf{r}_{B/A}$ $\left\{\begin{array}{l}\text{angular velocity effect caused} \\ \text{by rotation of } x, y, z \text{ frame}\end{array}\right.$

(plus)

$(\mathbf{v}_{B/A})_{xyz}$ $\left\{\begin{array}{l}\text{relative velocity of } B \\ \text{with respect to } A\end{array}\right.$ $\left.\begin{array}{l}\text{motion of } B \text{ observed} \\ \text{from the } x, y, z \text{ frame}\end{array}\right\}$

Acceleration. The acceleration of B, observed from the X, Y, Z coordinate system, may be expressed in terms of its motion measured with respect to the rotating or moving system of coordinates by taking the time derivative of Eq. 16–24, i.e.,

$$\frac{d\mathbf{v}_B}{dt} = \frac{d\mathbf{v}_A}{dt} + \frac{d\boldsymbol{\Omega}}{dt} \times \mathbf{r}_{B/A} + \boldsymbol{\Omega} \times \frac{d\mathbf{r}_{B/A}}{dt} + \frac{d(\mathbf{v}_{B/A})_{xyz}}{dt}$$

$$\mathbf{a}_B = \mathbf{a}_A + \dot{\boldsymbol{\Omega}} \times \mathbf{r}_{B/A} + \boldsymbol{\Omega} \times \frac{d\mathbf{r}_{B/A}}{dt} + \frac{d(\mathbf{v}_{B/A})_{xyz}}{dt} \quad (16\text{–}25)$$

Here $\dot{\boldsymbol{\Omega}} = d\boldsymbol{\Omega}/dt$ is the angular acceleration of the x, y, z coordinate system. For planar motion $\dot{\boldsymbol{\Omega}}$ is always perpendicular to the plane of motion, and therefore $\dot{\boldsymbol{\Omega}}$ measures *only the change in magnitude* of $\boldsymbol{\Omega}$. The derivative $d\mathbf{r}_{B/A}/dt$ in Eq. 16–25 is defined by Eq. 16–23, so that

$$\boldsymbol{\Omega} \times \frac{d\mathbf{r}_{B/A}}{dt} = \boldsymbol{\Omega} \times (\mathbf{v}_{B/A})_{xyz} + \boldsymbol{\Omega} \times (\boldsymbol{\Omega} \times \mathbf{r}_{B/A}) \quad (16\text{–}26)$$

Finding the time derivative of $(\mathbf{v}_{B/A})_{xyz} = (v_{B/A})_x \mathbf{i} + (v_{B/A})_y \mathbf{j}$,

$$\frac{d(\mathbf{v}_{B/A})_{xyz}}{dt} = \left[\frac{d(v_{B/A})_x}{dt} \mathbf{i} + \frac{d(v_{B/A})_y}{dt} \mathbf{j} \right] + \left[(v_{B/A})_x \frac{d\mathbf{i}}{dt} + (v_{B/A})_y \frac{d\mathbf{j}}{dt} \right]$$

The two terms in the first set of brackets represent the components of acceleration of point B as measured by an observer attached to the moving coordinate system. These terms will be denoted by $(\mathbf{a}_{B/A})_{xyz}$. The terms in the second set of brackets can be simplified using Eqs. 16–22.

$$\frac{d(\mathbf{v}_{B/A})_{xyz}}{dt} = (\mathbf{a}_{B/A})_{xyz} + \boldsymbol{\Omega} \times (\mathbf{v}_{B/A})_{xyz}$$

Substituting this and Eq. 16–26 into Eq. 16–25 and rearranging terms,

$$\boxed{\mathbf{a}_B = \mathbf{a}_A + \dot{\boldsymbol{\Omega}} \times \mathbf{r}_{B/A} + \boldsymbol{\Omega} \times (\boldsymbol{\Omega} \times \mathbf{r}_{B/A}) + 2\boldsymbol{\Omega} \times (\mathbf{v}_{B/A})_{xyz} + (\mathbf{a}_{B/A})_{xyz}}$$

$$(16\text{–}27)$$

where

$$\mathbf{a}_B = \text{acceleration of } B, \text{ measured from the } X, Y, Z \text{ reference}$$

$$\mathbf{a}_A = \text{acceleration of the origin } A \text{ of the } x, y, z \text{ reference, measured from the } X, Y, Z \text{ reference}$$

$$(\mathbf{a}_{B/A})_{xyz}, (\mathbf{v}_{B/A})_{xyz} = \text{relative acceleration and relative velocity of "} B \text{ with respect to } A \text{," as measured by an observer attached to the } rotating\ x, y, z \text{ reference}$$

$$\dot{\boldsymbol{\Omega}}, \boldsymbol{\Omega} = \text{angular acceleration and angular velocity of the } x, y, z \text{ reference, measured from the } X, Y, Z \text{ reference}$$

$$\mathbf{r}_{B/A} = \text{relative position of "} B \text{ with respect to } A \text{"}$$

If Eq. 16–27 is compared with Eq. 16–18, written in the form $\mathbf{a}_B = \mathbf{a}_A + \dot{\boldsymbol{\Omega}} \times \mathbf{r}_{B/A} + \boldsymbol{\Omega} \times (\boldsymbol{\Omega} \times \mathbf{r}_{B/A})$, which is valid for a translating frame of reference, it can be seen that the difference between the equations is represented by the terms $2\boldsymbol{\Omega} \times (\mathbf{v}_{B/A})_{xyz}$ and $(\mathbf{a}_{B/A})_{xyz}$. In particular, $2\boldsymbol{\Omega} \times (\mathbf{v}_{B/A})_{xyz}$ is called the *Coriolis acceleration*, named after the French engineer G. C. Coriolis, who was the first to determine it. This term represents the difference in the acceleration of B as measured from nonrotating and rotating x, y, z axes. As indicated by the vector cross product, the Coriolis acceleration will *always* be perpendicular to both $\boldsymbol{\Omega}$ and $(\mathbf{v}_{B/A})_{xyz}$. It is an important component of the acceleration which must be considered whenever rotating reference frames are used. This often occurs, for example, when studying the accelerations and forces which act on rockets, long-range projectiles, or other bodies having motions whose measurements are significantly affected by the rotation of the earth.

The following interpretation of the terms in Eq. 16–27 may be useful when applying this equation to the solution of problems.

$\mathbf{a}_B$	{absolute acceleration of B	} motion of B observed from the X, Y, Z frame
	(equals)	
$\mathbf{a}_A$	{ absolute acceleration of origin of x, y, z frame	
	(plus)	motion of
$\dot{\boldsymbol{\Omega}} \times \mathbf{r}_{B/A}$	{ angular acceleration effect caused by rotation of x, y, z frame	x, y, z frame observed from the X, Y, Z frame
	(plus)	
$\boldsymbol{\Omega} \times (\boldsymbol{\Omega} \times \mathbf{r}_{B/A})$	{ angular velocity effect caused by rotation of x, y, z frame	
	(plus)	
$2\boldsymbol{\Omega} \times (\mathbf{v}_{B/A})_{xyz}$	{ combined effect of B moving relative to x, y, z coordinates and rotation of x, y, z frame	} interacting motion
	(plus)	
$(\mathbf{a}_{B/A})_{xyz}$	{ relative acceleration of B with respect to A	} motion of B observed from the x, y, z frame

PROCEDURE FOR ANALYSIS

Equations 16–24 and 16–27 can be applied to the solution of problems involving the planar motion of particles or rigid bodies using the following procedure.

Coordinate Axes.

- Choose an appropriate location for the origin and proper orientation of the axes for both the X, Y, Z and moving x, y, z reference frames.
- Most often solutions are easily obtained if at the instant considered:
  (1) the origins are coincident
  (2) the corresponding axes are collinear
  (3) the corresponding axes are parallel
- The moving frame should be selected fixed to the body or device along which the relative motion occurs.

Kinematic Equations.

- After defining the origin A of the moving reference and specifying the moving point B, Eqs. 16–24 and 16–27 should be written in symbolic form

$$\mathbf{v}_B = \mathbf{v}_A + \boldsymbol{\Omega} \times \mathbf{r}_{B/A} + (\mathbf{v}_{B/A})_{xyz}$$

$$\mathbf{a}_B = \mathbf{a}_A + \dot{\boldsymbol{\Omega}} \times \mathbf{r}_{B/A} + \boldsymbol{\Omega} \times (\boldsymbol{\Omega} \times \mathbf{r}_{B/A}) + 2\boldsymbol{\Omega} \times (\mathbf{v}_{B/A})_{xyz} + (\mathbf{a}_{B/A})_{xyz}$$

- The Cartesian components of all these vectors may be expressed along either the X, Y, Z axes or the x, y, z axes. The choice is arbitrary provided a consistent set of unit vectors is used.
- Motion of the moving reference, is expressed by $\mathbf{v}_A$, $\mathbf{a}_A$, $\boldsymbol{\Omega}$, and $\dot{\boldsymbol{\Omega}}$; and motion of B with respect to the moving reference, is expressed by $\mathbf{r}_{B/A}$, $(\mathbf{v}_{B/A})_{xyz}$, and $(\mathbf{a}_{B/A})_{xyz}$.

The rotation of the dumping bin of the truck about point C is operated by the extension of the hydraulic cylinder AB. To determine the rotation of the bin due to this extension, we can use the equations of relative motion and fix the x, y axes to the cylinder so that the relative motion of the cylinder's extension occurs along the y axis.

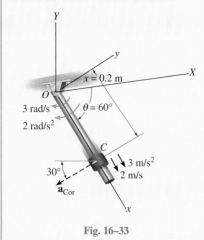

Fig. 16–33

At the instant $\theta = 60°$, the rod in Fig. 16–33 has an angular velocity of 3 rad/s and an angular acceleration of 2 rad/s². At this same instant, the collar C is traveling outward along the rod such that when $x = 0.2$ m the velocity is 2 m/s and the acceleration is 3 m/s², both measured relative to the rod. Determine the Coriolis acceleration and the velocity and acceleration of the collar at this instant.

Solution

Coordinate Axes.   The origin of both coordinate systems is located at point O, Fig. 16–33. Since motion of the collar is reported relative to the rod, the moving x, y, z frame of reference is *attached* to the rod.

Kinematic Equations.

$$\mathbf{v}_C = \mathbf{v}_O + \mathbf{\Omega} \times \mathbf{r}_{C/O} + (\mathbf{v}_{C/O})_{xyz} \tag{1}$$

$$\mathbf{a}_C = \mathbf{a}_O + \dot{\mathbf{\Omega}} \times \mathbf{r}_{C/O} + \mathbf{\Omega} \times (\mathbf{\Omega} \times \mathbf{r}_{C/O}) + 2\mathbf{\Omega} \times (\mathbf{v}_{C/O})_{xyz} + (\mathbf{a}_{C/O})_{xyz} \tag{2}$$

It will be simpler to express the data in terms of $\mathbf{i}, \mathbf{j}, \mathbf{k}$ component vectors rather than $\mathbf{I}, \mathbf{J}, \mathbf{K}$ components. Hence,

Motion of moving reference	*Motion of C with respect to moving reference*
$\mathbf{v}_O = \mathbf{0}$	$\mathbf{r}_{C/O} = \{0.2\mathbf{i}\}$ m
$\mathbf{a}_O = \mathbf{0}$	$(\mathbf{v}_{C/O})_{xyz} = \{2\mathbf{i}\}$ m/s
$\mathbf{\Omega} = \{-3\mathbf{k}\}$ rad/s	$(\mathbf{a}_{C/O})_{xyz} = \{3\mathbf{i}\}$ m/s²
$\dot{\mathbf{\Omega}} = \{-2\mathbf{k}\}$ rad/s²	

From Eq. 2, the Coriolis acceleration is defined as

$$\mathbf{a}_{\text{Cor}} = 2\mathbf{\Omega} \times (\mathbf{v}_{C/O})_{xyz} = 2(-3\mathbf{k}) \times (2\mathbf{i}) = \{-12\mathbf{j}\} \text{ m/s}^2 \quad Ans.$$

This vector is shown dashed in Fig. 16–33. If desired, it may be resolved into $\mathbf{I}, \mathbf{J}$ components acting along the X and Y axes, respectively.

The velocity and acceleration of the collar are determined by substituting the data into Eqs. 1 and 2 and evaluating the cross products, which yields

$$\mathbf{v}_C = \mathbf{v}_O + \mathbf{\Omega} \times \mathbf{r}_{C/O} + (\mathbf{v}_{C/O})_{xyz}$$
$$= 0 + (-3\mathbf{k}) \times (0.2\mathbf{i}) + 2\mathbf{i}$$
$$= \{2\mathbf{i} - 0.6\mathbf{j}\} \text{ m/s} \qquad\qquad Ans.$$

$$\mathbf{a}_C = \mathbf{a}_O + \dot{\mathbf{\Omega}} \times \mathbf{r}_{C/O} + \mathbf{\Omega} \times (\mathbf{\Omega} \times \mathbf{r}_{C/O}) + 2\mathbf{\Omega} \times (\mathbf{v}_{C/O})_{xyz} + (\mathbf{a}_{C/O})_{xyz}$$
$$= 0 + (-2\mathbf{k}) \times (0.2\mathbf{i}) + (-3\mathbf{k}) \times [(-3\mathbf{k}) \times (0.2\mathbf{i})] + 2(-3\mathbf{k}) \times (2\mathbf{i}) + 3\mathbf{i}$$
$$= 0 - 0.4\mathbf{j} - 1.80\mathbf{i} - 12\mathbf{j} + 3\mathbf{i}$$
$$= \{1.20\mathbf{i} - 12.4\mathbf{j}\} \text{ m/s}^2 \qquad\qquad Ans.$$

EXAMPLE 16.20

The rod AB, shown in Fig. 16–34, rotates clockwise such that it has an angular velocity $\omega_{AB} = 3$ rad/s and angular acceleration $\alpha_{AB} = 4$ rad/s^2 when $\theta = 45°$. Determine the angular motion of rod DE at this instant. The collar at C is pin connected to AB and slides over rod DE.

Solution

Coordinate Axes. The origin of both the fixed and moving frames of reference is located at D, Fig. 16–34. Furthermore, the x, y, z reference is attached to and rotates with rod DE so that the relative motion of the collar is easy to follow.

Kinematic Equations.

$$\mathbf{v}_C = \mathbf{v}_D + \mathbf{\Omega} \times \mathbf{r}_{C/D} + (\mathbf{v}_{C/D})_{xyz} \qquad (1)$$

$$\mathbf{a}_C = \mathbf{a}_D + \dot{\mathbf{\Omega}} \times \mathbf{r}_{C/D} + \mathbf{\Omega} \times (\mathbf{\Omega} \times \mathbf{r}_{C/D}) + 2\mathbf{\Omega} \times (\mathbf{v}_{C/D})_{xyz} + (\mathbf{a}_{C/D})_{xyz} \qquad (2)$$

All vectors will be expressed in terms of $\mathbf{i}, \mathbf{j}, \mathbf{k}$ components.

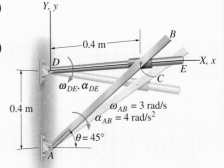

Fig. 16–34

Motion of moving reference	Motion of C with respect to moving reference
$\mathbf{v}_D = \mathbf{0}$	$\mathbf{r}_{C/D} = \{0.4\mathbf{i}\}$ m
$\mathbf{a}_D = \mathbf{0}$	$(\mathbf{v}_{C/D})_{xyz} = (v_{C/D})_{xyz}\mathbf{i}$
$\mathbf{\Omega} = -\omega_{DE}\mathbf{k}$	$(\mathbf{a}_{C/D})_{xyz} = (a_{C/D})_{xyz}\mathbf{i}$
$\dot{\mathbf{\Omega}} = -\alpha_{DE}\mathbf{k}$	

Motion of C: Since the collar moves along a *circular path*, its velocity and acceleration can be determined using Eqs. 16–9 and 16–14.

$$\mathbf{v}_C = \boldsymbol{\omega}_{AB} \times \mathbf{r}_{C/A} = (-3\mathbf{k}) \times (0.4\mathbf{i} + 0.4\mathbf{j}) = \{1.2\mathbf{i} - 1.2\mathbf{j}\} \text{ m/s}$$

$$\mathbf{a}_C = \boldsymbol{\alpha}_{AB} \times \mathbf{r}_{C/A} - \omega_{AB}^2 \mathbf{r}_{C/A}$$
$$= (-4\mathbf{k}) \times (0.4\mathbf{i} + 0.4\mathbf{j}) - (3)^2(0.4\mathbf{i} + 0.4\mathbf{j}) = \{-2\mathbf{i} - 5.2\mathbf{j}\} \text{ m/s}^2$$

Substituting the data into Eqs. 1 and 2, we have

$$\mathbf{v}_C = \mathbf{v}_D + \mathbf{\Omega} \times \mathbf{r}_{C/D} + (\mathbf{v}_{C/D})_{xyz}$$
$$1.2\mathbf{i} - 1.2\mathbf{j} = \mathbf{0} + (-\omega_{DE}\mathbf{k}) \times (0.4\mathbf{i}) + (v_{C/D})_{xyz}\mathbf{i}$$
$$1.2\mathbf{i} - 1.2\mathbf{j} = \mathbf{0} - 0.4\omega_{DE}\mathbf{j} + (v_{C/D})_{xyz}\mathbf{i}$$
$$(v_{C/D})_{xyz} = 1.2 \text{ m/s}$$
$$\omega_{DE} = 3 \text{ rad/s} \downarrow \qquad \qquad Ans.$$

$$\mathbf{a}_C = \mathbf{a}_D + \dot{\mathbf{\Omega}} \times \mathbf{r}_{C/D} + \mathbf{\Omega} \times (\mathbf{\Omega} \times \mathbf{r}_{C/D}) + 2\mathbf{\Omega} \times (\mathbf{v}_{C/D})_{xyz} + (\mathbf{a}_{C/D})_{xyz}$$
$$-2\mathbf{i} - 5.2\mathbf{j} = \mathbf{0} + (-\alpha_{DE}\mathbf{k}) \times (0.4\mathbf{i}) + (-3\mathbf{k}) \times [(-3\mathbf{k}) \times (0.4\mathbf{i})]$$
$$+2(-3\mathbf{k}) \times (1.2\mathbf{i}) + (a_{C/D})_{xyz}\mathbf{i}$$
$$-2\mathbf{i} - 5.2\mathbf{j} = -0.4\alpha_{DE}\mathbf{j} - 3.6\mathbf{i} - 7.2\mathbf{j} + (a_{C/D})_{xyz}\mathbf{i}$$
$$(a_{C/D})_{xyz} = 1.6 \text{ m/s}^2$$
$$\alpha_{DE} = -5 \text{ rad/s}^2 = 5 \text{ rad/s}^2 \uparrow \qquad \qquad Ans.$$

Two planes A and B are flying at the same elevation and have the motions shown in Fig. 16–35. Determine the velocity and acceleration of A as measured by the pilot of B.

Solution

Coordinate Axes. Since the relative motion of A with respect to the pilot in B is being sought, the x, y, z axes are attached to plane B, Fig. 16–35. At the *instant* considered, the origin B coincides with the origin of the fixed X, Y, Z frame.

Kinematic Equations.

$$\mathbf{v}_A = \mathbf{v}_B + \mathbf{\Omega} \times \mathbf{r}_{A/B} + (\mathbf{v}_{A/B})_{xyz} \tag{1}$$

$$\mathbf{a}_A = \mathbf{a}_B + \dot{\mathbf{\Omega}} \times \mathbf{r}_{A/B} + \mathbf{\Omega} \times (\mathbf{\Omega} \times \mathbf{r}_{A/B}) + 2\mathbf{\Omega} \times (\mathbf{v}_{A/B})_{xyz} + (\mathbf{a}_{A/B})_{xyz} \tag{2}$$

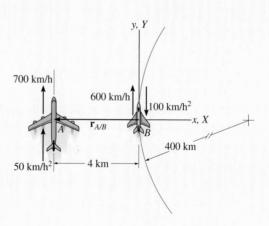

Fig. 16–35

Motion of moving reference:

$$\mathbf{v}_B = \{600\mathbf{j}\} \text{ km/h}$$

$$(a_B)_n = \frac{v_B^2}{\rho} = \frac{(600)^2}{400} = 900 \text{ km/h}^2$$

$$\mathbf{a}_B = (\mathbf{a}_B)_n + (\mathbf{a}_B)_t = \{900\mathbf{i} - 100\mathbf{j}\} \text{ km/h}^2$$

$$\Omega = \frac{v_B}{\rho} = \frac{600 \text{ km/h}}{400 \text{ km}} = 1.5 \text{ rad/h} \downarrow \qquad \mathbf{\Omega} = \{-1.5\mathbf{k}\} \text{ rad/h}$$

$$\dot{\Omega} = \frac{(a_B)_t}{\rho} = \frac{100 \text{ km/h}^2}{400 \text{ km}} = 0.25 \text{ rad/h}^2 \uparrow \qquad \dot{\mathbf{\Omega}} = \{0.25\mathbf{k}\} \text{ rad/h}^2$$

Motion of A with respect to moving reference:

$$\mathbf{r}_{A/B} = \{-4\mathbf{i}\} \text{ km} \quad (\mathbf{v}_{A/B})_{xyz} = ? \quad (\mathbf{a}_{A/B})_{xyz} = ?$$

Substituting the data into Eqs. 1 and 2, realizing that $\mathbf{v}_A = \{700\mathbf{j}\}$ km/h and $\mathbf{a}_A = \{50\mathbf{j}\}$ km/h^2, we have

$$\mathbf{v}_A = \mathbf{v}_B + \mathbf{\Omega} \times \mathbf{r}_{A/B} + (\mathbf{v}_{A/B})_{xyz}$$

$$700\mathbf{j} = 600\mathbf{j} + (-1.5\mathbf{k}) \times (-4\mathbf{i}) + (\mathbf{v}_{A/B})_{xyz}$$

$$(\mathbf{v}_{A/B})_{xyz} = \{94\mathbf{j}\} \text{ km/h} \qquad\qquad\qquad \textit{Ans.}$$

$$\mathbf{a}_A = \mathbf{a}_B + \dot{\mathbf{\Omega}} \times \mathbf{r}_{A/B} + \mathbf{\Omega} \times (\mathbf{\Omega} \times \mathbf{r}_{A/B}) + 2\mathbf{\Omega} \times (\mathbf{v}_{A/B})_{xyz} + (\mathbf{a}_{A/B})_{xyz}$$

$$50\mathbf{j} = (900\mathbf{i} - 100\mathbf{j}) + (0.25\mathbf{k}) \times (-4\mathbf{i})$$

$$+ (-1.5\mathbf{k}) \times [(-1.5\mathbf{k}) \times (-4\mathbf{i})] + 2(-1.5\mathbf{k}) \times (94\mathbf{j}) + (\mathbf{a}_{A/B})_{xyz}$$

$$(\mathbf{a}_{A/B})_{xyz} = \{-1191\mathbf{i} + 151\mathbf{j}\} \text{ km/h}^2 \qquad\qquad \textit{Ans.}$$

The solution of this problem should be compared with that of Example 12.26.

DESIGN PROJECTS

16–1D. DESIGN OF A BELT TRANSMISSION SYSTEM

The wheel A is used in a textile mill and must be rotated counterclockwise at 4 rad/s. This can be done using a motor which is mounted on the platform at the location shown. If the shaft B on the motor can rotate clockwise 50 rad/s, design a method for transmitting the rotation from B to A. Use a series of belts and pulleys as a basis for your design. A belt drives the rotation of wheel A by wrapping around its outer surface, and a pulley can be attached to the shaft of the motor as well as anywhere else. Do not let the length of any belt be longer than 6 ft. Submit a drawing of your design and the calculations of the kinematics. Also, determine the total cost of materials if any belt costs $2.50, and any pulley costs $2r$, where r is the radius of the pulley in inches.

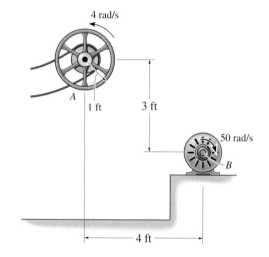

Prob. 16–1D

16–2D. DESIGN OF AN OSCILLATING LINK MECHANISM

The operation of a sewing machine requires the 200-mm-long bar to oscillate back-and-forth through an angle of 60° every 0.2 seconds. A motor having a drive shaft which turns at 40 rad/s is available to provide the necessary power. Specify the location of the motor and design a mechanism required to perform the motion. Submit a drawing of your design, showing the placement of the motor, and compute the velocity and acceleration of the end A of the link as a function of its angle of rotation $0° \le \theta \le 60°$.

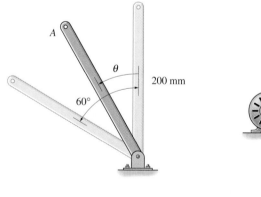

Prob. 16–2D

16–3D. DESIGN OF A RETRACTABLE AIRCRAFT LANDING-GEAR MECHANISM

The nose wheel of a small plane is attached to member *AB*, which is pinned to the aircraft frame at *B*. Design a mechanism that will allow the wheel to be fully retracted forward; i.e., rotated clockwise 90°, in $t \leq 4$ seconds. Use a hydraulic cylinder which has a closed length of 1.25 ft and, if needed, a fully extended length of 2 ft. Make sure that your design holds the wheel in a stable position when the wheel is on the ground. Show plots of the angular velocity and angular acceleration of *AB* versus its angular position $0° \leq \theta \leq 90°$.

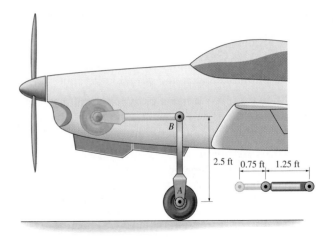

Prob. 16–3D

16–4D. DESIGN OF A SAW LINK MECHANISM

The saw blade in a lumber mill is required to remain in the horizontal position and undergo a complete back-and-forth motion in 2 seconds. An electric motor, having a shaft rotation of 50 rad/s, is available to power the saw and can be located anywhere. Design a mechanism that will transfer the rotation of the motor's shaft to the saw blade. Submit drawings of your design and calculations of the kinematics of the saw blade. Include a plot of the velocity and acceleration of the saw blade as a function of it horizontal position. Note that to cut through the log the blade must be allowed to move freely downward as well as back and forth.

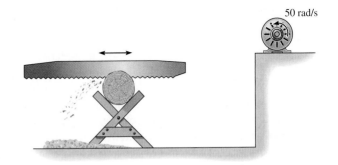

Prob. 16–4D

CHAPTER REVIEW

- *Rigid-Body Planar Motion.* A rigid body undergoes three types of planar motion: translation, rotation about a fixed axis, and general plane motion.

- *Translation.* When a body has rectilinear translation, all the particles of the body travel along straight-line paths. If the paths have the same radius of curvature, then curvilinear translation occurs. Provided we know the motion of one of the particles, then the motion of all of the others is also known.

- *Rotation about a Fixed Axis.* For this type of motion, all of the particles move along circular paths. Here, all line segments in the body undergo the same angular displacement, angular velocity, and angular acceleration. The differential relationships between these kinematic quantities are

$$\omega = d\theta/dt \qquad \alpha = d\omega/dt \qquad \alpha\, d\theta = \omega\, d\omega$$

If the angular acceleration is constant, $\alpha = \alpha_c$, then these equations can be integrated and become

$$\omega = \omega_0 + \alpha_c t$$
$$\theta = \theta_0 + \omega_0 t + \tfrac{1}{2}\alpha_c t^2$$
$$\omega^2 = \omega_0^2 + 2\alpha_c(\theta - \theta_0)$$

Once the angular motion of the body is known, then the velocity of any particle a distance r from the axis of rotation is

$$v = \omega r \qquad \text{or} \qquad \mathbf{v} = \boldsymbol{\omega} \times \mathbf{r}$$

The acceleration of the particle has two components. The tangential component accounts for the change in the magnitude of the velocity

$$a_t = \alpha r \qquad \text{or} \qquad \mathbf{a}_t = \boldsymbol{\alpha} \times \mathbf{r}$$

The normal component accounts for the change in the velocity direction

$$a_n = \omega^2 r \qquad \text{or} \qquad \mathbf{a}_n = -\omega^2 \mathbf{r}$$

- *General Plane Motion.* When a body undergoes general plane motion, it simultaneously translates and rotates. A typical example is a wheel that rolls without slipping. There are several methods for analyzing this motion.

Absolute Motion Analysis. If the motion of a point on a body or the angular motion of a line is known, then it may be possible to relate this motion to that of another point or line using an absolute motion analysis. To do so, linear position coordinates s or angular position coordinates θ are established (measured from a fixed point or line). These position coordinates are then related using the geometry of the body. The time derivative of this equation gives the relationship between the velocities and/or the angular velocities. A second time derivative relates the accelerations and/or the angular accelerations.

Relative Velocity Analysis. General plane motion can also be analyzed using a relative-motion analysis between two points A and B. This method considers the motion in parts: first a translation of the selected base point A, then a relative "rotation" of the body about point A, measured from a translating axis. The velocities of the two points A and B are then related using

$$\mathbf{v}_B = \mathbf{v}_A + \mathbf{v}_{B/A}$$

This equation can be applied in Cartesian vector form, written as

$$\mathbf{v}_B = \mathbf{v}_A + \boldsymbol{\omega} \times \mathbf{r}_{B/A}$$

In a similar manner, for acceleration,

$$\mathbf{a}_B = \mathbf{a}_A + (\mathbf{a}_{B/A})_t + (\mathbf{a}_{B/A})_n$$

or

$$\mathbf{a}_B = \mathbf{a}_A + \boldsymbol{\alpha} \times \mathbf{r}_{B/A} - \omega^2 \mathbf{r}_{B/A}$$

Since the relative motion is viewed as circular motion about the base point, point B will have a velocity $\mathbf{v}_{B/A}$, that is tangent to the circle. It also has two components of acceleration, $(\mathbf{a}_{B/A})_t$, and $(\mathbf{a}_{B/A})_n$. It is important to also realize that $\mathbf{a}_A$ and $\mathbf{a}_B$ may have two components if these points move along curved paths.

Instantaneous Center of Zero Velocity. If the base point A is selected as having zero velocity, then the relative velocity equation becomes

$$\mathbf{v}_B = \boldsymbol{\omega} \times \mathbf{r}_{B/A}$$

In this case, motion appears as if the body is rotating about an instantaneous axis.

The instantaneous center of rotation (*IC*) can be established provided the directions of the velocities of any two points on the body are known. Since the radial line r will always be perpendicular to each velocity, then the *IC* is at the point of intersection of these two radial lines. Its measured location is determined from the geometry of the body. Once it is established, then the velocity of any point P on the body can be determined from $v = \omega r$, where r extends from the *IC* to point P.

- *Relative Motion Using Rotating Axes.* Problems that involve connected members that slide relative to one another, or points not located on the same body, can be analyzed using a relative-motion analysis referenced from a rotating frame. The equations of relative motion are

$$\mathbf{v}_B = \mathbf{v}_A + \boldsymbol{\Omega} \times \mathbf{r}_{B/A} + (\mathbf{v}_{B/A})_{xyz}$$
$$\mathbf{a}_B = \mathbf{a}_A + \dot{\boldsymbol{\Omega}} \times \mathbf{r}_{B/A} + \boldsymbol{\Omega} \times (\boldsymbol{\Omega} \times \mathbf{r}_{B/A}) + 2\boldsymbol{\Omega} \times (\mathbf{v}_{B/A})_{xyz} + (\mathbf{a}_{B/A})_{xyz}$$

In particular, the term $2\boldsymbol{\Omega} \times (\mathbf{v}_{B/A})_{xyz}$ is called the Coriolis acceleration.

The forces acting on this dragster as it begins to accelerate are quite severe
and must be accounted for in the design of its structure.

Planar Kinetics of a Rigid Body: Force and Acceleration

- To introduce the methods used to determine the mass moment of inertia of a body.
- To develop the planar kinetic equations of motion for a symmetric rigid body.
- To discuss applications of these equations to bodies undergoing translation, rotation about a fixed axis, and general plane motion.

17.1 Moment of Inertia

Since a body has a definite size and shape, an applied non-concurrent force system may cause the body to both translate and rotate. The translational aspects of the motion were studied in Chapter 13 and are governed by the equation $\mathbf{F} = m\mathbf{a}$. It will be shown in Sec. 17.2 that the rotational aspects, caused by a moment $\mathbf{M}$, are governed by an equation of the form $\mathbf{M} = I\boldsymbol{\alpha}$. The symbol I in this equation is termed the moment of inertia. By comparison, the *moment of inertia* is a measure of the resistance of a body to *angular acceleration* $(\mathbf{M} = I\boldsymbol{\alpha})$ in the same way that *mass* is a measure of the body's resistance to *acceleration* $(\mathbf{F} = m\mathbf{a})$.

The flywheel on the engine of this tractor has a large moment of inertia about its axis of rotation. Once it is set into motion, it will be difficult to stop, and this in turn will prevent the engine from stalling and instead it will allow it to maintain a constant power.

We define the *moment of inertia* as the integral of the "second moment" about an axis of all the elements of mass dm which compose the body.* For example, the body's moment of inertia about the z axis in Fig. 17–1 is

$$I = \int_m r^2 \, dm. \tag{17–1}$$

Fig. 17–1

Here the "moment arm" r is the perpendicular distance from the z axis to the arbitrary element dm. Since the formulation involves r, the value of I is different for each axis about which it is computed. In the study of planar kinetics, the axis which is generally chosen for analysis passes through the body's mass center G and is always perpendicular to the plane of motion. The moment of inertia computed about this axis will be denoted as I_G. Realize that because r is squared in Eq. 17–1, the mass moment of inertia is always a *positive* quantity. Common units used for its measurement are kg·m² or slug·ft².

If the body consists of material having a variable density, $\rho = \rho(x, y, z)$, the elemental mass dm of the body may be expressed in terms of its density and volume as $dm = \rho \, dV$. Substituting dm into Eq. 17–1, the body's moment of inertia is then computed using *volume elements* for integration; i.e.,

$$I = \int_V r^2 \rho \, dV \tag{17–2}$$

*Another property of the body, which measures the symmetry of the body's mass with respect to a coordinate system, is the product of inertia. This property applies to the three-dimensional motion of a body and will be discussed in Chapter 21.

In the special case of ρ being a *constant*, this term may be factored out of the integral, and the integration is then purely a function of geometry,

$$I = \rho \int_V r^2 \, dV \qquad (17\text{–}3)$$

When the elemental volume chosen for integration has infinitesimal dimensions in all three directions, e.g., $dV = dx\, dy\, dz$, Fig. 17–2a, the moment of inertia of the body must be determined using "triple integration." The integration process can, however, be simplified to a *single integration* provided the chosen elemental volume has a differential size or thickness in only *one direction*. Shell or disk elements are often used for this purpose.

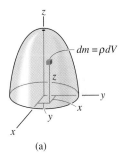

(a)

PROCEDURE FOR ANALYSIS

For integration, we will consider only symmetric bodies having surfaces which are generated by revolving a curve about an axis. An example of such a body which is generated about the z axis is shown in Fig. 17–2a. Two types of differential elements can be chosen.

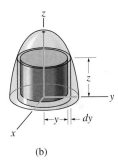

(b)

Shell Element.

- If a *shell element* having a height z, radius $r = y$, and thickness dy is chosen for integration, Fig. 17–2b, then the volume is $dV = (2\pi y)(z)\, dy$.
- This element may be used in Eq. 17–2 or 17–3 for determining the moment of inertia I_z of the body about the z axis, since the *entire element*, due to its "thinness," lies at the *same* perpendicular distance $r = y$ from the z axis (see Example 17.1).

Disk Element.

- If a disk element having a radius y and a thickness dz is chosen for integration, Fig. 17–2c, then the volume is $dV = (\pi y^2)\, dz$.
- This element is *finite* in the radial direction, and consequently its parts *do not* all lie at the *same radial distance* r from the z axis. As a result, Eq. 17–2 or 17–3 *cannot* be used to determine I_z directly. Instead, to perform the integration it is first necessary to determine the moment of inertia *of the element* about the z axis and then integrate this result (see Example 17.2).

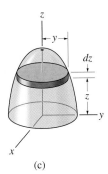

(c)

Fig. 17–2

EXAMPLE 17.1

Determine the moment of inertia of the cylinder shown in Fig. 17–3a about the z axis. The density of the material, ρ, is constant.

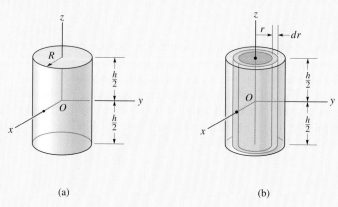

(a)   (b)

Fig. 17–3

Solution

Shell Element. This problem may be solved using the *shell element* in Fig. 17–3b and single integration. The volume of the element is $dV = (2\pi r)(h)\, dr$, so that its mass is $dm = \rho\, dV = \rho(2\pi hr\, dr)$. Since the *entire element* lies at the same distance r from the z axis, the moment of inertia *of the element* is

$$dI_z = r^2\, dm = \rho 2\pi hr^3\, dr$$

Integrating over the entire region of the cylinder yields

$$I_z = \int_m r^2\, dm = \rho 2\pi h \int_0^R r^3\, dr = \frac{\rho\pi}{2}R^4 h$$

The mass of the cylinder is

$$m = \int_m dm = \rho 2\pi h \int_0^R r\, dr = \rho\pi h R^2$$

so that

$$I_z = \frac{1}{2}mR^2 \qquad\qquad Ans.$$

E X A M P L E **17.2**

A solid is formed by revolving the shaded area shown in Fig. 17–4a about the y axis. If the density of the material is 5 slug/ft³, determine the moment of inertia about the y axis.

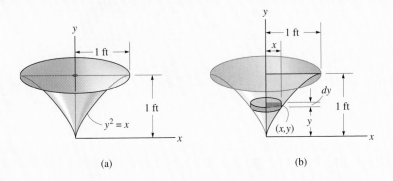

(a) (b)

Fig. 17–4

Solution

Disk Element. The moment of inertia will be computed using a *disk element*, as shown in Fig. 17–4b. Here the element intersects the curve at the arbitrary point (x, y) and has a mass

$$dm = \rho \, dV = \rho(\pi x^2) \, dy$$

Although all portions of the element are *not* located at the same distance from the y axis, it is still possible to determine the moment of inertia dI_y *of the element* about the y axis. In the preceding example it was shown that the moment of inertia of a cylinder about its longitudinal axis is $I = \frac{1}{2}mR^2$, where m and R are the mass and radius of the cylinder. Since the height of the cylinder is not involved in this formula, the cylinder itself can be thought of as a disk. Thus, for the disk element in Fig. 17–4b, we have

$$dI_y = \frac{1}{2}(dm)x^2 = \frac{1}{2}[\rho(\pi x^2) \, dy]x^2$$

Substituting $x = y^2$, $\rho = 5$ slug/ft³, and integrating with respect to y, from $y = 0$ to $y = 1$ ft, yields the moment of inertia for the entire solid.

$$I_y = \frac{\pi(5)}{2} \int_0^1 x^4 \, dy = \frac{\pi(5)}{2} \int_0^1 y^8 \, dy = 0.873 \text{ slug} \cdot \text{ft}^2 \qquad Ans.$$

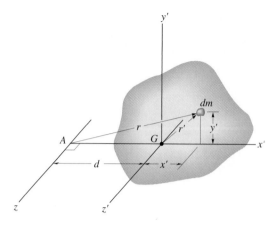

Fig. 17–5

Parallel-Axis Theorem. If the moment of inertia of the body about an axis passing through the body's mass center is known, then the moment of inertia about any other *parallel axis* may be determined by using the *parallel-axis theorem.* This theorem can be derived by considering the body shown in Fig. 17–5. The z' axis passes through the mass center G, whereas the corresponding *parallel z axis* lies at a constant distance d away. Selecting the differential element of mass dm, which is located at point (x', y'), and using the Pythagorean theorem, $r^2 = (d + x')^2 + y'^2$, we can express the moment of inertia of the body about the z axis as

$$I = \int_m r^2 \, dm = \int_m [(d + x')^2 + y'^2] \, dm$$

$$= \int_m (x'^2 + y'^2) \, dm + 2d \int_m x' \, dm + d^2 \int_m dm$$

Since $r'^2 = x'^2 + y'^2$, the first integral represents I_G. The second integral equals *zero*, since the z' axis passes through the body's mass center, i.e., $\int x' \, dm = \overline{x}' \int dm = 0$ since $\overline{x}' = 0$. Finally, the third integral represents the total mass m of the body. Hence, the moment of inertia about the

z axis can be written as

$$I = I_G + md^2 \qquad (17\text{--}4)$$

where

I_G = moment of inertia about the z' axis passing through the mass center G

m = mass of the body

d = perpendicular distance between the parallel axes

Radius of Gyration.   Occasionally, the moment of inertia of a body about a specified axis is reported in handbooks using the *radius of gyration, k.* This value has units of length, and when it and the body's mass m are known, the body's moment of inertia is determined from the equation

$$I = mk^2 \quad \text{or} \quad k = \sqrt{\frac{I}{m}} \qquad (17\text{--}5)$$

Note the *similarity* between the definition of k in this formula and r in the equation $dI = r^2\, dm$, which defines the moment of inertia of an elemental mass dm of the body about an axis.

Composite Bodies.   If a body is constructed of a number of simple shapes such as disks, spheres, and rods, the moment of inertia of the body about any axis z can be determined by adding algebraically the moments of inertia of all the composite shapes computed about the z axis. Algebraic addition is necessary since a composite part must be considered as a negative quantity if it has already been counted as a piece of another part—for example, a "hole" subtracted from a solid plate. The parallel-axis theorem is needed for the calculations if the center of mass of each composite part does not lie on the z axis. For the calculation, then, $I = \Sigma(I_G + md^2)$. Here I_G for each of the composite parts is computed by integration or can be determined from a table, such as the one given on the inside back cover of this book.

EXAMPLE 17.3

If the plate shown in Fig. 17–6a has a density of 8000 kg/m³ and a thickness of 10 mm, determine its moment of inertia about an axis directed perpendicular to the page and passing through point O.

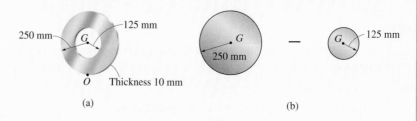

(a)   (b)

Fig. 17–6

Solution

The plate consists of two composite parts, the 250-mm-radius disk *minus* a 125-mm-radius disk, Fig. 17–6b. The moment of inertia about O can be determined by computing the moment of inertia of each of these parts about O and then adding the results *algebraically*. The calculations are performed by using the parallel-axis theorem in conjunction with the data listed in the table on the inside back cover.

Disk. The moment of inertia of a disk about the centroidal axis perpendicular to the plane of the disk is $I_G = \frac{1}{2}mr^2$. The mass center of the disk is located at a distance of 0.25 m from point O. Thus,

$$m_d = \rho_d V_d = 8000 \text{ kg/m}^3[\pi(0.25 \text{ m})^2(0.01 \text{ m})] = 15.71 \text{ kg}$$

$$(I_d)_O = \frac{1}{2}m_d r_d^2 + m_d d^2$$

$$= \frac{1}{2}(15.71 \text{ kg})(0.25 \text{ m})^2 + (15.71 \text{ kg})(0.25 \text{ m})^2$$

$$= 1.473 \text{ kg} \cdot \text{m}^2$$

Hole. For the 125-mm-radius disk (hole), we have

$$m_h = \rho_h V_h = 8000 \text{ kg/m}^3[\pi(0.125 \text{ m})^2(0.01 \text{ m})] = 3.93 \text{ kg}$$

$$(I_h)_O = \frac{1}{2}m_h r_h^2 + m_h d^2$$

$$= \frac{1}{2}(3.93 \text{ kg})(0.125 \text{ m})^2 + (3.93 \text{ kg})(0.25 \text{ m})^2$$

$$= 0.276 \text{ kg} \cdot \text{m}^2$$

The moment of inertia of the plate about point O is therefore

$$I_O = (I_d)_O - (I_h)_O$$
$$= 1.473 \text{ kg} \cdot \text{m}^2 - 0.276 \text{ kg} \cdot \text{m}^2$$
$$= 1.20 \text{ kg} \cdot \text{m}^2 \qquad\qquad \textit{Ans.}$$

E X A M P L E 17.4

The pendulum in Fig. 17–7 is suspended from point O and consists of two thin rods, each having a weight of 10 lb. Determine the pendulum's moment of inertia about an axis passing through (a) the pin at O, and (b) the mass center G of the pendulum.

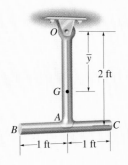

Solution

Part (a).   Using the table on the inside back cover, the moment of inertia of rod OA about an axis perpendicular to the page and passing through the end point O of the rod is $I_O = \frac{1}{3}ml^2$. Hence,

$$(I_{OA})_O = \frac{1}{3}ml^2 = \frac{1}{3}\left(\frac{10 \text{ lb}}{32.2 \text{ ft/s}^2}\right)(2 \text{ ft})^2 = 0.414 \text{ slug} \cdot \text{ft}^2$$

Fig. 17–7

This same value is obtained using $I_G = \frac{1}{12}ml^2$ and the parallel-axis theorem.

$$(I_{OA})_O = \frac{1}{12}ml^2 + md^2 = \frac{1}{12}\left(\frac{10 \text{ lb}}{32.2 \text{ ft/s}^2}\right)(2 \text{ ft})^2 + \left(\frac{10 \text{ lb}}{32.2 \text{ ft/s}^2}\right)(1 \text{ ft})^2$$

$$= 0.414 \text{ slug} \cdot \text{ft}^2$$

For rod BC we have

$$(I_{BC})_O = \frac{1}{12}ml^2 + md^2 = \frac{1}{12}\left(\frac{10 \text{ lb}}{32.2 \text{ ft/s}^2}\right)(2 \text{ ft})^2 + \left(\frac{10 \text{ lb}}{32.2 \text{ ft/s}^2}\right)(2 \text{ ft})^2$$

$$= 1.346 \text{ slug} \cdot \text{ft}^2$$

The moment of inertia of the pendulum about O is therefore

$$I_O = 0.414 + 1.346 = 1.76 \text{ slug} \cdot \text{ft}^2 \qquad\qquad Ans.$$

Part (b).   The mass center G will be located relative to the pin at O. Assuming this distance to be $\bar{y}$, Fig. 17–7, and using the formula for determining the mass center, we have

$$\bar{y} = \frac{\Sigma \tilde{y}m}{\Sigma m} = \frac{1(10/32.2) + 2(10/32.2)}{(10/32.2) + (10/32.2)} = 1.50 \text{ ft}$$

The moment of inertia I_G may be computed in the same manner as I_O, which requires successive applications of the parallel-axis theorem to transfer the moments of inertia of rods OA and BC to G. A more direct solution, however, involves using the result for I_O, i.e.,

$$I_O = I_G + md^2; \quad 1.76 \text{ slug} \cdot \text{ft}^2 = I_G + \left(\frac{20 \text{ lb}}{32.2 \text{ ft/s}^2}\right)(1.50 \text{ ft})^2$$

$$I_G = 0.362 \text{ slug} \cdot \text{ft}^2 \qquad\qquad Ans.$$

17.2 Planar Kinetic Equations of Motion

In the following analysis we will limit our study of planar kinetics to rigid bodies which, along with their loadings, are considered to be *symmetrical with respect to a fixed reference plane.** In this case the path of motion of each particle of the body is a plane curve parallel to a fixed reference plane. Since the motion of the body may be viewed within the reference plane, all the forces (and couple moments) acting on the body can then be projected onto the plane. An example of an arbitrary body of this type is shown in Fig. 17–8a. Here the *inertial frame of reference x, y, z* has its origin *coincident* with the arbitrary point P in the body. By definition, *these axes do not rotate and are either fixed or translate with constant velocity.*

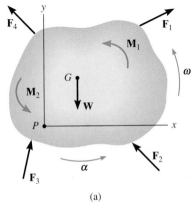

(a)

Fig. 17–8

Equation of Translational Motion.   The external forces shown on the body in Fig. 17–8a represent the effect of gravitational, electrical, magnetic, or contact forces between adjacent bodies. Since this force system has been considered previously in Sec. 13.3 for the analysis of a system of particles, the resulting Eq. 13–6 may be used here, in which case

$$\Sigma \mathbf{F} = m\mathbf{a}_G$$

This equation is referred to as the *translational equation of motion* for the mass center of a rigid body. It states that *the sum of all the external forces acting on the body is equal to the body's mass times the acceleration of its mass center G.*

For motion of the body in the *x–y* plane, the translational equation of motion may be written in the form of two independent scalar equations, namely,

$$\Sigma F_x = m(a_G)_x$$
$$\Sigma F_y = m(a_G)_y$$

*By doing this, the rotational equation of motion reduces to a rather simplified form. The more general case of body shape and loading is considered in Chapter 21.

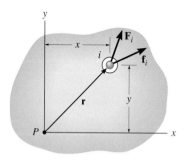

Particle free-body diagram

(b)

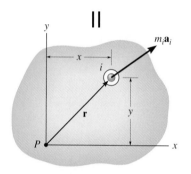

Particle kinetic diagram

(c)

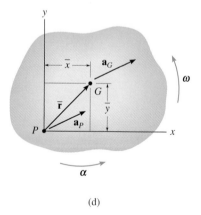

(d)

Fig. 17–8

Equation of Rotational Motion. We will now determine the effects caused by the moments of the external force system computed about an axis perpendicular to the plane of motion (the z axis) and passing through point P. As shown on the free-body diagram of the ith particle, Fig. 17–8b, $\mathbf{F}_i$, represents the *resultant external force* acting on the particle, and $\mathbf{f}_i$ is the *resultant of the internal forces* caused by interactions with adjacent particles. If the particle has a mass m_i and at the instant considered its acceleration is $\mathbf{a}_i$, then the kinetic diagram is constructed as shown in Fig. 17–8c. If moments of the forces acting on the particle are summed about point P, we require

$$\mathbf{r} \times \mathbf{F}_i + \mathbf{r} \times \mathbf{f}_i = \mathbf{r} \times m_i\mathbf{a}_i$$

or

$$(\mathbf{M}_P)_i = \mathbf{r} \times m_i\mathbf{a}_i$$

The moments about P can be expressed in terms of the acceleration of point P, Fig. 17–8d. If the body has an angular acceleration $\boldsymbol{\alpha}$ and angular velocity $\boldsymbol{\omega}$, then using Eq. 16–18 we have

$$(\mathbf{M}_P)_i = m_i\mathbf{r} \times (\mathbf{a}_P + \boldsymbol{\alpha} \times \mathbf{r} - \omega^2\mathbf{r})$$
$$= m_i[\mathbf{r} \times \mathbf{a}_P + \mathbf{r} \times (\boldsymbol{\alpha} \times \mathbf{r}) - \omega^2(\mathbf{r} \times \mathbf{r})]$$

The last term is zero, since $\mathbf{r} \times \mathbf{r} = \mathbf{0}$. Expressing the vectors with Cartesian components and carrying out the cross-product operations yields

$$(M_P)_i\mathbf{k} = m_i\{(x\mathbf{i} + y\mathbf{j}) \times [(a_P)_x\mathbf{i} + (a_P)_y\mathbf{j}]$$
$$+ (x\mathbf{i} + y\mathbf{j}) \times [\alpha\mathbf{k} \times (x\mathbf{i} + y\mathbf{j})]\}$$
$$(M_P)_i\mathbf{k} = m_i[-y(a_P)_x + x(a_P)_y + \alpha x^2 + \alpha y^2]\mathbf{k}$$
$$\zeta(M_P)_i = m_i[-y(a_P)_x + x(a_P)_y + \alpha r^2]$$

Letting $m_i \rightarrow dm$ and integrating with respect to the entire mass m of the body, we obtain the resultant moment equation

$$\zeta\Sigma M_P = -\left(\int_m y\,dm\right)(a_P)_x + \left(\int_m x\,dm\right)(a_P)_y + \left(\int_m r^2\,dm\right)\alpha$$

Here ΣM_P represents only the moment of the *external forces* acting on the body about point P. The resultant moment of the internal forces is zero, since for the entire body these forces occur in equal and opposite collinear pairs and thus the moment of each pair of forces about P cancels. The integrals in the first and second terms on the right are used to locate the body's center of mass G with respect to P, since $\bar{y}m = \int y\,dm$ and $\bar{x}m = \int x\,dm$, Fig. 17–8d. Also, the last integral represents the body's moment of inertia computed about the z axis, i.e., $I_P = \int r^2\,dm$. Thus,

$$\zeta\Sigma M_P = -\bar{y}m(a_P)_x + \bar{x}m(a_P)_y + I_P\alpha \qquad (17\text{–}6)$$

It is possible to reduce this equation to a simpler form if point P coincides with the mass center G for the body. If this is the case, then $\bar{x} = \bar{y} = 0$, and therefore*

$$\Sigma M_G = I_G\alpha \qquad (17\text{--}7)$$

This rotational equation of motion states that the sum of the moments of all the external forces computed about the body's mass center G is equal to the product of the moment of inertia of the body about an axis passing through G and the body's angular acceleration.

Equation 17–6 can also be rewritten in terms of the x and y components of $\mathbf{a}_G$ and the body's moment of inertia I_G. If point G is located at point $(\bar{x}, \bar{y})$, Fig. 17–8d, then by the parallel-axis theorem, $I_p = I_G + m(\bar{x}^2 + \bar{y}^2)$. Substituting into Eq. 17–6 and rearranging terms, we get

$$\curvearrowright\Sigma M_P = \bar{y}m[-(a_P)_x + \bar{y}\alpha] + \bar{x}m[(a_P)_y + \bar{x}\alpha] + I_G\alpha \qquad (17\text{--}8)$$

From the kinematic diagram of Fig. 17–8d, $\mathbf{a}_P$ can be expressed in terms of $\mathbf{a}_G$ as

$$\mathbf{a}_G = \mathbf{a}_P + \boldsymbol{\alpha} \times \bar{\mathbf{r}} - \omega^2\bar{\mathbf{r}}$$
$$(a_G)_x\mathbf{i} + (a_G)_y\mathbf{j} = (a_P)_x\mathbf{i} + (a_P)_y\mathbf{j} + \alpha\mathbf{k} \times (\bar{x}\mathbf{i} + \bar{y}\mathbf{j}) - \omega^2(\bar{x}\mathbf{i} + \bar{y}\mathbf{j})$$

Carrying out the cross product and equating the respective $\mathbf{i}$ and $\mathbf{j}$ components yields the two scalar equations

$$(a_G)_x = (a_P)_x - \bar{y}\alpha - \bar{x}\omega^2$$
$$(a_G)_y = (a_P)_y + \bar{x}\alpha - \bar{y}\omega^2$$

From these equations, $[-(a_P)_x + \bar{y}\alpha] = [-(a_G)_x - \bar{x}\omega^2]$ and $[(a_P)_y + \bar{x}\alpha] = [(a_G)_y + \bar{y}\omega^2]$. Substituting these results into Eq. 17–8 and simplifying gives

$$\curvearrowright\Sigma M_P = -\bar{y}m(a_G)_x + \bar{x}m(a_G)_y + I_G\alpha \qquad (17\text{--}9)$$

This important result indicates that when moments of the external forces shown on the free-body diagram are summed about point P, Fig. 17–8e, they are equivalent to the sum of the "kinetic moments" of the components of ma_G about P plus the "kinetic moment" of I_Gα, Fig. 17–8f. In other words, when the "kinetic moments," $\Sigma(\mathcal{M}_k)_P$, are computed, Fig. 17–8f, the vectors $m(\mathbf{a}_G)_x$ and $m(\mathbf{a}_G)_y$ are treated as sliding vectors; that is, they can act at any point along their line of action. In a similar manner, $I_G\alpha$ can be treated as a free vector and can therefore act at any point. It is important to keep in mind that ma_G and I_Gα are not the same as a force or a couple moment. Instead, they are caused by the external effects of forces and couple moments acting on the body. With this in mind we can therefore write Eq. 17–9 in a more general form as

$$\Sigma M_P = \Sigma(\mathcal{M}_k)_P \qquad (17\text{--}10)$$

*It also reduces to this same simple form $\Sigma M_P = I_P\alpha$ if point P is a *fixed point* (see Eq. 17–16) or the acceleration of point P is directed along the line PG.

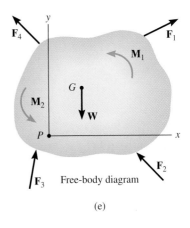

Free-body diagram

(e)

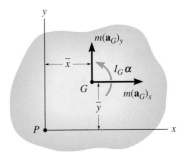

Kinetic diagram

(f)

Fig. 17–8

General Application of the Equations of Motion. To summarize this analysis, *three* independent scalar equations may be written to describe the general plane motion of a symmetrical rigid body.

$$\Sigma F_x = m(a_G)_x$$
$$\Sigma F_y = m(a_G)_y$$
$$\Sigma M_G = I_G\alpha \quad \text{or} \quad \Sigma M_P = \Sigma(\mathcal{M}_k)_P \qquad (17\text{--}11)$$

When applying these equations, one should *always* draw a free-body diagram, Fig. 17–8e, in order to account for the terms involved in ΣF_x, ΣF_y, ΣM_G, or ΣM_P. In some problems it may also be helpful to draw the *kinetic diagram* for the body. This diagram graphically accounts for the terms $m(\mathbf{a}_G)_x$, $m(\mathbf{a}_G)_y$, and $I_G\alpha$, and it is especially convenient when used to determine the components of $m\mathbf{a}_G$ and the moment terms in $\Sigma(\mathcal{M}_k)_P$.*

17.3 Equations of Motion: Translation

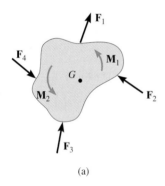

(a)

Fig. 17–9

When a rigid body undergoes a *translation*, Fig. 17–9a, all the particles of the body have the *same acceleration*, so that $\mathbf{a}_G = \mathbf{a}$. Furthermore, $\alpha = \mathbf{0}$, in which case the rotational equation of motion applied at point G reduces to a simplified form, namely, $\Sigma M_G = 0$. Application of this and the translational equations of motion will now be discussed for each of the two types of translation.

Rectilinear Translation. When a body is subjected to *rectilinear translation*, all the particles of the body (slab) travel along parallel straight-line paths. The free-body and kinetic diagrams are shown in Fig. 17–9b. Since $I_G\alpha = \mathbf{0}$, only $m\mathbf{a}_G$ is shown on the kinetic diagram. Hence, the equations of motion which apply in this case become

$$\boxed{\begin{aligned} \Sigma F_x &= m(a_G)_x \\ \Sigma F_y &= m(a_G)_y \\ \Sigma M_G &= 0 \end{aligned}} \qquad (17\text{--}12)$$

The last equation requires that the sum of the moments of all the external forces (and couple moments) computed about the body's center of mass be equal to zero. It is possible, of course, to sum moments about other points on or off the body, in which case the moment of $m\mathbf{a}_G$ must be

---

*For this reason, the kinetic diagram will be used in the solution of an example problem whenever $\Sigma M_P = \Sigma(\mathcal{M}_k)_P$ is applied.

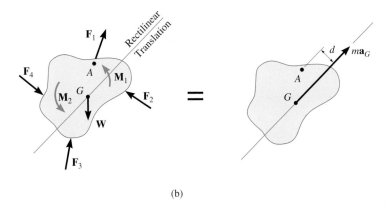

(b)

Fig. 17–9

taken into account. For example, if point A is chosen, which lies at a perpendicular distance d from the line of action of $m\mathbf{a}_G$, the following moment equation applies:

$$\zeta +\Sigma M_A = \Sigma(\mathcal{M}_k)_A; \qquad \Sigma M_A = (ma_G)d$$

Here the sum of moments of the external forces and couple moments about A (ΣM_A, free-body diagram) equals the moment of $m\mathbf{a}_G$ about A ($\Sigma(\mathcal{M}_k)_A$, kinetic diagram).

Curvilinear Translation. When a rigid body is subjected to *curvilinear translation*, all the particles of the body travel along *parallel curved paths*. For analysis, it is often convenient to use an inertial coordinate system having an origin which is coincident with the body's mass center at the instant considered, and axes which are oriented in the normal and tangential directions to the path of motion, Fig. 17–9c. The three scalar equations of motion are then

$$\boxed{\begin{aligned} \Sigma F_n &= m(a_G)_n \\ \Sigma F_t &= m(a_G)_t \\ \Sigma M_G &= 0 \end{aligned}} \qquad (17\text{–}13)$$

Here $(a_G)_t$ and $(a_G)_n$ represent, respectively, the magnitudes of the tangential and normal components of acceleration of point G.

If the moment equation $\Sigma M_G = 0$ is replaced by a moment summation about the arbitrary point B, Fig. 17–9c, it is necessary to account for the moments, $\Sigma(\mathcal{M}_k)_B$, of the two components $m(\mathbf{a}_G)_n$ and $m(\mathbf{a}_G)_t$ about this point. From the kinetic diagram, h and e represent the perpendicular distances (or "moment arms") from B to the lines of action of the components. The required moment equation therefore becomes

$$\zeta +\Sigma M_B = \Sigma(\mathcal{M}_k)_B; \qquad \Sigma M_B = e[m(a_G)_t] - h[m(a_G)_n]$$

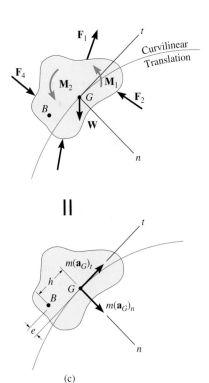

(c)

The free-body and kinetic diagrams for this boat and trailer are first drawn in order to apply the equations of motion. Here the forces on the free-body diagram cause the effect shown on the kinetic diagram. If moments are summed about the mass center, G, then $\Sigma M_G = 0$. However, if moments are summed about point B then $\curvearrowright + \Sigma M_B = ma_G(d)$.

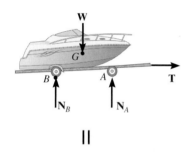

PROCEDURE FOR ANALYSIS

Kinetic problems involving rigid-body *translation* can be solved using the following procedure.

Free-Body Diagram.

• Establish the x, y or n, t inertial coordinate system and draw the free-body diagram in order to account for all the external forces and couple moments that act on the body.

• The direction and sense of the acceleration of the body's mass center $\mathbf{a}_G$ should be established.

• Identify the unknowns in the problem.

• If it is decided that the rotational equation of motion $\Sigma M_P = \Sigma(\mathcal{M}_k)_P$ is to be used in the solution, then consider drawing the kinetic diagram, since it graphically accounts for the components $m(\mathbf{a}_G)_x$, $m(\mathbf{a}_G)_y$ or $m(\mathbf{a}_G)_t$, $m(\mathbf{a}_G)_n$ and is therefore convenient for "visualizing" the terms needed in the moment sum $\Sigma(\mathcal{M}_k)_P$.

Equations of Motion.

• Apply the three equations of motion in accordance with the established sign convention.

• To simplify the analysis, the moment equation $\Sigma M_G = 0$ can be replaced by the more general equation $\Sigma M_P = \Sigma(\mathcal{M}_k)_P$, where point P is usually located at the intersection of the lines of action of as many unknown forces as possible.

• If the body is in contact with a *rough surface* and slipping occurs, use the frictional equation $F = \mu_k N$. Remember, $\mathbf{F}$ always acts on the body so as to oppose the motion of the body relative to the surface it contacts.

Kinematics.

• Use kinematics if the velocity and position of the body are to be determined.

• For *rectilinear translation* with *variable acceleration*, use
$a_G = dv_G/dt \quad a_G ds_G = v_G dv_G \quad v_G = ds_G/dt$

• For *rectilinear translation* with *constant acceleration*, use

$$v_G = (v_G)_0 + a_G t \qquad v_G^2 = (v_G)_0^2 + 2a_G[s_G - (s_G)_0]$$
$$s_G = (s_G)_0 + (v_G)_0 t + \tfrac{1}{2}a_G t^2$$

• For *curvilinear translation*, use $(a_G)_n = v_G^2/\rho = \omega^2\rho$, $(a_G)_t = dv_G/dt$, $(a_G)_t ds_G = v_G dv_G$, $(a_G)_t = \alpha\rho$

EXAMPLE 17.5

The car shown in Fig. 17–10a has a mass of 2 Mg and a center of mass at G. Determine the car's acceleration if the "driving" wheels in the back are always slipping, whereas the front wheels freely rotate. Neglect the mass of the wheels. The coefficient of kinetic friction between the wheels and the road is $\mu_k = 0.25$.

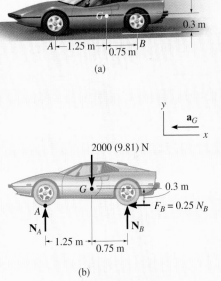

(a)

Solution I

Free-Body Diagram. As shown in Fig. 17–10b, the rear-wheel frictional force $\mathbf{F}_B$ pushes the car forward, and since *slipping occurs*, $F_B = 0.25N_B$. The frictional forces acting on the *front wheels* are *zero*, since these wheels have negligible mass.* There are three unknowns in the problem, N_A, N_B, and a_G. Here we will sum moments about the mass center. The car (point G) is assumed to accelerate to the left, i.e., in the negative x direction, Fig. 17–10b.

Equations of Motion.

$$\xrightarrow{+} \Sigma F_x = m(a_G)_x; \qquad -0.25N_B = -(2000 \text{ kg})a_G \qquad (1)$$

$$+\uparrow \Sigma F_y = m(a_G)_y; \qquad N_A + N_B - 2000(9.81) \text{ N} = 0 \qquad (2)$$

$$\curvearrowright{+} \Sigma M_G = 0; \quad -N_A(1.25 \text{ m}) - 0.25N_B(0.3 \text{ m}) + N_B(0.75 \text{ m}) = 0 \quad (3)$$

Solving,

$$a_G = 1.59 \text{ m/s}^2 \leftarrow \qquad\qquad Ans.$$

$$N_A = 6.88 \text{ kN}$$

$$N_B = 12.7 \text{ kN}$$

Solution II

Free-Body and Kinetic Diagrams. If the "moment" equation is applied about point A, then the unknown N_A will be eliminated from the equation. To "visualize" the moment of $m\mathbf{a}_G$ about A, we will include the kinetic diagram as part of the analysis, Fig. 17–10c.

Equation of Motion. We require

$$\curvearrowright{+} \Sigma M_A = \Sigma(\mathcal{M}_k)_A; \qquad N_B(2 \text{ m}) - 2000(9.81) \text{ N}(1.25 \text{ m}) =$$
$$(2000 \text{ kg})a_G(0.3 \text{ m})$$

Solving this and Eq. 1 for a_G leads to a simpler solution than that obtained from Eqs. 1 to 3.

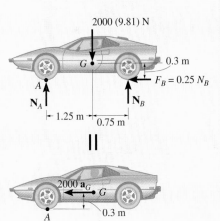

(c)

Fig. 17–10

*With negligible wheel mass, $I\alpha = 0$ and the frictional force at A required to turn the wheel is zero. If the wheels' mass were included, then the problem solution for this case would be more involved, since a general-plane-motion analysis of the wheels would have to be considered (see Sec. 17.5).

EXAMPLE 17.6

The motorcycle shown in Fig. 17–11a has a mass of 125 kg and a center of mass at G_1, while the rider has a mass of 75 kg and a center of mass at G_2. Determine the minimum coefficient of static friction between the wheels and the pavement in order for the rider to do a "wheely," i.e., lift the front wheel off the ground as shown in the photo. What acceleration is necessary to do this? Neglect the mass of the wheels and assume that the front wheel is free to roll.

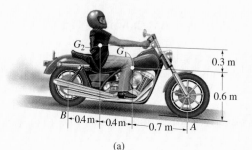

(a)

Solution

Free-Body and Kinetic Diagrams. In this problem we will consider both the motorcycle and the rider as the "system" to be analyzed. It is possible first to determine the location of the center of mass for this "system" by using the equations $\bar{x} = \Sigma \tilde{x}m/\Sigma m$ and $\bar{y} = \Sigma \tilde{y}m/\Sigma m$. Here, however, we will consider the separate weight and mass of each of its *component parts* as shown on the free-body and kinetic diagrams, Fig. 17–11b. Both parts move with the *same* acceleration and we have assumed that the front wheel is *about* to leave the ground, so that the normal reaction $N_A \approx 0$. The three unknowns in the problem are N_B, F_B, and a_G.

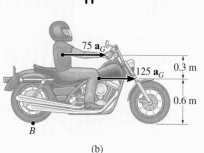

(b)

Fig. 17–11

Equations of Motion.

$$\xrightarrow{+} \Sigma F_x = m(a_G)_x; \qquad F_B = (75 \text{ kg} + 125 \text{ kg})a_G \qquad (1)$$

$$+\uparrow \Sigma F_y = m(a_G)_y; \qquad N_B - 735.75 \text{ N} - 1226.25 \text{ N} = 0 \qquad (2)$$

$$\zeta + \Sigma M_B = \Sigma(\mathcal{M}_k)_B; \quad -(735.75 \text{ N})(0.4 \text{ m}) - (1226.25 \text{ N})(0.8 \text{ m}) =$$
$$-(75 \text{ kg } a_G)(0.9 \text{ m}) - (125 \text{ kg } a_G)(0.6 \text{ m})$$

Solving,

$$a_G = 8.95 \text{ m/s}^2 \rightarrow \qquad Ans.$$
$$N_B = 1962 \text{ N}$$
$$F_B = 1790 \text{ N}$$

Thus the minimum coefficient of static friction is

$$(\mu_s)_{min} = \frac{F_B}{N_B} = \frac{1790 \text{ N}}{1962 \text{ N}} = 0.912 \qquad Ans.$$

EXAMPLE 17.7

A uniform 50-kg crate rests on a horizontal surface for which the coefficient of kinetic friction is $\mu_k = 0.2$. Determine the crate's acceleration if a force of $P = 600\ \text{N}$ is applied to the crate as shown in Fig. 17–12a.

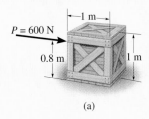

$P = 600\ \text{N}$

1 m

0.8 m

1 m

(a)

Solution

Free-Body Diagram. The force **P** can cause the crate either to slide or to tip over. As shown in Fig. 17–12b, it is assumed that the crate slides, so that $F = \mu_k N_C = 0.2 N_C$. Also, the resultant normal force $\mathbf{N}_C$ acts at O, a distance x (where $0 < x \le 0.5\ \text{m}$) from the crate's center line.* The three unknowns are N_C, x, and a_G.

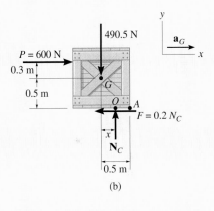

Equations of Motion

$$\xrightarrow{+}\ \Sigma F_x = m(a_G)_x;\qquad 600\ \text{N} - 0.2N_C = (50\ \text{kg})a_G \qquad (1)$$

$$+\uparrow \Sigma F_y = m(a_G)_y;\qquad N_C - 490.5\ \text{N} = 0 \qquad (2)$$

$$\zeta + \Sigma M_G = 0;\qquad -600\ \text{N}(0.3\ \text{m}) + N_C(x) - 0.2N_C(0.5\ \text{m}) = 0 \qquad (3)$$

Solving, we obtain

$$N_C = 490\ \text{N}$$
$$x = 0.467\ \text{m}$$
$$a_G = 10.0\ \text{m/s}^2 \rightarrow \qquad\qquad Ans.$$

Since $x = 0.467\ \text{m} < 0.5\ \text{m}$, indeed the crate slides as originally assumed. If the solution had given a value of $x > 0.5\ \text{m}$, the problem would have to be reworked with the assumption that tipping occurred. If this were the case, $\mathbf{N}_C$ would act at the *corner point A* and $F \le 0.2N_C$.

Fig. 17–12

---

*The line of action of $\mathbf{N}_C$ does not necessarily pass through the mass center $G\ (x = 0)$, since $\mathbf{N}_C$ must counteract the tendency for tipping caused by **P**. See Sec. 8.1 of *Engineering Mechanics: Statics.*

EXAMPLE 17.8

The 100-kg beam BD shown in Fig. 17–13a is supported by two rods having negligible mass. Determine the force created in each rod if at the instant $\theta = 30°$ and $\omega = 6$ rad/s.

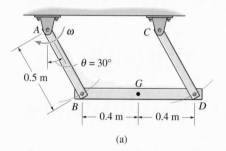

(a)

Solution

Free-Body Diagram. The beam moves with *curvilinear translation* since points B and D and the center of mass G all move along circular

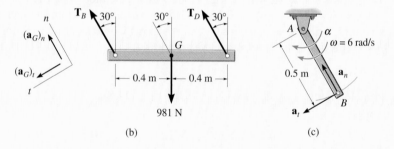

(b)   (c)

Fig. 17–13

paths, each path having the same radius of 0.5 m. Using normal and tangential coordinates, the free-body diagram for the beam is shown in Fig. 17–13b. Because of the *translation*, G has the *same* motion as the pin at B, which is connected to both the rod and the beam. By studying the angular motion of rod AB, Fig. 17–13c, note that the tangential component of acceleration acts downward to the left due to the clockwise direction of α. Furthermore, the normal component of acceleration is *always* directed toward the center of curvature (toward point A for rod AB). Since the angular velocity of AB is 6 rad/s, then

$$(a_G)_n = \omega^2 r = (6 \text{ rad/s})^2 (0.5 \text{ m}) = 18 \text{ m/s}^2$$

The three unknowns are T_B, T_D, and $(a_G)_t$. The directions of $(\mathbf{a}_G)_n$ and $(\mathbf{a}_G)_t$ have been established, and are indicated on the coordinate axes.

Equations of Motion.

$+\nwarrow \Sigma F_n = m(a_G)_n;\quad T_B + T_D - 981 \cos 30° \text{ N} = 100 \text{ kg}(18 \text{ m/s}^2)$ (1)

$+\swarrow \Sigma F_t = m(a_G)_t;\qquad\qquad 981 \sin 30° = 100 \text{ kg}(a_G)_t$ (2)

$\zeta + \Sigma M_G = 0;\quad -(T_B \cos 30°)(0.4 \text{ m}) + (T_D \cos 30°)(0.4 \text{ m}) = 0$ (3)

Simultaneous solution of these three equations gives

$$T_B = T_D = 1.32 \text{ kN } ^{\nwarrow}\!|^{30°} \qquad\qquad Ans.$$
$$(a_G)_t = 4.90 \text{ m/s}^2$$

17.4 Equations of Motion: Rotation About a Fixed Axis

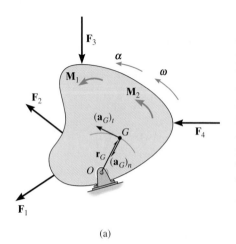

(a)

Fig. 17–14

Consider the rigid body (or slab) shown in Fig. 17–14a, which is constrained to rotate in the vertical plane about a fixed axis perpendicular to the page and passing through the pin at O. The angular velocity and angular acceleration are caused by the external force and couple moment system acting on the body. Because the body's center of mass G moves in a *circular path*, the acceleration of this point is represented by its tangential and normal components. The *tangential component of acceleration* has a *magnitude* of $(a_G)_t = \alpha r_G$ and must act in a *direction* which is *consistent* with the body's angular acceleration α. The *magnitude* of the *normal component of acceleration* is $(a_G)_n = \omega^2 r_G$. This component is *always directed* from point G to O, regardless of the direction of ω.

The free-body and kinetic diagrams for the body are shown in Fig. 17–14b. The weight of the body, $W = mg$, and the pin reaction $\mathbf{F}_O$ are included on the free-body diagram since they represent external forces acting on the body. The two components $m(\mathbf{a}_G)_t$ and $m(\mathbf{a}_G)_n$, shown on the kinetic diagram, are associated with the tangential and normal acceleration components of the body's mass center. These vectors act in the same *direction* as the acceleration components and have *magnitudes* of $m(a_G)_t$ and $m(a_G)_n$. The $I_G\boldsymbol{\alpha}$ vector acts in the same *direction* as $\boldsymbol{\alpha}$ and has a *magnitude* of $I_G\alpha$, where I_G is the body's moment of inertia calculated about an axis which is perpendicular to the page and passing through G. From the derivation given in Sec. 17.2, the equations of motion which apply to the body may be written in the form

$$\Sigma F_n = m(a_G)_n = m\omega^2 r_G$$
$$\Sigma F_t = m(a_G)_t = m\alpha r_G \qquad (17\text{–}14)$$
$$\Sigma M_G = I_G\alpha$$

The moment equation may be replaced by a moment summation about any arbitrary point P on or off the body provided one accounts for the moments $\Sigma(\mathcal{M}_k)_P$ produced by $I_G\boldsymbol{\alpha}$, $m(\mathbf{a}_G)_t$, and $m(\mathbf{a}_G)_n$ about the point. In many problems it is convenient to sum moments about the pin at O in order to eliminate the *unknown* force $\mathbf{F}_O$. From the kinetic diagram, Fig. 17–14b, this requires

$$\zeta + \Sigma M_O = \Sigma(\mathcal{M}_k)_O; \qquad \Sigma M_O = r_G m(a_G)_t + I_G\alpha \qquad (17\text{–}15)$$

Note that the moment of $m(\mathbf{a}_G)_n$ is not included in the summation since the line of action of this vector passes through O. Substituting $(a_G)_t = r_G\alpha$, we may rewrite the above equation as $\zeta + \Sigma M_O = (I_G + mr_G^2)\alpha$. From the parallel-axis theorem, $I_O = I_G + md^2$, and therefore the term in parentheses represents the *moment of inertia of the body about the fixed axis of rotation passing through O.** Consequently, we can write the three equations of motion for the body as

$$\Sigma F_n = m(a_G)_n = m\omega^2 r_G$$
$$\Sigma F_t = m(a_G)_t = m\alpha r_G \qquad (17\text{–}16)$$
$$\Sigma M_O = I_O\alpha$$

For applications, one should remember that "$I_O\alpha$" accounts for the "moment" of *both* $m(\mathbf{a}_G)_t$ *and* $I_G\boldsymbol{\alpha}$ about point O, Fig. 17–14b. In other words, $\Sigma M_O = \Sigma(\mathcal{M}_k)_O = I_O\alpha$, as indicated by Eqs. 17–15 and 17–16.

*The result $\Sigma M_O = I_O\alpha$ can also be obtained *directly* from Eq. 17–6 by selecting point P to coincide with O, realizing that $(a_P)_x = (a_P)_y = 0$.

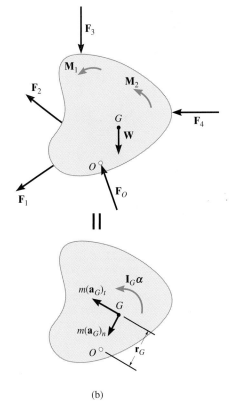

(a)

(b)

Fig. 17–14

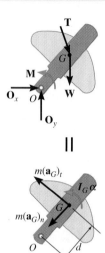

The crank on the oil-pumping rig undergoes rotation about a fixed axis which is caused by a driving torque **M** of the motor. The loadings shown on the free-body diagram cause the effects shown on the kinetic diagram. If moments are summed about the mass center, G, then $\Sigma M_G = I_G \alpha$. However, if moments are summed about point O, noting that $(a_G)_t = \alpha d$, then $\curvearrowright + \Sigma M_O = I_G \alpha + m(a_G)_t d + m(a_G)_n(0) = (I_G + md^2)\alpha = I_O \alpha$.

PROCEDURE FOR ANALYSIS

Kinetic problems which involve the rotation of a body about a fixed axis can be solved using the following procedure.

Free-Body Diagram.

- Establish the inertial x, y or n, t coordinate system and specify the direction and sense of the accelerations $(a_G)_n$ and $(a_G)_t$ and the angular acceleration α of the body. Recall that $(a_G)_t$ must act in a direction which is in accordance with α, whereas $(a_G)_n$ always acts toward the axis of rotation, point O.

- Draw the free-body diagram to account for all the external forces and couple moments that act on the body.

- Compute the moment of inertia I_G or I_O.

- Identify the unknowns in the problem.

- If it is decided that the rotational equation of motion $\Sigma M_P = \Sigma(\mathcal{M}_k)_P$ is to be used, i.e., P is a point other than G or O, then consider drawing the kinetic diagram in order to help "visualize" the "moments" developed by the components $m(a_G)_n$, $m(a_G)_t$, and $I_G\alpha$ when writing the terms for the moment sum $\Sigma(\mathcal{M}_k)_P$.

Equations of Motion.

- Apply the three equations of motion in accordance with the established sign convention.

- If moments are summed about the body's mass center, G, then $\Sigma M_G = I_G\alpha$, since $(ma_G)_t$ and $(ma_G)_n$ create no moment about G.

- If moments are summed about the pin support O on the axis of rotation, then $(ma_G)_n$ creates no moment about G, and it can be shown that $\Sigma M_O = I_O\alpha$.

Kinematics.

- Use kinematics if a complete solution cannot be obtained strictly from the equations of motion.

- If the *angular acceleration is variable*, use

$$\alpha = \frac{d\omega}{dt} \qquad \alpha\, d\theta = \omega\, d\omega \qquad \omega = \frac{d\theta}{dt}$$

- If the *angular acceleration is constant*, use

$$\omega = \omega_0 + \alpha_c t$$
$$\theta = \theta_0 + \omega_0 t + \tfrac{1}{2}\alpha_c t^2$$
$$\omega^2 = \omega_0^2 + 2\alpha_c(\theta - \theta_0)$$

EXAMPLE 17.9

The 30-kg uniform disk shown in Fig. 17–15a is pin supported at its center. If it starts from rest, determine the number of revolutions it must make to attain an angular velocity of 20 rad/s. Also, what are the reactions at the pin? The disk is acted upon by a constant force $F = 10$ N, which is applied to a cord wrapped around its periphery, and a constant couple moment $M = 5$ N·m. Neglect the mass of the cord in the calculation.

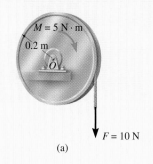

(a)

Solution

Free-Body Diagram. Fig. 17–15b. Note that the mass center is not subjected to an acceleration; however, the disk has a clockwise angular acceleration.

The moment of inertia of the disk about the pin is

$$I_O = \tfrac{1}{2}mr^2 = \frac{1}{2}(30 \text{ kg})(0.2 \text{ m})^2 = 0.6 \text{ kg} \cdot \text{m}^2$$

The three unknowns are O_x, O_y, and α.

Equations of Motion.

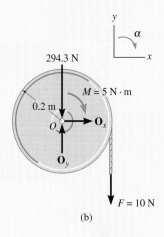

(b)

Fig. 17–15

$\xrightarrow{+} \Sigma F_x = m(a_G)_x;$ $\qquad O_x = 0$ $\qquad$ *Ans.*

$+\uparrow \Sigma F_y = m(a_G)_y;$ $\quad O_y - 294.3 \text{ N} - 10 \text{ N} = 0$

$\qquad\qquad\qquad O_y = 304 \text{ N}$ $\qquad$ *Ans.*

$\zeta + \Sigma M_O = I_O \alpha;$ $\quad -10 \text{ N}(0.2 \text{ m}) - 5 \text{ N} \cdot \text{m} = -(0.6 \text{ kg} \cdot \text{m}^2)\alpha$

$\qquad\qquad\qquad \alpha = 11.7 \text{ rad/s}^2 \zeta$

Kinematics. Since α is constant and is clockwise, the number of radians the disk must turn to obtain a clockwise angular velocity of 20 rad/s is

$\zeta +$ $\qquad\qquad \omega^2 = \omega_0^2 + 2\alpha_c(\theta - \theta_0)$

$\qquad (-20 \text{ rad/s})^2 = 0 + 2(-11.7 \text{ rad/s}^2)(\theta - 0)$

$\qquad\qquad \theta = -17.1 \text{ rad} = 17.1 \text{ rad} \zeta$

Hence,

$$\theta = 17.1 \text{ rad}\left(\frac{1 \text{ rev}}{2\pi \text{ rad}}\right) = 2.73 \text{ rev} \zeta \qquad \textit{Ans.}$$

E X A M P L E **17.10**

The 20-kg slender rod shown in Fig. 17–16a is rotating in the vertical plane, and at the instant shown it has an angular velocity of $\omega = 5$ rad/s. Determine the rod's angular acceleration and the horizontal and vertical components of reaction at the pin at this instant.

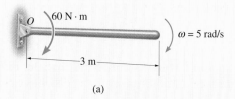

(a)

Solution

Free-Body and Kinetic Diagrams. Fig. 17–16b. As shown on the kinetic diagram, point G moves in a circular path and so has two components of acceleration. It is important that the tangential component $a_t = \alpha r_G$ act downward since it must be in accordance with the angular acceleration α of the rod. The three unknowns are O_n, O_t, and α.

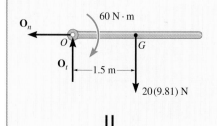

$\|$

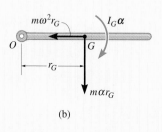

(b)

Fig. 17–16

Equations of Motion.

$$\xleftrightarrow{\pm} \Sigma F_n = m\omega^2 r_G; \qquad O_n = (20 \text{ kg})(5 \text{ rad/s})^2(1.5 \text{ m})$$

$$+\downarrow \Sigma F_t = m\alpha r_G; \qquad -O_t + 20(9.81) \text{ N} = (20 \text{ kg})(\alpha)(1.5 \text{ m})$$

$$\gamma+\Sigma M_G = I_G\alpha; \quad O_t(1.5 \text{ m}) + 60 \text{ N} \cdot \text{m} = [\tfrac{1}{12}(20 \text{ kg})(3 \text{ m})^2]\alpha$$

Solving

$$O_n = 750 \text{ N} \qquad O_t = 19.0 \text{ N} \qquad \alpha = 5.90 \text{ rad/s}^2 \qquad \textit{Ans.}$$

A more direct solution to this problem would be to sum moments about point O to eliminate $\mathbf{O}_n$ and $\mathbf{O}_t$ and obtain a *direct solution* for α. Here,

$$\gamma+\Sigma M_O = \Sigma(\mathcal{M}_k)_O; \quad 60 \text{ N} \cdot \text{m} + 20(9.81) \text{ N}(1.5 \text{ m}) =$$
$$[\tfrac{1}{12}(20 \text{ kg})(3 \text{ m})^2]\alpha + [20 \text{ kg}(\alpha)(1.5 \text{ m})](1.5 \text{ m})$$
$$\alpha = 5.90 \text{ rad/s}^2 \qquad \textit{Ans.}$$

Also, since $I_O = \tfrac{1}{3}ml^2$ for a slender rod, we can apply

$$\gamma+\Sigma M_O = I_O\alpha; \quad 60 \text{ N} \cdot \text{m} + 20(9.81) \text{ N}(1.5 \text{ m}) = [\tfrac{1}{3}(20 \text{ kg})(3 \text{ m})^2]\alpha$$
$$\alpha = 5.90 \text{ rad/s}^2 \qquad \textit{Ans.}$$

By comparison, the last equation provides the simplest solution for α and *does not* require use of the kinetic diagram.

EXAMPLE 17.11

The drum shown in Fig. 17–17a has a mass of 60 kg and a radius of gyration $k_O = 0.25$ m. A cord of negligible mass is wrapped around the periphery of the drum and attached to a block having a mass of 20 kg. If the block is released, determine the drum's angular acceleration.

(a)

Solution I

Free-Body Diagram. Here we will consider the drum and block separately, Fig. 17–17b. Assuming the block accelerates *downward* at **a**, it creates a *counterclockwise* angular acceleration **α** of the drum. The moment of inertia of the drum is

$$I_O = mk_O^2 = (60 \text{ kg})(0.25 \text{ m})^2 = 3.75 \text{ kg} \cdot \text{m}^2$$

There are five unknowns, namely O_x, O_y, T, a, and α.

Equations of Motion. Applying the translational equations of motion $\Sigma F_x = m(a_G)_x$ and $\Sigma F_y = m(a_G)_y$ to the drum is of no consequence to the solution, since these equations involve the unknowns O_x and O_y. Thus, for the drum and block, respectively,

$$\zeta + \Sigma M_O = I_O \alpha; \qquad T(0.4 \text{ m}) = (3.75 \text{ kg} \cdot \text{m}^2)\alpha \qquad (1)$$

$$+\uparrow \Sigma F_y = m(a_G)_y; \quad -20(9.81) \text{ N} + T = -20a \qquad (2)$$

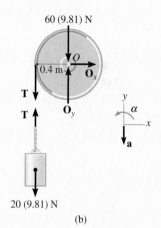

60 (9.81) N

0.4 m

T

T

20 (9.81) N

(b)

Kinematics. Since the point of contact A between the cord and drum has a tangential component of acceleration **a**, Fig. 17–17a, then

$$\zeta + a = \alpha r; \qquad a = \alpha(0.4) \qquad (3)$$

Solving the above equations,

$$T = 106 \text{ N}$$

$$a = 4.52 \text{ m/s}^2$$

$$\alpha = 11.3 \text{ rad/s}^2 \zeta \qquad \qquad Ans.$$

Solution II

Free-Body and Kinetic Diagrams. The cable tension T can be eliminated from the analysis by considering the drum and block as a *single system*, Fig. 17–17c. The kinetic diagram is shown since moments will be summed about point O.

Equations of Motion. Using Eq. 3 and applying the moment equation about O to eliminate the unknowns O_x and O_y, we have

$$\zeta + \Sigma M_O = \Sigma(\mathcal{M}_k)_O; \quad 20(9.81) \text{ N}(0.4 \text{ m}) =$$

$$(3.75 \text{ kg} \cdot \text{m}^2)\alpha + [20 \text{ kg}(0.4 \text{ m } \alpha)](0.4 \text{ m})$$

$$\alpha = 11.3 \text{ rad/s}^2 \zeta \qquad \qquad Ans.$$

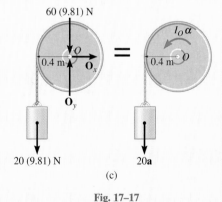

60 (9.81) N

0.4 m

$I_O \alpha$

=

0.4 m

20 (9.81) N

20a

(c)

Fig. 17–17

Note: If the block were *removed* and a force of 20(9.81) N were applied to the cord, show that $\alpha = 20.9$ rad/s^2 and explain the reason for the difference in the results.

EXAMPLE 17.12

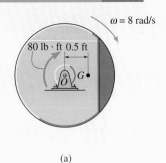

$\omega = 8$ rad/s

80 lb · ft 0.5 ft

(a)

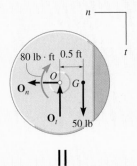

n

t

80 lb · ft 0.5 ft

O_n

O_t 50 lb

||

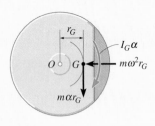

r_G

$I_G\alpha$

$m\omega^2 r_G$

$m\alpha r_G$

(b)

Fig. 17–18

The unbalanced 50-lb flywheel shown in Fig. 17–18a has a radius of gyration of $k_G = 0.6$ ft about an axis passing through its mass center G. If it has a clockwise angular velocity of 8 rad/s at the instant shown, determine the horizontal and vertical components of reaction at the pin O.

Solution

Free-Body and Kinetic Diagrams. Since G moves in a circular path, it will have both normal and tangential components of acceleration. Also, since α, which is caused by the flywheel's weight, acts clockwise, the tangential component of acceleration will act downward. Why? The vectors $m(a_G)_t = m\alpha r_G$, $m(a_G)_n = m\omega^2 r_G$, and $I_G\alpha$ are shown on the kinematic diagram in Fig. 17–18b. Here, the moment of inertia of the flywheel about its mass center is determined from the radius of gyration and the flywheel's mass; i.e., $I_G = mk_G^2 = (50 \text{ lb}/32.2 \text{ ft/s}^2)(0.6 \text{ ft})^2 = 0.559 \text{ slug} \cdot \text{ft}^2$.

The three unknowns are O_n, O_t, and α.

Equations of Motion.

$$\xleftarrow{+} \Sigma F_n = m\omega^2 r_G; \qquad O_n = \left(\frac{50 \text{ lb}}{32.2 \text{ ft/s}^2}\right)(8 \text{ rad/s})^2(0.5 \text{ ft}) \qquad (1)$$

$$+\downarrow \Sigma F_t = m\alpha r_G; \quad -O_t + 50 \text{ lb} = \left(\frac{50 \text{ lb}}{32.2 \text{ ft/s}^2}\right)(\alpha)(0.5 \text{ ft}) \qquad (2)$$

$$\zeta + \Sigma M_G = I_G\alpha; \qquad 80 \text{ lb} \cdot \text{ft} + O_t(0.5 \text{ ft}) = (0.559 \text{ slug} \cdot \text{ft}^2)\alpha \qquad (3)$$

Solving, $\alpha = 111 \text{ rad/s}^2 \qquad O_n = 49.7 \text{ lb} \qquad O_t = -36.1 \text{ lb} \qquad$ *Ans.*

Moments can also be summed about point O in order to eliminate $\mathbf{O}_n$ and $\mathbf{O}_t$ and thereby obtain a *direct solution* for $\boldsymbol{\alpha}$, Fig. 17–18b. This can be done in one of *two* ways, i.e., by using either $\Sigma M_O = \Sigma(\mathcal{M}_k)_O$ or $\Sigma M_O = I_O\alpha$. If the first of these equations is applied, we have

$$\zeta + \Sigma M_O = \Sigma(\mathcal{M}_k)_O; \ 80 \text{ lb} \cdot \text{ft} + 50 \text{ lb}(0.5 \text{ ft}) =$$

$$(0.559 \text{ slug} \cdot \text{ft}^2)\alpha + \left[\left(\frac{50 \text{ lb}}{32.2 \text{ ft/s}^2}\right)\alpha(0.5 \text{ ft})\right](0.5 \text{ ft})$$

$$105 = 0.947\alpha \qquad (4)$$

If $\Sigma M_O = I_O\alpha$ is applied, then by the parallel-axis theorem the moment of inertia of the flywheel about O is

$$I_O = I_G + mr_G^2 = 0.559 + \left(\frac{50}{32.2}\right)(0.5)^2 = 0.947 \text{ slug} \cdot \text{ft}^2$$

Hence, from the free-body diagram, Fig. 17–18b, we require

$$\zeta + \Sigma M_O = I_O\alpha; \quad 80 \text{ lb} \cdot \text{ft} + 50 \text{ lb}(0.5 \text{ ft}) = (0.947 \text{ slug} \cdot \text{ft}^2)\alpha$$

which is the same as Eq. 4. Solving for α and substituting into Eq. 2 yields the answer for O_t obtained previously.

EXAMPLE 17.13

The slender rod shown in Fig. 17–19a has a mass m and length l and is released from rest when $\theta = 0°$. Determine the horizontal and vertical components of force which the pin at A exerts on the rod at the instant $\theta = 90°$.

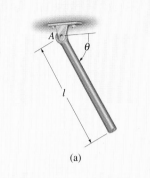

(a)

Solution

Free-Body Diagram. The free-body diagram for the rod is shown when the rod is in the general position θ, Fig. 17–19b. For convenience, the force components at A are shown acting in the n and t directions. Note that α acts clockwise.

The moment of inertia of the rod about point A is $I_A = \frac{1}{3}ml^2$.

Equations of Motion. Moments will be summed about A in order to eliminate the reactive forces there.*

$$+\nwarrow\Sigma F_n = m\omega^2 r_G; \quad A_n - mg \sin\theta = m\omega^2(l/2) \tag{1}$$

$$+\swarrow\Sigma F_t = m\alpha r_G; \quad A_t + mg \cos\theta = m\alpha(l/2) \tag{2}$$

$$\curvearrowright+\Sigma M_A = I_A\alpha; \quad mg \cos\theta(l/2) = (\tfrac{1}{3}ml^2)\alpha \tag{3}$$

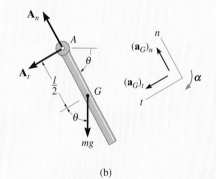

(b)

Fig. 17–19

Kinematics. For a given angle θ there are four unknowns in the above three equations: A_n, A_t, ω, and α. As shown by Eq. 3, α is *not constant*; rather, it depends on the position θ of the rod. The necessary fourth equation is obtained using kinematics, where α and ω can be related to θ by the equation

$$(\curvearrowright+) \qquad\qquad \omega\, d\omega = \alpha\, d\theta \tag{4}$$

Note that the positive clockwise direction for this equation *agrees* with that of Eq. 3. This is important since we are seeking a simultaneous solution.

In order to solve for ω at $\theta = 90°$, eliminate α from Eqs. 3 and 4, which yields

$$\omega\, d\omega = (1.5\, g/l) \cos\theta\, d\theta$$

Since $\omega = 0$ at $\theta = 0°$, we have

$$\int_0^\omega \omega\, d\omega = (1.5\, g/l) \int_{0°}^{90°} \cos\theta\, d\theta$$

$$\omega^2 = 3\, g/l$$

Substituting this value into Eq. 1 with $\theta = 90°$ and solving Eqs. 1 to 3 yields

$$\alpha = 0 \qquad A_t = 0 \qquad A_n = 2.5\, mg \qquad\qquad Ans.$$

---

*If $\Sigma M_A = \Sigma(\mathcal{M}_k)_A$ is used, one must account for the moments of $I_G\alpha$ and $m(\mathbf{a}_G)_t$ about A. Here, however, we have used $\Sigma M_A = I_A\alpha$.

17.5 Equations of Motion: General Plane Motion

The rigid body (or slab) shown in Fig. 17–20a is subjected to general plane motion caused by the externally applied force and couple-moment system. The free-body and kinetic diagrams for the body are shown in Fig. 17–20b. If an x and y inertial coordinate system is chosen as shown, the three equations of motion may be written as

$$\begin{aligned} \Sigma F_x &= m(a_G)_x \\ \Sigma F_y &= m(a_G)_y \\ \Sigma M_G &= I_G \alpha \end{aligned}$$

$$(17\text{–}17)$$

In some problems it may be convenient to sum moments about some point P other than G. This is usually done in order to eliminate unknown forces from the moment summation. When used in this more general sense, the three equations of motion become

$$\begin{aligned} \Sigma F_x &= m(a_G)_x \\ \Sigma F_y &= m(a_G)_y \\ \Sigma M_P &= \Sigma (\mathcal{M}_k)_P \end{aligned}$$

$$(17\text{–}18)$$

Here $\Sigma(\mathcal{M}_k)_P$ represents the moment sum of $I_G\alpha$ and $m\mathbf{a}_G$ (or its components) about P as determined by the data on the kinetic diagram.

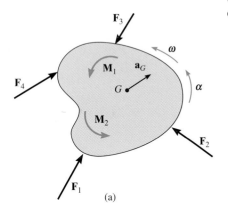

(a)

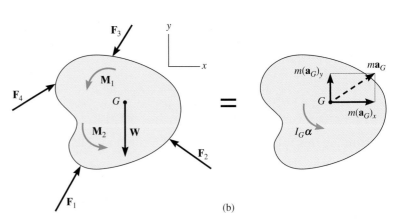

(b)

Fig. 17–20

Frictional Rolling Problems. There is a class of planar kinetics problems which deserves special mention. These problems involve wheels, cylinders, or bodies of similar shape, which roll on a *rough* plane surface. Because of the applied loadings, it may not be known if the body *rolls without slipping*, or if it *slides as it rolls*. For example, consider the homogeneous disk shown in Fig. 17–21a, which has a mass m and is subjected to a known horizontal force $\mathbf{P}$. The free-body diagram is shown in Fig. 17–21b. Since $\mathbf{a}_G$ is directed to the right and α is clockwise, we have

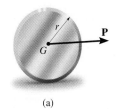

(a)

$$\xrightarrow{+} \Sigma F_x = m(a_G)_x; \qquad P - F = ma_G \qquad (17\text{–}19)$$
$$+\uparrow \Sigma F_y = m(a_G)_y; \qquad N - mg = 0 \qquad (17\text{–}20)$$
$$\zeta + \Sigma M_G = I_G\alpha; \qquad Fr = I_G\alpha \qquad (17\text{–}21)$$

A fourth equation is needed since these *three equations* contain *four unknowns: F, N, α, and a_G.*

No Slipping. If the frictional force F is great enough to allow the disk to roll without slipping, then a_G may be related to α by the kinematic equation,*

$$(\zeta+) \qquad\qquad a_G = \alpha r \qquad\qquad (17\text{–}22)$$

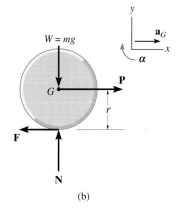

(b)

Fig. 17–21

The four unknowns are determined by *solving simultaneously* Eqs. 17–19 to 17–22. When the solution is obtained, the assumption of no slipping must be *checked*. Recall that no slipping occurs provided $F \le \mu_s N$, where μ_s is the coefficient of static friction. If the inequality is satisfied, the problem is solved. However, if $F > \mu_s N$, the problem must be *reworked*, since then the disk slips as it rolls.

Slipping. In the case of slipping, α and a_G are *independent of one another* so that Eq. 17–22 does not apply. Instead, the magnitude of the frictional force is related to the magnitude of the normal force using the coefficient of kinetic friction μ_k, i.e.,

$$F = \mu_k N \qquad\qquad (17\text{–}23)$$

In this case Eqs. 17–19 to 17–21 and 17–23 are used for the solution. It is important to keep in mind that whenever Eq. 17–22 or 17–23 is applied, it is necessary to have consistency in the directional sense of the vectors. In the case of Eq. 17–22, $\mathbf{a}_G$ must be directed to the right when $\boldsymbol{\alpha}$ is clockwise, since the rolling motion requires it. And in Eq. 17–23, $\mathbf{F}$ must be directed to the left to prevent the assumed slipping motion to the right, Fig. 17–21b. On the other hand, if these equations are *not used* for the solution, these vectors can have *any* assumed directional sense. Then if the calculated numerical value of these quantities is negative, the vectors act in their opposite sense of direction. Examples 17.15 and 17.16 illustrate these concepts numerically.

*See Example 16.3 or 16.14.

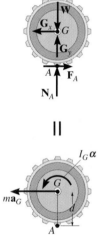

As the soil compactor moves forward, the roller has general plane motion. The forces shown on the roller's free-body diagram cause the effects shown on the kinetic diagram. If moments are summed about the mass center, G, then $\Sigma M_G = I_G \alpha$. However, if moments are summed about point A then $\downarrow + \Sigma M_A = I_G \alpha + (m a_G)d$.

PROCEDURE FOR ANALYSIS

Kinetic problems involving general plane motion of a rigid body can be solved using the following procedure.

Free-Body Diagram.

- Establish the x, y inertial coordinate system and draw the free-body diagram for the body.
- Specify the direction and sense of the acceleration of the mass center, $\mathbf{a}_G$, and the angular acceleration $\boldsymbol{\alpha}$ of the body.
- Compute the moment of inertia I_G.
- Identify the unknowns in the problem.
- If it is decided that the rotational equation of motion $\Sigma M_P = \Sigma(\mathcal{M}_k)_P$ is to be used, then consider drawing the kinetic diagram in order to help "visualize" the "moments" developed by the components $m(\mathbf{a}_G)_x$, $m(\mathbf{a}_G)_y$, and $I_G\boldsymbol{\alpha}$ when writing the terms in the moment sum $\Sigma(\mathcal{M}_k)_P$.

Equations of Motion.

- Apply the three equations of motion in accordance with the established sign convention.
- When friction is present, there is the possibility for motion with no slipping or tipping. Each possibility for motion should be considered.

Kinematics.

- Use kinematics if a complete solution cannot be obtained strictly from the equations of motion.
- If the body's motion is *constrained* due to its supports, additional equations may be obtained by using $\mathbf{a}_B = \mathbf{a}_A + \mathbf{a}_{B/A}$, which relates the accelerations of any two points A and B on the body.
- When a wheel, disk, cylinder, or ball *rolls without slipping*, then $a_G = \alpha r$.

E X A M P L E 17.14

The spool in Fig. 17–22a has a mass of 8 kg and a radius of gyration of $k_G = 0.35$ m. If cords of negligible mass are wrapped around its inner hub and outer rim as shown, determine the spool's angular acceleration.

Solution I

Free-Body Diagram. Fig. 17–22b. The 100-N force causes $\mathbf{a}_G$ to act upward. Also, α acts clockwise, since the spool winds around the cord at A.

There are three unknowns T, a_G, and α. The moment of inertia of the spool about its mass center is

$$I_G = mk_G^2 = 8\ \text{kg}(0.35\ \text{m})^2 = 0.980\ \text{kg} \cdot \text{m}^2$$

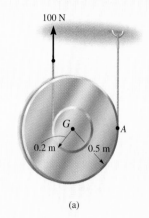

(a)

Equations of Motion.

$$+\uparrow \Sigma F_y = m(a_G)_y; \qquad T + 100\ \text{N} - 78.48\ \text{N} = (8\ \text{kg})a_G \qquad (1)$$

$$\zeta + \Sigma M_G = I_G\alpha; \quad 100\ \text{N}(0.2\ \text{m}) - T(0.5\ \text{m}) = (0.980\ \text{kg} \cdot \text{m}^2)\alpha \quad (2)$$

Kinematics. A complete solution is obtained if kinematics is used to relate a_G to α. In this case the spool "rolls without slipping" on the cord at A. Hence, we can use the results of Example 16.3 or 16.14, so that

$$(\zeta +)a_G = \alpha r, \qquad\qquad a_G = 0.5\alpha \qquad (3)$$

Solving Eqs. 1 to 3, we have

$$\alpha = 10.3\ \text{rad/s}^2 \qquad\qquad Ans.$$
$$a_G = 5.16\ \text{m/s}^2$$
$$T = 19.8\ \text{N}$$

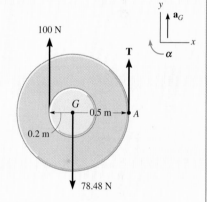

(b)

Solution II

Equations of Motion. We can eliminate the unknown T by summing moments about point A. From the free-body and kinetic diagrams Figs. 17–22b and 17–22c, we have

$$\zeta + \Sigma M_A = \Sigma(\mathcal{M}_k)_A; \qquad 100\ \text{N}(0.7\ \text{m}) - 78.48\ \text{N}(0.5\ \text{m})$$
$$= (0.980\ \text{kg} \cdot \text{m}^2)\alpha + [(8\ \text{kg})a_G](0.5\ \text{m})$$

Using Eq. (3),

$$\alpha = 10.3\ \text{rad/s}^2 \qquad\qquad Ans.$$

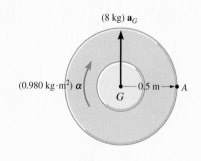

(c)

Fig. 17–22

EXAMPLE 17.15

The 50-lb wheel shown in Fig. 17–23a has a radius of gyration $k_G = 0.70$ ft. If a 35-lb·ft couple moment is applied to the wheel, determine the acceleration of its mass center G. The coefficients of static and kinetic friction between the wheel and the plane at A are $\mu_s = 0.3$ and $\mu_k = 0.25$, respectively.

Solution

Free-Body Diagram. By inspection of Fig. 17–23b, it is seen that the couple moment causes the wheel to have a clockwise angular acceleration of $\boldsymbol{\alpha}$. As a result, the acceleration of the mass center, $\mathbf{a}_G$, is directed to the right. The moment of inertia is

$$I_G = mk_G^2 = \frac{50\ \text{lb}}{32.2\ \text{ft/s}^2}(0.70\ \text{ft})^2 = 0.761\ \text{slug}\cdot\text{ft}^2$$

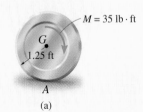

The unknowns are N_A, F_A, a_G, and α.

Equations of Motion.

$$\xrightarrow{+}\ \Sigma F_x = m(a_G)_x; \qquad F_A = \frac{50\ \text{lb}}{32.2\ \text{ft/s}^2}a_G \tag{1}$$

$$+\uparrow \Sigma F_y = m(a_G)_y; \qquad N_A - 50\ \text{lb} = 0 \tag{2}$$

$$\zeta+\Sigma M_G = I_G\alpha; \qquad 35\ \text{lb}\cdot\text{ft} - 1.25\ \text{ft}(F_A) = (0.761\ \text{slug}\cdot\text{ft}^2)\alpha \tag{3}$$

A fourth equation is needed for a complete solution.

Kinematics (No Slipping). If this assumption is made, then

$$(\zeta+) \qquad\qquad a_G = (1.25\ \text{ft})\alpha \tag{4}$$

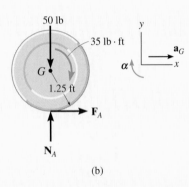

(b)

Fig. 17–23

Solving Eqs. 1 to 4,

$$N_A = 50.0\ \text{lb} \qquad F_A = 21.3\ \text{lb}$$
$$\alpha = 11.0\ \text{rad/s}^2 \qquad a_G = 13.7\ \text{ft/s}^2$$

The original assumption of no slipping requires $F_A \le \mu_s N_A$. However, since 21.3 lb > 0.3(50 lb) = 15 lb, the wheel slips as it rolls.

(Slipping). Equation 4 is not valid, and so $F_A = \mu_k N_A$, or

$$F_A = 0.25 N_A \tag{5}$$

Solving Eqs. 1 to 3 and 5 yields

$$N_A = 50.0\ \text{lb} \qquad F_A = 12.5\ \text{lb}$$
$$\alpha = 25.5\ \text{rad/s}^2$$
$$a_G = 8.05\ \text{ft/s}^2 \rightarrow \qquad\qquad Ans.$$

E X A M P L E 17.16

The uniform slender pole shown in Fig. 17–24a has a mass of 100 kg and a moment of inertia $I_G = 75 \text{ kg} \cdot \text{m}^2$. If the coefficients of static and kinetic friction between the end of the pole and the surface are $\mu_s = 0.3$ and $\mu_k = 0.25$, respectively, determine the pole's angular acceleration at the instant the 400-N horizontal force is applied. The pole is originally at rest.

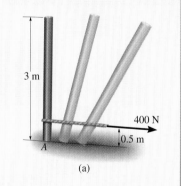

(a)

Solution

Free-Body Diagram. Figure 17–24b. The path of motion of the mass center G will be along an unknown curved path having a radius of curvature ρ, which is initially parallel to the y axis. There is no normal or y component of acceleration since the pole is originally at rest, i.e., $\mathbf{v}_G = \mathbf{0}$, so that $(a_G)_y = v_G^2/\rho = 0$. We will assume the mass center accelerates to the right and that the pole has a clockwise angular acceleration of $\boldsymbol{\alpha}$. The unknowns are N_A, F_A, a_G, and α.

Equations of Motion.

$$\xrightarrow{+} \Sigma F_x = m(a_G)_x; \qquad 400 \text{ N} - F_A = (100 \text{ kg})a_G \qquad (1)$$

$$+\uparrow \Sigma F_y = m(a_G)_y; \qquad N_A - 981 \text{ N} = 0 \qquad (2)$$

$$\zeta + \Sigma M_G = I_G\alpha; \quad F_A(1.5 \text{ m}) - 400 \text{ N}(1 \text{ m}) = (75 \text{ kg} \cdot \text{m}^2)\alpha \qquad (3)$$

A fourth equation is needed for a complete solution.

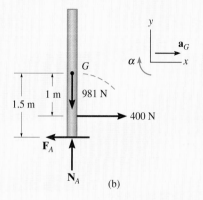

(b)

Fig. 17–24

Kinematics (No Slipping). In this case point A acts as a "pivot" so that indeed, if α is clockwise, then a_G is directed to the right.

$$\zeta + a_G = \alpha r_{AG}; \qquad a_G = (1.5 \text{ m})\alpha \qquad (4)$$

Solving Eqs. 1 to 4 yields

$$N_A = 981 \text{ N} \qquad F_A = 300 \text{ N}$$
$$a_G = 1 \text{ m/s}^2 \qquad \alpha = 0.667 \text{ rad/s}^2$$

Testing the original assumption of no slipping requires $F_A \leq \mu_s N_A$. However, 300 N > 0.3(981 N) = 294 N. (Slips at A.)

(Slipping). For this case Eq. 4 does *not* apply. Instead the frictional equation $F_A = \mu_k N_A$ is used. Hence,

$$F_A = 0.25N_A \qquad (5)$$

Solving Eqs. 1 to 3 and 5 simultaneously yields

$$N_A = 981 \text{ N} \qquad F_A = 245 \text{ N} \qquad a_G = 1.55 \text{ m/s}^2$$
$$\alpha = -0.428 \text{ rad/s}^2 = 0.428 \text{ rad/s}^2 \zeta \qquad \qquad Ans.$$

E X A M P L E 17.17

The 30-kg wheel shown in Fig. 17–25a has a mass center at G and a radius of gyration $k_G = 0.15$ m. If the wheel is originally at rest and released from the position shown, determine its angular acceleration. No slipping occurs.

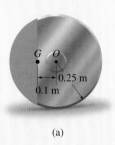

(a)

Solution

Free-Body and Kinetic Diagrams. The two unknowns $\mathbf{F}_A$ and $\mathbf{N}_A$ shown on the free-body diagram, Fig. 17–25b, can be eliminated from the analysis by summing moments about point A. The kinetic diagram accompanies the solution in order to illustrate application of $\Sigma(\mathcal{M}_k)_A$. Since point G moves along a curved path, the two components $m(\mathbf{a}_G)_x$ and $m(\mathbf{a}_G)_y$ are shown on the kinetic diagram, Fig. 17–25b.

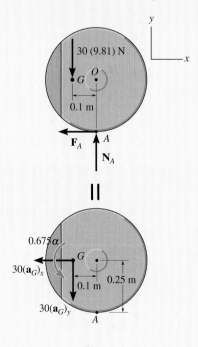

(b)

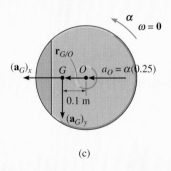

(c)

Fig. 17–25

The moment of inertia is

$$I_G = mk_G^2 = 30(0.15)^2 = 0.675 \text{ kg} \cdot \text{m}^2$$

There are five unknowns, N_A, F_A, $(a_G)_x$, $(a_G)_y$, and α.

Equation of Motion. Applying the rotational equation of motion about point A, to eliminate N_A, and F_A, we have

$$\zeta + \Sigma M_A = \Sigma(\mathcal{M}_k)_A; \quad 30(9.81) \text{ N}(0.1 \text{ m}) =$$
$$(0.675 \text{ kg} \cdot \text{m}^2)\alpha + (30 \text{ kg})(a_G)_x(0.25 \text{ m}) + (30 \text{ kg})(a_G)_y(0.1 \text{ m}) \quad (1)$$

There are three unknowns in this equation: $(a_G)_x$, $(a_G)_y$, and α.

Kinematics. Using kinematics, $(a_G)_x$, $(a_G)_y$ will be related to α. As shown in Fig. 17–25c, these vectors must have the same directional sense as the corresponding vectors on the kinetic diagram since we are seeking a simultaneous solution with Eq. 1. Since no slipping occurs, $a_O = \alpha r = \alpha(0.25 \text{ m})$, directed to the left, Fig. 17–25c. Also, $\omega = 0$, since the wheel is originally at rest. Applying the acceleration equation to point O (base point) and point G, we have

$$\mathbf{a}_G = \mathbf{a}_O + \boldsymbol{\alpha} \times \mathbf{r}_{G/O} - \omega^2 \mathbf{r}_{G/O}$$
$$-(a_G)_x\mathbf{i} - (a_G)_y\mathbf{j} = -\alpha(0.25)\mathbf{i} + (\alpha\mathbf{k}) \times (-0.1\mathbf{i}) - \mathbf{0}$$

Expanding and equating the respective $\mathbf{i}$ and $\mathbf{j}$ components, we have

$$(a_G)_x = \alpha(0.25) \quad (2)$$
$$(a_G)_y = \alpha(0.1) \quad (3)$$

Solving Eqs. 1 to 3 yields

$$\alpha = 10.3 \text{ rad/s}^2 \curvearrowright \qquad Ans.$$
$$(a_G)_x = 2.58 \text{ m/s}^2$$
$$(a_G)_y = 1.03 \text{ m/s}^2$$

As an exercise, show that $F_A = 77.4$ N and $N_A = 263$ N.

DESIGN PROJECTS

17-1D. DESIGN OF A DYNAMOMETER

In order to test the dynamic strength of cables, an instrument called a *dynamometer* must be used that will measure the tension in a cable when it hoists a very heavy object with accelerated motion. Design such an instrument, based on the use of single or multiple springs so that it can be used on the cable supporting the 300-kg pipe that is given an upward acceleration of 2 m/s². Submit a drawing and explain how your dynamometer operates.

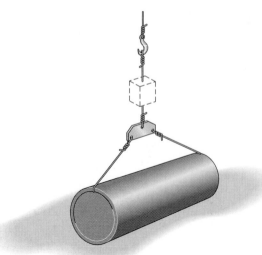

Prob. 17–1D

17-2D. DESIGN OF A SMALL ELEVATOR BRAKE

A small household elevator is operated using a hoist. For safety purposes it is necessary to install a braking mechanism which will automatically engage in case the cable fails during operation. Design the braking mechanism using steel members and springs. The elevator and its contents are assumed to have a mass of 300 kg, and it travels at 2.5 m/s. The maximum allowable deceleration to stop the motion is to be 4 m/s². Assume the coefficient of kinetic friction between any steel members and the walls of the elevator shaft is $\mu_k = 0.3$. The gap between the elevator frame and each wall of the shaft is 50 mm. Submit a scale drawing of your design along with a force analysis to show that your design will arrest the motion as required. Discuss the safety and reliability of the mechanism.

Prob. 17–2D

17-3D. SAFETY PERFORMANCE OF A BICYCLE

One of the most common accidents one can have on a bicycle is to flip over the handle bars. Obtain the necessary measurements of a standard-size bicycle and its mass and center of mass. Consider yourself as the rider, with center of mass at your navel. Perform an experiment to determine the coefficient of kinetic friction between the wheels and the pavement. With this data, calculate the possibility of flipping over when (a) only the rear brakes are applied, (b) only the front brakes are applied, and (c) both front and rear brakes are applied simultaneously. What effect does the height of the seat have on these results? Suggest a way to improve the bicycle's design, and write a report on the safety of cycling based on this analysis.

Prob. 17–3D

CHAPTER REVIEW

- **Moment of Inertia.** The moment of inertia is a measure of the resistance of a body to a change in its angular velocity. It is defined by $I = \int r^2 \, dm$ and will be different for each axis about which it is computed. For a body having an axial symmetry, the integration is usually performed using disk or shell elements.

  Many bodies are composed of simple shapes. If this is the case, then tabular values of I can be used, such as the ones given on the inside back cover of this book. To obtain the moment of inertia of a composite body about any specified axis, the moment of inertia of each part is determined about the axis and the results are added together. Doing this often requires use of the parallel-axis theorem $I = I_G + md^2$. Handbooks may also report values of the radius of gyration k for the body. If the body's mass is known, then the mass moment of inertia is determined from $I = mk^2$.

- **Planar Equations of Motion.** The equation which defines the translational motion of a rigid body is $\Sigma \mathbf{F} = m\mathbf{a}_G$. Here, $\mathbf{a}_G$ is the acceleration of the body's mass center.

  The equation which describes the rotational motion of the body is determined by taking the moments of all of the particles in the body about an axis. When the axis passes through the mass center, the result becomes $\Sigma M_G = I_G \alpha$. If moments are taken about some arbitrary point P, then we get $\Sigma M_P = \Sigma(\mathcal{M}_k)_P$. The sum on the right side represents the moments of the kinetic vectors $m\mathbf{a}_G$ and $I_G\alpha$ about point P.

  In order to account for all of the terms in these equations, application should always be accompanied with a free-body diagram, and for some problems, it may also be convenient to draw the kinetic diagram.

- **Translation.** Here $I_G\boldsymbol{\alpha} = \mathbf{0}$ since $\boldsymbol{\alpha} = \mathbf{0}$. If the body undergoes rectilinear translation, use an inertial x–y axis, in which case the equations of motion are

$$\Sigma F_x = m(a_G)_x$$
$$\Sigma F_y = m(a_G)_y$$
$$\Sigma M_G = 0$$

For curvilinear translation, use inertial n–t axes so that

$$\Sigma F_n = m(a_G)_n$$
$$\Sigma F_t = m(a_G)_t$$
$$\Sigma M_G = 0$$

- **Rotation About a Fixed Axis.** For fixed axis rotation, the kinetic vector $m(\mathbf{a}_G)_n$ produces no moment about the axis of rotation, and so the rotational equation of motion reduces to a simplified form about point O. The equations of motion are
$$\Sigma F_n = m\omega^2 r_G \qquad \Sigma F_t = m\alpha r_G$$
$$\Sigma M_G = I_G\alpha \quad \text{or} \quad \Sigma M_O = I_O\alpha$$

- **General Plane Motion.** For general plane motion, we have
$$\Sigma F_x = m(a_G)_x \qquad \Sigma F_y = m(a_G)_y$$
$$\Sigma M_G = I_G\alpha \quad \text{or} \quad \Sigma M_P = \Sigma(\mathcal{M}_k)_P$$

If the body is constrained by its supports, then additional equations of kinematics can be obtained by using $\mathbf{a}_B = \mathbf{a}_A + \mathbf{a}_{B/A}$ to relate the accelerations of any two points A and B on the body.

The principle of work and energy plays an important role in the motion of the
draw works used to lift pipe on this drilling rig.

18

Planar Kinetics of a Rigid Body: Work and Energy

CHAPTER OBJECTIVES

- To develop formulations for the kinetic energy of a body, and define the various ways a force and couple do work.
- To apply the principle of work and energy to solve rigid-body planar kinetic problems that involve force, velocity, and displacement.
- To show how the conservation of energy can be used to solve rigid-body planar kinetic problems.

18.1 Kinetic Energy

In this chapter we will apply work and energy methods to problems involving force, velocity, and displacement related to the planar motion of a rigid body. Before doing this, however, it will first be necessary to develop a means of obtaining the body's kinetic energy when the body is subjected to translation, rotation about a fixed axis, or general plane motion.

To do this we will consider the rigid body shown in Fig. 18–1, which is represented here by a *slab* moving in the inertial x–y reference plane. An arbitrary ith particle of the body, having a mass dm, is located at r from the arbitrary point P. If at the *instant* shown the particle has a velocity v_i, then the particle's kinetic energy is $T_i = \frac{1}{2} dm\, v_i^2$.

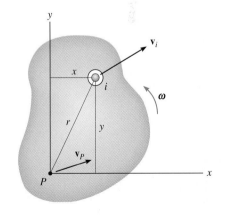

Fig. 18–1

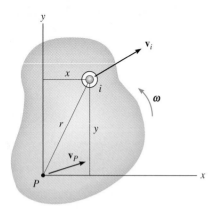

Fig. 18–1

The kinetic energy of the entire body is determined by writing similar expressions for each particle of the body and integrating the results, i.e.,

$$T = \frac{1}{2} \int_m dm\, v_i^2$$

This equation may also be expressed in terms of the velocity of point P. If the body has an angular velocity ω, then from Fig. 18–1 we have

$$v_i = v_P + v_{i/P}$$
$$= (v_P)_x \mathbf{i} + (v_P)_y \mathbf{j} + \omega \mathbf{k} \times (x\mathbf{i} + y\mathbf{j})$$
$$= [(v_P)_x - \omega y]\mathbf{i} + [(v_P)_y + \omega x]\mathbf{j}$$

The square of the magnitude of v_i is thus

$$v_i \cdot v_i = v_i^2 = [(v_P)_x - \omega y]^2 + [(v_P)_y + \omega x]^2$$
$$= (v_P)_x^2 - 2(v_P)_x \omega y + \omega^2 y^2 + (v_P)_y^2 + 2(v_P)_y \omega x + \omega^2 x^2$$
$$= v_P^2 - 2(v_P)_x \omega y + 2(v_P)_y \omega x + \omega^2 r^2$$

Substituting into the equation of kinetic energy yields

$$T = \frac{1}{2}\left(\int_m dm\right)v_P^2 - (v_P)_x \omega\left(\int_m y\, dm\right) + (v_P)_y \omega\left(\int_m x\, dm\right) + \frac{1}{2}\omega^2\left(\int_m r^2 dm\right)$$

The first integral on the right represents the entire mass m of the body. Since $\bar{y}m = \int y\, dm$ and $\bar{x}m = \int x\, dm$, the second and third integrals locate the body's center of mass G with respect to P. The last integral represents the body's moment of inertia I_P, computed about the z axis passing through point P. Thus,

$$T = \frac{1}{2}mv_P^2 - (v_P)_x \omega \bar{y}m + (v_P)_y \omega \bar{x}m + \frac{1}{2}I_P \omega^2 \qquad (18\text{–}1)$$

As a special case, if point P coincides with the mass center G for the body, then $\bar{y} = \bar{x} = 0$, and therefore

$$T = \frac{1}{2}mv_G^2 + \frac{1}{2}I_G\omega^2 \qquad (18\text{–}2)$$

Here I_G is the moment of inertia for the body about an axis which is perpendicular to the plane of motion and passes through the mass center. Both terms on the right side are *always positive*, since the velocities are squared. Furthermore, it may be verified that these terms have units of length times force, common units being m $\cdot$ N or ft$\cdot$ lb. Recall, however, that in the SI system the unit of energy is the joule (J), where $1\,\text{J} = 1\,\text{m} \cdot \text{N}$.

Translation. When a rigid body of mass m is subjected to either rectilinear or curvilinear *translation*, the kinetic energy due to rotation is zero, since $\boldsymbol{\omega} = 0$. From Eq. 18–2, the kinetic energy of the body is therefore

$$T = \tfrac{1}{2}mv_G^2 \qquad\qquad (18\text{–}3)$$

where v_G is the magnitude of the translational velocity $\mathbf{v}$ at the instant considered, Fig. 18–2.

Rotation About a Fixed Axis. When a rigid body is *rotating about a fixed axis* passing through point O, Fig. 18–3, the body has both *translational* and *rotational* kinetic energy as defined by Eq. 18–2, i.e.,

$$T = \tfrac{1}{2}mv_G^2 + \tfrac{1}{2}I_G\omega^2 \qquad\qquad (18\text{–}4)$$

The body's kinetic energy may also be formulated by noting that $v_G = r_G\omega$, in which case $T = \tfrac{1}{2}(I_G + mr_G^2)\omega^2$. By the parallel-axis theorem, the terms inside the parentheses represent the moment of inertia I_O of the body about an axis perpendicular to the plane of motion and passing through point O. Hence,*

$$T = \tfrac{1}{2}I_O\omega^2 \qquad\qquad (18\text{–}5)$$

From the derivation, this equation will give the same result as Eq. 18–4, since it accounts for *both* the translational and rotational kinetic energies of the body.

General Plane Motion. When a rigid body is subjected to general plane motion, Fig. 18–4, it has an angular velocity $\boldsymbol{\omega}$ and its mass center has a velocity v_G. Hence, the kinetic energy is defined by Eq. 18–2, i.e.,

$$T = \tfrac{1}{2}mv_G^2 + \tfrac{1}{2}I_G\omega^2 \qquad\qquad (18\text{–}6)$$

Here it is seen that the total kinetic energy of the body consists of the *scalar* sum of the body's *translational* kinetic energy, $\tfrac{1}{2}mv_G^2$, and *rotational* kinetic energy about its mass center, $\tfrac{1}{2}I_G\omega^2$.

Because energy is a scalar quantity, the total kinetic energy for a system of *connected* rigid bodies is the sum of the kinetic energies of all its moving parts. Depending on the type of motion, the kinetic energy of *each body* is found by applying Eq. 18–2 or the alternative forms mentioned above.

---

*The similarity between this derivation and that of $\Sigma M_O = I_O\alpha$, Eq. 17–16, should be noted. Also note that the same result can be obtained directly from Eq. 18–1 by selecting point P at O, realizing that $v_O = 0$.

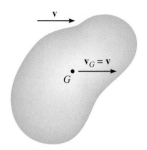

Translation

Fig. 18–2

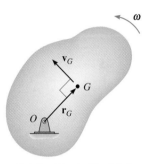

Rotation About a Fixed Axis

Fig. 18–3

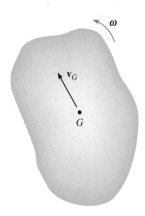

General Plane Motion

Fig. 18–4

The total kinetic energy of this soil compactor consists of the kinetic energy of the body or frame of the machine due to its translation, and the translational and rotational kinetic energies of the roller and the wheels due to their general plane motion. Here we exclude the additional kinetic energy developed by the moving parts of the engine and drive train.

E X A M P L E 18.1

The system of three elements shown in Fig. 18–5a consists of a 6-kg block B, a 10-kg disk D, and a 12-kg cylinder C. If no slipping occurs, determine the total kinetic energy of the system at the instant shown.

Solution

In order to compute the kinetic energy of the disk and cylinder, it is first necessary to determine ω_D, ω_C, and v_G, Fig. 18–5a. From the *kinematics* of the disk,

$$v_B = r_D\omega_D; \qquad 0.8\,\text{m/s} = (0.1\,\text{m})\omega_D \qquad \omega_D = 8\,\text{rad/s}$$

Since the cylinder rolls without slipping, the instantaneous center of zero velocity is at the point of contact with the ground, Fig. 18–5b, hence,

$$v_E = r_{E/IC}\omega_C; \qquad 0.8\,\text{m/s} = (0.2\,\text{m})\omega_C \qquad \omega_C = 4\,\text{rad/s}$$
$$v_G = r_{G/IC}\omega_C; \qquad v_G = (0.1\,\text{m})(4\,\text{rad/s}) = 0.4\,\text{m/s}$$

Block

$$T_B = \tfrac{1}{2}m_Bv_B^2 = \tfrac{1}{2}(6\,\text{kg})(0.8\,\text{m/s})^2 = 1.92\,\text{J}$$

Disk

$$T_D = \tfrac{1}{2}I_D\omega_D^2 = \tfrac{1}{2}(\tfrac{1}{2}m_Dr_D^2)\omega_D^2$$
$$= \tfrac{1}{2}\tfrac{1}{2}(10\,\text{kg})(0.1\,\text{m})^2](8\,\text{rad/s})^2 = 1.60\,\text{J}$$

Cylinder

$$T_C = \tfrac{1}{2}mv_G^2 + \tfrac{1}{2}I_G\omega_C^2 = \tfrac{1}{2}mv_G^2 + \tfrac{1}{2}(\tfrac{1}{2}m_Cr_C^2)\omega_C^2$$
$$= \tfrac{1}{2}(12\,\text{kg})(0.4\,\text{m/s})^2 + \tfrac{1}{2}\tfrac{1}{2}(12\,\text{kg})(0.1\,\text{m})^2](4\,\text{rad/s})^2 = 1.44\,\text{J}$$

The total kinetic energy of the system is therefore

$$T = T_B + T_D + T_C$$
$$= 1.92\,\text{J} + 1.60\,\text{J} + 1.44\,\text{J} = 4.96\,\text{J} \qquad \textit{Ans.}$$

Fig. 18–5

(a)

(b)

18.2   The Work of a Force

Several types of forces are often encountered in planar kinetics problems involving a rigid body. The work of each of these forces has been presented in Sec. 14.1 and is listed below as a summary.

Work of a Variable Force.   If an external force **F** acts on a rigid body, the work done by the force when it moves along the path s, Fig. 18–6, is defined as

$$U_F = \mathbf{F} \cdot \mathbf{r} = \int_s F \cos\theta \, ds \qquad (18\text{–}7)$$

Here θ is the angle between the "tails" of the force vector and the differential displacement. In general, the integration must account for the variation of the force's direction and magnitude.

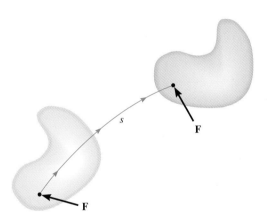

Fig. 18–6

Work of a Constant Force.   If an external force $\mathbf{F}_c$ acts on a rigid body, Fig. 18–7, and maintains a constant magnitude F_c and constant direction θ, while the body undergoes a translation s, Eq. 18–7 can be integrated so that the work becomes

$$U_{F_c} = (F_c \cos\theta)s \qquad (18\text{–}8)$$

Here $F_c \cos\theta$ represents the magnitude of the component of force in the direction of displacement.

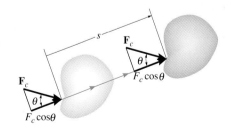

Fig. 18–7

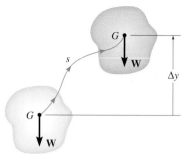

Fig. 18–8

Work of a Weight. The weight of a body does work only when the body's center of mass G undergoes a *vertical displacement* Δy. If this displacement is *upward*, Fig. 18–8, the work is negative, since the weight and displacement are in opposite directions.

$$U_W = -W\Delta y \qquad (18\text{–}9)$$

Likewise, if the displacement is *downward* $(-\Delta y)$ the work becomes *positive*. In both cases the elevation change is considered to be small so that $\mathbf{W}$, which is caused by gravitation, is constant.

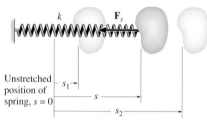

Fig. 18–9

Work of a Spring Force. If a linear elastic spring is attached to a body, the spring force $F_s = ks$ *acting on the body* does work when the spring either stretches or compresses from s_1 to a *further* position s_2. In both cases the work will be *negative* since the *displacement of the body* is in the opposite direction to the force, Fig. 18–9. The work done is

$$U_s = -(\tfrac{1}{2}ks_2^2 - \tfrac{1}{2}ks_1^2) \qquad (18\text{–}10)$$

where $|s_2| > |s_1|$.

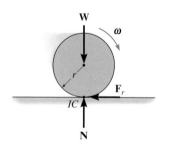

Fig. 18–10

Forces That Do No Work. There are some external forces that do no work when the body is displaced. These forces can act either at *fixed points* on the body, or they can have a direction *perpendicular to their displacement*. Examples include the reactions at a pin support about which a body rotates, the normal reaction acting on a body that moves along a fixed surface, and the weight of a body when the center of gravity of the body moves in a *horizontal plane*, Fig. 18–10. A rolling resistance force F_r acting on a round body as it *rolls without slipping* over a rough surface also does no work, Fig. 18–10.* This is because, during any *instant of time* dt, F_r acts at a point on the body which has *zero velocity* (instantaneous center, IC), and so the work done by the force on the point is zero. In other words, the point is not displaced in the direction of the force during this instant. Since F_r contacts successive points for only an instant, the work of F_r will be zero.

*The work done by the frictional force *when the body slips* has been discussed in Sec. 14.3.

18.3   The Work of a Couple

When a body subjected to a couple undergoes general plane motion, the two couple forces do work *only* when the body undergoes a *rotation*. To show this, consider the body in Fig. 18–11a, which is subjected to a couple moment $M = Fr$. Any general differential displacement of the body can be considered as a translation plus rotation. When the body *translates*, such that the *component of displacement* along the line of action of the forces is ds_t, Fig. 18–11b, clearly the "positive" work of one force *cancels* the "negative" work of the other. If the body undergoes a differential rotation $d\theta$ about an axis which is perpendicular to the plane of the couple and intersects the plane at point O, Fig. 18–11c, then each force undergoes a displacement $ds_\theta = (r/2)\, d\theta$ in the direction of the force. Hence, the total work done is

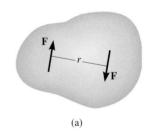

(a)

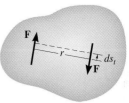

Translation
(b)

$$dU_M = F\left(\frac{r}{2}d\theta\right) + F\left(\frac{r}{2}d\theta\right) = (Fr)\, d\theta$$

$$= M\, d\theta$$

Here the line of action of $d\theta$ is parallel to the line of action of M. This is *always the case for general plane motion*, since **M** and $d\boldsymbol{\theta}$ are perpendicular to the plane of motion. Furthermore, the resultant work is *positive* when **M** and $d\boldsymbol{\theta}$ have the *same sense of direction* and *negative* if these vectors have an *opposite sense of direction*.

When the body rotates in the plane through a finite angle θ measured in radians, from θ_1 to θ_2, the work of a couple is

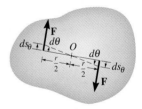

Rotation
(c)

Fig. 18–11

$$U_M = \int_{\theta_1}^{\theta_2} M\, d\theta \qquad (18\text{–}11)$$

If the couple moment **M** has a *constant magnitude*, then

$$U_M = M(\theta_2 - \theta_1) \qquad (18\text{–}12)$$

Here the work is *positive* provided **M** and $(\boldsymbol{\theta}_2 - \boldsymbol{\theta}_1)$ are in the same direction.

E X A M P L E 18.2

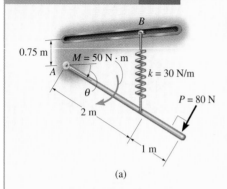

(a)

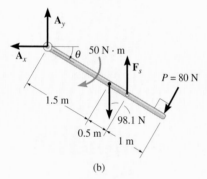

(b)

Fig. 18–12

The bar shown in Fig. 18–12a has a mass of 10 kg and is subjected to a couple moment of $M = 50\,\text{N} \cdot \text{m}$ and a force of $P = 80\,\text{N}$, which is always applied perpendicular to the end of the bar. Also, the spring has an unstretched length of 0.5 m and remains in the vertical position due to the roller guide at B. Determine the total work done by all the forces acting on the bar when it has rotated downward from $\theta = 0°$ to $\theta = 90°$.

Solution

First the free-body diagram of the bar is drawn in order to account for all the forces that act on it, Fig. 18–12b.

Weight W. Since the weight $10(9.81)\,\text{N} = 98.1\,\text{N}$ is displaced downward 1.5 m, the work is

$$U_W = 98.1\,\text{N}\,(1.5\,\text{m}) = 147.2\,\text{J}$$

Why is the work positive?

Couple Moment M. The couple moment rotates through an angle of $\theta = \pi/2\,\text{rad}$. Hence

$$U_M = 50\,\text{N} \cdot \text{m}\,(\pi/2) = 78.5\,\text{J}$$

Spring Force F_s. When $\theta = 0°$ the spring is stretched $(0.75\,\text{m} - 0.5) = 0.25\,\text{m}$, and when $\theta = 90°$, the stretch is $(2\,\text{m} + 0.75\,\text{m}) - 0.5\,\text{m} = 2.25\,\text{m}$. Thus

$$U_s = -\tfrac{1}{2}(30\,\text{N}/\text{m})(2.25\,\text{m})^2 - \tfrac{1}{2}(30\,\text{N}/\text{m})(0.25\,\text{m})^2] = -75.0\,\text{J}$$

By inspection the spring does negative work on the bar since F_s acts in the opposite direction to displacement. This checks with the result.

Force P. As the bar moves downward, the force is displaced through a distance of $(\pi/2)(3\,\text{m}) = 4.712\,\text{m}$. The work is positive. Why?

$$U_P = 80\,\text{N}\,(4.712\,\text{m}) = 377.0\,\text{J}$$

Pin Reactions. Forces A_x and A_y do no work since they are not displaced.

Total Work. The work of all the forces when the bar is displaced is thus

$$U = 147.2\,\text{J} + 78.5\,\text{J} - 75.0\,\text{J} + 377.0\,\text{J} = 528\,\text{J} \qquad Ans.$$

18.4 Principle of Work and Energy

By applying the principle of work and energy developed in Sec. 14.2 to each of the particles of a rigid body and adding the results algebraically, since energy is a scalar, the principle of work and energy for a rigid body becomes

$$T_1 + \Sigma U_{1-2} = T_2 \qquad (18\text{--}13)$$

This equation states that the body's initial translational *and* rotational kinetic energy, plus the work done by all the external forces and couple moments acting on the body as the body moves from its initial to its final position, is equal to the body's final translational *and* rotational kinetic energy. Note that the work of the body's *internal forces* does not have to be considered since the body is rigid. These forces occur in equal but opposite collinear pairs, so that when the body moves, the work of one force cancels that of its counterpart. Furthermore, since the body is rigid, *no relative movement* between these forces occurs, so that no internal work is done.

When several rigid bodies are pin connected, connected by inextensible cables, or in mesh with one another, Eq. 18–13 may be applied to the entire system of connected bodies. In all these cases the internal forces, which hold the various members together, do no work and hence are eliminated from the analysis.

The work of the torque or moment developed by the driving gears on the two motors is transformed into kinetic energy of rotation of the drum of the mixer.

PROCEDURE FOR ANALYSIS

The principle of work and energy is used to solve kinetic problems that involve *velocity, force*, and *displacement*, since these terms are involved in the formulation. For application, it is suggested that the following procedure be used.

Kinetic Energy (Kinematic Diagrams).

- The kinetic energy of a body is made up of two parts. Kinetic energy of translation is referenced to the velocity of the mass center, $T = \frac{1}{2}mv_G^2$, and kinetic energy of rotation is determined from knowing the moment of inertia about the mass center, $T = \frac{1}{2}I_G\omega^2$. In the special case of rotation about a fixed axis, these two kinetic energies are combined and can be expressed as $T = \frac{1}{2}I_O\omega^2$, where I_O is the moment of inertia about the axis of rotation.

- *Kinematic diagrams* for velocity may be useful for determining v_G and ω or for establishing a *relationship* between v_G and ω.*

Work (Free-Body Diagram).

- Draw a free-body diagram of the body when it is located at an intermediate point along the path in order to account for all the forces and couple moments which do work on the body as it moves along the path.

- A force does work when it moves through a displacement in the direction of the force.

- Forces that are functions of displacement must be integrated to obtain the work. Graphically, the work is equal to the area under the force–displacement curve.

- The work of a weight is the product of its magnitude and the vertical displacement, $U_W = Wy$. It is positive when the weight moves downwards.

- The work of a spring is of the form $U_s = \frac{1}{2}ks^2$, where k is the spring stiffness and s is the stretch or compression of the spring.

- The work of a couple is the product of the couple moment and the angle in radians through which it rotates.

- Since *algebraic addition* of the work terms is required, it is important that the proper sign of each term be specified. Specifically, work is *positive* when the force (couple moment) is in the *same direction* as its displacement (rotation); otherwise, it is negative.

Principle of Work and Energy.

- Apply the principle of work and energy, $T_1 + \Sigma U_{1-2} = T_2$. Since this is a scalar equation, it can be used to solve for only one unknown when it is applied to a single rigid body.

*A brief review of Secs. 16.5 to 16.7 may prove helpful in solving problems, since computations for kinetic energy require a kinematic analysis of velocity.

EXAMPLE 18.3

The 30-kg disk shown in Fig. 18–13a is pin supported at its center. Determine the number of revolutions it must make to attain an angular velocity of 20 rad/s starting from rest. It is acted upon by a constant force $F = 10\,\text{N}$, which is applied to a cord wrapped around its periphery, and a constant couple moment $M = 5\,\text{N} \cdot \text{m}$. Neglect the mass of the cord in the calculation.

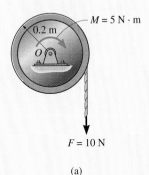

$M = 5\,\text{N} \cdot \text{m}$

0.2 m

$F = 10\,\text{N}$

(a)

Solution

Kinetic Energy. Since the disk rotates about a fixed axis, the kinetic energy can be computed using $T = \frac{1}{2}I_O\omega^2$, where the moment of inertia is $I_O = \frac{1}{2}mr^2$. Initially, the disk is at rest, so that

$$T_1 = 0$$
$$T_2 = \frac{1}{2}I_O\omega_2^2 = \frac{1}{2}[\frac{1}{2}(30\,\text{kg})(0.2\,\text{m})^2](20\,\text{rad/s})^2 = 120\,\text{J}$$

Work (Free-Body Diagram). As shown in Fig. 18–13b, the pin reactions O_x and O_y and the weight (294.3 N) do no work, since they are not displaced. The *couple moment*, having a constant magnitude, does positive work $U_M = M\theta$ as the disk *rotates* through a clockwise angle of θ rad, and the *constant force* **F** does positive work $U_{F_c} = Fs$ as the cord *moves* downward $s = \theta r = \theta(0.2\,\text{m})$.

Principle of Work and Energy

$$\{T_1\} + \{\Sigma U_{1-2}\} = \{T_2\}$$
$$\{T_1\} + \{M\theta + Fs\} = \{T_2\}$$
$$\{0\} + \{(5\,\text{N}\cdot\text{m})\theta + (10\,\text{N})\theta(0.2\,\text{m})\} = \{120\,\text{J}\}$$

$$\theta = 17.1\,\text{rad} = 17.1\,\text{rad}\left(\frac{1\,\text{rev}}{2\pi\,\text{rad}}\right) = 2.73\,\text{rev} \qquad \textit{Ans.}$$

This problem has also been solved in Example 17.9. Compare the two methods of solution and note that since force, velocity, and displacement θ are involved, a work-energy approach yields a more direct solution.

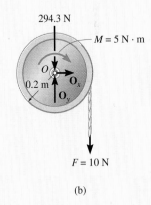

294.3 N

$M = 5\,\text{N} \cdot \text{m}$

O_x

0.2 m

O_y

$F = 10\,\text{N}$

(b)

Fig. 18–13

E X A M P L E   **18.4**

The 700-kg pipe is equally suspended from the two tines of the fork lift shown in the photo. It is undergoing a swinging motion such that when $\theta = 30°$ it is momentarily at rest. Determine the normal and frictional forces acting on each tine which are needed to support the pipe at the instant $\theta = 0°$. Measurements of the pipe and the suspender are shown in Fig. 18–14a. Neglect the mass of the suspender and the thickness of the pipe.

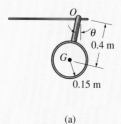

(a)

Fig. 18–14

Solution
We must use the equations of motion to find the forces on the tines since these forces do no work. Before doing this, however, we will apply the principle of work and energy to determine the angular velocity of the pipe when $\theta = 0°$.

Kinetic Energy (Kinematic Diagram). Since the pipe is originally at rest, then

$$T_1 = 0$$

The final kinetic energy may be computed with reference to either the fixed point O or the center of mass G. For the calculation we will consider the pipe to be a thin ring so that $I_G = mr^2$. If point G is considered, we have

$$T_2 = \tfrac{1}{2}m(v_G)_2^2 + \tfrac{1}{2}I_G\omega_2^2$$
$$= \tfrac{1}{2}(700 \text{ kg})\,[(0.4 \text{ m})\omega_2]^2 + \tfrac{1}{2}\,[700 \text{ kg}(0.15 \text{ m})^2]\omega_2^2$$
$$= 63.875\omega_2^2$$

If point O is considered then the parallel-axis theorem must be used to determine I_O. Hence,

$$T_2 = \tfrac{1}{2}I_O\omega_2^2 = \tfrac{1}{2}\,[700 \text{ kg}(0.15 \text{ m})^2 + 700 \text{ kg}(0.4 \text{ m})^2]\omega_2^2$$
$$= 63.875\omega_2^2$$

Work (Free-Body Diagram). Fig. 18–14*b*. The normal and frictional forces on the tines do no work since they do not move as the pipe swings. The weight, centered at *G*, does positive work since the weight moves downward through a vertical distance $\Delta y = 0.4$m $-$ $0.4 \cos 30°$ m $= 0.05359$ m .

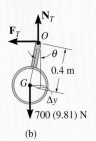

(b)

Principle of Work and Energy

$$\{T_1\} + \{\Sigma U_{1-2}\} = \{T_2\}$$

$$\{0\} + \{700(9.81)\text{ N}(0.05359\text{ m})\} = \{63.875\omega_2^2\}$$

$$\omega_2 = 2.40 \text{ rad/s}$$

Equations of Motion. Referring to the free-body and kinetic diagrams shown in Fig. 18–14*c*, and using the result for ω_2, we have

$$\underleftarrow{+} \Sigma F_t = m(a_G)_t; \qquad F_T = 700(a_G)_t$$

$$+\uparrow \Sigma F_n = m(a_G)_n; \quad N_T - 700(9.81) \text{ N} = 700 \text{ kg}(2.40 \text{ rad/s})^2(0.4 \text{ m})$$

$$\overset{\curvearrowleft}{+} \Sigma M_O = I_O\alpha; \qquad 0 = [700 \text{ kg}(0.15 \text{ m})^2 + 700 \text{ kg}(0.4 \text{ m})^2]\alpha$$

Since $(a_G)_t = 0.4\alpha$, then

$$\alpha = 0, (a_G)_t = 0$$

$$F_T = 0$$

$$N_T = 8.48 \text{ kN}$$

There are two tines used to support the load, therefore

$$F'_T = 0 \qquad\qquad\qquad Ans.$$

$$N'_T = \frac{8.48 \text{ kN}}{2} = 4.24 \text{ kN} \qquad\qquad\qquad Ans.$$

Due to the swinging motion the tines are subjected to a *greater* normal force than would be the case if the load was static, in which case $N'_T = 700(9.81)\text{N}/2 = 3.43$ kN .

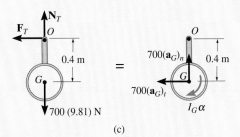

(c)

E X A M P L E 18.5

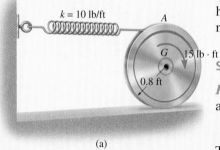

$k = 10$ lb/ft

(a)

The wheel shown in Fig. 18–15a weighs 40 lb and has a radius of gyration $k_G = 0.6$ ft about its mass center G. If it is subjected to a clockwise couple moment of 15 lb·ft and rolls from rest without slipping, determine its angular velocity after its center G moves 0.5 ft. The spring has a stiffness $k = 10$ lb/ft and is initially unstretched when the couple moment is applied.

Solution

Kinetic Energy (Kinematic Diagram). Since the wheel is initially at rest,

$$T_1 = 0$$

The kinematic diagram of the wheel when it is in the final position is shown in Fig. 18–15b. Hence, the final kinetic energy is

$$T_2 = \tfrac{1}{2}m(v_G)_2^2 + \tfrac{1}{2}I_G\omega_2^2$$

$$= \frac{1}{2}\left(\frac{40\text{ lb}}{32.2\text{ ft/s}^2}\right)(v_G)_2^2 + \frac{1}{2}\left[\frac{40\text{ lb}}{32.2\text{ ft/s}^2}(0.6\text{ ft})^2\right]\omega_2^2$$

The velocity of the mass center can be related to the angular velocity from the instantaneous center of zero velocity (IC), i.e., $(v_G)_2 = 0.8\omega_2$. Substituting into the above equation and simplifying, we have

$$T_2 = 0.621\omega_2^2$$

Work (Free-Body Diagram). As shown in Fig. 18–15c, only the spring force $\mathbf{F}_s$ and the couple moment do work. The normal force does not move along its line of action and the frictional force does *no work*, since the wheel does not slip as it rolls.

The work of $\mathbf{F}_s$ may be computed using $U_s = -\tfrac{1}{2}ks^2$. Here the work is negative since $\mathbf{F}_s$ is in the opposite direction to displacement. Since the wheel does not slip when the center G moves 0.5 ft, then the wheel rotates $\theta = s_G/r_{G/IC} = 0.5\text{ ft}/0.8\text{ ft} = 0.625$ rad, Fig. 18–15b. Hence, the spring stretches $s_A = \theta r_{A/IC} = 0.625\text{ rad}(1.6\text{ ft}) = 1$ ft.

Principle of Work and Energy

$$\{T_1\} + \{\Sigma U_{1-2}\} = \{T_2\}$$
$$\{T_1\} + \{M\theta - \tfrac{1}{2}ks^2\} = \{T_2\}$$
$$\{0\} + \left\{15\text{ lb}\cdot\text{ft}(0.625\text{ rad}) - \frac{1}{2}(10\text{ lb/ft})(1\text{ ft})^2\right\} = \{0.621\omega_2^2\,\text{ft}\cdot\text{lb}\}$$
$$\omega_2 = 2.65\text{ rad/s} \qquad\qquad Ans.$$

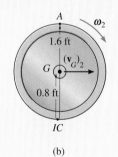

(b)

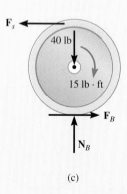

(c)

Fig. 18–15

EXAMPLE 18.6

The 10-kg rod shown in Fig. 18–16a is constrained so that its ends move along the grooved slots. The rod is initially at rest when $\theta = 0°$. If the slider block at B is acted upon by a horizontal force $P = 50\,N$, determine the angular velocity of the rod at the instant $\theta = 45°$. Neglect friction and the mass of blocks A and B.

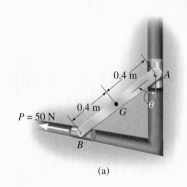

Solution

Why can the principle of work and energy be used to solve this problem?

Kinetic Energy (Kinematic Diagrams). Two kinematic diagrams of the rod, when it is in the initial position 1 and final position 2, are shown in Fig. 18–16b. When the rod is in position 1, $T_1 = 0$ since $(v_G)_1 = \omega_1 = 0$. In position 2 the angular velocity is ω_2 and the velocity of the mass center is $(v_G)_2$. Hence, the kinetic energy is

$$T_2 = \tfrac{1}{2}m(v_G)_2^2 + \tfrac{1}{2}I_G\omega_2^2$$
$$= \tfrac{1}{2}(10\,kg)(v_G)_2^2 + \tfrac{1}{2}[\tfrac{1}{12}(10\,kg)(0.8\,m)^2]\omega_2^2$$
$$= 5(v_G)_2^2 + 0.267(\omega_2)^2$$

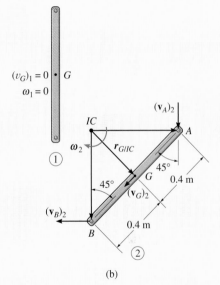

The two unknowns $(v_G)_2$ and ω_2 may be related from the instantaneous center of zero velocity for the rod. Fig. 18–16b. It is seen that as A moves downward with a velocity $(v_A)_2$, B moves horizontally to the left with a velocity $(v_B)_2$. Knowing these directions, the IC is determined as shown in the figure. Hence,

$$(v_G)_2 = r_{G/IC}\omega_2 = (0.4\tan 45°\,m)\omega_2$$
$$= 0.4\omega_2$$

Therefore,

$$T_2 = 0.8\omega_2^2 + 0.267\omega_2^2 = 1.067\omega_2^2$$

Work (Free-Body Diagram). Fig. 18–16c. The normal forces N_A and N_B do no work as the rod is displaced. Why? The 98.1-N weight is displaced a vertical distance of $\Delta y = (0.4 - 0.4\cos 45°)\,m$; whereas the 50-N force moves a horizontal distance of $s = (0.8\sin 45°)\,m$. Both of these forces do positive work. Why?

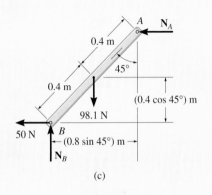

Principle of Work and Energy

$$\{T_1\} + \{\Sigma U_{1-2}\} = \{T_2\}$$
$$\{T_1\} + \{W\Delta y + Ps\} = \{T_2\}$$

$$\{0\} + \{98.1\,N(0.4\,m - 0.4\cos 45°\,m) + 50\,N(0.8\sin 45°\,m)\}$$
$$= \{1.067\omega_2^2\,J\}$$

Solving for ω_2 gives

$$\omega_2 = 6.11\,rad/s \qquad\qquad Ans.$$

Fig. 18–16

18.5 Conservation of Energy

When a force system acting on a rigid body consists only of *conservative forces*, the conservation of energy theorem may be used to solve a problem which otherwise would be solved using the principle of work and energy. This theorem is often easier to apply since the work of a conservative force is *independent of the path* and depends only on the initial and final positions of the body. It was shown in Sec. 14.5 that the work of a conservative force may be expressed as the difference in the body's potential energy measured from an arbitrarily selected reference or datum.

Gravitational Potential Energy.   Since the total weight of a body can be considered concentrated at its center of gravity, the *gravitational potential energy* of the body is determined by knowing the height of the body's center of gravity above or below a horizontal datum. Measuring y_G as *positive upward*, the gravitational potential energy of the body is thus

$$V_g = W y_G \qquad (18\text{--}14)$$

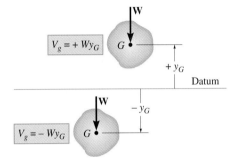

Gravitational potential energy

Fig. 18–17

Here the potential energy is *positive* when y_G is positive, since the weight has the ability to do *positive work* when the body is moved back to the datum, Fig. 18–17. Likewise, if the body is located *below* the datum $(-y_G)$, the gravitational potential energy is *negative*, since the weight does *negative work* when the body is returned to the datum.

Elastic Potential Energy.   The force developed by an elastic spring is also a conservative force. The *elastic potential energy* which a spring imparts to an attached body when the spring is elongated or compressed from an initial undeformed position $(s = 0)$ to a final position s, Fig. 18–18, is

$$V_e = +\tfrac{1}{2} k s^2 \qquad (18\text{--}15)$$

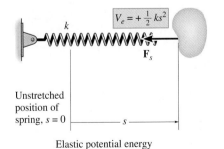

Elastic potential energy

Fig. 18–18

In the deformed position, the spring force acting *on the body* always has the capacity for doing positive work when the spring is returned back to its original undeformed position (see Sec. 14.5).

Conservation of Energy. In general, if a body is subjected to both gravitational and elastic forces, the total *potential energy* is expressed as a potential function V represented as the algebraic sum

$$V = V_g + V_e \qquad (18\text{--}16)$$

Here measurement of V depends on the location of the body with respect to the selected datum.

Realizing that the work of conservative forces can be written as a difference in their potential energies, i.e., $(\Sigma U_{1-2})_{\text{cons}} = V_1 - V_2$, Eq. 14–16, we can rewrite the principle of work and energy for a rigid body as

$$T_1 + V_1 + (\Sigma U_{1-2})_{\text{noncons}} = T_2 + V_2 \qquad (18\text{--}17)$$

Here $(\Sigma U_{1-2})_{\text{noncons}}$ represents the work of the nonconservative forces such as friction. If this term is zero, then

$$T_1 + V_1 = T_2 + V_2 \qquad (18\text{--}18)$$

This equation is referred to as the conservation of mechanical energy. It states that the *sum* of the potential and kinetic energies of the body remains *constant* when the body moves from one position to another. It also applies to a system of smooth, pin-connected rigid bodies, bodies connected by inextensible cords, and bodies in mesh with other bodies. In all these cases the forces acting at the points of contact are *eliminated* from the analysis, since they occur in equal but opposite collinear pairs and each pair of forces moves through an equal distance when the system undergoes a displacement.

It is important to remember that only problems involving conservative force systems may be solved by using Eq. 18–18. As stated in Sec. 14.5, friction or other drag-resistant forces, which depend on velocity or acceleration, are nonconservative. The work of such forces is transformed into thermal energy used to heat up the surfaces of contact, and consequently this energy is dissipated into the surroundings and may not be recovered. Therefore, problems involving frictional forces can be solved by using either the principle of work and energy written in the form of Eq. 18–17, if it applies, or the equations of motion.

The torsional springs located at the top of the garage door wind up as the door is lowered. When the door is raised, the potential energy stored in the springs is then transferred into gravitational potential energy of the door's weight, thereby making it easy to open.

PROCEDURE FOR ANALYSIS

The conservation of energy equation is used to solve problems involving *velocity, displacement,* and *conservative force systems*. For application it is suggested that the following procedure be used.

Potential Energy.

- Draw two diagrams showing the body located at its initial and final positions along the path.

- If the center of gravity, G, is subjected to a *vertical displacement*, establish a fixed horizontal datum from which to measure the body's gravitational potential energy V_g.

- Data pertaining to the elevation y_G of the body's center of gravity from the datum and the extension or compression of any connecting springs can be determined from the problem geometry and listed on the two diagrams.

- Recall that the potential energy $V = V_g + V_e$. Here $V_g = W y_G$, which can be positive or negative, and $V_e = \frac{1}{2}ks^2$, which is always positive.

Kinetic Energy.

- The kinetic energy of the body consists of two parts, namely translational kinetic energy, $T = \frac{1}{2}mv_G^2$, and rotational kinetic energy, $T = \frac{1}{2}I_G\omega^2$.

- Kinematic diagrams for velocity may be useful for determining v_G and ω for establishing a *relationship* between these quantities.

Conservation of Energy.

- Apply the conservation of energy equation $T_1 + V_1 = T_2 + V_2$.

E X A M P L E 18.7

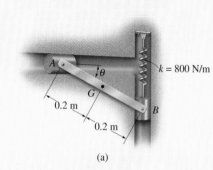

(a)

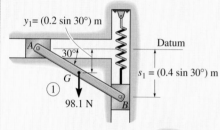

$y_1 = (0.2 \sin 30°)$ m

Datum

$s_1 = (0.4 \sin 30°)$ m

98.1 N

①

$s_2 = 0$

98.1 N

②

(b)

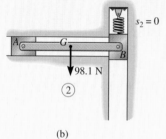

$(v_G)_2$

ω_2

$r_{G/IC}$

0.2 m

(c)

Fig. 18–19

The 10-kg rod *AB* shown in Fig. 18–19*a* is confined so that its ends move in the horizontal and vertical slots. The spring has a stiffness of $k = 800$ N/m and is unstretched when $\theta = 0°$. Determine the angular velocity of *AB* when $\theta = 0°$, if the rod is released from rest when $\theta = 30°$. Neglect the mass of the slider blocks.

Solution

Potential Energy. The two diagrams of the rod, when it is located at its initial and final positions, are shown in Fig. 18–19*b*. The datum, used to measure the gravitational potential energy, is placed in line with the rod when $\theta = 0°$.

When the rod is in position 1, the center of gravity *G* is located *below the datum* so that the gravitational potential energy is *negative*. Furthermore, (positive) elastic potential energy is stored in the spring, since it is stretched a distance of $s_1 = (0.4 \sin 30°)$ m. Thus,

$$V_1 = -Wy_1 + \tfrac{1}{2}ks_1^2$$
$$= -98.1\,\text{N}(0.2 \sin 30°\,\text{m}) + \tfrac{1}{2}(800\,\text{N/m})(0.4 \sin 30°\,\text{m})^2 = 6.19\,\text{J}$$

When the rod is in position 2, the potential energy of the rod is zero, since the spring is unstretched, $s_2 = 0$, and the center of gravity *G* is located at the datum. Thus,

$$V_2 = 0$$

Kinetic Energy. The rod is released from rest from position 1, thus $(v_G)_1 = 0$ and $\boldsymbol{\omega}_1 = 0$, and

$$T_1 = 0$$

In position 2, the angular velocity is $\boldsymbol{\omega}_2$ and the rod's mass center has a velocity of $(v_G)_2$. Thus,

$$T_2 = \tfrac{1}{2}m(v_G)_2^2 + \tfrac{1}{2}I_G\omega_2^2$$
$$= \tfrac{1}{2}(10\,\text{kg})(v_G)_2^2 + \tfrac{1}{2}[\tfrac{1}{12}(10\,\text{kg})(0.4\,\text{m})^2]\omega_2^2$$

Using *kinematics*, $(v_G)_2$ can be related to $\boldsymbol{\omega}_2$ as shown in Fig. 18–19*c*. At the instant considered, the instantaneous center of zero velocity (*IC*) for the rod is at point *A*; hence, $(v_G)_2 = (r_{G/IC})\omega_2 = (0.2)\omega_2$. Substituting into the above expression and simplifying, we get

$$T_2 = 0.267\omega_2^2$$

Conservation of Energy

$$\{T_1\} + \{V_1\} = \{T_2\} + \{V_2\}$$
$$\{0\} + \{6.19\} = \{0.267\omega_2^2\} + \{0\}$$
$$\omega_2 = 4.82\,\text{rad/s} \qquad\qquad\qquad Ans.$$

EXAMPLE 18.8

The disk shown in Fig. 18–20a has a weight of 30 lb and a radius of gyration of $k_G = 0.6$ ft, and it is attached to a spring which has a stiffness $k = 2$ lb/ft and an unstretched length of 1 ft. If the disk is released from rest in the position shown and rolls without slipping, determine its angular velocity at the instant G moves 3 ft to the left.

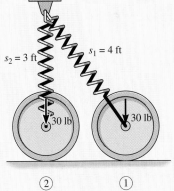

(a)

Solution

Potential Energy. Two diagrams of the disk, when it is located in its initial and final positions, are shown in Fig. 18–20b. A gravitational datum is not needed here since the weight is not displaced vertically. From the problem geometry the spring is stretched $s_1 = (\sqrt{3^2 + 4^2} - 1) = 4$ ft and $s_2 = (4 - 1) = 3$ ft in the initial and final positions, respectively. Hence,

$$V_1 = \tfrac{1}{2}ks_1^2 = \tfrac{1}{2}(2 \text{ lb/ft})(4 \text{ ft})^2 = 16 \text{ J}$$
$$V_2 = \tfrac{1}{2}ks_2^2 = \tfrac{1}{2}(2 \text{ lb/ft})(3 \text{ ft})^2 = 9 \text{ J}$$

Kinetic Energy. The disk is released from rest so that $(v_G)_1 = 0$, $\omega_1 = 0$, and

$$T_1 = 0$$

In the final position,

$$T_2 = \frac{1}{2}m(v_G)_2^2 + \frac{1}{2}I_G\omega_2^2$$

$$= \frac{1}{2}\left(\frac{30 \text{ lb}}{32.2 \text{ ft/s}^2}\right)(v_G)_2^2 + \frac{1}{2}\left[\left(\frac{30 \text{ lb}}{32.2 \text{ ft/s}^2}\right)(0.6 \text{ ft})^2\right]\omega_2^2$$

Since the disk rolls without slipping, $(v_G)_2$ can be related to ω_2 from the instantaneous center of zero velocity, Fig. 18–20c, i.e., $(v_G)_2 = (0.75 \text{ ft})\omega_2$. Substituting and simplifying yields

$$T_2 = 0.430\omega_2^2$$

Conservation of Energy

$$\{T_1\} + \{V_1\} = \{T_2\} + \{V_2\}$$
$$\{0\} + \{16\} = \{0.430\omega_2^2\} + \{9\}$$
$$\omega_2 = 4.04 \text{ rad/s} \qquad\qquad Ans.$$

(b)

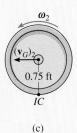

(c)

Fig. 18–20

EXAMPLE 18.9

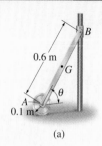

(a)

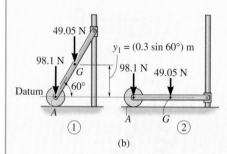

(b)

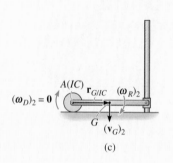

(c)

Fig. 18–21

The 10-kg homogeneous disk shown in Fig. 18–21a is attached to a uniform 5-kg rod AB. If the assembly is released from rest when $\theta = 60°$, determine the angular velocity of the rod when $\theta = 0°$. Assume that the disk rolls without slipping. Neglect friction along the guide and the mass of the collar at B.

Solution

Potential Energy. Two diagrams for the rod and disk, when they are located at their initial and final positions, are shown in Fig. 18–21b. For convenience the datum passes through point A.

When the system is in position 1, the rod's weight has positive potential energy. Thus,

$$V_1 = W_R y_1 = 49.05\,\text{N}\,(0.3\sin 60°\,\text{m}) = 12.74\,\text{J}$$

When the system is in position 2, both the weight of the rod and the weight of the disk have zero potential energy. Why? Thus,

$$V_2 = 0$$

Kinetic Energy. Since the entire system is at rest at the initial position,

$$T_1 = 0$$

In the final position the rod has an angular velocity $(\omega_R)_2$ and its mass center has a velocity $(v_G)_2$, Fig. 18–21c. Since the rod is *fully extended* in this position, the disk is momentarily at rest, so $(\omega_D)_2 = 0$ and $(v_A)_2 = 0$. For the rod $(v_G)_2$ can be related to $(\omega_R)_2$ from the instantaneous center of zero velocity, which is located at point A, Fig. 18–21c. Hence, $(v_G)_2 = r_{G/IC}(\omega_R)_2$ or $(v_G)_2 = 0.3(\omega_R)_2$. Thus,

$$T_2 = \frac{1}{2}m_R(v_G)_2^2 + \frac{1}{2}I_G(\omega_R)_2^2 + \frac{1}{2}m_D(v_A)_2^2 + \frac{1}{2}I_A(\omega_D)_2^2$$

$$= \frac{1}{2}(5\,\text{kg})\,[0.3\,\text{m}\,(\omega_R)_2]^2 + \frac{1}{2}\left[\frac{1}{12}(5\,\text{kg})(0.6\,\text{m})^2\right](\omega_R)_2^2 + 0 + 0$$

$$= 0.3(\omega_R)_2^2$$

Conservation of Energy

$$\{T_1\} + \{V_1\} = \{T_2\} + \{V_2\}$$
$$\{0\} + \{12.74\} = \{0.3(\omega_R)_2^2\} + \{0\}$$
$$(\omega_R)_2 = 6.52\,\text{rad/s} \qquad\qquad Ans.$$

CHAPTER REVIEW

- **Kinetic Energy.** The kinetic energy of a rigid body that undergoes planar motion can be referenced to its mass center. For a translating body $T = \frac{1}{2}mv_G^2$. If the body is rotating about a fixed axis through point O, then its mass center has a velocity, and the body also has an angular velocity. Therefore $T = \frac{1}{2}mv_G^2 + \frac{1}{2}I_G\omega^2$. Using $v_G = \omega r$ and the parallel-axis theorem, we can also determine the kinetic energy relative to point O. We have $T = \frac{1}{2}I_O\omega^2$.

  For general plane motion, the kinetic energy is simply $T = \frac{1}{2}mv_G^2 + \frac{1}{2}I_G\omega^2$, the scalar sum of its translational and rotational kinetic energies.

- **Work of a Force and a Couple Moment.** A force does work when it undergoes a displacement ds in the direction of the force. The work is $U = \int F\,ds$. If the force is constant and is in the direction of its displacement Δs, then $U = F\Delta s$. If a weight $\mathbf{W}$ is displaced downward by Δy, then $U = W\Delta y$. If a force $\mathbf{F}$ stretches a spring a distance s, then $U = \frac{1}{2}ks^2$. The frictional and normal forces that act on a cylinder or sphere that rolls without slipping will do no work, since the normal force does not displace and the frictional force acts on successive points on the surface of the body. A couple moment will do work when it undergoes a rotation θ in the direction of the couple moment. If this moment is constant, then $U = M\theta$.

- **Principle of Work and Energy.** Problems that involve velocity, force, and displacement can be solved using the principle of work and energy.

$$T_1 + \Sigma U_{1-2} = T_2$$

Here, the kinetic energy is the sum of both its rotational and translational parts. For application, a free-body diagram should be drawn in order to account for the work of all of the forces and couple moments that act on the body as it moves along the path.

- **Conservation of Energy.** If a rigid body is subjected only to conservative forces, then the conservation of energy equation can be used to solve the problem. This equation requires that the sum of the potential and kinetic energies of the body remains the same at any two points along the path, that is,

$$T_1 + V_1 = T_2 + V_2$$

Here, the potential energy is the sum of its gravitational and elastic potential energies.

$$V = V_g + V_e$$

In particular, the gravitational potential energy will be positive if the body's center of mass is located above a datum. If it is below the datum, then it will be negative. The elastic potential energy is always positive, regardless if the spring is stretched or compressed.

The launching of this weather satellite requires application of impulse and momentum principles to accurately predict its orbital angular motion and proper orientation.

Planar Kinetics of a Rigid Body: Impulse and Momentum

- To develop formulations for the linear and angular momentum of a body.
- To apply the principles of linear and angular impulse and momentum to solve rigid-body planar kinetic problems that involve force, velocity, and time.
- To discuss application of the conservation of momentum.
- To analyze the mechanics of eccentric impact.

19.1 Linear and Angular Momentum

In this chapter we will use the principles of linear and angular impulse and momentum to solve problems involving force, velocity, and time as related to the planar motion of a rigid body. Before doing this, however, it will first be necessary to formalize the methods for obtaining the body's linear and angular momentum when the body is subjected to translation, rotation about a fixed axis, and general plane motion. Here we will assume the body is symmetric with respect to an inertial x–y reference plane.

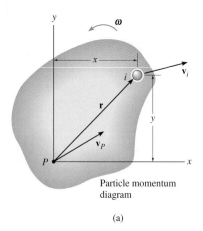

Particle momentum
diagram

(a)

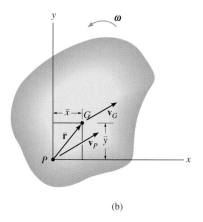

(b)

Fig. 19–1

Linear Momentum.

The linear momentum of a rigid body is determined by summing vectorially the linear momenta of all the particles of the body, i.e., $\mathbf{L} = \Sigma m_i \mathbf{v}_i$. Since $\Sigma m_i \mathbf{v}_i = m \mathbf{v}_G$ (see Sec. 15.2) we can also write

$$\mathbf{L} = m \mathbf{v}_G \qquad (19\text{–}1)$$

This equation states that the body's linear momentum is a vector quantity having a *magnitude* $m v_G$, which is commonly measured in units of kg·m/s or slug·ft/s, and a *direction* defined by $\mathbf{v}_G$, the velocity of the body's mass center.

Angular Momentum.

Consider the body in Fig. 19–1a, which is subjected to general plane motion. At the instant shown, the arbitrary point P has a velocity $\mathbf{v}_P$, and the body has an angular velocity $\boldsymbol{\omega}$. If the velocity of the ith particle of the body is to be determined, Fig. 19–1a, then

$$\mathbf{v}_i = \mathbf{v}_P + \mathbf{v}_{i/P} = \mathbf{v}_P + \boldsymbol{\omega} \times \mathbf{r}$$

The angular momentum of particle i about point P is equal to the "moment" of the particle's linear momentum about P, Fig. 19–1a. Thus,

$$(\mathbf{H}_P)_i = \mathbf{r} \times m_i \mathbf{v}_i$$

Expressing $\mathbf{v}_i$ in terms of $\mathbf{v}_P$ and using Cartesian vectors, we have

$$(H_P)_i \mathbf{k} = m_i (x\mathbf{i} + y\mathbf{j}) \times [(v_P)_x \mathbf{i} + (v_P)_y \mathbf{j} + \omega \mathbf{k} \times (x\mathbf{i} + y\mathbf{j})]$$
$$(H_P)_i = -m_i y (v_P)_x + m_i x (v_P)_y + m_i \omega r^2$$

Letting $m_i \rightarrow dm$ and integrating over the entire mass m of the body, we obtain

$$H_P = -\left(\int_m y \, dm \right)(v_P)_x + \left(\int_m x \, dm \right)(v_P)_y + \left(\int_m r^2 \, dm \right)\omega$$

Here H_P represents the angular momentum of the body about an axis (the z axis) perpendicular to the plane of motion and passing through point P. Since $\bar{y} m = \int y \, dm$ and $\bar{x} m = \int x \, dm$, the integrals for the first and second terms on the right are used to locate the body's center of mass G with respect to P, Fig. 19–1b. Also, the last integral represents the body's moment of inertia computed about the z axis, i.e., $I_P = \int r^2 \, dm$. Thus,

$$H_P = -\bar{y} m (v_P)_x + \bar{x} m (v_P)_y + I_P \omega \qquad (19\text{–}2)$$

This equation reduces to a simpler form if point P coincides with the mass center G for the body,* in which case $\bar{x} = \bar{y} = 0$. Hence,

---

*It also reduces to the same simple form, $H_P = I_P \omega$, if point P is a *fixed point* (see Eq. 19–9) or the velocity of point P is directed along the line PG.

$$H_G = I_G\omega \qquad (19\text{–}3)$$

This equation states that the angular momentum of the body computed about G is equal to the product of the moment of inertia of the body about an axis passing through G and the body's angular velocity. It is important to realize that $\mathbf{H}_G$ is a vector quantity having a *magnitude* $I_G\omega$, which is commonly measured in units of kg·m²/s or slug·ft²/s, and a *direction* defined by ω, which is always perpendicular to the plane of motion.

Equation 19–2 can also be rewritten in terms of the x and y components of the velocity of the body's mass center, $(v_G)_x$ and $(v_G)_y$, and the body's moment of inertia I_G. Since point G is located at coordinates $(\bar{x}, \bar{y})$, then by the parallel-axis theorem, $I_P = I_G + m(\bar{x}^2 + \bar{y}^2)$. Substituting into Eq. 19–2 and rearranging terms, we have

$$H_P = \bar{y}m[-(v_P)_x + \bar{y}\omega] + \bar{x}m[(v_P)_y + \bar{x}\omega] + I_G\omega \qquad (19\text{–}4)$$

From the kinematic diagram of Fig. 19–1b, $\mathbf{v}_G$ can be expressed in terms of $\mathbf{v}_P$ as

$$\mathbf{v}_G = \mathbf{v}_p + \boldsymbol{\omega} \times \bar{\mathbf{r}}$$
$$(v_G)_x\mathbf{i} + (v_G)_y\mathbf{j} = (v_p)_x\mathbf{i} + (v_p)_y\mathbf{j} + \omega\mathbf{k} \times (\bar{x}\mathbf{i} + \bar{y}\mathbf{j})$$

Carrying out the cross product and equating the respective $\mathbf{i}$ and $\mathbf{j}$ components yields the two scalar equations

$$(v_G)_x = (v_P)_x - \bar{y}\omega$$
$$(v_G)_y = (v_P)_y + \bar{x}\omega$$

Substituting these results into Eq. 19–4 yields

$$H_P = -\bar{y}m(v_G)_x + \bar{x}m(v_G)_y + I_G\omega \qquad (19\text{–}5)$$

With reference to Fig. 19–1c, *this result indicates that when the angular momentum of the body is computed about point P, it is equivalent to the moment of the linear momentum* $m\mathbf{v}_G$ *or its components* $m(\mathbf{v}_G)_x$ *and* $m(\mathbf{v}_G)_y$ *about P plus the angular momentum* $I_G\omega$. Note that since ω is a free vector, $\mathbf{H}_G$ *can act at any point on the body* provided it preserves its same magnitude and direction. Furthermore, since angular momentum is equal to the "moment" of the linear momentum, the *line of action of* $\mathbf{L}$ *must pass through the body's mass center G* in order to preserve the correct magnitude of $\mathbf{H}_P$ when "moments" are computed about P, Fig. 19–1c. As a result of this analysis, we will now consider three types of motion.

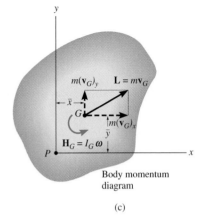

Body momentum diagram

(c)

Fig. 19–1

Translation. When a rigid body of mass m is subjected to either rectilinear or curvilinear *translation*, Fig. 19–2a, its mass center has a velocity of $\mathbf{v}_G = \mathbf{v}$ and $\boldsymbol{\omega} = \mathbf{0}$ for the body. Hence, the linear momentum and the angular momentum computed about G become

$$\boxed{\begin{array}{c} L = mv_G \\ H_G = 0 \end{array}} \qquad (19\text{–}6)$$

If the angular momentum is computed about any other point A on or off the body, Fig. 19–2a, the "moment" of the linear momentum $\mathbf{L}$ must be computed about the point. Since d is the "moment arm" as shown in the figure, then in accordance with Eq. 19–5, $H_A = (d)(mv_G)\,\text{\Large\char"21BB}$.

Rotation About a Fixed Axis. When a rigid body is *rotating about a fixed axis* passing through point O, Fig. 19–2b, the linear momentum and the angular momentum computed about G are

$$\boxed{\begin{array}{c} L = mv_G \\ H_G = I_G\omega \end{array}} \qquad (19\text{–}7)$$

It is sometimes convenient to compute the angular momentum of the body about point O. In this case it is necessary to account for the "moment" of *both* $\mathbf{L}$ and $\mathbf{H}_G$ about O. Noting that $\mathbf{L}$ (or $\mathbf{v}_G$) is always *perpendicular to* $\mathbf{r}_G$, we have

$$\zeta+ \qquad\qquad H_O = I_G\omega + r_G(mv_G) \qquad (19\text{–}8)$$

This equation may be *simplified* by first substituting $v_G = r_G\omega$, in which case $H_O = (I_G + mr_G^2)\omega$, and, by the parallel-axis theorem, noting that the terms inside the parentheses represent the moment of inertia I_O of the body about an axis perpendicular to the plane of motion and passing through point O. Hence,*

$$\boxed{H_O = I_O\boldsymbol{\omega}} \qquad (19\text{–}9)$$

For the calculation, then, either Eq. 19–8 or 19–9 can be used.

---

*The similarity between this derivation and that of Eq. 17–16 ($\Sigma M_O = I_O\alpha$) and Eq. 18–5 ($T = \frac{1}{2}I_O\omega^2$) should be noted. Also note that the same result can be obtained from Eq. 19–2 by selecting point P at O, realizing that $(v_O)_x = (v_O)_y = 0$.

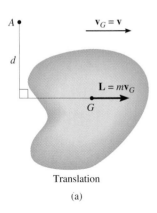

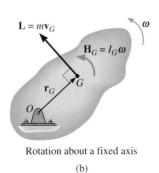

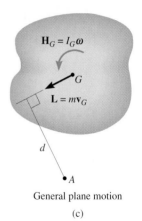

Translation
(a)

Rotation about a fixed axis
(b)

General plane motion
(c)

Fig. 19–2

General Plane Motion. When a rigid body is subjected to general plane motion, Fig. 19–2c, the linear momentum and the angular momentum computed about G become

$$\boxed{\begin{aligned} L &= mv_G \\ H_G &= I_G\boldsymbol{\omega} \end{aligned}}$$ (19–10)

If the angular momentum is computed about a point A located either on or off the body, Fig. 19–2c, it is necessary to find the moments of *both* **L** and $\mathbf{H}_G$ about this point. In this case,

$\zeta +$

$$H_A = I_G\omega + (d)(mv_G)$$

Here d is the moment arm, as shown in the figure.

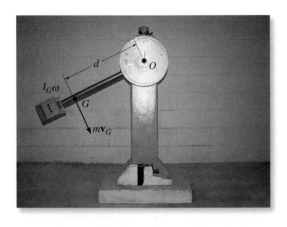

As the pendulum swings downward, its angular momentum about point O can be determined by computing the moment of $I_G\boldsymbol{\omega}$ and $m\mathbf{v}_G$ about O. This is $H_O = I_G\omega + (mv_G)d$. Since $v_G = \omega d$, then

$$H_O = I_G\omega + m(\omega d)d = (I_G + md^2)\omega = I_O\omega.$$

E X A M P L E 19.1

At a given instant the 10-kg disk and 5-kg bar have the motions shown in Fig. 19–3a. Determine their angular momenta about point G and about point B for the disk and about G and about the IC for the bar at this instant.

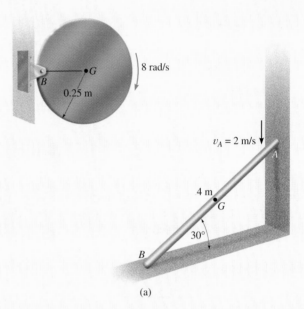

(a)

Solution

Disk.   Since the disk is *rotating about a fixed axis* (through point B), then $v_G = (8 \text{ rad/s})(0.25 \text{ m}) = 2 \text{ m/s}$, Fig. 19–3b. Hence

$$\zeta + H_G = I_G\omega = [\tfrac{1}{2}(10 \text{ kg})(0.25 \text{ m})^2](8 \text{ rad/s}) = 2.50 \text{ kg} \cdot \text{m}^2/\text{s} \quad Ans.$$

$$\zeta + H_B = I_G\omega + (mv_G)r_G = 2.50 \text{ kg} \cdot \text{m}^2/\text{s} + (10 \text{ kg})(2 \text{ m/s})(0.25 \text{ m})$$
$$= 7.50 \text{ kg} \cdot \text{m}^2/\text{s}\downarrow \quad Ans.$$

Also, from the table on the inside back cover, $I_B = (3/2)mr^2$, so that

$$\zeta + H_B = I_B\omega = [\tfrac{3}{2}(10 \text{ kg})(0.25 \text{ m})^2](8 \text{ rad/s}) = 7.50 \text{ kg} \cdot \text{m}^2/\text{s}\downarrow \quad Ans.$$

Bar.   The bar undergoes *general plane motion*. The IC is established in Fig. 19–3c, so that $\omega = (2 \text{ m/s})/(3.464 \text{ m}) = 0.5774 \text{ rad/s}$ and $v_G = (0.5774 \text{ rad/s})(2 \text{ m}) = 1.155 \text{ m/s}$. Thus,

$$\zeta + H_G = I_G\omega = [\tfrac{1}{12}(5 \text{ kg})(4 \text{ m})^2](0.5774 \text{ rad/s}) = 3.85 \text{ kg} \cdot \text{m}^2/\text{s}\downarrow Ans.$$

Moments of $I_G\omega$ and mv_G about the IC yield

$$\zeta + H_{IC} = I_G\omega + d(mv_G) = 3.85 \text{ kg} \cdot \text{m}^2/\text{s} + (2 \text{ m})(5 \text{ kg})(1.155 \text{ m/s})$$
$$= 15.4 \text{ kg} \cdot \text{m}^2/\text{s}\downarrow \quad Ans.$$

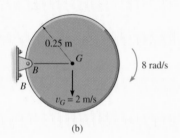

(b)

(c)

Fig. 19–3

19.2 Principle of Impulse and Momentum

Like the case for particle motion, the principle of impulse and momentum for a rigid body is developed by *combining* the equation of motion with kinematics. The resulting equation will allow a *direct solution to problems involving force, velocity, and time.*

Principle of Linear Impulse and Momentum. The equation of translational motion for a rigid body can be written as $\Sigma \mathbf{F} = m\mathbf{a}_G = m \, (d\mathbf{v}_G/dt)$. Since the mass of the body is constant,

$$\Sigma \mathbf{F} = \frac{d}{dt}(m\mathbf{v}_G)$$

Multiplying both sides by dt and integrating from $t = t_1$, $\mathbf{v}_G = (\mathbf{v}_G)_1$ to $t = t_2$, $\mathbf{v}_G = (\mathbf{v}_G)_2$ yields

$$\Sigma \int_{t_1}^{t_2} \mathbf{F} \, dt = m(\mathbf{v}_G)_2 - m(\mathbf{v}_G)_1 \qquad (19\text{–}11)$$

This equation is referred to as the *principle of linear impulse and momentum.* It states that the sum of all the impulses created by the *external force system* which acts on the body during the time interval t_1 to t_2 is equal to the change in the linear momentum of the body during the time interval, Fig. 19–4.

Principle of Angular Impulse and Momentum. If the body has *general plane motion* we can write $\Sigma M_G = I_G \alpha = I_G(d\omega/dt)$. Since the moment of inertia is constant,

$$\Sigma M_G = \frac{d}{dt}(I_G \omega)$$

Multiplying both sides by dt and integrating from $t = t_1$, $\omega = \omega_1$ to $t = t_2$, $\omega = \omega_2$ gives

$$\Sigma \int_{t_1}^{t_2} M_G \, dt = I_G \omega_2 - I_G \omega_1 \qquad (19\text{–}12)$$

In a similar manner, for *rotation about a fixed axis* passing through point O, Eq. 17–16 ($\Sigma M_O = I_O \alpha$) when integrated becomes

$$\Sigma \int_{t_1}^{t_2} M_O \, dt = I_O \omega_2 - I_O \omega_1 \qquad (19\text{–}13)$$

Equations 19–12 and 19–13 are referred to as the *principle of angular impulse and momentum.* Both equations state that the sum of the angular impulses acting on the body during the time interval t_1 to t_2 is equal to the change in the body's angular momentum during this time interval. In particular, the angular impulse considered is determined by integrating the moments about point G or O of all the external forces and couple moments applied to the body.

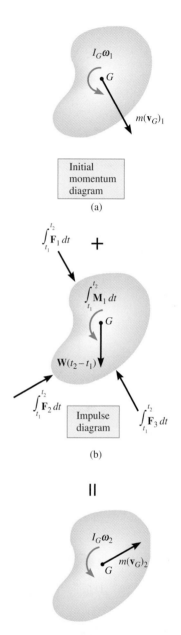

Initial momentum diagram

(a)

+

Impulse diagram

(b)

‖

Final momentum diagram

(c)

Fig. 19–4

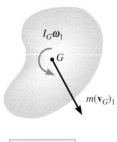

Initial
momentum
diagram

(a)

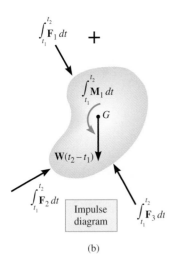

Impulse
diagram

(b)

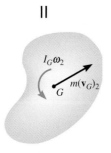

Final
momentum
diagram

(c)

Fig. 19–4

To summarize the preceding concepts, if motion is occurring in the *x–y* plane, using impulse and momentum principles the following *three scalar equations* may be written which describe the *planar motion* of the body:

$$m(v_{Gx})_1 + \Sigma \int_{t_1}^{t_2} F_x \, dt = m(v_{Gx})_2$$

$$m(v_{Gy})_1 + \Sigma \int_{t_1}^{t_2} F_y \, dt = m(v_{Gy})_2 \qquad (19\text{–}14)$$

$$I_G \omega_1 + \Sigma \int_{t_1}^{t_2} M_G \, dt = I_G \omega_2$$

The first two of these equations represent the principle of linear impulse and momentum in the *x–y* plane, Eq. 19–11, and the third equation represents the principle of angular impulse and momentum about the *z* axis, which passes through the body's mass center *G*, Eq. 19–12.

The terms in Eqs. 19–14 can be graphically accounted for by drawing a set of impulse and momentum diagrams for the body, Fig. 19–4. Note that the linear momenta mv_G are applied at the body's mass center, Figs. 19–4*a* and 19–4*c*; whereas the angular momenta $I_G \omega$ are free vectors, and therefore, like a couple moment, they may be applied at any point on the body. When the impulse diagram is constructed, Fig. 19–4*b*, the forces **F** and moment **M** vary with time, and are indicated by the integrals. However, if **F** and **M** are *constant* from t_1 to t_2, integration of the impulses yields $\mathbf{F}(t_2 - t_1)$ and $\mathbf{M}(t_2 - t_1)$, respectively. Such is the case for the body's weight **W**, Fig. 19–4*b*.

Equations 19–14 may also be applied to an entire system of connected bodies rather than to each body separately. Doing this eliminates the need to include reactive impulses which occur at the connections since they are *internal* to the system. The resultant equations may be written in symbolic form as

$$\left(\Sigma \begin{matrix} \text{syst. linear} \\ \text{momentum} \end{matrix} \right)_{x1} + \left(\Sigma \begin{matrix} \text{syst. linear} \\ \text{impulse} \end{matrix} \right)_{x(1\text{–}2)} = \left(\Sigma \begin{matrix} \text{syst. linear} \\ \text{momentum} \end{matrix} \right)_{x2}$$

$$\left(\Sigma \begin{matrix} \text{syst. linear} \\ \text{momentum} \end{matrix} \right)_{y1} + \left(\Sigma \begin{matrix} \text{syst. linear} \\ \text{impulse} \end{matrix} \right)_{y(1\text{–}2)} = \left(\Sigma \begin{matrix} \text{syst. linear} \\ \text{momentum} \end{matrix} \right)_{y2}$$

$$\left(\Sigma \begin{matrix} \text{syst. angular} \\ \text{momentum} \end{matrix} \right)_{O1} + \left(\Sigma \begin{matrix} \text{syst. angular} \\ \text{impulse} \end{matrix} \right)_{O(1\text{–}2)} = \left(\Sigma \begin{matrix} \text{syst. angular} \\ \text{momentum} \end{matrix} \right)_{O2}$$

$$(19\text{–}15)$$

As indicated, the system's angular momentum and angular impulse must be computed with respect to the *same fixed reference point O* for all the bodies of the system.

PROCEDURE FOR ANALYSIS

Impulse and momentum principles are used to solve kinetic problems that involve *velocity, force*, and *time* since these terms are involved in the formulation.

Free-Body Diagram.

- Establish the x, y, z inertial frame of reference and draw the free-body diagram in order to account for all the forces and couple moments that produce impulses on the body.

- The direction and sense of the initial and final velocity of the body's mass center, $\mathbf{v}_G$, and the body's angular velocity $\boldsymbol{\omega}$ should be established. If any of these motions is unknown, assume that the sense of its components is in the direction of the positive inertial coordinates.

- Compute the moment of inertia I_G or I_O.

- As an alternative procedure, draw the impulse and momentum diagrams for the body or system of bodies. Each of these diagrams represents an outlined shape of the body which graphically accounts for the data required for each of the three terms in Eqs. 19–14 or 19–15, Fig. 19–4. These diagrams are particularly helpful in order to visualize the "moment" terms used in the principle of angular impulse and momentum, if application is about a point other than the body's mass center G or a fixed point O.

Principle of Impulse and Momentum.

- Apply the three scalar equations of impulse and momentum.

- The angular momentum of a rigid body rotating about a fixed axis is the moment of $m\mathbf{v}_G$ plus $I_G\boldsymbol{\omega}$ about the axis. This can be shown to be equal to $H_O = I_O\boldsymbol{\omega}$, where I_O is the moment of inertia of the body about the axis.

- All the forces acting on the body's free-body diagram will create an impulse; however, some of these forces will do no work.

- Forces that are functions of time must be integrated to obtain the impulse. Graphically, the impulse is equal to the area under the force-time curve.

- The principle of angular impulse and momentum is often used to eliminate unknown impulsive forces that are parallel or pass through a common axis, since the moment of these forces is zero about this axis.

Kinematics.

- If more than three equations are needed for a complete solution, it may be possible to relate the velocity of the body's mass center to the body's angular velocity using *kinematics*. If the motion appears to be complicated, kinematic (velocity) diagrams may be helpful in obtaining the necessary relation.

EXAMPLE 19.2

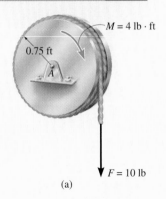

0.75 ft

$M = 4$ lb · ft

A

$F = 10$ lb

(a)

The 20-lb disk shown in Fig. 19–5a is assumed to be uniform and is pin supported at its center. If it is acted upon by a constant couple moment of 4 lb · ft and a force of 10 lb which is applied to a cord wrapped around its periphery, determine the angular velocity of the disk two seconds after starting from rest. Also, what are the force components of reaction at the pin?

Solution

Since angular velocity, force, and time are involved in the problems we will apply the principles of impulse and momentum to the solution.

Free-Body Diagram. Fig. 19–5b. The disk's mass center does not move; however, the loading causes the disk to rotate clockwise.
  The moment of inertia of the disk about its fixed axis of rotation is

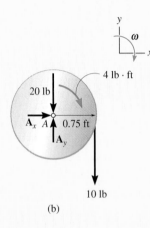

y

ω

x

4 lb · ft

20 lb

A_x A 0.75 ft

A_y

10 lb

(b)

Fig. 19–5

$$I_A = \frac{1}{2}mr^2 = \frac{1}{2}\left(\frac{20\text{ lb}}{32.2\text{ ft/s}^2}\right)(0.75\text{ ft})^2 = 0.175\text{ slug}\cdot\text{ft}^2$$

Principle of Impulse and Momentum.

$(\xrightarrow{+})$
$$m(v_{Ax})_1 + \Sigma\int_{t_1}^{t_2} F_x\,dt = m(v_{Ax})_2$$
$$0 + A_x(2\text{ s}) = 0$$

$(+\uparrow)$
$$m(v_{Ay})_1 + \Sigma\int_{t_1}^{t_2} F_y\,dt = m(v_{Ay})_2$$
$$0 + A_y(2\text{ s}) - 20\text{ lb}(2\text{ s}) - 10\text{ lb}(2\text{ s}) = 0$$

$(\curvearrowright+)$
$$I_A\omega_1 + \Sigma\int_{t_1}^{t_2} M_A\,dt = I_A\omega_2$$
$$0 + 4\text{ lb}\cdot\text{ft}(2\text{ s}) + [10\text{ lb}(2\text{ s})](0.75\text{ ft}) = 0.175\omega_2$$

Solving these equations yields

$$A_x = 0 \qquad\qquad\qquad Ans.$$
$$A_y = 30\text{ lb} \qquad\qquad Ans.$$
$$\omega_2 = 132\text{ rad/s}\,\downarrow \qquad Ans.$$

EXAMPLE 19.3

The 100-kg spool shown in Fig. 19–6a has a radius of gyration $k_G = 0.35$ m. A cable is wrapped around the central hub of the spool, and a horizontal force having a variable magnitude of $P = (t + 10)$ N is applied, where t is in seconds. If the spool is initially at rest, determine its angular velocity in 5 s. Assume that the spool rolls without slipping at A.

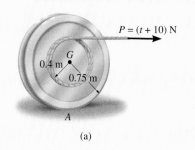

(a)

Solution

Free-Body Diagram. By inspection of the free-body diagram, Fig. 19–6b, the *variable* force P will cause the friction force F_A to be variable, and thus the impulses created by both P and F_A must be determined by integration. The force **P** causes the mass center to have a velocity $\mathbf{v}_G$ to the right, and the spool has a clockwise angular velocity $\boldsymbol{\omega}$.

The moment of inertia of the spool about its mass center is

$$I_G = mk_G^2 = 100 \text{ kg}(0.35 \text{ m})^2 = 12.25 \text{ kg} \cdot \text{m}^2$$

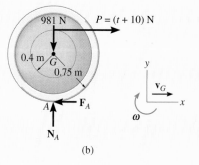

(b)

Principle of Impulse and Momentum.

$(\xrightarrow{+})$
$$m(v_G)_1 + \Sigma \int F_x \, dt = m(v_G)_2$$

$$0 + \int_0^{5 \text{ s}} (t + 10) \text{ N } dt - \int F_A \, dt = 100 \text{ kg}(v_G)_2$$

$$62.5 - \int F_A \, dt = 100(v_G)_2 \qquad (1)$$

$(\curvearrowleft+)$
$$I_G \omega_1 + \Sigma \int M_G \, dt = I_G \omega_2$$

$$0 + \left[\int_0^{5 \text{ s}} (t + 10) \text{ N } dt\right](0.4 \text{ m}) + \left(\int F_A \, dt\right)(0.75 \text{ m}) = (12.25 \text{ kg} \cdot \text{m}^2)\omega_2$$

$$25 + \left(\int F_A \, dt\right)(0.75) = 12.25\omega_2 \qquad (2)$$

Kinematics. Since the spool does not slip, the instantaneous center of zero velocity is at point A, Fig. 19–6c. Hence, the velocity of G can be expressed in terms of the spool's angular velocity as $(v_G)_2 = (0.75 \text{ m})\omega_2$. Substituting this into Eq. 1 and eliminating the unknown impulse $\int F_A \, dt$ between Eqs. 1 and 2, we obtain

$$\omega_2 = 1.05 \text{ rad/s} \downarrow \qquad \textit{Ans.}$$

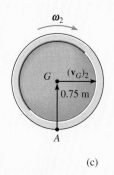

(c)

Fig. 19–6

Note: A more direct solution can be obtained by applying the principle of angular impulse and momentum about point A. As an exercise, do this and show that one obtains the same result.

E X A M P L E **19.4**

The block shown in Fig. 19–7a has a mass of 6 kg. It is attached to a cord which is wrapped around the periphery of a 20-kg disk that has a moment of inertia $I_A = 0.40 \text{ kg} \cdot \text{m}^2$. If the block is initially moving downward with a speed of 2 m/s, determine its speed in 3 s. Neglect the mass of the cord in the calculation.

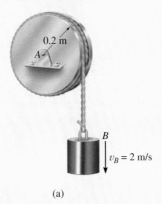

(a)

Solution I

Free-Body Diagram. The free-body diagrams of the block and disk are shown in Fig. 19–7b. All the forces are *constant* since the weight of the block causes the motion. The downward motion of the block, v_B, causes $\boldsymbol{\omega}$ of the disk to be clockwise.

Principle of Impulse and Momentum. We can eliminate $\mathbf{A}_x$ and $\mathbf{A}_y$ from the analysis by applying the principle of angular impulse and momentum about point A. Hence

Disk

$$(\curvearrowright +) \qquad I_A \omega_1 + \Sigma \int M_A \, dt = I_A \omega_2$$

$$0.40 \text{ kg} \cdot \text{m}^2 (\omega_1) + T(3 \text{ s})(0.2 \text{ m}) = (0.4 \text{ kg} \cdot \text{m}^2)\omega_2$$

Block

$$(+\uparrow) \qquad m_B(v_B)_1 + \Sigma \int F_y \, dt = m_B(v_B)_2$$

$$-6 \text{ kg}(2 \text{ m/s}) + T(3 \text{ s}) - 58.86 \text{ N}(3 \text{ s}) = -6 \text{ kg}(v_B)_2$$

Kinematics. Since $\omega = v_B/r$, then $\omega_1 = (2 \text{ m/s})/(0.2 \text{ m}) = 10 \text{ rad/s}$ and $\omega_2 = (v_B)_2/0.2 \text{ m} = 5(v_B)_2$. Substituting and solving the equations simultaneously for $(v_B)_2$ yields

$$(v_B)_2 = 13.0 \text{ m/s} \downarrow \qquad\qquad Ans.$$

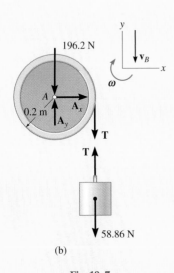

Fig. 19–7

Solution II

Impulse and Momentum Diagrams. We can obtain $(v_B)_2$ *directly* by considering the *system* consisting of the block, the cord, and the disk. The impulse and momentum diagrams have been drawn to clarify application of the principle of angular impulse and momentum about point A, Fig. 19–7c.

Principle of Angular Impulse and Momentum. Realizing that $\omega_1 = 10$ rad/s and $\omega_2 = 5(v_B)_2$, we have

$$(\gamma+) \quad \left(\sum {\text{syst. angular} \atop \text{momentum}}\right)_{A1} + \left(\sum {\text{syst. angular} \atop \text{impulse}}\right)_{A(1-2)} = \left(\sum {\text{syst. angular} \atop \text{momentum}}\right)_{A2}$$

$$6 \text{ kg}(2 \text{ m/s})(0.2 \text{ m}) + 0.4 \text{ kg} \cdot \text{m}^2(10 \text{ rad/s}) + 58.86 \text{ N}(3 \text{ s})(0.2 \text{ m})$$
$$= 6 \text{ kg}(v_B)_2(0.2 \text{ m}) + 0.40 \text{ kg} \cdot \text{m}^2[5(v_B)_2(0.2 \text{ m})]$$
$$(v_B)_2 = 13.0 \text{ m/s} \downarrow \qquad\qquad Ans.$$

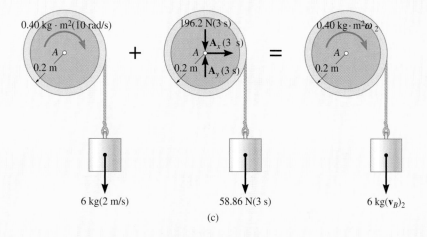

(c)

Fig. 19–7

EXAMPLE 19.5

The Charpy impact test is used in materials testing to determine the energy absorption characteristics of a material during impact. The test is performed using the pendulum shown in Fig. 19–8a, which has a mass m, mass center at G, and a radius of gyration k_G about G. Determine the distance r_P from the pin at A to the point P where the impact with the specimen S should occur so that the horizontal force at the pin is essentially zero during the impact. For the calculation, assume the specimen absorbs all the pendulum's kinetic energy gained during the time it falls and thereby stops the pendulum from swinging when $\theta = 0°$.

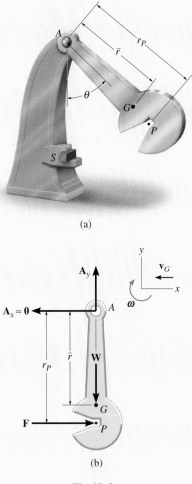

(a)

(b)

Fig. 19–8

Solution

Free-Body Diagram. As shown on the free-body diagram, Fig. 19–8*b*, the conditions of the problem require the horizontal impulse at *A* to be zero. Just before impact, the pendulum has a clockwise angular velocity ω_1, and the mass center of the pendulum is moving to the left at $(v_G)_1 = \bar{r}\omega_1$.

Principle of Impulse and Momentum. We will apply the principle of angular impulse and momentum about point *A*. Thus,

$$(\curvearrowright+) \qquad I_A\omega_1 + \Sigma M_A \, dt = I_A\omega_2$$

$$I_A\omega_1 - \left(\int F \, dt\right)r_p = 0$$

$$(\xrightarrow{+}) \qquad m(v_G)_1 + \Sigma F \, dt = m(v_G)_2$$

$$-m(\bar{r}\omega_1) + \int F \, dt = 0$$

Eliminating the impulse $\int F \, dt$ and substituting $I_A = mk_G^2 + m\bar{r}^2$ yields

$$[mk_G^2 + m\bar{r}^2]\omega_1 - m(\bar{r}\omega_1)r_P = 0$$

Factoring out $m\omega_1$ and solving for r_P, we obtain

$$r_P = \bar{r} + \frac{k_G^2}{\bar{r}} \qquad\qquad \textit{Ans.}$$

The point *P*, so defined, is called the *center of percussion*. By placing the striking point at *P*, the force developed at the pin will be minimized. Many sports rackets, clubs, etc. are designed so that collision with the object being struck occurs at the center of percussion. As a consequence, no "sting" or little sensation occurs in the hand of the player.

19.3 Conservation of Momentum

Conservation of Linear Momentum. If the sum of all the *linear impulses* acting on a system of connected rigid bodies is *zero*, the linear momentum of the system is constant, or conserved. Consequently, the first two of Eqs. 19–15 reduce to the form

$$\left(\sum \begin{array}{c} \text{syst. linear} \\ \text{momentum} \end{array} \right)_1 = \left(\sum \begin{array}{c} \text{syst. linear} \\ \text{momentum} \end{array} \right)_2 \qquad (19\text{–}16)$$

This equation is referred to as the *conservation of linear momentum*.

Without inducing appreciable errors in the calculations, it may be possible to apply Eq. 19–16 in a specified direction for which the linear impulses are small or *nonimpulsive*. Specifically, nonimpulsive forces occur when small forces act over very short periods of time. Typical examples include the force of a slightly deformed spring, the initial contact with soft ground, and in some cases the weight of the body.

Conservation of Angular Momentum. The angular momentum of a system of connected rigid bodies is conserved about the system's center of mass G, or a fixed point O, when the sum of all the angular impulses created by the external forces acting on the system is zero or appreciably small (nonimpulsive) when computed about these points. The third of Eqs. 19–15 then becomes

$$\left(\sum \begin{array}{c} \text{syst. angular} \\ \text{momentum} \end{array} \right)_{O1} = \left(\sum \begin{array}{c} \text{syst. angular} \\ \text{momentum} \end{array} \right)_{O2} \qquad (19\text{–}17)$$

This equation is referred to as the *conservation of angular momentum*. In the case of a single rigid body, Eq. 19–17 applied to point G becomes $(I_G\omega)_1 = (I_G\omega)_2$. To illustrate an application of this equation, consider a swimmer who executes a somersault after jumping off a diving board. By tucking his arms and legs in close to his chest, he *decreases* his body's moment of inertia and thus *increases* his angular velocity ($I_G\omega$ must be constant). If he straightens out just before entering the water, his body's moment of inertia is *increased*, and his angular velocity *decreases*. Since the weight of his body creates a linear impulse during the time of motion, this example also illustrates how the angular momentum of a body can be conserved and yet the linear momentum is *not*. Such cases occur whenever the external forces creating the linear impulse pass through either the center of mass of the body or a fixed axis of rotation.

Provided the initial linear or angular velocity of the body is known, the conservation of linear or angular momentum is used to determine the respective final linear or angular velocity of the body *just after* the time period considered. Furthermore, by applying these equations to a *system* of bodies, the internal impulses acting within the system, which may be unknown, are eliminated from the analysis, since they occur in equal but opposite collinear pairs. If it is necessary to determine an *internal impulsive force* acting on only one body of a system of connected bodies, the body must be *isolated* (free-body diagram) and the principle of linear or angular impulse and momentum applied *to the body*. After the impulse $\int F\, dt$ is calculated; then, provided the time Δt for which the impulse acts is known, the *average impulsive force* F_{avg} can be determined from $F_{avg} = (\int F\, dt)/\Delta t$.

PROCEDURE FOR ANALYSIS

The conservation of linear or angular momentum should be applied using the following procedure.

Free-Body Diagram.

- Establish the x, y inertial frame of reference and draw the free-body diagram for the body or system of bodies during the time of impact. From this diagram classify each of the applied forces as being either "impulsive" or "nonimpulsive."

- By inspection of the free-body diagram, the *conservation of linear momentum* applies in a given direction when *no* external impulsive forces act on the body or system in that direction; whereas the *conservation of angular momentum* applies about a fixed point O or at the mass center G of a body or system of bodies when all the external impulsive forces acting on the body or system create zero moment (or zero angular impulse) about O or G.

- As an alternative procedure, draw the impulse and momentum diagrams for the body or system of bodies. These diagrams are particularly helpful in order to visualize the "moment" terms used in the conservation of angular momentum equation, when it has been decided that angular momenta are to be computed about a point other than the body's mass center G.

Conservation of Momentum.

- Apply the conservation of linear or angular momentum in the appropriate directions.

Kinematics.

- If the motion appears to be complicated, kinematic (velocity) diagrams may be helpful in obtaining the necessary kinematic relations.

E X A M P L E 19.6

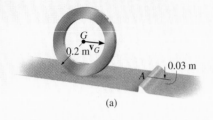

(a)

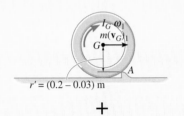

$+$

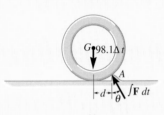

$\parallel$

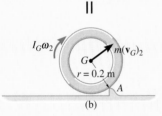

(b)

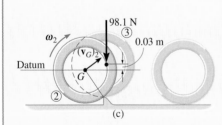

(c)

Fig. 19–9

The 10-kg wheel shown in Fig. 19–9a has a moment of inertia $I_G = 0.156 \text{ kg} \cdot \text{m}^2$. Assuming that the wheel does not slip or rebound, determine the minimum velocity $\mathbf{v}_G$ it must have to just roll over the obstruction at A.

Solution

Impulse and Momentum Diagrams. Since no slipping or rebounding occurs, the wheel essentially *pivots* about point A during contact. This condition is shown in Fig. 19–9b, which indicates, respectively, the momentum of the wheel *just before impact*, the impulses given to the wheel *during impact*, and the momentum of the wheel *just after impact*. Only two impulses (forces) act on the wheel. By comparison, the force at A is much greater than that of the weight, and since the time of impact is very short, the weight can be considered nonimpulsive. The impulsive force $\mathbf{F}$ at A has both an unknown magnitude and an unknown direction θ. To eliminate this force from the analysis, note that angular momentum about A is essentially *conserved* since $(98.1\Delta t)d \approx 0$.

Conservation of Angular Momentum. With reference to Fig. 19–9b,

$(\curvearrowright +)$
$$(H_A)_1 = (H_A)_2$$
$$r'm(v_G)_1 + I_G\omega_1 = rm(v_G)_2 + I_G\omega_2$$
$$(0.2 \text{ m} - 0.03 \text{ m})(10 \text{ kg})(v_G)_1 + (0.156 \text{ kg} \cdot \text{m}^2)(\omega_1) =$$
$$(0.2 \text{ m})(10 \text{ kg})(v_G)_2 + (0.156 \text{ kg} \cdot \text{m}^2)(\omega_2)$$

Kinematics. Since no slipping occurs, in general $\omega = v_G/r = v_G/0.2 \text{ m} = 5v_G$. Substituting this into the above equation and simplifying yields

$$(v_G)_2 = 0.892(v_G)_1 \qquad (1)$$

Conservation of Energy.* In order to roll over the obstruction, the wheel must pass position 3 shown in Fig. 19–9c. Hence, if $(v_G)_2$ [or $(v_G)_1$] is to be a minimum, it is necessary that the kinetic energy of the wheel at position 2 be equal to the potential energy at position 3. Constructing the datum through the center of gravity, as shown in the figure, and applying the conservation of energy equation, we have

$$\{T_2\} + \{V_2\} = \{T_3\} + \{V_3\}$$
$$\{\tfrac{1}{2}(10 \text{ kg})(v_G)_2^2 + \tfrac{1}{2}(0.156 \text{ kg} \cdot \text{m}^2)\omega_2^2\} + \{0\} =$$
$$\{0\} + \{(98.1 \text{ N})(0.03 \text{ m})\}$$

Substituting $\omega_2 = 5(v_G)_2$ and Eq. 1 into this equation, and solving,

$$(v_G)_1 = 0.729 \text{ m/s} \rightarrow \qquad \textit{Ans.}$$

*This principle *does not* apply during *impact*, since energy is *lost* during the collision; however, just after impact, position 2, it can be used.

EXAMPLE 19.7

The 5-kg slender rod shown in Fig. 19–10a is pinned at O and is initially at rest. If a 4-g bullet is fired into the rod with a velocity of 400 m/s, as shown in the figure, determine the angular velocity of the rod just after the bullet becomes embedded in it.

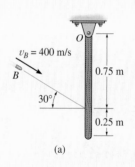

(a)

Solution

Impulse and Momentum Diagrams. The impulse which the bullet exerts on the rod can be eliminated from the analysis, and the angular velocity of the rod just after impact can be determined by considering the bullet and rod as a single system. To clarify the principles involved, the impulse and momentum diagrams are shown in Fig. 19–10b. The momentum diagrams are drawn *just before and just after impact.* During impact, the bullet and rod exchange equal but *opposite internal impulses* at A. As shown on the impulse diagram, the impulses that are external to the system are due to the reactions at O and the weights of the bullet and rod. Since the time of impact, Δt, is very short, the rod moves only a slight amount, and so the "moments" of the weight impulses about point O are essentially zero. Therefore angular momentum is conserved about this point.

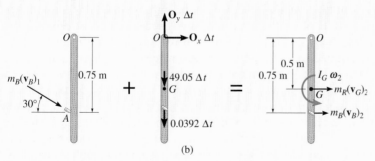

(b)

Conservation of Angular Momentum. From Fig. 19–10b, we have

$$(\downarrow +) \qquad\qquad \Sigma(H_O)_1 = \Sigma(H_O)_2$$

$$m_B(v_B)_1 \cos 30°(0.75 \text{ m}) = m_B(v_B)_2(0.75 \text{ m}) + m_R(v_G)_2(0.5 \text{ m}) + I_G\omega_2$$

$$(0.004 \text{ kg})(400 \cos 30°\text{m/s})(0.75 \text{ m}) =$$

$$(0.004 \text{ kg})(v_B)_2(0.75 \text{ m}) + (5 \text{ kg})(v_G)_2(0.5 \text{ m}) + [\tfrac{1}{12}(5 \text{ kg})(1 \text{ m})^2]\omega_2$$

or

$$1.039 = 0.003(v_B)_2 + 2.50(v_G)_2 + 0.417\omega_2 \qquad (1)$$

Kinematics. Since the rod is pinned at O, from Fig. 19–10c we have

$$(v_G)_2 = (0.5 \text{ m})\omega_2 \qquad (v_B)_2 = (0.75 \text{ m})\omega_2$$

Substituting into Eq. 1 and solving yields

$$\omega_2 = 0.623 \text{ rad/s} \qquad\qquad Ans.$$

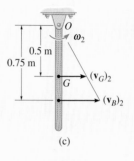

(c)

Fig. 19–10

19.4 Eccentric Impact

An example of eccentric impact occurring between this bowling ball and pin.

The concepts involving central and oblique impact of particles have been presented in Sec. 15.4. We will now expand this treatment and discuss the eccentric impact of two bodies. *Eccentric impact* occurs when the line connecting the *mass centers* of the two bodies *does not* coincide with the line of impact.* This type of impact often occurs when one or both of the bodies are constrained to rotate about a fixed axis. Consider, for example, the collision at *C* between the two bodies *A* and *B*, shown in Fig. 19–11*a*. It is assumed that just before collision *B* is rotating counterclockwise with an angular velocity $(\boldsymbol{\omega}_B)_1$, and the velocity of the contact point *C* located on *A* is $(\mathbf{u}_A)_1$. Kinematic diagrams for both bodies just before collision are shown in Fig. 19–11*b*. Provided the bodies are smooth, the *impulsive forces* they exert on each other *are directed along the line of impact*. Hence, the component of velocity of point *C* on body. *B*, which is directed along the line of impact, is $(v_B)_1 = (\omega_B)_1 r$, Fig. 19–11*b*. Likewise, on body *A* the component of velocity $(\mathbf{u}_A)_1$ along the line of impact is $(\mathbf{v}_A)_1$. In order for a collision to occur, $(v_A)_1 > (v_B)_1$.

During the impact an equal but opposite impulsive force **P** is exerted between the bodies which *deforms* their shapes at the point of contact. The resulting impulse is shown on the impulse diagrams for both bodies, Fig. 19–11*c*. Note that the impulsive force created at point *C* on the rotating body creates impulsive pin reactions at *O*. On these diagrams it is assumed that the impact creates forces which are much larger than the nonimpulsive weights of the bodies, which are not shown. When the deformation at point *C* is a maximum, *C* on both the bodies moves with a common velocity **v** along the line of impact, Fig. 19–11*d*. A period of *restitution* then occurs in which the bodies tend to regain their original shapes. The restitution phase creates an equal but opposite impulsive force **R** acting between the bodies as shown on the impulse diagram, Fig. 19–11*e*. After restitution the bodies move apart such that point *C* on body *B* has a velocity $(\mathbf{v}_B)_2$ and point *C* on body *A* has a velocity $(\mathbf{u}_A)_2$, Fig. 19–11*f*, where $(v_B)_2 > (v_A)_2$.

In general, a problem involving the impact of two bodies requires determining the *two unknowns* $(v_A)_2$ and $(v_B)_2$, assuming $(v_A)_1$ and $(v_B)_1$ are known (or can be determined using kinematics, energy methods, the equations of motion, etc.). To solve this problem, two equations must be written. The *first equation* generally involves application of *the conservation of angular momentum to the two bodies*. In the case of both bodies *A* and *B*, we can state that angular momentum is conserved about point *O* since the impulses at *C* are internal to the system and the impulses at *O* create zero moment (or zero angular impulse) about point

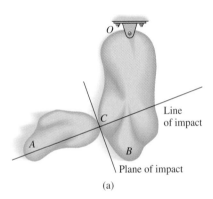

Line
of impact

Plane of impact

(a)

Fig. 19–11

---

*When these lines coincide, central impact occurs and the problem can be analyzed as discussed in Sec. 15.4.

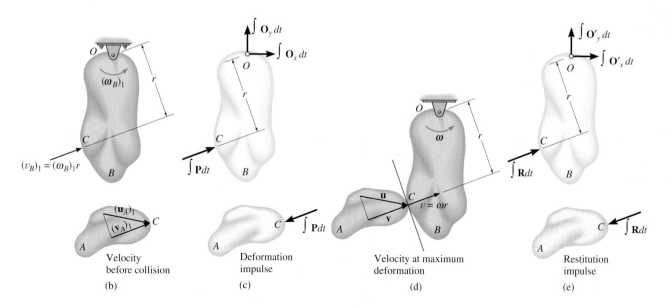

(b) Velocity before collision

(c) Deformation impulse

(d) Velocity at maximum deformation

(e) Restitution impulse

O. The *second equation* is obtained using the definition of the *coefficient of restitution, e*, which is a ratio of the restitution impulse to the deformation impulse. To establish a useful form of this equation we must first apply the principle of angular impulse and momentum about point O to bodies B and A separately. Combining the results, we then obtain the necessary equation. Proceeding in this manner, the principle of impulse and momentum applied to body B from the time just before the collision to the instant of maximum deformation, Figs. 19–11b, 19–11c, and 19–11d, becomes

$$(\zeta+) \qquad I_O(\omega_B)_1 + r\int P\,dt = I_O\omega \qquad (19\text{–}18)$$

Here I_O is the moment of inertia of body B about point O. Similarly, applying the principle of angular impulse and momentum from the instant of maximum deformation to the time just after the impact, Figs. 19–11d, 19–11e, and 19–11f, yields

$$(\zeta+) \qquad I_O\omega + r\int R\,dt = I_O(\omega_B)_2 \qquad (19\text{–}19)$$

Solving Eqs. 19–18 and 19–19 for $\int P\,dt$ and $\int R\,dt$, respectively, and formulating e, we have

$$e = \frac{\displaystyle\int R\,dt}{\displaystyle\int P\,dt} = \frac{r(\omega_B)_2 - r\omega}{r\omega - r(\omega_B)_1} = \frac{(v_B)_2 - v}{v - (v_B)_1}$$

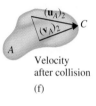

(f) Velocity after collision

Fig. 19–11

In the same manner, we may write an equation which relates the magnitudes of velocity $(v_A)_1$ and $(v_A)_2$ of body A. The result is

$$e = \frac{v - (v_A)_2}{(v_A)_1 - v}$$

Combining the above equations by eliminating the common velocity v yields the desired result, i.e.,

$(+\nearrow)$ $$e = \frac{(v_B)_2 - (v_A)_2}{(v_A)_1 - (v_B)_1}$$ (19–20)

This equation is identical to Eq. 15–11, which was derived for the central impact between two particles. Equation 19–20 states that the coefficient of restitution is equal to the ratio of the relative velocity of *separation* of the points of contact (C) *just after impact* to the relative velocity at which the points *approach* one another *just* before impact. In deriving this equation, we assumed that the points of contact for both bodies move up and to the right *both* before and after impact. If motion of any one of the contacting points occurs down and to the left, the velocity of this point is considered a negative quantity in Eq. 19–20.

As stated previously, when Eq. 19–20 is used in conjunction with the conservation of angular momentum for the bodies, it provides a useful means of obtaining the velocities of two colliding bodies just after collision.

During impact the columns on many highway signs are intended to break out of their supports and easily collapse at their joints. This is shown by the slotted connections at their base and the breaks at the column's midsection. The mechanics of eccentric impact is used in the design of these structures.

EXAMPLE 19.8

The 10-lb slender rod is suspended from the pin at A, Fig. 19–12a. If a 2-lb ball B is thrown at the rod and strikes its center with a horizontal velocity of 30 ft/s, determine the angular velocity of the rod just after impact. The coefficient of restitution is $e = 0.4$.

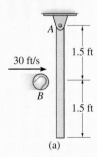

(a)

Solution

Conservation of Angular Momentum. Consider the ball and rod as a system, Fig. 19–12b. Angular momentum is conserved about point A since the impulsive force between the rod and ball is *internal*. Also, the *weights* of the ball and rod are *nonimpulsive*. Noting the directions of the velocities of the ball and rod just after impact as shown on the kinematic diagram, Fig. 19–12c, we require

$$(\downarrow+) \qquad (H_A)_1 = (H_A)_2$$

$$m_B(v_B)_1(1.5 \text{ ft}) = m_B(v_B)_2(1.5 \text{ ft}) + m_R(v_G)_2(1.5 \text{ ft}) + I_G\omega_2$$

$$\left(\frac{2 \text{ lb}}{32.2 \text{ ft/s}^2}\right)(30 \text{ ft/s})(1.5 \text{ ft}) = \left(\frac{2 \text{ lb}}{32.2 \text{ ft/s}^2}\right)(v_B)_2(1.5 \text{ ft}) +$$

$$\left(\frac{10 \text{ lb}}{32.2 \text{ ft/s}^2}\right)(v_G)_2(1.5 \text{ ft}) + \left[\frac{1}{12}\left(\frac{10 \text{ lb}}{32.2 \text{ ft/s}^2}\right)(3 \text{ ft})^2\right]\omega_2$$

Since $(v_G)_2 = 1.5\omega_2$ then

$$2.795 = 0.09317(v_B)_2 + 0.9317\omega_2 \qquad (1)$$

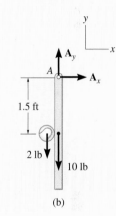

(b)

Coefficient of Restitution. With reference to Fig. 19–12c, we have

$$(\xrightarrow{+}) \qquad e = \frac{(v_G)_2 - (v_B)_2}{(v_B)_1 - (v_G)_1} \qquad 0.4 = \frac{(1.5 \text{ ft})\omega_2 - (v_B)_2}{30 \text{ ft/s} - 0}$$

$$12.0 = 1.5\omega_2 - (v_B)_2$$

Solving,

$$(v_B)_2 = -6.52 \text{ ft/s} = 6.52 \text{ ft/s} \leftarrow$$

$$\omega_2 = 3.65 \text{ rad/s} \qquad\qquad\qquad\qquad \textit{Ans.}$$

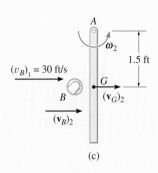

(c)

Fig. 19–12

CHAPTER REVIEW

- *Linear and Angular Momentum.* The linear momentum of a rigid body can be referenced to the velocity of its mass center G. The result is $\mathbf{L} = m\mathbf{v}_G$. By summing the moments of the linear momenta of all of the particles of the body about an axis passing through G, it can be shown that the angular momentum for the body about G is $\mathbf{H}_G = I_G\boldsymbol{\omega}$. If the angular momentum is to be determined about an axis other than the one passing through the mass center, then the angular momentum is determined by summing vector $\mathbf{H}_G$ and the moment of vector $\mathbf{L}_G$ about this axis.

  Translation. If the body is translating, then $\omega = 0$, and so

  $$L = mv_G \qquad H_G = 0$$

  Rotation About a Fixed Axis. Here

  $$L = mv_G \qquad H_G = I_G\omega$$

  Since $v_G = \omega r$, then, using the parallel-axis theorem, the angular momentum about the axis of rotation becomes

  $$H_O = I_O\omega$$

  General Plane Motion. The linear momentum and the angular momentum about point G for this case are

  $$L = mv_G \qquad H_G = I_G\omega$$

- *Principle of Impulse and Momentum.* The principles of linear and angular impulse and momentum for a rigid body are

  $$m(\mathbf{v}_G)_1 + \Sigma \int_{t_1}^{t_2} \mathbf{F}\, dt = m(\mathbf{v}_G)_2$$

  $$I_G\omega_1 + \Sigma \int_{t_1}^{t_2} M_G\, dt = I_G\omega_2$$

  Before applying these equations, it is important to establish the x, y, z inertial coordinate system. The free-body diagram for the body should also be drawn in order to account for all of the forces and couple moments that produce impulses on the body.

- *Conservation of Momentum.* Provided the sum of the linear impulses acting on a system of connected rigid bodies is zero in a particular direction, then the linear momentum for the system is conserved in this direction. Conservation of angular momentum occurs if the impulses pass through an axis or are parallel to it. Momentum is also conserved if the external forces are small and thereby create nonimpulsive forces on the system.

  A free-body diagram should accompany any application in order to classify the forces as impulsive or nonimpulsive and to determine an axis about which the angular momentum may be conserved.

- *Eccentric Impact.* If the line of impact does not coincide with the line connecting the mass centers of two colliding bodies, then eccentric impact will occur. If the motion of the bodies just after the impact is to be determined, then it is necessary to consider a conservation of momentum equation for the system and use the coefficient of restitution equation.

Planar Kinematics and Kinetics of a Rigid Body

Having presented the various topics in planar kinematics and kinetics in Chapters 16 through 19, we will now summarize these principles and provide an opportunity for applying them to the solution of various types of problems.

Kinematics.   Here we are interested in studying the geometry of motion, without concern for the forces which cause the motion. Before solving a planar kinematics problem, it is *first* necessary to *classify the motion* as being either rectilinear or curvilinear translation, rotation about a fixed axis, or general plane motion. In particular, problems involving general plane motion can be solved either with reference to a fixed axis (absolute motion analysis) or using translating or rotating frames of reference (relative motion analysis). The choice generally depends upon the type of constraints and the problem's geometry. In all cases, application of the necessary equations may be clarified by drawing a kinematic diagram. Remember that the *velocity* of a point is always *tangent* to its path of motion, and the *acceleration* of a point can have *components* in the *n-t* directions when the path is *curved*.

Translation.   When the body moves with rectilinear or curvilinear translation, *all* the points on the body have the *same motion*.

$$\mathbf{v}_B = \mathbf{v}_A \qquad \mathbf{a}_B = \mathbf{a}_A$$

Rotation About a Fixed Axis.

Angular Motion.

Variable Angular Acceleration. Provided a mathematical relationship is given between *any two* of the *four* variables θ, ω, α, and t, then a *third* variable can be determined by solving one of the following equations which relate all three variables.

$$\omega = \frac{d\theta}{dt} \qquad \alpha = \frac{d\omega}{dt} \qquad \alpha \, d\theta = \omega \, d\omega$$

Constant Angular Acceleration. The following equations apply when it is *absolutely certain* that the angular acceleration is constant.

$$\theta = \theta_0 + \omega_0 t + \tfrac{1}{2}\alpha_c t^2 \qquad \omega = \omega_0 + \alpha_c t \qquad \omega^2 = \omega_0^2 + 2\alpha_c(\theta - \theta_0)$$

Motion of Point P.

Once $\boldsymbol{\omega}$ and $\boldsymbol{\alpha}$ have been determined, then the circular motion of point P can be specified using the following scalar or vector equations.

$$v = \omega r \qquad\qquad \mathbf{v} = \boldsymbol{\omega} \times \mathbf{r}$$

$$a_t = \alpha r \quad a_n = \omega^2 r \qquad \mathbf{a} = \boldsymbol{\alpha} \times \mathbf{r} + \boldsymbol{\omega} \times (\boldsymbol{\omega} \times \mathbf{r})$$

General Plane Motion—Relative-Motion Analysis.   Recall that when *translating axes* are placed at the "base point" A, the *relative motion* of point B with respect to A is simply *circular motion of B about A.* The following equations apply to two points A and B located on the *same* rigid body.

$$\mathbf{v}_B = \mathbf{v}_A + \mathbf{v}_{B/A} = \mathbf{v}_A + (\boldsymbol{\omega} \times \mathbf{r}_{B/A})$$

$$\mathbf{a}_B = \mathbf{a}_A + \mathbf{a}_{B/A} = \mathbf{a}_A + \boldsymbol{\alpha} \times \mathbf{r}_{B/A} + \boldsymbol{\omega} \times (\boldsymbol{\omega} \times \mathbf{r}_{B/A})$$

Rotating and translating axes are often used to analyze the motion of rigid bodies which are connected together by collars or slider blocks.

$$\mathbf{v}_B = \mathbf{v}_A + \boldsymbol{\Omega} \times \mathbf{r}_{B/A} + (\mathbf{v}_{B/A})_{xyz}$$

$$\mathbf{a}_B = \mathbf{a}_A + \dot{\boldsymbol{\Omega}} \times \mathbf{r}_{B/A} + \boldsymbol{\Omega} \times (\boldsymbol{\Omega} \times \mathbf{r}_{B/A}) + 2\boldsymbol{\Omega} \times (\mathbf{v}_{B/A})_{xyz} + (\mathbf{a}_{B/A})_{xyz}$$

Kinetics.   To analyze the forces which cause the motion we must use the principles of kinetics. When applying the necessary equations, it is important to first establish the inertial coordinate system and define the positive directions of the axes. The *directions* should be the *same* as those selected when writing any equations of kinematics provided *simultaneous solution* of equations becomes necessary.

Equations of Motion.   These equations are used to determine accelerated motions or forces causing the motion. If used to determine position, velocity, or time of motion, then kinematics will have to be considered for part of the solution. Before applying the equations of motion, *always draw a free-body diagram* in order to identify all the forces acting on the body. Also, establish the directions of the acceleration of the mass center and the angular acceleration of the body. (A kinetic diagram may also be drawn in order to represent $m\mathbf{a}_G$ and $I_G\boldsymbol{\alpha}$ graphically. This diagram is particularly convenient for resolving $m\mathbf{a}_G$ into components and for identifying the terms in the moment sum $\Sigma(\mathcal{M}_k)_P$.)

The three equations of motion are

$$\Sigma F_x = m(a_G)_x$$
$$\Sigma F_y = m(a_G)_y$$
$$\Sigma M_G = I_G\alpha \quad \text{or} \quad \Sigma M_P = \Sigma(\mathcal{M}_k)_P$$

In particular, if the body is *rotating about a fixed axis*, moments may also be summed about point O on the axis, in which case

$$\Sigma M_O = \Sigma(\mathcal{M}_k)_O = I_O\alpha$$

Work and Energy. *The equation of work and energy is used to solve problems involving force, velocity, and displacement.* Before applying this equation, *always draw a free-body diagram* of the body in order to identify the forces which do work. Recall that the kinetic energy of the body is due to translational motion of the mass center, $\mathbf{v}_G$, *and* rotational motion of the body, $\boldsymbol{\omega}$.

$$T_1 + \Sigma U_{1-2} = T_2$$

where

$$T = \tfrac{1}{2}mv_G^2 + \tfrac{1}{2}I_G\omega^2$$

$$U_F = \int F\cos\theta\, ds \qquad \text{(variable force)}$$

$$U_{F_c} = F_c\cos\theta(s_2 - s_1) \quad \text{(constant force)}$$
$$U_W = -W\,\Delta y \qquad\qquad \text{(weight)}$$
$$U_s = -(\tfrac{1}{2}ks_2^2 - \tfrac{1}{2}ks_1^2) \quad \text{(spring)}$$
$$U_M = M\theta \qquad\qquad\qquad \text{(constant couple moment)}$$

If the forces acting on the body are *conservative forces*, then apply the *conservation of energy equation.* This equation is easier to use than the equation of work and energy, since it applies only at *two points* on the path and *does not* require calculation of the work done by a force as the body moves along the path.

$$T_1 + V_1 = T_2 + V_2$$

where

$$V_g = Wy \quad \text{(gravitational potential energy)}$$
$$V_e = \tfrac{1}{2}ks^2 \text{ (elastic potential energy)}$$

Impulse and Momentum.   *The principles of linear and angular impulse and momentum are used to solve problems involving force, velocity, and time.* Before applying the equations, *draw a free-body diagram* in order to identify all the forces which cause linear and angular impulses on the body. Also, establish the directions of the velocity of the mass center and the angular velocity of the body just before and just after the impulses are applied. (As an alternative procedure, the impulse and momentum diagrams may accompany the solution in order to graphically account for the terms in the equations. These diagrams are particularly advantageous when computing the angular impulses and angular momenta about a point other than the body's mass center.)

$$m(\mathbf{v}_G)_1 + \Sigma \int \mathbf{F}\, dt = m(\mathbf{v}_G)_2$$

$$(\mathbf{H}_G)_1 + \Sigma \int \mathbf{M}_G\, dt = (\mathbf{H}_G)_2$$

or

$$(\mathbf{H}_O)_1 + \Sigma \int \mathbf{M}_O\, dt = (\mathbf{H}_O)_2$$

Conservation of Momentum.   If nonimpulsive forces or no impulsive forces act on the body in a particular direction, or if the motions of several bodies are involved in the problem, then consider applying the conservation of linear or angular momentum for the solution. Investigation of the free-body diagram (or the impulse diagram) will aid in determining the directions for which the impulsive forces are zero, or axes about which the impulsive forces cause zero angular momentum. For these cases,

$$m(\mathbf{v}_G)_1 = m(\mathbf{v}_G)_2$$

$$(\mathbf{H}_O)_1 = (\mathbf{H}_O)_2$$

The problems that follow involve application of all the above concepts. They are presented in *random order* so that practice may be gained at identifying the various types of problems and developing the skills necessary for their solution.

REVIEW PROBLEMS

R2-1. At a given instant, the wheel is rotating with the angular motions shown. Determine the acceleration of the collar at A at this instant.

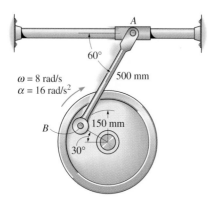

Prob. R2–1

R2-2. The hoisting gear A has an initial angular velocity of 60 rad/s and a constant deceleration of 1 rad/s². Determine the velocity and deceleration of the block which is being hoisted by the hub on gear B when $t = 3$ s.

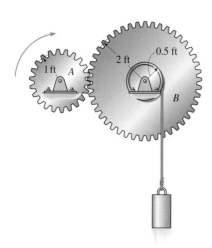

Prob. R2–2

R2-3. The rod is bent into the shape of a sine curve and is forced to rotate about the y axis by connecting the spindle S to a motor. If the rod starts from rest in the position shown and a motor drives it for a short time with an angular acceleration $\alpha = (1.5e^t)$ rad/s², where t is in seconds, determine the magnitudes of the angular velocity and angular displacement of the rod when $t = 3$ s. Locate the point on the rod which has the greatest velocity and acceleration, and compute the magnitudes of the velocity and acceleration of this point when $t = 3$ s. The curve defining the rod is $z = 0.25 \sin(\pi y)$, where the argument for the sine is given in radians when y is in meters.

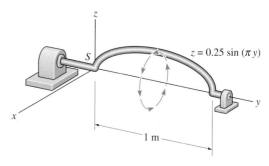

Prob. R2–3

*R2-4. A cord is wrapped around the inner spool of the gear. If it is pulled with a constant velocity v, determine the velocity and acceleration of points A and B. The gear rolls on the fixed gear rack.

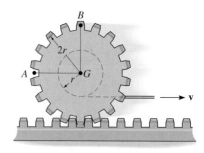

Prob. R2–4

R2-5. A 7-kg automobile tire is released from rest at A on the incline and rolls without slipping to point B, where it then travels in free flight. Determine the maximum height h the tire attains. The radius of gyration of the tire about its mass center is $k_G = 0.3$ m.

R2-7. The uniform connecting rod BC has a mass of 3 kg and is pin-connected at its end points. Determine the vertical forces which the pins exert on the ends B and C of the rod at the instant (a) $\theta = 0°$, and (b) $\theta = 90°$. The crank AB is turning with a constant angular velocity $\omega_{AB} = 5$ rad/s.

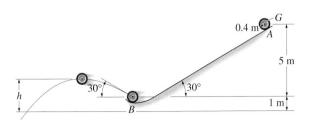

Prob. R2–5

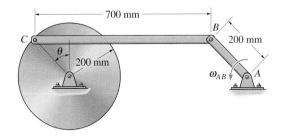

Prob. R2–7

R2-6. The link OA is pinned at O and rotates because of the sliding action of rod R along the horizontal groove. If R starts from rest when $\theta = 0°$ and has a constant acceleration $a_R = 60$ mm/s^2 to the right, determine the angular velocity and angular acceleration of OA when $t = 2$ s.

*R2-8. The tire has a mass of 9 kg and a radius of gyration $k_O = 225$ mm. If it is released from rest and rolls down the plane without slipping, determine the speed of its center O when $t = 3$ s.

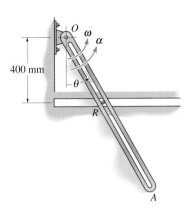

Prob. R2–6

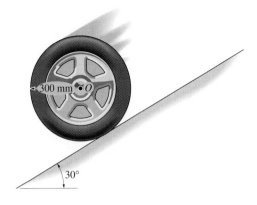

Prob. R2–8

R2-9. The double pendulum consists of two rods. Rod *AB* has a constant angular velocity of 3 rad/s, and rod *BC* has a constant angular velocity of 2 rad/s. Both of these absolute motions are measured counterclockwise. Determine the velocity and acceleration of point *C* at the instant shown.

R2-11. If the ball has a weight of 15 lb and is thrown onto a *rough surface* so that its center has a velocity of 6 ft/s parallel to the surface, determine the amount of backspin, ω, the ball must be given so that it stops spinning at the same instant that its forward velocity is zero. It is not necessary to know the coefficient of kinetic friction at *A* for the calculation.

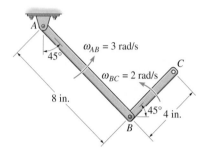

Prob. R2–9

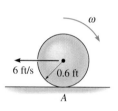

Prob. R2–11

R2-10. The spool and wire wrapped around its core have a mass of 20 kg and a centroidal radius of gyration $k_G = 250$ mm. If the coefficient of kinetic friction at the ground is $\mu_B = 0.1$, determine the angular acceleration of the spool when the 30-N · m couple moment is applied.

*R2-12. Blocks *A* and *B* weigh 50 and 10 lb, respectively. If *P* = 100 lb, determine the normal force exerted by block *A* on block *B*. Neglect friction and the weights of the pulleys, cord, and bars of the triangular frame.

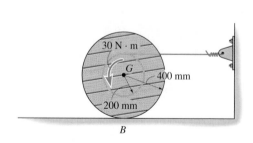

Prob. R2–10

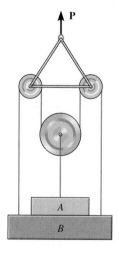

Prob. R2–12

R2-13. Determine the velocity and acceleration of rod R for any angle θ of cam C if the cam rotates with a constant angular velocity ω. The pin connection at O does not cause an interference with the motion of A on C.

R2-15. A tape having a thickness s wraps around the wheel which is turning at a constant rate ω. Assuming the unwrapped portion of tape remains horizontal, determine the acceleration of point P on the tape when the radius is r. *Hint:* Since $v_p = \omega r$, take the time derivative and note that $dr/dt = \omega(s/2\pi)$.

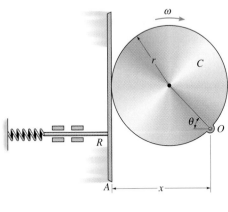

Prob. R2–13

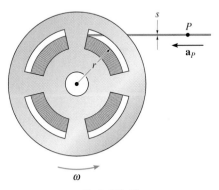

Prob. R2–15

R2-14. The uniform plate weighs 40 lb and is supported by a roller at A. If a horizontal force $F = 70$ lb is suddenly applied to the roller, determine the acceleration of the center of the roller at the instant the force is applied. The plate has a moment of inertia about its center of mass of $I_G = 0.414$ slug·ft². Neglect the weight of the roller.

*R2-16. The 15-lb cylinder is initially at rest on a 5-lb plate. If a couple moment $M = 40$ lb·ft is applied to the cylinder, determine the angular acceleration of the cylinder and the time needed for the end B of the plate to travel 3 ft and strike the wall. Assume the cylinder does not slip on the plate, and neglect the mass of the rollers under the plate.

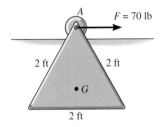

Prob. R2–14

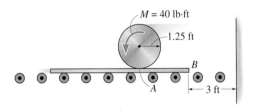

Prob. R2–16

R2-17. The wheel barrow and its contents have a mass of 40 kg and a mass center at G, excluding the wheel. The wheel has a mass of 2 kg and a radius of gyration $k_O = 0.120$ m. If the wheelbarrow is released from rest from the position shown, determine its speed after it travels 4 m down the incline. The coefficient of kinetic friction between the incline and A is $\mu_A = 0.3$. The wheels roll without slipping at B.

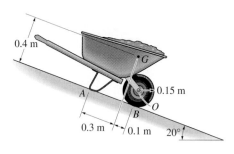

Prob. R2-17

R2-18. The drum of mass m, radius r, and radius of gyration k_O rolls along an inclined plane for which the coefficient of static friction is μ. If the drum is released from rest, determine the maximum angle θ for the incline so that it rolls without slipping.

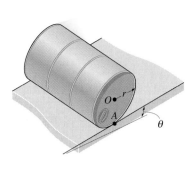

Prob. R2-18

R2-19. The 20-lb solid ball is cast on the floor such that it has a backspin $\omega = 15$ rad/s and its center has an initial horizontal velocity $v_G = 20$ ft/s. If the coefficient of kinetic friction between the floor and the ball is $\mu_A = 0.3$, determine the distance it travels before it stops spinning.

***R2-20.** Determine the backspin ω which should be given to the 20-lb ball so that when its center is given an initial horizontal velocity $v_G = 20$ ft/s it stops spinning and translating at the same instant. The coefficient of kinetic friction is $\mu_A = 0.3$.

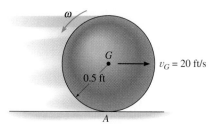

Probs. R2-19/20

R2-21. A 20-kg roll of paper, originally at rest, is pin-supported at its ends to bracket AB. The roll rests against a wall for which the coefficient of kinetic friction at C is $\mu_C = 0.3$. If a force of 40 N is applied uniformly to the end of the sheet, determine the initial angular acceleration of the roll and the tension in the bracket as the paper unwraps. For the calculation, treat the roll as a cylinder.

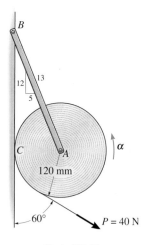

Prob. R2-21

R2-22. Compute the velocity of rod R for any angle θ of the cam C if the cam rotates with a constant angular velocity ω. The pin connection at O does not cause an interference with the motion of A on C.

*R2-24. The pendulum consists of a 30-lb sphere and a 10-lb slender rod. Compute the reaction at the pin O just after the cord AB is cut.

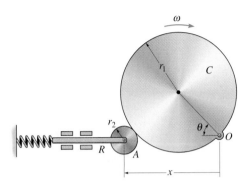

Prob. R2–22

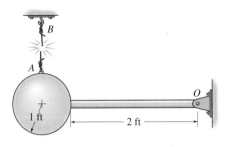

Prob. R2–24

R2-23. The assembly weighs 10 lb and has a radius of gyration $k_G = 0.6$ ft about its center of mass G. The kinetic energy of the assembly is 31 ft-lb when it is in the position shown. If it is rolling counterclockwise on the surface without slipping, determine its linear momentum at this instant.

R2-25. The board rests on the surface of two drums. At the instant shown, it has an acceleration of 0.5 m/s^2 to the right, while at the same instant points on the outer rim of each drum have an acceleration with a magnitude of 3 m/s^2. If the board does not slip on the drums, determine its speed due to the motion.

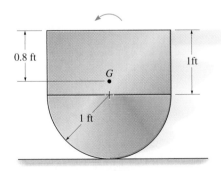

Prob. R2–23

Prob. R2–25

R2-26. The center of the pulley is being lifted vertically with an acceleration of 4 m/s² at the instant it has a velocity of 2 m/s. If the cable does not slip on the pulley's surface, determine the accelerations of the cylinder B and point C on the pulley.

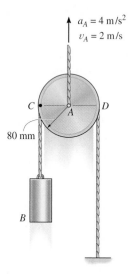

Prob. R2–26

R2-27. At the instant shown, two forces act on the 30-lb slender rod which is pinned at O. Determine the magnitude of force $\mathbf{F}$ and the initial angular acceleration of the rod so that the horizontal reaction which the *pin exerts on the rod* is 5 lb directed to the right.

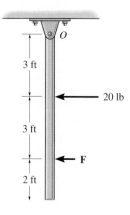

Prob. R2–27

***R-28.** The tub of the mixer has a weight of 70 lb and a radius of gyration $k_G = 1.3$ ft about its center of gravity. If a constant torque $M = 60$ lb·ft is applied to the dumping wheel, determine the angular velocity of the tub when it has rotated $\theta = 90°$. Originally the tub is at rest when $\theta = 0°$.

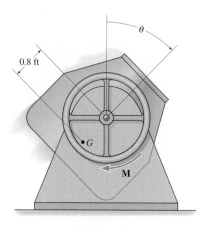

Prob. R2–28

R2-29. The spool has a weight of 30 lb and a radius of gyration $k_O = 0.65$ ft. If a force of 40 lb is applied to the cord at A, determine the angular velocity of the spool in $t = 3$ s starting from rest. Neglect the mass of the pulley and cord.

R2-30. Solve Prob. R2–29 if a 40-lb block is suspended from the cord at A, rather than applying the 40-lb force.

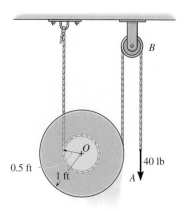

Probs. R2–29/30

R2-31. The dresser has a weight of 80 lb and is pushed along the floor. If the coefficient of static friction at A and B is $\mu_s = 0.3$ and the coefficient of kinetic friction is $\mu_k = 0.2$, determine the smallest horizontal force P needed to cause motion. If this force is increased slightly, determine the acceleration of the dresser. Also, what are the normal reactions at A and B when it begins to move?

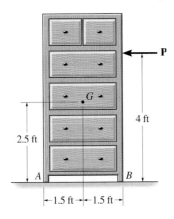

Prob. R2–31

*R2-32. When the crank on the Chinese windlass is turning, the rope on shaft A unwinds while that on shaft B winds up. Determine the speed at which the block lowers if the crank is turning with an angular velocity $\omega = 4$ rad/s. What is the angular velocity of the pulley at C? The rope segments on each side of the pulley are both parallel and vertical, and the rope does not slip on the pulley.

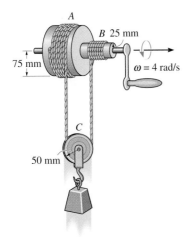

Prob. R2–32

R2-33. The semicircular disk has a mass of 50 kg and is released from rest from the position shown. The coefficients of static and kinetic friction between the disk and the beam are $\mu_s = 0.5$ and $\mu_k = 0.3$, respectively. Determine the initial reactions at the pin A and roller B, used to support the beam. Neglect the mass of the beam for the calculation.

R2-34. The semicircular disk has a mass of 50 kg and is released from rest from the position shown. The coefficients of static and kinetic friction between the disk and the beam are $\mu_s = 0.2$ and $\mu_k = 0.1$, respectively. Determine the initial reactions at the pin A and roller B used to support the beam. Neglect the mass of the beam for the calculation.

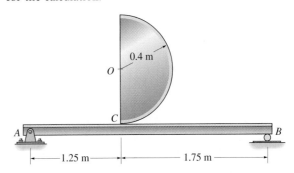

Probs. R2–33/34

R2-35. The cylinder having a mass of 5 kg is initially at rest when it is placed in contact with the wall B and the rotor at A. If the rotor always maintains a constant clockwise angular velocity $\omega = 6$ rad/s, determine the initial angular acceleration of the cylinder. The coefficient of kinetic friction at the contacting surfaces B and C is $\mu_k = 0.2$.

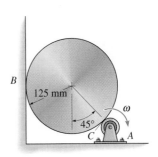

Prob. R2–35

*R2-36. The truck carries the 800-lb crate which has a center of gravity at G_c. Determine the largest acceleration of the truck so that the crate will not slip or tip on the truck bed. The coefficient of static friction between the crate and the truck is $\mu_s = 0.6$.

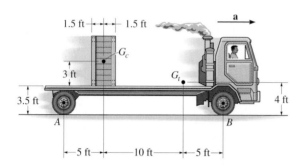

Prob. R2-36

R2-37. The truck has a weight of 8000 lb and center of gravity at G_t. It carries the 800-lb crate, which has a center of gravity at G_c. Determine the normal reaction at *each* of its four tires if it accelerates at $a = 0.5 \text{ ft/s}^2$. Also, what is the frictional force acting between the crate and the truck, and between *each* of the rear tires and the road? Assume that power is delivered only to the rear tires. The front tires are free to roll. Neglect the mass of the tires. The crate does not slip or tip on the truck.

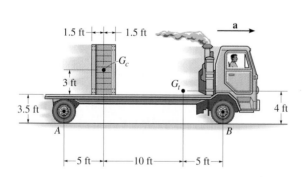

Prob. R2-37

R2-38. Spool B is at rest and spool A is rotating at 6 rad/s when the slack in the cord connecting them is taken up. Determine the angular velocity of each spool immediately after the cord is jerked tight. The spools A and B have weights and radii of gyration $W_A = 30$ lb, $k_A = 0.8$ ft and $W_B = 15$ lb, $k_B = 0.6$ ft, respectively.

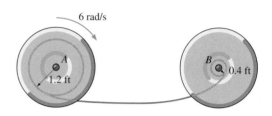

Prob. R2-38

R2-39. The two 3-lb rods EF and HI are fixed (welded) to the link AC at E. Determine the internal axial force E_x, shear force E_y, and moment M_E, which the bar AC exerts on FE at E if at the instant $\theta = 30°$ link AB has an angular velocity $\omega = 5$ rad/s and an angular acceleration $\alpha = 8$ rad/s^2 as shown.

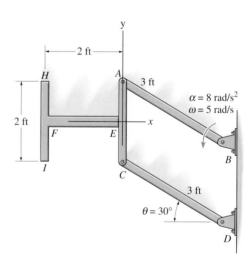

Prob. R2-39

*R2-40. The dragster has a mass of 1500 kg and a center of mass at G. If the coefficient of kinetic friction between the rear wheels and the pavement is $\mu_k = 0.6$, determine if it is possible for the driver to lift the front wheels, A, off the ground while the rear wheels are slipping. If so, what acceleration is necessary to do this? Neglect the mass of the wheels and assume that the front wheels are free to roll.

R2-42. The 1.6-Mg car shown has been "raked" by increasing the height $h = 0.2$ m of its center of mass. This was done by raising the springs on the rear axle. If the coefficient of kinetic friction between the rear wheels and the ground is $\mu_k = 0.3$, show that the car can accelerate slightly faster than its counterpart for which $h = 0$. Neglect the mass of the wheels and driver and assume the front wheels at B are free to roll while the rear wheels slip.

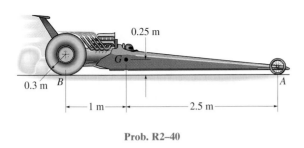

Prob. R2–40

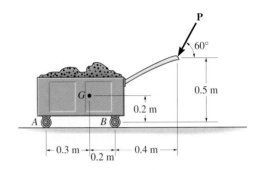

Prob. R2–42

R2-41. The dragster has a mass of 1500 kg and a center of mass at G. If no slipping occurs, determine the friction force F_B which must be applied to *each* of the rear wheels B in order to develop an acceleration $a = 6$ m/s^2. What are the normal reactions of *each* wheel on the ground? Neglect the mass of the wheels and assume that the front wheels are free to roll.

R2-43. The handcart has a mass of 200 kg and center of mass at G. Determine the normal reactions at *each* of the wheels at A and B if a force $P = 50$ N is applied to the handle. Neglect the mass and rolling resistance of the wheels.

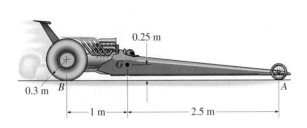

Prob. R2–41

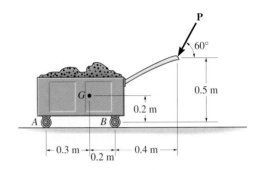

Prob. R2–43

*R2-44. If bar AB has an angular velocity $\omega_{AB} = 6$ rad/s, determine the velocity of the slider block C at the instant shown.

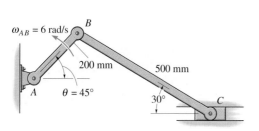

Prob. R2–44

R2-45. The disk is rotating at a constant rate $\omega = 4$ rad/s, and as it falls freely, its center has an acceleration of 32.2 ft/s². Determine the acceleration of points A and B on the rim of the disk at the instant shown.

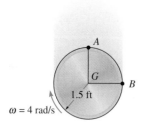

Prob. R2–45

R2-46. The 80-lb cylinder is attached to the 10-lb slender rod which is pinned from point A. At the instant $\theta = 30°$ it has an angular velocity $\omega_1 = 1$ rad/s as shown. Determine the largest angle θ to which the rod swings before it momentarily stops.

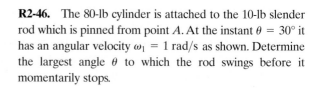

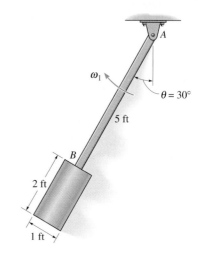

Prob. R2–46

R2-47. The bicycle and rider have a mass of 80 kg with center of mass located at G. If the coefficient of kinetic friction at the rear tire is $\mu_B = 0.8$, determine the normal reactions at the tires A and B, and the deceleration of the rider, when the rear wheel locks for braking. What is the normal reaction at the rear wheel when the bicycle is traveling at constant velocity and the brakes are not applied? Neglect the mass of the wheels.

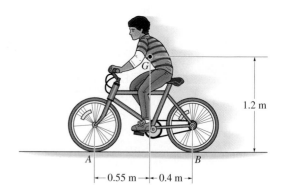

Prob. R2–47

*R2-48. At the instant shown, link AB has an angular velocity $\omega_{AB} = 2$ rad/s and an angular acceleration $\alpha_{AB} = 6$ rad/s². Determine the acceleration of the pin at C and the angular acceleration of link CB at this instant, when $\theta = 60°$.

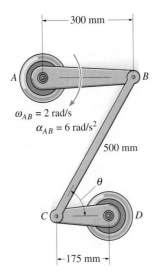

Prob. R2–48

R2-49. The spool has a mass of 60 kg and a radius of gyration $k_G = 0.3$ m. If it is released from rest, determine how far it descends down the smooth plane before it attains an angular velocity $\omega = 6$ rad/s. Neglect friction and the mass of the cord which is wound around the central core.

R2-50. Solve Prob. R2–49 if the plane is rough, such that the coefficient of kinetic friction at A is $\mu_A = 0.2$.

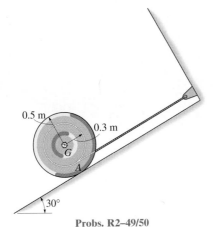

Probs. R2–49/50

R2-51. The gear rack has a mass of 6 kg, and the gears each have a mass of 4 kg and a radius of gyration $k = 30$ mm at their centers. If the rack is originally moving downward at 2 m/s, when $s = 0$, determine the speed of the rack when $s = 600$ mm. The gears are free to turn about their centers, A and B.

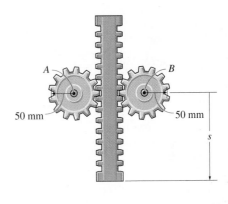

Prob. R2–51

*R2-52. The car has a mass of 1.50 Mg and a mass center at G. Determine the maximum acceleration it can have if (a) power is supplied only to the rear wheels, (b) power is supplied only to the front wheels. Neglect the mass of the wheels in the calculation, and assume that the wheels that do not receive power are free to roll. Also, assume that slipping of the powered wheels occurs, where the coefficient of kinetic friction is $\mu_k = 0.3$.

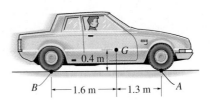

Prob. R2–52

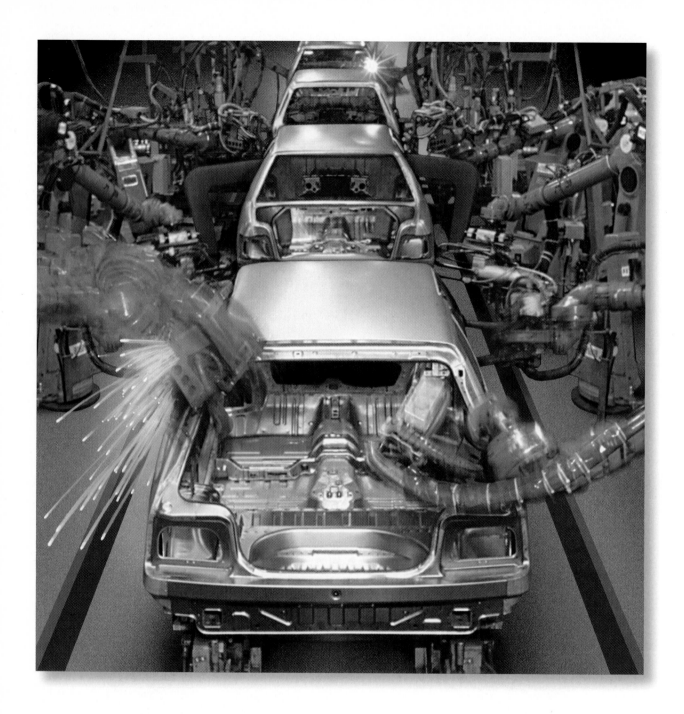

The three-dimensional motion of these industrial robots used in the manufacturing of automobiles must be accurately specified.

CHAPTER

20

Three-Dimensional Kinematics of a Rigid Body

CHAPTER OBJECTIVES

- To analyze the kinematics of a body subjected to rotation about a fixed axis and general plane motion.
- To provide a relative-motion analysis of a rigid body using translating and rotating axes.

20.1 Rotation About a Fixed Point

When a rigid body rotates about a fixed point, the distance r from the point to a particle P located on the body is the *same* for *any position* of the body. Thus, the path of motion for the particle lies on the *surface of a sphere* having a radius r and centered at the fixed point. Since motion along this path occurs only from a series of rotations made during a finite time interval, we will first develop a familiarity with some of the properties of rotational displacements.

The boom can rotate up and down, and because it is hinged at a point on the vertical axis about which it turns it is subjected to rotation about a fixed point.

Euler's Theorem. Euler's theorem states that two "component" rotations about different axes passing through a point are equivalent to a single resultant rotation about an axis passing through the point. If more than two rotations are applied, they can be combined into pairs, and each pair can be further reduced to combine into one rotation.

Finite Rotations. If component rotations used in Euler's theorem are *finite*, it is important that the *order* in which they are applied be maintained. This is because finite rotations do *not* obey the law of vector addition, and hence they cannot be classified as vector quantities. To show this, consider the two finite rotations $\boldsymbol{\theta}_1 + \boldsymbol{\theta}_2$ applied to the block in Fig. 20–1a. Each rotation has a magnitude of 90° and a direction defined by the right-hand rule, as indicated by the arrow. The resultant orientation of the block is shown at the right. When these two rotations are applied in the order $\boldsymbol{\theta}_2 + \boldsymbol{\theta}_1$, as shown in Fig. 20–1b, the resultant position of the block is *not* the same as it is in Fig. 20–1a. Consequently, *finite rotations* do not obey the commutative law of addition ($\boldsymbol{\theta}_1 + \boldsymbol{\theta}_2 \neq \boldsymbol{\theta}_2 + \boldsymbol{\theta}_1$), and therefore *they cannot be classified as vectors*. If smaller, yet finite, rotations had been used to illustrate this point, e.g., 10° instead of 90°, the *resultant* orientation of the block after each combination of rotations would also be different; however, in this case, the difference is only a small amount.

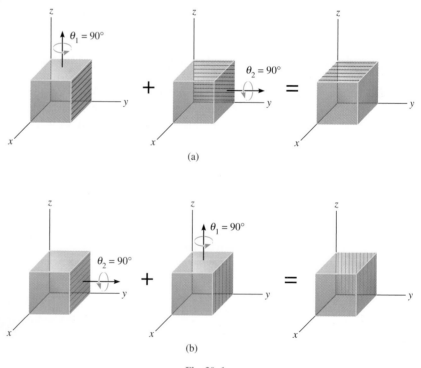

Fig. 20–1

Infinitesimal Rotations. When defining the angular motions of a body subjected to three-dimensional motion, only rotations which are *infinitesimally small* will be considered. *Such rotations may be classified as vectors, since they can be added vectorially in any manner.* To show this, let us for purposes of simplicity consider the rigid body itself to be a sphere which is allowed to rotate about its central fixed point O, Fig. 20–2a. If we impose two infinitesimal rotations $d\boldsymbol{\theta}_1 + d\boldsymbol{\theta}_2$ on the body, it is seen that point P moves along the path $d\boldsymbol{\theta}_1 \times \mathbf{r} + d\boldsymbol{\theta}_2 \times \mathbf{r}$ and ends up at P'. Had the two successive rotations occurred in the order $d\boldsymbol{\theta}_2 + d\boldsymbol{\theta}_1$, then the resultant displacements of P would have been $d\boldsymbol{\theta}_2 \times \mathbf{r} + d\boldsymbol{\theta}_1 \times \mathbf{r}$. Since the vector cross product obeys the distributive law, by comparison $(d\boldsymbol{\theta}_1 + d\boldsymbol{\theta}_2) \times \mathbf{r} = (d\boldsymbol{\theta}_2 + d\boldsymbol{\theta}_1) \times \mathbf{r}$. Here infinitesimal rotations $d\boldsymbol{\theta}$ are vectors, since these quantities have both a magnitude and direction for which the order of (vector) addition is not important, i.e., $d\boldsymbol{\theta}_1 + d\boldsymbol{\theta}_2 = d\boldsymbol{\theta}_2 + d\boldsymbol{\theta}_1$. Furthermore, as shown in Fig. 20–2a, the two "component" rotations $d\boldsymbol{\theta}_1$ and $d\boldsymbol{\theta}_2$ are equivalent to a single resultant rotation $d\boldsymbol{\theta} = d\boldsymbol{\theta}_1 + d\boldsymbol{\theta}_2$, a consequence of Euler's theorem.

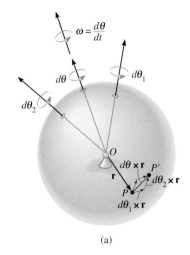

(a)

Angular Velocity. If the body is subjected to an angular rotation $d\boldsymbol{\theta}$ about a fixed point, the angular velocity of the body is defined by the time derivative,

$$\boldsymbol{\omega} = \dot{\boldsymbol{\theta}} \qquad (20\text{–}1)$$

The line specifying the direction of $\boldsymbol{\omega}$, which is collinear with $d\boldsymbol{\theta}$, is referred to as the *instantaneous axis of rotation*, Fig. 20–2b. In general, this axis changes direction during each instant of time. Since $d\boldsymbol{\theta}$ is a vector quantity, so too is $\boldsymbol{\omega}$, and it follows from vector addition that if the body is subjected to two component angular motions, $\boldsymbol{\omega}_1 = \dot{\boldsymbol{\theta}}_1$ and $\boldsymbol{\omega}_2 = \dot{\boldsymbol{\theta}}_2$, the resultant angular velocity is $\boldsymbol{\omega} = \boldsymbol{\omega}_1 + \boldsymbol{\omega}_2$.

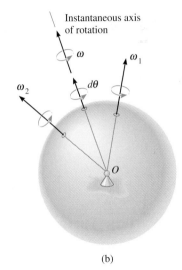

(b)

Fig. 20–2

Angular Acceleration. The body's angular acceleration is determined from the time derivative of the angular velocity, i.e.,

$$\boldsymbol{\alpha} = \dot{\boldsymbol{\omega}} \qquad (20\text{–}2)$$

For motion about a fixed point, $\boldsymbol{\alpha}$ must account for a change in *both* the magnitude and direction of $\boldsymbol{\omega}$, so that, in general, $\boldsymbol{\alpha}$ is not directed along the instantaneous axis of rotation, Fig. 20–3.

As the direction of the instantaneous axis of rotation (or the line of action of $\boldsymbol{\omega}$) changes in space, the locus of points defined by the axis generates a fixed *space cone*. If the change in this axis is viewed with respect to the rotating body, the locus of the axis generates a *body cone*,

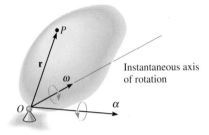

Fig. 20–3

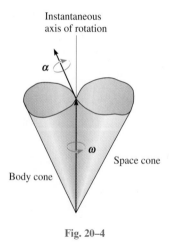

Fig. 20–4

Fig. 20–4. At any given instant, these cones are tangent along the instantaneous axis of rotation, and when the body is in motion, the body cone appears to roll either on the inside or the outside surface of the fixed space cone. Provided the paths defined by the open ends of the cones are described by the head of the ω vector, α must act tangent to these paths at any given instant, since the time rate of change of ω is equal to α. Fig. 20–4.

Velocity. Once ω is specified, the velocity of any point P on a body rotating about a fixed point can be determined using the same methods as for a body rotating about a fixed axis (Sec. 16.3). Hence, by the cross product,

$$\boxed{\mathbf{v} = \boldsymbol{\omega} \times \mathbf{r}} \tag{20–3}$$

Here $\mathbf{r}$ defines the position of P measured from the fixed point O, Fig. 20–3.

Acceleration. If ω and α are known at a given instant, the acceleration of any point P on the body can be obtained by time differentiation of Eq. 20–3, which yields

$$\boxed{\mathbf{a} = \boldsymbol{\alpha} \times \mathbf{r} + \boldsymbol{\omega} \times (\boldsymbol{\omega} \times \mathbf{r})} \tag{20–4}$$

The form of this equation is the same as that developed in Sec. 16.3, which defines the acceleration of a point located on a body subjected to rotation about a fixed axis.

*20.2 The Time Derivative of a Vector Measured from Either a Fixed or Translating-Rotating System

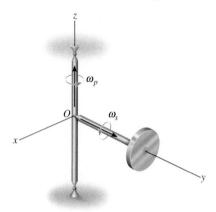

Fig. 20–5

In many types of problems involving the motion of a body about a fixed point, the angular velocity ω is specified in terms of its component angular motions. For example, the disk in Fig. 20–5 spins about the horizontal y axis at ω_s while it rotates or precesses about the vertical z axis at ω_p. Therefore, its resultant angular velocity is $\omega = \omega_s + \omega_p$. If the angular acceleration $\alpha = \dot{\omega}$ of such a body is to be determined, it is sometimes easier to compute the time derivative of ω by using a coordinate system which has a *rotation* defined by one or more of the components of ω.* For this reason, and for other uses later, an equation will presently be derived that relates the time derivative of any vector $\mathbf{A}$ defined from a translating-rotating reference to its time derivative defined from a fixed reference.

---

*In the case of the spinning disk, Fig. 20–5, the x, y, z axes may be given an angular velocity of ω_p.

Consider the x, y, z axes of the moving frame of reference to have an angular velocity Ω which is measured from the fixed X, Y, Z axes, Fig. 20–6a. In the following discussion, it will be convenient to express vector $\mathbf{A}$ in terms of its $\mathbf{i}$, $\mathbf{j}$, $\mathbf{k}$ components, which define the directions of the moving axes. Hence,

$$\mathbf{A} = A_x\mathbf{i} + A_y\mathbf{j} + A_z\mathbf{k}$$

In general, the time derivative of $\mathbf{A}$ must account for the change in both the vector's magnitude and direction. However, if this derivative is taken *with respect to the moving frame of reference*, only a change in the magnitudes of the components of $\mathbf{A}$ must be accounted for, since the directions of the components do not change with respect to the moving reference. Hence,

$$(\dot{\mathbf{A}})_{xyz} = \dot{A}_x\mathbf{i} + \dot{A}_y\mathbf{j} + \dot{A}_z\mathbf{k} \qquad (20\text{–}5)$$

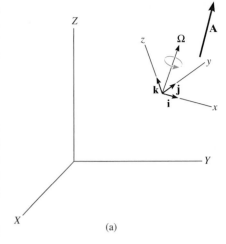

(a)

When the time derivative of $\mathbf{A}$ is taken *with respect to the fixed frame of reference*, the *directions* of $\mathbf{i}$, $\mathbf{j}$, and $\mathbf{k}$ change only on account of the *rotation* Ω of the axes and not their translation. Hence, in general,

$$\dot{\mathbf{A}} = \dot{A}_x\mathbf{i} + \dot{A}_y\mathbf{j} + \dot{A}_z\mathbf{k} + A_x\dot{\mathbf{i}} + A_y\dot{\mathbf{j}} + A_z\dot{\mathbf{k}}$$

The time derivatives of the unit vectors will now be considered. For example, $\dot{\mathbf{i}} = d\mathbf{i}/dt$ represents only a change in the *direction* of $\mathbf{i}$ with respect to time, since $\mathbf{i}$ has a fixed magnitude of 1 unit. As shown in Fig. 20–6b, the change, $d\mathbf{i}$, is *tangent to the path* described by the arrowhead of $\mathbf{i}$ as $\mathbf{i}$ moves due to the rotation Ω. Accounting for both the magnitude and direction of $d\mathbf{i}$, we can therefore define $\dot{\mathbf{i}}$ using the cross product, $\dot{\mathbf{i}} = \Omega \times \mathbf{i}$. In general,

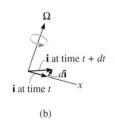

(b)

Fig. 20–6

$$\dot{\mathbf{i}} = \Omega \times \mathbf{i} \qquad \dot{\mathbf{j}} = \Omega \times \mathbf{j} \qquad \dot{\mathbf{k}} = \Omega \times \mathbf{k}$$

These formulations were also developed in Sec. 16.8, regarding planar motion of the axes. Substituting the results into the above equation and using Eq. 20–5 yields

$$\boxed{\dot{\mathbf{A}} = (\dot{\mathbf{A}})_{xyz} + \Omega \times \mathbf{A}} \qquad (20\text{–}6)$$

This result is rather important, and it will be used throughout Sec. 20.4 and Chapter 21. It states that the time derivative of *any vector* $\mathbf{A}$ as observed from the fixed X, Y, Z frame of reference is equal to the time rate of change of $\mathbf{A}$ as observed from the x, y, z translating-rotating frame of reference, Eq. 20–5, plus $\Omega \times \mathbf{A}$, the change of $\mathbf{A}$ caused by the rotation of the x, y, z frame. As a result, Eq. 20–6 should always be used whenever Ω produces a change in the direction of $\mathbf{A}$ as seen from the X, Y, Z reference. If this change does not occur, i.e., $\Omega = \mathbf{0}$, then $\dot{\mathbf{A}} = (\dot{\mathbf{A}})_{xyz}$, and so the time rate of change of $\mathbf{A}$ as observed from both coordinate systems will be the *same*.

E X A M P L E 20.1

The disk shown in Fig. 20–7a is spinning about its horizontal axis with a constant angular velocity $\omega_s = 3$ rad/s, while the horizontal platform on which the disk is mounted is rotating about the vertical axis at a constant rate $\omega_p = 1$ rad/s. Determine the angular acceleration of the disk and the velocity and acceleration of point A on the disk when it is in the position shown.

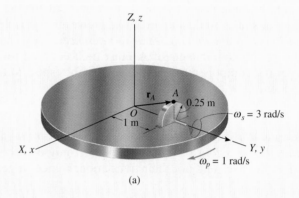

(a)

Fig. 20–7

Solution

Point O represents a fixed point of rotation for the disk if one considers a hypothetical extension of the disk to this point. To determine the velocity and acceleration of point A, it is first necessary to determine the resultant angular velocity ω and angular acceleration α of the disk, since these vectors are used in Eqs. 20–3 and 20–4.

Angular Velocity. The angular velocity, which is measured from X, Y, Z, is simply the vector addition of the two component motions. Thus,

$$\omega = \omega_s + \omega_p = \{3\mathbf{j} - 1\mathbf{k}\} \text{ rad/s}$$

At first glance, it may not appear that the disk is actually rotating with this angular velocity, since it is generally more difficult to imagine the resultant of angular motions in comparison with linear motions. To further understand the angular motion, consider the disk as being replaced by a cone (a body cone), which is rolling over the stationary space cone, Fig. 20–7b. The instantaneous axis of rotation is along the line of contact of the cones. This axis defines the direction of the resultant ω, which has components ω_s and ω_p.

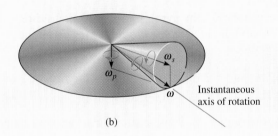

(b)

Angular Acceleration. Since the magnitude of $\boldsymbol{\omega}$ is constant, only a change in its direction, as seen from a fixed reference, creates the angular acceleration $\boldsymbol{\alpha}$ of the disk. One way to obtain $\boldsymbol{\alpha}$ is to compute the time derivative of *each of the two components* of $\boldsymbol{\omega}$ using Eq. 20–6. At the instant shown in Fig. 20–7a, imagine the fixed X, Y, Z and a rotating x, y, z frame to be coincident. If the rotating x, y, z frame is chosen to have an angular velocity of $\boldsymbol{\Omega} = \boldsymbol{\omega}_p = \{-1\mathbf{k}\}$ rad/s, then $\boldsymbol{\omega}_s$ will *always* be directed along the y (not Y) axis, and the time rate of change of $\boldsymbol{\omega}_s$ as seen from x, y, z is *zero*; i.e., $(\dot{\boldsymbol{\omega}}_s)_{xyz} = \mathbf{0}$ (the magnitude and direction of $\boldsymbol{\omega}_s$ is constant). Thus, by Eq. 20–6,

$$\dot{\boldsymbol{\omega}}_s = (\dot{\boldsymbol{\omega}}_s)_{xyz} + \boldsymbol{\omega}_p \times \boldsymbol{\omega}_s = \mathbf{0} + (-1\mathbf{k}) \times (3\mathbf{j}) = \{3\mathbf{i}\} \text{ rad/s}^2$$

By the same choice of axes rotation, $\boldsymbol{\Omega} = \boldsymbol{\omega}_p$, or even with $\boldsymbol{\Omega} = \mathbf{0}$, the time derivative $(\dot{\boldsymbol{\omega}}_p)_{xyz} = \mathbf{0}$, since $\boldsymbol{\omega}_p$ has a constant magnitude and direction. Hence,

$$\dot{\boldsymbol{\omega}}_p = (\dot{\boldsymbol{\omega}}_p)_{xyz} + \boldsymbol{\omega}_p \times \boldsymbol{\omega}_p = \mathbf{0} + \mathbf{0} = \mathbf{0}$$

The angular acceleration of the disk is therefore

$$\boldsymbol{\alpha} = \dot{\boldsymbol{\omega}} = \dot{\boldsymbol{\omega}}_s + \dot{\boldsymbol{\omega}}_p = \{3\mathbf{i}\} \text{ rad/s}^2 \qquad \textit{Ans.}$$

Velocity and Acceleration. Since $\boldsymbol{\omega}$ and $\boldsymbol{\alpha}$ have been determined, the velocity and acceleration of point A can be computed using Eqs. 20–3 and 20–4. Realizing that $\mathbf{r}_A = \{1\mathbf{j} + 0.25\mathbf{k}\}$ m, Fig. 20–7a, we have

$$\mathbf{v}_A = \boldsymbol{\omega} \times \mathbf{r}_A = (3\mathbf{j} - 1\mathbf{k}) \times (1\mathbf{j} + 0.25\mathbf{k}) = \{1.75\mathbf{i}\} \text{ m/s} \qquad \textit{Ans.}$$

$$\mathbf{a}_A = \boldsymbol{\alpha} \times \mathbf{r}_A + \boldsymbol{\omega} \times (\boldsymbol{\omega} \times \mathbf{r}_A)$$

$$= (3\mathbf{i}) \times (1\mathbf{j} + 0.25\mathbf{k}) + (3\mathbf{j} - 1\mathbf{k}) \times [(3\mathbf{j} - 1\mathbf{k}) \times (1\mathbf{j} + 0.25\mathbf{k})]$$

$$= \{-2.50\mathbf{j} - 2.25\mathbf{k}\} \text{ m/s}^2 \qquad \textit{Ans.}$$

E X A M P L E **20.2**

At the instant $\theta = 60°$, the gyrotop in Fig. 20–8 has three components of angular motion directed as shown and having magnitudes defined as:

spin: $\omega_s = 10$ rad/s, increasing at the rate of 6 rad/s^2

nutation: $\omega_n = 3$ rad/s, increasing at the rate of 2 rad/s^2

precession: $\omega_p = 5$ rad/s, increasing at the rate of 4 rad/s^2

Determine the angular velocity and angular acceleration of the top.

Solution

Angular Velocity. The top is rotating about the fixed point O. If the fixed and rotating frames are coincident at the instant shown, then the angular velocity can be expressed in terms of **i**, **j**, **k** components, appropriate to the x, y, z frame; i.e.,

$$\boldsymbol{\omega} = -\omega_n\mathbf{i} + \omega_s \sin\theta\,\mathbf{j} + (\omega_p + \omega_s\cos\theta)\mathbf{k}$$
$$= -3\mathbf{i} + 10\sin 60°\mathbf{j} + (5 + 10\cos 60°)\mathbf{k}$$
$$= \{-3\mathbf{i} + 8.66\mathbf{j} + 10\mathbf{k}\}\ \text{rad/s} \qquad Ans.$$

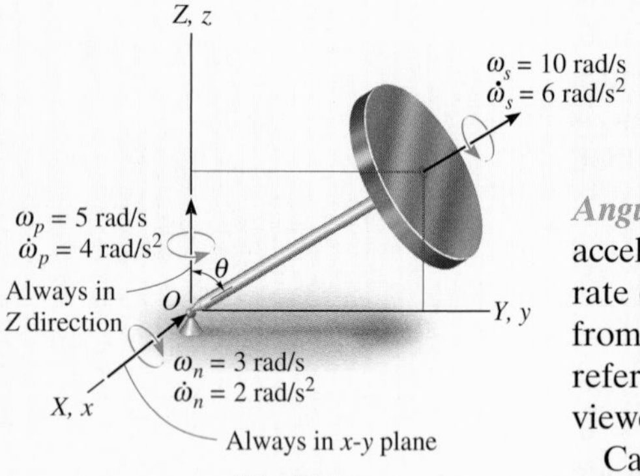

Angular Acceleration. As in the solution of Example 20.1, the angular acceleration $\boldsymbol{\alpha}$ will be determined by investigating separately the time rate of change of *each of the angular velocity components* as observed from the fixed X, Y, Z reference. We will choose an $\boldsymbol{\Omega}$ for the x, y, z reference so that the component of $\boldsymbol{\omega}$ which is being considered is viewed as having a *constant direction* when observed from x, y, z.

Careful examination of the motion of the top reveals that $\boldsymbol{\omega}_s$ has a *constant direction* relative to x, y, z if these axes rotate at $\boldsymbol{\Omega} = \boldsymbol{\omega}_n + \boldsymbol{\omega}_p$. Thus,

$$\dot{\boldsymbol{\omega}}_s = (\dot{\boldsymbol{\omega}}_s)_{xyz} + (\boldsymbol{\omega}_n + \boldsymbol{\omega}_p) \times \boldsymbol{\omega}_s$$
$$= (6\sin 60°\mathbf{j} + 6\cos 60°\mathbf{k}) + (-3\mathbf{i} + 5\mathbf{k}) \times (10\sin 60°\mathbf{j} + 10\cos 60°\mathbf{k})$$
$$= \{-43.30\mathbf{i} + 20.20\mathbf{j} - 22.98\mathbf{k}\}\ \text{rad/s}^2$$

Since $\boldsymbol{\omega}_n$ *always* lies in the fixed X-Y plane, this vector has a *constant direction* if the motion is viewed from axes x, y, z having a rotation of $\boldsymbol{\Omega} = \boldsymbol{\omega}_p$ (not $\boldsymbol{\Omega} = \boldsymbol{\omega}_s + \boldsymbol{\omega}_p$). Thus,

$$\dot{\boldsymbol{\omega}}_n = (\dot{\boldsymbol{\omega}}_n)_{xyz} + \boldsymbol{\omega}_p \times \boldsymbol{\omega}_n = -2\mathbf{i} + (5\mathbf{k}) \times (-3\mathbf{i}) = \{-2\mathbf{i} - 15\mathbf{j}\}\ \text{rad/s}^2$$

Finally, the component $\boldsymbol{\omega}_p$ is *always directed* along the Z axis so that here it is not necessary to think of x, y, z as rotating, i.e., $\boldsymbol{\Omega} = \mathbf{0}$. Expressing the data in terms of the **i**, **j**, **k** components, we therefore have

$$\dot{\boldsymbol{\omega}}_p = (\dot{\boldsymbol{\omega}}_p)_{xyz} + \mathbf{0} \times \boldsymbol{\omega}_p = \{4\mathbf{k}\}\ \text{rad/s}^2$$

Thus, the angular acceleration of the top is

$$\boldsymbol{\alpha} = \dot{\boldsymbol{\omega}}_s + \dot{\boldsymbol{\omega}}_n + \dot{\boldsymbol{\omega}}_p = \{-45.3\mathbf{i} + 5.20\mathbf{j} - 19.0\mathbf{k}\}\ \text{rad/s}^2 \quad Ans.$$

$\omega_s = 10$ rad/s
$\dot{\omega}_s = 6$ rad/s^2

$\omega_p = 5$ rad/s
$\dot{\omega}_p = 4$ rad/s^2
Always in Z direction

$\omega_n = 3$ rad/s
$\dot{\omega}_n = 2$ rad/s^2

Always in x-y plane

Fig. 20–8

20.3 General Motion

Shown in Fig. 20–9 is a rigid body subjected to general motion in three dimensions for which the angular velocity is $\boldsymbol{\omega}$ and the angular acceleration is $\boldsymbol{\alpha}$. If point A has a known motion of $\mathbf{v}_A$ and $\mathbf{a}_A$, the motion of any other point B may be determined by using a relative-motion analysis. In this section a *translating coordinate system* will be used to define the relative motion, and in the next section a reference that is both rotating and translating will be considered.

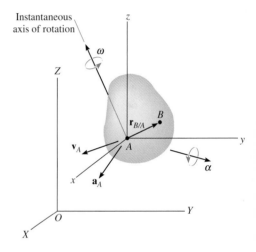

Fig. 20–9

If the origin of the translating coordinate system x, y, z $(\boldsymbol{\Omega} = \mathbf{0})$ is located at the "base point" A, then, at the instant shown, the motion of the body may be regarded as the sum of an instantaneous translation of the body having a motion of $\mathbf{v}_A$ and $\mathbf{a}_A$ and a rotation of the body about an instantaneous axis passing through the base point. Since the body is rigid, the motion of point B measured by an observer located at A is the same as *motion of the body about a fixed point*. This relative motion occurs about the instantaneous axis of rotation and is defined by $\mathbf{v}_{B/A} = \boldsymbol{\omega} \times \mathbf{r}_{B/A}$, Eq. 20–3, and $\mathbf{a}_{B/A} = \boldsymbol{\alpha} \times \mathbf{r}_{B/A} + \boldsymbol{\omega} \times (\boldsymbol{\omega} \times \mathbf{r}_{B/A})$, Eq. 20–4. For translating axes the relative motions are related to absolute motions by $\mathbf{v}_B = \mathbf{v}_A + \mathbf{v}_{B/A}$ and $\mathbf{a}_B = \mathbf{a}_A + \mathbf{a}_{B/A}$, Eqs. 16–15 and 16–17, so that the absolute velocity and acceleration of point B can be determined from the equations

$$\boxed{\mathbf{v}_B = \mathbf{v}_A + \boldsymbol{\omega} \times \mathbf{r}_{B/A}} \qquad (20\text{–}7)$$

and

$$\boxed{\mathbf{a}_B = \mathbf{a}_A + \boldsymbol{\alpha} \times \mathbf{r}_{B/A} + \boldsymbol{\omega} \times (\boldsymbol{\omega} \times \mathbf{r}_{B/A})} \qquad (20\text{–}8)$$

These two equations are identical to those describing the general plane motion of a rigid body, Eqs. 16–16 and 16–18. However, difficulty in application arises for three-dimensional motion, because $\boldsymbol{\alpha}$ measures the change in *both* the magnitude and direction of $\boldsymbol{\omega}$. (Recall that, for general plane motion, $\boldsymbol{\alpha}$ and $\boldsymbol{\omega}$ are always parallel or perpendicular to the plane of motion, and therefore $\boldsymbol{\alpha}$ measures only a change in the magnitude of $\boldsymbol{\omega}$.) In some problems the constraints or connections of a body will require that the directions of the angular motions or displacement paths of points on the body be defined. As illustrated in the following example, this information is useful for obtaining some of the terms in the above equations.

E X A M P L E   **20.3**

One end of the rigid bar CD shown in Fig. 20–10a slides along the horizontal member AB, and the other end slides along the vertical member EF. If the collar at C is moving towards B at a speed of 3 m/s, determine the velocity of the collar at D and the angular velocity of the bar at the instant shown. The bar is connected to the collars at its end points by ball-and-socket joints.

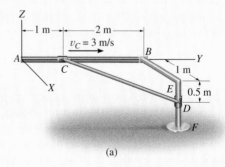

(a)

Solution

Bar CD is subjected to general motion. Why? The velocity of point D on the bar may be related to the velocity of point C by the equation

$$\mathbf{v}_D = \mathbf{v}_C + \boldsymbol{\omega} \times \mathbf{r}_{D/C}$$

The fixed and translating frames of reference are assumed to coincide at the instant considered, Fig. 20–10b. We have

$$\mathbf{v}_D = -v_D\mathbf{k} \qquad \mathbf{v}_C = \{3\mathbf{j}\} \text{ m/s}$$
$$\mathbf{r}_{D/C} = \{1\mathbf{i} + 2\mathbf{j} - 0.5\mathbf{k}\} \text{ m} \qquad \boldsymbol{\omega} = \omega_x\mathbf{i} + \omega_y\mathbf{j} + \omega_z\mathbf{k}$$

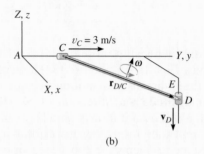

(b)

Fig. 20–10

Substituting these quantities into the above equation gives

$$-v_D\mathbf{k} = 3\mathbf{j} + \begin{vmatrix} \mathbf{i} & \mathbf{j} & \mathbf{k} \\ \omega_x & \omega_y & \omega_z \\ 1 & 2 & -0.5 \end{vmatrix}$$

Expanding and equating the respective $\mathbf{i}$, $\mathbf{j}$, $\mathbf{k}$ components yields

$$-0.5\omega_y - 2\omega_z = 0 \qquad (1)$$

$$0.5\omega_x + 1\omega_z + 3 = 0 \qquad (2)$$

$$2\omega_x - 1\omega_y + v_D = 0 \qquad (3)$$

These equations contain four unknowns.* A fourth equation can be written if the direction of $\boldsymbol{\omega}$ is specified. In particular, any component of $\boldsymbol{\omega}$ acting along the bar's axis has no effect on moving the collars. This is because the bar is *free to rotate* about its axis. Therefore, if $\boldsymbol{\omega}$ is specified as acting *perpendicular* to the axis of the bar, then $\boldsymbol{\omega}$ must have a unique magnitude to satisfy the above equations. Perpendicularity is guaranteed provided the dot product of $\boldsymbol{\omega}$ and $\mathbf{r}_{D/C}$ is zero (see Eq. C–14 of Appendix C). Hence,

$$\boldsymbol{\omega} \cdot \mathbf{r}_{D/C} = (\omega_x\mathbf{i} + \omega_y\mathbf{j} + \omega_z\mathbf{k}) \cdot (1\mathbf{i} + 2\mathbf{j} - 0.5\mathbf{k}) = 0$$

$$1\omega_x + 2\omega_y - 0.5\omega_z = 0 \qquad (4)$$

Solving Eqs. 1 through 4 simultaneously yields

$$\omega_x = -4.86 \text{ rad/s} \quad \omega_y = 2.29 \text{ rad/s} \quad \omega_z = -0.571 \text{ rad/s} \quad \textit{Ans.}$$

$$v_D = 12.0 \text{ m/s} \downarrow \qquad\qquad\qquad \textit{Ans.}$$

*Although this is the case, the magnitude of $\mathbf{v}_D$ can be obtained. For example, solve Eqs. 1 and 2 for ω_y and ω_x in terms of ω_z and substitute into Eq. 3. It will be noted that ω_z will *cancel out*, which will allow a solution for v_D.

*20.4 Relative-Motion Analysis Using Translating and Rotating Axes

The most general way to analyze the three-dimensional motion of a rigid body requires the use of x, y, z axes that both translate and rotate relative to a second frame X, Y, Z. This analysis will also provide a means for determining the motions of two points A and B located on separate members of a mechanism, and for determining the relative motion of one particle with respect to another when one or both particles are moving along *rotating paths*.

As shown in Fig. 20–11, the locations of points A and B are specified relative to the X, Y, Z frame of reference by position vectors $\mathbf{r}_A$ and $\mathbf{r}_B$. The base point A represents the origin of the x, y, z coordinate system, which is translating and rotating with respect to X, Y, Z. At the instant considered, the velocity and acceleration of point A are $\mathbf{v}_A$ and $\mathbf{a}_A$, respectively, and the angular velocity and angular acceleration of the x, y, z axes are $\mathbf{\Omega}$ and $\dot{\mathbf{\Omega}} = d\mathbf{\Omega}/dt$, respectively. All these vectors are *measured* with respect to the X, Y, Z frame of reference, although they may be expressed in Cartesian component form along either set of axes.

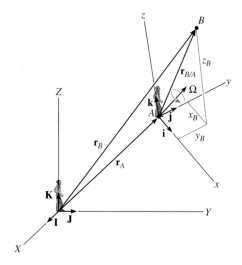

Fig. 20–11

Position. If the position of "B with respect to A" is specified by the *relative-position vector* $\mathbf{r}_{B/A}$, Fig. 20–11, then, by vector addition,

$$\boxed{\mathbf{r}_B = \mathbf{r}_A + \mathbf{r}_{B/A}} \qquad (20\text{–}9)$$

where

$\qquad \mathbf{r}_B$ = position of B

$\qquad \mathbf{r}_A$ = position of the origin A

$\qquad \mathbf{r}_{B/A}$ = relative position of "B with respect to A"

Velocity. The velocity of point B measured from X, Y, Z is determined by taking the time derivative of Eq. 20–9, which yields

$$\dot{\mathbf{r}}_B = \dot{\mathbf{r}}_A + \dot{\mathbf{r}}_{B/A}$$

The first two terms represent $\mathbf{v}_B$ and $\mathbf{v}_A$. The last term is evaluated by applying Eq. 20–6, since $\mathbf{r}_{B/A}$ is measured between two points in a rotating reference. Hence,

$$\dot{\mathbf{r}}_{B/A} = (\dot{\mathbf{r}}_{B/A})_{xyz} + \mathbf{\Omega} \times \mathbf{r}_{B/A} = (\mathbf{v}_{B/A})_{xyz} + \mathbf{\Omega} \times \mathbf{r}_{B/A} \quad (20\text{–}10)$$

Here $(\mathbf{v}_{B/A})_{xyz}$ is the relative velocity of B with respect to A measured from x, y, z. Thus,

$$\boxed{\mathbf{v}_B = \mathbf{v}_A + \mathbf{\Omega} \times \mathbf{r}_{B/A} + (\mathbf{v}_{B/A})_{xyz}} \qquad (20\text{–}11)$$

where

$\qquad \mathbf{v}_B$ = velocity of B

$\qquad \mathbf{v}_A$ = velocity of the origin A of the x, y, z frame of reference

$(\mathbf{v}_{B/A})_{xyz}$ = relative velocity of "B with respect to A" as measured by an observer attached to the rotating x, y, z frame of reference

$\qquad \mathbf{\Omega}$ = angular velocity of the x, y, z frame of reference

$\qquad \mathbf{r}_{B/A}$ = relative position of "B with respect to A"

Acceleration. The acceleration of point B measured from X, Y, Z is determined by taking the time derivative of Eq. 20–11, which yields

$$\dot{\mathbf{v}}_B = \dot{\mathbf{v}}_A + \dot{\boldsymbol{\Omega}} \times \mathbf{r}_{B/A} + \boldsymbol{\Omega} \times \dot{\mathbf{r}}_{B/A} + \frac{d}{dt}(\mathbf{v}_{B/A})_{xyz}$$

The time derivatives defined in the first and second terms represent $\mathbf{a}_B$ and $\mathbf{a}_A$ respectively. The fourth term is evaluated using Eq. 20–10, and the last term is evaluated by applying Eq. 20–6, which yields

$$\frac{d}{dt}(\mathbf{v}_{B/A})_{xyz} = (\dot{\mathbf{v}}_{B/A})_{xyz} + \boldsymbol{\Omega} \times (\mathbf{v}_{B/A})_{xyz}$$

$$= (\mathbf{a}_{B/A})_{xyz} + \boldsymbol{\Omega} \times (\mathbf{v}_{B/A})_{xyz}$$

Here $(\mathbf{a}_{B/A})_{xyz}$ is the relative acceleration of B with respect to A measured from x, y, z. Substituting this result and Eq. 20–10 into the above equation and simplifying, we have

$$\boxed{\mathbf{a}_B = \mathbf{a}_A + \dot{\boldsymbol{\Omega}} \times \mathbf{r}_{B/A} + \boldsymbol{\Omega} \times (\boldsymbol{\Omega} \times \mathbf{r}_{B/A}) + 2\boldsymbol{\Omega} \times (\mathbf{v}_{B/A})_{xyz} + (\mathbf{a}_{B/A})_{xyz}}$$

$$(20\text{–}12)$$

where

$$\mathbf{a}_B = \text{acceleration of } B$$

$$\mathbf{a}_A = \text{acceleration of the origin } A \text{ of the } x, y, z \text{ frame of reference}$$

$$(\mathbf{a}_{B/A})_{xyz}, (\mathbf{v}_{B/A})_{xyz} = \text{relative acceleration and relative velocity of "}B \text{ with respect to } A\text{" as measured by an observer attached to the rotating } x, y, z \text{ frame of reference}$$

$$\dot{\boldsymbol{\Omega}}, \boldsymbol{\Omega} = \text{angular acceleration and angular velocity of the } x, y, z \text{ frame of reference}$$

$$\mathbf{r}_{B/A} = \text{relative position of "}B \text{ with respect to } A\text{"}$$

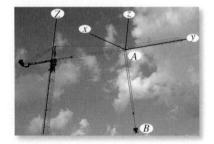

Complicated spatial motion of the concrete bucket B occurs due to the rotation of the boom about the Z axis, motion of the carriage A along the boom, and extension and swinging of the cable AB. A translating-rotating x, y, z coordinate system can be established on the carriage, and a relative-motion analysis can then be applied to study this motion.

Equations 20–11 and 20–12 are identical to those used in Sec. 16.8 for analyzing relative plane motion.* In that case, however, application is simplified since $\boldsymbol{\Omega}$ and $\dot{\boldsymbol{\Omega}}$ have a *constant direction* which is always perpendicular to the plane of motion. For three-dimensional motion, $\dot{\boldsymbol{\Omega}}$ must be computed by using Eq. 20–6, since $\dot{\boldsymbol{\Omega}}$ depends on the change in *both* the magnitude and direction of $\boldsymbol{\Omega}$.

*Refer to Sec. 16.8 for an interpretation of the terms.

PROCEDURE FOR ANALYSIS

Three-dimensional motion of particles or rigid bodies can be analyzed with Eqs. 20–11 and 20–12 by using the following procedure.

Coordinate Axes.

- Select the location and orientation of the X, Y, Z and x, y, z coordinate axes. Most often solutions are easily obtained if at the instant considered:

  (1) the origins are *coincident*
  (2) the axes are collinear
  (3) the axes are parallel

- If several components of angular velocity are involved in a problem, the calculations will be reduced if the x, y, z axes are selected such that only one component of angular velocity is observed in this frame (Ω_{xyz}) and the frame rotates with Ω defined by the other components of angular velocity.

Kinematic Equations.

- After the origin of the moving reference, A, is defined and the moving point B is specified, Eqs. 20–11 and 20–12 should be written in symbolic form as

$$\mathbf{v}_B = \mathbf{v}_A + \Omega \times \mathbf{r}_{B/A} + (\mathbf{v}_{B/A})_{xyz}$$

$$\mathbf{a}_B = \mathbf{a}_A + \dot{\Omega} \times \mathbf{r}_{B/A} + \Omega \times (\Omega \times \mathbf{r}_{B/A}) + 2\Omega \times (\mathbf{v}_{B/A})_{xyz} + (\mathbf{a}_{B/A})_{xyz}$$

- If $\mathbf{r}_A$ and Ω appear to *change direction* when observed from the fixed X, Y, Z reference use a set of primed reference axes, x', y', z' having $\Omega' = \Omega$ and Eq. 20–6 to determine $\dot{\Omega}$ and the motion $\mathbf{v}_A$ and $\mathbf{a}_A$ of the origin of the moving x, y, z axes.

- If $(\mathbf{r}_{B/A})_{xyz}$ and Ω_{xyz} appear to *change direction* as observed from x, y, z, then use a set of primed reference axes x', y', z' having $\Omega' = \Omega_{xyz}$ and Eq. 20–6 to determine $\dot{\Omega}_{xyz}$ and the relative motion $(\mathbf{v}_{B/A})_{xyz}$ and $(\mathbf{a}_{B/A})_{xyz}$.

- After the final forms of $\dot{\Omega}$, $\mathbf{v}_A$, $\mathbf{a}_A$, $\dot{\Omega}_{xyz}$, $(\mathbf{v}_{B/A})_{xyz}$, and $(\mathbf{a}_{B/A})_{xyz}$ are obtained, numerical problem data may be substituted and the kinematic terms evaluated. The components of all these vectors may be selected either along the X, Y, Z axes or along x, y, z. The choice is arbitrary, provided a consistent set of unit vectors is used.

E X A M P L E 20.4

A motor and attached rod AB have the angular motions shown in Fig. 20–12. A collar C on the rod is located 0.25 m from A and is moving downward along the rod with a velocity of 3 m/s and an acceleration of 2 m/s². Determine the velocity and acceleration of C at this instant.

Solution

Coordinate Axes. The origin of the fixed X, Y, Z reference is chosen at the center of the platform, and the origin of the moving, x, y, z frame at point A, Fig. 20–12. Since the collar is subjected to two components of angular motion, $\boldsymbol{\omega}_p$ and $\boldsymbol{\omega}_M$, it will be viewed as having an angular velocity of $\Omega_{xyz} = \boldsymbol{\omega}_M$ in x, y, z. The x, y, z axes will be attached to the platform so that $\Omega = \boldsymbol{\omega}_p$.

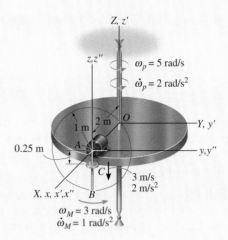

Fig. 20–12

Kinematic Equations.   Equations 20–11 and 20–12, applied to points C and A, become

$$\mathbf{v}_C = \mathbf{v}_A + \mathbf{\Omega} \times \mathbf{r}_{C/A} + (\mathbf{v}_{C/A})_{xyz}$$

$$\mathbf{a}_C = \mathbf{a}_A + \dot{\mathbf{\Omega}} \times \mathbf{r}_{C/A} + \mathbf{\Omega} \times (\mathbf{\Omega} \times \mathbf{r}_{C/A}) + 2\mathbf{\Omega} \times (\mathbf{v}_{C/A})_{xyz} + (\mathbf{a}_{C/A})_{xyz}$$

Motion of A.

Here $\mathbf{r}_A$ changes direction relative to XYZ.

To find the time derivatives of $\mathbf{r}_A$ we will use a set of x', y', z' axes coincident with the X, Y, Z axes that rotate at $\mathbf{\Omega}' = \mathbf{\Omega} = \boldsymbol{\omega}_p$. Thus

$\mathbf{\Omega} = \boldsymbol{\omega}_p = \{5\mathbf{k}\}$ rad/s ($\mathbf{\Omega}$ does not change direction relative to XYZ.)

$\dot{\mathbf{\Omega}} = \dot{\boldsymbol{\omega}}_p = \{2\mathbf{k}\}$ rad/s^2

$\mathbf{r}_A = \{2\mathbf{i}\}$ m

$\mathbf{v}_A = \dot{\mathbf{r}}_A = (\dot{\mathbf{r}}_A)_{x'y'z'} + \boldsymbol{\omega}_p \times \mathbf{r}_A = 0 + 5\mathbf{k} \times 2\mathbf{i} = \{10\mathbf{j}\}$ m/s

$\mathbf{a}_A = \ddot{\mathbf{r}}_A = [(\ddot{\mathbf{r}}_A)_{x'y'z'} + \boldsymbol{\omega}_p \times (\dot{\mathbf{r}}_A)_{x'y'z'}] + \dot{\boldsymbol{\omega}}_p \times \mathbf{r}_A + \boldsymbol{\omega}_p \times \dot{\mathbf{r}}_A$

$\qquad = [0 + 0] + 2\mathbf{k} \times 2\mathbf{i} + 5\mathbf{k} \times 10\mathbf{j} = \{-50\mathbf{i} + 4\mathbf{j}\}$ m/s^2

Motion of C with Respect to A.

Here $(\mathbf{r}_{C/A})_{xyz}$ changes direction relative to $x\,y\,z$. To find the time derivatives of $(\mathbf{r}_{C/A})_{xyz}$ use a set of x'', y'', z'' axes that rotate at $\mathbf{\Omega}'' = \mathbf{\Omega}_{xyz} = \boldsymbol{\omega}_M$. Thus

$\mathbf{\Omega}_{xyz} = \boldsymbol{\omega}_M = \{3\mathbf{i}\}$ rad/s ($\mathbf{\Omega}_{xyz}$ does not change direction relative to xyz.)

$\dot{\mathbf{\Omega}}_{xyz} = \dot{\boldsymbol{\omega}}_M = \{1\mathbf{i}\}$ rad/s^2

$(\mathbf{r}_{C/A})_{xyz} = \{-0.25\mathbf{k}\}$ m

$(\mathbf{v}_{C/A})_{xyz} = (\dot{\mathbf{r}}_{C/A})_{xyz} = (\dot{\mathbf{r}}_{C/A})_{xyz} + \boldsymbol{\omega}_M \times (\mathbf{r}_{C/A})_{xyz}$

$\qquad = -3\mathbf{k} + [3\mathbf{i} \times (-0.25\mathbf{k})] = \{0.75\mathbf{j} - 3\mathbf{k}\}$ m/s

$(\mathbf{a}_{C/A})_{xyz} = (\ddot{\mathbf{r}}_{C/A})_{xyz} = [(\ddot{\mathbf{r}}_{C/A})_{xyz} + \boldsymbol{\omega}_M \times (\dot{\mathbf{r}}_{C/A})_{xyz}] + \dot{\boldsymbol{\omega}}_M \times (\mathbf{r}_{C/A})_{xyz} + \boldsymbol{\omega}_M \times (\mathbf{r}_{C/A})_{xyz}$

$\qquad = [-2\mathbf{k} + 3\mathbf{i} \times (-3\mathbf{k})] + (1\mathbf{i}) \times (-0.25\mathbf{k}) + (3\mathbf{i}) \times (0.75\mathbf{j} - 3\mathbf{k})$

$\qquad = \{18.25\mathbf{j} + 0.25\mathbf{k}\}$ m/s^2

Motion of C.

$\mathbf{v}_C = \mathbf{v}_A + \mathbf{\Omega} \times \mathbf{r}_{C/A} + (\mathbf{v}_{C/A})_{xyz}$

$\qquad = 10\mathbf{j} + [5\mathbf{k} \times (-0.25\mathbf{k})] + (0.75\mathbf{j} - 3\mathbf{k})$

$\qquad = \{10.8\mathbf{j} - 3\mathbf{k}\}$ m/s *Ans.*

$\mathbf{a}_C = \mathbf{a}_A + \dot{\mathbf{\Omega}} \times \mathbf{r}_{C/A} + \mathbf{\Omega} \times (\mathbf{\Omega} \times \mathbf{r}_{C/A}) + 2\mathbf{\Omega} \times (\mathbf{v}_{C/A})_{xyz} + (\mathbf{a}_{C/A})_{xyz}$

$\qquad = (-50\mathbf{i} + 4\mathbf{j}) + [2\mathbf{k} \times (-0.25\mathbf{k})] + 5\mathbf{k} \times [5\mathbf{k} \times (-0.25\mathbf{k})]$

$\qquad + 2[5\mathbf{k} \times (0.75\mathbf{j} - 3\mathbf{k})] + (18.25\mathbf{j} + 0.25\mathbf{k})$

$\qquad = \{-57.5\mathbf{i} + 22.2\mathbf{j} + 0.25\mathbf{k}\}$ m/s^2 *Ans.*

EXAMPLE 20.5

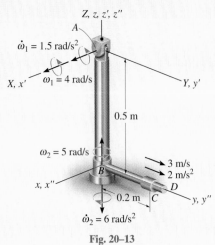

Fig. 20–13

The pendulum shown in Fig. 20–13 consists of two rods; AB is pin supported at A and swings only in the Y–Z plane, whereas a bearing at B allows the attached rod BD to spin about rod AB. At a given instant, the rods have the angular motions shown. Also, a collar C, located 0.2 m from B, has a velocity of 3 m/s and an acceleration of 2 m/s² along the rod. Determine the velocity and acceleration of the collar at this instant.

Solution I

Coordinate Axes. The origin of the fixed X, Y, Z frame will be chosen at A. Motion of the collar is conveniently observed from B, so the origin of the x, y, z frame is located at this point. We will choose $\mathbf{\Omega} = \boldsymbol{\omega}_1$ and $\mathbf{\Omega}_{xyz} = \boldsymbol{\omega}_2$.

Kinematic Equations.

$$\mathbf{v}_C = \mathbf{v}_B + \mathbf{\Omega} \times \mathbf{r}_{C/B} + (\mathbf{v}_{C/B})_{xyz}$$

$$\mathbf{a}_C = \mathbf{a}_B + \dot{\mathbf{\Omega}} \times \mathbf{r}_{C/B} + \mathbf{\Omega} \times (\mathbf{\Omega} \times \mathbf{r}_{C/B}) + 2\mathbf{\Omega} \times (\mathbf{v}_{C/B})_{xyz} + (\mathbf{a}_{C/B})_{xyz}$$

Motion of B. To find the time derivatives of $\mathbf{r}_B$ let the x', y', z' axes rotate with $\mathbf{\Omega} = \boldsymbol{\omega}_1$. Then

$$\mathbf{\Omega} = \boldsymbol{\omega}_1 = \{4\mathbf{i}\} \text{ rad/s} \qquad \dot{\mathbf{\Omega}} = \dot{\boldsymbol{\omega}}_1 = \{1.5\mathbf{i}\} \text{ rad/s}^2$$

$$\mathbf{r}_B = \{-0.5\mathbf{k}\} \text{ m}$$

$$\mathbf{v}_B = \dot{\mathbf{r}}_B = (\dot{\mathbf{r}}_B)_{x'y'z'} + \boldsymbol{\omega}_1 \times \mathbf{r}_B = 0 + 4\mathbf{i} \times (-0.5\mathbf{k}) = \{2\mathbf{j}\} \text{ m/s}$$

$$\mathbf{a}_B = \ddot{\mathbf{r}}_B = [(\ddot{\mathbf{r}}_B)_{x'y'z'} + \boldsymbol{\omega}_1 \times (\dot{\mathbf{r}}_B)_{x'y'z'}] + \dot{\boldsymbol{\omega}}_1 \times \mathbf{r}_B + \boldsymbol{\omega}_1 \times \dot{\mathbf{r}}_B$$

$$= [0 + 0] + 1.5\mathbf{i} \times (-0.5\mathbf{k}) + 4\mathbf{i} \times 2\mathbf{j} = \{0.75\mathbf{j} + 8\mathbf{k}\} \text{ m/s}^2$$

Motion of C with Respect to B.

To find the time derivatives of $(\mathbf{r}_{C/B})_{xyz}$, let the x'', y'', z'' axes rotate with $\mathbf{\Omega}_{xyz} = \boldsymbol{\omega}_2$. Then

$$\mathbf{\Omega}_{xyz} = \boldsymbol{\omega}_2 = \{5\mathbf{k}\} \text{ rad/s} \qquad \dot{\mathbf{\Omega}}_{xyx} = \dot{\boldsymbol{\omega}}_2 = \{-6\mathbf{k}\} \text{ rad/s}^2$$

$$(\mathbf{r}_{C/B})_{xyz} = \{0.2\mathbf{j}\} \text{ m}$$

$$(\mathbf{v}_{C/B})_{xyz} = (\dot{\mathbf{r}}_{C/B})_{xyz} = (\dot{\mathbf{r}}_{C/B})_{x''y''z''} + \boldsymbol{\omega}_2 \times (\mathbf{r}_{C/B})_{xyz} = 3\mathbf{j} + 5\mathbf{k} \times 0.2\mathbf{j} = \{-1\mathbf{i} + 3\mathbf{j}\} \text{ m/s}$$

$$(\mathbf{a}_{C/B})_{xyz} = (\ddot{\mathbf{r}}_{C/B})_{xyz} = [(\ddot{\mathbf{r}}_{C/B})_{x''y''z''} + \boldsymbol{\omega}_2 \times (\dot{\mathbf{r}}_{C/B})_{x''y''z''}] + \dot{\boldsymbol{\omega}}_2 \times (\mathbf{r}_{C/B})_{xyz} + \boldsymbol{\omega}_2 \times (\dot{\mathbf{r}}_{C/B})_{xyz}$$

$$= (2\mathbf{j} + 5\mathbf{k} \times 3\mathbf{j}) + (-6\mathbf{k} \times 0.2\mathbf{j}) + [5\mathbf{k} \times (-1\mathbf{i} + 3\mathbf{j})]$$

$$= \{-28.8\mathbf{i} - 3\mathbf{j}\} \text{ m/s}^2$$

Motion of C.

$$\mathbf{v}_C = \mathbf{v}_B + \mathbf{\Omega} \times \mathbf{r}_{C/B} + (\mathbf{v}_{C/B})_{xyz} = 2\mathbf{j} + 4\mathbf{i} \times 0.2\mathbf{j} + (-1\mathbf{i} + 3\mathbf{j})$$

$$= \{-1\mathbf{i} + 5\mathbf{j} + 0.8\mathbf{k}\} \text{ m/s} \qquad\qquad Ans.$$

$$\mathbf{a}_C = \mathbf{a}_B + \dot{\mathbf{\Omega}} \times \mathbf{r}_{C/B} + \mathbf{\Omega} \times (\mathbf{\Omega} \times \mathbf{r}_{C/B}) + 2\mathbf{\Omega} \times (\mathbf{v}_{C/B})_{xyz} + (\mathbf{a}_{C/B})_{xyz}$$

$$= (0.75\mathbf{j} + 8\mathbf{k}) + (1.5\mathbf{i} \times 0.2\mathbf{j}) + [4\mathbf{i} \times (4\mathbf{i} \times 0.2\mathbf{j})]$$

$$\quad + 2[4\mathbf{i} \times (-1\mathbf{i} + 3\mathbf{j})] + (-28.8\mathbf{i} - 3\mathbf{j})$$

$$= \{-28.8\mathbf{i} - 5.45\mathbf{j} + 32.3\mathbf{k}\} \text{ m/s}^2 \qquad\qquad Ans.$$

Solution II

Coordinate Axes. Here we will let the x, y, z axes rotate at

$$\boldsymbol{\Omega} = \boldsymbol{\omega}_1 + \boldsymbol{\omega}_2 = \{4\mathbf{i} + 5\mathbf{k}\} \text{ rad/s}$$

Then $\boldsymbol{\Omega}_{xyz} = \mathbf{0}$.

Motion of B.

From the constraints of the problem $\boldsymbol{\omega}_1$ does not change direction relative to X, Y, Z; however, the direction of $\boldsymbol{\omega}_2$ is changed by $\boldsymbol{\omega}_1$. Thus, to obtain $\dot{\boldsymbol{\Omega}}$ consider x', y', z' axes coincident with the X, Y, Z axes at A, such that $\dot{\boldsymbol{\Omega}} = \boldsymbol{\omega}_1$. Then taking the derivative of its components,

$$\dot{\boldsymbol{\Omega}} = \dot{\boldsymbol{\omega}}_1 + \dot{\boldsymbol{\omega}}_2 = [(\dot{\boldsymbol{\omega}}_1)_{x'y'z'} + \boldsymbol{\omega}_1 \times \boldsymbol{\omega}_1] + [(\dot{\boldsymbol{\omega}}_2)_{x'y'z'} + \boldsymbol{\omega}_1 \times \boldsymbol{\omega}_2]$$
$$= [1.5\mathbf{i}+0] + [-6\mathbf{k}+4\mathbf{i} \times 5\mathbf{k}] = \{1.5\mathbf{i}-20\mathbf{j}-6\mathbf{k}\} \text{ rad/s}^2$$

Also, $\boldsymbol{\omega}_1$ changes the direction of $\mathbf{r}_B$ so that the time derivatives of $\mathbf{r}_B$ can be computed using the primed axes defined above. Hence,

$$\mathbf{v}_B = \dot{\mathbf{r}}_B = (\dot{\mathbf{r}}_B)_{x'y'z'} + \boldsymbol{\omega}_1 \times \mathbf{r}_B$$
$$= 0 + 4\mathbf{i} \times (-0.5\mathbf{k}) = \{2\mathbf{j}\} \text{ m/s}$$
$$\mathbf{a}_B = \ddot{\mathbf{r}}_B = [(\ddot{\mathbf{r}}_B)_{x'y'z'} + \boldsymbol{\omega}_1 \times (\dot{\mathbf{r}}_B)_{x'y'z'}] + \dot{\boldsymbol{\omega}}_1 \times \mathbf{r}_B + \boldsymbol{\omega}_1 \times \dot{\mathbf{r}}_B$$

$$= [0 + 0]+1.5\mathbf{i}\times(-0.5\mathbf{k})+4\mathbf{i} \times 2\mathbf{j} = \{0.75\mathbf{j}+8\mathbf{k}\} \text{ m/s}^2$$

Motion of C with Respect to B.

$$\boldsymbol{\Omega}_{xyz} = \mathbf{0}$$
$$\dot{\boldsymbol{\Omega}}_{xyz} = \mathbf{0}$$
$$(\mathbf{r}_{C/B})_{xyz} = \{0.2\mathbf{j}\} \text{ m}$$
$$(\mathbf{v}_{C/B})_{xyz} = \{3\mathbf{j}\} \text{ m/s}$$
$$(\mathbf{a}_{C/B})_{xyz} = \{2\mathbf{j}\} \text{ m/s}^2$$

Motion of C.

$$\mathbf{v}_C = \mathbf{v}_B + \boldsymbol{\Omega} \times \mathbf{r}_{C/B} + (\mathbf{v}_{C/B})_{xyz}$$
$$= 2\mathbf{j} + [(4\mathbf{i} + 5\mathbf{k}) \times (0.2\mathbf{j})] + 3\mathbf{j}$$
$$= \{-1\mathbf{i} + 5\mathbf{j} + 0.8\mathbf{k}\} \text{ m/s} \qquad\qquad Ans.$$
$$\mathbf{a}_C = \mathbf{a}_B + \dot{\boldsymbol{\Omega}} \times \mathbf{r}_{C/B} + \boldsymbol{\Omega} \times (\boldsymbol{\Omega} \times \mathbf{r}_{C/B}) + 2\boldsymbol{\Omega} \times (\mathbf{v}_{C/B})_{xyz} + (\mathbf{a}_{C/B})_{xyz}$$
$$= (0.75\mathbf{j} + 8\mathbf{k}) + [(1.5\mathbf{i} - 20\mathbf{j} - 6\mathbf{k}) \times (0.2\mathbf{j})]$$
$$+ (4\mathbf{i} + 5\mathbf{k}) \times [(4\mathbf{i} + 5\mathbf{k}) \times 0.2\mathbf{j}] + 2[(4\mathbf{i} + 5\mathbf{k}) \times 3\mathbf{j}] + 2\mathbf{j}$$
$$= \{-28.8\mathbf{i} - 5.45\mathbf{j} + 32.3\mathbf{k}\} \text{ m/s}^2 \qquad\qquad Ans.$$

CHAPTER REVIEW

- *Rotation About a Fixed Point.* When a body rotates about a fixed point O, then points on the body follow a path that lies on the surface of a sphere. Infinitesimal rotations are vector quantities, whereas finite rotations are not.

Since the angular acceleration is a time rate of change in the angular velocity, then we must account for both the magnitude and directional changes of $\boldsymbol{\omega}$ when finding its derivative. To do this, the angular velocity is often specified in terms of its component motions, such that some of these components will remain constant relative to rotating x, y, z axes. If this is the case, then the time derivative relative to the fixed axis can be determined from Eq. 20–6, that is,

$$\dot{\mathbf{A}} = (\dot{\mathbf{A}})_{xyz} + \boldsymbol{\Omega} \times \mathbf{A}$$

Once $\boldsymbol{\omega}$ and $\boldsymbol{\alpha}$ are known, then the velocity and acceleration of point P can be determined from

$$\mathbf{a} = \boldsymbol{\alpha} \times \mathbf{r} + \boldsymbol{\omega} \times (\boldsymbol{\omega} \times \mathbf{r})$$

$$\mathbf{v} = \boldsymbol{\omega} \times \mathbf{r}$$

- *General Motion.* If the body undergoes general motion, then the motion of a point B in the body can be related to the motion of another point A using a relative motion analysis, along with translating axes at A. The relationships are

$$\mathbf{v}_B = \mathbf{v}_A + \boldsymbol{\omega} \times \mathbf{r}_{B/A}$$

$$\mathbf{a}_B = \mathbf{a}_A + \boldsymbol{\alpha} \times \mathbf{r}_{B/A} + \boldsymbol{\omega} \times (\boldsymbol{\omega} \times \mathbf{r}_{B/A})$$

- *Relative Motion Analysis Using Translating and Rotating Axes.* The motion of two points A and B on a body, a series of connected bodies, or located on two different paths, can be related using a relative motion analysis with rotating and translating axes at A. The relationships are

$$\mathbf{v}_B = \mathbf{v}_A + \boldsymbol{\Omega} \times \mathbf{r}_{B/A} + (\mathbf{v}_{B/A})_{xyz}$$

$$\mathbf{a}_B = \mathbf{a}_A + \dot{\boldsymbol{\Omega}} \times \mathbf{r}_{B/A} + \boldsymbol{\Omega} \times (\boldsymbol{\Omega} \times \mathbf{r}_{B/A}) + 2\boldsymbol{\Omega} \times (\mathbf{v}_{B/A})_{xyz} + (\mathbf{a}_{B/A})_{xyz}$$

When applying these equations, it is important to account for both the magnitude and directional changes of $\mathbf{r}_A, (\mathbf{r}_{B/A})_{xyz}, \boldsymbol{\Omega}$, and $\boldsymbol{\Omega}_{xyz}$ when taking their time derivatives to find $\mathbf{v}_A, \mathbf{a}_A, (\mathbf{v}_{B/A})_{xyz}, (\mathbf{a}_{B/A})_{xyz}$, and $\dot{\boldsymbol{\Omega}}$, and $\dot{\boldsymbol{\Omega}}_{xyz}$. To do this properly, one must use Eq. 20–6.

The design of amusement-park rides requires a force analysis that depends on
their three-dimensional motion.

Three-Dimensional Kinetics of a Rigid Body

- To introduce the methods for finding the moments of inertia and products of inertia of a body about various axes.
- To show how to apply the principles of work and energy and linear and angular momentum to a rigid body having three-dimensional motion.
- To develop and apply the equations of motion in three dimensions.
- To study the motion of a gyroscope and torque-free motion.

*21.1 Moments and Products of Inertia

When studying the planar kinetics of a body, it was necessary to introduce the moment of inertia I_G, which was computed about an axis perpendicular to the plane of motion and passing through the body's mass center G. For the kinetic analysis of three-dimensional motion it will sometimes be necessary to calculate six inertial quantities. These terms, called the moments and products of inertia, describe in a particular way the distribution of mass for a body relative to a given coordinate system that has a specified orientation and point of origin.

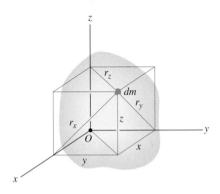

Fig. 21–1

Moment of Inertia. Consider the rigid body shown in Fig. 21–1. The *moment of inertia* for a differential element dm of the body about any one of the three coordinate axes is defined as the product of the mass of the element and the square of the shortest distance from the axis to the element. For example, as noted in the figure, $r_x = \sqrt{y^2 + z^2}$, so that the mass moment of inertia of dm about the x axis is

$$dI_{xx} = r_x^2 \, dm = (y^2 + z^2) \, dm$$

The moment of inertia I_{xx} for the body is determined by integrating this expression over the entire mass of the body. Hence, for each of the axes, we may write

$$
\begin{aligned}
I_{xx} &= \int_m r_x^2 \, dm = \int_m (y^2 + z^2) \, dm \\
I_{yy} &= \int_m r_y^2 \, dm = \int_m (x^2 + z^2) \, dm \\
I_{zz} &= \int_m r_z^2 \, dm = \int_m (x^2 + y^2) \, dm
\end{aligned}
\tag{21–1}
$$

Here it is seen that the moment of inertia is *always a positive quantity*, since it is the summation of the product of the mass dm, which is always positive, and the distances squared.

Product of Inertia. The *product of inertia* for a differential element dm is defined with respect to a set of *two orthogonal planes* as the product of the mass of the element and the perpendicular (or shortest) distances from the planes to the element. For example, this distance is x to the y–z plane and it is y to the x–z plane, Fig. 21–1. The product of inertia dI_{xy} for the element dm is therefore

$$dI_{xy} = xy \, dm$$

Note also that $dI_{yx} = dI_{xy}$. By integrating over the entire mass, the product of inertia of the body for each combination of planes may be expressed as

$$
\begin{aligned}
I_{xy} &= I_{yx} = \int_m xy \, dm \\
I_{yz} &= I_{zy} = \int_m yz \, dm \\
I_{xz} &= I_{zx} = \int_m xz \, dm
\end{aligned}
\tag{21–2}
$$

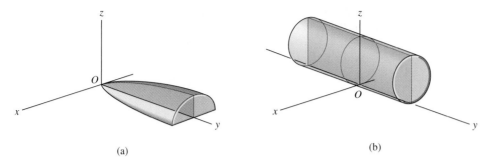

(a)

(b)

Fig. 21–2

Unlike the moment of inertia, which is always positive, the product of inertia may be positive, negative, or zero. The result depends on the signs of the two defining coordinates, which vary independently from one another. In particular, if either one or both of the orthogonal planes are *planes of symmetry* for the mass, the *product of inertia* with respect to these planes will be *zero*. In such cases, elements of mass will occur in *pairs* located on each side of the plane of symmetry. On one side of the plane the product of inertia for the element will be positive, while on the other side the product of inertia for the corresponding element will be negative, the sum therefore yielding zero. Examples of this are shown in Fig. 21–2. In the first case, Fig. 21–2a, the y–z plane is a plane of symmetry, and hence $I_{xy} = I_{xz} = 0$. Calculation of I_{yz} will yield a *positive* result, since all elements of mass are located using only positive y and z coordinates. For the cylinder, with the coordinate axes located as shown in Fig. 21–2b, the x–z and y–z planes are both planes of symmetry. Thus, $I_{xy} = I_{yz} = I_{zx} = 0$.

Parallel-Axis and Parallel-Plane Theorems.   The techniques of integration which are used to determine the moment of inertia of a body were described in Sec. 17.1. Also discussed were methods to determine the moment of inertia of a composite body, i.e., a body that is composed of simpler segments, as tabulated on the inside back cover. In both of these cases the *parallel-axis theorem* is often used for the calculations. This theorem, which was developed in Sec. 17.1, is used to transfer the moment of inertia of a body from an axis passing through its mass center G to a parallel axis passing through some other point. In this regard, if G has coordinates x_G, y_G, z_G defined from the x, y, z axes, Fig. 21–3, then

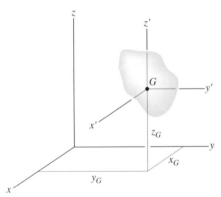

Fig. 21–3

the parallel-axis equations used to calculate the moments of inertia about the x, y, z axes are

$$
\begin{aligned}
I_{xx} &= (I_{x'x'})_G + m(y_G^2 + z_G^2) \\
I_{yy} &= (I_{y'y'})_G + m(x_G^2 + z_G^2) \\
I_{zz} &= (I_{z'z'})_G + m(x_G^2 + y_G^2)
\end{aligned}
\qquad (21\text{–}3)
$$

The products of inertia of a composite body are computed in the same manner as the body's moments of inertia. Here, however, the *parallel-plane theorem* is important. This theorem is used to transfer the products of inertia of the body from a set of three orthogonal planes passing through the body's mass center to a corresponding set of three parallel planes passing through some other point O. Defining the perpendicular distances between the planes as x_G, y_G and z_G, Fig. 21–3, the parallel-plane equations can be written as

$$
\begin{aligned}
I_{xy} &= (I_{x'y'})_G + mx_G y_G \\
I_{yz} &= (I_{y'z'})_G + my_G z_G \\
I_{zx} &= (I_{z'x'})_G + mz_G x_G
\end{aligned}
\qquad (21\text{–}4)
$$

The derivation of these formulas is similar to that given for the parallel-axis equation, Sec. 17.1.

Inertia Tensor. The inertial properties of a body are completely characterized by nine terms, six of which are independent of one another. This set of terms is defined using Eqs. 21–1 and 21–2 and can be written as

$$
\begin{pmatrix}
I_{xx} & -I_{xy} & -I_{xz} \\
-I_{yx} & I_{yy} & -I_{yz} \\
-I_{zx} & -I_{zy} & I_{zz}
\end{pmatrix}
$$

This array is called an *inertia tensor*. It has a unique set of values for a body when it is computed for each location of the origin O and orientation of the coordinate axes.

In general, for point O we can specify a unique axes inclination for which the products of inertia for the body are zero when computed with respect to these axes. When this is done, the inertia tensor is said to be "diagonalized" and may be written in the simplified form

$$
\begin{pmatrix}
I_x & 0 & 0 \\
0 & I_y & 0 \\
0 & 0 & I_z
\end{pmatrix}
$$

The dynamics of the space shuttle as it orbits the earth can be predicted only if its moments and products of inertia are known.

Here $I_x = I_{xx}$, $I_y = I_{yy}$, and $I_z = I_{zz}$ are termed the *principal moments of inertia* for the body, which are computed from the *principal axes of inertia*. Of these three principal moments of inertia, one will be a maximum and another a minimum of the body's moment of inertia.

Mathematical determination of the directions of principal axes of inertia will not be discussed here (see Prob. 21–21). There are many cases, however, in which the principal axes may be determined by inspection. From the previous discussion it was noted that if the coordinate axes are oriented such that *two* of the three orthogonal planes containing the axes are planes of *symmetry* for the body, then all the products of inertia for the body are zero with respect to the coordinate planes, and hence the coordinate axes are principal axes of inertia. For example, the x, y, z axes shown in Fig. 21–2b represent the principal axes of inertia for the cylinder at point O.

Moment of Inertia About an Arbitrary Axis.

Consider the body shown in Fig. 21–4, where the nine elements of the inertia tensor have been computed for the x, y, z axes having an origin at O. Here we wish to determine the moment of inertia of the body about the Oa axis, for which the direction is defined by the unit vector $\mathbf{u}_a$. By definition $I_{Oa} = \int b^2\, dm$, where b is the *perpendicular distance* from dm to Oa. If the position of dm is located using $\mathbf{r}$, then $b = r\sin\theta$, which represents the *magnitude* of the cross product $\mathbf{u}_a \times \mathbf{r}$. Hence, the moment of inertia can be expressed as

$$I_{Oa} = \int_m |(\mathbf{u}_a \times \mathbf{r})|^2\, dm = \int_m (\mathbf{u}_a \times \mathbf{r}) \cdot (\mathbf{u}_a \times \mathbf{r})\, dm$$

Provided $\mathbf{u}_a = u_x\mathbf{i} + u_y\mathbf{j} + u_z\mathbf{k}$ and $\mathbf{r} = x\mathbf{i} + y\mathbf{j} + z\mathbf{k}$, so that $\mathbf{u}_a \times \mathbf{r} = (u_y z - u_z y)\mathbf{i} + (u_z x - u_x z)\mathbf{j} + (u_x y - u_y x)\mathbf{k}$, then, after substituting and performing the dot-product operation, we can write the moment of inertia as

$$I_{Oa} = \int_m [(u_y z - u_z y)^2 + (u_z x - u_x z)^2 + (u_x y - u_y x)^2]\, dm$$

$$= u_x^2 \int_m (y^2 + z^2)\, dm + u_y^2 \int_m (z^2 + x^2)\, dm + u_z^2 \int_m (x^2 + y^2)\, dm$$

$$- 2u_x u_y \int_m xy\, dm - 2u_y u_z \int_m yz\, dm - 2u_z u_x \int_m zx\, dm$$

Recognizing the integrals to be the moments and products of inertia of the body, Eqs. 21–1 and 21–2, we have

$$\boxed{I_{Oa} = I_{xx}u_x^2 + I_{yy}u_y^2 + I_{zz}u_z^2 - 2I_{xy}u_x u_y - 2I_{yz}u_y u_z - 2I_{zx}u_z u_x} \quad (21\text{–}5)$$

Thus, if the inertia tensor is specified for the x, y, z axes, the moment of inertia of the body about the inclined Oa axis can be found by using Eq. 21–5. For the calculation, the direction cosines u_x, u_y, u_z of the axes must be determined. These terms specify the cosines of the coordinate direction angles α, β, γ made between the positive Oa axis and the positive x, y, z axes, respectively (see Appendix C).

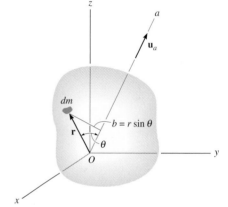

Fig. 21–4

E X A M P L E 21.1

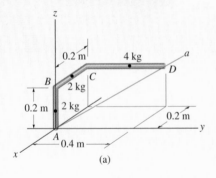

(a)

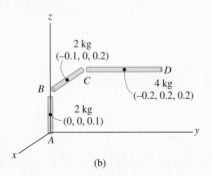

(b)

Fig. 21–5

Determine the moment of inertia of the bent rod shown in Fig. 21–5a about the Aa axis. The mass of each of the three segments is shown in the figure.

Solution

Before applying Eq. 21–5, it is first necessary to determine the moments and products of inertia of the rod about the x, y, z axes. This is done using the formula for the moment of inertia of a slender rod, $I = \frac{1}{12}ml^2$, and the parallel-axis and parallel-plane theorems, Eqs. 21–3 and 21–4. Dividing the rod into three parts and locating the mass center of each segment, Fig. 21–5b, we have

$$I_{xx} = [\tfrac{1}{12}(2)(0.2)^2 + 2(0.1)^2] + [0 + 2(0.2)^2]$$
$$+ [\tfrac{1}{12}(4)(0.4)^2 + 4((0.2)^2 + (0.2)^2)] = 0.480 \text{ kg} \cdot \text{m}^2$$
$$I_{yy} = [\tfrac{1}{12}(2)(0.2)^2 + 2(0.1)^2] + [\tfrac{1}{12}(2)(0.2)^2 + 2((-0.1)^2 + (0.2)^2)]$$
$$+ [0 + 4((-0.2)^2 + (0.2)^2)] = 0.453 \text{ kg} \cdot \text{m}^2$$
$$I_{zz} = [0 + 0] + [\tfrac{1}{12}(2)(0.2)^2 + 2(0.1)^2] + [\tfrac{1}{12}(4)(0.4)^2 + 4((-0.2)^2$$
$$+ (0.2)^2)] = 0.400 \text{ kg} \cdot \text{m}^2$$
$$I_{xy} = [0 + 0] + [0 + 0] + [0 + 4(-0.2)(0.2)] = -0.160 \text{ kg} \cdot \text{m}^2$$
$$I_{yz} = [0 + 0] + [0 + 0] + [0 + 4(0.2)(0.2)] = 0.160 \text{ kg} \cdot \text{m}^2$$
$$I_{zx} = [0 + 0] + [0 + 2(0.2)(-0.1)] + [0 + 4(0.2)(-0.2)] = -0.200 \text{ kg} \cdot \text{m}^2$$

The Aa axis is defined by the unit vector

$$\mathbf{u}_{Aa} = \frac{\mathbf{r}_D}{r_D} = \frac{-0.2\mathbf{i} + 0.4\mathbf{j} + 0.2\mathbf{k}}{\sqrt{(-0.2)^2 + (0.4)^2 + (0.2)^2}} = -0.408\mathbf{i} + 0.816\mathbf{j} + 0.408\mathbf{k}$$

Thus,

$$u_x = -0.408 \qquad u_y = 0.816 \qquad u_z = 0.408$$

Substituting these results into Eq. 21–5 yields

$$I_{Aa} = I_{xx}u_x^2 + I_{yy}u_y^2 + I_{zz}u_z^2 - 2I_{xy}u_xu_y - 2I_{yz}u_yu_z - 2I_{zx}u_zu_x$$
$$= 0.480(-0.408)^2 + (0.453)(0.816)^2 + 0.400(0.408)^2$$
$$- 2(-0.160)(-0.408)(0.816) - 2(0.160)(0.816)(0.408)$$
$$- 2(-0.200)(0.408)(-0.408)$$
$$= 0.169 \text{ kg} \cdot \text{m}^2 \qquad\qquad Ans.$$

*21.2 Angular Momentum

In this section we will develop the necessary equations used to determine the angular momentum of a rigid body about an arbitrary point. This formulation will provide a means for developing both the principle of impulse and momentum and the equations of rotational motion for a rigid body.

Consider the rigid body in Fig. 21–6, which has a mass m and center of mass at G. The X, Y, Z coordinate system represents an inertial frame of reference, and hence, its axes are fixed or translate with a constant velocity. The angular momentum as measured from this reference will be computed relative to the arbitrary point A. The position vectors $\mathbf{r}_A$ and $\boldsymbol{\rho}_A$ are drawn from the origin of coordinates to point A and from A to the ith particle of the body. If the particle's mass is m_i, the angular momentum about point A is

$$(\mathbf{H}_A)_i = \boldsymbol{\rho}_A \times m_i \mathbf{v}_i$$

where $\mathbf{v}_i$ represents the particle's velocity measured from the X, Y, Z coordinate system. If the body has an angular velocity $\boldsymbol{\omega}$ at the instant considered, $\mathbf{v}_i$ may be related to the velocity of A by applying Eq. 20–7, i.e.,

$$\mathbf{v}_i = \mathbf{v}_A + \boldsymbol{\omega} \times \boldsymbol{\rho}_A$$

Thus,

$$(\mathbf{H}_A)_i = \boldsymbol{\rho}_A \times m_i(\mathbf{v}_A + \boldsymbol{\omega} \times \boldsymbol{\rho}_A)$$
$$= (\boldsymbol{\rho}_A m_i) \times \mathbf{v}_A + \boldsymbol{\rho}_A \times (\boldsymbol{\omega} \times \boldsymbol{\rho}_A) m_i$$

Summing all the particles of the body requires an integration, and since $m_i \rightarrow dm$, we have

$$\mathbf{H}_A = \left(\int_m \boldsymbol{\rho}_A \, dm \right) \times \mathbf{v}_A + \int_m \boldsymbol{\rho}_A \times (\boldsymbol{\omega} \times \boldsymbol{\rho}_A) \, dm \qquad (21\text{–}6)$$

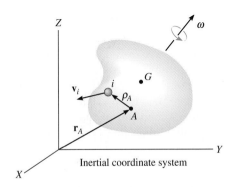

Inertial coordinate system

Fig. 21–6

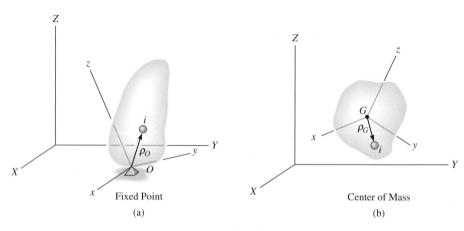

Fig. 21–7

Fixed Point O. If A becomes a *fixed point O* in the body, Fig. 21–7a, then $\mathbf{v}_A = \mathbf{0}$ and Eq. 21–6 reduces to

$$\mathbf{H}_O = \int_m \boldsymbol{\rho}_O \times (\boldsymbol{\omega} \times \boldsymbol{\rho}_O)\, dm \tag{21–7}$$

Center of Mass G. If A is located at the *center of mass G* of the body, Fig. 21–7b, then $\int_m \boldsymbol{\rho}_A\, dm = \mathbf{0}$ and

$$\mathbf{H}_G = \int_m \boldsymbol{\rho}_G \times (\boldsymbol{\omega} \times \boldsymbol{\rho}_G)\, dm \tag{21–8}$$

Arbitrary Point A. In general, A may be some point other than O or G, Fig. 21–7c, in which case Eq. 21–6 may nevertheless be simplified to the following form (see Prob. 21–22).

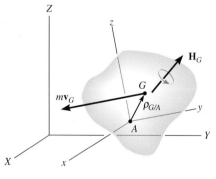

Arbitrary Point

(c)

$$\mathbf{H}_A = \boldsymbol{\rho}_{G/A} \times m\mathbf{v}_G + \mathbf{H}_G \tag{21–9}$$

Here the angular momentum consists of two parts—the moment of the linear momentum $m\mathbf{v}_G$ of the body about point A added (vectorially) to the angular momentum $\mathbf{H}_G$. Equation 21–9 may also be used for computing the angular momentum of the body about a fixed point O; the results, of course, will be the same as those computed using the more convenient Eq. 21–7.

Rectangular Components of H. To make practical use of Eqs. 21–7 through 21–9, the angular momentum must be expressed in terms of its scalar components. For this purpose, it is convenient to choose a second set of x, y, z axes having an arbitrary orientation relative to the X, Y, Z axes, Fig. 21–7, and for a general formulation, note that Eqs. 21–7 and 21–8 are both of the form

$$\mathbf{H} = \int_m \boldsymbol{\rho} \times (\boldsymbol{\omega} \times \boldsymbol{\rho})\, dm$$

Expressing $\mathbf{H}$, $\boldsymbol{\rho}$, and $\boldsymbol{\omega}$ in terms of x, y, and z components, we have

$$H_x\mathbf{i} + H_y\mathbf{j} + H_z\mathbf{k} = \int_m (x\mathbf{i} + y\mathbf{j} + z\mathbf{k}) \times [(\omega_x\mathbf{i} + \omega_y\mathbf{j} + \omega_z\mathbf{k})$$
$$\times (x\mathbf{i} + y\mathbf{j} + z\mathbf{k})]\, dm$$

Expanding the cross products and combining terms yields

$$H_x\mathbf{i} + H_y\mathbf{j} + H_z\mathbf{k} = \left[\omega_x \int_m (y^2 + z^2)\, dm - \omega_y \int_m xy\, dm - \omega_z \int_m xz\, dm\right]\mathbf{i}$$
$$+ \left[-\omega_x \int_m xy\, dm + \omega_y \int_m (x^2 + z^2)\, dm - \omega_z \int_m yz\, dm\right]\mathbf{j}$$
$$+ \left[-\omega_x \int_m zx\, dm - \omega_y \int_m yz\, dm + \omega_z \int_m (x^2 + y^2)\, dm\right]\mathbf{k}$$

Equating the respective $\mathbf{i}$, $\mathbf{j}$, $\mathbf{k}$ components and recognizing that the integrals represent the moments and products of inertia, we obtain

$$\begin{aligned}
H_x &= I_{xx}\omega_x - I_{xy}\omega_y - I_{xz}\omega_z \\
H_y &= -I_{yx}\omega_x + I_{yy}\omega_y - I_{yz}\omega_z \\
H_z &= -I_{zx}\omega_x - I_{zy}\omega_y + I_{zz}\omega_z
\end{aligned} \tag{21–10}$$

These three equations represent the scalar form of the $\mathbf{i}$, $\mathbf{j}$, $\mathbf{k}$ components of $\mathbf{H}_O$ or $\mathbf{H}_G$ (given in vector form by Eqs. 21–7 and 21–8). The angular momentum of the body about the arbitrary point A, other than the fixed point O or the center of mass G, may also be expressed in scalar form. Here it is necessary to use Eq. 21–9 to represent $\boldsymbol{\rho}_{G/A}$ and $\mathbf{v}_G$ as Cartesian vectors, carry out the cross-product operation, and substitute the components, Eqs. 21–10, for $\mathbf{H}_G$.

Equations 21–10 may be simplified further if the x, y, z coordinate axes are oriented such that they become *principal axes of inertia* for the body at the point. When these axes are used, the products of inertia $I_{xy} = I_{yz} = I_{zx} = 0$, and if the principal moments of inertia about the x, y, z axes are represented as $I_x = I_{xx}$, $I_y = I_{yy}$, and $I_z = I_{zz}$, the three components of angular momentum become

$$H_x = I_x\omega_x \qquad H_y = I_y\omega_y \qquad H_z = I_z\omega_z \tag{21–11}$$

The motion of the astronaut is controlled by use of small directional jets attached to his or her space suit. The impulses these jets provide must be carefully specified in order to prevent tumbling and loss of orientation.

Principle of Impulse and Momentum. Now that the formulation of the angular momentum for a body has been developed, the *principle of impulse and momentum*, as discussed in Sec. 19.2, may be used to solve kinetic problems which involve *force, velocity, and time*. For this case, the following two vector equations are available:

$$m(\mathbf{v}_G)_1 + \Sigma \int_{t_1}^{t_2} \mathbf{F}\, dt = m(\mathbf{v}_G)_2 \qquad (21\text{–}12)$$

$$(\mathbf{H}_O)_1 + \Sigma \int_{t_1}^{t_2} \mathbf{M}_O\, dt = (\mathbf{H}_O)_2 \qquad (21\text{–}13)$$

In three dimensions each vector term can be represented by three scalar components, and therefore a total of *six scalar equations* can be written. Three equations relate the linear impulse and momentum in the x, y, z directions, and the other three equations relate the body's angular impulse and momentum about the x, y, z axes. Before applying Eqs. 21–12 and 21–13 to the solution of problems, the material in Secs. 19.2 and 19.3 should be reviewed.

---

*21.3 Kinetic Energy

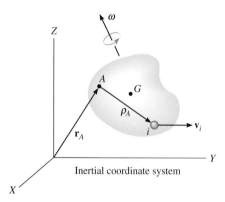

Fig. 21–8

In order to apply the principle of work and energy to the solution of problems involving general rigid body motion, it is first necessary to formulate expressions for the kinetic energy of the body. In this regard, consider the rigid body shown in Fig. 21–8, which has a mass m and center of mass at G. The kinetic energy of the ith particle of the body having a mass m_i and velocity $\mathbf{v}_i$, measured relative to the inertial X, Y, Z frame of reference, is

$$T_i = \tfrac{1}{2} m_i v_i^2 = \tfrac{1}{2} m_i (\mathbf{v}_i \cdot \mathbf{v}_i)$$

Provided the velocity of an arbitrary point A in the body is known, $\mathbf{v}_i$ may be related to $\mathbf{v}_A$ by the equation $\mathbf{v}_i = \mathbf{v}_A + \boldsymbol{\omega} \times \boldsymbol{\rho}_A$, where $\boldsymbol{\omega}$ is the angular velocity of the body, measured from the X, Y, Z coordinate system, and $\boldsymbol{\rho}_A$ is a position vector drawn from A to i. Using this expression for $\mathbf{v}_i$, the kinetic energy for the particle may be written as

$$T_i = \tfrac{1}{2} m_i (\mathbf{v}_A + \boldsymbol{\omega} \times \boldsymbol{\rho}_A) \cdot (\mathbf{v}_A + \boldsymbol{\omega} \times \boldsymbol{\rho}_A)$$
$$= \tfrac{1}{2}(\mathbf{v}_A \cdot \mathbf{v}_A)\, m_i + \mathbf{v}_A \cdot (\boldsymbol{\omega} \times \boldsymbol{\rho}_A)\, m_i + \tfrac{1}{2}(\boldsymbol{\omega} \times \boldsymbol{\rho}_A) \cdot (\boldsymbol{\omega} \times \boldsymbol{\rho}_A)\, m_i$$

The kinetic energy for the entire body is obtained by summing the kinetic energies of all the particles of the body. This requires an integration, and since $m_i \rightarrow dm$, we get

$$T = \tfrac{1}{2} m (\mathbf{v}_A \cdot \mathbf{v}_A) + \mathbf{v}_A \cdot \left(\boldsymbol{\omega} \times \int_m \boldsymbol{\rho}_A\, dm \right) + \tfrac{1}{2} \int_m (\boldsymbol{\omega} \times \boldsymbol{\rho}_A) \cdot (\boldsymbol{\omega} \times \boldsymbol{\rho}_A)\, dm$$

The last term on the right may be rewritten using the vector identity $\mathbf{a} \times \mathbf{b} \cdot \mathbf{c} = \mathbf{a} \cdot \mathbf{b} \times \mathbf{c}$, where $\mathbf{a} = \boldsymbol{\omega}$, $\mathbf{b} = \boldsymbol{\rho}_A$, and $\mathbf{c} = \boldsymbol{\omega} \times \boldsymbol{\rho}_A$. The final result is

$$T = \tfrac{1}{2}m(\mathbf{v}_A \cdot \mathbf{v}_A) + \mathbf{v}_A \cdot \left(\boldsymbol{\omega} \times \int_m \boldsymbol{\rho}_A \, dm \right)$$

$$+ \tfrac{1}{2}\boldsymbol{\omega} \cdot \int_m \boldsymbol{\rho}_A \times (\boldsymbol{\omega} \times \boldsymbol{\rho}_A) \, dm \quad (21\text{--}14)$$

This equation is rarely used because of the computations involving the integrals. Simplification occurs, however, if the reference point A is either a fixed point O or the center of mass G.

Fixed Point O. If A is a *fixed point* O in the body, Fig. 21–7a, then $\mathbf{v}_A = \mathbf{0}$, and using Eq. 21–7, we can express Eq. 21–14 as

$$T = \tfrac{1}{2}\boldsymbol{\omega} \cdot \mathbf{H}_O$$

If the x, y, z axes represent the principal axes of inertia for the body, then $\boldsymbol{\omega} = \omega_x\mathbf{i} + \omega_y\mathbf{j} + \omega_z\mathbf{k}$ and $\mathbf{H}_O = I_x\omega_x\mathbf{i} + I_y\omega_y\mathbf{j} + I_z\omega_z\mathbf{k}$. Substituting into the above equation and performing the dot product operations yields

$$T = \tfrac{1}{2}I_x\omega_x^2 + \tfrac{1}{2}I_y\omega_y^2 + \tfrac{1}{2}I_z\omega_z^2 \quad (21\text{--}15)$$

Center of Mass G. If A is located at the *center of mass* G of the body, Fig. 21–7b, then $\int \boldsymbol{\rho}_A \, dm = \mathbf{0}$ and, using Eq. 21–8, we can write Eq. 21–14 as

$$T = \tfrac{1}{2}mv_G^2 + \tfrac{1}{2}\boldsymbol{\omega} \cdot \mathbf{H}_G$$

In a manner similar to that for a fixed point, the last term on the right side may be represented in scalar form, in which case

$$T = \tfrac{1}{2}mv_G^2 + \tfrac{1}{2}I_x\omega_x^2 + \tfrac{1}{2}I_y\omega_y^2 + \tfrac{1}{2}I_z\omega_z^2 \quad (21\text{--}16)$$

Here it is seen that the kinetic energy consists of two parts; namely, the translational kinetic energy of the mass center, $\tfrac{1}{2}mv_G^2$, and the body's rotational kinetic energy.

Principle of Work and Energy. Having formulated the kinetic energy for a body, the *principle of work and energy* may be applied to solve kinetics problems which involve *force, velocity, and displacement*. For this case only one scalar equation can be written for each body, namely,

$$T_1 + \Sigma U_{1\text{--}2} = T_2 \quad (21\text{--}17)$$

Before applying this equation, the material in Chapter 18 should be reviewed.

EXAMPLE 21.2

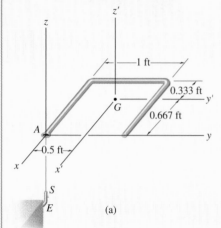

(a)

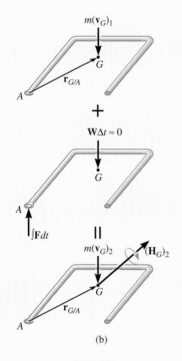

$m(\mathbf{v}_G)_1$

$\mathbf{r}_{G/A}$

A

$+$

$\mathbf{W}\Delta t \approx 0$

G

A

$\int \mathbf{F} dt$

$||$

$m(\mathbf{v}_G)_2$ $(\mathbf{H}_G)_2$

G

$\mathbf{r}_{G/A}$

A

(b)

Fig. 21–9

The rod in Fig. 21–9a has a weight of 1.5 lb/ft. Determine its angular velocity just after the end A falls onto the hook at E. The hook provides a permanent connection for the rod due to the spring-lock mechanism S. Just before striking the hook the rod is falling downward with a speed $(v_G)_1 = 10$ ft/s.

Solution

The principle of impulse and momentum will be used since impact occurs.

Impulse and Momentum Diagrams. Fig. 21–9b. During the short time Δt, the impulsive force $\mathbf{F}$ acting at A changes the momentum of the rod. (The impulse created by the rod's weight $\mathbf{W}$ during this time is small compared to $\int \mathbf{F} \, dt$, so that it is neglected, i.e., the weight is a nonimpulsive force.) Hence, the angular momentum of the rod is *conserved* about point A since the moment of $\int \mathbf{F} \, dt$ about A is zero.

Conservation of Angular Momentum. Equation 21–9 must be used for computing the angular momentum of the rod, since A does not become a *fixed point* until *after* the impulsive interaction with the hook. Thus, with reference to Fig. 21–9b, $(\mathbf{H}_A)_1 = (\mathbf{H}_A)_2$, or

$$\mathbf{r}_{G/A} \times m(\mathbf{v}_G)_1 = \mathbf{r}_{G/A} \times m(\mathbf{v}_G)_2 + (\mathbf{H}_G)_2 \qquad (1)$$

From Fig. 21–9a, $\mathbf{r}_{G/A} = \{-0.667\mathbf{i} + 0.5\mathbf{j}\}$ ft. Furthermore, the primed axes are principal axes of inertia for the rod because $I_{x'y'} = I_{x'z'} = I_{z'y'} = 0$. Hence, from Eqs. 21–11, $(\mathbf{H}_G)_2 = I_{x'}\omega_x\mathbf{i} + I_{y'}\omega_y\mathbf{j} + I_{z'}\omega_z\mathbf{k}$. The principal moments of inertia are $I_{x'} = 0.0272$ slug·ft^2, $I_{y'} = 0.0155$ slug·ft^2, $I_{z'} = 0.0427$ slug·ft^2 (see Prob. 21–13). Substituting into Eq. 1, we have

$$(-0.667\mathbf{i} + 0.5\mathbf{j}) \times \left[\left(\frac{4.5}{32.2}\right)(-10\mathbf{k})\right] = (-0.667\mathbf{i} + 0.5\mathbf{j}) \times \left[\left(\frac{4.5}{32.2}\right)(-v_G)_2\mathbf{k}\right]$$
$$+ 0.0272\omega_x\mathbf{i} + 0.0155\omega_y\mathbf{j} + 0.0427\omega_z\mathbf{k}$$

Expanding and equating the respective $\mathbf{i}$, $\mathbf{j}$, and $\mathbf{k}$ components yields

$$-0.699 = -0.0699(v_G)_2 + 0.0272\omega_x \qquad (2)$$
$$-0.932 = -0.0932(v_G)_2 + 0.0155\omega_y \qquad (3)$$
$$0 = 0.0427\omega_z \qquad (4)$$

Kinematics. There are four unknowns in the above equations; however, another equation may be obtained by relating $\boldsymbol{\omega}$ to $(\mathbf{v}_G)_2$ using *kinematics*. Since $\omega_z = 0$ (Eq. 4) and after impact the rod rotates about the fixed point A, Eq. 20–3 may be applied, in which case $(\mathbf{v}_G)_2 = \boldsymbol{\omega} \times \mathbf{r}_{G/A}$, or

$$-(v_G)_2\mathbf{k} = (\omega_x\mathbf{i} + \omega_y\mathbf{j}) \times (-0.667\mathbf{i} + 0.5\mathbf{j})$$
$$-(v_G)_2 = 0.5\omega_x + 0.667\omega_y \qquad (5)$$

Solving Eqs. 2 through 5 simultaneously yields

$$(\mathbf{v}_G)_2 = \{-8.41\mathbf{k}\} \text{ ft/s} \qquad \boldsymbol{\omega} = \{-4.09\mathbf{i} - 9.55\mathbf{j}\} \text{ rad/s} \quad \textit{Ans.}$$

EXAMPLE 21.3

A 5-N·m torque is applied to the vertical shaft CD shown in Fig. 21–10a, which allows the 10-kg gear A to turn freely about CE. Assuming that gear A starts from rest, determine the angular velocity of CD after it has turned two revolutions. Neglect the mass of shaft CD and axle CE and assume that gear A can be approximated by a thin disk. Gear B is fixed.

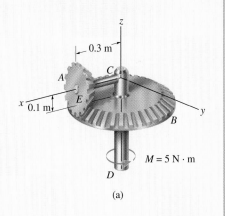

Solution

The principle of work and energy may be used for the solution. Why?

Work. If shaft CD, the axle CE, and gear A are considered as a system of connected bodies, only the applied torque $\mathbf{M}$ does work. For two revolutions of CD, this work is $\Sigma U_{1-2} = (5\,\text{N} \cdot \text{m})(4\pi\,\text{rad}) = 62.83\,\text{J}$.

Kinetic Energy. Since the gear is initially at rest, its initial kinetic energy is zero. A kinematic diagram for the gear is shown in Fig. 21–10b. If the angular velocity of CD is taken as $\boldsymbol{\omega}_{CD}$, then the angular velocity of gear A is $\boldsymbol{\omega}_A = \boldsymbol{\omega}_{CD} + \boldsymbol{\omega}_{CE}$. The gear may be imagined as a portion of a massless extended body which is rotating about the *fixed point C.* The instantaneous axis of rotation for this body is along line CH, because both points C and H on the body (gear) have zero velocity and must therefore lie on this axis. This requires that the components $\boldsymbol{\omega}_{CD}$ and $\boldsymbol{\omega}_{CE}$ be related by the equation $\omega_{CD}/0.1\,\text{m} = \omega_{CE}/0.3\,\text{m}$ or $\omega_{CE} = 3\omega_{CD}$. Thus,

$$\boldsymbol{\omega}_A = -\omega_{CE}\mathbf{i} + \omega_{CD}\mathbf{k} = -3\omega_{CD}\mathbf{i} + \omega_{CD}\mathbf{k} \qquad (1)$$

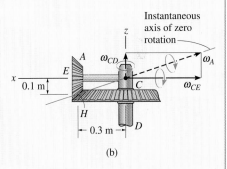

Fig. 21–10

The x, y, z axes in Fig. 21–10a represent *principal axes of inertia* at C for the gear. Since point C is a fixed point of rotation, Eq. 21–15 may be applied to determine the kinetic energy, i.e.,

$$T = \tfrac{1}{2}I_x\omega_x^2 + \tfrac{1}{2}I_y\omega_y^2 + \tfrac{1}{2}I_z\omega_z^2 \qquad (2)$$

Using the parallel-axis theorem, the moments of inertia of the gear about point C are as follows:

$$I_x = \tfrac{1}{2}(10\,\text{kg})(0.1\,\text{m})^2 = 0.05\,\text{kg} \cdot \text{m}^2$$

$$I_y = I_z = \tfrac{1}{4}(10\,\text{kg})(0.1\,\text{m})^2 + 10\,\text{kg}(0.3\,\text{m})^2 = 0.925\,\text{kg} \cdot \text{m}^2$$

Since $\omega_x = -3\omega_{CD}$, $\omega_y = 0$, $\omega_z = \omega_{CD}$, Eq. 2 becomes

$$T_A = \tfrac{1}{2}(0.05)(-3\omega_{CD})^2 + 0 + \tfrac{1}{2}(0.925)(\omega_{CD})^2 = 0.6875\omega_{CD}^2$$

Principle of Work and Energy. Applying the principle of work and energy, we obtain

$$T_1 + \Sigma U_{1-2} = T_2$$
$$0 + 62.83 = 0.6875\omega_{CD}^2$$
$$\omega_{CD} = 9.56\,\text{rad/s} \qquad\qquad Ans.$$

*21.4 Equations of Motion

Having become familiar with the techniques used to describe both the inertial properties and the angular momentum of a body, we can now write the equations which describe the motion of the body in their most useful forms.

Equations of Translational Motion. The *translational motion* of a body is defined in terms of the acceleration of the body's mass center, which is measured from an inertial X, Y, Z reference. The equation of translational motion for the body can be written in vector form as

$$\Sigma \mathbf{F} = m\mathbf{a}_G \qquad (21\text{-}18)$$

or by the three scalar equations

$$\boxed{\begin{aligned} \Sigma F_x &= m(a_G)_x \\ \Sigma F_y &= m(a_G)_y \\ \Sigma F_z &= m(a_G)_z \end{aligned}} \qquad (21\text{-}19)$$

Here, $\Sigma \mathbf{F} = \Sigma F_x \mathbf{i} + \Sigma F_y \mathbf{j} + \Sigma F_z \mathbf{k}$ represents the sum of all the external forces acting on the body.

Equations of Rotational Motion. In Sec. 15.6, we developed Eq. 15–17, namely,

$$\Sigma \mathbf{M}_O = \dot{\mathbf{H}}_O \qquad (21\text{-}20)$$

which states that the sum of the moments about a fixed point O of all the external forces acting on a system of particles (contained in a rigid body) is equal to the time rate of change of the total angular momentum of the body about point O. When moments of the external forces acting on the particles are summed about the system's *mass center G*, one again obtains the same simple form of Eq. 21–20, relating the moment summation $\Sigma \mathbf{M}_G$ to the angular momentum $\mathbf{H}_G$. To show this, consider the system of particles in Fig. 21–11, where X, Y, Z represents an inertial frame of reference and the x, y, z axes, with origin at G, *translate* with respect to this frame. In general, G is *accelerating*, so by definition the translating frame is *not* an inertial reference. The angular momentum of the ith particle with respect to this frame is, however,

$$(\mathbf{H}_i)_G = \mathbf{r}_{i/G} \times m_i \mathbf{v}_{i/G}$$

where $\mathbf{r}_{i/G}$ and $\mathbf{v}_{i/G}$ represent the relative position and relative velocity of the ith particle with respect to G. Taking the time derivative we have

$$(\dot{\mathbf{H}}_i)_G = \dot{\mathbf{r}}_{i/G} \times m_i \mathbf{v}_{i/G} + \mathbf{r}_{i/G} \times m_i \dot{\mathbf{v}}_{i/G}$$

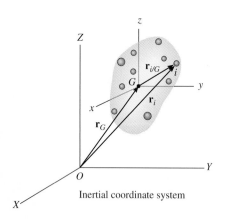

Inertial coordinate system

Fig. 21–11

By definition, $\mathbf{v}_{i/G} = \dot{\mathbf{r}}_{i/G}$. Thus, the first term on the right side is zero since the cross product of equal vectors is zero. Also, $\mathbf{a}_{i/G} = \dot{\mathbf{v}}_{i/G}$, so that

$$(\dot{\mathbf{H}}_i)_G = (\mathbf{r}_{i/G} \times m_i \mathbf{a}_{i/G})$$

Similar expressions can be written for the other particles of the body. When the results are summed, we get

$$\dot{\mathbf{H}}_G = \Sigma(\mathbf{r}_{i/G} \times m_i \mathbf{a}_{i/G})$$

Here $\dot{\mathbf{H}}_G$ is the time rate of change of the total angular momentum of the body computed relative to point G.

The relative acceleration for the ith particle is defined by the equation $\mathbf{a}_{i/G} = \mathbf{a}_i - \mathbf{a}_G$, where $\mathbf{a}_i$ and $\mathbf{a}_G$ represent, respectively, the accelerations of the ith particle and point G measured with respect to the *inertial frame of reference*. Substituting and expanding, using the distributive property of the vector cross product, yields

$$\dot{\mathbf{H}}_G = \Sigma(\mathbf{r}_{i/G} \times m_i \mathbf{a}_i) - (\Sigma m_i \mathbf{r}_{i/G}) \times \mathbf{a}_G$$

By definition of the mass center, the sum $(\Sigma m_i \mathbf{r}_{i/G}) = (\Sigma m_i)\bar{\mathbf{r}}$ is equal to zero, since the position vector $\bar{\mathbf{r}}$ relative to G is zero. Hence, the last term in the above equation is zero. Using the equation of motion, the product $m_i \mathbf{a}_i$ may be replaced by the resultant *external force* $\mathbf{F}_i$ acting on the ith particle. Denoting $\Sigma \mathbf{M}_G = (\Sigma \mathbf{r}_{i/G} \times \mathbf{F}_i)$, the final result may be written as

$$\Sigma \mathbf{M}_G = \dot{\mathbf{H}}_G \qquad (21\text{–}21)$$

The rotational equation of motion for the body will now be developed from either Eq. 21–20 or 21–21. In this regard, the scalar components of the angular momentum $\mathbf{H}_O$ or $\mathbf{H}_G$ are defined by Eqs. 21–10 or, if principal axes of inertia are used either at point O or G, by Eqs. 21–11. If these components are computed about x, y, z axes that are *rotating* with an angular velocity $\boldsymbol{\Omega}$, which may be *different* from the body's angular velocity $\boldsymbol{\omega}$, then the time derivative $\dot{\mathbf{H}} = d\mathbf{H}/dt$, as used in Eqs. 21–20 and 21–21, must account for the rotation of the x, y, z axes as measured from the inertial X, Y, Z axes. Hence, the time derivative of $\mathbf{H}$ must be determined from Eq. 20-6, in which case Eqs. 21–20 and 21–21 become

$$\Sigma \mathbf{M}_O = (\dot{\mathbf{H}}_O)_{xyz} + \boldsymbol{\Omega} \times \mathbf{H}_O$$
$$\Sigma \mathbf{M}_G = (\dot{\mathbf{H}}_G)_{xyz} + \boldsymbol{\Omega} \times \mathbf{H}_G \qquad (21\text{–}22)$$

Here $(\dot{\mathbf{H}})_{xyz}$ is the time rate of change of $\mathbf{H}$ measured from the x, y, z reference.

There are three ways in which one can define the motion of the x, y, z axes. Obviously, motion of this reference should be chosen to yield the simplest set of moment equations for the solution of a particular problem.

***x, y, z* Axes Having Motion $\Omega = 0$.** If the body has general motion, the *x, y, z* axes may be chosen with origin at *G*, such that the axes only *translate* relative to the inertial *X, Y, Z* frame of reference. Doing this would certainly simplify Eq. 21–22, since $\Omega = 0$. However, the body may have a rotation ω about these axes, and therefore the moments and products of inertia of the body would have to be expressed as *functions of time*. In most cases this would be a difficult task, so that such a choice of axes has restricted value.

***x, y, z* Axes Having Motion $\Omega = \omega$.** The *x, y, z* axes may be chosen such that they are *fixed in and move with the body*. The moments and products of inertia of the body relative to these axes will be *constant* during the motion. Since $\Omega = \omega$, Eqs. 21–22 become

$$\Sigma \mathbf{M}_O = (\dot{\mathbf{H}}_O)_{xyz} + \boldsymbol{\omega} \times \mathbf{H}_O$$

$$\Sigma \mathbf{M}_G = (\dot{\mathbf{H}}_G)_{xyz} + \boldsymbol{\omega} \times \mathbf{H}_G \tag{21–23}$$

We may express each of these vector equations as three scalar equations using Eqs. 21–10. Neglecting the subscripts *O* and *G* yields

$$\Sigma M_x = I_{xx}\dot{\omega}_x - (I_{yy} - I_{zz})\omega_y\omega_z - I_{xy}(\dot{\omega}_y - \omega_z\omega_x) - I_{yz}(\omega_y^2 - \omega_z^2)$$
$$- I_{zx}(\dot{\omega}_z + \omega_x\omega_y)$$

$$\Sigma M_y = I_{yy}\dot{\omega}_y - (I_{zz} - I_{xx})\omega_z\omega_x - I_{yz}(\dot{\omega}_z - \omega_x\omega_y) - I_{zx}(\omega_z^2 - \omega_x^2)$$
$$- I_{xy}(\dot{\omega}_x + \omega_y\omega_z) \tag{21–24}$$

$$\Sigma M_z = I_{zz}\dot{\omega}_z - (I_{xx} - I_{yy})\omega_x\omega_y - I_{zx}(\dot{\omega}_x - \omega_y\omega_z) - I_{xy}(\omega_x^2 - \omega_y^2)$$
$$- I_{yz}(\dot{\omega}_y + \omega_z\omega_x)$$

Notice that for a rigid body symmetric with respect to the x-y reference plane, and undergoing general plane motion in this plane, $I_{xz} = I_{yz} = 0$, and $\omega_x = \omega_y = d\omega_x/dt = d\omega_y/dt = 0$. Equations 21–24 reduce to the form $\Sigma M_x = \Sigma M_y = 0$, and $\Sigma M_z = I_{zz}\alpha_z$ (where $\alpha_z = \dot{\omega}_z$), which is essentially the third of Eqs. 17–16 or 17–17 depending on the choice of point *O* or *G* for summing moments.

If the *x, y,* and *z* axes are chosen as *principal axes of inertia*, the products of inertia are zero, $I_{xx} = I_x$, etc., and Eqs. 21–24 reduce to the from

$$\boxed{\begin{aligned} \Sigma M_x &= I_x\dot{\omega}_x - (I_y - I_z)\omega_y\omega_z \\ \Sigma M_y &= I_y\dot{\omega}_y - (I_z - I_x)\omega_z\omega_x \\ \Sigma M_z &= I_z\dot{\omega}_z - (I_x - I_y)\omega_x\omega_y \end{aligned}} \tag{21–25}$$

This set of equations is known historically as the *Euler equations of motion*, named after the Swiss mathematician Leonhard Euler, who first developed them. They apply *only* for moments summed about either point *O* or *G*.

When applying these equations it should be realized that $\dot{\omega}_x$, $\dot{\omega}_y$, $\dot{\omega}_z$ represent the time derivatives of the magnitudes of the x, y, z components of $\boldsymbol{\omega}$ as observed from x, y, z. Since the x, y, z axes are rotating at $\boldsymbol{\Omega} = \boldsymbol{\omega}$, then, from Eq. 20–6, it should be noted that $\dot{\boldsymbol{\omega}} = (\dot{\boldsymbol{\omega}})_{xyz} + \boldsymbol{\omega} \times \boldsymbol{\omega}$. Since $\boldsymbol{\omega} \times \boldsymbol{\omega} = \mathbf{0}$, $\dot{\boldsymbol{\omega}} = (\dot{\boldsymbol{\omega}})_{xyz}$. This important result indicates that the required time derivative of $\boldsymbol{\omega}$ can be obtained either by first finding the components of $\boldsymbol{\omega}$ along the x, y, z axes when these axes are oriented in a general position *and then* taking the time derivative of the magnitudes of these components, i.e., $(\dot{\boldsymbol{\omega}})_{xyz}$, or by finding the time derivative of $\boldsymbol{\omega}$ with respect to the X, Y, Z axes, i.e., $\dot{\boldsymbol{\omega}}$, and then determining the components $\dot{\omega}_x$, $\dot{\omega}_y$, $\dot{\omega}_z$. In practice, it is generally easier to compute $\dot{\omega}_x$, $\dot{\omega}_y$, $\dot{\omega}_z$ on the basis of finding $\dot{\boldsymbol{\omega}}$. See Example 21–5.

x, y, z Axes Having Motion $\Omega \neq \omega$.

To simplify the calculations for the time derivative of $\boldsymbol{\omega}$, it is often convenient to choose the x, y, z axes having an angular velocity $\boldsymbol{\Omega}$ which is different from the angular velocity $\boldsymbol{\omega}$ of the body. This is particularly suitable for the analysis of spinning tops and gyroscopes which are *symmetrical* about their spinning axes.* When this is the case, the moments and products of inertia remain constant during the motion.

Equations 21–22 are applicable for such a set of chosen axes. Each of these two vector equations may be reduced to a set of three scalar equations which are derived in a manner similar to Eqs. 21–25, i.e.,

$$\begin{aligned}
\Sigma M_x &= I_x \dot{\omega}_x - I_y \Omega_z \omega_y + I_z \Omega_y \omega_z \\
\Sigma M_y &= I_y \dot{\omega}_y - I_z \Omega_x \omega_z + I_x \Omega_z \omega_x \\
\Sigma M_z &= I_z \dot{\omega}_z - I_x \Omega_y \omega_x + I_y \Omega_x \omega_y
\end{aligned} \tag{21–26}$$

Here Ω_x, Ω_y, Ω_z represent the x, y, z components of $\boldsymbol{\Omega}$, measured from the inertial frame of reference, and $\dot{\omega}_x$, $\dot{\omega}_y$, $\dot{\omega}_z$ must be determined relative to the x, y, z axes that have the rotation $\boldsymbol{\Omega}$. See Example 21–6.

Any one of these sets of moment equations, Eqs. 21–24, 21–25, or 21–26, represents a series of three first-order nonlinear differential equations. These equations are "coupled," since the angular-velocity components are present in all the terms. Success in determining the solution for a particular problem therefore depends upon what is unknown in these equations. Difficulty certainly arises when one attempts to solve for the unknown components of $\boldsymbol{\omega}$, given the external moments as functions of time. Further complications can arise if the moment equations are coupled to the three scalar equations of translational motion, Eqs. 21–19. This can happen because of the existence of kinematic constraints which

---

*A detailed discussion of such devices is given in Sec. 21.5.

relate the rotation of the body to the translation of its mass center, as in the case of a hoop which rolls without slipping. Problems necessitating the simultaneous solution of differential equations generally require application of numerical methods with the aid of a computer. In many engineering problems, however, one is required to determine the applied moments acting on the body, given information about the motion of the body. Fortunately, many of these types of problems have direct solutions, so that there is no need to resort to computer techniques.

PROCEDURE FOR ANALYSIS

Problems involving the three-dimensional motion of a rigid body can be solved using the following procedure.

Free-Body Diagram

- Draw a *free-body diagram* of the body at the instant considered and specify the x, y, z coordinate system. The origin of this reference must be located either at the body's mass center G, or at point O, considered fixed in an inertial reference frame and located either in the body or on a massless extension of the body.

- Unknown reactive forces can be shown having a positive sense of direction.

- Depending on the nature of the problem, decide what type of rotational motion Ω the x, y, z coordinate system should have, i.e., $\Omega = \mathbf{0}$, $\Omega = \boldsymbol{\omega}$, or $\Omega \neq \boldsymbol{\omega}$. When choosing, one should keep in mind that the moment equations are simplified when the axes move in such a manner that they represent principal axes of inertia for the body at all times.

- Compute the necessary moments and products of inertia for the body relative to the x, y, z axes.

Kinematics

- Determine the x, y, z components of the body's angular velocity and compute the time derivatives of $\boldsymbol{\omega}$.

- Note that if $\Omega = \boldsymbol{\omega}$ then, $\dot{\boldsymbol{\omega}} = (\dot{\boldsymbol{\omega}})_{xyz}$, and we can either find the components of $\boldsymbol{\omega}$ along the x, y, z axes when the axes are oriented in a general position, and then take the time derivative of the magnitudes of these components, $(\dot{\boldsymbol{\omega}})_{xyz}$; or we can find the time derivative of $\boldsymbol{\omega}$ with respect to the X, Y, Z axes, $\dot{\boldsymbol{\omega}}$, and then determine the components $\dot{\omega}_x, \dot{\omega}_y, \dot{\omega}_z$.

Equations of Motion

- Apply either the two vector equations 21–18 and 21–22, or the six scalar component equations appropriate for the x, y, z coordinate axes chosen for the problem.

EXAMPLE **21.4**

The gear shown in Fig. 21–12a has a mass of 10 kg and is mounted at an angle of 10° with a rotating shaft having negligible mass. If $I_z = 0.1$ kg·m², $I_x = I_y = 0.05$ kg·m², and the shaft is rotating with a constant angular velocity of $\omega = 30$ rad/s, determine the reactions that the bearing supports A and B exert on the shaft at the instant shown.

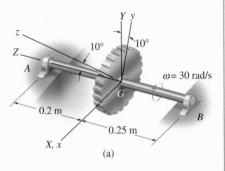

(a)

Solution

Free-Body Diagram. Fig. 21–12b. The origin of the x, y, z coordinate system is located at the gear's center of mass G, which is also a fixed point. The axes are fixed in and rotate with the gear, since these axes will then always represent the principal axes of inertia for the gear. Hence $\Omega = \omega$.

Kinematics. As shown in Fig. 21–12c, the angular velocity ω of the gear is constant in magnitude and is always directed along the axis of the shaft AB. Since this vector is measured from the X, Y, Z inertial frame of reference, for any position of the x, y, z axes,

$$\omega_x = 0 \qquad \omega_y = -30 \sin 10° \qquad \omega_z = 30 \cos 10°$$

These components remain constant for any general orientation of the x, y, z axes, and so $\dot{\omega}_x = \dot{\omega}_y = \dot{\omega}_z = 0$. Also note that since $\Omega = \omega$, then $\dot{\omega} = (\dot{\omega})_{xyz}$ and we can find these time derivatives relative to the X, Y, Z axes. In this regard ω has a constant magnitude and direction $(+Z)$ and so $\dot{\omega} = 0$, and so $\dot{\omega}_x = \dot{\omega}_y = \dot{\omega}_z = 0$. Furthermore, since G is a fixed point, $(a_G)_x = (a_G)_y = (a_G)_z = 0$.

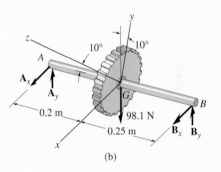

(b)

Equations of Motion. Applying Eqs. 21–25 ($\Omega = \omega$) yields

$$\Sigma M_x = I_x \dot{\omega}_x - (I_y - I_z)\omega_y\omega_z$$
$$-(A_Y)(0.2) + (B_Y)(0.25) = 0 - (0.05 - 0.1)(-30 \sin 10°)(30 \cos 10°)$$
$$-0.2A_Y + 0.25B_Y = -7.70 \qquad (1)$$

$$\Sigma M_y = I_y \dot{\omega}_y - (I_z - I_x)\omega_z\omega_x$$
$$A_X(0.2) \cos 10° - B_X(0.25) \cos 10° = 0 + 0$$
$$A_X = 1.25B_X \qquad (2)$$

$$\Sigma M_z = I_z \dot{\omega}_z - (I_x - I_y)\omega_x\omega_y$$
$$A_X(0.2) \sin 10° - B_X(0.25) \sin 10° = 0 + 0$$
$$A_X = 1.25B_X$$

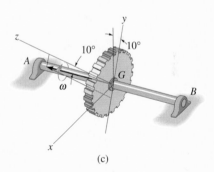

(c)

Fig. 21–12

Applying Eqs. 21–19, we have

$$\Sigma F_X = m(a_G)_X; \qquad A_X + B_X = 0 \qquad (3)$$
$$\Sigma F_Y = m(a_G)_Y; \qquad A_Y + B_Y - 98.1 = 0 \qquad (4)$$
$$\Sigma F_Z = m(a_G)_Z; \qquad 0 = 0$$

Solving Eqs. 1 through 4 simultaneously gives

$$A_X = B_X = 0 \qquad A_Y = 71.6 \text{ N} \qquad B_Y = 26.5 \text{ N} \qquad \qquad Ans.$$

EXAMPLE 21.5

The airplane shown in Fig. 21–13*a* is in the process of making a steady *horizontal* turn at the rate of ω_p. During this motion, the airplane's propeller is spinning at the rate of ω_s. If the propeller has two blades, determine the moments which the propeller shaft exerts on the propeller at the instant the blades are in the vertical position. For simplicity, assume the blades to be a uniform slender bar having a moment of inertia I about an axis perpendicular to the blades and passing through their center, and having zero moment of inertia about a longitudinal axis.

(a)

Solution

Free-Body Diagram. Fig. 21–13*b*. The effect of the connecting shaft on the propeller is indicated by the resultants $\mathbf{F}_R$ and $\mathbf{M}_R$. (The propeller's weight is assumed to be negligible.) The x, y, z axes will be taken fixed to the propeller, since these axes always represent the principal axes of inertia for the propeller. Thus, $\mathbf{\Omega} = \boldsymbol{\omega}$. The moments of inertia I_x and I_y are equal ($I_x = I_y = I$) and $I_z = 0$.

Kinematics. The angular velocity of the propeller observed from the X, Y, Z axes, coincident with the x, y, z axes, Fig. 21–13*c*, is $\boldsymbol{\omega} = \boldsymbol{\omega}_s + \boldsymbol{\omega}_p = \omega_s\mathbf{i} + \omega_p\mathbf{k}$, so that the x, y, z components of $\boldsymbol{\omega}$ are

$$\omega_x = \omega_s \qquad \omega_y = 0 \qquad \omega_z = \omega_p$$

Since $\mathbf{\Omega} = \boldsymbol{\omega}$, then $\dot{\boldsymbol{\omega}} = (\dot{\boldsymbol{\omega}})_{xyz}$. The time derivative of $\dot{\boldsymbol{\omega}}$ can be determined with respect to the fixed X, Y, Z axes. To do this, Eq. 20–6 must be used since $\boldsymbol{\omega}$ is changing direction relative to X, Y, Z. Since $\boldsymbol{\omega} = \boldsymbol{\omega}_s + \boldsymbol{\omega}_p$, then $\dot{\boldsymbol{\omega}} = \dot{\boldsymbol{\omega}}_s + \dot{\boldsymbol{\omega}}_p$. The time rate of change of each of these components relative to the X, Y, Z axes can be obtained by using a third coordinate system x', y', z', which has an angular velocity $\mathbf{\Omega}' = \boldsymbol{\omega}_p$ and is coincident with the X, Y, Z axes at the instant shown. Thus

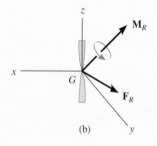

(b)

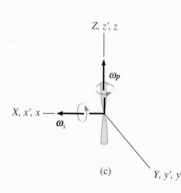

(c)

Fig. 21–13

$$\dot{\boldsymbol{\omega}} = (\dot{\boldsymbol{\omega}})_{x'y'z'} + \boldsymbol{\omega}_p \times \boldsymbol{\omega}$$

$$= (\dot{\boldsymbol{\omega}}_s)_{x'y'z'} + (\dot{\boldsymbol{\omega}}_p)_{x'y'z'} + \boldsymbol{\omega}_p \times (\boldsymbol{\omega}_s + \boldsymbol{\omega}_p)$$

$$= \mathbf{0} + \mathbf{0} + \boldsymbol{\omega}_p \times \boldsymbol{\omega}_s + \boldsymbol{\omega}_p \times \boldsymbol{\omega}_p$$

$$= \mathbf{0} + \mathbf{0} + \omega_p \mathbf{k} \times \omega_s \mathbf{i} + \mathbf{0} = \omega_p \omega_s \mathbf{j}$$

Since the X, Y, Z axes are coincident with the x, y, z axes at the instant shown, the components of $\dot{\boldsymbol{\omega}}$ along $x\,y\,z$ are therefore

$$\dot{\omega}_x = 0 \qquad \dot{\omega}_y = \omega_p \omega_s \qquad \dot{\omega}_z = 0$$

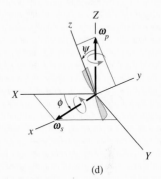

(d)

These same results can also be determined by direct calculation of $(\dot{\boldsymbol{\omega}})_{xyz}$; however, this will involve a bit more work. To do this, it will be necessary to view the propeller (or the x, y, z axes) in some *general position* such as shown in Fig. 21–13d. Here the plane has turned through an angle ϕ (phi) and the propeller has turned through an angle ψ (psi) relative to the plane. Notice that $\boldsymbol{\omega}_p$ is always directed along the fixed Z axis and $\boldsymbol{\omega}_s$ follows the x axis. Thus the general components of $\boldsymbol{\omega}$ are

$$\omega_x = \omega_s \qquad \omega_y = \omega_p \sin \psi \qquad \omega_z = \omega_p \cos \psi$$

Since ω_s and ω_p are constant, the time derivatives of these components become

$$\dot{\omega}_x = 0 \qquad \dot{\omega}_y = \omega_p \cos \psi \,\dot{\psi} \qquad \dot{\omega}_z = -\omega_p \sin \psi \,\dot{\psi}$$

But $\phi = \psi = 0°$ and $\dot{\psi} = \omega_s$ at the instant considered. Thus,

$$\omega_x = \omega_s \qquad \omega_y = 0 \qquad \omega_z = \omega_p$$

$$\dot{\omega}_x = 0 \qquad \dot{\omega}_y = \omega_p \omega_s \qquad \dot{\omega}_z = 0$$

which are the same results as those obtained previously.

Equations of Motion. Using Eqs. 21–25, we have

$$\Sigma M_x = I_x \dot{\omega}_x - (I_y - I_z)\omega_y \omega_z = I(0) - (I - 0)(0)\omega_p$$

$$M_x = 0 \qquad\qquad\qquad Ans.$$

$$\Sigma M_y = I_y \dot{\omega}_y - (I_z - I_x)\omega_z \omega_x = I(\omega_p \omega_s) - (0 - I)\omega_p \omega_s$$

$$M_y = 2I\omega_p \omega_s \qquad\qquad Ans.$$

$$\Sigma M_z = I_z \dot{\omega}_z - (I_x - I_y)\omega_x \omega_y = 0(0) - (I - I)\omega_s(0)$$

$$M_z = 0 \qquad\qquad\qquad Ans.$$

E X A M P L E   21.6

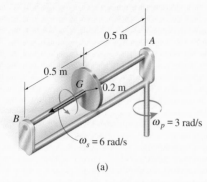

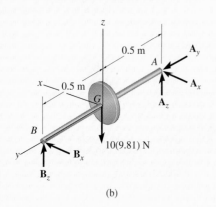

(a)

(b)

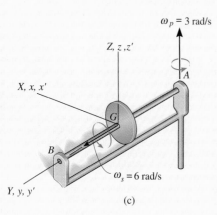

(c)

Fig. 21–14

The 10-kg flywheel (or thin disk) shown in Fig. 21–14a rotates (spins) about the shaft at a constant angular velocity of $\omega_s = 6$ rad/s. At the same time, the shaft is rotating (precessing) about the bearing at A with an angular velocity of $\omega_p = 3$ rad/s. If A is a thrust bearing and B is a journal bearing, determine the components of force reaction at each of these supports due to the motion.

Solution I

Free-Body Diagram. Fig. 21–14b. The origin of the x, y, z coordinate system is located at the center of mass G of the flywheel. Here we will let these coordinates have an angular velocity of $\Omega = \omega_p = \{3\mathbf{k}\}$ rad/s. Although the wheel spins relative to these axes, the moments of inertia *remain constant,* *i.e.,

$$I_x = I_z = \tfrac{1}{4}(10\text{ kg})(0.2\text{ m})^2 = 0.1 \text{ kg} \cdot \text{m}^2$$
$$I_y = \tfrac{1}{2}(10\text{ kg})(0.2\text{ m})^2 = 0.2 \text{ kg} \cdot \text{m}^2$$

Kinematics. From the coincident inertial X, Y, Z frame of reference, Fig. 21–41c, the flywheel has an angular velocity of $\omega = \{6\mathbf{j} + 3\mathbf{k}\}$ rad/s, so that

$$\omega_x = 0, \quad \omega_y = 6 \text{ rad/s}, \quad \omega_z = 3 \text{ rad/s}$$

The time derivative of ω must be determined relative to the x, y, z axes. In this case both ω_p and ω_s do not change and so

$$\dot{\omega}_x = 0, \quad \dot{\omega}_y = 0, \quad \dot{\omega}_z = 0$$

Equations of Motion. Applying Eqs. 21–26 ($\Omega \neq \omega$) yields

$$\Sigma M_x = I_x \dot{\omega}_x - I_y \Omega_z \omega_y + I_z \Omega_y \omega_z$$
$$-A_z(0.5) + B_z(0.5) = 0 - (0.2)(3)(6) + 0 = -3.6$$
$$\Sigma M_y = I_y \dot{\omega}_y - I_z \Omega_x \omega_z + I_x \Omega_z \omega_x$$
$$0 = 0 - 0 + 0$$
$$\Sigma M_z = I_z \dot{\omega}_z - I_x \Omega_y \omega_x + I_y \Omega_x \omega_y$$
$$A_x(0.5) - B_x(0.5) = 0 - 0 + 0$$

Applying Eqs. 21–19, we have

$$\Sigma F_X = m(a_G)_X; \quad A_x + B_x = 0$$
$$\Sigma F_Y = m(a_G)_Y; \quad A_y = -10(0.5)(3)^2$$
$$\Sigma F_Z = m(a_G)_Z; \quad A_z + B_z - 10(9.81) = 0$$

Solving these equations, we obtain

$$A_x = 0 \qquad A_y = -45.0 \text{ N} \qquad A_z = 52.6 \text{ N} \qquad Ans.$$
$$B_x = 0 \qquad\qquad\qquad\qquad\qquad B_z = 45.4 \text{ N} \qquad Ans.$$

Note that if the precession ω_p had not occurred, the reactions at A and B would be equal to 49.05 N. In this case, however, the difference in the reactions is caused by the "gyroscopic moment" created whenever a spinning body precesses about another axis. We will study this effect in detail in the next section.

*This would not be true for the propeller in Example 21–5.

Solution II

This example can also be solved using Euler's equations of motion, Eqs. 21–25. In this case $\Omega = \omega = \{6\mathbf{j} + 3\mathbf{k}\}$ rad/s, and the time derivative $(\dot{\omega})_{xyz}$ can be conveniently obtained with reference to the fixed X, Y, Z axes since $\dot{\omega} = (\dot{\omega})_{xyz}$. This calculation can be performed by choosing x', y', z' axes to have an angular velocity of $\Omega' = \omega_p$, Fig. 21–14c, so that

$$\dot{\omega} = (\dot{\omega})_{x'y'z'} + \omega_p \times \omega = 0 + 3\mathbf{k} \times (6\mathbf{j} + 3\mathbf{k}) = \{-18\mathbf{i}\} \text{ rad/s}^2$$

$$\dot{\omega}_x = -18 \text{ rad/s} \quad \dot{\omega}_y = 0 \quad \dot{\omega}_z = 0$$

The moment equations then become

$$\Sigma M_x = I_x \dot{\omega}_x - (I_y - I_z)\omega_y \omega_z$$
$$-A_x(0.5) + B_z(0.5) = 0.1(-18) - (0.2 - 0.1)(6)(3) = -3.6$$
$$\Sigma M_y = I_y \dot{\omega}_y - (I_z - I_x)\omega_z \omega_x$$
$$0 = 0 - 0$$
$$\Sigma M_z = I_z \dot{\omega}_z - (I_x - I_y)\omega_x \omega_y$$
$$A_x(0.5) - B_x(0.5) = 0 - 0$$

The solution then proceeds as before.

*21.5 Gyroscopic Motion

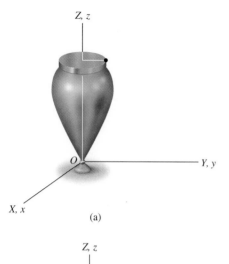

(a)

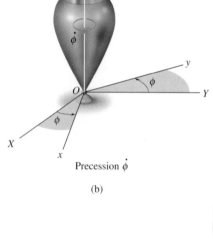

Precession $\dot{\phi}$

(b)

In this section the equations defining the motion of a body (top) which is symmetrical with respect to an axis and moving about a fixed point lying on the axis will be developed. These equations will then be applied to study the motion of a particularly interesting device, the gyroscope.

The body's motion will be analyzed using *Euler angles* ϕ, θ, ψ (phi, theta, psi). To illustrate how these angles define the position of a body, reference is made to the top shown in Fig. 21–15a. The top is attached to point O and has an orientation relative to the fixed X, Y, Z axes at some instant of time as shown in Fig. 21–15d. To define this final position, a second set of x, y, z axes will be needed. For purposes of discussion, assume that this reference is fixed in the top. Starting with the X, Y, Z and x, y, z axes in coincidence, Fig. 21–15a, the final position of the top is determined using the following three steps:

1. Rotate the top about the Z (or z) axis through an angle ϕ ($0 \leq \phi < 2\pi$), Fig. 21–15b.

2. Rotate the top about the x axis through an angle θ ($0 \leq \theta \leq \pi$), Fig. 21–15c.

3. Rotate the top about the z axis through an angle ψ ($0 \leq \psi < 2\pi$) to obtain the final position, Fig. 20-15d.

The sequence of these three angles, ϕ, θ, then ψ, must be maintained, since finite rotations are *not vectors* (see Fig. 20–1). Although this is the case, the differential rotations $d\phi$, $d\theta$, and $d\psi$ are vectors, and thus the angular velocity $\boldsymbol{\omega}$ of the top can be expressed in terms of the time derivatives of the Euler angles. The angular-velocity components $\dot{\phi}$, $\dot{\theta}$, and $\dot{\psi}$ are known as the *precession*, *nutation*, and *spin*, respectively.

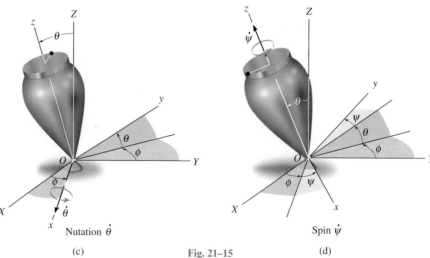

Nutation $\dot{\theta}$

(c)

Spin $\dot{\psi}$

(d)

Fig. 21–15

Their positive directions are shown in Fig. 21–16. It is seen that these vectors are not all perpendicular to one another; however, $\boldsymbol{\omega}$ of the top can still be expressed in terms of these three components.

In our case the body (top) is symmetric with respect to the z or spin axis. If we consider the top to be oriented so that at the instant considered the spin angle $\psi = 0$ and *the x, y, z axes follow the motion of the body only in nutation and precession*, i.e., $\boldsymbol{\Omega} = \boldsymbol{\omega}_p + \boldsymbol{\omega}_n$, then the nutation and spin are always directed along the x and z axes, respectively, Fig. 21–16. Hence, the angular velocity of the body is specified only in terms of the Euler angle θ, i.e.,

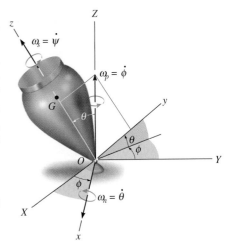

$$\boldsymbol{\omega} = \omega_x \mathbf{i} + \omega_y \mathbf{j} + \omega_z \mathbf{k}$$
$$= \dot{\theta}\mathbf{i} + (\dot{\phi} \sin \theta)\mathbf{j} + (\dot{\phi} \cos \theta + \dot{\psi})\mathbf{k} \qquad (21\text{--}27)$$

Since motion of the axes is not affected by the spin component,

$$\boldsymbol{\Omega} = \Omega_x \mathbf{i} + \Omega_y \mathbf{j} + \Omega_z \mathbf{k}$$
$$= \dot{\theta}\mathbf{i} + (\dot{\phi} \sin \theta)\mathbf{j} + (\dot{\phi} \cos \theta)\mathbf{k} \qquad (21\text{--}28)$$

Fig. 21–16

The x, y, z axes in Fig. 21–16 represent *principal axes of inertia* of the body for *any* spin of the body about these axes. Hence, the moments of inertia are constant and will be represented as $I_{xx} = I_{yy} = I$ and $I_{zz} = I_z$. Since $\boldsymbol{\Omega} \neq \boldsymbol{\omega}$, Eqs. 21–26 are used to establish the rotational equations of motion. Substituting into these equations the respective angular-velocity components defined by Eqs. 21–27 and 21–28, their corresponding time derivatives, and the moment of inertia components yields

$$\Sigma M_x = I(\ddot{\theta} - \dot{\phi}^2 \sin \theta \cos \theta) + I_z \dot{\phi} \sin \theta (\dot{\phi} \cos \theta + \dot{\psi})$$
$$\Sigma M_y = I(\ddot{\phi} \sin \theta + 2\dot{\phi}\dot{\theta} \cos \theta) - I_z \dot{\theta}(\dot{\phi} \cos \theta + \dot{\psi}) \qquad (21\text{--}29)$$
$$\Sigma M_z = I_z(\ddot{\psi} + \ddot{\phi} \cos \theta - \dot{\phi}\dot{\theta} \sin \theta)$$

Each moment summation applies only at the fixed point O or the center of mass G of the body. Since the equations represent a coupled set of nonlinear second-order differential equations, in general a closed-form solution may not be obtained. Instead, the Euler angles ϕ, θ, and ψ may be obtained graphically as functions of time using numerical analysis and computer techniques.

A special case, however, does exist for which simplification of Eqs. 21–29 is possible. Commonly referred to as *steady precession*, it occurs when the nutation angle θ, precession $\dot{\phi}$, and spin $\dot{\psi}$ all remain *constant*. Equations 21–29 then reduce to the form

$$\boxed{\Sigma M_x = -I\dot{\phi}^2 \sin \theta \cos \theta + I_z \dot{\phi} \sin \theta (\dot{\phi} \cos \theta + \dot{\psi})} \qquad (21\text{--}30)$$

$$\Sigma M_y = 0$$
$$\Sigma M_z = 0$$

Equation 21–30 may be further simplified by noting that, from Eq. 21–27, $\omega_z = \dot\phi \cos\theta + \dot\psi$, so that

$$\Sigma M_x = -I\dot\phi^2 \sin\theta \cos\theta + I_z\dot\phi \sin\theta\omega_z$$

or

$$\boxed{\Sigma M_x = \dot\phi \sin\theta(I_z\omega_z - I\dot\phi \cos\theta)} \qquad (21\text{–}31)$$

It is interesting to note what effects the spin $\dot\psi$ has on the moment about the x axis. In this regard consider the spinning rotor shown in Fig. 21–17. Here $\theta = 90°$, in which case Eq. 21–30 reduces to the form

$$\Sigma M_x = I_z\dot\phi\dot\psi$$

or

$$\boxed{\Sigma M_x = I_z\Omega_y\omega_z} \qquad (21\text{–}32)$$

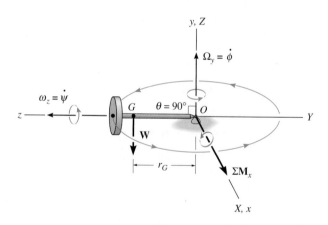

Fig. 21–17

From the figure it is seen that vectors $\Sigma\mathbf{M}_x$, $\mathbf{\Omega}_y$, and $\boldsymbol{\omega}_z$ all act along their respective *positive axes* and therefore are mutually perpendicular. Instinctively, one would expect the rotor to fall down under the influence of gravity! However, this is not the case at all, provided the product $I_z\Omega_y\omega_z$ is correctly chosen to counterbalance the moment $\Sigma M_x = Wr_G$ of the rotor's weight about O. This unusual phenomenon of rigid-body motion is often referred to as the *gyroscopic effect*.

Perhaps a more intriguing demonstration of the gyroscopic effect comes from studying the action of a *gyroscope*, frequently referred to as a *gyro*. A gyro is a rotor which spins at a very high rate about its axis of symmetry. This rate of spin is considerably greater than its precessional rate of rotation about the vertical axis. Hence, for all practical purposes, the angular momentum of the gyro can be assumed directed among its axis of spin. Thus, for the gyro rotor shown in Fig. 21–18, $\omega_z \gg \Omega_y$, and the magnitude of the angular momentum about point O, as determined from Eqs. 21–11, reduces to the form $H_O = I_z\omega_z$. Since both the magnitude and direction of $\mathbf{H}_O$ are constant as observed from x, y, z, direct application of Eq. 21–22 yields

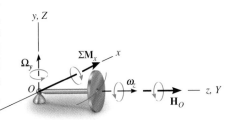

$$\boxed{\Sigma\mathbf{M}_x = \mathbf{\Omega}_y \times \mathbf{H}_O} \qquad (21\text{–}33)$$

Fig. 21–18

Using the right-hand rule applied to the cross product, it is seen that $\mathbf{\Omega}_y$ always swings $\mathbf{H}_O$ (or $\boldsymbol{\omega}_z$) toward the sense of $\Sigma\mathbf{M}_x$. In effect, the *change in direction* of the gyro's angular momentum, $d\mathbf{H}_O$, is equivalent to the angular impulse caused by the gyro's weight about O, i.e., $d\mathbf{H}_O = \Sigma\mathbf{M}_x\, dt$, Eq. 21–20. Also, since $H_O = I_z\omega_z$ and $\Sigma\mathbf{M}_x$, $\mathbf{\Omega}_y$, and H_O are mutually perpendicular, Eq. 21–33 reduces to Eq. 21–32.

When a gyro is mounted in gimbal rings, Fig. 21–19, it becomes *free* of external moments applied to its base. Thus, in theory, its angular momentum $\mathbf{H}$ will never precess but, instead, maintain its same fixed orientation along the axis of spin when the base is rotated. This type of gyroscope is called a *free gyro* and is useful as a gyrocompass when the spin axis of the gyro is directed north. In reality, the gimbal mechanism is never completely free of friction, so such a device is useful only for the local navigation of ships and aircraft. The gyroscopic effect is also useful as a means of stabilizing both the rolling motion of ships at sea and the trajectories of missiles and projectiles. Furthermore, this effect is of significant importance in the design of shafts and bearings for rotors which are subjected to forced precessions.

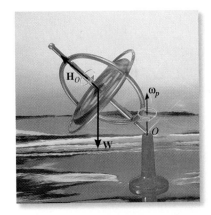

The spinning of the gyro within the frame of this toy gyroscope produces angular momentum $\mathbf{H}_O$ which is changing direction as the frame precesses $\boldsymbol{\omega}_p$ about the vertical axis. The gyroscope will not fall down since the moment of its weight $\mathbf{W}$ about the support is balanced by the change in the direction of $\mathbf{H}_O$.

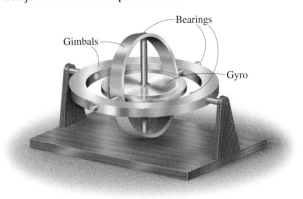

Fig. 21–19

EXAMPLE 21.7

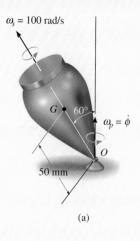

$\omega_s = 100$ rad/s

G 60°

$\omega_p = \dot{\phi}$

O

50 mm

(a)

Fig. 21–20

The top shown in Fig. 21–20a has a mass of 0.5 kg and is precessing about the vertical axis at a constant angle of $\theta = 60°$. If it spins with an angular velocity $\omega_s = 100$ rad/s, determine the precessional velocity ω_p. Assume that the axial and transverse moments of inertia of the top are $0.45(10^{-3})$ kg·m² and $1.20(10^{-3})$ kg·m², respectively, measured with respect to the fixed point O.

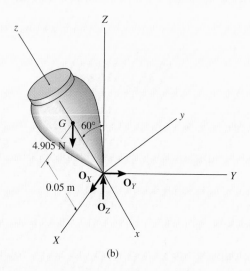

(b)

Solution

Equation 21–30 will be used for the solution since the motion is *steady precession*. As shown on the free-body diagram, Fig. 21–20b, the coordinate axes are established in the usual manner, that is, with the positive z axis in the direction of spin, the positive Z axis in the direction of precession, and the positive x axis in the direction of the moment ΣM_x (refer to Fig. 21–16). Thus,

$$\Sigma M_x = -I\dot{\phi}^2 \sin\theta \cos\theta + I_z\dot{\phi}\sin\theta(\dot{\phi}\cos\theta + \dot{\psi})$$

$$4.905\ \text{N}(0.05\ \text{m})\sin 60° = -[1.20(10^{-3})\ \text{kg·m}^2\ \dot{\phi}^2]\sin 60°\cos 60°$$
$$+ [0.45(10^{-3})\ \text{kg·m}^2]\ \dot{\phi}\sin 60°(\dot{\phi}\cos 60° + 100\ \text{rad/s})$$

or

$$\dot{\phi}^2 - 120.0\dot{\phi} + 654.0 = 0 \qquad (1)$$

Solving this quadratic equation for the precession gives

$$\dot{\phi} = 114\ \text{rad/s} \qquad \text{(high precession)} \qquad Ans.$$

and

$$\dot{\phi} = 5.72\ \text{rad/s} \qquad \text{(low precession)} \qquad Ans.$$

In reality, low precession of the top would generally be observed, since high precession would require a larger kinetic energy.

E X A M P L E **21.8**

The 1-kg disk shown in Fig. 21–21a is spinning about its axis with a constant angular velocity $\omega_D = 70$ rad/s. The block at B has a mass of 2 kg, and by adjusting its position s one can change the precession of the disk about its supporting pivot at O. Determine the position s which will enable the disk to have a constant precessional velocity $\omega_p = 0.5$ rad/s about the pivot. Neglect the weight of the shaft.

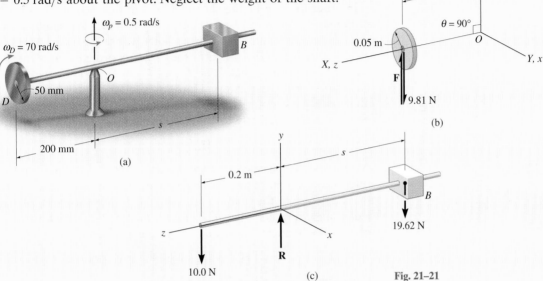

Fig. 21–21

Solution

The free-body diagram of the disk is shown in Fig. 21–21b, where **F** represents the force reaction of the shaft on the disk. The origin for both the x, y, z and X, Y, Z coordinate systems is located at point O, which represents a *fixed point* for the disk. (Although point O does not lie on the disk, imagine a massless extension of the disk to this point.) In the conventional sense, the Z axis is chosen along the axis of precession, and the z axis is along the axis of spin, so that $\theta = 90°$. Since the precession is *steady*, Eq. 21–31 may be used for the solution. This equation reduces to

$$\Sigma M_x = \dot{\phi} I_z \omega_z$$

which is the same as Eq. 21–32. Substituting the required data gives

$$9.81 \text{ N}(0.2 \text{ m}) - F(0.2 \text{ m}) = 0.5 \text{ rad/s}[\tfrac{1}{2}(1 \text{ kg})(0.05 \text{ m})^2](-70 \text{ rad/s})$$
$$F = 10.0 \text{ N}$$

As shown on the free-body diagram of the shaft and block B, Fig. 21–21c, summing moments about the x axis requires

$$(19.62 \text{ N})s = (10.0 \text{ N})(0.2 \text{ m})$$
$$= 0.102 \text{ m} = 102 \text{ mm} \qquad \qquad Ans.$$

*21.6 Torque-Free Motion

When the only external force acting on a body is caused by gravitation, the general motion of the body is referred to as *torque-free motion*. This type of motion is characteristic of planets, artificial satellites, and projectiles—provided the effects of air friction are neglected.

In order to describe the characteristics of this motion, the distribution of the body's mass will be assumed *axisymmetric*. The satellite shown in Fig. 21–22 is an example of such a body, where the z axis represents an axis of symmetry. The origin of the x, y, z coordinates is located at the mass center G, such that $I_{zz} = I_z$ and $I_{xx} = I_{yy} = I$ for the body. Since gravitation is the only external force present, the summation of moments about the mass center is zero. From Eq. 21–21, this requires the angular momentum of the body to be constant, i.e.,

$$\mathbf{H}_G = \text{constant}$$

At the instant considered, it will be assumed that the inertial frame of reference is oriented so that the positive Z axis is directed along $\mathbf{H}_G$ and the y axis lies in the plane formed by the z and Z axes, Fig. 21–22. The Euler angle formed between Z and z is θ, and therefore, with this choice of axes the angular momentum may be expressed as

$$\mathbf{H}_G = H_G \sin\theta\, \mathbf{j} + H_G \cos\theta\, \mathbf{k}$$

Furthermore, using Eqs. 21–11, we have

$$\mathbf{H}_G = I\omega_x \mathbf{i} + I\omega_y \mathbf{j} + I_z\omega_z \mathbf{k}$$

where ω_x, ω_y, ω_z represent the x, y, z components of the body's angular velocity. Equating the respective $\mathbf{i}$, $\mathbf{j}$, and $\mathbf{k}$ components of the above two equations yields

$$\omega_x = 0 \qquad \omega_y = \frac{H_G \sin\theta}{I} \qquad \omega_z = \frac{H_G \cos\theta}{I_z} \qquad (21\text{–}34)$$

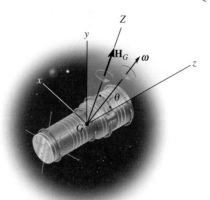

Fig. 21–22

or

$$\boldsymbol{\omega} = \frac{H_G \sin \theta}{I}\mathbf{j} + \frac{H_G \cos \theta}{I_z}\mathbf{k} \qquad (21\text{--}35)$$

In a similar manner, equating the respective **i**, **j**, **k** components of Eq. 21–27 to those of Eq. 21–34, we obtain

$$\dot{\theta} = 0$$

$$\dot{\phi} \sin \theta = \frac{H_G \sin \theta}{I}$$

$$\dot{\phi} \cos \theta + \dot{\psi} = \frac{H_G \cos \theta}{I_z}$$

Solving, we get

$$\theta = \text{const}$$

$$\dot{\phi} = \frac{H_G}{I}$$

$$\dot{\psi} = \frac{I - I_z}{II_z}H_G \cos \theta \qquad (21\text{--}36)$$

Thus, for torque-free motion of an axisymmetrical body, the angle θ formed between the angular-momentum vector and the spin of the body remains constant. Furthermore, the angular momentum $\mathbf{H}_G$, precession $\dot{\phi}$, and spin $\dot{\psi}$ for the body remain constant at all times during the motion. Eliminating H_G from the second and third of Eqs. 21–36 yields the following relationship between the spin and precession:

$$\dot{\psi} = \frac{I - I_z}{I_z}\dot{\phi} \cos \theta \qquad (21\text{--}37)$$

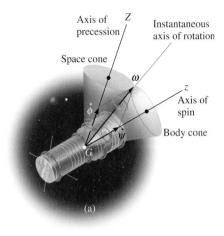

(a)

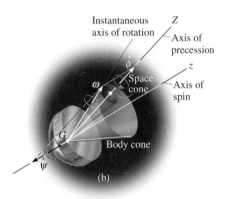

(b)

Fig. 21–23

As shown in Fig. 21–23a, the body precesses about the Z axis, which is fixed in direction, while it spins about the z axis. These two components of angular motion may be studied by using a simple cone model, introduced in Sec. 20.1. The *space cone* defining the precession is fixed from rotating, since the precession has a fixed direction, while the *body cone* rotates around the space cone's outer surface without slipping. On this basis, an attempt should be made to imagine the motion. The interior angle of each cone is chosen such that the resultant angular velocity of the body is directed along the line of contact of the two cones. This line of contact represents the instantaneous axis of rotation for the body cone, and hence the angular velocity of both the body cone and the body must be directed along this line. Since the spin is a function of the moments of inertia I and I_z of the body, Eq. 21–36, the cone model in Fig. 21–23a is satisfactory for describing the motion, provided $I > I_z$. Torque-free motion which meets these requirements is called *regular precession*. If $I < I_z$, the spin is negative and the precession positive. This motion is represented by the satellite motion shown in Fig. 21–23b ($I < I_z$). The cone model may again be used to represent the motion; however, to preserve the correct vector addition of spin and precession to obtain the angular velocity $\boldsymbol{\omega}$, the inside surface of the body cone must roll on the outside surface of the (fixed) space cone. This motion is referred to as *retrograde precession*.

Satellites are often given a spin before they are launched. If their angular momentum is not collinear with the axis of spin they will exhibit precession. In the photo on the left regular precession will occur since $I > I_z$, and in the photo on the right, retrograde precession will occur since $I < I_z$.

EXAMPLE 21.9

The motion of a football is observed using a slow-motion projector. From the film, the spin of the football is seen to be directed 30° from the horizontal, as shown in Fig. 21–24a. Also, the football is precessing about the vertical axis at a rate $\dot{\phi} = 3$ rad/s. If the ratio of the axial to transverse moments of inertia of the football is $\frac{1}{3}$, measured with respect to the center of mass, determine the magnitude of the football's spin and its angular velocity. Neglect the effect of air resistance.

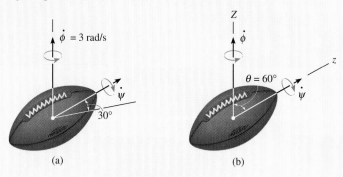

(a)   (b)

Fig. 21–24

Solution

Since the weight of the football is the only force acting, the motion is torque-free. In the conventional sense, if the z axis is established along the axis of spin and the Z axis along the precession axis, as shown in Fig. 21–24b, then the angle $\theta = 60°$. Applying Eq. 21–37, the spin is

$$\dot{\psi} = \frac{I - I_z}{I_z}\dot{\phi}\cos\theta = \frac{I - \frac{1}{3}I}{\frac{1}{3}I}(3)\cos 60°$$

$$= 3 \text{ rad/s} \qquad\qquad Ans.$$

Using Eqs. 21–34, where $H_G = \dot{\phi}I$ (Eq. 21–36), we have

$$\omega_x = 0$$

$$\omega_y = \frac{H_G \sin\theta}{I} = \frac{3I \sin 60°}{I} = 2.60 \text{ rad/s}$$

$$\omega_z = \frac{H_G \cos\theta}{I_z} = \frac{3I \cos 60°}{\frac{1}{3}I} = 4.50 \text{ rad/s}$$

Thus,

$$\omega = \sqrt{(\omega_x)^2 + (\omega_y)^2 + (\omega_z)^2}$$

$$= \sqrt{(0)^2 + (2.60)^2 + (4.50)^2}$$

$$= 5.20 \text{ rad/s} \qquad\qquad Ans.$$

CHAPTER REVIEW

- **Moments and Products of Inertia.** A body has six components of inertia for any specified x, y, z axes. Three of these are moments of inertia about each of the axes, I_x, I_y, I_z, and three are products of inertia, each defined from two orthogonal planes, I_{xy}, I_{yz}, I_{xz}. If either one or both of the planes are planes of symmetry, then the product of inertia with respect to these planes will be zero.

  The moments and products of inertia can be determined by direct integration, or by using tabulated values. If these quantities are to be determined with respect to axes or planes that do not pass through the mass center, then parallel-axis and parallel-plane theorems must be used.

  Provided the six components of inertia are known, then the moment of inertia about any axis can be determined using the transformation equation

$$I_{Oa} = I_{xx}u_x^2 + I_{yy}u_y^2 + I_{zz}u_z^2 - 2I_{xy}u_x u_y - 2I_{yz}u_y u_z - 2I_{zx}u_z u_x$$

- **Principal Moments of Inertia.** At any point on or off the body, the x, y, z axes can be oriented so that the products of inertia will be zero. The resulting moments of inertia are called the principal moments of inertia, one of which will be a maximum and the other a minimum moment of inertia for the body.

- **Principle of Impulse and Momentum.** The angular momentum for a body can be determined about any point A using the equation

$$\mathbf{H}_A = \boldsymbol{\rho}_{G/A} \times m\mathbf{v}_G + \mathbf{H}_G$$

If it is to be determined using principal axes of inertia, with origin located at the body's mass center or at a fixed point, then the components of momentum become

$$H_x = I_{xx}\omega_x - I_{xy}\omega_y - I_{xz}\omega_z$$

$$H_y = -I_{yx}\omega_x + I_{yy}\omega_y - I_{yz}\omega_z$$

$$H_z = -I_{zx}\omega_x - I_{zy}\omega_y + I_{zz}\omega_z$$

Once the linear and angular momentum for the body have been formulated, then the principle of impulse and momentum can be used to solve problems that involve force, velocity, and time. These equations are

$$m(\mathbf{v}_G)_1 + \Sigma \int_{t_1}^{t_2} \mathbf{F}\, dt = m(\mathbf{v}_G)_2$$

$$(\mathbf{H}_O)_1 + \Sigma \int_{t_1}^{t_2} \mathbf{M}_O\, dt = (\mathbf{H}_O)_2$$

- **Principle of Work and Energy.** The kinetic energy for a body is usually determined relative to a fixed point or the body's mass center. Provided the axes are principal axes of inertia, then for a fixed point,

$$T = \tfrac{1}{2}I_x\omega_x^2 + \tfrac{1}{2}I_y\omega_y^2 + \tfrac{1}{2}I_z\omega_z^2$$

And relative to the mass center,

$$T = \tfrac{1}{2}mv_G^2 + \tfrac{1}{2}I_x\omega_x^2 + \tfrac{1}{2}I_y\omega_y^2 + \tfrac{1}{2}I_z\omega_z^2$$

These formulations can be used with the principle of work and energy to solve problems that involve force, velocity, and displacement. This equation is

$$T_1 + \Sigma U_{1-2} = T_2$$

• **Equations of Motion.** There are three scalar equations of translational motion for a rigid body that moves in three dimensions. They are

$$\Sigma F_x = m(a_G)_x$$
$$\Sigma F_y = m(a_G)_y$$
$$\Sigma F_z = m(a_G)_z$$

The three scalar equations of rotational motion depend upon the location of the x, y, z reference. Most often, these axes are oriented so that the axes are principal axes of inertia. If the axes are fixed in and move with the rotation $\boldsymbol{\omega}$ of the body, then the equations are referred to as the Euler equations of motion. They are

$$\Sigma M_x = I_x \dot{\omega}_x - (I_y - I_z)\omega_y\omega_z$$
$$\Sigma M_y = I_y \dot{\omega}_y - (I_z - I_x)\omega_z\omega_x$$
$$\Sigma M_z = I_z \dot{\omega}_z - (I_x - I_y)\omega_x\omega_y$$

If the axes have a rotation $\Omega \neq \omega$, then the equations become

$$\Sigma M_x = I_x \dot{\omega}_x - I_y\Omega_z\omega_y + I_z\Omega_y\omega_z$$
$$\Sigma M_y = I_y \dot{\omega}_y - I_z\Omega_x\omega_z + I_x\Omega_z\omega_x$$
$$\Sigma M_z = I_z \dot{\omega}_z - I_x\Omega_y\omega_x + I_y\Omega_x\omega_y$$

A free-body diagram should always accompany the application of these equations.

• **Gyroscopic Motion.** The angular motion of a gyroscope is best described using the changes in motion of the three Euler angles. These angular velocity components are the precession $\dot{\phi}$, the nutation $\dot{\theta}$, and the spin $\dot{\psi}$. If $\ddot{\psi} = 0$ and $\dot{\phi}$ and $\dot{\theta}$ are constant, then the motion is referred to as steady precession. In this case, the equations of rotational motion become

$$\Sigma M_x = -I\dot{\phi}^2\sin\theta\cos\theta + I_z\dot{\phi}\sin\theta(\dot{\phi}\cos\theta + \dot{\psi})$$
$$\Sigma M_y = 0$$
$$\Sigma M_z = 0$$

It is the spin of a gyro rotor that is responsible for holding the rotor from falling downward, and instead causing it to process about a vertical axis. This phenomenon is called the gyroscopic effect.

• **Torque-Free Motion.** A body that is only subjected to a gravitational force will have no moments on it about its mass center, and so the motion is described as torque-free motion. The angular momentum for the body will remain constant. This causes the body to have both a spin and a precession. The behavior depends upon the size of the moment of inertia of a symmetric body about the spin axis I_z versus that about a perpendicular axis I. If $I > I_z$, then regular precession occurs. If $I < I_z$, then the motion is referred to as retrograde precession.

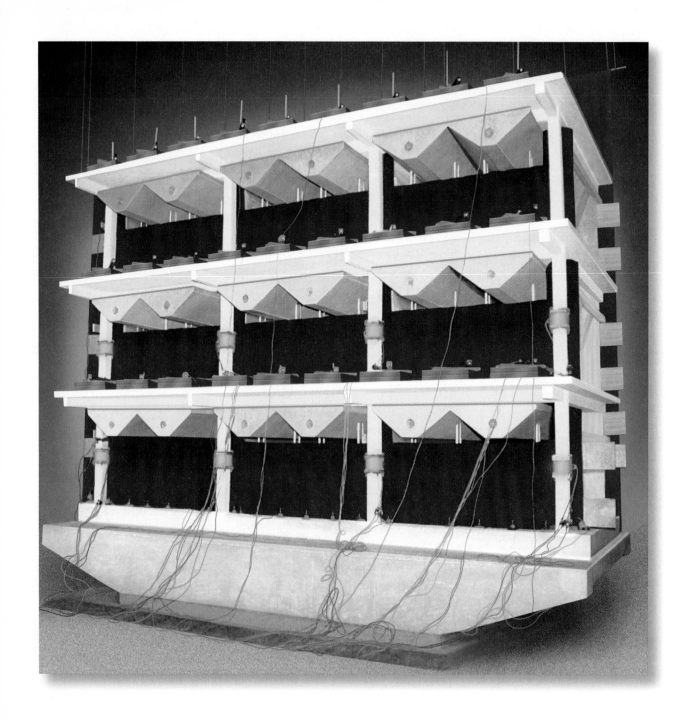

The analysis of vibrations plays an important role in the study
of the behavior of structures subjected to earthquakes.

Vibrations

- To discuss undamped one-degree-of-freedom vibration of a rigid body using the equation of motion and energy methods.
- To study the analysis of undamped forced vibration and viscous damped forced vibration.
- To introduce the concept of electrical circuit analogs to study vibrational motion.

*22.1   Undamped Free Vibration

A *vibration* is the periodic motion of a body or system of connected bodies displaced from a position of equilibrium. In general, there are two types of vibration, free and forced. *Free vibration* occurs when the motion is maintained by gravitational or elastic restoring forces, such as the swinging motion of a pendulum or the vibration of an elastic rod. *Forced vibration* is caused by an external periodic or intermittent force applied to the system. Both of these types of vibration may be either damped or undamped. *Undamped* vibrations can continue indefinitely because frictional effects are neglected in the analysis. Since in reality both internal and external frictional forces are present, the motion of all vibrating bodies is actually *damped*.

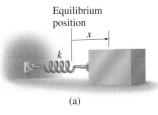

Equilibrium position

(a)

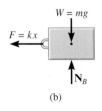

(b)

Fig. 22–1

The simplest type of vibrating motion is undamped free vibration, represented by the model shown in Fig. 22–1a. The block has a mass m and is attached to a spring having a stiffness k. Vibrating motion occurs when the block is released from a displaced position x so that the spring pulls on the block. The block will attain a velocity such that it will proceed to move out of equilibrium when $x = 0$, and provided the supporting surface is smooth, oscillation will continue indefinitely.

The time-dependent path of motion of the block may be determined by applying the equation of motion to the block when it is in the displaced position x. The free-body diagram is shown in Fig. 22–1b. The elastic restoring force $F = kx$ is always directed toward the equilibrium position, whereas the acceleration $\mathbf{a}$ is assumed to act in the direction of *positive displacement*. Noting that $a = d^2x/dt^2 = \ddot{x}$, we have

$$\xrightarrow{+} \Sigma F_x = ma_x; \qquad\qquad -kx = m\ddot{x}$$

Note that the acceleration is proportional to the block's displacement. Motion described in this manner is called *simple harmonic motion*. Rearranging the terms into a "standard form" gives

$$\ddot{x} + \omega_n^2 x = 0 \qquad\qquad (22\text{–}1)$$

The constant ω_n is called the *circular frequency* or *natural frequency*, expressed in rad/s, and in this case

$$\omega_n = \sqrt{\dfrac{k}{m}} \qquad\qquad (22\text{–}2)$$

Equation 22–1 may also be obtained by considering the block to be suspended and measuring the displacement y from the block's *equilibrium position*, Fig. 22–2a. When the block is in equilibrium, the spring exerts an upward force of $F = W = mg$ on the block. Hence, when the block is displaced a distance y downward from this position, the magnitude of the spring force is $F = W + ky$, Fig. 22–2b. Applying the equation of motion gives

$$+\downarrow \Sigma F_y = ma_y; \qquad\qquad -W - ky + W = m\ddot{y}$$

or

$$\ddot{y} + \omega_n^2 y = 0$$

which is the same form as Eq. 22–1, where ω_n is defined by Eq. 22–2.

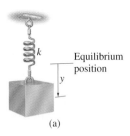

Equilibrium position

(a)

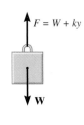

(b)

Fig. 22–2

Equation 22–1 is a homogeneous, second-order, linear, differential equation with constant coefficients. It can be shown, using the methods of differential equations, that the general solution is

$$x = A \sin \omega_n t + B \cos \omega_n t \qquad (22\text{–}3)$$

where A and B represent two constants of integration. The block's velocity and acceleration are determined by taking successive time derivatives, which yields

$$v = \dot{x} = A\omega_n \cos \omega_n t - B\omega_n \sin \omega_n t \qquad (22\text{–}4)$$

$$a = \ddot{x} = -A\omega_n^2 \sin \omega_n t - B\omega_n^2 \cos \omega_n t \qquad (22\text{–}5)$$

When Eqs. 22–3 and 22–5 are substituted into Eq. 22–1, the differential equation is indeed satisfied, showing that Eq. 22–3 is the solution to Eq. 22–1.

The integration constants A and B in Eq. 22–3 are generally determined from the initial conditions of the problem. For example, suppose that the block in Fig. 22–1a has been displaced a distance x_1 to the right from its equilibrium position and given an initial (positive) velocity $\mathbf{v}_1$ directed to the right. Substituting $x = x_1$ at $t = 0$ into Eq. 22–3 yields $B = x_1$. Since $v = v_1$ at $t = 0$, using Eq. 22–4 we obtain $A = v_1/\omega_n$. If these values are substituted into Eq. 22–3, the equation describing the motion becomes

$$x = \frac{v_1}{\omega_n} \sin \omega_n t + x_1 \cos \omega_n t \qquad (22\text{–}6)$$

Equation 22–3 may also be expressed in terms of simple sinusoidal motion. Let

$$A = C \cos \phi \qquad (22\text{–}7)$$

and

$$B = C \sin \phi \qquad (22\text{–}8)$$

where C and ϕ are new constants to be determined in place of A and B. Substituting into Eq. 22–3 yields

$$x = C \cos \phi \sin \omega_n t + C \sin \phi \cos \omega_n t$$

Since $\sin(\theta + \phi) = \sin \theta \cos \phi + \cos \theta \sin \phi$, then

$$x = C \sin(\omega_n t + \phi) \qquad (22\text{–}9)$$

If this equation is plotted on an x-versus-$\omega_n t$ axis, the graph shown in Fig. 22–3 is obtained. The maximum displacement of the block from its

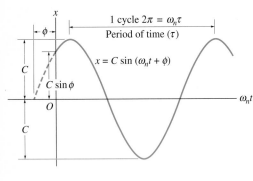

Fig. 22–3

equilibrium position is defined as the *amplitude* of vibration. From either the figure or Eq. 22–9 the amplitude is C. The angle ϕ is called the *phase angle* since it represents the amount by which the curve is displaced from the origin when $t = 0$. The constants C and ϕ are related to A and B by Eqs. 22–7 and 22–8. Squaring and adding these two equations, the amplitude becomes

$$C = \sqrt{A^2 + B^2} \qquad (22\text{–}10)$$

If Eq. 22–8 is divided by Eq. 22–7, the phase angle is

$$\phi = \tan^{-1}\frac{B}{A} \qquad (22\text{–}11)$$

Note that the sine curve, Eq. 22–9, completes one *cycle* in time $t = \tau$ (tau) when $\omega_n\tau = 2\pi$, or

$$\tau = \frac{2\pi}{\omega_n} \qquad (22\text{–}12)$$

This length of time is called a *period*, Fig. 22–3. Using Eq. 22–2, the period may also be represented as

$$\tau = 2\pi\sqrt{\frac{m}{k}} \qquad (22\text{–}13)$$

The *frequency* f is defined as the number of cycles completed per unit of time, which is the reciprocal of the period:

$$f = \frac{1}{\tau} = \frac{\omega_n}{2\pi} \qquad (22\text{–}14)$$

or

$$f = \frac{1}{2\pi}\sqrt{\frac{k}{m}} \qquad (22\text{–}15)$$

The frequency is expressed in cycles/s. This ratio of units is called a *hertz* (Hz), where 1 Hz = 1 cycle/s = 2π rad/s.

When a body or system of connected bodies is given an initial displacement from its equilibrium position and released, it will vibrate with the *natural frequency*, ω_n. Provided the body has a single degree of freedom, that is, it requires only one coordinate to specify completely the position of the system at any time, then the vibrating motion of the body will have the same characteristics as the simple harmonic motion of the block and spring just presented. Consequently, the body's motion is described by a differential equation of the same "standard form" as Eq. 22–1, i.e.,

$$\ddot{x} + \omega_n^2 x = 0 \qquad (22\text{–}16)$$

Hence, if the natural frequency ω_n of the body is known, the period of vibration τ, natural frequency f, and other vibrating characteristics of the body can be established using Eqs. 22–3 through 22–15.

IMPORTANT POINTS

- Free vibration occurs when the motion is maintained by gravitational or elastic restoring forces.
- The amplitude is the maximum displacement of the body.
- The period is the time required to complete one cycle.
- The frequency is the number of cycles completed per unit of time, where 1 Hz = 1 cycle/s.
- Only one position-coordinate system is needed to describe the location of a one-degree-of-freedom system.

PROCEDURE FOR ANALYSIS

As in the case of the block and spring, the natural frequency ω_n of a rigid body or system of connected rigid bodies having a single degree of freedom can be determined using the following procedure:

Free-Body Diagram.

- Draw the free-body diagram of the body when the body is displaced by a *small amount* from its equilibrium position.
- Locate the body with respect to its equilibrium position by using an appropriate *inertial coordinate q*. The acceleration of the body's mass center $\mathbf{a}_G$ or the body's angular acceleration $\boldsymbol{\alpha}$ should have a sense which is in the *positive direction* of the position coordinate.
- If the rotational equation of motion $\Sigma M_P = \Sigma(\mathcal{M}_k)_P$ is to be used, then it may be beneficial to also draw the kinetic diagram since it graphically accounts for the components $m(\mathbf{a}_G)_x$, $m(\mathbf{a}_G)_y$, and $I_G\boldsymbol{\alpha}$, and thereby makes it convenient for visualizing the terms needed in the moment sum $\Sigma(\mathcal{M}_k)_P$.

Equation of Motion.

- Apply the equation of motion to relate the elastic or gravitational *restoring* forces and couple moments acting on the body to the body's accelerated motion.

Kinematics.

- Using kinematics, express the body's accelerated motion in terms of the second time derivative of the position coordinate, $\ddot{q}$.
- Substitute the result into the equation of motion and determine ω_n by rearranging the terms so that the resulting equation is of the "standard form," $\ddot{q} + \omega_n^2 q = 0$.

E X A M P L E 22.1

(a)

(b)

Fig. 22–4

Determine the period of vibration for the simple pendulum shown in Fig. 22–4a. The bob has a mass m and is attached to a cord of length l. Neglect the size of the bob.

Solution

Free-Body Diagram. Motion of the system will be related to the position coordinate $(q =)\ \theta$, Fig. 22–4b. When the bob is displaced by an angle θ, the *restoring force* acting on the bob is created by the weight component $mg \sin \theta$. Furthermore, $\mathbf{a}_t$ acts in the direction of *increasing s* (or θ).

Equation of Motion. Applying the equation of motion in the *tangential direction*, since it involves the restoring force, yields

$$+\nearrow \Sigma F_t = ma_t; \qquad -mg \sin \theta = ma_t \qquad (1)$$

Kinematics. $a_t = d^2s/dt^2 = \ddot{s}$. Furthermore, s may be related to θ by the equation $s = l\theta$, so that $a_t = l\ddot{\theta}$. Hence, Eq. 1 reduces to

$$\ddot{\theta} + \frac{g}{l}\sin \theta = 0 \qquad (2)$$

The solution of this equation involves the use of an elliptic integral. For *small displacements*, however, $\sin \theta \approx \theta$, in which case

$$\ddot{\theta} + \frac{g}{l}\theta = 0 \qquad (3)$$

Comparing this equation with Eq. 22–16 ($\ddot{x} + \omega_n^2 x = 0$), it is seen that $\omega_n = \sqrt{g/l}$. From Eq. 22–12, the period of time required for the bob to make one complete swing is therefore

$$\tau = \frac{2\pi}{\omega_n} = 2\pi\sqrt{\frac{l}{g}} \qquad \textit{Ans.}$$

This interesting result, originally discovered by Galileo Galilei through experiment, indicates that the period depends only on the length of the cord and not on the mass of the pendulum bob or the angle θ.

The solution of Eq. 3 is given by Eq. 22–3, where $\omega_n = \sqrt{g/l}$ and θ is substituted for x. Like the block and spring, the constants A and B in this problem may be determined if, for example, one knows the displacement and velocity of the bob at a given instant.

E X A M P L E **22.2**

The 10-kg rectangular plate shown in Fig. 22–5*a* is suspended at its center from a rod having a torsional stiffness $k = 1.5\ \text{N}\cdot\text{m/rad}$. Determine the natural period of vibration of the plate when it is given a small angular displacement θ in the plane of the plate.

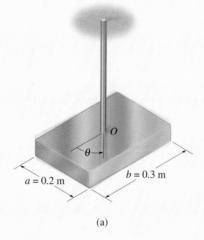

(a)

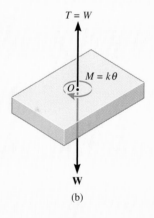

(b)

Fig. 22–5

Solution

Free-Body Diagram. Fig. 22–5*b*. Since the plate is displaced in its own plane, the torsional *restoring* moment created by the rod is $M = k\theta$. This moment acts in the direction opposite to the angular displacement θ. The angular acceleration $\ddot{\theta}$ acts in the direction of *positive* θ.

Equation of Motion.

$$\Sigma M_O = I_O \alpha; \qquad -k\theta = I_O\ddot{\theta}$$

or

$$\ddot{\theta} + \frac{k}{I_O}\theta = 0$$

Since this equation is in the "standard form," the natural frequency is $\omega_n = \sqrt{k/I_O}$.

From the table on the inside back cover, the moment of inertia of the plate about an axis coincident with the rod is $I_O = \frac{1}{12}m(a^2 + b^2)$. Hence,

$$I_O = \frac{1}{12}(10\ \text{kg})[(0.2\ \text{m})^2 + (0.3\ \text{m})^2] = 0.108\ \text{kg}\cdot\text{m}^2$$

The natural period of vibration is therefore,

$$\tau = \frac{2\pi}{\omega_n} = 2\pi\sqrt{\frac{I_O}{k}} = 2\pi\sqrt{\frac{0.108}{1.5}} = 1.69\ \text{s} \qquad \textit{Ans.}$$

EXAMPLE 22.3

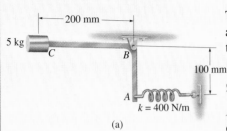

(a)

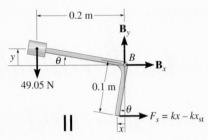

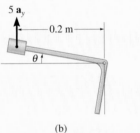

(b)

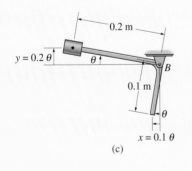

(c)

Fig. 22–6

The bent rod shown in Fig. 22–6a has a negligible mass and supports a 5-kg collar at its end. Determine the natural period of vibration for the system.

Solution

Free-Body and Kinetic Diagrams. Fig. 22–6b. Here the rod is displaced by a small amount θ from the equilibrium position. Since the spring is subjected to an initial compression of x_{st} for equilibrium, then when the displacement $x > x_{st}$ the spring exerts a force of $F_s = kx - kx_{st}$ on the rod. To obtain the "standard form," Eq. 22–16, $5\mathbf{a}_y$ acts *upward*, which is in accordance with positive θ displacement.

Equation of Motion. Moments will be summed about point B to eliminate the unknown reaction at this point. Since θ is small,

$$\zeta + \Sigma M_B = \Sigma(\mathcal{M}_k)_B;$$
$$kx(0.1\text{ m}) - kx_{st}(0.1\text{ m}) + 49.05\text{ N}(0.2\text{ m}) = -(5\text{ kg})a_y(0.2\text{ m})$$

The second term on the left side, $-kx_{st}(0.1\text{ m})$, represents the moment created by the spring force which is necessary to hold the collar in *equilibrium*, i.e., at $x = 0$. Since this moment is equal and opposite to the moment $49.05(0.2)$ created by the weight of the collar, these two terms cancel in the above equation, so that

$$kx(0.1) = -5a_y(0.2) \qquad (1)$$

Kinematics. The positions of the spring and the collar may be related to the angle θ, Fig. 22–6c. Since θ is small, $x = (0.1\text{ m})\theta$ and $y = (0.2\text{ m})\theta$. Therefore, $a_y = \ddot{y} = 0.2\ddot{\theta}$. Substituting into Eq. 1 yields

$$400(0.1\theta)0.1 = -5(0.2\ddot{\theta})0.2$$

Rewriting this equation in the "standard form" gives

$$\ddot{\theta} + 20\theta = 0$$

Compared with $\ddot{x} + \omega_n^2 x = 0$ (Eq. 22–16), we have

$$\omega_n^2 = 20 \qquad \omega_n = 4.47\text{ rad/s}$$

The natural period of vibration is therefore

$$\tau = \frac{2\pi}{\omega_n} = \frac{2\pi}{4.47} = 1.40\text{ s} \qquad \textit{Ans.}$$

E X A M P L E 22.4

A 10-lb block is suspended from a cord that passes over a 15-lb disk, as shown in Fig. 22–7a. The spring has a stiffness $k = 200$ lb/ft. Determine the natural period of vibration for the system.

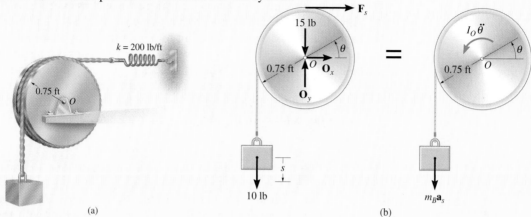

(a)

(b)

Solution

Free-Body and Kinetic Diagrams. Fig. 22–7b. The *system* consists of the disk, which undergoes a rotation defined by the angle θ, and the block, which translates by an amount s. The vector $I_O\ddot{\theta}$ acts in the direction of *positive* θ, and consequently $m_B\mathbf{a}_s$ acts downward in the direction of *positive* s.

Equation of Motion. Summing moments about point O to eliminate the reactions $\mathbf{O}_x$ and $\mathbf{O}_y$, realizing that $I_O = \frac{1}{2}mr^2$, yields

$$\zeta + \Sigma M_O = \Sigma(\mathcal{M}_k)_O;$$
$$10 \text{ lb}(0.75 \text{ ft}) - F_s(0.75 \text{ ft})$$
$$= \frac{1}{2}\left(\frac{15 \text{ lb}}{32.2 \text{ ft/s}^2}\right)(0.75 \text{ ft})^2\ddot{\theta} + \left(\frac{10 \text{ lb}}{32.2 \text{ ft/s}^2}\right)a_s(0.75 \text{ ft}) \quad (1)$$

Kinematics. As shown on the kinematic diagram in Fig. 22–7c, a small positive displacement θ of the disk causes the block to lower by an amount $s = 0.75\theta$; hence, $a_s = \ddot{s} = 0.75\ddot{\theta}$. When $\theta = 0°$, the spring force required for *equilibrium* of the disk is 10 lb, acting to the right. For position θ, the spring force is $F_s = (200 \text{ lb/ft})(0.75\theta \text{ ft}) + 10 \text{ lb}$. Substituting these results into Eq. 1 and simplifying yields

$$\ddot{\theta} + 368\theta = 0$$

Hence,

$$\omega_n^2 = 368 \qquad \omega_n = 19.2 \text{ rad/s}$$

Therefore, the natural period of vibration is

$$T = \frac{2\pi}{\omega_n} = \frac{2\pi}{19.2} = 0.328 \text{ s} \qquad \textit{Ans.}$$

(c)

Fig. 22–7

*22.2 Energy Methods

The simple harmonic motion of a body, discussed in the previous section, is due only to gravitational and elastic restoring forces acting on the body. Since these types of forces are *conservative*, it is also possible to use the conservation of energy equation to obtain the body's natural frequency or period of vibration. To show how to do this, consider again the block and spring in Fig. 22–8. When the block is displaced an arbitrary amount x from the equilibrium position, the kinetic energy is $T = \frac{1}{2}mv^2 = \frac{1}{2}m\dot{x}^2$ and the potential energy is $V = \frac{1}{2}kx^2$. By the conservation of energy equation, Eq. 14–21, it is necessary that

$$T + V = \text{constant}$$

$$\tfrac{1}{2}m\dot{x}^2 + \tfrac{1}{2}kx^2 = \text{constant} \tag{22–17}$$

The differential equation describing the *accelerated motion* of the block can be obtained by *differentiating* this equation with respect to time; i.e.,

$$m\dot{x}\ddot{x} + kx\dot{x} = 0$$
$$\dot{x}(m\ddot{x} + kx) = 0$$

Since the velocity $\dot{x}$ is not *always* zero in a vibrating system,

$$\ddot{x} + \omega_n^2 x = 0 \qquad \omega_n = \sqrt{k/m}$$

which is the same as Eq. 22–1.

If the energy equation is written for a *system of connected bodies*, the natural frequency or the equation of motion can also be determined by time differentiation. Here it is *not necessary* to dismember the system to account for reactive and connective forces which do no work.

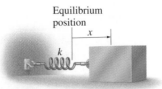

Equilibrium
position

x

k

Fig. 22–8

The suspension of a railroad car consists of a set of springs which are mounted between the frame of the car and the wheel truck. This will give the car a natural frequency of vibration which can be determined.

PROCEDURE FOR ANALYSIS

The circular or natural frequency ω_n of a body or system of connected bodies can be determined by applying the conservation of energy equation using the following procedure.

Energy Equation.

- Draw the body when it is displaced by a *small amount* from its equilibrium position and define the location of the body from its equilibrium position by an appropriate position coordinate q.

- Formulate the equation of energy for the body, $T + V = $ constant, in terms of the position coordinate.

- In general, the kinetic energy must account for both the body's translational and rotational motion, $T = \frac{1}{2}mv_G^2 + \frac{1}{2}I_G\omega_n^2$, Eq. 18–2.

- The potential energy is the sum of the gravitational and elastic potential energies of the body, $V = V_g + V_e$, Eq. 18–16. In particular, V_g should be measured from a datum for which $q = 0$ (equilibrium position).

Time Derivative.

- Take the time derivative of the energy equation using the chain rule of calculus and factor out the common terms. The resultant differential equation represents the equation of motion for the system. The value of ω_n is obtained after rearranging the terms in the "standard form," $\ddot{q} + \omega_n^2 q = 0$.

EXAMPLE 22.5

(a)

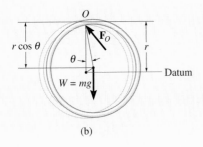

(b)

Fig. 22–9

The thin hoop shown in Fig. 22–9a is supported by a peg at O. Determine the natural period of oscillation for small amplitudes of swing. The hoop has a mass m.

Solution

Energy Equation. A diagram of the hoop when it is displaced a small amount $(q =)$ θ from the equilibrium position is shown in Fig. 22–9b. Using the table on the inside back cover and the parallel-axis theorem to determine I_O, we can express the kinetic energy as

$$T = \tfrac{1}{2} I_O \omega_n{}^2 = \tfrac{1}{2}[mr^2 + mr^2]\dot{\theta}^2 = mr^2\dot{\theta}^2$$

If a horizontal datum is placed through the center of gravity of the hoop when $\theta = 0$, then the center of gravity moves upward $r(1 - \cos \theta)$ in the displaced position. For *small angles*, $\cos \theta$ may be replaced by the first two terms of its power series expansion, $\cos \theta = 1 - \theta^2/2 + \cdots$. Therefore, the potential energy is

$$V = mgr\left[1 - \left(1 - \frac{\theta^2}{2}\right)\right] = mgr\frac{\theta^2}{2}$$

The total energy in the system is

$$T + V = mr^2\dot{\theta}^2 + mgr\frac{\theta^2}{2}$$

Time Derivative.

$$mr^2 2\dot{\theta}\ddot{\theta} + mgr\theta\dot{\theta} = 0$$

$$mr\dot{\theta}(2r\ddot{\theta} + g\theta) = 0$$

Since $\dot{\theta}$ is not always equal to zero, from the terms in parentheses,

$$\ddot{\theta} + \frac{g}{2r}\theta = 0$$

Hence,

$$\omega_n = \sqrt{\frac{g}{2r}}$$

so that

$$\tau = \frac{2\pi}{\omega_n} = 2\pi\sqrt{\frac{2r}{g}} \qquad \textit{Ans.}$$

E X A M P L E **22.6**

A 10-kg block is suspended from a cord wrapped around a 5-kg disk, as shown in Fig. 22–10a. If the spring has a stiffness $k = 200$ N/m, determine the natural period of vibration for the system.

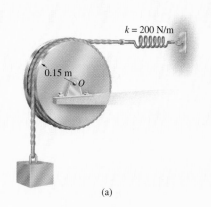

$k = 200$ N/m

0.15 m

O

(a)

Solution

Energy Equation. A diagram of the block and disk when they are displaced by respective amounts s and θ from the equilibrium position is shown in Fig. 22–10b. Since $s = (0.15$ m$)\theta$, the kinetic energy of the system is

$$T = \tfrac{1}{2}m_b v_b^2 + \tfrac{1}{2}I_O \omega_d^2$$
$$= \tfrac{1}{2}(10 \text{ kg})[(0.15 \text{ m})\dot{\theta}]^2 + \tfrac{1}{2}[\tfrac{1}{2}(5 \text{ kg})(0.15 \text{ m})^2](\dot{\theta})^2$$
$$= 0.141(\dot{\theta})^2$$

Establishing the datum at the equilibrium position of the block and realizing that the spring stretches s_{st} for equilibrium, we can write the potential energy as

$$V = \tfrac{1}{2}k(s_{st} + s)^2 - Ws$$
$$= \tfrac{1}{2}(200 \text{ N/m})[s_{st} + (0.15 \text{ m})\theta]^2 - 98.1 \text{ N}[(0.15 \text{ m})\theta]$$

The total energy for the system is, therefore,

$$T + V = 0.141(\dot{\theta})^2 + 100(s_{st} + 0.15\theta)^2 - 14.72\theta$$

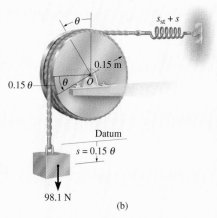

$s_{st} + s$

0.15 m

θ

O

0.15 θ

Datum

$s = 0.15\ \theta$

Time Derivative.

$$0.281(\dot{\theta})\ddot{\theta} + 200(s_{st} + 0.15\theta)0.15\dot{\theta} - 14.72\dot{\theta} = 0$$

Since $s_{st} = 98.1/200 = 0.4905$ m, the above equation reduces to the "standard form"

98.1 N

(b)

Fig. 22–10

$$\ddot{\theta} + 16\theta = 0$$

so that

$$\omega_n = \sqrt{16} = 4 \text{ rad/s}$$

Thus,

$$\tau = \frac{2\pi}{\omega_n} = \frac{2\pi}{4} = 1.57 \text{ s} \qquad\qquad Ans.$$

*22.3 Undamped Forced Vibration

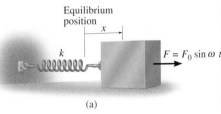

Equilibrium position

x

k

$F = F_0 \sin \omega t$

(a)

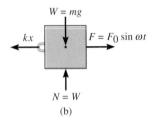

$W = mg$

kx

$F = F_0 \sin \omega t$

$N = W$

(b)

Fig. 22–11

Undamped forced vibration is considered to be one of the most important types of vibrating motion in engineering work. The principles which describe the nature of this motion may be used to analyze the forces which cause vibration in many types of machines and structures.

Periodic Force. The block and spring shown in Fig. 22–11a provide a convenient model which represents the vibrational characteristics of a system subjected to a periodic force $F = F_0 \sin \omega_0 t$. This force has an amplitude of F_0 and a *forcing frequency* ω_0. The free-body diagram for the block when it is displaced a distance x is shown in Fig. 22–11b. Applying the equation of motion yields

$$\xrightarrow{+} \Sigma F_x = ma_x; \qquad F_0 \sin \omega t - kx = m\ddot{x}$$

or

$$\ddot{x} + \frac{k}{m}x = \frac{F_0}{m}\sin \omega_0 t \qquad (22\text{–}18)$$

This equation is referred to as a nonhomogeneous second-order differential equation. The general solution consists of a complementary solution, x_c, *plus* a particular solution, x_p.

The *complementary solution* is determined by setting the term on the right side of Eq. 22–18 equal to zero and solving the resulting homogeneous equation, which is equivalent to Eq. 22–1. The solution is defined by Eq. 22–3, i.e.,

$$x_c = A \sin \omega_n t + B \cos \omega_n t \qquad (22\text{–}19)$$

where ω_n is the natural frequency, $\omega_n = \sqrt{k/m}$, Eq. 22–2.

Since the motion is periodic, the *particular solution* of Eq. 22–18 may be determined by assuming a solution of the form

$$x_p = C \sin \omega_0 t \qquad (22\text{–}20)$$

where C is a constant. Taking the second time derivative and substituting into Eq. 22–18 yields

$$-C\omega^2 \sin \omega_0 t + \frac{k}{m}(C \sin \omega_0 t) = \frac{F_0}{m}\sin \omega_0 t$$

Factoring out $\sin \omega t$ and solving for C gives

$$C = \frac{F_0/m}{(k/m) - \omega^2} = \frac{F_0/k}{1 - (\omega/\omega_n)^2} \qquad (22\text{–}21)$$

Substituting into Eq. 22–20, we obtain the particular solution

$$\boxed{x_p = \frac{F_0/k}{1 - (\omega_0/\omega_n)^2} \sin \omega_0 t} \qquad (22\text{–}22)$$

Shaker tables provide forced vibration and are used to separate out granular materials.

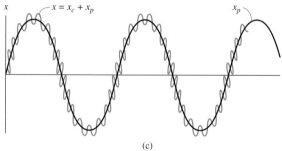

(a)

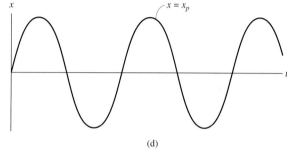

(b)

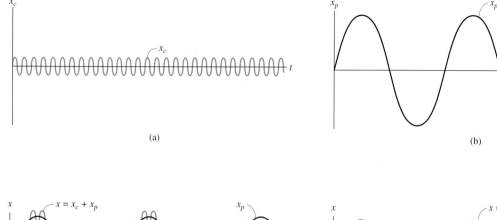

(c)

(d)

Fig. 22–12

The *general solution* is therefore

$$x = x_c + x_p = A \sin \omega_n + B \cos \omega_n + \frac{F_0/k}{1 - (\omega/\omega_n)^2} \sin \omega t \quad (22\text{–}23)$$

Here x describes two types of vibrating motion of the block. The *complementary solution* x_c defines the *free vibration*, which depends on the circular frequency $\omega_n = \sqrt{k/m}$ and the constants A and B, Fig. 22–12a. Specific values for A and B are obtained by evaluating Eq. 22–23 at a given instant when the displacement and velocity are known. The *particular solution* x_p describes the *forced vibration* of the block caused by the applied force $F = F_0 \sin \omega t$, Fig. 22–12b. The resultant vibration x is shown in Fig. 22–12c. Since all vibrating systems are subject to *friction*, the free vibration, x_c, will in time dampen out. For this reason the free vibration is referred to as *transient*, and the forced vibration is called *steady-state*, since it is the only vibration that remains, Fig. 22–12d.

From Eq. 22–21 it is seen that the *amplitude* of forced vibration depends on the *frequency ratio* ω/ω_n. If the *magnification factor* MF is defined as the ratio of the amplitude of steady-state vibration, $(x_p)_{max}$, to the static deflection F_0/k, which would be produced by the amplitude of the periodic force F_0, then, from Eq. 22–22,

The soil compactor operates by forced vibration developed by an internal motor. It is important that the forcing frequency not be close to the natural frequency of vibration, which is determined when the motor is turned off; otherwise resonance will occur and the machine will become uncontrollable.

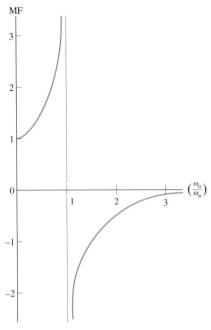

Fig. 22–13

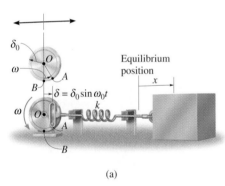

(a)

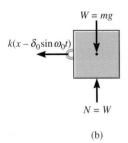

(b)

Fig. 22–14

$$MF = \frac{(x_p)_{max}}{F_0/k} = \frac{1}{1 - (\omega/\omega_n)^2} \qquad (22-24)$$

This equation is graphed in Fig. 22–13. Note that for $\omega_0 \approx 0$, the MF ≈ 1. In this case, because of the very low frequency $\omega \ll \omega_n$, the magnitude of the force **F** changes slowly and so the vibration of the block will be in phase with the applied force **F**. If the force or displacement is applied with a frequency close to the natural frequency of the system, i.e., $\omega/\omega \approx 1$, the amplitude of vibration of the block becomes extremely large. This occurs because the force **F** is applied to the block so that it always follows the motion of the block. This condition is called *resonance*, and in practice, resonating vibrations can cause tremendous stress and rapid failure of parts.* When the cyclic force $F_0 \sin \omega t$ is applied at high frequencies ($\omega > \omega_n$), the value of the MF becomes negative, indicating that the motion of the block is out of phase with the force. Under these conditions, as the block is displaced to the right, the force acts to the left, and vice versa. For extremely high frequencies ($\omega \gg \omega_n$) the block remains almost stationary, and hence the MF is approximately zero.

Periodic Support Displacement. Forced vibrations can also arise from the periodic excitation of the support of a system. The model shown in Fig. 22–14a represents the periodic vibration of a block which is caused by harmonic movement $\delta = \delta_0 \sin \omega_0 t$ of the support. The free-body diagram for the block in this case is shown in Fig. 22–14b. The coordinate x is measured from the point of zero displacement of the support, i.e., when the radial line OA coincides with OB, Fig. 22–14a. Therefore, general displacement of the spring is $(x - \delta_0 \sin \omega t)$. Applying the equation of motion yields

$$\xrightarrow{+} F_x = ma_x; \qquad -k(x - \delta_0 \sin \omega t) = m\ddot{x}$$

or

$$\ddot{x} + \frac{k}{m}x = \frac{k\delta_0}{m} \sin \omega t \qquad (22-25)$$

By comparison, this equation is identical to the form of Eq. 22–18, *provided F_0 is replaced by $k\delta_0$.* If this substitution is made into the solutions defined by Eqs. 22–21 to 22–23, the results are appropriate for describing the motion of the block when subjected to the support displacement $\delta = \delta_0 \sin \omega t$.

*A swing has a natural period of vibration, as determined in Example 22.1. If someone pushes on the swing only when it reaches its highest point, neglecting drag or wind resistance, resonance will occur since the natural and forcing frequencies will be equal.

E X A M P L E 22.7

The instrument shown in Fig. 22–15 is rigidly attached to a platform P, which in turn is supported by *four* springs, each having a stiffness $k = 800 \text{ N/m}$. Initially the platform is at rest when the floor is subjected to a displacement $\delta = 10 \sin(8t)$ mm, where t is in seconds. If the instrument is constrained to move vertically and the total mass of the instrument and platform is 20 kg, determine the vertical displacement y of the platform as a function of time, measured from the equilibrium position. What floor vibration is required to cause resonance?

Solution

Since the induced vibration is caused by the displacement of the supports, the motion is described by Eq. 22–23, with F_0 replaced by $k\delta_0$, i.e.,

$$y = A \sin \omega_n t + B \cos \omega_n t + \frac{\delta_0}{1 - (\omega/\omega_n)^2} \sin \omega t \qquad (1)$$

Here $\delta = \delta_0 \sin \omega t = 10 \sin(8t)$ mm, so that

$$\delta_0 = 10 \text{ mm} \qquad \omega_0 = 8 \text{ rad/s}$$

$$\omega_n = \sqrt{\frac{k}{m}} = \sqrt{\frac{4(800 \text{ N/m})}{20 \text{ kg}}} = 12.6 \text{ rad/s}$$

Fig. 22–15

From Eq. 22–22, with $k\delta_0$ replacing F_0, the amplitude of vibration caused by the floor displacement is

$$(y_{\omega_n})_{\max} = \frac{\delta_0}{1 - (\omega_0/\omega_n)^2} = \frac{10}{1 - [(8 \text{ rad/s})/(12.6 \text{ rad/s})]^2} = 16.7 \text{ mm} \quad (2)$$

Hence, Eq. 1 and its time derivative become

$$y = A \sin(12.6t) + B \cos(12.6t) + 16.7 \sin(8t)$$
$$\dot{y} = A(12.6) \cos(12.6t) - B(12.6) \sin(12.6t) + 133.3 \cos(8t)$$

The constants A and B are evaluated from these equations. Since $y = 0$ and $\dot{y} = 0$ at $t = 0$, then

$$0 = 0 + B + 0 \qquad\qquad B = 0$$
$$0 = A(12.6) - 0 + 133.3 \quad A = -10.5$$

The vibrating motion is therefore described by the equation

$$y = -10.5 \sin(12.6t) + 16.7 \sin(8t) \qquad\qquad \textit{Ans.}$$

Resonance will occur when the amplitude of vibration caused by the floor displacement approaches infinity. From Eq. 2, this requires

$$\omega = \omega_n = 12.6 \text{ rad/s} \qquad\qquad \textit{Ans.}$$

*22.4 Viscous Damped Free Vibration

The vibration analysis considered thus far has not included the effects of friction or damping in the system, and as a result, the solutions obtained are only in close agreement with the actual motion. Since all vibrations die out in time, the presence of damping forces should be included in the analysis.

In many cases damping is attributed to the resistance created by the substance, such as water, oil, or air, in which the system vibrates. Provided the body moves slowly through this substance, the resistance to motion is directly proportional to the body's speed. The type of force developed under these conditions is called a *viscous damping force*. The magnitude of this force is expressed by an equation of the form

$$F = c\dot{x} \qquad (22\text{–}26)$$

where the constant c is called the *coefficient of viscous damping* and has units of $N \cdot s/m$ or $lb \cdot s/ft$.

The vibrating motion of a body or system having viscous damping may be characterized by the block and spring shown in Fig. 22–16a. The effect of damping is provided by the *dashpot* connected to the block on the right side. Damping occurs when the piston P moves to the right or left within the enclosed cylinder. The cylinder contains a fluid, and the motion of the piston is retarded since the fluid must flow around or through a small hole in the piston. The dashpot is assumed to have a coefficient of viscous damping c.

If the block is displaced a distance x from its equilibrium position, the resulting free-body diagram is shown in Fig. 22–16b. Both the spring force kx and the damping force $c\dot{x}$ oppose the forward motion of the block, so that applying the equation of motion yields

$$\xrightarrow{+} \Sigma F_x = ma_x; \qquad -kx - c\dot{x} = m\ddot{x}$$

or

$$m\ddot{x} + c\dot{x} + kx = 0 \qquad (22\text{–}27)$$

This linear, second-order, homogeneous, differential equation has solutions of the form

$$x = e^{\lambda t}$$

where e is the base of the natural logarithm and λ (lambda) is a constant. The value of λ may be obtained by substituting this solution into Eq. 22–27, which yields

$$m\lambda^2 e^{\lambda t} + c\lambda e^{\lambda t} + ke^{\lambda t} = 0$$

or

$$e^{\lambda t}(m\lambda^2 + c\lambda + k) = 0$$

Equilibrium position

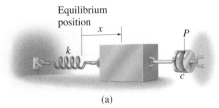

(a)

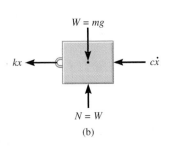

(b)

Fig. 22–16

Since $e^{\lambda t}$ is never zero, a solution is possible provided

$$m\lambda^2 + c\lambda + k = 0$$

Hence, by the quadratic formula, the two values of λ are

$$\lambda_1 = -\frac{c}{2m} + \sqrt{\left(\frac{c}{2m}\right)^2 - \frac{k}{m}}$$

$$\lambda_2 = -\frac{c}{2m} - \sqrt{\left(\frac{c}{2m}\right)^2 - \frac{k}{m}} \qquad (22\text{–}28)$$

The general solution of Eq. 22–27 is therefore a linear combination of exponentials which involves both of these roots. There are three possible combinations of λ_1 and λ_2 which must be considered. Before discussing these combinations, however, we will first define the *critical damping coefficient* c_c as the value of c which makes the radical in Eqs. 22–28 equal to zero; i.e.,

$$\left(\frac{c_c}{2m}\right)^2 - \frac{k}{m} = 0$$

or

$$c_c = 2m\sqrt{\frac{k}{m}} = 2m\omega_n \qquad (22\text{–}29)$$

Here the value of ω_n is the natural frequency $\omega_n = \sqrt{k/m}$, Eq. 22–2.

Overdamped System.   When $c > c_c$, the roots λ_1 and λ_2 are both real. The general solution of Eq. 22–27 may then be written as

$$x = Ae^{\lambda_1 t} + Be^{\lambda_2 t} \qquad (22\text{–}30)$$

Motion corresponding to this solution is *nonvibrating*. The effect of damping is so strong that when the block is displaced and released, it simply creeps back to its original position without oscillating. The system is said to be *overdamped*.

Critically Damped System.   If $c = c_c$, then $\lambda_1 = \lambda_2 = -c_c/2m = -\omega_n$. This situation is known as *critical damping*, since it represents a condition where c has the smallest value necessary to cause the system to be nonvibrating. Using the methods of differential equations, it may be shown that the solution to Eq. 22–27 for critical damping is

$$x = (A + Bt)e^{-\omega_n t} \qquad (22\text{–}31)$$

Underdamped System. Most often $c < c_c$, in which case the system is referred to as *underdamped*. In this case the roots λ_1 and λ_2 are complex numbers, and it may be shown that the general solution of Eq. 22–27 can be written as

$$x = D[e^{-(c/2m)t} \sin(\omega_d t + \phi)] \tag{22–32}$$

where D and ϕ are constants generally determined from the initial conditions of the problem. The constant ω_d is called the *damped natural frequency* of the system. It has a value of

$$\omega_d = \sqrt{\frac{k}{m} - \left(\frac{c}{2m}\right)^2} = \omega_n \sqrt{1 - \left(\frac{c}{c_c}\right)^2} \tag{22–33}$$

where the ratio c/c_c is called the *damping factor*.

The graph of Eq. 22–32 is shown in Fig. 22–17. The initial limit of motion, D, diminishes with each cycle of vibration, since motion is confined within the bounds of the exponential curve. Using the damped natural frequency ω_d, the period of damped vibration may be written as

$$\tau_d = \frac{2\pi}{\omega_d} \tag{22–34}$$

Since $\omega_d < \omega_n$, Eq. 22–33, the period of damped vibration, τ_d, will be greater than that of free vibration, $\tau = 2\pi/\omega_n$.

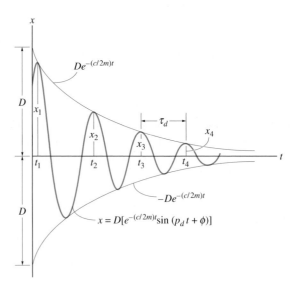

Fig. 22–17

*22.5   Viscous Damped Forced Vibration

The most general case of single-degree-of-freedom vibrating motion occurs when the system includes the effects of forced motion and induced damping. The analysis of this particular type of vibration is of practical value when applied to systems having significant damping characteristics.

If a dashpot is attached to the block and spring shown in Fig. 22–11a, the differential equation which describes the motion becomes

$$m\ddot{x} + c\dot{x} + kx = F_0 \sin \omega t \qquad (22\text{--}35)$$

A similar equation may be written for a block and spring having a periodic support displacement, Fig. 22–14a, which includes the effects of damping. In that case, however, F_0 is replaced by $k\delta_0$. Since Eq. 22–35 is nonhomogeneous, the general solution is the sum of a complementary solution, x_c, and a particular solution, x_p. The complementary solution is determined by setting the right side of Eq. 22–35 equal to zero and solving the homogeneous equation, which is equivalent to Eq. 22–27. The solution is therefore given by Eq. 22–30, 22–31, or 22–32, depending on the values of λ_1 and λ_2. Because all systems contain friction, however, this solution will dampen out with time. Only the particular solution, which describes the *steady-state vibration* of the system, will remain. Since the applied forcing function is harmonic, the steady-state motion will also be harmonic. Consequently, the particular solution will be of the form

$$x_p = A' \sin \omega t + B' \cos \omega t \qquad (22\text{--}36)$$

The constants A' and B' are determined by taking the necessary time derivatives and substituting them into Eq. 22–35, which after simplification yields

$$(-A'm\omega^2 - cB'\omega + kA') \sin \omega t +$$

$$(-B'm\omega^2 + cA'\omega + kB') \cos \omega t = F_0 \sin \omega t$$

Since this equation holds for all time, the constant coefficients of $\sin \omega t$ and $\cos \omega t$ may be equated; i.e.,

$$-A'm\omega^2 - cB'\omega + kA' = F_0$$

$$-B'm\omega^2 + cA'\omega + kB' = 0$$

Solving for A' and B', realizing that $\omega_n^2 = k/m$, yields

$$A' = \frac{(F_0/m)(\omega_n^2 - \omega^2)}{(\omega_n^2 - \omega^2)^2 + (c\omega/m)^2}$$
$$B' = \frac{-F_0(c\omega/m^2)}{(\omega_n^2 - \omega^2)^2 + (c\omega/m)^2}$$

(22–37)

It is also possible to express Eq. 22–36 in a form similar to Eq. 22–9,

$$x_p = C' \sin(\omega t - \phi')$$

(22–38)

in which case the constants C' and ϕ' are

$$C' = \frac{F_0/k}{\sqrt{[1 - (\omega/\omega_n)^2]^2 + [2(c/c_c)(\omega/\omega_n)]^2}}$$

(22–39)

$$\phi' = \tan^{-1}\left[\frac{2(c/c_c)(\omega/\omega_n)}{1 - (\omega/\omega_n)^2}\right]$$

The angle ϕ' represents the phase difference between the applied force and the resulting steady-state vibration of the damped system.

The *magnification factor* MF has been defined in Sec. 22.3 as the ratio of the amplitude of deflection caused by the forced vibration to the deflection caused by a static force $\mathbf{F}_0$. From Eq. 22–38, the forced vibration has an amplitude of C'; thus,

$$MF = \frac{C'}{F_0/k} = \frac{1}{\sqrt{[1 - (\omega/\omega_n)^2]^2 + [2(c/c_c)(\omega/\omega_n)]^2}}$$

(22–40)

The MF is plotted in Fig. 22–18 versus the frequency ratio ω/ω_n for various values of the damping factor c/c_c. It can be seen from this graph that the magnification of the amplitude increases as the damping factor decreases. Resonance obviously occurs only when the damping factor is zero and the frequency ratio equals 1.

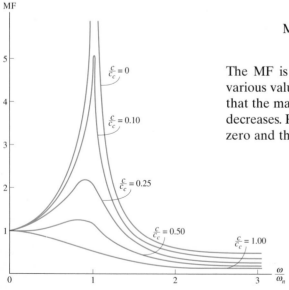

Fig. 22–18

E X A M P L E   22.8

The 30-kg electric motor shown in Fig. 22–19 is supported by *four* springs, each spring having a stiffness of 200 N/m. If the rotor R is unbalanced such that its effect is equivalent to a 4-kg mass located 60 mm from the axis of rotation, determine the amplitude of vibration when the rotor is turning at $\omega = 10$ rad/s. The damping factor is $c/c_c = 0.15$.

Fig. 22–19

Solution

The periodic force which causes the motor to vibrate is the centrifugal force due to the unbalanced rotor. This force has a constant magnitude of

$$F_0 = ma_n = mr\omega^2 = 4 \text{ kg}(0.06 \text{ m})(10 \text{ rad/s})^2 = 24 \text{ N}$$

Since $F = F_0 \sin \omega t$, where $\omega = 10$ rad/s, then

$$F = 24 \sin 10t$$

The stiffness of the entire system of four springs is $k = 4(200 \text{ N/m}) = 800 \text{ N/m}$. Therefore, the natural frequency of vibration is

$$\omega_n = \sqrt{\frac{k}{m}} = \sqrt{\frac{800 \text{ N/m}}{30 \text{ kg}}} = 5.16 \text{ rad/s}$$

Since the damping factor is known, the steady-state amplitude may be determined from the first of Eqs. 22–39, i.e.,

$$C' = \frac{F_0/k}{\sqrt{[1 - (\omega/\omega_n)^2]^2 + [2(c/c_c)(\omega/\omega_n)]^2}}$$

$$= \frac{24/800}{\sqrt{[1 - (10/5.16)^2]^2 + [2(0.15)(10/5.16)]^2}}$$

$$= 0.0107 \text{ m} = 10.7 \text{ mm} \qquad\qquad Ans.$$

*22.6 Electrical Circuit Analogs

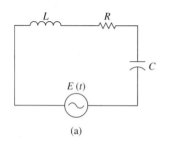

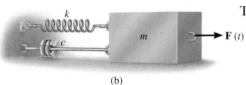

Fig. 22–20

The characteristics of a vibrating mechanical system may be represented by an electric circuit. Consider the circuit shown in Fig. 22–20a, which consists of an inductor L, a resistor R, and a capacitor C. When a voltage $E(t)$ is applied, it causes a current of magnitude i to flow through the circuit. As the current flows past the inductor the voltage drop is $L(di/dt)$, when it flows across the resistor the drop is Ri, and when it arrives at the capacitor the drop is $(1/C)\int i\,dt$. Since current cannot flow past a capacitor, it is only possible to measure the charge q acting on the capacitor. The charge may, however, be related to the current by the equation $i = dq/dt$. Thus, the voltage drops which occur across the inductor, resistor, and capacitor may be written as $L\,d^2q/dt^2$, $R\,dq/dt$, and q/C, respectively. According to Kirchhoff's voltage law, the applied voltage balances the sum of the voltage drops around the circuit. Therefore,

$$L\frac{d^2q}{dt^2} + R\frac{dq}{dt} + \frac{1}{C}q = E(t) \qquad (22\text{–}41)$$

Consider now the model of a single-degree-of-freedom mechanical system, Fig. 22–20b, which is subjected to both a general forcing function $F(t)$ and damping. The equation of motion for this system was established in the previous section and can be written as

$$m\frac{d^2x}{dt^2} + c\frac{dx}{dt} + kx = F(t) \qquad (22\text{–}42)$$

By comparison, it is seen that Eqs. 22–41 and 22–42 have the same form, and hence mathematically the problem of analyzing an electric circuit is the same as that of analyzing a vibrating mechanical system. The analogs between the two equations are given in Table 22–1.

This analogy has important application to experimental work, for it is much easier to simulate the vibration of a complex mechanical system using an electric circuit, which can be constructed on an analog computer, than to make an equivalent mechanical spring-and-dashpot model.

TABLE 22–1 • Electrical–Mechanical Analogs

Electrical		Mechanical	
Electric charge	q	Displacement	x
Electric current	i	Velocity	dx/dt
Voltage	$E(t)$	Applied force	$F(t)$
Inductance	L	Mass	m
Resistance	R	Viscous damping coefficient	c
Reciprocal of capacitance	$1/C$	Spring stiffness	k

CHAPTER REVIEW

- *Undamped Free Vibration.*   A body has free vibration provided gravitational or elastic restoring forces cause the motion. This motion is undamped when friction forces are neglected. The periodic motion of an undamped, freely vibrating body can be studied by displacing the body from the equilibrium position and then applying the equation of motion along the path. For a one-degree-of-freedom system, the resulting differential equation can be written in the form $\ddot{x} + \omega_n^2 x = 0$. Here ω_n is the circular or natural frequency. If it is known, then the cycle time is $\tau = 2\pi/\omega_n$. Also, the frequency, or number of cycles completed per unit of time, is

$$f = \omega_n/2\pi$$

- *Energy Methods.*   Provided the restoring forces acting on the body are gravitational and elastic, then conservation of energy can also be used to determine its simple harmonic motion. To do this, the body is displaced a small amount from its equilibrium position, and an expression for its kinetic and potential energy is written. The time derivative of this equation can then be rearranged in the standard form $\ddot{x} + \omega_n^2 x = 0$. Knowing ω_n, the other properties of the motion can then be obtained.

- *Undamped Forced Vibration.*   When the equation of motion is applied to a body, which is subjected to a periodic force or a periodic support displacement having a frequency ω, then the displacement consists of a complementary solution and a particular solution. The complementary solution is caused by the free vibration and can be neglected. The particular solution is caused by the forced vibration. Resonance will occur if the natural period of vibration, ω_n, is equal to the forcing frequency, ω. This should be avoided, since the motion will become unbounded.

- *Viscous Damped Free Vibration.*   A viscous damping force is caused by a fluid drag on the system as it vibrates. If the motion is slow, this drag force is then proportional to the velocity, that is $F = c\dot{x}$. Here c is the coefficient of viscous damping. By comparing its value to the critical damping coefficient $c_c = 2m\omega_n$, we can specify the type of vibration that occurs. If $c > c_c$, it is an overdamped system; if $c = c_c$, it is a critically damped system; if $c < c_c$, it is an underdamped system.

- *Viscous Damped Forced Vibration.*   The most general type of vibration for a one-degree-of-freedom system occurs when the system is damped and subjected to periodic forced motion. The solution provides insight as to how the damping factor, c/c_c, and the frequency ratio, ω/ω_n, influence the vibration. Resonance is avoided provided $c/c_c \neq 0$ and $\omega/\omega_n \neq 1$.

- *Electrical Circuit Analogs.*   The vibrating motion of a complex system can be studied by modeling it as an electrical circuit. This is possible since the differential equations that govern the behavior of each system are the same.

APPENDIX

A

Mathematical Expressions

Quadratic Formula

If $ax^2 + bx + c = 0$, then $x = \dfrac{-b \pm \sqrt{b^2 - 4ac}}{2a}$

Hyperbolic Functions

$$\sinh x = \frac{e^x - e^{-x}}{2}, \quad \cosh x = \frac{e^x + e^{-x}}{2}, \quad \tanh x = \frac{\sinh x}{\cosh x}$$

Trigonometric Identities

$$\sin \theta = \frac{A}{C}, \quad \csc \theta = \frac{C}{A}$$

$$\cos \theta = \frac{B}{C}, \quad \sec \theta = \frac{C}{B}$$

$$\tan \theta = \frac{A}{B}, \quad \cot \theta = \frac{B}{A}$$

$$\sin^2 \theta + \cos^2 \theta = 1$$

$$\sin(\theta \pm \phi) = \sin \theta \cos \phi \pm \cos \theta \sin \phi$$

$$\sin 2\theta = 2 \sin \theta \cos \theta$$

$$\cos(\theta \pm \phi) = \cos \theta \cos \phi \mp \sin \theta \sin \phi$$

$$\cos 2\theta = \cos^2 \theta - \sin^2 \theta$$

$$\cos \theta = \pm\sqrt{\frac{1 + \cos 2\theta}{2}}, \quad \sin \theta = \pm\sqrt{\frac{1 - \cos 2\theta}{2}}$$

$$\tan \theta = \frac{\sin \theta}{\cos \theta}$$

$$1 + \tan^2 \theta = \sec^2 \theta \qquad 1 + \cot^2 \theta = \csc^2 \theta$$

Power-Series Expansions

$$\sin x = x - \frac{x^3}{3!} + \cdots \qquad \sinh x = x + \frac{x^3}{3!} + \cdots$$

$$\cos x = 1 - \frac{x^2}{2!} + \cdots \qquad \cosh x = 1 + \frac{x^2}{2!} + \cdots$$

Derivatives

$$\frac{d}{dx}(u^n) = nu^{n-1}\frac{du}{dx}$$

$$\frac{d}{dx}(uv) = u\frac{dv}{dx} + v\frac{du}{dx}$$

$$\frac{d}{dx}\left(\frac{u}{v}\right) = \frac{v\dfrac{du}{dx} - u\dfrac{dv}{dx}}{v^2}$$

$$\frac{d}{dx}(\cot u) = -\csc^2 u\frac{du}{dx}$$

$$\frac{d}{dx}(\sec u) = \tan u \sec u\frac{du}{dx}$$

$$\frac{d}{dx}(\csc u) = -\csc u \cot u\frac{du}{dx}$$

$$\frac{d}{dx}(\sin u) = \cos u\frac{du}{dx}$$

$$\frac{d}{dx}(\cos u) = -\sin u\frac{du}{dx}$$

$$\frac{d}{dx}(\tan u) = \sec^2 u\frac{du}{dx}$$

$$\frac{d}{dx}(\sinh u) = \cosh u\frac{du}{dx}$$

$$\frac{d}{dx}(\cosh u) = \sinh u\frac{du}{dx}$$

Integrals

$$\int x^n \, dx = \frac{x^{n+1}}{n+1} + C, \quad n \neq -1$$

$$\int \frac{dx}{a+bx} = \frac{1}{b}\ln(a+bx) + C$$

$$\int \frac{dx}{a+bx^2} = \frac{1}{2\sqrt{-ba}}\ln\left[\frac{a+x\sqrt{-ab}}{a-x\sqrt{-ab}}\right] + C, \quad ab < 0$$

$$\int \frac{x \, dx}{a+bx^2} = \frac{1}{2b}\ln(bx^2+a) + C,$$

$$\int \frac{x^2 \, dx}{a+bx^2} = \frac{x}{b} - \frac{a}{b\sqrt{ab}}\tan^{-1}\frac{x\sqrt{ab}}{a} + C, \quad ab > 0$$

$$\int \frac{dx}{a^2-x^2} = \frac{1}{2a}\ln\left[\frac{a+x}{a-x}\right] + C, \quad a^2 > x^2$$

$$\int \sqrt{a+bx} \, dx = \frac{2}{3b}\sqrt{(a+bx)^3} + C$$

$$\int x\sqrt{a+bx} \, dx = \frac{-2(2a-3bx)\sqrt{(a+bx)^3}}{15b^2} + C$$

$$\int x^2\sqrt{a+bx} \, dx = \frac{2(8a^2-12abx+15b^2x^2)\sqrt{(a+bx)^3}}{105b^3} + C$$

$$\int \sqrt{a^2-x^2} \, dx = \frac{1}{2}\left[x\sqrt{a^2-x^2} + a^2\sin^{-1}\frac{x}{a}\right] + C, \quad a > 0$$

$$\int x\sqrt{a^2-x^2} \, dx = -\frac{1}{3}\sqrt{(a^2-x^2)^3} + C$$

$$\int x^2\sqrt{a^2-x^2} \, dx = -\frac{x}{4}\sqrt{(a^2-x^2)^3}$$

$$\qquad + \frac{a^2}{8}\left(x\sqrt{a^2-x^2} + a^2\sin^{-1}\frac{x}{a}\right) + C, \, a > 0$$

$$\int \sqrt{x^2 \pm a^2} \, dx = \frac{1}{2}\left[x\sqrt{x^2 \pm a^2} \pm a^2\ln(x + \sqrt{x^2 \pm a^2})\right] + C$$

$$\int x\sqrt{x^2 \pm a^2} \, dx = \frac{1}{3}\sqrt{(x^2 \pm a^2)^3} + C$$

$$\int x^2\sqrt{x^2 \pm a^2} \, dx = \frac{x}{4}\sqrt{(x^2 \pm a^2)^3} \mp \frac{a^2}{8}x\sqrt{x^2 \pm a^2}$$

$$\qquad - \frac{a^4}{8}\ln(x + \sqrt{x^2 \pm a^2}) + C$$

$$\int \frac{dx}{\sqrt{a+bx}} = \frac{2\sqrt{a+bx}}{b} + C$$

$$\int \frac{x \, dx}{\sqrt{x^2 \pm a^2}} = \sqrt{x^2 \pm a^2} + C$$

$$\int \frac{dx}{\sqrt{a+bx+cx^2}} = \frac{1}{\sqrt{c}}\ln\left[\sqrt{a+bx+cx^2}\right.$$

$$\qquad \left. + x\sqrt{c} + \frac{b}{2\sqrt{c}}\right] + C, \, c > 0$$

$$\qquad = \frac{1}{\sqrt{-c}}\sin^{-1}\left(\frac{-2cx-b}{\sqrt{b^2-4ac}}\right) + C, \, c > 0$$

$$\int \sin x \, dx = -\cos x + C$$

$$\int \cos x \, dx = \sin x + C$$

$$\int x\cos(ax) \, dx = \frac{1}{a^2}\cos(ax) + \frac{x}{a}\sin(ax) + C$$

$$\int x^2\cos(ax) \, dx = \frac{2x}{a^2}\cos(ax)$$

$$\qquad + \frac{a^2x^2-2}{a^3}\sin(ax) + C$$

$$\int e^{ax} \, dx = \frac{1}{a}e^{ax} + C$$

$$\int xe^{ax} \, dx = \frac{e^{ax}}{a^2}(ax-1) + C$$

$$\int \sinh x \, dx = \cosh x + C$$

$$\int \cosh x \, dx = \sinh x + C$$

Numerical and Computer Analysis

Occasionally the application of the laws of mechanics will lead to a system of equations for which a closed-form solution is difficult or impossible to obtain. When confronted with this situation, engineers will often use a numerical method which in most cases can be programmed on a microcomputer or "programmable" pocket calculator. Here we will briefly present a computer program for solving a set of linear algebraic equations, and three numerical methods which can be used to solve an algebraic or transcendental equation, evaluate a definite integral, and solve an ordinary differential equation. Application of each method will be explained by example, and an associated computer program written in Microsoft BASIC, which is designed to run on most personal computers, is provided.* A text on numerical analysis should be consulted for further discussion regarding a check of the accuracy of each method and the inherent errors that can develop from the methods.

B.1 Linear Algebraic Equations

Application of the equations of static equilibrium or the equations of motion sometimes requires solving a set of linear algebraic equations. The computer program listed in Fig. B–1 can be used for this purpose. It is based on the method of a Gaussian elimination and can solve at

---

*Similar types of programs can be written or purchased for programmable pocket calculators.

```
1  PRINT"Linear system of equations":PRINT
2  DIM A(10,11)
3  INPUT"Input number of equations : ",N
4  PRINT
5  PRINT"A  coefficients"
6  FOR I = 1 TO N
7  FOR J = 1 TO N
8  PRINT "A(";I;",";J;
9  INPUT")=",A(I,J)
10 NEXT J
11 NEXT I
12 PRINT
13 PRINT"B  coefficients"
14 FOR I = 1 TO N
15 PRINT "B(";I;
16 INPUT")=",A(I,N+1)
17 NEXT I
18 GOSUB 25
19 PRINT
20 PRINT"Unknowns"
21 FOR I = 1 TO N
22 PRINT "X(";I;")=";A(I,N+1)
23 NEXT I
24 END
25 REM Subroutine Guassian
26 FOR M=1 TO N
27 NP=M
28 BG=ABS(A(M,M))
29 FOR I = M TO N
30 IF ABS(A(I,M))<=BG THEN 33
31 BG=ABS(A(I,M))
32 NP=I
33 NEXT I
34 IF NP=M THEN 40
35 FOR I = M TO N+1
36 TE=A(M,I)
37 A(M,I)=A(NP,I)
38 A(NP,I)=TE
39 NEXT I
40 FOR I = M+1 TO N
41 FC=A(I,M)/A(M,M)
42 FOR J = M+1 TO N+1
43 A(I,J)=A(I,J)-FC*A(M,J)
44 NEXT J
45 NEXT I
46 NEXT M
47 A(N,N+1)=A(N,N+1)/A(N,N)
48 FOR I = N-1 TO 1 STEP -1
49 SM=0
50 FOR J=I+1 TO N
51 SM=SM+A(I,J)*A(J,N+1)
52 NEXT J
53 A(I,N+1)=(A(I,N+1)-SM)/A(I,I)
54 NEXT I
55 RETURN
```

Fig. B–1

most 10 equations with 10 unknowns. To do so, the equations should first be written in the following general format:

$$A_{11}x_1 + A_{12}x_2 + \cdots + A_{1n}x_n = B_1$$
$$A_{21}x_1 + A_{22}x_2 + \cdots + A_{2n}x_n = B_2$$
$$\vdots$$
$$A_{n1}x_1 + A_{n2}x_2 + \cdots + A_{nn}x_n = B_n$$

The "A" and "B" coefficients are "called" for when running the program. The output presents the unknowns $x_1, \ldots, x_n$.

EXAMPLE B.1

Solve the two equations

$$3x_1 + x_2 = 4$$
$$2x_1 - x_2 = 10$$

Solution

When the program begins to run, it first calls for the number of equations (2); then the A coefficients in the sequence $A_{11} = 3$, $A_{12} = 1$, $A_{21} = 2$, $A_{22} = -1$; and finally the B coefficients $B_1 = 4$, $B_2 = 10$. The output appears as

Unknowns

$X(1) = 2.8$ *Ans.*

$X(2) = -4.4$ *Ans.*

B.2 Simpson's Rule

Simpson's rule is a numerical method that can be used to determine the area under a curve given as a graph or as an explicit function $y = f(x)$. Likewise, it can be used to find the value of a definite integral which involves the function $y = f(x)$. To do so, the area must be subdivided into an *even number* of strips or intervals having a width h. The curve between three consecutive ordinates is approximated by a parabola, and the entire area or definite integral is then determined from the formula

$$\int_{x_0}^{x_n} f(x)\, dx \approx \frac{h}{3}[y_0 + 4(y_1 + y_3 + \cdots + y_{n-1})$$
$$+ 2(y_2 + y_4 + \cdots + y_{n-2}) + y_n] \qquad \text{(B–1)}$$

The computer program for this equation is given in Fig. B–2. For its use, we must first specify the function (on line 6 of the program). The upper and lower limits of the integral and the number of intervals are called for when the program is executed. The value of the integral is then given as the output.

```
1 PRINT"Simpson's rule":PRINT
2 PRINT" To execute this program :":PRINT
3 PRINT"    1- Modify right-hand side of the equation given below,
4 PRINT"       then press RETURN key"
5 PRINT"    2- Type   RUN 6":PRINT:EDIT 6
6 DEF FNF(X)=LOG(X)
7 PRINT:INPUT" Enter Lower Limit = ",A
8 INPUT" Enter Upper Limit = ",B
9 INPUT" Enter Number (even) of Intervals = ",N%
10 H=(B-A)/N%:AR=FNF(A):X=A+H
11 FOR J%=2 TO N%
12 K=2*(2-J%+2*INT(J%/2))
13 AR=AR+K*FNF(X)
14 X=X+H:NEXT J%
15 AR=H*(AR+FNF(B))/3
16 PRINT" Integral = ",AR
17 END
```

Fig. B–2

E X A M P L E **B.2**

Evaluate the definite integral

$$\int_2^5 \ln x \, dx$$

Solution

The interval $x_0 = 2$ to $x_6 = 5$ will be divided into six equal parts ($n = 6$), each having a width $h = (5 - 2)/6 = 0.5$. We then compute $y = f(x) = \ln x$ at each point of subdivision.

n	x_n	y_n
0	2	0.693
1	2.5	0.916
2	3	1.099
3	3.5	1.253
4	4	1.386
5	4.5	1.504
6	5	1.609

Thus, Eq. B–1 becomes

$$\int_2^5 \ln x \, dx \approx \frac{0.5}{3}[0.693 + 4(0.916 + 1.253 + 1.504)$$

$$+ \, 2(1.099 + 1.386) + 1.609]$$

$$\approx 3.66 \qquad\qquad\qquad Ans.$$

This answer is equivalent to the exact answer to three significant figures. Obviously, accuracy to a greater number of significant figures can be improved by selecting a smaller interval h (or larger n).

Using the computer program, we first specify the function $\ln x$, line 6 in Fig. B–2. During execution, the program input requires the upper and lower limits 2 and 5 and the number of intervals $n = 6$. The output appears as

$$\text{Integral} = 3.66082 \qquad\qquad Ans.$$

B.3 The Secant Method

The secant method is used to find the real roots of an algebraic or transcendental equation $f(x) = 0$. The method derives its name from the fact that the formula used is established from the slope of the secant line to the graph $y = f(x)$. This slope is $[f(x_n) - f(x_{n-1})]/(x_n - x_{n-1})$, and the secant formula is

$$x_{n+1} = x_n - f(x_n)\left[\frac{x_n - x_{n-1}}{f(x_n) - f(x_{n-1})}\right] \tag{B-2}$$

For application it is necessary to provide two initial guesses, x_0 and x_1, and thereby evaluate x_2 from Eq. B–2 ($n = 1$). One then proceeds to reapply Eq. B–2 with x_1 and the calculated value of x_2 and obtain x_3 ($n = 2$), etc., until the value $x_{n+1} \approx x_n$. One can see this will occur if x_n is approaching the root of the function $f(x) = 0$, since the correction term on the right of Eq. B–2 will tend toward zero. In particular, the larger the slope, the smaller the correction to x_n, and the faster the root will be found. On the other hand, if the slope is very small in the neighborhood of the root, the method leads to large corrections for x_n, and convergence to the root is slow and may even lead to a failure to find it. In such cases other numerical techniques must be used for solution.

A computer program based on Eq. B–2 is listed in Fig. B–3. We must first specify the function on line 7 of the program. When the program is executed, two initial guesses, x_0 and x_1, must be entered in order to approximate the solution. The output specifies the value of the root. If it cannot be determined, this is so stated.

```
1 PRINT"Secant method":PRINT
2 PRINT" To execute this program :":PRINT
3 PRINT"    1) Modify right hand side of the equation given below,"
4 PRINT"       then press RETURN key."
5 PRINT"    2) Type  RUN 7"
6 PRINT:EDIT 7
7 DEF FNF(X)=.5*SIN(X)-2*COS(X)+1.3
8 INPUT"Enter point #1 =",X
9 INPUT"Enter point #2 =",X1
10 IF X=X1 THEN 14
11 EP=.00001:TL=2E-20
12 FP=(FNF(X1)-FNF(X))/(X1-X)
13 IF ABS(FP)>TL THEN 15
14 PRINT"Root can not be found.":END
15 DX=FNF(X1)/FP
16 IF ABS(DX)>EP THEN 19
17 PRINT "Root = ";X1;"      Function evaluated at this root = ";FNF(X1)
18 END
19 X=X1:X1=X1-DX
20 GOTO 12
```

Fig. B–3

E X A M P L E   B.3

Determine the root of the equation

$$f(x) = 0.5 \sin x - 2 \cos x + 1.30 = 0$$

Solution
Guesses of the initial roots will be $x_0 = 45°$ and $x_1 = 30°$. Applying
Eq. B–2,

$$x_2 = 30° - (-0.1821)\frac{(30° - 45°)}{(-0.1821 - 0.2393)} = 36.48°$$

Using this value in Eq. B–2, along with $x_1 = 30°$, we have

$$x_3 = 36.48° - (-0.0108)\frac{36.48° - 30°}{(-0.0108 + 0.1821)} = 36.89°$$

Repeating the process with this value and $x_2 = 36.48°$ yields

$$x_4 = 36.89° - (0.0005)\left[\frac{36.89° - 36.48°}{(0.0005 + 0.0108)}\right] = 36.87°$$

Thus $x = 36.9°$ is appropriate to three significant figures.

If the problem is solved using the computer program, first we specify
the function, line 7 in Fig. B–3. During execution, the first and second
guesses must be entered in radians. Choosing these to be 0.8 rad and
0.5 rad, the result appears as

Root = 0.6435022

Function evaluated at this root = 1.66893E-06

This result converted from radians to degrees is therefore

$$x = 36.9° \qquad\qquad Ans.$$

B.4 Runge-Kutta Method

The Runge-Kutta method is used to solve an ordinary differential equation. It consists of applying a set of formulas which are used to find specific values of y for corresponding incremental values h in x. The formulas given in general form are as follows:

First-Order Equation.   To integrate $\dot{x} = f(t, x)$ step by step, use

$$x_{i+1} = x_i + \frac{1}{6}(k_1 + 2k_2 + 2k_3 + k_4) \tag{B-3}$$

where

$$
\begin{aligned}
k_1 &= hf(t_i, x_i) \\
k_2 &= hf\left(t_i + \frac{h}{2}, x_i + \frac{k_1}{2}\right) \\
k_3 &= hf\left(t_i + \frac{h}{2}, x_i + \frac{k_2}{2}\right) \\
k_4 &= hf(t_i + h, x_i + k_3)
\end{aligned} \tag{B-4}
$$

Second-Order Equation.   To integrate $\ddot{x} = f(t, x, \dot{x})$, use

$$
\begin{aligned}
x_{i+1} &= x_i + h\left[\dot{x}_i + \frac{1}{6}(k_1 + k_2 + k_3)\right] \\
\dot{x}_{i+1} &= \dot{x}_i + \frac{1}{6}(k_1 + 2k_2 + 2k_3 + k_4)
\end{aligned} \tag{B-5}
$$

where

$$
\begin{aligned}
k_1 &= hf(t_i, x_i, \dot{x}_i) \\
k_2 &= hf\left(t_i + \frac{h}{2}, x_i + \frac{h}{2}\dot{x}_i, \dot{x}_i + \frac{k_1}{2}\right) \\
k_3 &= hf\left(t_i + \frac{h}{2}, x_i + \frac{h}{2}\dot{x}_i + \frac{h}{4}k_1, \dot{x}_i + \frac{k_2}{2}\right) \\
k_4 &= hf\left(t_i + h, x_i + h\dot{x}_i + \frac{h}{2}k_2, \dot{x}_i + k_3\right)
\end{aligned} \tag{B-6}
$$

To apply these equations, one starts with initial values $t_i = t_0$, $x_i = x_0$ and $\dot{x}_i = \dot{x}_0$ (for the second-order equation). Choosing an increment h for t_0, the four constants k are computed, and these results are substituted into Eq. B-3 or B-5 in order to compute $x_{i+1} = x_1$, $\dot{x}_{i+1} = x_1$, corresponding to $t_{i+1} = t_1 = t_0 + h$. Repeating this process using t_1, x_1, $\dot{x}_1$ and h, values for $x_2, \dot{x}_2$ and $t_2 = t_1 + h$ are then computed, etc.

Computer programs which solve first- and second-order differential equations by this method are listed in Figs. B–4 and B–4, respectively. In order to use these programs, the operator specifies the function $\dot{x} = f(t, x)$ or $\ddot{x} = f(t, x, \dot{x})$ (line 7), the initial values t_0, x_0, $\dot{x}_0$ (for second-order equation), the final time t_n, and the step size h. The output gives the values of t, x, and $\dot{x}$ for each time increment until t_n is reached.

```
1 PRINT"Runge-Kutta Method for 1-st order Differential Equation":PRINT
2 PRINT" To execute this program :":PRINT
3 PRINT"   1) Modify right hand side of the equation given below,"
4 PRINT"      then Press RETURN key"
5 PRINT"   2) Type  RUN 7"
6 PRINT:EDIT 7
7 DEF FNF(T,X)=5*T+X
8 CLS:PRINT" Initial Conditions":PRINT
9 INPUT"Input  t   = ",T
10 INPUT"       x   = ",X
11 INPUT"Final  t   = ",T1
12 INPUT"step size  = ",H:PRINT
13 PRINT"        t            x"
14 IF T>=T1+H THEN 23
15 PRINT USING"######.#####";T;X
16 K1=H*FNF(T,X)
17 K2=H*FNF(T+.5*H,X+.5*K1)
18 K3=H*FNF(T+.5*H,X+.5*K2)
19 K4=H*FNF(T+H,X+K3)
20 T=T+H
21 X=X+(K1+K2+K2+K3+K3+K4)/6
22 GOTO 14
23 END
```

Fig. B–4

```
1 PRINT"Runge-Kutta Method for 2-nd order Differential Equation":PRINT
2 PRINT" To execute this program :":PRINT
3 PRINT"   1) Modify right hand side of the equation given below,"
4 PRINT"      then Press RETURN key"
5 PRINT"   2) Type  RUN 7"
6 PRINT:EDIT 7
7 DEF FNF(T,X,XD)=
8 INPUT"Input  t   = ",T
9 INPUT"       x   = ",X
10 INPUT"     dx/dt = ",XD
11 INPUT"Final  t   = ",T1
12 INPUT"step size  = ",H:PRINT
13 PRINT"        t         x          dx/dt"
14 IF T>=T1+H THEN 24
15 PRINT USING"######.#####";T;X;XD
16 K1=H*FNF(T,X,XD)
17 K2=H*FNF(T+.5*H,X+.5*H*XD,XD+.5*K1)
18 K3=H*FNF(T+.5*H,X+(.5*H)*(XD+.5*K1),XD+.5*K2)
19 K4=H*FNF(T+H,X+H*XD+.5*H*K2,XD+K3)
20 T=T+H
21 X=X+H*XD+H*(K1+K2+K3)/6
22 XD=XD+(K1+K2+K2+K3+K3+K4)/6
23 GOTO 14
24 END
```

Fig. B–5

E X A M P L E B.4

Solve the differential equations $\dot{x} = 5t + x$. Obtain the results for two steps using time increments of $h = 0.02$ s. At $t_0 = 0$, $x_0 = 0$.

Solution

This is a first-order equation, so Eqs. B–3 and B–4 apply. Thus, for $t_0 = 0$, $x_0 = 0$, $h = 0.02$, we have

$$k_1 = 0.02(0 + 0) = 0$$
$$k_2 = 0.02[5(0.01) + 0] = 0.001$$
$$k_3 = 0.02[5(0.01) + 0.0005] = 0.00101$$
$$k_4 = 0.02[5(0.02) + 0.00101] = 0.00202$$
$$x_1 = 0 + \tfrac{1}{6}[0 + 2(0.001) + 2(0.00101) + 0.00202] = 0.00101$$

Using the values $t_1 = 0 + 0.02 = 0.02$ and $x_1 = 0.00101$ with $h = 0.02$, the value for x_2 is now computed from Eqs. B–3 and B–4.

$$k_1 = 0.02[5(0.02) + 0.00101] = 0.00202$$
$$k_2 = 0.02[5(0.03) + 0.00202] = 0.00304$$
$$k_3 = 0.02[5(0.03) + 0.00253] = 0.00305$$
$$k_4 = 0.02[5(0.04) + 0.00406] = 0.00408$$
$$x_2 = 0.00101 + \tfrac{1}{6}[0.00202 + 2(0.00304) + 2(0.00305) + 0.00408]$$
$$= 0.00405 \qquad\qquad\qquad Ans.$$

To solve this problem using the computer program in Fig. B–4, the function is first entered on line 7, then the data $t_0 = 0$, $x_0 = 0$, $t_n = 0.04$, and $h = 0.02$ is specified. The results appear as

t	x
0.00000	0.00000
0.02000	0.00101
0.04000	0.00405

Ans.

Vector Analysis

The following discussion provides a brief review of vector analysis. A more detailed treatment of these topics is given in *Engineering Mechanics: Statics*.

Vector. A vector, **A**, is a quantity which has magnitude and direction, and adds according to the parallelogram law. As shown in Fig. C–1, **A** = **B** + **C**, where **A** is the *resultant vector* and **B** and **C** are *component vectors*.

Unit Vector. A unit vector, $\mathbf{u}_A$, has a magnitude of one "dimensionless" unit and acts in the same direction as **A**. It is determined by dividing **A** by its magnitude A, i.e,

$$\mathbf{u}_A = \frac{\mathbf{A}}{A} \qquad\qquad (C\text{--}1)$$

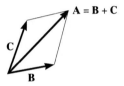

Fig. C–1

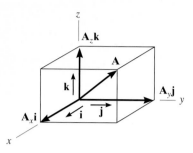

Fig. C–2

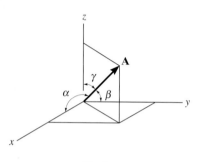

Fig. C–3

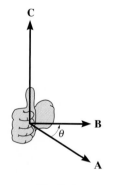

Fig. C–4

Cartesian Vector Notation. The directions of the positive x, y, z axes are defined by the Cartesian unit vectors $\mathbf{i}$, $\mathbf{j}$, $\mathbf{k}$, respectively.

As shown in Fig. C–2, vector $\mathbf{A}$ is formulated by the addition of its x, y, z components as

$$\mathbf{A} = A_x\mathbf{i} + A_y\mathbf{j} + A_z\mathbf{k} \qquad \text{(C–2)}$$

The *magnitude* of $\mathbf{A}$ is determined from

$$A = \sqrt{A_x^2 + A_y^2 + A_z^2} \qquad \text{(C–3)}$$

The *direction* of $\mathbf{A}$ is defined in terms of its *coordinate direction angles*, α, β, γ, measured from the *tail* of $\mathbf{A}$ to the *positive x, y, z* axes, Fig. C–3. These angles are determined from the *direction cosines* which represent the $\mathbf{i}$, $\mathbf{j}$, $\mathbf{k}$ components of the unit vector $\mathbf{u}_A$; i.e., from Eqs. C–1 and C–2.

$$\mathbf{u}_A = \frac{A_x}{A}\mathbf{i} + \frac{A_y}{A}\mathbf{j} + \frac{A_z}{A}\mathbf{k} \qquad \text{(C–4)}$$

so that the direction cosines are

$$\cos\alpha = \frac{A_x}{A} \qquad \cos\beta = \frac{A_y}{A} \qquad \cos\gamma = \frac{A_z}{A} \qquad \text{(C–5)}$$

Hence, $\mathbf{u}_A = \cos\alpha\,\mathbf{i} + \cos\beta\,\mathbf{j} + \cos\gamma\,\mathbf{k}$, and using Eq. C–3, it is seen that

$$\cos^2\alpha + \cos^2\beta + \cos^2\gamma = 1 \qquad \text{(C–6)}$$

The Cross Product. The cross product of two vectors $\mathbf{A}$ and $\mathbf{B}$, which yields the resultant vector $\mathbf{C}$, is written as

$$\mathbf{C} = \mathbf{A} \times \mathbf{B} \qquad \text{(C–7)}$$

and reads $\mathbf{C}$ equals $\mathbf{A}$ "cross" $\mathbf{B}$. The *magnitude* of $\mathbf{C}$ is

$$C = AB \sin\theta \qquad \text{(C–8)}$$

where θ is the angle made between the *tails* of $\mathbf{A}$ and $\mathbf{B}$ ($0° \le \theta \le 180°$). The *direction* of $\mathbf{C}$ is determined by the right-hand rule, whereby the fingers of the right hand are curled *from* $\mathbf{A}$ *to* $\mathbf{B}$ and the thumb points in the direction of $\mathbf{C}$, Fig. C–4. This vector is perpendicular to the plane containing vectors $\mathbf{A}$ and $\mathbf{B}$.

The vector cross product is *not* commutative, i.e., $\mathbf{A} \times \mathbf{B} \ne \mathbf{B} \times \mathbf{A}$. Rather,

$$\mathbf{A} \times \mathbf{B} = -\mathbf{B} \times \mathbf{A} \qquad \text{(C–9)}$$

The distributive law is valid; i.e.,

$$\mathbf{A} \times (\mathbf{B} + \mathbf{D}) = \mathbf{A} \times \mathbf{B} + \mathbf{A} \times \mathbf{D} \qquad \text{(C–10)}$$

And the cross product may be multiplied by a scalar m in any manner; i.e.,

$$m(\mathbf{A} \times \mathbf{B}) = (m\mathbf{A}) \times \mathbf{B} = \mathbf{A} \times (m\mathbf{B}) = (\mathbf{A} \times \mathbf{B})m \qquad \text{(C–11)}$$

Equation C–7 can be used to find the cross product of any pair of Cartesian unit vectors. For example, to find $\mathbf{i} \times \mathbf{j}$, the magnitude is $(i)(j) \sin 90° = (1)(1)(1) = 1$, and its direction $+\mathbf{k}$ is determined from the right-hand rule, applied to $\mathbf{i} \times \mathbf{j}$, Fig. C–2. A simple scheme shown in Fig. C–5 may be helpful in obtaining this and other results when the need arises. If the circle is constructed as shown, then "crossing" two of the unit vectors in a *counterclockwise* fashion around the circle yields a *positive* third unit vector, e.g., $\mathbf{k} \times \mathbf{i} = \mathbf{j}$. Moving *clockwise*, a *negative* unit vector is obtained, e.g., $\mathbf{i} \times \mathbf{k} = -\mathbf{j}$.

If $\mathbf{A}$ and $\mathbf{B}$ are expressed in Cartesian component form, then the cross product, Eq. C–7, may be evaluated by expanding the determinant

Fig. C–5

$$\mathbf{C} = \mathbf{A} \times \mathbf{B} = \begin{vmatrix} \mathbf{i} & \mathbf{j} & \mathbf{k} \\ A_x & A_y & A_z \\ B_x & B_y & B_z \end{vmatrix} \qquad \text{(C–12)}$$

which yields

$$\mathbf{C} = (A_y B_z - A_z B_y)\mathbf{i} - (A_x B_z - A_z B_x)\mathbf{j} + (A_x B_y - A_y B_x)\mathbf{k}$$

Recall that the cross product is used in statics to define the moment of a force $\mathbf{F}$ about point O, in which case

$$\mathbf{M}_O = \mathbf{r} \times \mathbf{F} \qquad \text{(C–13)}$$

where $\mathbf{r}$ is a position vector directed from point O to *any point* on the line of action of $\mathbf{F}$.

The Dot Product.   The dot product of two vectors $\mathbf{A}$ and $\mathbf{B}$, which yields a scalar, is defined as

$$\mathbf{A} \cdot \mathbf{B} = AB \cos \theta \qquad \text{(C–14)}$$

and reads $\mathbf{A}$ "dot" $\mathbf{B}$. The angle θ is formed between the *tails* of $\mathbf{A}$ and $\mathbf{B}$ ($0° \le \theta \le 180°$).

The dot product is commutative; i.e.,

$$\mathbf{A} \cdot \mathbf{B} = \mathbf{B} \cdot \mathbf{A} \qquad \text{(C–15)}$$

The distributive law is valid; i.e.,

$$\mathbf{A} \cdot (\mathbf{B} + \mathbf{D}) = \mathbf{A} \cdot \mathbf{B} + \mathbf{A} \cdot \mathbf{D} \qquad \text{(C–16)}$$

And scalar multiplication can be performed in any manner, i.e.,

$$m(\mathbf{A} \cdot \mathbf{B}) = (m\mathbf{A}) \cdot \mathbf{B} = \mathbf{A} \cdot (m\mathbf{B}) = (\mathbf{A} \cdot \mathbf{B})m \qquad \text{(C–17)}$$

Using Eq. C–14, the dot product between any two Cartesian vectors can be determined. For example, $\mathbf{i} \cdot \mathbf{i} = (1)(1) \cos 0° = 1$ and $\mathbf{i} \cdot \mathbf{j} = (1)(1) \cos 90° = 0$.

If $\mathbf{A}$ and $\mathbf{B}$ are expressed in Cartesian component form, then the dot product, Eq. C–14, can be determined from

$$\boxed{\mathbf{A} \cdot \mathbf{B} = A_x B_x + A_y B_y + A_z B_z} \qquad \text{(C–18)}$$

The dot product may be used to determine the *angle θ formed between two vectors*. From Eq. C–14,

$$\theta = \cos^{-1}\left(\frac{\mathbf{A} \cdot \mathbf{B}}{AB}\right) \qquad \text{(C–19)}$$

It is also possible to find the *component of a vector in a given direction* using the dot product. For example, the magnitude of the component (or projection) of vector $\mathbf{A}$ in the direction of $\mathbf{B}$, Fig. C–6, is defined by $A \cos \theta$. From Eq. C–14, this magnitude is

$$A \cos \theta = \mathbf{A} \cdot \frac{\mathbf{B}}{B} = \mathbf{A} \cdot \mathbf{u}_B \qquad \text{(C–20)}$$

where $\mathbf{u}_B$ represents a unit vector acting in the direction of $\mathbf{B}$, Fig. C–6.

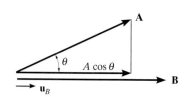

Fig. C–6

Differentiation and Integration of Vector Functions.

The rules for differentiation and integration of the sums and products of scalar functions also apply to vector functions. Consider, for example, the two vector functions $\mathbf{A}(s)$ and $\mathbf{B}(s)$. Provided these functions are smooth and continuous for all s, then

$$\frac{d}{ds}(\mathbf{A} + \mathbf{B}) = \frac{d\mathbf{A}}{ds} + \frac{d\mathbf{B}}{ds} \qquad \text{(C–21)}$$

$$\int (\mathbf{A} + \mathbf{B})\, ds = \int \mathbf{A}\, ds + \int \mathbf{B}\, ds \qquad \text{(C–22)}$$

For the cross product,

$$\frac{d}{ds}(\mathbf{A} \times \mathbf{B}) = \left(\frac{d\mathbf{A}}{ds} \times \mathbf{B}\right) + \left(\mathbf{A} \times \frac{d\mathbf{B}}{ds}\right) \qquad \text{(C–23)}$$

Similarly, for the dot product,

$$\frac{d}{ds}(\mathbf{A} \cdot \mathbf{B}) = \frac{d\mathbf{A}}{ds} \cdot \mathbf{B} + \mathbf{A} \cdot \frac{d\mathbf{B}}{ds} \qquad \text{(C–24)}$$

Review for the Fundamentals of Engineering Examination

The Fundamentals of Engineering (FE) exam is given semiannually by the National Council of Engineering Examiners (NCEE) and is one of the requirements for obtaining a Professional Engineering License. A portion of this exam contains problems in dynamics, and this appendix provides a review of the subject matter most often asked on this exam. Before solving any of the problems, you should review the sections indicated in each chapter in order to become familiar with the boldfaced definitions and the procedures used to solve the various types of problems. Also, review the example problems in these sections.

The following problems are arranged in the same sequence as the topics in each chapter. Besides helping as preparation for the FE exam, these problems also provide additional examples for general practice of the subject matter. Partial solutions and answers to *all the problems* are given at the back of this appendix.

Chapter 12—Review Sections 12.1, 12.4–12.6, 12.8–12.9

D-1. The position of a particle is $s = (0.5t^3 + 4t)$ ft, where t is in seconds. Determine the velocity and the acceleration of the particle when $t = 3$ s.

D-2. After traveling a distance of 100 m, a particle reaches a velocity of 30 m/s, starting from rest. Determine its constant acceleration.

D-3. A particle moves in a straight line such that $s = (12t^3 + 2t^2 + 3t)$ m, where t is in seconds. Determine the velocity and acceleration of the particle when $t = 2$ s.

D-4. A particle moves along a straight line such that $a = (4t^2 - 2)$ m/s^2, where t is in seconds. When $t = 0$, the particle is located 2 m to the left of the origin, and when $t = 2$ s, it is 20 m to the left of the origin. Determine the position of the particle when $t = 4$ s.

D-5. Determine the speed at which the basketball at A must be thrown at the angle of $30°$ so that it makes it to the basket at B.

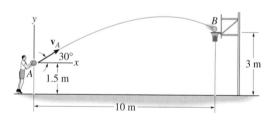

Prob. D–5

D-6. A particle moves with curvilinear motion in the x–y plane such that the y component of motion is described by the equation $y = (7t^3)$ m, where t is in seconds. If the particle starts from rest at the origin when $t = 0$, and maintains a *constant* acceleration in the x direction of 12 m/s^2, determine the particle's speed when $t = 2$ s.

D-7. Water is sprayed at an angle of $90°$ from the slope at 20 m/s. Determine the range R.

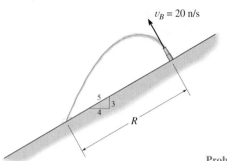

Prob. D–7

D-8. An automobile is traveling with a *constant speed* along a horizontal circular curve that has a radius of $\rho = 250$ m. If the magnitude of acceleration is $a = 1.5$ m/s^2, determine the speed at which the automobile is traveling.

D-9. A boat is traveling along a circular path having a radius of 30 m. Determine the magnitude of the boat's acceleration if at a given instant the boat's speed is $v = 6$ m/s and the rate of increase in speed is $\dot{v} = 2$ m/s^2.

D-10. A train travels along a horizontal circular curve that has a radius of 600 m. If the speed of the train is uniformly increased from 40 km/h to 60 km/h in 5 s, determine the magnitude of the acceleration at the instant the speed of the train is 50 km/h.

D-11. At a given instant, the automobile has a speed of 25 m/s and an acceleration of 3 m/s^2 acting in the direction shown. Determine the radius of curvature of the path and the rate of increase of the automobile's speed.

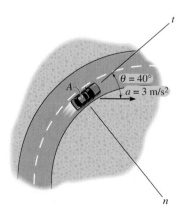

Prob. D–11

D-12. At the instant shown, cars A and B are traveling at the speeds shown. If B is accelerating at 1200 km/h^2 while A maintains a constant speed, determine the velocity and acceleration of A with respect to B.

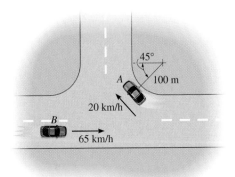

Prob. D–12

D-13. Determine the speed of point P on the cable in order to lift the platform at 2 m/s.

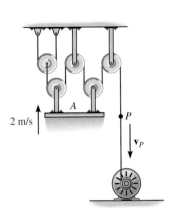

Prob. D–13

Chapter 13—Review Sections 13.1–13.5

D-14. The effective weight of a man in an elevator varies between 130 lb and 170 lb while he is riding in the elevator. When the elevator is at rest the man weighs 153 lb. Determine how fast the elevator car can accelerate, going up and going down.

D-15. Neglecting friction and the mass of the pulley and cord, determine the acceleration at which the 4-kg block B will descend. What is the tension in the cord? Block A has a mass of 2 kg.

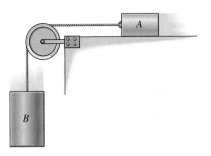

Prob. D–15

D-16. The blocks are suspended over a pulley by a rope. Neglecting the mass of the rope and the pulley, determine the acceleration of both blocks and the tension in the rope.

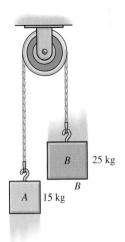

Prob. D–16

D-17. Block *B* rests upon a smooth surface. If the coefficients of static and kinetic friction between *A* and *B* are $\mu_s = 0.4$ and $\mu_k = 0.3$, respectively, determine the acceleration of each block if $P = 6$ lb.

D-19. Determine the maximum speed that the jeep can travel over the crest of the hill and not lose contact with the road.

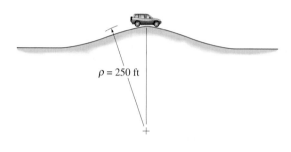

Prob. D–19

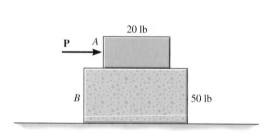

Prob. D–17

D-18. The block rests at a distance of 2 m from the center of the platform. If the coefficient of static friction between the block and the platform is $\mu_s = 0.3$, determine the maximum speed which the block can attain before it begins to slip. Assume the angular motion of the disk is slowly increasing.

D-20. A pilot weighs 150 lb and is traveling at a constant speed of 120 ft/s. Determine the normal force he exerts on the seat of the plane when he is upside down at *A*. The loop has a radius of curvature of 400 ft.

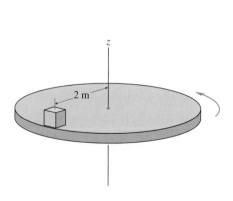

Prob. D–18

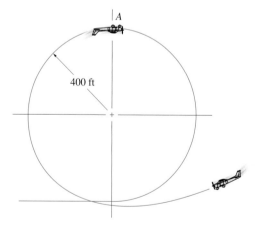

Prob. D–20

D-21. The sports car is traveling along a 30° banked road having a radius of curvature of $\rho = 500$ ft. If the coefficient of static friction between the tires and the road is $\mu_s = 0.2$, determine the maximum safe speed for travel so no slipping occurs. Neglect the size of the car.

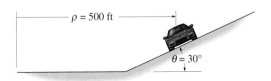

Prob. D–21

D-22. The 5-lb pendulum bob B is released from rest when $\theta = 0°$. Determine the tension in string BC immediately after it is released and when the pendulum reaches point D, where $\theta = 90°$.

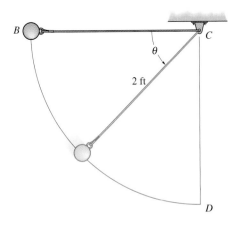

Prob. D–22

Chapter 14—Review All Sections

D-23. A 15 000-lb freight car is pulled along a horizontal track. If the car starts from rest and attains a velocity of 40 ft/s after traveling a distance of 300 ft, determine the total work done on the car by the towing force in this distance if the rolling frictional force between the car and track is 80 lb.

D-24. The 20-lb block resting on the 30° inclined plane is acted upon by a 40-lb force. If the block's initial velocity is 5 ft/s down the plane, determine its velocity after it has traveled 10 ft down the plane. The coefficient of kinetic friction between the block and the plane is $\mu_k = 0.2$.

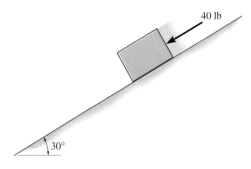

Prob. D–24

D-25. The 3-kg block is subjected to the action of the two forces shown. If the block starts from rest, determine the distance it has moved when it attains a velocity of 10 m/s. The coefficient of kinetic friction between the block and the surface is $\mu_k = 0.2$.

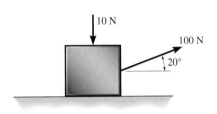

Prob. D–25

D-26. The 6-lb ball is to be fired from rest using a spring having a stiffness of $k = 40$ lb/ft. Determine how far the spring must be compressed so that when the ball reaches a height of 8 ft it has a velocity of 6 ft/s.

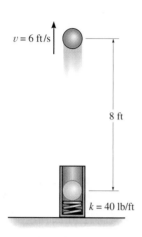

Prob. D–26

D-27. The 2-kg pendulum bob is released from rest when it is at A. Determine the speed of the bob and the tension in the cord when the bob passes through its lowest position, B.

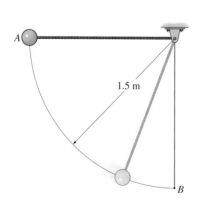

Prob. D–27

D-28. The 5-lb collar is released from rest at A and travels along the frictionless guide. Determine the speed of the collar when it strikes the stop B. The spring has an unstretched length of 0.5 ft.

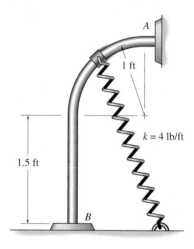

Prob. D–28

D-29. The 2-kg collar is given a downward velocity of 4 m/s when it is at A. If the spring has an unstretched length of 1 m and a stiffness of $k = 30$ N/m, determine the velocity of the collar at $s = 1$ m.

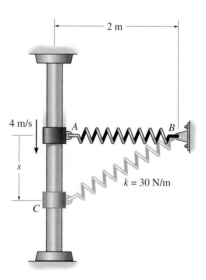

Prob. D–29

Chapter 15—Review Sections 15.1–15.4

D-30. A 30-ton engine exerts a constant horizontal force of $40(10^3)$ lb on a train having three cars that have a total weight of 250 tons. If the rolling resistance is 10 lb per ton for both the engine and cars, determine how long it takes to increase the speed of the train from 20 ft/s to 30 ft/s. What is the driving force exerted by the engine wheels on the tracks?

D-31. A 5-kg block is moving up a 30° inclined plane with an initial velocity of 3 m/s. If the coefficient of kinetic friction between the block and the plane is $\mu_k = 0.3$, determine how long a 100-N horizontal force must act on the block in order to increase the velocity of the block to 10 m/s up the plane.

D-32. The 10-lb block A attains a velocity of 1 ft/s in 5 seconds, starting from rest. Determine the tension in the cord and the coefficient of kinetic friction between block A and the horizontal plane. Neglect the weight of the pulley. Block B has a weight of 8 lb.

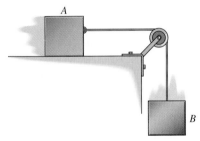

Prob. D–32

D-33. Determine the velocity of each block 10 seconds after the blocks are released from rest. Neglect the mass of the pulleys.

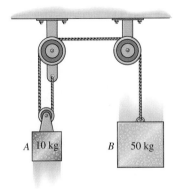

Prob. D–33

D-34. The two blocks have a coefficient of restitution of $e = 0.5$. If the surface is smooth, determine the velocity of each block after impact.

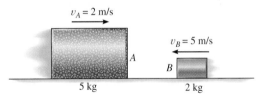

Prob. D–34

D-35. A 6-kg disk A has an initial velocity of $(v_A)_1 = 20$ m/s and strikes head-on disk B that has a mass of 24 kg and is originally at rest. If the collision is perfectly elastic, determine the speed of each disk after the collision and the impulse which disk A imparts to disk B.

D-36. Blocks A and B weigh 5 lb and 10 lb, respectively. After striking block B, A slides 2 in. to the right, and B slides 3 in. to the right. If the coefficient of kinetic friction between the blocks and the surface is $\mu_k = 0.2$, determine the coefficient of restitution between the blocks. Block B is originally at rest.

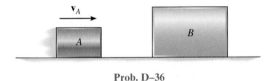

Prob. D–36

D-37. Disk A weighs 2 lb and is sliding on the smooth horizontal plane with a velocity of 3 ft/s. Disk B weighs 11 lb and is initially at rest. If after the impact A has a velocity of 1 ft/s, directed along the positive x axis, determine the speed of disk B after impact.

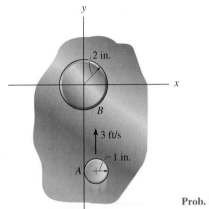

Prob. D–37

Chapter 16—Review Sections 16.3, 16.5–16.7

D-38. If gear A is rotating clockwise with an angular velocity of $\omega_A = 3$ rad/s, determine the angular velocities of gears B and C. Gear B is one unit, having radii of 2 in. and 5 in.

Prob. D–38

D-39. The spin drier of a washing machine has a constant angular acceleration of 2 rev/s², starting from rest. Determine how many turns it makes in 10 seconds and its angular velocity when $t = 5$ s.

D-40. Starting from rest, point P on the cord has a constant acceleration of 20 ft/s². Determine the angular acceleration and angular velocity of the disk after it has completed 10 revolutions. How many revolutions will the disk turn after it has completed 10 revolutions and P continues to move downward for 4 seconds longer?

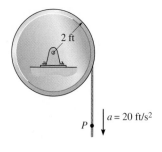

Prob. D–40

D-41. The center of the wheel has a velocity of 3 m/s. At the same time, it is slipping and has a clockwise angular velocity of $\omega = 2$ rad/s. Determine the velocity of point A at the instant shown.

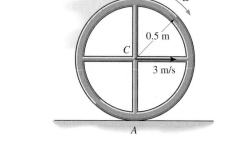

Prob. D–41

D-42. A cord is wrapped around the inner core of the gear and it is pulled with a constant velocity of 2 ft/s. Determine the velocity of the center of the gear, C.

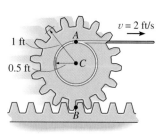

Prob. D–42

D-43. The center of the wheel is moving to the right with a speed of 2 m/s. If no slipping occurs at the ground, A, determine the velocity of point B at the instant shown.

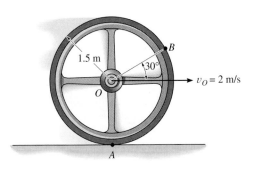

Prob. D–43

D-45. Determine the angular velocity of link AB at the instant shown.

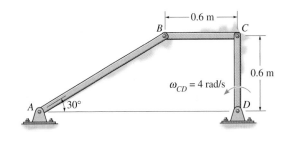

Prob. D–45

D-44. If the velocity of the slider block at B is 2 ft/s to the left, compute the velocity of the block at A and the angular velocity of the rod at the instant shown.

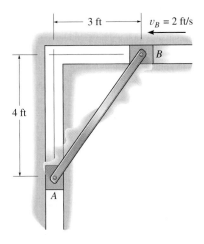

Prob. D–44

D-46. When the slider block C is in the position shown, the link AB has a clockwise angular velocity of 2 rad/s. Determine the velocity of block C at this instant.

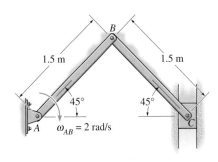

Prob. D–46

D-47. The center of the pulley is being lifted vertically with an acceleration of 3 m/s², and at the instant shown its velocity is 2 m/s. Determine the accelerations of points A and B. Assume that the rope does not slip on the pulley's surface.

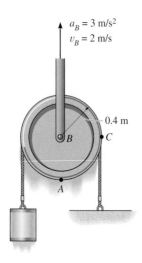

Prob. D–47

D-48. At a given instant, the slider block A has the velocity and deceleration shown. Determine the acceleration of block B and the angular acceleration of the link at this instant.

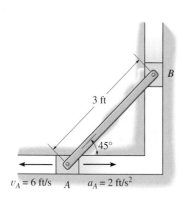

Prob. D–48

Chapter 17—Review All Sections

D-49. The 3500-lb car has a center of mass located at G. Determine the normal reactions of both front and both rear wheels on the road and the acceleration of the car if it is rolling freely down the incline. Neglect the weight of the wheels.

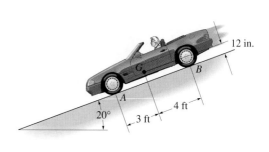

Prob. D–49

D-50. The 20-lb link AB is pinned to a moving frame at A and held in a vertical position by means of a string BC which can support a maximum tension of 10 lb. Determine the maximum acceleration of the link without breaking the string. What are the corresponding components of reaction at the pin A?

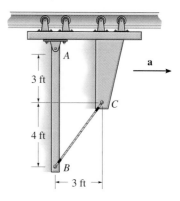

Prob. D–50

D-51. The 50-lb triangular plate is released from rest. Determine the initial angular acceleration of the plate and the horizontal and vertical components of reaction at B. The moment of inertia of the plate about the pinned axis B is $I_B = 2.30$ slug·ft².

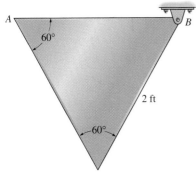

Prob. D–51

D-52. The 20-kg slender rod is pinned at O. Determine the reaction at O just after the cable is cut.

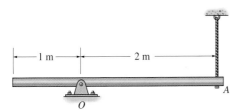

Prob. D–52

D-53. The 20-kg wheel has a radius of gyration of $K_G = 0.8$ m. Determine the angular acceleration of the wheel if no slipping occurs.

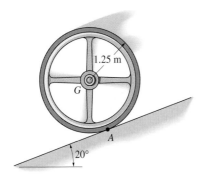

Prob. D–53

D-54. The 15-kg wheel has a wire wrapped around its inner hub and is released from rest on the inclined plane, for which the coefficient of kinetic friction is $\mu_k = 0.1$. If the centroidal radius of gyration of the wheel is $k_G = 0.8$ m, determine the angular acceleration of the wheel.

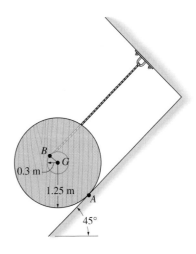

Prob. D–54

D-55. The 2-kg gear is at rest on the surface of a gear rack. If the rack is suddenly given an acceleration of 5 m/s², determine the initial angular acceleration of the gear. The radius of gyration of the gear is $k_G = 0.3$ m.

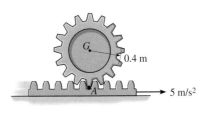

Prob. D–55

Solutions and Answers

D-1. $v = \dfrac{ds}{dt} = 1.5t^2 + 4|_{t=3} = 17.5 \text{ ft/s}$ *Ans.*

$a = \dfrac{dv}{dt} = 3t|_{t=3} = 9 \text{ ft/s}^2$ *Ans.*

D-2. $(30)^2 = (0)^2 + 2a(100 - 0)$
$a = 4.5 \text{ m/s}^2$ *Ans.*

D-3. $v = \dfrac{ds}{dt} = 36t^2 + 4t + 3|_{t=2} = 155 \text{ m/s}$ *Ans.*

$a = \dfrac{dv}{dt} = 72t + 4|_{t=2} = 148 \text{ m/s}^2$ *Ans.*

D-4. $v = \int (4t^2 - 2)\, dt$

$v = \dfrac{4}{3}t^3 - 2t + C_1$

$s = \int \left(\dfrac{4}{3}t^3 - 2t + C_1\right) dt$

$s = \dfrac{1}{3}t^4 - t^2 + C_1 t + C_2$

$t = 0, s = -2, C_2 = -2$
$t = 2, s = -20, C_1 = -9.67$
$t = 4, s = 28.7 \text{ m}$ *Ans.*

D-5. $\xrightarrow{+} s = s_0 + v_0 t$
$10 = 0 + v_A \cos 30° \, t$

$+\uparrow s = s_0 + v_0 t + \dfrac{1}{2}a_c t^2$

$3 = 1.5 + v_A \sin 30° \, t + \dfrac{1}{2}(-9.81)t^2$

$t = 0.933, v_A = 12.4 \text{ m/s}$ *Ans.*

D-6. $v = v_0 + a_c t$
$v_x = 0 + 12(2) = 24 \text{ m/s}$

$v_y = \dfrac{dy}{dt} = 21t^2|_{t=2} = 84 \text{ m/s}$

$v = \sqrt{(24)^2 + (84)^2} = 87.4 \text{ m/s}$ *Ans.*

D-7. $(\xrightarrow{+}) s = s_0 + v_0 t$
$R\left(\dfrac{4}{5}\right) = 0 + 20\left(\dfrac{3}{5}\right)t$

$(+\uparrow) s = s_0 + v_0 t + \dfrac{1}{2}a_c t^2$

$-R\left(\dfrac{3}{5}\right) = 0 + 20\left(\dfrac{4}{5}\right)t + \dfrac{1}{2}(-9.81)t^2$

$t = 5.10 \text{ s}$
$R = 76.5 \text{ m}$ *Ans.*

D-8. $a_t = 0$

$a_n = a = 1.5 = \dfrac{v^2}{250}; \ v = 19.4 \text{ m/s}$ *Ans.*

D-9. $a_t = 2 \text{ m/s}^2$

$a_n = \dfrac{v^2}{\rho} = \dfrac{(6)^2}{30} = 1.20 \text{ m/s}^2$

$a = \sqrt{(2)^2 + (1.20)^2} = 2.33 \text{ m/s}^2$ *Ans.*

D-10. $a_t = \dfrac{\Delta v}{\Delta t} = \dfrac{60 - 40}{[(5 - 0)/3600]} = 14\,400 \text{ km/h}^2$

$a_n = \dfrac{v^2}{\rho} = \dfrac{(50)^2}{0.6} = 4167 \text{ km/h}^2$

$a = \sqrt{(14.4)^2 + (4.167)^2}\, 10^3 = 15.0(10^3) \text{ km/h}^2$
   Ans.

D-11. $a_t = 3 \cos 40° = 2.30 \text{ m/s}^2$ *Ans.*

$a_n = \dfrac{v^2}{\rho}; \ 3 \sin 40° = \dfrac{(25)^2}{\rho}, \rho = 324 \text{ m}$ *Ans.*

D-12. $\mathbf{v}_A = \mathbf{v}_B + \mathbf{v}_{A/B}$
$-20 \cos 45° \, \mathbf{i} + 20 \sin 45° \, \mathbf{j} = 65\mathbf{i} + \mathbf{v}_{A/B}$
$\mathbf{v}_{A/B} = -79.14\mathbf{i} + 14.14\mathbf{j}$
$\mathbf{v}_{A/B} = \sqrt{(-79.14)^2 + (14.14)^2} = 80.4 \text{ km/h}$ *Ans.*
$\mathbf{a}_A = \mathbf{a}_B + \mathbf{a}_{A/B}$
$\dfrac{(20)^2}{0.1} \cos 45° \, \mathbf{i} + \dfrac{(20)^2}{0.1} \sin 45° \, \mathbf{j} = 1200\mathbf{i} + \mathbf{a}_{A/B}$
$\mathbf{a}_{A/B} = 1628\mathbf{i} + 2828\mathbf{j}$
$a_{A/B} = \sqrt{(1628)^2 + (2828)^2} = 3.26(10^3) \text{ km/h}^2$ *Ans.*

D-13. $4s_A + s_P = l$
$v_P = -4v_A = -4(-2) = 8 \text{ m/s}$ *Ans.*

D-14. $+\uparrow \Sigma F_y = ma_y; \ 170 - 153 = \dfrac{153}{32.2}a$
$a = 3.58 \text{ ft/s}^2 \ \uparrow$ *Ans.*

$+\downarrow \Sigma F_y = ma_y; \ 153 - 130 = \dfrac{153}{32.2}a'$
$a' = 4.84 \text{ ft/s}^2 \ \downarrow$ *Ans.*

D-15. Block *B*:
$+\downarrow \Sigma F_y = ma_y; \ 4(9.81) - T = 4a$
Block *A*:
$\xrightarrow{+} \Sigma F_x = ma_x; \ T = 2a$
$T = 13.1 \text{ N}, \ a = 6.54 \text{ m/s}^2$ *Ans.*

D-16. Block *A*:

$+\downarrow \Sigma F_y = ma_y;\ 15(9.81) - T = -15a$

Block *B*:

$+\downarrow \Sigma F_y = ma_y;\ 25(9.81) - T = 25a$

$a = 2.45 \text{ m/s}^2,\ T = 184 \text{ N}$ *Ans.*

D-17. Blocks *A* and *B*:

$\xrightarrow{+} \Sigma F_x = ma_x;\ 6 = \dfrac{70}{32.2}a;\ a = 2.76 \text{ ft/s}^2$

Check if slipping occurs between *A* and *B*.

$\xrightarrow{+} \Sigma F_x = ma_x;\ 6 - F = \dfrac{20}{32.2}(2.76);$

$F = 4.29 \text{ lb} < 0.4(20) = 8 \text{ lb}$

$a_A = a_B = 2.76 \text{ m/s}^2$ *Ans.*

D-18. $\Sigma F_n = m\dfrac{v^2}{\rho};\ (0.3)m(9.81) = m\dfrac{v^2}{2}$

$v = 2.43 \text{ m/s}$ *Ans.*

D-19. $+\downarrow \Sigma F_n = ma_n;\ m(32.2) = m\left(\dfrac{v^2}{250}\right)$

$v = 89.7 \text{ ft/s}$ *Ans.*

D-20. $+\downarrow \Sigma F_n = ma_n;\ 150 + N_p = \dfrac{150}{32.2}\left(\dfrac{(120)^2}{400}\right)$

$N_p = 17.7 \text{ lb}$ *Ans.*

D-21. $\xleftarrow{+} \Sigma F_n = ma_n;\ N_c \sin 30° + 0.2\,N_c \cos 30° = m\dfrac{v^2}{500}$

$+\uparrow \Sigma F_b = 0;$

$\qquad N_c \cos 30° - 0.2\,N_c \sin 30° - m(32.2) = 0$

$v = 119 \text{ ft/s}$ *Ans.*

D-22. At *B*:

$\xrightarrow{+} \Sigma F_n = ma_n,\ T = \left(\dfrac{5}{32.2}\right)\left(\dfrac{(0)^2}{2}\right) = 0$ *Ans.*

In the general position,

$+\nearrow \Sigma F_n = ma_n;\ T - 5 \sin \theta = \left(\dfrac{5}{32.2}\right)\left(\dfrac{v^2}{2}\right)$

$\searrow + \Sigma F_t = ma_t;\ 5 \cos \theta = \dfrac{5}{32.2}\left(\dfrac{v\,dv}{2\,d\theta}\right)$

$\displaystyle\int_0^{90°} 64.4 \cos \theta\, d\theta = \int_0^v v\,dv$

$v = 11.3 \text{ ft/s},$

When $\theta = 90°;$

$T = 15 \text{ lb}$ *Ans.*

D-23. $T_1 + \Sigma U_{1-2} = T_2$

$0 + U_{1-2} - 80(300) = \dfrac{1}{2}\left(\dfrac{15000}{32.2}\right)(40)^2$

$U_{1-2} = 397(10^3) \text{ ft} \cdot \text{lb}$ *Ans.*

D-24. $T_1 + \Sigma U_{1-2} = T_2$

$\dfrac{1}{2}\left(\dfrac{20}{32.2}\right)(5)^2 + 40(10) -$

$(0.2)(20 \cos 30°)(10) + 20(10 \sin 30°) = \dfrac{1}{2}\left(\dfrac{20}{32.2}\right)v^2$

$v = 39.0 \text{ ft/s}$ *Ans.*

D-25. $+\uparrow \Sigma F_y = ma_y;\ N_b + 100 \sin 20° - 10 - 3(9.81) = 0$

$N_b = 5.23 \text{ N}$

$T_1 + \Sigma U_{1-2} = T_2$

$0 + (100 \cos 20°)d - 0.2(5.23)d = \dfrac{1}{2}(3)(10)^2$

$d = 1.61 \text{ m}$ *Ans.*

D-26. $T_1 + V_1 = T_2 + V_2$

$0 + \dfrac{1}{2}(40)(x)^2 = \dfrac{1}{2}\left(\dfrac{6}{32.2}\right)(6)^2 + 6(8)$

$x = 1.60 \text{ ft}$ *Ans.*

D-27. $T_A + V_A = T_B + V_B$

$0 + 2(9.81)(1.5) = \dfrac{1}{2}(2)(v_B)^2 + 0$

$v_B = 5.42 \text{ m/s}$ *Ans.*

$+\uparrow \Sigma F_n = ma_n;\ T - 2(9.81) = 2\left(\dfrac{(5.42)^2}{1.5}\right)$

$T = 58.9 \text{ N}$ *Ans.*

D-28. $T_A + V_A = T_B + V_B$

$0 + \dfrac{1}{2}(4)(2.5 - 0.5)^2 + 5(2.5)$

$\qquad = \dfrac{1}{2}\left(\dfrac{5}{32.2}\right)v_B^2 + \dfrac{1}{2}(4)(1 - 0.5)^2$

$v_B = 16.0 \text{ ft/s}$ *Ans.*

D-29. $T_1 + V_1 = T_2 + V_2$

$\dfrac{1}{2}(2)(4)^2 + \dfrac{1}{2}(30)(2 - 1)^2$

$\qquad = \dfrac{1}{2}(2)(v)^2 - 2(9.81)(1) + \dfrac{1}{2}(30)(\sqrt{5} - 1)^2$

$v = 5.26 \text{ m/s}$ *Ans.*

D-30. Three cars:

$$\xrightarrow{+} mv_1 + \Sigma \int F\, dt = mv_2$$

$$\frac{250(2000)}{32.2}(20) + 40(10^3)(t) - 10(250)t$$

$$= \frac{250(2000)}{32.2}(30)$$

$t = 4.14$ s   *Ans.*

Engine:

$$\xrightarrow{+} mv_1 + \Sigma \int F\, dt = mv_2$$

$$\frac{30(2000)}{32.2}(20) + F(4.14) - 10(30)(4.14)$$

$$= \frac{30(2000)}{32.2}(30)$$

$F = 4800$ lb   *Ans.*

D-31. $+\nwarrow \Sigma F_y = 0;$

$N_b - 5(9.81)\cos 30° - 100\sin 30° = 0$

$N_b = 92.48$ N

$+\nearrow mv_1 + \Sigma \int F\, dt = mv_2$

$5(3) + (100\cos 30°)t - 5(9.81)\sin 30°(t) -$
$\qquad\qquad 0.3(92.48)t = 5(10)$

$t = 1.02$ s   *Ans.*

D-32. Block B:

$(+\downarrow) mv_1 + \int F\, dt = mv_2$

$$0 + 8(5) - T(5) = \frac{8}{32.2}(1)$$

$T = 7.95$ lb   *Ans.*

Block A:

$(\xrightarrow{+}) mv_1 + \int F\, dt = mv_2$

$$0 + 7.95(5) - \mu_k(10)(5) = \frac{10}{32.2}(1)$$

$\mu_k = 0.789$ *Ans.*

D-33. $2s_A + s_B = l$

$2v_A = -v_B$

$+\downarrow m(v_A)_1 + \int F\, dt = m(v_A)_2$

$0 + 10(9.81)(10) - 2T(10) = 10(v_A)_2$

$+\downarrow m(v_B)_1 + \int F\, dt = m(v_B)_2$

$0 + 50(9.81)(10) - T(10) = 50(v_B)_2$

$T = 70.1$ N

$(v_A)_2 = -42.0$ m/s $= 42.0$ m/s $\uparrow$ *Ans.*

$(v_B)_2 = 84.1$ m/s $\downarrow$ *Ans.*

D-34. $\xrightarrow{+} \Sigma mv_1 = \Sigma mv_2;\; 5(2) - 2(5) = 5(v_A)_2 + 2(v_B)_2$

$$\xrightarrow{+} e = \frac{(v_B)_2 - (v_A)_2}{(v_A)_1 - (v_B)_1};\; 0.5 = \frac{(v_B)_2 - (v_A)_2}{2 - (-5)}$$

$(v_A)_2 = -1$ m/s $= 1$ m/s $\leftarrow$ *Ans.*

$(v_B)_2 = 2.5$ m/s $\rightarrow$ *Ans.*

D-35. $\xrightarrow{+} \Sigma mv_1 = \Sigma mv_2;\; 6(20) + 0 = 6(v_A)_2 + 24(v_B)_2$

$$\xrightarrow{+} e = \frac{(v_B)_2 - (v_A)_2}{(v_A)_1 - (v_B)_1};\; 1 = \frac{(v_B)_2 - (v_A)_2}{20 - 0}$$

$(v_A)_2 = -12$ m/s $= 12$ m/s $\leftarrow$ *Ans.*

$(v_B)_2 = 8$ m/s $\rightarrow$ *Ans.*

Disk A:

$mv_1 + \int F\, dt = mv_2;$

$6(20) - \int F\, dt = -6(12)$

$\int F\, dt = 192$ N·s   *Ans.*

D-36. After collision: $T_1 + \Sigma U_{1-2} = T_2$

$$\frac{1}{2}\left(\frac{5}{32.2}\right)(v_A)_2^2 - 0.2(5)\left(\frac{2}{12}\right) = 0$$

$(v_A)_2 = 1.465$ ft/s

$$\frac{1}{2}\left(\frac{10}{32.2}\right)(v_B)_2^2 - 0.2(10)\left(\frac{3}{12}\right) = 0$$

$(v_B)_2 = 1.794$ ft/s

$\Sigma mv_1 = \Sigma mv_2$

$$\frac{5}{32.2}(v_A)_1 + 0 = \frac{5}{32.2}(1.465) + \frac{10}{32.2}(1.794)$$

$(v_A)_1 = 5.054$

$$e = \frac{(v_B)_2 - (v_A)_2}{(v_A)_1 - (v_B)_1} = \frac{1.794 - 1.465}{5.054 - 0} = 0.0652 \; Ans.$$

D-37. $\Sigma m(v_x)_1 = \Sigma m(v_x)_2$

$$0 + 0 = \frac{2}{32.2}(1) + \frac{11}{32.2}(v_{Bx})_2$$

$(v_{Bx})_2 = -0.1818$ ft/s

$\Sigma m(v_y)_1 = \Sigma m(v_y)_2$

$$\frac{2}{32.2}(3) + 0 = 0 + \frac{11}{32.2}(v_{By})_2$$

$(v_{By})_2 = 0.545$ ft/s

$(v_B)_2 = \sqrt{(-0.1818)^2 + (0.545)^2} = 0.575$ ft/s *Ans.*

D-38. $\omega_B(5) = 3(4)$

$\omega_B = 2.40$ rad/s $\uparrow$ *Ans.*

$\omega_C(3) = 2.40(2)$

$\omega_C = 1.60$ rad/s $\downarrow$ *Ans.*

D-39. $\theta = \theta_0 + \omega_0 t + \dfrac{1}{2}\alpha_c t^2$

$\theta = 0 + 0 + \dfrac{1}{2}(2)(10)^2 = 100$ rev   *Ans.*

$\omega = \omega_0 + \alpha_c t$

$\omega = 0 + 2(5) = 10$ rev/s   *Ans.*

D-40. $\alpha = \dfrac{a_t}{r} = \dfrac{20}{2} = 10 \text{ rad/s}^2$ *Ans.*

$10 \text{ rev} = 20\pi \text{ rad}$

$\omega^2 = \omega_0^2 + 2\,\alpha_c(\theta - \theta_0)$

$\omega^2 = 0 + 2(10)(20\pi - 0)$

$\omega = 35.4 \text{ rad/s}$ Ans.

$\theta = \theta_0 + \omega_0 t + \dfrac{1}{2}\alpha_c t^2$

$\theta = 0 + 35.4(4) + \dfrac{1}{2}(10)(4)^2 = 222 \text{ rad}$

$\theta = \dfrac{222}{2\pi} = 35.3 \text{ rev.}$ *Ans.*

D-41. $\mathbf{v}_A = \mathbf{v}_C + \boldsymbol{\omega} \times \mathbf{r}_{A/C}$

$\mathbf{v}_A\mathbf{i} = 3\mathbf{i} + (-2\mathbf{k}) \times (-0.5\mathbf{j})$

$v_A = 2 \text{ m/s} \rightarrow$ *Ans.*

D-42. $\mathbf{v}_A = \mathbf{v}_B + \boldsymbol{\omega} \times \mathbf{r}_{A/B}$

$2\mathbf{i} = 0 + (-\omega\mathbf{k}) \times (1.5\mathbf{j})$

$\omega = 1.33 \text{ rad/s} \downarrow$

$\mathbf{v}_C = \mathbf{v}_B + \boldsymbol{\omega} \times \mathbf{r}_{C/B}$

$\mathbf{v}_C\mathbf{i} = 0 + (-1.33\mathbf{k}) \times (1\mathbf{j})$

$v_C = 1.33 \text{ ft/s} \rightarrow$ *Ans.*

D-43. $\mathbf{v}_O = \mathbf{v}_A + \boldsymbol{\omega} \times \mathbf{r}_{O/A}$

$2\mathbf{i} = 0 + (-\omega\mathbf{k}) \times 1.5\mathbf{j}$

$\omega = 1.33 \text{ rad/s} \downarrow$

$\mathbf{v}_B = \mathbf{v}_O + \boldsymbol{\omega} \times \mathbf{r}_{B/O}$

$\mathbf{v}_B = 2\mathbf{i} + (-1.33\mathbf{k}) \times (1.5\cos 30°\,\mathbf{i} + 1.5\sin 30°\,\mathbf{j})$

$\mathbf{v}_B = \{3\mathbf{i} - 1.73\mathbf{j}\} \text{ m/s}$ *Ans.*

D-44. $\mathbf{v}_A = \mathbf{v}_B + \boldsymbol{\omega} \times \mathbf{r}_{A/B}$

$-v_A\mathbf{j} = -2\mathbf{i} + \omega\mathbf{k} \times (-3\mathbf{i} - 4\mathbf{j})$

$v_A = 1.5 \text{ ft/s} \downarrow$ *Ans.*

$\omega = 0.5 \text{ rad/s} \nwarrow$ *Ans.*

D-45. $v_C = 4(0.6) = 2.4 \text{ m/s} \leftarrow$

$\mathbf{v}_B = \mathbf{v}_C + \boldsymbol{\omega} \times \mathbf{r}_{B/C}$

$-v_B \sin 30°\,\mathbf{i} + v_B \cos 30°\,\mathbf{j}$

$\qquad = -2.4\mathbf{i} + (\omega_{BC}\mathbf{k}) \times (-0.6\mathbf{i})$

$v_B = 4.80 \text{ m/s} \nwarrow \ \omega_{BC} = 6.93 \text{ rad/s} \downarrow$

$\omega_{AB} = \dfrac{4.80}{0.6/\sin 30°} = 4 \text{ rad/s} \nwarrow$ *Ans.*

D-46. $v_B = 2(1.5) = 3 \text{ m/s} \searrow$

$\mathbf{v}_C = \mathbf{v}_B + \boldsymbol{\omega} \times \mathbf{r}_{C/B}$

$-v_C\mathbf{j} = 3\cos 45°\,\mathbf{i} - 3\sin 45°\,\mathbf{j}$

$\qquad\quad + (\omega\mathbf{k}) \times (1.5\cos 45°\,\mathbf{i} - 1.5\sin 45°\,\mathbf{j})$

$v_C = 4.24 \text{ m/s} \downarrow$ *Ans.*

$\omega = 2 \text{ rad/s} \downarrow$

D-47. $\omega = \dfrac{v_B}{r_{B/C}} = \dfrac{2}{0.4} = 5 \text{ rad/s} \downarrow$

$\alpha = \dfrac{a_B}{r_{B/C}} = \dfrac{3}{0.4} = 7.5 \text{ rad/s}^2 \downarrow$

$\mathbf{a}_A = \mathbf{a}_B + \boldsymbol{\alpha} \times \mathbf{r}_{A/B} - \omega^2 \mathbf{r}_{A/B}$

$\mathbf{a}_A = 3\mathbf{j} + (-7.5\mathbf{k}) \times (-0.4\mathbf{j}) - (5)^2(-0.4\mathbf{j})$

$\mathbf{a}_A = \{-3\mathbf{i} + 13\mathbf{j}\} \text{ m/s}^2$ *Ans.*

$a_B = 3 \text{ m/s}^2$ *Ans.*

D-48. $\mathbf{v}_B = \mathbf{v}_A + \boldsymbol{\omega} \times \mathbf{r}_{B/A}$

$-v_B\mathbf{j} = -6\mathbf{i} + \omega\mathbf{k} \times (3\cos 45°\,\mathbf{i} + 3\sin 45°\,\mathbf{j})$

$\omega = 2.828 \text{ rad/s} \downarrow,\ v_B = 6 \text{ ft/s} \downarrow$

$\mathbf{a}_B = \mathbf{a}_A + \boldsymbol{\alpha} \times \mathbf{r}_{B/A} - \omega^2 \mathbf{r}_{B/A}$

$-a_B\mathbf{j} = 2\mathbf{i} + (\alpha\mathbf{k}) \times (3\cos 45°\,\mathbf{i} + 3\sin 45°\,\mathbf{j})$

$\qquad\qquad -(2.828)^2 (3\cos 45°\,\mathbf{i} + 3\sin 45°\,\mathbf{j})$

$a_B = 31.9 \text{ ft/s}^2 \downarrow$ *Ans.*

$\alpha = 7.06 \text{ rad/s}^2 \downarrow$ *Ans.*

D-49. $+\swarrow \Sigma F_x = m(a_G)_x;\ 3500 \sin 20° = \dfrac{3500}{32.2} a_G$

$a_G = 11.0 \text{ ft/s}^2$ *Ans.*

$+\nwarrow \Sigma F_y = m(a_G)_y;\ N_A + N_B - 3500 \cos 20° = 0$

$\downarrow + \Sigma M_G = 0;\ N_B(4) - N_A(3) = 0$

$N_A = 1879 \text{ lb}$ *Ans.*

$N_B = 1410 \text{ lb}$ *Ans.*

D-50. $\downarrow + \Sigma M_A = \Sigma(M_k)_A;\ 10\left(\dfrac{3}{5}\right)(7) = \dfrac{20}{32.2} a(3.5)$

$a = 19.32 \text{ ft/s}^2$ *Ans.*

$\xrightarrow{+} \Sigma F_x = m(a_G)_x;\ A_x + 10\left(\dfrac{3}{5}\right) = \dfrac{20}{32.2}(19.32)$

$A_x = 6 \text{ lb}$ *Ans.*

$+\uparrow \Sigma F_y = m(a_G)_y;\ A_y - 20 + 10\left(\dfrac{4}{5}\right) = 0$

$A_y = 12 \text{ lb}$ *Ans.*

D-51. $\zeta + \Sigma M_B = I_B\alpha;\ 50(1) = 2.30\alpha$

$\alpha = 21.7\ \text{rad/s}^2$ *Ans.*

$a_G = \left(\dfrac{1}{\cos 30°}\right)(21.7) = 25.1\ \text{m/s}^2$

$\overset{+}{\to} \Sigma F_x = m(a_G)_x;$

$B_x = \left(\dfrac{50}{32.2}\right)25.1 \sin 30° = 19.5\ \text{lb}$ *Ans.*

$+\uparrow \Sigma F_y = m(a_G)_y;$

$\qquad B_y - 50 = -\left(\dfrac{50}{32.2}\right)25.1 \cos 30°.$

$B_y = 16.2\ \text{lb}$ *Ans.*

D-52. $\zeta + \Sigma M_O = I_O\alpha;$

$20(9.81)(0.5) = \left[\dfrac{1}{12}(20)(3)^2 + 20(0.5)^2\right]\alpha$

$\alpha = 4.90\ \text{rad/s}^2$ *Ans.*

D-53. No slipping, so that

$a_G = 1.25\alpha$

$\zeta + \Sigma M_A = \Sigma(M_k)_A;\ 20(9.81)\sin 20°(1.25)$

$\qquad\qquad = 20(0.8)^2\alpha + (20a_G)(1.25)$

$\alpha = 1.90\ \text{rad/s}^2$ *Ans.*

$a_G = 2.38\ \text{m/s}^2$

D-54. $+\nwarrow \Sigma F_y = 0;\ N_A - 15(9.81)\cos 45° = 0;$

$N_A = 104.1\ \text{N}$

$a_G = \alpha(0.3)$

$\zeta + \Sigma M_B = \Sigma(M_k)_B;\ (0.1)(104.1)(1.55) -$

$\qquad 15(9.81)\sin 45°(0.3) = -15(0.8)^2\alpha - 15(a_G)(0.3)$

$a_G = 0.413\ \text{m/s}^2$

$\alpha = 1.38\ \text{rad/s}^2$ *Ans.*

D-55. $\overset{+}{\to} \Sigma F_x = m(a_G)_x;\ F_A = 2a_G$

$\zeta + \Sigma M_G = I_G\alpha;\ F_A(0.4) = 2(0.3)^2\alpha$

$\mathbf{a}_A = \mathbf{a}_G + \boldsymbol{\alpha} \times \mathbf{r}_{A/G} - \omega^2 \mathbf{r}_{A/G}$

$5\mathbf{i} = a_G\mathbf{i} + (\alpha\mathbf{k}) \times (-0.4\mathbf{j}) - \mathbf{0}$

$5 = a_G + 0.4\alpha$

$F_A = 3.60\ \text{N}$

$a_G = 1.80\ \text{m/s}^2$

$\alpha = 8\ \text{rad/s}^2$ *Ans.*

Answers to Selected Problems

Review Problems 1

R1–1. $t_{min} = 23.8$ s
R1–2. $y = 0.0208x^2 + 0.333x$
R1–3. $v_B = 55.2$ ft/s, $v_A = 27.6$ ft/s
R1–5. $v = 38.5$ ft/s
R1–6. **a)** $t = 1.07$ s, **b)** $v_{A/B} = 5.93$ m/s $\rightarrow$
R1–7. $t = 5$ s
R1–9. $a = 0.343$ m/s^2, $a = 0.436$ m/s^2
R1–10. $y = 3\ln\left(\dfrac{\sqrt{x^2 + 4} + x}{2}\right)$
R1–11. $v_B = 3.33$ ft/s$\uparrow$, $v_{B/C} = 13.3$ ft/s$\uparrow$
R1–13. $s = 1.84$ m
R1–14. $s = 980$ m
R1–15. $P = 946$ kW
R1–17. $v = 4.38$ m/s
R1–18. $x = 7.85$ in.
R1–19. $v_B = 10.4$ m/s
R1–21. $(v_B)_2 = 0.575$ ft/s, $71.5°$ ⬎,
$\Delta T = -0.192$ ft · lb
R1–22. $a = 29.0$ m/s^2
R1–23. $v_B = 0.5$ m/s $\uparrow$
R1–25. $(v_A)_{min} = 0.838$ m/s,
$(v_A)_{max} = 1.76$ m/s
R1–26. $v = 2.68$ ft/s
R1–27. $v = 25.4$ ft/s
R1–29. $a = 322$ m/s^2, $26.6°$ ⬏
R1–30. $F = 38.5$ N
R1–31. $s = 20.2$ m
R1–33. $N_s = 240$ lb
R1–34. $v_B = 47.8$ ft/s
R1–35. $e = 0.901$
R1–37. $v_2 = 13.8$ ft/s
R1–38. $T = 158$ N
R1–39. $s = 20.0$ ft
R1–41. $v_A = 26.8$ ft/s $\downarrow$
R1–42. $t' = 27.3$ s
R1–43. $P_i = 1.80$ hp
R1–45. **a)** $a_B = a_A = 2.76$ ft/s^2,
b) $a_A = 70.8$ ft/s^2, $a_B = 3.86$ ft/s^2
R1–46. $s = 0.933$ m
R1–47. $v_2 = 31.7$ m/s
R1–49. $v_2 = 21.5$ ft/s

R1–50. $a_A = 0.755$ m/s^2, $a_B = 1.51$ m/s^2,
$T_A = 90.6$ N,
$T_B = 45.3$ N
R1–51. $v = 5.38$ ft/s

Review Problems 2

R2–1. $\alpha_A = 12.5$ m/s $\leftarrow$
R2–2. $v_W = 14.2$ ft/s, $a_W = 0.25$ ft/s^2
R2–3. $\omega = 28.6$ rad/s, $\theta = 24.1$ rad,
$v_P = 7.16$ m/s,
$a_P = 205$ m/s^2
R2–5. $h = 1.80$ m
R2–6. $\omega = 0.275$ rad/s, $\alpha = 0.0922$ rad/s^2
R2–7. **a)** $C_y = 7.22$ N, $B_y = 7.22$ N,
b) $C_y = 14.7$ N, $B_y = 14.7$ N
R2–9. $v_C = 25.3$ in./s, $63.4°$ ⬏;
$a_C = 73.8$ in./s^2, $32.5°$ ⬊
R2–10. $\alpha = 8.89$ rad/s^2
R2–11. $\omega = 25.0$ rad/s
R2–13. $v = -r\omega \sin\theta$, $a = -r\omega^2 \cos\theta$
R2–14. $a_A = 282$ ft/s^2
R2–15. $a = \dfrac{s}{2\pi}\omega^2$
R2–17. $v = 4.78$ m/s
R2–18. $\theta = \tan^{-1}\left[\dfrac{\mu(k_O^2 + r^2)}{k_O^2}\right]$
R2–19. $d = 5.75$ ft
R2–21. $T = 218$ N, $\alpha = 21.0$ rad/s^2
R2–22.
$$v = -r_1\omega \sin\theta - \dfrac{r_1^2\omega \sin 2\theta}{2\sqrt{(r_1 + r_2)^2 - (r_1 \sin\theta)^2}}$$
R2–23. $L = 3.92$ slug · ft/s
R2–25. $v_B = 0.860$ m/s
R2–26. $a_B = 8.00$ m/s^2 $\uparrow$
R2–27. $\alpha = 12.1$ rad/s^2, $F = 30.0$ lb
R2–29. $\omega = 215$ rad/s
R2–30. $\omega = 39.5$ rad/s
R2–31. $P = 24$ lb, $a_G = 3.22$ ft/s^2,
$N_B = 14.8$ lb,
$N_A = 65.3$ lb
R2–33. $B_y = 180$ N, $A_y = 252$ N, $A_x = 139$ N
R2–34. $B_y = 143$ N, $A_y = 200$ N, $A_x = 34.3$ N
R2–35. 14.2 rad/s^2

R2–37. $F_C = 12.4$ lb, $N_B = 3.09$ kip,
$N_A = 1.31$ kip, $F_A = 68.3$ lb
R2–38. $\omega_A = 1.70$ rad/s, $\omega_B = 5.10$ rad/s
R2–39. $E_x = 9.87$ lb, $E_y = 4.86$ lb,
$M_E = 7.29$ lb · ft
R2–41. $F_B' = 4.50$ kN, $N_A = 1.78$ kN,
$N_B' = 5.58$ kN
R2–42. 1.41 m/s^2, 1.38 m/s^2
R2–43. $N_A' = 383$ N, $N_B' = 620$ N
$N_B' = 5.58$ kN
R2–45. $\alpha_A = 56.2$ ft/s^2 $\downarrow$, $\alpha_B = 40.2$ ft/s^2,
$53.3°$ ⬈
R2–46. $39.3°$
R2–47. 2.26 m/s^2, $N_B = 226$ N, $N_A = 559$ N,
$N_B = 454$ N
R2–49. 0.661 m
R2–50. 0.859 m
R2–51. 3.46 m/s

Index

Geometric Properties of Line and Area Elements

Centroid Location	Centroid Location	Area Moment of Inertia

$L = 2\theta r$

$\dfrac{r\sin\theta}{\theta}$

Circular arc segment

$A = \theta r^2$

$\dfrac{2}{3}\dfrac{r\sin\theta}{\theta}$

Circular sector area

$I_x = \tfrac{1}{4}r^4(\theta - \tfrac{1}{2}\sin 2\theta)$

$I_x = \tfrac{1}{4}r^4(\theta + \tfrac{1}{2}\sin 2\theta)$

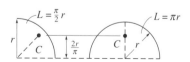

$L = \tfrac{\pi}{2}r$ $\qquad L = \pi r$

$\dfrac{2r}{\pi}$

Quarter and semicircle arcs

$A = \tfrac{1}{4}\pi r^2$

$\dfrac{4r}{3\pi}$

$\dfrac{4r}{3\pi}$

Quarter circle area

$I_x = \tfrac{1}{16}\pi r^4$

$I_y = \tfrac{1}{16}\pi r^4$

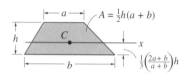

$A = \tfrac{1}{2}h(a+b)$

$\tfrac{1}{3}\left(\dfrac{2a+b}{a+b}\right)h$

Trapezoidal area

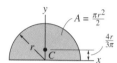

$A = \dfrac{\pi r^2}{2}$

$\dfrac{4r}{3\pi}$

Semicircular area

$I_x = \tfrac{1}{8}\pi r^4$

$I_y = \tfrac{1}{8}\pi r^4$

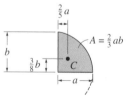

$\tfrac{2}{5}a$

$A = \tfrac{2}{3}ab$

$\tfrac{3}{8}b$

Semiparabolic area

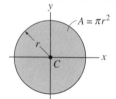

$A = \pi r^2$

Circular area

$I_x = \tfrac{1}{4}\pi r^4$

$I_y = \tfrac{1}{4}\pi r^4$

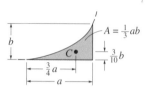

$A = \tfrac{1}{3}ab$

$\tfrac{3}{10}b$

$\tfrac{3}{4}a$

Exparabolic area

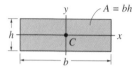

$A = bh$

b

Rectangular area

$I_x = \tfrac{1}{12}bh^3$

$I_y = \tfrac{1}{12}hb^3$

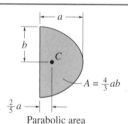

$A = \tfrac{4}{3}ab$

$\tfrac{2}{5}a$

Parabolic area

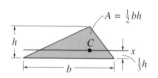

$A = \tfrac{1}{2}bh$

$\tfrac{1}{3}h$

Triangular area

$I_x = \tfrac{1}{36}bh^3$

Center of Gravity and Mass Moment of Inertia of Homogeneous Solids

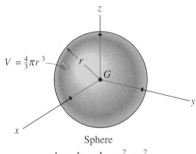

Sphere

$$I_{xx} = I_{yy} = I_{zz} = \tfrac{2}{5}mr^2$$

$$V = \tfrac{4}{3}\pi r^3$$

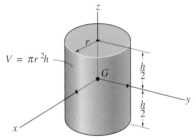

Cylinder

$$V = \pi r^2 h$$

$$I_{xx} = I_{yy} = \tfrac{1}{12}m(3r^2 + h^2) \quad I_{zz} = \tfrac{1}{2}mr^2$$

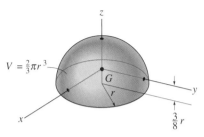

Hemisphere

$$V = \tfrac{2}{3}\pi r^3$$

$$I_{xx} = I_{yy} = 0.259mr^2 \quad I_{zz} = \tfrac{2}{5}mr^2$$

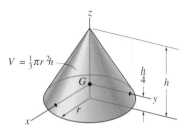

Cone

$$V = \tfrac{1}{3}\pi r^2 h$$

$$I_{xx} = I_{yy} = \tfrac{3}{80}m(4r^2 + h^2) \quad I_{zz} = \tfrac{3}{10}mr^2$$

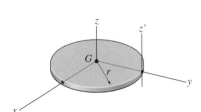

Thin Circular disk

$$I_{xx} = I_{yy} = \tfrac{1}{4}mr^2 \quad I_{zz} = \tfrac{1}{2}mr^2 \quad I_{z'z'} = \tfrac{3}{2}mr^2$$

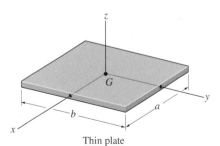

Thin plate

$$I_{xx} = \tfrac{1}{12}mb^2 \quad I_{yy} = \tfrac{1}{12}ma^2 \quad I_{zz} = \tfrac{1}{12}m(a^2 + b^2)$$

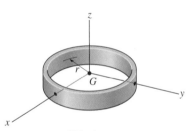

Thin ring

$$I_{xx} = I_{yy} = \tfrac{1}{2}mr^2 \quad I_{zz} = mr^2$$

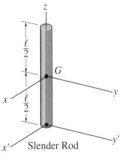

Slender Rod

$$I_{xx} = I_{yy} = \tfrac{1}{12}m\ell^2 \quad I_{x'x'} = I_{y'y'} = \tfrac{1}{3}m\ell^2 \quad I_{z'z'} = 0$$